Fossil Energy

This volume collects selected topical entries from the Encyclopedia of Sustainability Science and Technology (ESST). ESST addresses the grand challenges for science and engineering today. It provides unprecedented, peer-reviewed coverage of sustainability science and technology with contributions from nearly 1,000 of the world's leading scientists and engineers, who write on more than 600 separate topics in 38 sections. ESST establishes a foundation for the research, engineering, and economics supporting the many sustainability and policy evaluations being performed in institutions worldwide.

Ripudaman Malhotra
Editor

# Fossil Energy

## Selected Entries from the Encyclopedia of Sustainability Science and Technology

Springer

*Editor*
Ripudaman Malhotra
Chemistry & Chemical Engineering Laboratory
Pure and Applied Physical Sciences Division
SRI International
Menlo Park, CA, USA

This book consists of selections from the Encyclopedia of Sustainability Science and Technology edited by Robert A. Meyers, originally published by Springer Science +Business Media New York in 2012.

ISBN 978-1-4899-8536-1 ISBN 978-1-4614-5722-0 (eBook)
DOI 10.1007/978-1-4614-5722-0
Springer New York Heidelberg Dordrecht London



Printed on acid-free paper

Springer is part of Springer Science+Business Media (www.springer.com)

# Contents

# Chapter 1
# Fossil Energy, Introduction

**Ripudaman Malhotra**

To many in the sustainability community, fossil energy is an anathema. Continued use of fossil resources – oil, coal, and natural gas – poses threats to the environment through the emission of pollutants and greenhouse gases. The fact that they are a limited or exhaustible resource means that in the future we could either run out of them or their extraction will get progressively harder to a point that it takes more energy to extract them than would be derived from their use. Using fossil energy is clearly not sustainable, and the world has to look to renewable resources for long-term survival.

This is an encyclopedia about the science and technology of sustainability. The word, *sustainability* shares its root with *sustenance*. Any discussion of sustainability must therefore include discussions of sustenance, and in the context of modern society, sustenance stems from the use of energy. We derive energy from a number of different sources. Annual global consumption of energy is currently on the order of 500 exajoules (EJ), about 85% of which comes from fossil resources. From 500 EJ/year, the global energy demand is expected to rise to somewhere between 1,000 and 1,500 EJ/year by the middle of this century. The drivers for this increased demand are already in place. Large segments of China, India, and Brazil are poised to increase their standard of living – and the concomitant energy demand – substantially. If the average energy consumption in the Asia-Pacific region were to reach the current global average, which is about half of what is consumed in Europe and Japan, the demand would increase by an amount equal to the total energy consumption in the North America region. The challenge of achieving a sustainable future is in being able to balance the energy requirements for the

This chapter was originally published as part of the Encyclopedia of Sustainability Science and Technology edited by Robert A. Meyers. DOI:10.1007/978-1-4419-0851-3

R. Malhotra (✉)
Chemistry and Chemical Engineering Laboratory, Pure and Applied Physical Sciences Division, SRI International, 333 Ravenswood Avenue, Menlo Park, CA 94025, USA
e-mail: ripudaman.malhotra@sri.com

R. Malhotra (ed.), *Fossil Energy: Selected Entries from the Encyclopedia of Sustainability Science and Technology*, DOI 10.1007/978-1-4614-5722-0_1,

billions of people so they can lead healthy and productive lives against the need to preserve the environment by not running into its limits in its ability to supply the resources or act as a sink for the waste [1]. The entries in this section examine the current status, assess the resource potential of the various fossil fuels, examine the technologies for using them, and review the environmental impact of their use.

Petroleum and natural gas have similar origin and often occur together in geologic formations, and their global distribution and production is discussed together in the entry by McCabe (▸Oil and Natural Gas: Global Resources). This entry makes clear the distinction between reserves and resources. Reserves represent only that fraction of the resource base that can be economically recovered using current technology. These are not fixed quantities as both technology and economics change over time. Global annual production (and consumption) of oil in 2010 was 31 billion barrels and of natural gas was around 120 trillion cubic feet. In energy units, they correspond to 180 EJ of oil and 120 EJ of natural gas [2]. The current reserves are estimated at 1,236 billion barrels of conventional oil (7,500 EJ) and 6,545 tcf of natural gas (6,500 EJ). The current reserves to production ratio (R/P) is about 40 for oil and about 55 for natural gas. The R/P ratio has often been mistakenly taken as the time to exhaustion, but new discoveries as well as advances in technology add to the reserves. In the case of oil, for example, the R/P ratio has stayed around 40–50 years for more than 60 years even with the steadily increasing oil consumption. In addition, there are also unconventional accumulations of these hydrocarbon resources and extracting them requires development of new technologies. In the case of oil, the unconventional resources are oil sands, oil shale, and heavy oil. Unconventional resources of natural gas are coal bed methane, tight gas, shale gas, and gas hydrates. These unconventional resources are vast and have the potential of more than doubling our resource endowment.

Exploration and production of oil from sedimentary deposits and oil sands is the subject of an entry by Speight (Petroleum and Oil Sands Exploration and Production). The processes for recovering oil could be a simple matter of drilling into the formation with the oil flowing to the surface under its own pressure, or it may require injection of gases, fluids, and surfactants to coax it to flow. In extreme cases, it may even require underground combustion of a portion of the resource to release the oil. Speight describes the chemical and physical factors that govern the flow of oil and the technology options currently available. In a different entry, Speight (▸Petroleum Refining and Environmental Control and Environmental Effects) provides and account of the different processes such as distillation, catalytic cracking, hydrotreating, reforming, and deasphalting used in the refining of crude oil. This entry also deals with environmental effects of the gaseous, liquid, and solid effluents from these processes. Production of oil from shale is principally achieved by retorting of shale, or other thermal processes including in situ pyrolysis. Oja and Suuberg (Oil Shale Processing, Chemistry and Technology) detail the chemistry and technology of these processes in the entry.

Gas hydrates represent a particularly noteworthy resource and are reviewed in the entry by Patil (Alaska Gas Hydrate Research and Field Studies). They are naturally occurring ice-like substances in which methane or other light gases are trapped in the cage structures formed by water molecules. They are stable under certain conditions of temperature and pressure, and can be found at many places around the world at depths of several hundred meters below the seabed and in the permafrost in the polar region. Global estimates of gas hydrates are on the order of 500,000 tcf, about a hundred times the proved reserves of natural gas. Apart from being a potentially very important source for energy in the future, the gas hydrates are also important from the perspective of climate change. A general warming could release substantial amounts of methane, which – being a potent greenhouse gas – would reinforce any global warming.

Oil remains the prized fuel. It has a high energy density and as a liquid it is well suited for transportation. The transportation sector relies on oil for over 90% of its energy needs (the remainder being mostly furnished by coal – via electricity) [3]. Given the importance of liquid fuels, there is considerable interest in converting coal and natural gas, the other hydrocarbon resources, into oil. In a pair of entries, Araso and Smith (Gas to Liquid Technologies and Coal to Liquids Technologies) provide an overview of the various processes for converting natural gas and coal to liquids. These reviews cover the basic chemistry and catalytic technologies behind the different approaches. The entry on Gas to Liquid Technologies deals with steam reforming of methane, autothermal reforming, and partial oxidation approaches for producing syngas, including strategies for managing the heat and mass transfer through the use of different reactor technologies. The entry next covers the conversion of syngas into liquids by Fischer-Tropsch (FT) synthesis. The entry on Coal to Liquids Technologies reviews the different approaches like pyrolysis, direct liquefaction, coprocessing with petroleum, and indirect liquefaction that first converts coal into syngas and then uses FT or other conversion processes to make liquid fuels.

The internal combustion engines that power these vehicles covert only about 20–30% of the energy in the fuel to motive power. Increasing the efficiency of the engines used in transportation represents a significant opportunity to reduce future demand for oil, as well as reduce the carbon footprint of the cars, trucks, and planes. Jacobs (Internal Combustion Engines, Developments in) provides a detailed account of the thermodynamics of the different kinds of engines and of the various technologies under development for improving the efficiency of the internal combustion engines. These include strategies such as engine downsizing, turbocharging, better controls, variable geometry engines, variable valve timing, homogeneous charge compression engines, and waste heat recovery.

Höök (Coal and Peat: Global Resources and Future Supply) reviews the global coal and peat resources, which are substantially larger than conventional oil and gas resources. Peat and coal are also distributed more evenly around the globe. The proved reserves of coal are estimated at between 800 billion and a trillion metric tons, representing roughly 20,000 EJ. Most coal is derived from trees and ferns that grew some 250–300 million years ago during the carboniferous period. Aerobic bacteria generally decompose most plant matter into $CO_2$, but when a tree fell into a swamp and was buried with limited exposure to oxygen, it could be partially preserved. That phenomenon occurs even today and is evidenced in the formation of peat (very young coal) in bogs. Unlike coal, peat has a low calorific content and its commercial use as an energy resource is limited. However, peat is important from global carbon cycle. Peat bogs store vast amounts of carbon, but the product of anaerobic decomposition, methane, contributes to greenhouse gases in the atmosphere such that under steady state conditions they comprise only a slight sink of carbon. The situation changes with fires as peat bogs become major source of carbon in the atmosphere. Human activities, such as drainage of the area for agriculture, exacerbate the situation, as more peat gets exposed and the swamp is not wet-enough to squelch the fire. Estimates of the amounts of carbon released from peat are on the order of several billion tons, the same order of magnitude as from fuels used in transportation or power production.

As coal resources are vast, they are likely to continue to contribute substantially to global energy mix. However, their use also presents a number of challenges, beginning with mining, and through coal preparation and use. Mining operations result in excavation of billions of tons of earth, and depending on the nature of the waste rock and how it is handled it can lead to contamination of air, water, and soils systems. The wide-ranging nature of environmental impact necessitates a commensurately broad portfolio of technologies for the remediation, restoration and reclamation. Chattopadhyay and Chattopadhyay (Mining Industries and Their Sustainable Management) provide an overview of the environmental impact of coal mining, and mining processes in general, as well as review the range of the chemical and biochemical strategies to mitigate the impact.

The run-of-mine coal is often laden with noncombustible minerals such as shale and clays that reduce its heating value. Cleaning and washing of coals removes many of these impurities and upgrades the coal into a marketable resource, improve the performance of power plants, and also reduce potential of harmful emissions and dust. Luttrell and Honaker (Coal Preparation) review these cleaning and preparation operations, and point out the importance of cheap preparation in extending the resource base.

Electricity is one of the most useful forms of energy; its importance to the way we live cannot be over emphasized. It can be used to perform all kinds of useful functions and services that people desire, and at the point of use it produces no pollution. Electricity is a secondary source of energy, as it must be produced from primary sources such as coal, natural gas, oil, nuclear, hydro and, of late, increasingly from wind and solar. In 2010, the global production of energy was 21,325 trillion kWh, which is equivalent to 77 EJ. However, since 68% of electricity is derived from burning fossil fuels at an average efficiency of 38%, the total primary energy the world consumed as electricity amounts to 135 EJ, or roughly 27% of the total energy consumed. Electricity demand in the world is projected to rise at a rate greater than for total energy, and meeting that need for additional sources for producing electricity is critical human welfare.

Several entries in this section are devoted to the production of electrical power. As the fuel with the lowest levelized cost of electricity, coal remains the dominant fuel for producing electricity, and furnishes over half of the electricity. It results in the emission of about 900 g $CO_2$ per kWh, highest of any fossil resource.

The development of coal power in future is going to be constrained by the concerns of $CO_2$ emissions, and Beer ($CO_2$ Reduction and Coal-Based Electricity Generation) discusses the different technology options for increasing the efficiency of coal-fired power plants and reducing the $CO_2$ emissions. Ultra supercritical steam cycles and combined heat and power applications are relatively straightforward to apply, and have the potential to reduce the largest tonnage of emissions.

Of course, there are other pollutants also that are emitted when coal is used for power productions. Chief among them are oxides of sulfur and nitrogen and toxic metals such as mercury, selenium, and arsenic, as well as carcinogenic PAH-containing soot aerosols. Moyeda (Pulverized Coal-Fired Boilers and Pollution Control) reviews the technologies for scrubbing the stack gas emissions. The deployment of scrubbers for the nitrogen and sulfur oxides following the Clean Air and Water Act in the 1980s greatly improved the quality of air and water systems.

Natural gas is the other major fuel for producing electric power. When used in a gas-fired combined-cycle mode, it produces only 380 g $CO_2$/kWh. Smith and Gülen (Natural Gas Power) review the technology for power generation from natural gas. They discuss the different sources of natural gas, provide a brief history followed by detailed thermodynamics of gas turbines, and modern power plants designs that incorporate combined cycle for maximum overall efficiency. They also review emissions from NGCC plants, notably $NO_x$, and strategies for minimizing them. The entry concludes with a discussion of power plant economics.

Carbon capture and sequestration (CCS) as a strategy for stabilizing atmospheric $CO_2$ levels is receiving considerable attention. Friedman ($CO_2$ Capture and Sequestration) provides an overview of the state of technology and the different embodiments of CCS. These cover various strategies for first capturing the $CO_2$, which may be performed pre- or post-combustion, or from the atmosphere. He discusses the energy requirements under these different scenarios. The entry also reviews the geochemistry of various options for sequestering the captured $CO_2$, either in deep saline aquifers or in other geological formations.

## Bibliography

1. Meadows D, Randers J, Meadows D (2004) Limits to growth, the 30-year update. Chelsea Green Publishing, White River Junction, Vermount
2. BP statistical review of world energy, June 2011 From www.BP.com
3. Energy Information Administration, Annual energy review 2010. http://www.eia.gov/emeu/aer/pecss_diagram.html

# Chapter 2
# Oil and Natural Gas: Global Resources

**Peter J. McCabe**

## Glossary

| | |
|---|---|
| Coal bed methane (CBM) | Natural gas (methane) that can be extracted from coal beds. Synonymous with coal bed gas and coal seam gas (CSG). |
| Conventional oil and gas | Oil and natural gas that occur in the subsurface and that can be produced using conventional methods of well drilling. |
| Gas hydrates | Accumulations of natural gas that are trapped in ice-like crystalline solids consisting of gas molecules surrounded by cages of water molecules. Hydrates are stable at certain temperatures and pressures within some sea-floor sediments and within permafrost in polar regions. Synonymous with gas clathrates. |
| Oil sands | Sandstones that are naturally impregnated with bitumen, a highly viscous form of petroleum. Synonymous with bituminous sands and tar sands. |
| Oil shale | A rock that contains significant amounts of solid organic chemical compounds (kerogen) that can generate oil when heated. |
| Reserves | The discovered, but not yet produced, amounts of oil or gas that could be extracted profitably with existing technology under present economic conditions. |
| Resources | The amounts of oil and gas that have been discovered plus the estimated amount that remains to be discovered. |

---

This chapter was originally published as part of the Encyclopedia of Sustainability Science and Technology edited by Robert A. Meyers. DOI:10.1007/978-1-4419-0851-3

P.J. McCabe (✉)
School of Earth, Environmental and Biological Sciences, Queensland University of Technology, 2 George Street, Brisbane, QLD 4000, Australia
e-mail: Peter.McCabe@csiro.au

R. Malhotra (ed.), *Fossil Energy: Selected Entries from the Encyclopedia of Sustainability Science and Technology*, DOI 10.1007/978-1-4614-5722-0_2,

| | |
|---|---|
| Shale gas | Natural gas that is produced from shale. |
| Tight gas | Natural gas that is extracted from rocks with very low porosity and permeability and which is, therefore, relatively difficult to produce. |
| Unconventional oil and gas | Oil and natural gas accumulations that require extraction techniques that allow easier flow of oil and gas to a well (for example, hydraulic fracturing to open pathways or in situ heating to reduce viscosity) or by processing after mining. |

## Definition of the Subject

Crude oil and natural gas (mostly methane but including some longer-chain hydrocarbons) have been used by humans for thousands of years for a variety of purposes including lighting, heating, and medicinal uses. However, use was limited by access to natural seeps of oil and gas and the available technologies to extract and store the products. The earliest known oil wells were drilled in China in the fourth century using bamboo. It was not, however, until the mid-nineteenth century that large-scale production began, when metal piping allowed deeper drilling into hard rock. Early commercial production began in Poland and Romania and this was followed rapidly by drilling successes in the Russian Empire, in what is now Azerbaijan, and in the United States. Development of the process of fractional distillation at this time fuelled the demand for crude oil, which could now be economically refined into kerosene for use in oil lamps. The development of the internal combustion engine also created a demand for oil and this greatly expanded with Karl Benz's invention of the gasoline-powered automobile, patented in 1886. The transport sector became the dominant user of oil and currently accounts for 61.4% of oil consumption [1]. Other major uses include lubrication, chemical feedstock, and domestic heating. Production has risen dramatically over the last century (Fig. 2.1) and production of crude oil today is approximately 73 million barrels per day (a "barrel" is 42 US gallons).

The United States dominated world oil production for the first half of the twentieth century, accounting for over 50% of the world's annual production until 1960. After World War II demand for oil increased at a rapid rate and production increased fivefold between 1950 and 1972. As the world economy rapidly grew, international trade in oil was facilitated by the development of supertankers. The dominance of the United States decreased as several other regions increased their share of production, particularly the Middle East and the Soviet Union. The Organization of the Petroleum Exporting Countries (OPEC) was formed in 1960 and by 1972 was responsible for over 50% of the world's oil production. The cartel restricted the supply of oil during the 1970s, resulting in a fivefold increase in prices. The resulting fall in demand and oversupply led to

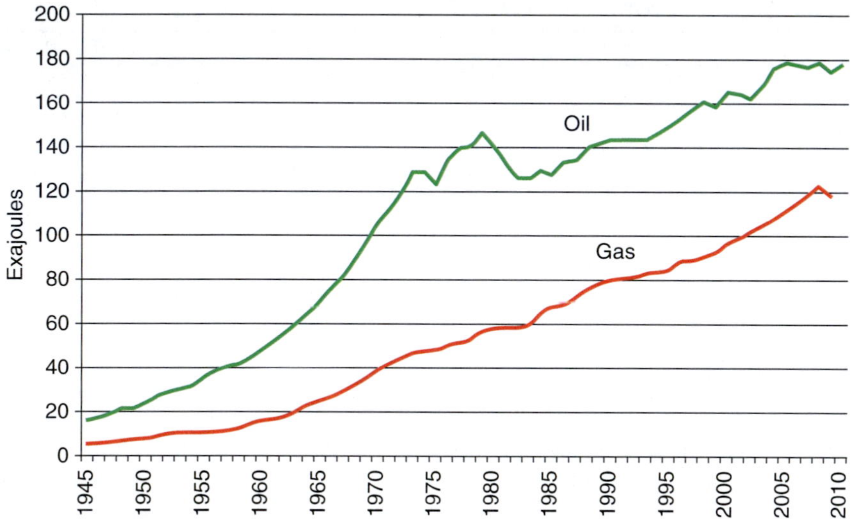


**Fig. 2.1** World production of crude oil and natural gas since World War II expressed in units of energy for comparative purposes. Based on EIA data [2]

**Table 2.1** Major producers of crude oil. From [1]

| Country | % of world total |
|---|---|
| Russia | 12.9 |
| Saudi Arabia | 11.8 |
| USA | 8.3 |
| Iran | 5.4 |
| China | 5.0 |
| Canada | 4.0 |
| Mexico | 3.8 |
| Venezuela | 3.3 |
| Kuwait | 3.2 |
| U.A.E. | 3.1 |
| Rest of world | 39.2 |

a price collapse in the early 1980s. OPEC has subsequently never exceeded 42% of world production [2] but the rapid economic expansion of many countries in the first decade of the twenty-first century resulted in relatively high prices again as demand increased substantially. Today 60% of the world's oil production is from six countries, of which five belong to OPEC (Table 2.1).

Although natural gas had long been used in limited amounts for illumination, much of the natural gas that was originally discovered in association with oil was vented or flared because there was no way of commercializing the gas. Reliable pipelines to transport gas were not developed until after World War II. Natural gas is now used primarily for electrical generation in power plants, domestic heating, manufacturing fertilizers, and for other industrial purposes. World production of

**Table 2.2** Major producers of natural gas. From [1]

| Country | % of world total |
|---|---|
| USA | 19.2 |
| Russia | 19.0 |
| Canada | 5.1 |
| Iran | 4.6 |
| Norway | 3.4 |
| China | 2.9 |
| Qatar | 2.9 |
| Algeria | 2.6 |
| Netherlands | 2.5 |
| Indonesia | 2.5 |
| Rest of world | 35.3 |

natural gas grew steadily in the 30 years after World War II, though at a slower rate than oil production. Production was dominated by the United States, which produced over 50% of the world's annual production until 1973. Since the early 1970s world gas production has risen at a similar rate to oil production (Fig. 2.1). While the market for natural gas was once limited to the reach of pipelines, natural gas can now be transported globally using liquefied natural gas (LNG) tankers. This requires removal of impurities and cooling of the gas to approximately −160°C. LNG now accounts for 6% of the global market. Today ten countries produce 65% of the world's natural gas (Table 2.2).

Together oil and natural gas currently comprise 54.3% of the world's total annual consumption of energy and, as such, are a major foundation of the global economy. In particular, the transportation of food, raw materials, and goods to market, as well as public and private transport systems, are all heavily dependent on oil and natural gas: they fuel over 96% of the transport sector. On the negative side they are a major source of $CO_2$ emissions, though cleaner than coal per unit of power generated. The successful development of cleaner and economically viable alternative fuels is essential in the long run but the world's economy today is heavily dependent on the supply of oil and natural gas. This raises the question as to how much oil and natural gas remain in the ground. How long will oil and gas resources be available for mankind?

## Introduction

Predictions of future shortages of oil began shortly after commercial production started in the late nineteenth century. In the first half of the twentieth century, there was national concern about imminent and irreversible shortages of oil on at least six occasions [3]. In the 1950s M. King Hubbert, a geophysicist at Shell Development Company in Houston, developed a model of a cycle of production of finite nonrenewable resources that aimed to predict future production from analysis of production-to-date and estimates of the amount of remaining resource in the ground [4, 5]. In a series of publications Hubbert predicted that US oil production would peak by 1975 at the latest: actual peak production was in 1970. The peak in US production gave credence to Hubbert's methodology, which is now commonly referred to as the "Hubbert curve" [6].

Fears arose of global depletion of fossil energy resources. In 1972 a report for the Club of Rome, a global think tank, examined known reserves of oil and gas and predicted the time of total depletion under various scenarios [7]. Assuming exponential growth in consumption, the report predicted depletion of known reserves of oil by 1992 and natural gas by 1994. Even under what was considered an optimistic scenario wherein reserves could be increased fivefold by new exploration, oil was predicted to be totally depleted by 2022 and gas by 2021. These predictions, though influential at the time, proved unfounded largely because consumption did not increase at an exponential rate.

In 1962 Hubbert predicted that world oil production would peak in 2000 and, again, in 1969 predicted the peak would be between 1990 and 2000 [5]. As the turning point of the millennium approached predictions of "peak oil" increased. Most of the predictions suggested a peak in the first years of the twenty-first century. For example, in 1988 Campbell, using the Hubbert curve analysis, predicted a peak between 2000 and 2005 [8]. Actual production between 2006 and 2010 has been 4.7% higher than in the previous 5 years (Fig. 2.1). However, predictions of imminent "peak oil" persist and proponents point out that production has been relatively level for the last 6 years (to 2010). Others point to global political and economic reasons for the recent relatively flat production rate.

The basic problem with the Hubbert curve analysis is that estimates of remaining resources change through time. The amount of oil and gas on the planet is undoubtedly finite. The amount is large but much of it would be prohibitively expensive to produce using present-day technologies. Resource estimates therefore tend to focus on oil and gas accumulations that are anticipated to be economically viable for production, either at present or within the foreseeable future. Advances in technology, however, may make accumulations technically accessible at lower costs over time, thus increasing the size of estimated remaining resources. In addition, advances in geologic concepts and subsurface imaging through geophysical techniques have shown oil and gas resources to be substantially more extensive than was at one time thought. As a result, many predictions using the Hubbert curve analysis have been far less successful than Hubbert's prediction of US peak production. Hubbert [5] in 1962, for example, predicted US natural gas production would peak in 1975 but production in 2010 was at an all time high. Another example is Campbell's 1988 prediction [8] that underestimated US oil production in 2010 by 70%.

The following sections describe the various types of geologic occurrence of oil and gas and how scientific and technological advances have changed the perception of the size and economic viability of the resources over the last few decades. Oil and gas resources are often divided into conventional or unconventional resources.

The definition of what is conventional varies between authors but generally refers to those resources that can be produced using long-established methods of well-drilling where the oil and gas can be brought to the surface under its own pressure or by pumping without significant stimulation such as in situ fracturing of the host rock. Unconventional resources, by contrast, require more complex methods of extraction of the resource from the host rock. They include tight gas, oil sands, coal bed methane (CBM), shale gas, shale oil, and gas hydrates.

## Oil and Gas in Conventional Fields

Oil and gas is generated from organic-rich rocks by thermogenic or biogenic processes. When such "source rocks" are exposed over periods of geologic time to high temperatures the organic material breaks down releasing oil and gas that then migrate toward the surface due to buoyancy. The temperature required for thermal maturation of a source rock varies depending on the type of organic material but the minimum temperature for oil generation is approximately 50°C and for gas generation is 100°C. Most thermogenic oil and gas are generated at depths of 2–6 km. Methane may also be produced from source rocks by the biogenic breakdown of organic material in source rocks. Such biogenic gas is produced at lower temperatures and at shallower depths. Whether of thermogenic or biogenic origin, oil and gas can be trapped by buoyancy in "reservoirs" in porous rocks beneath an impermeable layer of rock typically at depths of over 1 km. The permeable nature of the host rock, high pressures related to depth of burial, and the concentration of oil and gas in discrete reservoirs allow relatively easy extraction of oil and gas by drilling wells. Until the 1990s almost all the world's oil and gas production was produced from such "conventional" fields. Natural gas in conventional fields may occur with ("associated gas") or without ("nonassociated gas") oil. In either case the gas may contain compounds that can be separated at the surface as liquids. Natural gas liquids (NGLs) such as propane, butane, and pentane are sometimes included in reporting oil production but are not insignificant: they comprise, for example, 3.5% of the total energy production of the United States.

Estimates of the world's original endowment of conventional oil and gas resources have tended to increase because of the advances in technology over time [9]. Advances in geologic concepts, drilling technologies, seismic imaging, and computer modeling using large datasets have progressively revolutionized the search for oil and gas over the last century. Likewise, advances in production technologies, pipeline construction, and tankers have all made resources available that at one time would have been considered economically unviable. Advances in science and technology allow more oil and gas to be found in old fields, within existing petroleum provinces, and in new frontier regions.

Existing oil and gas fields that have already been in production for years may, at first blush, seem unusual sites to look for further reserves. However, the

US Geological Survey (USGS) in 2000 estimated that 48% of the oil and 41% of the natural gas that remains to be added to reserves in the future lie within existing fields [10]. Increases in successive estimates of the estimated recoverable oil and gas are known as "reserve growth." On average, only 22% of the oil is currently recovered from fields worldwide [11]. There are many reasons for such a low percentage. Much of the oil within a reservoir rock will not easily move toward wells. Enhanced recovery methods such as gas reinjection, water-flooding, and flushing with polymers and surfactants can free up more of the oil. While such secondary and tertiary recovery methods are commonly used in mature oil fields in developed countries, they are not yet widely used in many other regions. Another reason for reserve growth is that there is rarely a single reservoir within an oil or gas field as most are split into numerous compartments, each containing amounts of oil/gas in varying amounts. Advances in seismic imaging over the last 20 years have allowed not only good visualization of these compartments but also the fluids within the compartments. At the same time, technological advances allow drilling with much greater precision permitting access to much smaller subsurface targets over time. While it will never be possible to economically drain all the oil and gas out of a field, recovery rates of over 70% may eventually be feasible in many areas [12].

Much oil and natural gas also remains to be found in undiscovered fields within existing petroleum provinces. The largest fields tend to be discovered early in the exploration history of a basin and progressively smaller ones are found over time. The infrastructure that is built to develop the larger fields allows fields to be developed that would otherwise be too small to justify the construction of pipelines, platforms, ports, and processing facilities. By analogy estimates can be made of the number and size of remaining undiscovered fields from the discovery history of a petroleum province. This methodology may underestimate the total resource if more than one type of oil and gas accumulation is present. For example, the discovery of giant oil and gas fields in the subsalt regions in offshore regions of the Gulf of Mexico added a new "play" in what was thought to be a mature oil province.

There is a strong possibility that further major accumulations of conventional oil and gas will also be found in frontier basins where little or no exploration has taken place. Most of the potential areas are in regions of deepwater or harsh climates. Over time technological advances have made exploration feasible in areas once thought inaccessible. This has been particularly true in the ability to drill to deeper depths. In 1960 any drilling in water depths over 20 m was regarded as "deepwater" but by 1980 oil fields had been found in over 300 m water depth, by 1990 in over 850 m water depth, and by 2010 in over 2,400 m water depth [13]. Total drilling depths also increased dramatically with a record set in 2009 with the discovery of the giant Tiber oil field in the Gulf of Mexico which is at 1,200 m water depth, with a total drilling depth of 10,680 m. In the last decade giant oil fields have been found in ultra-deepwater areas not only in the Gulf of Mexico but also offshore West Africa and Brazil. The giant Brazilian Tupi field lies not only in deepwater (>2,000 m) but also below a 2,000 m thick layer of salt. It is only recently that seismic technology has allowed an understanding of the structures below thick salts and drilling technologies have permitted drilling through such strata.

As in the past, technological advances will permit exploration for oil and gas in frontier regions currently regarded as inaccessible. Future exploration in deepwater will likely lead to more significant discoveries. Outside the Atlantic realm other regions with potential for deepwater exploration include offshore East Africa, the Great Australian Bight, and offshore New Zealand. There is a growing interest in the potential of the Arctic region where geologists believe there are substantial oil and gas resources but where the severe climate, ice floes, and icebergs all make exploration and development hazardous and in many areas either prohibitively expensive or impossible using currently available technologies [14].

## Tight Gas

In conventional fields gas moves easily through the permeable host rock and naturally moves toward wells and up to the surface because of pressure differences. By contrast, sandstones that have very low permeability require considerably more effort to produce the gas. These sandstones are generally thinner than those in conventional fields and have been buried to great depth. The pores in the sandstone, in which the gas is trapped, have been reduced in size by compaction and cementation during burial. These "tight sands" typically occur near the center of sedimentary basins and are sometimes referred to as "basin-centered gas" accumulations. Production of gas from tight sands requires extensive and complex drilling. Because each well produces relatively small volumes of natural gas, many wells must be drilled. Hydraulic fracturing of the host rock can increase the rate of flow of gas to a well.

Although basin-centered gas accumulations can occur over very wide geographic areas, production is generally from limited regions of a basin. The geologic nature of these so-called "sweet spots" is debated as to whether they are regions with enhanced permeability created by natural fractures in the rock or are buoyancy traps similar to conventional gas fields [15]. Resource estimates of basin-centered gas are typically very large but in some cases may be reevaluated as geologists develop a better understanding of the nature of sweet spots. Nevertheless, tight gas is undoubtedly a major resource and, historically, it has been the most important component of unconventional gas production in North America but relatively few tight gas fields have yet been developed outside of North America. Production has risen from 10% of total natural gas production in the United States in 1990 to 28% in 2009 [2, 16]. Given the magnitude and widespread nature of these accumulations in North America it seems likely that tight gas resources occur in many sedimentary basins worldwide. Exploration and development of such resources must compete with conventional gas resources that are much cheaper to produce, and which in many cases may lie at shallower depths within the same sedimentary basin.

However, it is probably only a matter of time before tight gas resources are developed in many regions of the world.

## Oil Sands

Oil sands are rocks that consist predominantly of sandstones which contain bitumen within the pore spaces that has been produced by the biodegradation of oils in the subsurface. Most oil reservoirs are sufficiently hot that biogenic activity is curtailed or absent. However, oil that migrates into shallow reservoir rocks may be altered completely to bitumen that is very viscous. Although found close to the surface, production of oil from oil sands is expensive. The bitumen must be heated before it will flow and commercial extraction requires large amounts of energy. Once extracted the bitumen must also be upgraded by purification and hydrogenation before it can be refined like conventional crude oil.

Although oil sands have long been recognized and used in limited ways, it is only within the last 40 years that commercial production has grown. Canadian oil sands production began in 1967 and production has grown steadily. By 2009 production from oil sands was equivalent to over 550 million barrels of oil, accounting for 49% of Canada's oil production in 2009 [17]. Though production was initially subsidized, production costs have fallen with technological advances and is now economically viable at $50/barrel and production is forecast to more than triple by 2025 [18]. Venezuela also has substantial oil sand deposits [19] and smaller accumulations are known in Russia and the Middle East.

## Coal Bed Methane

The presence of methane in coal beds had long been a hazard for underground coal mining and it is only recently that this gas has been seen as an economic resource. Modern mining techniques aimed at mitigating the potential for lethal build-up of methane in mines extracts the gas during or before mining as coal mine methane (CMM), which in some cases is used for power generation. More significant from an energy resource perspective is the natural gas extracted from coal beds by wells in regions where no mining is planned: this is known as coal bed methane (CBM) or coal seam gas (CSG).

Commercial production of CBM began in the United States in 1989 and annual production had grown to almost 2 trillion cubic feet (TCF) in 2009, which was 8% of total US gas production [2]. Early production was predominantly from bituminous coals in the San Juan and Raton Basins of New Mexico and Colorado. Production from these areas still accounts for almost 50% of US production. In 2000 production began from the lower-rank subbituminous and lignite coals of the Powder River Basin in Wyoming which by 2010 comprised 28% of US CBM production. Over the last decade, the technology and knowhow for CBM extraction has spread from the United States to other major coal-bearing regions, particularly Canada, Australia, China, and parts of Europe. Australia plans to export CBM gas as LNG [20].

CBM occurs at much shallower depths than most conventional gas fields: generally less than 900 m. Extraction of CBM, however, requires many closely spaced wells: over 20,000 wells had been drilled in the Powder River Basin [21] and over 17,000 in Alberta [22]. The gas is adsorbed onto the pore space in coals and must be released by depressurization, allowing the gas to flow toward the well along natural fractures known as "cleats" or by pathways produced by artificial hydraulic fracturing.

## Oil Shale and Shale Oil

Oil can be produced from some organic-rich fine-grained rocks that are normally referred to as "shales" even if the host rock is not strictly a shale by a geologic definition. Production from such sources has a long history. Production of oil from shales for illumination preceded the discovery of conventional petroleum resources in the mid-nineteenth century but for most of the twentieth century production of such oil was a very minor component of global petroleum production. However, in the first decade of the twenty-first century there has been a resurgence of interest in these resources because of advances in technology and rising energy prices.

"Oil shales" are rocks that contain significant amounts of solid organic chemical compounds (kerogen) that have not been buried deeply enough to allow for oil maturation. Production is generally done by mining the rock and heating it in a retort in a processing plant close to the mine where the oil and associated gases can be captured. The oil may also be extracted using in situ methods which require heating the subsurface rock by injection of hot fluids, gases, or steam, or by the use of heating elements. As the oil is expelled from the kerogen it can then be induced to flow toward conventional oil wells for extraction.

The leading producer of oil shale in the world is Estonia, where 90% of the power is generated from that source. By far the largest accumulations of oil shale, however, are in the United States, particularly in the Green River Formation of Colorado, Utah, and Wyoming, that were deposited within ancient saline lake systems some 40–50 million years ago. There are also major accumulations in Devonian–Mississippian black shales in the eastern United States that were deposited in marine environments over 350 million years ago. With the high oil prices of the late 1970s a number of pilot projects produced oil from the Green River Formation in Colorado but plans for major commercial exploitation were abandoned when prices fell in the early 1980s. Interest in the potential for production has been rekindled with the high oil prices in recent years. Other countries with significant oil shale accumulations include Australia, Brazil, the Democratic Republic of the Congo, and Russia [23].

"Shale oil" is oil that is trapped within a fine-grained rock. Extraction of the oil does not require heating but the low permeability of the rocks requires that the rock be artificially fractured in situ to allow flow toward a well. Once regarded only of scientific interest, recent advances in hydraulic fracturing have made shale oil economically viable in some areas. The best known shale oil accumulation is the Bakken Formation of the Williston Basin of Montana and North Dakota and adjoining parts of Canada [24]. From a geologic perspective, the Bakken is a petroleum source rock that reached maturation but, unlike most mature source rocks, the oil was never expelled to migrate to conventional traps.

## Shale Gas

Natural gas is present in some organic-rich mudrocks. Shale gas has been produced in small quantities for well over a century but recent advances in technology allow much higher rates of production. Shale beds are relatively thin and vertical wells provide access to a very limited volume of rock. It is now possible to drill at any angle and to deviate wells at depth to follow individual shale beds over great distances, allowing access to large volumes of gas-bearing shale. While production used to be primarily from naturally fractured shales, much recent production from shales is aided by hydraulic fracturing technologies. The fractures provide avenues for gas migration toward the wells.

The new technologies for shale gas production have revolutionized US gas production. Shale gas production was approximately 1% of total US production in 2000 but had risen to 14% of total production in 2009 and is forecast to be 45% of production by 2035 [2]. The potential of shale gas has been questioned by some [25]. Like tight gas, production is from sweet spots whose nature and lateral extent remains to be better defined. The dramatic rise in US production has, however, spurred interest in shale gas in many other parts of the world, especially in Canada, Central Europe, China, India, and Australia.

## Gas Hydrates

Gas hydrates (also called clathrates) are solid ice-like substances consisting of rigid cages of water molecules that enclose molecules of gas – mainly methane. Hydrates are stable in a restricted range of temperatures and pressures and occur in two regions: in polar regions, where they are associated with permafrost, and at shallow depths in sediment on the outer continental shelves, in water depths over 300 m [26]. Hydrates can pose an environmental problem because natural breakdown of accumulations can result in slumping of the seabed and outgassing of methane to the atmosphere. On the other hand, if appropriate technologies are developed, hydrates may be a future source of natural gas. Current technologies allow production only at costs substantially above the market prices that have prevailed over the last decade [27]. Production would require depressurization, thermal stimulation, or injection of inhibitors to destabilize the hydrate lattices, releasing the methane to flow toward wells.

## Estimated Volumes of Remaining Oil and Gas Resources

Proved reserves are the volume of known oil and gas accumulations that can be produced at a profit under existing economic and operating conditions. Estimates of world-proved reserves of oil and gas are compiled using best available information by several organizations including the International Energy Agency, IHS Energy, and BP. Proved reserves of oil and gas are not routinely reported in many countries. In fact, some of the largest producers consider reserves as state secrets. This lack of transparent reporting has led some to question the size of global reserves giving credence to predictions of imminent production declines [28]. However, though differing in detail the reserves estimated by different agencies are similar in estimated overall size. As of the end of 2009, BP [29] estimated globally proved reserves of 1,333 billion barrels of oil (BBO), equivalent to over 45 years of production at 2009 rates. With the exception of 237 billion barrels of oil in the Canadian oil sands, these reserves are entirely of conventional oil. BP also estimated globally proved reserves of natural gas at 6,621 trillion cubic feet (TCF), equivalent to 63 years of production. These reserves are predominantly conventional gas.

In addition to proved reserves, there are undiscovered resources and "contingent resources" that are currently noncommercial but could probably be produced under different economic conditions. The most recent comprehensive global estimate of undiscovered oil and gas resources was published by the USGS in 2000 [10]. This was a geologically based study of 128 geologic provinces that included the producing basins that accounted for more than 95% of the world's known oil and gas outside of the United States. The study also examined the discovery history of each province up to 1996. The report included probabilistic estimates of the volumes of conventional oil, gas, and natural gas liquids that might be added to proved reserves from new field discoveries in the studied provinces from the 1996 baseline. The assessment estimated that there was a 95% chance of discovering another 334 BBO of conventional oil and a 5% chance of discovering 1,107 BBO. The median estimate, the 50% chance, was that 607 BBO of oil might yet be found in the studied provinces. For conventional natural gas, the assessment estimated that there was a 95% chance of discovering another 2,299 TCF and a 5% chance of

**Table 2.3** Estimate of remaining oil resources including NGL (billion barrels). Sources and rationale for estimates are given in text. For comparison, annual production of oil was approximately 29.2 billion barrels in 2009

| Conventional | |
|---|---|
| Proved reserves | 1,236 |
| Reserve growth | 600 |
| Undiscovered fields | 600 |
| Total conventional | 2,436 |
| **Unconventional** | |
| Oil sands proved reserves | 237 |
| Oil sands resources (incl. reserves) | 820 |
| Oil shale resources | 2,826 |
| Total unconventional | 3,141 |

discovering 8,174 TCF. The median estimate was that 4,333 of natural gas might be found in the studied provinces.

More recent estimates by the USGS that have concentrated on specific regions of the world suggest that on balance their 2000 estimates are probably reasonable estimates of the global picture though there is need for revision in some areas. A 2010 study [30] estimates 17% less undiscovered oil and 20% less undiscovered natural gas than in the 2000 assessment in eight basins in Southeast Asia; however, there had been at least 10 years of discoveries between the two assessments. Furthermore, the 2010 assessment examined a number of additional basins in Southeast Asia resulting in an overall increase in estimated undiscovered oil of 71% and natural gas by 66%. The 2008 USGS assessment of the Arctic revised down the estimate of undiscovered oil by 48%, largely because of a reappraisal of the West Siberian Basin, but increased the estimate of natural gas by 21% by including a number of basins not in the 2000 assessment [14].

The USGS 2000 assessment [10] also estimated the amount of reserve growth that could be anticipated to 2025 for fields that had been discovered prior to 1996. The median estimate for reserve growth for conventional oil was 612 BBO and for conventional natural gas 3,305 trillion cubic feet. At the time of writing, the midpoint of the time interval has been reached over which reserve growth was predicted and a substantial amount of the predicted reserve growth has already occurred [31]. It seems unreasonable, however, to assume that reserve growth will cease in 2025 and interestingly King [32], using a different methodology and separate data from that used by the USGS, estimated future reserve growth of oil at between 200 to 1,000 BBO – almost exactly the same range that the USGS predicted would occur from 1996.

Table 2.3 provides an estimate of remaining oil resources at the end of 2007. The proved reserve numbers are from BP [29] and are slightly lower than those of EIA [2]. The estimated volume of oil in undiscovered fields is revised down from the median estimate in the USGS 2000 assessment because 16.5% of that volume had been discovered by 2007 [31]. However, it should be noted that this may be conservative as the 2000 assessment did not include the United States and many geologic basins that have relatively small volumes of discovered oil or where there has been little or no exploration. The reserve growth numbers are based on King [31] for crude oil, as discussed above, and 50% of the USGS 2000 estimated median reserve growth for NGL that was predicted to occur between 1996 and 2025. The oil sand reserve and resource estimates are for Canada [33] and Venezuela [19] only though there are smaller oil sand accumulations in several other countries. The shale oil estimate is from Dyni [23]. The estimated remaining conventional oil is equivalent to 79 years of annual production at 2009 rates. The estimated unconventional oil adds up to the equivalent of 125 years of annual production.

Global reserves of natural gas in 2010 were estimated at a little over 6,600 TCF by BP [29] and EIA [2]. Taking into account discoveries since 1996 [31] and revising down the median estimate of gas in undiscovered fields from the USGS 2000 assessment, and also assuming 50% of the estimated reserve growth has taken place, the remaining conventional natural gas resources are estimated at approximately 11,850 TCF, equivalent to about 87 years of annual production at 2009 rates. Estimating global abundance of unconventional gas resources is difficult because development of these resources is largely in its infancy, especially outside of North America. Recently published estimates of recoverable shale gas resources in the United States and Canada vary between 50 and 1,000 TCF [34]. Estimates for the rest of the world would likely have an even larger range. It is particularly important to discriminate between "in-place resources" and those that are economically recoverable. The USGS [35] estimated that the United States has 700 TCF of CBM gas in place but only 100 TCF that would be economically recoverable. As 20.5 TCF was produced over the following 13 years under a major drilling effort, 100 TCF may be an optimistic estimate of ultimate recovery. The USGS [35] also suggested that global in-place CBM gas resources may be as high as 7,500 TCF but recoverable resources may well be an order of magnitude lower. Australia, for example, has 9% of the world's coal resources [2] and an estimated 153 TCF of recoverable CBM gas, of which 90% is currently sub-economic or not yet proven by drilling [20]. While extraordinarily large estimates of unconventional gas resources (particularly for hydrates) should be regarded with some scepticism from an economic perspective, there is ample evidence that unconventional gas resources are abundant. In the United States, which has a wide variety of geologic basins, 50% of the gas produced in 2010 was unconventional and it is not unreasonable to suppose that globally the abundance of unconventional gas resources is of a similar scale to conventional gas resources.

## Future Directions

A strong case could be made that the estimates of remaining oil and gas presented above are on the low side because, as in the past, unanticipated technological advances may make additional oil and gas accumulations economically viable.

Forty years ago large-scale economic extraction of oil from oil sands, gas from gas shales, gas from coal beds, and gas from tight sandstones all seemed improbable, but all are now a significant part of world production. However, even if the estimates are somewhat optimistic, it is clear that there is a large volume of remaining recoverable oil and gas resources. How much of that resource will eventually be extracted remains to be seen.

There are clearly environmental costs to oil and gas extraction. Pursuing hard-to-get resources raises the risk of disasters such as the BP Deepwater Horizon spill in 2010 [13] that can profoundly influence public opinion. Ecosystems are impacted by large-scale mining of oil sands and oil shales and some in situ extraction of unconventional oil and gas has the potential to influence groundwater supplies. Large volumes of water can be produced during CBM development. The biggest concern, however, is in the consumption of oil and gas with associated rising levels of greenhouse gases in the atmosphere and predictions of global climate change. On the other hand, substitution of natural gas for coal in power plants, more efficient vehicles, and carbon capture and storage could also make substantial reductions in global greenhouse gas emissions.

How much of the remaining oil and gas resources will be produced will depend on competition over time from other energy sources, especially renewables and nuclear. It is sometimes assumed that the depletion of global oil resources will inevitably lead to high oil prices that will allow more expensive energy sources to become competitive. However, historically substitution of one energy source by another has taken place primarily by reduction in costs of the substituting energy source [9]. Rather than a result of long-term depletion of the resource, the high oil prices of the first decade of the twenty-first century arguably reflect more restrictions of supply related to conflicts and political problems coinciding with a rapid rise in demand, especially from developing countries. Oil and natural gas may lose global market share in the future either because technological advances permit cheaper production of alternate energy sources or because of political influence on the market, including restrictions on supplies of oil and natural gas or incentives against their consumption, such as carbon taxes. Nevertheless, given the rising global demand for energy, it seems likely that oil and gas will be a major part of the energy mix for most, if not all, of the twenty-first century.

## Bibliography

### *Primary Literature*

1. International Energy Agency (2010) Key world energy statistics. OECD/IEA, Paris
2. Energy Information Agency (2011) http://www.eia.doe.gov/. Accessed April 2011
3. Fanning LM (1950) A case history of oil-shortage scares. In: Fanning LM (ed) Our oil resources. McGraw-Hill, New York, pp 306–406

4. Hubbert MK (1956) Nuclear energy and the fossil fuels, Shell Development Company Publication 95. Shell Development Company, Houston, 40 p
5. Hubbert MK (1962) Energy resources: a report to the Committee on Natural Resources, National Academy of Sciences-National Research Council Publication 1000-D. National Academy of Sciences-National Research Council, Washington, 141 p
6. Deffeyes KS (2001) Hubbert's peak: the impending world oil shortage. Princeton University Press, Princeton, 208 p
7. Meadows DH, Meadows DL, Randers J, Behrens WW III (1972) The limits to growth. Universe Books, New York, 205 p
8. Campbell CJ (1988) The coming oil crisis. MultiScience, Brentwood, 210 p
9. McCabe PJ (1998) Energy resources – cornucopia or empty barrel? AAPG Bull 82:2110–2134
10. United States Geological Survey (2000) World petroleum assessment 2000 – description and results, U.S. Geological Survey Digital Data Series DDS-60. U.S. Geological Survey, Denver. http://pubs.usgs.gov/dds/dds-060/
11. Sandrea I, Sandrea R (2007) Global oil reserves-1: recovery factors leave vast target for EOR technologies. Oil Gas J 105(Nov 5):44–47
12. Sandrea I, Sandrea R (2007) Global oil reserves-2: recovery factors leave EOR plenty of room for growth. Oil Gas J 105(Nov 12):39–42
13. National Commission on the BP Deepwater Horizon Oil Spill and Offshore Drilling (2011) Deep water: the Gulf oil disaster and the future of offshore drilling. 380 p, https://s3.amazonaws.com/pdf_final/DEEPWATER_ReporttothePresident_FINAL.pdf
14. Gautier DL, Bird KJ, Charpentier RC, Grantz A, Houseknecht DW, Klett TR, Moore TE, Pitman JK, Schenk CJ, Schuenemeyer JH, Sørensen K, Tennyson ME, Valin ZC, Wandrey CJ (2009) Assessment of undiscovered oil and gas in the Arctic. Science 324:1175–1179
15. Shanley KW, Cluff RM, Robinson JW (2004) Factors controlling prolific gas production from low-permeability sandstone reservoirs: implications for resource assessment, prospect development, and risk analysis. AAPG Bull 88:1083–1121
16. Nehring R (2008) Growing and indispensable: the contribution of production from tight-gas sands to U.S. gas production. In: Cumella SP, Shanley KW, Camp WK (eds) Understanding, exploring, and developing tight-gas sands – 2005 Vail Hedberg conference, AAPG Hedberg Series, No. 3. American Association of Petroleum Geologists, Tulsa, pp 5–12
17. Canadian Association of Petroleum Producers (2011) Statistical handbook for Canada's upstream petroleum industry. Canadian Association of Petroleum Producers, 2011–9999, Calgary, 211 p
18. The Economist (2011) Muck and brass. Economist 398(8717):77–80
19. United States Geological Survey (2009) An estimate of recoverable heavy oil resources of the Orinoco oil belt, Venezuela, U.S. Geological Survey Fact Sheet 2009–3028. U.S. Geological Survey, Reston, 4 p
20. Australian Government (2010) Australian energy resource assessment. Department of Resources, Energy and Tourism, Canberra, 358 p
21. Swindell GS (2007) Powder River Basin coalbed methane wells – reserves and rates. Society of Petroleum Engineers Report SPE 107308
22. Energy Resources Conservation Board (2010) ST109: Alberta coalbed methane well locations. Energy Resources Conservation Board, Calgary, 426 p
23. Dyni JR (2006) Geology and resources of some world oil-shale deposits, U.S. Geological Survey Scientific Investigations Report 2005–5294. U.S. Geological Survey, Reston, 42 p
24. U.S. Geological Survey (2008) Assessment of undiscovered oil resources in the Devonian-Mississippian Bakken Formation, Williston Basin province, Montana and North Dakota, U.S Geological Survey Fact Sheet 2008–3021. U.S. Geological Survey, Reston
25. Klump E, Polson J (2009) Shale-gas skeptic's supply doubts draw wrath of Devon (update 2). http://www.bloomberg.com/apps/news?pid=newsarchive&sid=asEUlpJcuZB4. Accessed April 2011

26. Kvenvolden KA (1993) A primer on gas hydrates. In: Howell DG et al (eds) The future of energy gases, US Geological Survey Professional Paper 1570. U.S. G.P.O, Washington, DC, pp 279–291
27. Walsh MR, Hancock SH, Wilson SJ, Patil SL, Moridis GJ, Boswell R, Collett TS, Koh CA, Sloan ED (2009) Preliminary report on the commercial viability of gas production from natural gas hydrates. Energy Econ 31:815–823
28. Simmons MW (2005) Twilight in the desert. Wiley, Hoboken, 422 p
29. BP (2010) BP Statistical review of world energy, June 2010. http://bp.com/statisticalreview. Accessed April 2011
30. United States Geological Survey (2010) Assessment of undiscovered oil and gas resources of Southeast Asia, USGS Fact Sheet 2010–3015. U.S. Geological Survey, Reston
31. Gautier DL, McCabe PJ, Ogden J, Demayo TN (2010) Resources, reserves, and consumption of energy. In: Graedel TE, van der Voet E (eds) Linkages of sustainability, Strüngmann Forum Report. MIT, Cambridge, MA, pp 323–340
32. King KC (2007) Growth; are we underestimating recent discoveries?: abstracts, annual meeting. AAPG 2007:77
33. Government of Canada (2010) Oil sands – a strategic resource for Canada, North America and the world. Natural Resources Canada, Ottawa, http://www.nrcan-rncan.gc.ca/eneene/pdf/os-sb-eng.pdf
34. Mohr SH, Evans GM (2010) Shale gas changes N. American gas production projections. Oil Gas J 108(July 26):60–64
35. U.S. Geological Survey (1997) Coalbed methane – an untapped energy resource and an environmental concern, USGS Fact Sheet FS-019-97. U.S. Geological Survey, Denver, http://energy.usgs.gov/factsheets/Coalbed/coalmeth.html

# Chapter 3
# Petroleum and Oil Sands Exploration and Production

**James G. Speight**

## Glossary

| | |
|---|---|
| Bitumen | A semisolid to solid hydrocarbonaceous material found filling pores and crevices of sandstone, limestone, or argillaceous sediments such as tar sand. |
| Exploration | The search for petroleum using a variety of physical and spectrographic methods. |
| Hot-water process | The recovery of bitumen from tar sand by use of hot water whereby the bitumen floats and the sand sinks. |
| In situ conversion | Partial or complete conversion of heavy oil or tar sand bitumen in the reservoir or deposit as part of the recovery process. |
| Oil mining | The recovery of petroleum using a mining method whereby an underground chamber is produced by mining and the oil is allowed or encouraged to drain into the chamber. |
| Recovery | Recovery of petroleum at the surface using primary, secondary, and tertiary recovery methods. |
| Tar sand mining | Recovery of tar sand by mining (digging) tar sand from the formations at or close to the surface. |

This chapter was originally published as part of the Encyclopedia of Sustainability Science and Technology edited by Robert A. Meyers. DOI:10.1007/978-1-4419-0851-3

J.G. Speight (✉)
CD&W Inc., 2476 Overland Road, P.O. Box 1722 Laramie, WY 82073–1722, USA
e-mail: JamesSp8@aol.com

R. Malhotra (ed.), *Fossil Energy: Selected Entries from the Encyclopedia of Sustainability Science and Technology*, DOI 10.1007/978-1-4614-5722-0_3,

## Definition of the Subject

Exploration for petroleum is an essential part of petroleum technology. Depletion of reserves is continuing at a noticeable rate and other sources of hydrocarbons are required – these include heavy oil (a type of petroleum) and tar sand bitumen.

## Introduction

Petroleum occurs in the microscopic pores of sedimentary rocks that form a reservoir – typically, reservoir rock consists of sand, sandstone, limestone, or dolomite. However, not all of the pores in a rock will contain petroleum – some will be filled with water or brine that is saturated with minerals.

Both oil and gas have a low specific gravity relative to water and will thus float through the more porous sections of reservoir rock from their source area to the surface unless restrained by a trap. A trap is a reservoir that is overlain and underlain by dense impermeable cap rock or a zone of very low or no porosity that restrains migrating hydrocarbon. Reservoirs vary from being quite small to covering several thousands of acres, and range in thickness from a few inches to hundreds of feet or more.

In general, petroleum is extracted by drilling wells from an appropriate surface configuration into the hydrocarbon-bearing reservoir or reservoirs. Wells are designed to contain and control all fluid flow at all times throughout drilling and producing operations. The number of wells required is dependent on a combination of technical and economic factors used to determine the most likely range of recoverable reserves relative to a range of potential investment alternatives.

There are three phases for recovering oil from reservoirs (Fig. 3.1):

1. *Primary recovery* occurs as wells produce because of natural energy from expansion of gas and water within the producing formation, pushing fluids into the well bore and lifting the fluids to the surface.
2. *Secondary recovery* requires energy to be applied to lift fluids to surface – this may be accomplished by injecting gas down a hole to lift fluids to the surface, installation of a subsurface pump, or injecting gas or water into the formation itself.
3. *Tertiary recovery* occurs when a means is required to increase fluid mobility within the reservoir – this may be accomplished by introducing additional heat into the formation to lower the viscosity (thin the oil) and improve its ability to flow to the well bore. Heat may be introduced by either (1) injecting chemicals with water (*chemical flood*, *surfactant flood*), (2) injecting steam (*steam flood*), or (3) injecting oxygen to enable the ignition and combustion of oil within the reservoir (*fire flood*).

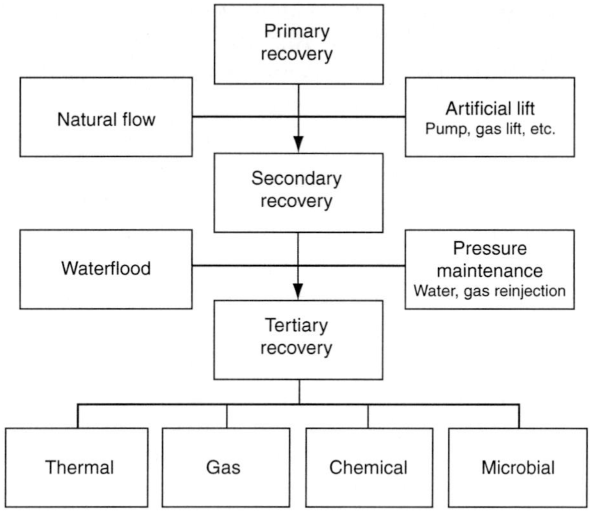


**Fig. 3.1** Methods for oil recovery

Production rates from reservoirs depend on a number of factors, such as reservoir geometry (primarily formation thickness and reservoir continuity), reservoir pressure, reservoir depth, rock type and permeability, fluid saturations and properties, extent of fracturing, number of wells and their locations, and the ratio of the permeability of the formation to the viscosity of the oil [1, 2].

The geological variability of reservoirs means that production profiles differ from field to field. Heavy oil reservoirs can be developed to significant levels of production and maintained for a period of time by supplementing natural drive force, while gas reservoirs normally decline more rapidly.

## Petroleum Exploration and Production

### *Exploration*

Exploration for petroleum originated in the latter part of the nineteenth century when geologists began to map land features that were favorable for the collection of oil in a reservoir. Of particular interest to geologists were outcrops that provided evidence of alternating layers of porous and impermeable rock. The porous rock (typically a sandstone, limestone, or dolomite) provides the reservoir for the petroleum while the impermeable rock (typically clay or shale) prevents migration of the petroleum from the reservoir.

By the early part of the twentieth century, most of the areas where surface structural characteristics offered the promise of oil had been investigated and the era of subsurface exploration for oil began in the early 1920s. New geological and geophysical techniques were developed for areas where the strata were not sufficiently exposed to permit surface mapping of the subsurface characteristics. In the 1960s, the development of geophysics provided methods for exploring below the surface of the earth.

The principles used are basically *magnetism* (*magnetometer*), *gravity* (*gravimeter*), and *sound waves* (*seismograph*). These techniques are based on the physical properties of materials that can be utilized for measurements and include those that are responsive to the methods of applied geophysics. Furthermore, the methods can be subdivided into those that focus on *gravitational properties*, *magnetic properties*, *seismic properties*, *electrical properties*, *electromagnetic properties*, *properties*, and *radioactive properties*. These geophysical methods can be subdivided into two principal groups: (1) those methods without depth control and (2) those methods having depth control.

In the first group of the measurements (those without depth control), the methods incorporate effects from both local and distant sources. For example, gravity measurements are affected by the variation in the radius of the earth with latitude. They are also affected by the elevation of the site relative to sea level, the thickness of the earth's crust, and the configuration and density of the underlying rocks, as well as by any abnormal mass variation that might be associated with a mineral deposit.

In the second group of measurements (those with depth control), seismic or electric energy is introduced into the ground and variations in transmissibility with distance are observed and interpreted in terms of geological quantities. Depths to geological horizons having marked differences in transmissibility can be computed on a quantitative basis and the physical nature of these horizons deduced.

However, geophysical exploration techniques cannot be applied indiscriminately. Knowledge of the geological parameters likely to be associated with the mineral or subsurface condition being studied is essential both in choosing the method to be applied and in interpreting the results obtained. Furthermore, not all the techniques described here may be suitable for petroleum exploration.

In petroleum exploration, terms as *geophysical borehole logging* can imply the use of one or more of the geophysical exploration techniques. This procedure involves drilling a well and using instruments to log or make measurements at various levels in the hole by such means as *gravity* (*density*), *electrical resistivity*, or *radioactivity*.

A basic rule of thumb in the upstream (or producing) sector of the oil and gas industry has been (and maybe still is in some circles of exploration technology) that the best place to find new crude oil or natural gas is near formations where it has already been found. The financial risk of doing so is far lower than that associated with drilling a rank wildcat hole in a prospective, but previously unproductive, area. On the other hand, there is a definite tradeoff between rewards for risk. The returns on drilling investment become ever leaner as more wells are drilled in a particular area because the natural distribution of oil and gas field volumes tends to be approximately log-geometric – there are only a few large fields, whereas there are a great many small ones [3].

Drilling does not end when production commences and continues after a field enters production. Extension wells must be drilled to define the boundaries of the crude oil pool. In-field wells are necessary to increase recovery rates, and service wells are used to reopen wells that have become clogged. Additionally, wells are often drilled at the same location but to different depths, to test other geological structures for the presence of crude oil.

Finally, the drilling job is complete when the drill bit penetrates the reservoir and the reservoir is evaluated to see whether the well represents the discovery of a *prospect* or whether it is a dry hole. If the hole is dry, it is plugged and abandoned.

At the stage when the prospect has been identified, reservoir evaluation is usually initiated by examining the cuttings from the well bore for evidence of hydrocarbons while the drill bit passes through a reservoir trap. The evaluation of these cuttings helps pinpoint the possible producing intervals in the well bore. At this time, a wire-line is lowered into the hole and an electric log is run to help define possible producing intervals, presence of hydrocarbons, and detailed information about the different formations throughout the well bore. Further tests (such as pressure tests, formation fluid recovery, and sidewall core analysis) can also be run on individual formations within the well bore.

If hydrocarbons are detected, the prospect becomes a *live prospect* and once the final depth has been reached, the well is completed to allow oil to flow into the casing in a controlled manner. First, a *perforating gun* is lowered into the well to the production depth. The gun has explosive charges to create holes in the casing through which oil can flow. After the casing has been perforated, a small-diameter pipe (*tubing*) is run into the hole as a conduit for oil and gas to flow up the well and a *packer* is run down the outside of the tubing. When the packer is set at the production level, it is expanded to form a seal around the outside of the tubing. Finally, a multivalve structure (the *Christmas tree*; Fig. 3.2) is installed at the top of the tubing and cemented to the top of the casing. The Christmas tree allows them to control the flow of oil from the well.

Finally, the development of an onshore shallow gas reservoir located among other established fields may be expected to incur relatively high cost and be nominally complex. A deep oil or gas reservoir located in more than 1 mile of water depth located miles away from other existing producing fields will push the limits of emerging technology at extreme costs.

Onshore developments may permit the phasing of facility investments as wells are drilled and production established to minimize economic risk. However, offshore projects may require 65% or more of the total planned investments to be made before production start-up, and impose significant economic risk.

As might be expected, the type of exploration technique employed depends upon the nature of the site. In other words, and as for many environmental operations, the recovery techniques applied to a specific site are dictated by the nature of the site and are, in fact, *site specific*. For example, in areas where little is known about the subsurface, preliminary reconnaissance techniques are necessary to identify potential reservoir systems that warrant further investigation. Techniques for reconnaissance that have been employed to make inferences about the subsurface

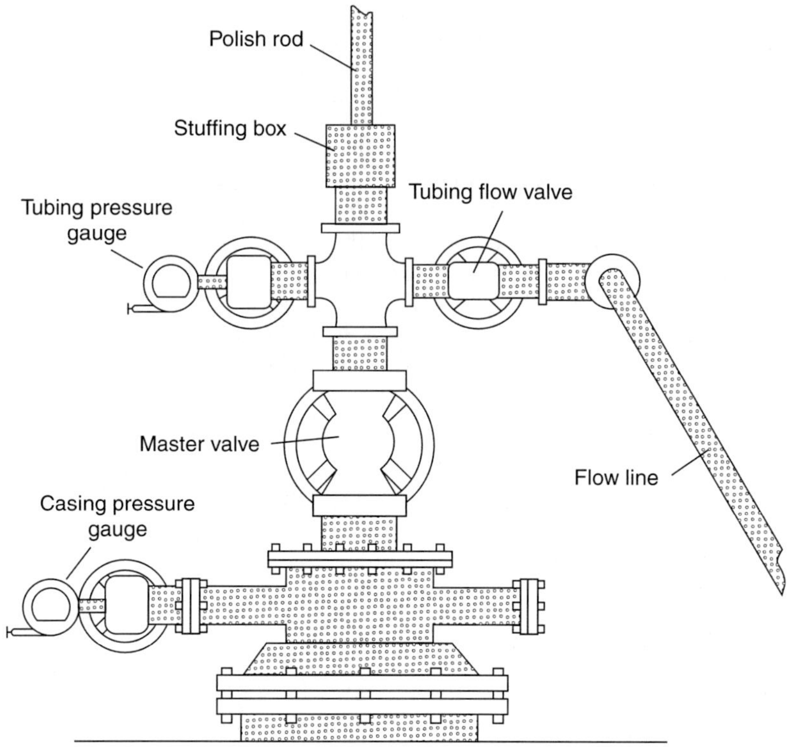


**Fig. 3.2** The Christmas tree

structure include *satellite and high-altitude imagery* and *magnetic and gravity surveys*.

## *Production*

Recovery, as applied in the petroleum industry, is the production of oil from a reservoir. There are several methods by which this can be achieved that range from recovery due to reservoir energy (i.e., the oil flows from the well hole without assistance) to enhanced recovery methods in which considerable energy must be added to the reservoir to produce the oil. However, the effect of the method on the oil and on the reservoir must be considered before application.

Generally, crude oil reservoirs sometimes exist with an overlying *gas cap*, in communication with aquifers, or both. The oil resides together with water and free gas in very small holes (pore spaces) and fractures. The size, shape, and degree of interconnection of the pores vary considerably from place to place in an individual reservoir. Below the oil layer, the sandstone is usually saturated with salt water.

The oil is released from this formation by drilling a well and puncturing the limestone layer on either side of the limestone dome or fold. If the peak of the formation is tapped, only the gas is obtained. If the penetration is made too far from the center, only salt water is obtained.

Therefore, in designing a recovery project, it is a general practice to locate injection and producing wells in a regular geometric pattern so that a symmetrical and interconnected network is formed and production can be maximized. However, the relative location of injectors and producers depends on: (1) reservoir geometry, (2) lithology, (3) reservoir depth, (4) porosity, (5) permeability, (6) continuity of reservoir rock properties, (7) magnitude and distribution of fluid saturations, and last, but certainly not least (8) fluid, i.e., oil, properties. Overall, the goal is to increase the mobility of the oil.

Once production begins, the performance of each well and reservoir is monitored and a variety of engineering techniques are used to progressively refine reserve recovery estimates over the producing life of the field. The total recoverable reserves are not known with complete certainty until the field has produced to depletion or its economic limit and abandonment.

Generally, the first stage in the extraction of crude oil is to drill a well into the underground reservoir. Often many wells (*multilateral wells*) will be drilled into the same reservoir, to ensure that the extraction rate will be economically viable. Also, some wells (*secondary wells*) may be used to pump water, steam, acids, or various gas mixtures into the reservoir to raise or maintain the reservoir pressure, and so maintain an economic extraction rate.

*Directional drilling* is also used to reach formations and targets not directly below the penetration point or drilling from shore to locations under water [4]. A controlled deviation may also be used from a selected depth in an existing hole to attain economy in drilling costs. Various types of tools are used in directional drilling along with instruments to help orient their position and measure the degree and direction of deviation; two such tools are the *whipstock* and the *knuckle joint*. The whipstock is a gradually tapered wedge with a chisel-shaped base that prevents rotation after it has been forced into the bottom of an open hole. As the bit moves down, it is deflected by the taper about 5° from the alignment of the existing hole.

If the underground pressure in the oil reservoir is sufficient, the oil will be forced to the surface under this pressure (*primary recovery*). Natural gas (*associated natural gas*) is often present, which also supplies needed underground pressure (*primary recovery*). In this situation, it is sufficient to place an arrangement of valves (the *Christmas tree* ; Fig. 3.2) on the well head to connect the well to a pipeline network for storage and processing.

For limestone reservoir rock, acid is pumped down the well and out the perforations. The acid dissolves channels in the limestone that lead oil into the well. For sandstone reservoir rock, a specially blended fluid containing *proppants* (sand, walnut shells, aluminum pellets) is pumped down the well and out the perforations. The pressure from this fluid makes small fractures in the sandstone that allow oil to flow into the well, while the proppants hold these fractures open. Once the oil is flowing, production equipment is set up to extract the oil from the well.

A well is always carefully controlled in its flush stage of production to prevent the potentially dangerous and wasteful *gusher*. This is actually dangerous condition, and is (hopefully) prevented by the blowout preventer and the pressure of the drilling mud. In most wells, acidizing or fracturing the well starts the oil flow.

Whatever the nature of the reservoir rock (sandstone or limestone), over the lifetime of the well the pressure will fall, and at some point, there will be insufficient underground pressure to force the oil to the surface. *Secondary oil recovery* uses various techniques to aid in recovering oil from depleted or low-pressure reservoirs. Sometimes pumps, such as beam pumps (*horsehead pumps*) and electrical submersible pumps, are used to bring the oil to the surface. Other secondary recovery techniques increase the reservoir's pressure by water injection, natural gas reinjection, and gas lift, which injects air, carbon dioxide, or some other gas into the reservoir.

Reservoir heterogeneity, such as fractures and faults, can cause reservoirs to drain inefficiently by conventional methods. Also, highly cemented or shale zones can produce barriers to the flow of fluids in reservoirs and lead to high residual oil saturation. Reservoirs containing crude oils with low API gravity often cannot be produced efficiently without application of *enhanced oil recovery* (EOR) methods because of the high viscosity of the crude oil.

Conventional primary and secondary recovery processes are ultimately expected to produce about one third of the original oil-in-place (OOIP), although recoveries from individual reservoirs can range from less than 5% to as high as 80% v/v of the original oil-in-place. This broad range of recovery efficiency is a result of variations in the properties of the specific rock and fluids involved from reservoir to reservoir as well as the kind and level of energy that drives the oil to producing wells, where it is captured.

Conventional oil production methods may be unsuccessful because the management of the reservoir was poor or because reservoir heterogeneity has prevented the recovery of crude oil in an economical manner. In some cases, the reservoir pressure may have been depleted prematurely by poor reservoir management practices to create reservoirs with low energy and high oil saturation.

Crude oil is also produced from offshore fields, usually from steel drilling platforms set on the ocean floor. In shallow, calm waters, these may be little more than a wellhead and workspace but the larger ocean rigs include the well equipment and processing equipment as well as crew quarters. Such platforms include the *floating tension leg platform* that is secured to the sea floor by giant cables and drill ships. Such platforms can hold a steady position above a sea floor well using constant, computer-controlled adjustments. In Arctic areas, islands may be built from dredged gravel and sand to provide platforms capable of resisting drifting ice fields.

## Primary Recovery Methods

Petroleum recovery usually starts with a formation pressure high enough to force crude oil into the well and sometimes to the surface through the tubing [5]. In this situation, it is sufficient to place the Christmas tree (Fig. 3.2) on the wellhead to connect the well to a pipeline network for storage and processing.

For a newly opened formation and under ideal conditions, the proportions of gas may be so high that the oil is, in fact, a solution of liquid in gas that leaves the reservoir rock so efficiently that a core sample will not show any obvious oil content. A general rough indication of this situation is a high ratio of gas to oil produced. This ratio may be zero for fields in which the rock pressure has been dissipated. The oil must be pumped out to as much as 50,000 $ft^3$ or more of gas per barrel of oil in the so-called condensate reservoirs, in which a very light crude oil (0.80 specific gravity or lighter) exists as vapor at high pressure and elevated temperature.

Crude oil moves out of the reservoir into the well by one or more of three processes. These processes are: *dissolved gas drive*, *gas cap drive*, and *water drive*. Early recognition of the type of drive involved is essential to the efficient development of an oil field.

In *dissolved gas drive* (*solution gas drive*) [2, 4], the propulsive force is the gas in solution in the oil, which tends to come out of solution because of the pressure release at the point of penetration of a well. Dissolved gas drive is the least efficient type of natural drive as it is difficult to control the gas-oil ratio and the bottom-hole pressure drops rapidly.

If gas overlies the oil beneath the top of the trap, it is compressed and can be utilized (*gas cap drive*) to drive the oil into wells situated at the bottom of the oil-bearing zone [2, 4]. By producing oil only from below the gas cap, it is possible to maintain a high gas-oil ratio in the reservoir until almost the very end of the life of the pool. If, however, the oil deposit is not systematically developed so that bypassing of the gas occurs, an undue proportion of oil is left behind.

Usually the gas in a gas cap (*associated natural gas*) contains methane and other hydrocarbons that may be separated out by compressing the gas. A well-known example is *natural gasoline* that was formerly referred to as *casinghead gasoline* or *natural gas gasoline*. However at high pressures, such as those existing in the deeper fields, the density of the gas increases and the density of the oil decreases until they form a single phase in the reservoir. These are the so-called retrograde condensate pools because a decrease (instead of an increase) in pressure brings about condensation of the liquid hydrocarbons. When this reservoir fluid is brought to the surface and the condensate is removed, a large volume of residual gas remains. In many cases, this gas is recycled by compression and injection back into the reservoir, thus maintaining adequate pressure within the gas cap, and condensation in the reservoir is prevented.

The most efficient propulsive force in driving oil into a well is natural *water drive*, in which the pressure of the water forces the lighter recoverable oil out of the reservoir into the producing wells [2, 4]. In anticlinal accumulations, the structurally lowest wells around the flanks of the dome are the first to come into water. Then the oil–water contact plane moves upward until only the wells at the top of the anticline are still producing oil; eventually these also must be abandoned as the water displaces the oil. The force behind the water drive may be hydrostatic pressure, the expansion of the reservoir water, or a combination of both. Water drive is also used in certain submarine fields.

*Gravity drive* is an important factor when oil columns of several thousands of feet exist. Furthermore, the last bit of recoverable oil is produced in many pools by gravity drainage of the reservoir. Another source of energy during the early stages of withdrawal from a reservoir containing undersaturated oil is the expansion of that oil as the pressure reduction brings the oil to the bubble point (the pressure and temperature at which the gas starts to come out of solution).

The recovery efficiency for primary production is generally low when liquid expansion and solution gas evolution are the driving mechanisms. Much higher recoveries are associated with reservoirs with water and gas cap drives and with reservoirs in which gravity effectively promotes drainage of the oil from the rock pores. The overall recovery efficiency is related to how the reservoir is delineated by production wells.

For primary recovery operations, no pumping equipment is required. If the reservoir energy is not sufficient to force the oil to the surface, then the well must be pumped. In either case, nothing is added to the reservoir to increase or maintain the reservoir energy or to sweep the oil toward the well. The rate of production from a flowing well tends to decline as the natural reservoir energy is expended. When a flowing well is no longer producing at an efficient rate, a pump is installed.

Two processes used to improve formation characteristics are *acidizing* and *fracturing*. *Acidizing* involves injecting an acid into a soluble formation, such as a carbonate, where it dissolves rock. This process enlarges the existing voids and increases permeability. *Hydraulic fracturing* (*fracking*) involves injecting a fluid into the formation under significant pressure that makes existing small fractures larger and creates new fractures.

Heavy oil and Tar sands (*oil sands*) have a shorter history of production and generally heavy oil reservoirs and tar sand deposits have only been subject to only one recovery technology. In the case of tar sands, primary and secondary recovery technologies, as defined for conventional oil, are not applicable because tar sand bitumen is not mobile at reservoir conditions [2, 4]. Therefore, tar sands developments generally start with a thermal recovery technology which would be considered a tertiary method or enhanced recovery method for conventional oil. However, as the development of heavy oil and tar sand technology matures, the concept of applying more than one recovery technology in a specific order is likely to also be applied to heavy oil reservoirs and tar sand deposits. In particular, in the Lloydminster area (Alberta, Canada), producers have already been investigating for several years the concept of follow-up recovery technologies once primary production is no longer economic.

## Secondary Recovery

Petroleum production is invariably accompanied by a decline in reservoir pressure and *primary recovery* comes to an end as the reservoir energy is reduced. At this stage, secondary recovery methods are applied to replace produce reservoir fluids and maintain (or increase) reservoir pressure.

*Secondary oil recovery* methods use various techniques to aid in recovering oil from depleted or low-pressure reservoirs. Sometimes pumps on the surface or submerged (electrical submersible pumps, ESPs) are used to bring the oil to the surface. Other secondary recovery techniques increase the reservoir's pressure by water injection and gas injection, which injects air or some other gas into the reservoir. In fact, the first method recommended for improving the recovery of oil was a pressure maintenance project which involved the reinjection of natural gas, and there are indications that gas injection was utilized for this purpose before 1900 [6, 7].

The most common follow-up, or *secondary recovery*, operations usually involve the application of pumping operations or of injection of materials into a well to encourage movement and recovery of the remaining petroleum. The pump, generally known as the *horsehead pump* (*pump jack*, *nodding donkey*, or *sucker rod pump*), provides mechanical lift to the fluids in the reservoir.

The up-and-down movement of the sucker rods forces the oil up the tubing to the surface. A walking beam powered by a nearby engine may supply this vertical movement, or it may be brought about through the use of a pump jack, which is connected to a central power source by means of pull rods. Depending on the size of the pump, it generally produces up to one third of a barrel of an oil–water emulsion at each stroke. The size of the pump is also determined by the depth and weight of the oil to be removed, with deeper extraction requiring more power to move the heavier lengths of polish rod.

There are also *secondary oil recovery* operations that involve the injection of water or gas into the reservoir. When water is used, the process is called a *waterflood*; when gas is used, it is called a *gas flood*. Separate wells are usually used for injection and production. The injected fluids maintain reservoir pressure or re-pressure the reservoir after primary depletion and displace a portion of the remaining crude oil to production wells.

During the withdrawal of fluids from a well, it is usual practice to maintain pressures in the reservoir at or near the original levels by pumping either gas or water into the reservoir as the hydrocarbons are withdrawn. This practice has the advantage of retarding the decline in the production of individual wells and considerably increasing the ultimate yield. It also may bring about the conservation of gas that otherwise would be wasted, and the disposal of brines that otherwise might pollute surface and near-surface potable waters.

In the *waterflooding process*, water is injected into a reservoir to obtain additional oil recovery through movement of reservoir oil to a producing well. Generally, the selection of an appropriate flooding pattern for the reservoir depends on the quantity and location of accessible wells. Frequently, producing wells can be converted to injection wells whereas in other circumstances, it may be necessary or advantageous to drill new injection wells.

The mobility of oil is the effective permeability of the rock to the oil divided by the viscosity of the oil.

**Table 3.1** The ratio of injectors to producers for various well patterns

| Pattern | Ratio of producing wells to injection wells | Drilling pattern required |
|---|---|---|
| Four spot | 2 | Equilateral triangle |
| Five spot | 1 | Square |
| Seven spot | 1/2 | Equilateral triangle |
| Inverted seven spot | 2 | Equilateral triangle |
| Nine spot | 1/3 | Square |
| Inverted nine spot | 3 | Square |
| Direct line drive | 1 | Rectangle |
| Staggered line drive | 1 | Offset lines of wells |

$$\lambda = k/\mu$$

where $\lambda$ is the mobility, mD/cP, $k$ is the effective permeability of reservoir rock to a given fluid, mD, and $\mu$ is the fluid viscosity, cP. Thus, the mobility ratio ($M$) is the mobility of the water divided by the mobility of oil:

$$M = K_{\mathrm{rw}}\mu_{\mathrm{o}}/K_{\mathrm{ro}}\mu_{\mathrm{w}}$$

where $K_{\mathrm{rw}}$ is the relative permeability to water, $K_{\mathrm{ro}}$ is the relative permeability to oil, $\mu_{\mathrm{o}}$ is the viscosity of the oil, and $\mu_{\mathrm{w}}$ is the viscosity of water.

The mobility ratio ($M$) refers that $K_{\mathrm{o}}$ is the mobility of oil ahead of the front (measured at $S_{\mathrm{wc}}$) while $K_{\mathrm{w}}$ is the mobility of water at average water saturation in the water-contacted portion of the reservoir.

The mobility ratio of a waterflood will remain constant before breakthrough, but will increase after water breakthrough corresponding to the increase in water saturation and relative permeability to water in the water-contacted portion of the reservoir. Furthermore, the mobility ratio at water breakthrough is the term that is of significance in describing relative mobility ratio, i.e., $M < 1$ indicates a favorable displacement as oil moves faster than water and $M = 1$ indicates a favorable displacement as both oil and water move at equal speed whereas $M > 1$ indicates an unfavorable displacement as water moves faster than oil.

Generally, the choice of pattern (Table 3.1) for waterflooding must be consistent with the existing wells. The objective is to select the proper pattern that will provide the injection fluid with the maximum possible contact with the crude oil to minimize bypassing by the water.

In a *four-spot pattern*, the distance between all like wells is constant. Any three injection wells form an equilateral triangle with a production well at the center. The four spot may be used when the injectivity is high or the heterogeneity is minimal.

In a *five-spot pattern*, the distance between all like wells is constant. Four injection wells form a square with a production well at the center. If existing wells were drilled on square patterns, five-spot patterns (as well as nine-spot patterns) are most commonly used since they allow easy conversion to a five-spot waterflood.

In the *seven-spot pattern*, the injection wells are located at the corner of a hexagon with a production well at its center. If the reservoir characteristics yield lower than preferred injection rates, either a seven-spot (or a nine-spot) pattern should be considered because there are more injection wells per pattern than producing wells.

In the *nine-spot pattern*, the arrangement is similar to that of the five spot but with an extra injection well drilled at the middle of each side of the square. The pattern essentially contains eight injectors surrounding one producer. If existing wells were drilled on square patterns, nine-spot patterns (as well as five-spot patterns) are most commonly used. If the reservoir characteristics yield lower injection rates than those desired, one should consider using either a nine-spot pattern (or a seven-spot pattern) where there are more injection wells per pattern than producing wells.

In the *inverted seven-spot pattern*, the arrangement is similar to the normal seven-spot pattern except where the position of the producer well was in the normal seven-spot pattern there is now an injector well. Likewise where the injector wells were in the normal seven-spot pattern, there are now producer wells. The inverted seven-spot pattern may be used when the injectivity is high or the heterogeneity is minimal.

In the *inverted nine-spot pattern*, the arrangement of the wells is similar to the normal nine-spot pattern except the position of the producer well in the normal nine-spot pattern is occupied by an injector well. Likewise where the positions of the injector wells were in the normal nine-spot, there are now producer wells. If the reservoir is fairly homogenous and the mobility ratio is unfavorable, the inverted nine-spot pattern may be promising.

In the *direct line-drive pattern*, the lines of injection and production are directly opposite to each other. If the injectivity is low or the heterogeneity is large, direct line drive is a good option. Anisotropic permeability, permeability trends, or oriented fracture systems favor line drive patterns.

In the *staggered line-drive pattern*, the wells are in lines as in the direct line, but the injectors and producers are no longer directly opposed but laterally displaced by a distance by a specified that is dependent upon the distance between wells of the same type and the distance between the lines of injector wells and producer wells. The staggered line-drive pattern is also effective for reservoirs there is anisotropic permeability or where permeability trends or oriented fracture systems.

Reservoir uniformities (and heterogeneity) also dictate the choice of pattern and mobility ratio has an important influence on pattern selection. If the ratio is unfavorable, the injectivity of an injector will exceed the productivity of a producer and water injection will supersede oil production. Hence, to balance the production with the water injection, more producers than injectors are required. On the other hand, if the mobility ratio is favorable, the injectivity is impaired, and the pattern should have more injectors than producers.

## Enhanced Oil Recovery

Traditional primary and secondary recovery methods typically recover less than half (sometimes less than one third) of the oil only one third of the original oil-in-place. It is at some point before secondary recovery ceases to remain feasible that enhanced oil recovery methods must be applied if further oil is to be recovered.

*Enhanced oil recovery* (*tertiary oil recovery*) is the incremental ultimate oil that can be recovered from a petroleum reservoir over oil that can be obtained by primary and secondary recovery methods [2, 4, 8, 9]. Enhanced oil recovery methods offer prospects for ultimately producing 30–60%, or more, of the reservoir's original oil-in-place.

Enhanced oil recovery processes use *thermal*, *chemical*, or *fluid phase behavior* effects to reduce or eliminate the capillary forces that trap oil within pores, to thin the oil or otherwise improve its mobility or to alter the mobility of the displacing fluids. In some cases, the effects of gravity forces, which ordinarily cause vertical segregation of fluids of different densities, can be minimized or even used to advantage. The various processes differ considerably in complexity, the physical mechanisms responsible for oil recovery, and the amount of experience that has been derived from field application. The degree to which the enhanced oil recovery methods are applicable in the future will depend on development of improved process technology. It will also depend on improved understanding of fluid chemistry, phase behavior, and physical properties, and also on the accuracy of geology and reservoir engineering in characterizing the physical nature of individual reservoirs [10].

For taxation purposes, the Internal Revenue Service of the United States has listed the projects that qualify as enhanced oil recovery projects [11] and are therefore available for a tax credit and these projects are:

1. *Thermal recovery methods*:
   Thermal methods of recovery reduce the viscosity of the crude oil by heat so that it flows more easily into the production well.

   (a) *Steam drive injection* – the continuous injection of steam into one set of wells (injection wells) or other injection source to effect oil displacement toward and production from a second set of wells (production wells).
   (b) *Cyclic steam injection* – the alternating injection of steam and production of oil with condensed steam from the same well or wells.
   (c) *In situ combustion* – the combustion of oil or fuel in the reservoir sustained by injection of air, oxygen-enriched air, oxygen, or supplemental fuel supplied from the surface to displace unburned oil toward producing wells. This process may include the concurrent, alternating, or subsequent injection of water.

   Steam-based methods are the most advanced of all enhanced oil recovery methods in terms of field experience and thus have the least uncertainty in estimating performance, provided that a good reservoir description is available.

Steam processes are most often applied in reservoirs containing heavy crude oil, usually in place of rather than following secondary or primary methods. Commercial application of steam processes has been underway since the early 1960s.

2. *Gas flood recovery methods*:

   (a) *Miscible fluid displacement* – the injection of gas (e.g., natural gas, enriched natural gas, a liquefied petroleum slug driven by natural gas, carbon dioxide, nitrogen, or flue gas) or alcohol into the reservoir at pressure levels such that the gas or alcohol and reservoir oil are miscible.
   (b) *Carbon dioxide–augmented waterflooding* – the injection of carbonated water, or water and carbon dioxide, to increase waterflood efficiency.
   (c) *Immiscible carbon dioxide displacement* – the injection of carbon dioxide into an oil reservoir to effect oil displacement under conditions in which miscibility with reservoir oil is not obtained; this process may include the concurrent, alternating, or subsequent injection of water.
   (d) *Immiscible nonhydrocarbon gas displacement* – the injection of nonhydrocarbon gas (e.g., nitrogen) into an oil reservoir, under conditions in which miscibility with reservoir oil is not obtained, to obtain a chemical or physical reaction (other than pressure) between the oil and the injected gas or between the oil and other reservoir fluids; this process may include the concurrent, alternating, or subsequent injection of water.

3. *Chemical flood recovery methods*:
   Three enhanced oil recovery processes involve the use of chemicals – surfactant/polymer, polymer, and alkaline flooding [12]. However, each reservoir has unique fluid and rock properties, and specific chemical systems must be designed for each individual application. The chemicals used, their concentrations in the slugs, and the slug sizes depend upon the specific properties of the fluids and the rocks involved and upon economic considerations.

   (a) *Surfactant flooding* is a multiple-slug process involving the addition of surface-active chemicals to water [13]. These chemicals reduce the capillary forces that trap the oil in the pores of the rock. The surfactant slug displaces the majority of the oil from the reservoir volume contacted, forming a flowing oil–water bank that is propagated ahead of the surfactant slug. The principal factors that influence the surfactant slug design are interfacial properties, slug mobility in relation to the mobility of the oil–water bank, the persistence of acceptable slug properties and slug integrity in the reservoir, and cost.
   (b) *Microemulsion flooding* also known as *surfactant-polymer flooding* involves injection of a surfactant system (e.g., a surfactant, hydrocarbon, cosurfactant, electrolyte, and water) to enhance the displacement of oil toward producing wells; and [2] *caustic flooding* – the injection of water that has been made chemically basic by the addition of alkali metal hydroxides, silicates, or other chemicals.
   (c) *Polymer-augmented waterflooding* – the injection of polymeric additives with water to improve the areal and vertical sweep efficiency of the reservoir

by increasing the viscosity and decreasing the mobility of the water injected; polymer-augmented waterflooding does not include the injection of polymers for the purpose of modifying the injection profile of the wellbore or the relative permeability of various layers of the reservoir, rather than modifying the water-oil mobility ratio.

Certain types of reservoirs, such as those with very viscous crude oils and some low-permeability carbonate (limestone, dolomite, or chert) reservoirs, respond poorly to conventional secondary recovery techniques. The viscosity (or the API gravity) of petroleum is an important factor that must be taken into account when heavy oil is recovered from a reservoir.

In these reservoirs, it is desirable to initiate *enhanced oil recovery* operations as early as possible. This may mean considerably abbreviating conventional secondary recovery operations or bypassing them altogether.

*Thermal methods* for oil recovery have found most use when the oil in the reservoir has a high viscosity. For example, heavy oil is usually highly viscous (hence the use of the adjective *heavy*), with a viscosity ranging from approximately 100 cP to several million centipoises at the reservoir conditions. In addition, oil viscosity is also a function of temperature and API gravity [2, 14]. Thus, for heavy crude oil samples with API gravity ranging from 4 to 21°API (1.04–0.928 kg/m$^3$):

$$\log \log(\mu\sigma + \alpha) = A - B\log(T + 460)$$

In this equation, $\mu\sigma$ is oil viscosity in cP, $T$ is temperature in °F, $A$ and $B$ are constants, and $\alpha$ is an empirical factor used to achieve a straight-line correlation at low viscosity. This equation is usually used to correlate kinematic viscosity in centistokes, in which case an $\alpha$ of 0.6–0.8 is suggested (dynamic viscosity in cP equals kinematic viscosity in cSt times density in g/mL).

An alternative equation for correlating viscosity data (where $a$ and $b$ are constants, and $T$ * is the absolute temperature) is:

$$\mu = a\mathrm{e}^{b/T^*}$$

*Thermal-enhanced oil recovery processes* add heat to the reservoir to reduce oil viscosity and/or to vaporize the oil. In both instances, the oil is made more mobile so that it can be more effectively driven to producing wells. In addition to adding heat, these processes provide a driving force (pressure) to move oil to producing wells.

*Steam drive injection* (*steam injection*) has been commercially applied since the early 1960s. The process occurs in two steps: (1) steam stimulation of production wells, that is, direct steam stimulation and (2) steam drive by steam injection to increase production from other wells (indirect steam stimulation).

When there is some natural reservoir energy, steam stimulation normally precedes steam drive. In steam stimulation, heat is applied to the reservoir by the injection of high-quality steam into the production well. This cyclic process, also called *huff and puff* or *steam soak*, uses the same well for both injection and production. The period of steam injection is followed by production of reduced viscosity oil and condensed steam (water). One mechanism that aids production of the oil is the flashing of hot water (originally condensed from steam injected under high pressure) back to steam as pressure is lowered when a well is put back on production.

*Cyclic steam injection* is the alternating injection of steam and production of oil with condensed steam from the same well or wells. Thus, steam generated at surface is injected in a well and the same well is subsequently put back on production.

A cyclic steam injection process includes three stages. The first stage is injection, during which a measured amount of steam is introduced into the reservoir. The second stage (*the soak period*) requires that the well be shut in for a period of time (usually several days) to allow uniform heat distribution to reduce the viscosity of the oil (alternatively, to raise the reservoir temperature above the pour point of the oil). Finally, during the third stage, the now-mobile oil is produced through the same well. The cycle is repeated until the flow of oil diminishes to a point of no returns.

The high gas mobility may limit recovery through its adverse effect on the sweep efficiency of the burning front. Because of the density contrast between air and reservoir liquids, the burning front tends to override the reservoir liquids. To date, combustion has been most effective for the recovery of viscous oils in moderately thick reservoirs in which reservoir dip and continuity provide effective gravity drainage or operational factors permit close well spacing.

Using combustion to stimulate oil production is regarded as attractive for deep reservoirs [15] and, in contrast to steam injection, usually involves no loss of heat. The duration of the combustion may be short (<30 days) or more prolonged (approximately 90 days), depending upon requirements. In addition, backflow of the oil through the hot zone must be prevented or coking occurs.

Both forward and reverse combustion methods have been used with some degree of success when applied to tar sand deposits. The forward-combustion process has been applied to the Orinoco deposits [16] and in the Kentucky sands [15]. The reverse combustion process has been applied to the Orinoco deposit [17] and the Athabasca [2, 4]. In tests such as these, it is essential to control the airflow and to mitigate the potential for spontaneous ignition [17]. A modified combustion approach has been applied to the Athabasca deposit [2, 4]. The technique involved a heat-up phase and a production (or blowdown phase) followed by a displacement phase using a fireflood-waterflood (COFCAW) process.

## Oil Sand Exploration and Production

Heavy oil and bitumen (the component of interest in tar sand) are often defined (loosely and incorrectly) in terms of API gravity. A more appropriate definition of bitumen, which sets it aside from heavy oil and conventional petroleum, is based on the definition offered by the US government as the *extremely viscous hydrocarbon which is not recoverable in its natural state by conventional oil well production methods including currently used enhanced recovery techniques* [2, 4].

By inference, conventional petroleum and heavy oil (recoverable by *conventional oil well production methods including currently used enhanced recovery techniques*) are different to tar sand bitumen. Be that as it may, in some stage of production, conventional petroleum (in the later stages of recovery) and heavy oil (in the earlier stages of recovery) may require the application of enhanced oil recovery methods for recovery.

### *Oil Mining*

Oil mining includes recovery of oil and/or heavy oil by drainage from reservoir beds to mine shafts or other openings driven into the rock or by drainage from the reservoir rock into mine openings driven outside the reservoir but connected with it by boreholes or mine wells.

Oil mining methods should be applied in reservoirs that have significant residual oil saturation and have reservoir or fluid properties that make production by conventional methods inefficient or impossible. The high well density in improved oil mining usually compensates for the inefficient production caused by reservoir heterogeneity.

However, close well spacing can also magnify the deleterious effects of reservoir heterogeneity. If a high-permeability streak exists with a lateral extent that is less than the inter-well spacing of conventional wells but is comparable to that of improved oil mining, the channeling is more unfavorable for the improved oil mining method.

### *Tar Sand Mining*

The bitumen occurring in tar sand deposits poses a major recovery problem. The material is notoriously immobile at formation temperatures and must therefore require some stimulation (usually by thermal means) in order to ensure recovery. Alternately, proposals have been noted which advocate bitumen recovery by solvent flooding or by the use of emulsifiers. There is no doubt that with time, one or more of these functions may come to fruition, but for the present, the two commercial operations rely on the mining technique.

The alternative to in situ processing is to mine tar sand, transport the mined material to a processing plant, extract the bitumen, and dispose of the waste sand. Such a procedure is often referred to as *oil mining*. This is the term applied to the surface or subsurface excavation of petroleum-bearing formations for subsequent removal of the heavy oil or bitumen by washing, flotation, or retorting treatments.

The tar sand mining method of recovery has received considerable attention since it was chosen as the technique of preference for the only two commercial bitumen recovery plants in operation in North America. In situ processes have been tested many times in the United States, Canada, and other parts of the world and are ready for commercialization. There are also conceptual schemes that are a combination of both mining (aboveground recovery) and in situ (non-mining recovery) methods.

Engineering a successful oil mining project must address a number of items because there must be sufficient recoverable resources, the project must be conducted safely, and the project should be engineered to maximize recovery within economic limits. The use of a reliable screening technique is necessary to locate viable candidates. Once the candidate is defined, this should be followed by an exhaustive literature search covering the local geology, drilling, production, completion, and secondary and tertiary recovery operations.

The reservoir properties, which can affect the efficiency of heavy oil or bitumen production by mining technology, can be grouped into three classes:

1. *Primary properties*, i.e., those properties that have an influence on the fluid flow and fluid storage properties and include rock and fluid properties, such as porosity, permeability, wettability, crude oil viscosity, and pour point
2. *Secondary properties*, i.e., those properties that significantly influence the primary properties, including pore size distribution, clay type, and content
3. *Tertiary properties*, i.e., those other properties that mainly influence oil production operation (fracture breakdown pressure, hardness, and thermal properties) and the mining operations (e.g., temperature, subsidence potential, and fault distribution)

There are also important rock mechanical parameters of the formation in which a tunnel is to be mined and from where all oil mining operations will be conducted. These properties are mostly related to the mining aspects of the operations, and not all are of equal importance in their influence on the mining technology. Their relative importance also depends on the individual reservoir.

Surface mining is the mining method that is currently being used by Suncor Energy and Syncrude Canada Limited to recover tar sand from the ground. Surface mining can be used in mineable tar sand areas which lie under 250 ft or less of overburden material. Less than 10% of the Athabasca Oil Sands deposit can be mined using the surface mining technique, as the other 90% of the deposit has more than 250 ft of overburden. This other 90% will have to be mined using different mining techniques.

There are two methods of mining currently in use in the Athabasca Oil Sands. Suncor Energy uses the truck and shovel method of mining whereas Syncrude uses the truck and shovel method of mining, as well as draglines and bucket-wheel reclaimers. These enormous draglines and bucket-wheels are being phased out and soon will be completely replaced with large trucks and shovels. The shovel scoops up the tar sand and dumps it into a heavy hauler truck. The heavy hauler truck takes the tar sand to a conveyor belt that transports the tar sand from the mine to the extraction plant. Presently, there are extensive conveyor belt systems that transport the mined tar sand from the recovery site to the extraction plant. With the development of new technologies, these conveyors are being phased out and replaced with hydrotransport technology.

Hydrotransport is a combination of ore transport and preliminary extraction. After the bituminous sands have been recovered using the truck and shovel method, it is mixed with water and caustic soda to form a slurry and is pumped along a pipeline to the extraction plant. The extraction process thus begins with the mixing of the water and agitation needed to initiate bitumen separation from the sand and clay.

Mine spoils need to be disposed of in a manner that assures physical stabilization. This means appropriate slope stability for the pile against not only gravity but also earthquake forces. Since return of the spoils to the mine excavations is seldom economical, the spoil pile must be designed as a permanent structure whose outline blends into the landscape. Straight, even lines in the pile must be avoided.

Underground mining options have also been proposed but for the moment have not been developed because of the fear of collapse of the formation onto any operation/equipment. This particular option should not, however, be rejected out-of-hand because a novel aspect or the requirements of the developer (which remove the accompanying dangers) may make such an option acceptable.

The tar sand recovered by mining is sent to the processing plant for separation of the bitumen from the sand prior to upgrading.

## *The Hot-Water Process*

The *hot-water process* is, to date, the only successful commercial process to be applied to bitumen recovery from mined tar sands in North America [18–22]. Many process options have been tested with varying degrees of success, and one of these options may even supersede the hot-water process.

The process utilizes (1) the film of water coats most of the mineral matter, which permits extraction by the *hot-water process*, (2) the linear and the nonlinear variation of bitumen density and water density, respectively, with temperature so that the bitumen that is heavier than water at room temperature becomes lighter than water at 80°C (180°F), and (3) natural surface-active materials (surfactants) in the tar sand which also contribute to freeing the bitumen from the sand.

In the process, the tar sand is introduced into a *conditioning* drum where the sand is heated and mixed with water to encourage agglomeration of the oil particles. Conditioning is carried out in a slowly rotating drum that contains a steam-sparging system for temperature control as well as mixing devices to assist in lump size reduction and a size ejector at the outlet end. The tar sand lumps are reduced in size by ablation and mixing action. The conditioned *pulp* has the following characteristics: (1) solids 60–85% and (2) pH 7.5–8.5.

Lumps of as-mined tar sand are reduced in size by ablation, and the conditioned pulp is screened through a double-layer vibrating screen. Water is then added to the screened material (to achieve more beneficial pumping conditions), and the pulp enters the separation cell through a central feed well and distributor. The bulk of the sand settles in the cell and is removed from the bottom as tailing, but the majority of the bitumen floats to the surface and is removed as froth. A middlings stream (mostly water with suspended fines and some bitumen) is withdrawn from approximately midway up the side of the cell wall. Part of the middlings is recycled to dilute the conditioning-drum effluent for pumping. Clays do not settle readily and generally accumulate in the middlings layer. High concentrations of clays increase the viscosity and can prevent normal operation in the separation cell.

The separation cell acts like two settlers – one on top of the other – and in the lower settler, the sand settles down, whereas in the upper settler, the bitumen floats. The bulk of the sand in the feed is removed from the bottom of the separation cell as tailings. A large portion of the feed bitumen floats to the surface of the separation cell and is removed as bituminous froth. A middlings stream consists mostly of water with some suspended fine minerals and bitumen particles, and a portion of the middlings may be returned for mixing with the conditioning-drum effluent in order to dilute the separation-cell feed for pumping. The remainder of the middlings is withdrawn from the separation cell to be rejected after processing in the scavenger cells.

The combined froth from the separation cell and scavenging operation contains an average of about 10% by weight mineral material and up to 40% by weight water. The dewatering and demineralizing is accomplished in two stages of centrifuging; in the first stage, the coarser mineral material is removed but much of the water remains. The feed then passes through a filter to remove any additional large-size mineral matter that would plug up the nozzles of the second stage centrifuges.

In the scavenging cell, froth flotation with air is usually employed to recover more bitumen. The scavenger froth is combined with the separation-cell froth to be further treated and upgraded to synthetic crude oil. Tailings from the scavenger cell join the separation-cell tailings stream and go to waste.

The bituminous froth from the hot-water process may be mixed with a hydrocarbon diluent, e.g., coker naphtha, and centrifuged. The Suncor process employs a two-stage centrifuging operation, and each stage consists of multiple centrifuges of conventional design installed in parallel. The bitumen product contains 1% by weight to 2% by weight mineral (dry bitumen basis) and 5% by weight to 15% by weight water (wet diluted basis). Syncrude also utilizes a centrifuge system with naphtha diluent.

One of the major problems that arises from the hot-water process is the disposal and control of the tailings. The fact is that each ton of tar sand in place has a volume of about 16 $ft^3$, which will generate about 22 $ft^3$ of tailings giving a volume gain on the order of 40%. If the mine produces about 200,000 t of tar sand per day, the volume expansion represents a considerable solids disposal problem. Tailings from the process consist of about 49–50% by weight of sand, 1% by weight of bitumen, and about 50% by weight of water. The average particle size of the sand is about 200 mm, and it is a suitable material for dike building. Accordingly, Suncor used this material to build the sand dike, but for fine sand, the sand must be well compacted.

Environmental regulations in Canada or the United States will not allow the discharge of tailings streams into (1) the river; (2) on to the surface; or (3) on to any area where contamination of groundwater domains or the river may be contaminated. The tailings streams is essentially high in clays and contains some bitumen, hence the current need for tailings ponds, where some settling of the clay occurs. In addition, an approach to acceptable reclamation of the tailings ponds will have to be accommodated at the time of site abandonment.

The structure of the dike may be stabilized on the upstream side by beaching. This gives a shallow slope but consumes sand during the season when it is impossible to build the dike. In remote areas such as the Fort McMurray (Alberta) site, the dike can only be built in above-freezing weather because (1) frozen water in the pores of the dike will create an unstable layer and (2) the vapor emanating from the water creates a fog, which can create a work hazard. The slope of the tailings dike is about 2.5:1 depending on the amount of fines in the material. It may be possible to build with 2:1 slopes with coarser material, but steeper slopes must be stabilized quickly by beaching. After discharge from the hot-water separation system, it is preferable that attempts be made to separate the sand, sludge, and water, hence, the tailings pond. The sand is used to build dikes and the runoff that contains the silt, clay, and water collects in the pond. Silt and some clay settle out to form sludge, and some of the water is recycled to the plant.

In summary, the hot-water separation process involves extremely complicated surface chemistry with interfaces among various combinations of solids (including both silica sand and aluminosilicate clays), water, bitumen, and air. The control of pH is critical with the preferred range being 8.0–8.5, which is achievable by use of any of the monovalent bases. Polyvalent cations must be excluded because they tend to flocculate the clays and thus raise the viscosity of the middlings in the separation cell.

## *Other Processes*

The issues arising from bitumen mining and bitumen recovery may be alleviated somewhat by the development of process options that require considerably less water in the sand/bitumen separation step. Such an option would allow a more gradual removal of the tailings ponds.

A *cold-water process* for bitumen separation from mined tar sand has also been recommended [23, 24]. The process uses a combination of cold water and solvent, and the first step usually involves disintegration of the tar sand charge that is mixed with water, diluent, and reagents. The diluent may be a petroleum distillate fraction such as aromatic naphtha or kerosene and is added in approximately a 1:1 weight ratio to the bitumen in the feed. The pH is maintained at 9–9.5 by the addition of wetting agents and approximately 0.77 kg of soda ash per ton of tar sand. The effluent is mixed with more water, and in a raked classifier, the sand is settled from the bulk of the remaining mixture. The water and oil overflow the classifier and are passed to thickeners where the oil is concentrated. Clay in the tar sand feed has a distinct effect on the process; it forms emulsions that are hard to break and are wasted with the underflow from the thickeners.

The *sand-reduction process* is a cold-water process without solvent. In the first step, the tar sand feedstock is mixed with water at approximately 20°C (68°F) in a screw conveyor in a ratio of 0.75–3 t per ton of tar sand (the lower range is preferred). The mixed pulp from the screw conveyor is discharged into a rotary-drum screen, which is submerged in a water-filled settling vessel. The bitumen forms agglomerates that are retained by an 840-μm (20-mesh) screen. These agglomerates settle and are withdrawn as oil product. The sand readily passes through the 840 μm (20 mesh) screen and is withdrawn as waste stream. The process is called sand reduction because its objective is the removal of sand from the tar sand to provide a feed suitable for a fluid coking process; ca 80% of sand is removed. Nominal composition of the oil product is 58% by weight (bitumen), 27% by weight mineral matter, and 15% by weight water.

The *spherical agglomeration process* resembles the sand-reduction process. Water is added to tar sands and the mixture is ball-milled. The bitumen forms dense agglomerates of 75% by weight to 87% by weight bitumen, 12% by weight to 25% by weight sand, and 1% by weight to 5% by weight water.

An *oleophilic sieve process* [25, 26] offers the potential for reducing tailings pond size because of a reduction in the water requirements. The process is based on the concept that when a mixture of an oil phase and an aqueous phase is passed through a sieve made from oleophilic materials, the aqueous phase and any hydrophilic solids pass through the sieve but the oil adheres to the sieve surface on contact. The sieve is in the form of a moving conveyor, the oil is captured in a recovery zone, and recovery efficiency is high.

An anhydrous *solvent extraction process* for bitumen recovery has been attempted and usually involves the use of a low-boiling hydrocarbon. The process generally involves up to four steps. In the mixer step, fresh tar sand is mixed with recycle solvent that contains some bitumen and small amounts of water and mineral and the solvent-to-bitumen weight ratio is adjusted to approximately 0.5. The drain step consists of a three-stage countercurrent wash. Settling and draining time is approximately 30 min for each stage. After each extraction step, a bed of sand is formed and the extract is drained through the bed until the interstitial pore volume of the bed is emptied. The last two steps of the process are devoted to solvent recovery solvent recovery from the bitumen and from the solids, which holds the key to the economic success the process.

Another aboveground method of separating bitumen from mined tar sand involves *direct heating of the tar sand* without previous separation of the bitumen [27]. Thus, the bitumen is not recovered as such but is an upgraded overhead product. In the process, the sand is crushed and introduced into a vessel, where it is contacted with either hot (spent) sand or with hot product gases that furnish part of the heat required for cracking and volatilization. The volatile products are passed out of the vessel and are separated into gases and (condensed) liquids. The coke that is formed as a result of the thermal decomposition of the bitumen remains on the sand, which is then transferred to a vessel for coke removal by burning in air. The hot flue gases can be used either to heat incoming tar sand or as refinery fuel. As expected, processes of this type yield an upgraded product but require various arrangements of pneumatic and mechanical equipment for solids movement around the refinery.

In *improved mining*, directional (horizontal or slant) wells are drilled into the reservoir from a mine in an underlying formation to drain oil by pressured depletion and gravity drainage. In the process of gravity drainage extraction of liquid crude oil, the wells are completed so that only the forces acting within the reservoir are used. A large number of closely spaced wells can be drilled into a reservoir from an underlying tunnel more economically than the same number of wells from the surface. In addition, only one pumping system is required in underground drainage, whereas at the surface, each well must have a pumping system. The objective of using a large number of wells is to produce each well slowly so that the gas–oil and water–oil interfaces move toward each other efficiently. By maintaining the reservoir pressures because of forces acting on the reservoir, it is then assured that the oil production is provided by the internal forces due to gravity (the buoyancy effect) and capillary effects.

Large vertical shafts sunk from the surface are generally the means through which underground openings can be excavated. These shafts are one means of access to offer an outlet for removal of excavated rock, provide sufficient opening for equipment, provide ventilation, and allow the removal of oil and gas products during later production. These requirements plus geological conditions and oil reservoir dimensions determine the shaft size. It is expected that an access shaft will range from 8 to 20 ft in diameter.

## *Non-mining Methods*

Whereas conventional crude oils may have a viscosity of several poise (at 40°C, 105°F), the tar sand bitumen has a viscosity of the order of 50,000–1,000,000 cP or more at formation temperatures (approximately 0–10°C, 32–50°F depending upon the season). This offers a formidable (but not insurmountable) obstacle to bitumen recovery.

In principle, the *non-mining recovery of bitumen from tar sand deposits* is an enhanced recovery technique and requires the injection of a fluid into the formation through an injection wall. This leads to the in situ displacement of the bitumen from the sand followed by bitumen production at the surface through an egress well (production well).

In tar sand deposits, it is often desirable to initiate *enhanced oil recovery* (EOR) operations as early as possible, which mean considerably abbreviating conventional secondary recovery operations or bypassing them altogether. Thermal floods using steam and controlled in situ combustion methods are also used. Thermal methods of recovery reduce the viscosity of the crude oil by heat so that it flows more easily into the production well [28].

The technologies applied to oil recovery involve different concepts, some of which can cause changes to the oil during production. Technologies such as alkaline flooding, microemulsion (micellar/emulsion) flooding, polymer-augmented waterflooding, and carbon dioxide miscible/immiscible flooding do not require or cause any change to the oil. The steaming technologies may cause some steam distillation that can augment the process when the steam-distilled material moves with the steam front and acts as a solvent for oil ahead of the steam front. Again, there is no change to the oil although there may be favorable compositional changes to the oil insofar as lighter fractions are recovered and heavier materials remain in the reservoir.

The technology where changes do occur involves combustion of the oil in situ. The concept of any combustion technology requires that the oil be partially combusted and that thermal decomposition occur to other parts of the oil. This is sufficient to cause irreversible chemical and physical changes to the oil to the extent that the product is markedly different to the oil-in-place, indicating upgrading of the bitumen during the process. Recognition of this phenomenon is essential before combustion technologies are applied to oil recovery.

*Thermal recovery methods* (Fig. 3.1) have found most use when heavy oil or bitumen has an extremely high viscosity under reservoir conditions [2, 4]. For example, bitumen is highly viscous, with a viscosity ranging up to a million centipoises or more at the reservoir conditions.

*Thermal-enhanced oil recovery processes* (i.e., cyclic steam injection, steam flooding, and in situ combustion) add heat to the reservoir to reduce oil viscosity and/or to vaporize the oil. In both instances, the oil is made more mobile so that it can be more effectively driven to producing wells. In addition to adding heat, these processes provide a driving force (pressure) to move oil to producing wells.

In the *modified* in situ *extraction* processes, combinations of in situ and mining techniques are used to access the reservoir. A portion of the reservoir rock must be removed to enable application of the in situ extraction technology. The most common method is to enter the reservoir through a large-diameter vertical shaft, excavate horizontal drifts from the bottom of the shaft, and drill injection and production wells horizontally from the drifts. Thermal extraction processes are then applied through the wells. When the horizontal wells are drilled at or near the base of the tar sand reservoir, the injected heat rises from the injection wells through the reservoir, and drainage of produced fluids to the production wells is assisted by gravity.

There are, however, several serious constraints that are particularly important and relate to bulk properties of the tar sand and the bitumen. In fact, both must be considered in the context of bitumen recovery by non-mining techniques. For example, the Canadian deposits are unconsolidated sands with a porosity ranging up to about 45% whereas other deposits may range from predominantly low-porosity, low-permeability consolidated sand to, in a few instances, unconsolidated sands. In addition, the bitumen properties are not conducive to fluid flow under deposit conditions. Nevertheless, where the general nature of the deposits prohibits the application of a mining technique, a non-mining method may be the only feasible bitumen recovery option.

Another general constraint to bitumen recovery by non-mining methods is the relatively low injectivity of tar sand formations. Thus, it is usually necessary to inject displacement or recovery finds at a pressure such that fracturing (parting) is achieved. Such a technique therefore changes the reservoir profile and introduces a series of channels through which fluids can flow from the injection well to the production well. On the other hand, the technique may be disadvantageous insofar as the fracture occurs along the path of least resistance, giving undesirable (i.e., inefficient) flow characteristics within the reservoir between the injection and production wells, leaving a large part of the reservoir relatively untouched by the displacement or recovery fluids.

Another general constraint to bitumen recovery by non-mining methods is the relatively low injectivity of tar sand formations. It is usually necessary to inject displacement/recovery fluids at a pressure such that fracturing (parting) is achieved. Such a technique, therefore, changes the reservoir profile and introduces a series of channels through which fluids can flow from the injection well to the production well. On the other hand, the technique may be disadvantageous insofar as the fracture occurs along the path of least resistance giving undesirable (i.e., inefficient) flow characteristics within the reservoir between the injection and production wells which leave a part of the reservoir relatively untouched by the displacement or recovery fluids.

## Steam-Based Processes

Steam-based processes are the most advanced of all enhanced oil recovery methods in terms of field experience and thus have the least uncertainty in estimating performance, provided that a good reservoir description is available. Steam processes are most often applied in reservoirs containing heavy oil, which is mobile at reservoir temperature. Commercial application of steam processes has been underway since the early 1960s.

*Steam drive injection* (*steam injection*) has been commercially applied since the early 1960s. The process occurs in two steps: (1) steam stimulation of production wells, that is, direct steam stimulation, and (2) steam drive by steam injection to increase production from other wells (i.e., indirect steam stimulation). Steam drive requires sufficient effective permeability (with the immobile bitumen in place) to allow injection of the steam at rates sufficient to raise the reservoir temperature to mobilize the bitumen and drive it to the production well.

*Cyclic steam injection* (also called *huff and puff* or *steam soak*) is the alternating injection of steam and production of oil with condensed steam from the same well or wells. This process is predominantly a vertical well process, with each well alternately injecting steam and producing heavy oil and steam condensate. In practice, steam is injected into the formation at greater than fracturing pressure followed by a *soak* period after which production is commenced. The heat injected warms the heavy oil and lowers its viscosity. A heated zone is created through which the warmed heavy oil can flow back into the well. This is a well-developed process; the major limitation is that less than 30% (usually less than 20%) of the initial oil-in-place can be recovered.

## Combustion Processes

In situ *combustion* (*fireflood*) is normally applied to reservoirs containing low-gravity oil but has been tested over perhaps the widest spectrum of conditions of any enhanced oil recovery process. In the process, heat is generated within the reservoir by injecting air and burning part of the crude oil. This reduces the oil viscosity and partially vaporizes the oil-in-place, and the oil is driven out of the reservoir by a combination of steam, hot water, and gas drive. Forward combustion involves movement of the hot front in the same direction as the injected air; reverse combustion involves movement of the hot front opposite to the direction of the injected air.

During the process, energy is generated in the formation by igniting bitumen in the formation and sustaining it in a state of combustion or partial combustion. The high temperatures generated decrease the viscosity of the oil and make it more mobile. Some cracking of the bitumen also occurs, and an upgraded product rather than bitumen itself is the fluid recovered from the production wells.

The relatively small portion of the oil that remains after the displacement mechanisms have acted becomes the fuel for the in situ combustion process. Production is obtained from wells offsetting the injection locations. In some applications, the efficiency of the total in situ combustion operation can be improved by alternating water and air injection. The injected water tends to improve the utilization of heat by transferring heat from the rock behind the combustion zone to the rock immediately ahead of the combustion zone.

The use of combustion to stimulate oil production is regarded as attractive for deep reservoirs. In contrast to steam injection, it usually involves no loss of heat. The duration of the combustion may be less than 30 days or much as 90 days depending on requirements. In addition, backflow of the oil through the hot zone must be prevented or coking will occur.

*Forward combustion* involves movement of the hot front in the same direction as the injected air while reverse combustion involves movement of the hot front opposite to the direction of the injected air. In forward combustion, the hydrocarbon products released from the zone of combustion move into a relatively cold portion of the formation. Thus, there is a definite upper limit of the viscosity of the liquids that can be recovered by a forward-combustion process. On the other hand, since the air passes through the hot formation before reaching the combustion zone, burning is complete; the formation is left completely cleaned of hydrocarbons.

*Reverse combustion* is particularly applicable to reservoirs with lower effective permeability (in contrast with forward combustion). It is more effective because the lower permeability would cause the reservoir to be plugged by the mobilized fluids ahead of a forward-combustion front. In the reverse combustion process, the vaporized and mobilized fluids move through the heated portion of the reservoir behind the combustion front. The reverse combustion partially cracks the bitumen, consumes a portion of the bitumen as fuel, and deposits residual coke on the sand grains. In the process, part of the bitumen will be consumed as fuel and part will be deposited on the sand grains as coke leaving 40–60% recoverable. This coke deposition serves as a cementing material, reducing movement and production of sand.

The addition of water or steam to an in situ combustion process can result in a significant increase in the overall efficiency of that process. Two major benefits may be derived. Heat transfer in the reservoir is improved because the steam and condensate have greater heat-carrying capacity than combustion gases and gaseous hydrocarbons. Sweep efficiency may also be improved because of the more favorable mobility ratio of steam-bitumen compared with gas-bitumen.

Applying a preheating phase before the bitumen recovery phase may significantly enhance the steam or combustion extraction processes. Preheating can be particularly beneficial if the saturation of highly viscous bitumen is sufficiently great as to lower the effective permeability to the point of production being precluded by reservoir plugging. Preheating partially mobilizes the bitumen by raising its temperature and lowering its viscosity. The result is a lower required pressure to inject steam or air and move the bitumen.

In the *fracture-assisted steam technology* (FAST) process, steam is injected rapidly into an induced horizontal fracture near the bottom of the reservoir to preheat the reservoir. This process has been applied successfully in three pilot projects in southwest Texas. Shell has accomplished the same preheating goal by injecting steam into a high-permeability bottom water zone in the Peace River (Alberta) field. Electrical heating of the reservoir by radio-frequency waves may also be an effective method.

In situ combustion has been field tested under a wide variety of reservoir conditions, but few projects have proven economical and advanced to commercial scale and the concept has been abandoned by many recovery operators. However, in situ combustion may make a comeback with a new concept. THAI (toe-to-heel air injection) (Fig. 3.3) is based on the geometry of horizontal wells that may solve the problems that have plagued conventional in situ combustion. The well geometry

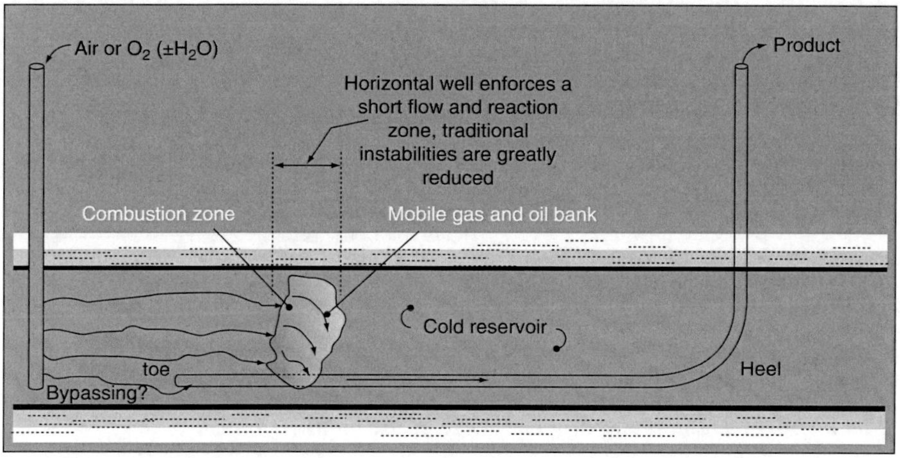


**Fig. 3.3** The THAI process

enforces a short flow path so that any instability issues associated with conventional combustion are reduced or even eliminated [2, 4].

In situ conversion, or underground refining, is a promising new technology to tap the extensive reservoirs of heavy oil and deposits of bitumen. The new technology [29, 30] features the injection of high-temperature, high-quality steam, and hot hydrogen into a formation containing heavy hydrocarbons to initiate conversion of the heavy hydrocarbons into lighter hydrocarbons. In effect, the heavy hydrocarbons undergo partial underground refining that converts them into a synthetic crude oil (or *syncrude*). The heavier portion of the syncrude is treated to provide the fuel and hydrogen required by the process, and the lighter portion is marketed as a conventional crude oil.

Thus, below ground, superheated steam and hot hydrogen are injected into a heavy oil or bitumen formation, which simultaneously produces the heavy oil or bitumen and converts it in situ (i.e., within the formation) into syncrude. Above ground, the heavier fraction of the syncrude is separated and treated on-site to produce the fuel and hydrogen required by the process, while the lighter fraction is sent to a conventional refinery to be made into petroleum products (United States Patent 6,016,867; United States Patent 6,016,868).

The potential advantages of an in situ process for bitumen and heavy oil include (1) leaving the carbon-forming precursors in the ground, (2) leaving the heavy metals in the ground, (3) reducing sand handling, and (4) bringing a partially upgraded product to the surface. The extent of the upgrading can, hopefully, be adjusted by adjusting the exposure of the bitumen of heavy oil to the underground thermal effects.

Finally, by all definitions, the quality of the bitumen from tar sand deposits is poor when considered as a refinery feedstock. As in any field in which primary recovery operations are followed by secondary or enhanced recovery operations and there is a change in product quality, such is also the case for tar sand recovery operations.

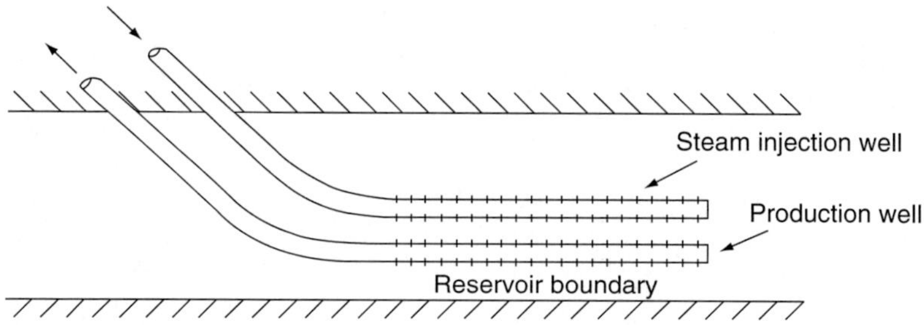


**Fig. 3.4** The SAGD process

Thus, product oils recovered by the thermal stimulation of tar sand deposits show some improvement in properties over those of the bitumen in-place.

In situ recovery processes (although less efficient in terms of bitumen recovery relative to mining operations) may have the added benefit of *leaving* some of the more obnoxious constituents (from the processing objective) in the ground.

## Other Processes

Many innovative concepts in heavy oil production have been developed in the last 10 years [2, 4]. However, there are varying degrees of success and all are dependent on the properties of the deposit. There is no panacea for bitumen recovery that can be applied on a worldwide basis.

Long *horizontal wells* with several multilateral branches have been used widely in the development of the heavy oils of Venezuela, where production rates as high as 2,000–2,500 bbl/day in some wells have been achieved through the use of aggregate horizontal lengths as large as 10,000 m in oil of 1,200–5,000 cP viscosity. Unfortunately, this technology can only achieve 8–15% v/v recovery of the oil and only from the best high-permeability zones.

*Inert gas injection* (IGI) is a technology for conventional oils in reservoirs where good vertical permeability exists, or where it can be created through propped hydraulic fracturing. It is generally viewed as a *top-down* process with nitrogen or methane injection through vertical wells at the top of the reservoirs, creating a gas–oil interface that is slowly displaced toward long horizontal production wells [2, 4]. As with all gravity drainage processes, it is essential to balance the injection and production volumes precisely so that the system does not become pressure driven, but remains in the gravity-dominated flow regime.

SAGD (steam-assisted gravity drainage) (Fig. 3.4) was developed first in Canada for reservoirs where the immobile bitumen occurs [31]. This process uses paired horizontal wells. Low-pressure steam continuously injected through the upper well creates a steam chamber along the walls of which the heated bitumen flows and is produced in the lower well.

In the process, a pair of horizontal wells that are separated vertically by about 15–20 ft are drilled at the bottom of a thick unconsolidated sandstone reservoir. Steam, perhaps along with a mixture of hydrocarbons that dissolve into the oil and help reduce its viscosity, is injected into the upper well. The heat reduces the oil viscosity to values as low as 1–10 cP (depending on temperature and initial conditions) and develops a *steam chamber* that grows vertically and laterally. The steam and gases rise because of their low density, and the oil and condensed water are removed through the lower well. The gases produced during SAGD tend to be methane with some carbon dioxide and traces of hydrogen sulfide.

The SAGD process, as for all gravity-driven processes, is extremely stable because the process zone grows only by gravity segregation, and there are no pressure-driven instabilities such as channeling, coning, and fracturing. SAGD seems to be relatively insensitive to shale streaks and similar horizontal barriers, even up to several meters thick (3–6 ft), that otherwise would restrict vertical flow rates. The combined processes of gravity segregation and shale thermal fracturing make SAGD so efficient that recovery ratios of 60–70% are claimed. Nevertheless, the process is not universally applicable to all reservoirs and deposits.

*Cold heavy oil production with sand* (CHOPS) is also used as a production approach in unconsolidated sandstones. The process results in the development of high-permeability channels (*wormholes*) in the adjacent low cohesive strength sands, facilitating the flow of oil foam that is caused by solution gas drive. Instead of blocking sand ingress by screens or gravel packs, sand is encouraged to enter the wellbore by aggressive perforation and swabbing strategies. Vertical or slightly inclined wells (vertical to 45°) are operated with rotary progressive cavity pumps (rather than reciprocating pumps) and old fields are converting to higher-capacity progressive cavity pumps, giving production boosts to old wells. Because massive sand production creates a large disturbed zone, the reservoir may be positively affected for later implementation of thermal processes.

Typically, a well placed on CHOPS production will initially produce a high percentage of sand, greater than 20% by volume of liquids. However, this generally drops after some weeks or months. The huge volumes of sand are disposed of by slurry fracture injection or salt cavern placement or by sand placement in a landfill in an environmentally acceptable manner. Obviously, the production of excessive amounts of sand is a cause for mechanical and environmental concern.

*Pressure pulsing technologies* (PPT) involves a radically new aspect of porous media mechanics discovered and developed into a production enhancement method in the period 1997–2003. The mechanism by which PPT works is to generate a porosity dilation wave (a fluid displacement wave similar to a tsunami); this generates pore-scale dilation and contraction so that oil and water flow into and out of pores, leading to periodic fluid accelerations in the pore throats. As the porosity dilation wave moves through the porous medium at a velocity of about 50–100 ft/s (40–80 m/s), the small expansion and contraction of the pores with the passage of each packet of wave energy helps unblock pore throats, increase the velocity of liquid flow, overcome part of the effects of capillary blockage, and reduce some of the negative effects of instability due to viscous fingering, coning, and permeability streak channeling.

PPT promises to be a major adjunct to a number of oil production processes, particularly all pressure-driven processes, where it will both accelerate flow rates as well as increasing oil recovery factors. It is also now used in environmental applications to help purge shallow aquifers of nonmiscible phases such as oil. The basis for its use in tar sand deposits is largely unknown and unproven.

*Vapor-assisted petroleum extraction* (VAPEX) is a new process in which the physics of the process are essentially the same as for SAGD and the configuration of wells is generally similar. The process involves the injection of vaporized solvents such as ethane or propane to create a vapor-chamber through which the oil flows due to gravity drainage [32–35]. The process can be applied in paired horizontal wells, single horizontal wells, or a combination of vertical and horizontal wells. The key benefits are significantly lower energy costs, potential for in situ upgrading, and application to thin reservoirs, with bottom water or reactive mineralogy.

Because of the slow diffusion of gases and liquids into viscous oils, this approach, used alone, perhaps will be suited only for less viscous oils although preliminary tests indicate that there are micro-mechanisms that act so that the VAPEX dilution process is not diffusion rate limited and the process may be suitable for the highly viscous tar sand bitumen [36, 37].

Nevertheless, VAPEX can undoubtedly be used in conjunction with SAGD methods. As with SAGD and IGI, a key factor is the generation of a three-phase system with a continuous gas phase so that as much of the oil as possible can be contacted by the gaseous phases, generating the thin oil film drainage mechanism. As with IGI, vertical permeability barriers are a problem, and must be overcome through hydraulic fracturing to create vertical permeable channels, or undercut by the lateral growth of the chamber beyond the lateral extent of the limited barrier, or "baffle." As with any solvent process, the loss of solvents in geological formations (such as by adsorption on clay and other minerals) drastically affects process economics and raises many serious environmental issues.

Hybrid approaches that involve the simultaneous use of several technologies are evolving and will see greater applications in the future. In addition to hybrid approaches, the new production technologies, along with older, pressure-driven technologies, will be used in successive phases to extract more oil from reservoirs, even from reservoirs that have been abandoned after primary exploitation. These hybrid approaches hold (on paper at least) the promise of significantly increasing recoverable reserves worldwide, not just in heavy oil cases.

*Microbial-enhanced oil recovery* (MEOR) processes involve use of reservoir microorganisms or specially selected natural bacterial to produce specific metabolic events that lead to enhanced oil recovery.

In microbial-enhanced oil recovery processes, microbial technology is exploited in oil reservoirs to improve recovery [38–40]. From a microbiologist's perspective, microbial-enhanced oil recovery processes are somewhat akin to in situ bioremediation processes. Injected nutrients, together with indigenous or added microbes, promote in situ microbial growth and/or generation of products which mobilize additional oil and move it to producing wells through reservoir repressurization, interfacial tension/oil viscosity reduction, and selective plugging of the most permeable zones [41, 42].

This technology requires consideration of the physicochemical properties of the reservoir in terms of salinity, pH, temperature, pressure, and nutrient availability [43, 44].

The microbial-enhanced oil recovery process may modify the immediate reservoir environment in a number of ways that could also damage the production hardware or the formation itself. Certain sulfate reducers can produce $H_2S$, which can corrode pipeline and other components of the recovery equipment, and considerable uncertainty still remains regarding process performance. In addition, conditions vary from reservoir to reservoir, which calls for reservoir-specific customization of the microbial-enhanced oil recovery process, and this alone has the potential to undermine microbial process economic viability. Even though microbes produce the necessary chemical reactions in situ, there is need for caution and astute observation of the effects of the microorganisms on the reservoir chemistry.

Finally, recent developments in *upgrading* of heavy oil and bitumen [2, 4] indicate that the near future could see a reduction of the differential cost of upgrading heavy oil. These processes are based on a better understanding of the issues of asphaltene solubility effects at high temperatures, incorporation of a catalyst that is chemically precipitated internally during the upgrading, and improving hydrogen addition or carbon rejection.

## Future Directions

With the current energy problems, the motivation for recovering as much as possible of the in-place reserves is greater than ever. There is a potentially uncomfortable and politically disastrous situation where the gap between energy requirements and available energy supplies is widening quickly.

Consequently, the search for new domestic supplies has shifted in large measure to increasingly hostile environments such as the Alaskan North Slope and the offshore waters along the outer continental shelves of the United States. These changes in production operations (not to mention the associated environmental disasters that often accompany such venture) have meant both significantly higher costs of production operations and fewer and fewer new commercial discoveries.

The importance of improving the rate of recovery from domestic petroleum reservoirs is underscored by the increasing difficulty of finding significant new reserves to meet the increasing demand for energy. One solution lies in greater emphasis on a multidisciplinary approach – on an intracompany basis and on a cooperative intercompany basis within the industry.

In the near term, a more immediate solution lies in improved application of existing technology as regards selection and quality control of materials, rigorous application of procedures, and the training and supervision of personnel. Part of the answer to the shortage, at least for the short term, has been to import more crude oil, but this is a less than ideal solution for a number of economic and political reasons.

In terms of petroleum recovery, steam-based processes will remain the processes of choice for the recovery of much of the oil in the ground over the next 2 or 3 decades. In some instances, fire flooding will be rejuvenated as the need to recovery of bitumen and residual oil becomes more important.

With the preponderance of heavier oils and tars and bitumen, it is likely that efforts will be made to emphasize partial (or full) upgrading in situ as an integral part of the recovery process. Any type of upgrading during recovery will enhance the quality of the recovered oil, leaving some of the undesirable constituents in the ground as thermal products. Enhancement of the quality of the recovered oil will facilitate upgrading of the oil in the refinery.

Biotechnology will play a more significant role in enhancing crude oil recovery from the depleted oil reservoirs to solve stagnant petroleum production. Such enhanced oil recovery processes (microbial-enhanced oil recovery, MEOR) involve stimulating indigenous reservoir microbes or injecting specially selected consortia of natural bacteria into the reservoir to produce specific metabolic events that lead to improved oil recovery. This also involves flooding with oil recovery agents produced ex situ by industrial or pilot scale fermentation. However, like all recovery processes, a complete evaluation and assessment of microbial from a scientific and engineering standpoint must be performed and must be based on economics, applicability, and the performance standards required to further improve the process efficiency.

Above all, there is the need to manage recovery operations in such a manner that environmental issues do not become issues!

## Bibliography

### *Primary Literature*

1. Taber JJ, Martin FD (1983) Technical screening guides for the enhanced recovery of oil. In: Proceedings of the 58th SPE annual technical conference and exhibition, San Francisco, 5–8 Oct 1983 (SPE 12069)
2. Speight JG (2009) Enhanced recovery methods for heavy oil and tar sands. Gulf Publishing, Houston
3. Drew LJ (1997) Undiscovered mineral and petroleum deposits: assessment & controversy. Plenum, New York (Chap. 3)
4. Speight JG (2007) The chemistry and technology of petroleum, 4th edn. CRC Press/Taylor and Francis, Boca Raton
5. Lake LW, Walsh MP (2004) Primary hydrocarbon recovery. Elsevier, Amsterdam
6. Craft BC, Hawkins MF (1959) Applied petroleum reservoir engineering. Prentice-Hall, Englewood Cliffs
7. Frick TC (1962) Petroleum production handbook, vol II. McGraw-Hill, New York
8. Lake LW (1989) Enhanced oil recovery. Prentice-Hall, Englewood Cliffs

9. Arnarnath A (1999) Enhanced oil recovery scoping study. Report No. TR-113836, Electric Power Research Institute, Palo Alto
10. Borchardt JK, Yen TF (1989) Oil field chemistry, Symposium series No. 396. American Chemical Society, Washington, DC
11. CFR 1.43-2 (2004) Internal revenue service, Department of the Treasury, Government of the United States, Washington, DC
12. OTA (1978) Enhanced oil recovery potential in the United States. Office of Technology Assessment, Washington, DC. NTIS order #PB-276594
13. Reed RL, Healy RN (1977) Some physicochemical aspects of microemulsion flooding: a review. In: Shah DO, Schechter RS (eds) Improved oil recovery by surfactant and polymer flooding. Academic, New York
14. Speight JG (2000) Desulfurization of heavy oils and residua, 2nd edn. Marcel Dekker, New York
15. Terwilliger PL (1975) Paper 5568. In: Proceedings of 50th annual fall meeting of the Society of Petroleum Engineers, Dallas. American Institute of Mechanical Engineers, Dallas
16. Terwilliger PL, Clay RR, Wilson LA, Gonzalez-Gerth E (1975) J Petrol Technol 27:9
17. Burger J (1978) Developments in petroleum science. In: Chilingarian GV, Yen TF (eds) Bitumens, asphalts and tar sands, vol 7. Elsevier, New York, p 191
18. Clark KA (1944) Hot-water separation of Alberta bituminous sand. Trans Can Inst Min Met 47:257
19. Carrigy MA (1963) Bulletin No. 14. Alberta Research Council, Edmonton
20. Carrigy MA (1963) The oil sands of Alberta. Information Series No. 45. Alberta Research Council, Edmonton
21. Fear JVD, Innes, ED (1967) In: Proceedings seventh world petroleum congress, Mexico, vol 3, p 549
22. Speight JG, Moschopedis SE (1978) Fuel Process Technol 1:261
23. Miller JC, Misra M (1982) Fuel Process Technol 6:27
24. Misra M, Aguilar R, Miller JD (1981) Sep Sci Technol 16(10):1523
25. Kruyer J (1982) In: Proceedings of the second international conference on heavy crude and tar sands, Caracas, 7–17 Feb 1982
26. Kruyer J (1983) Preprint No. 3d. Summer national meeting of the American Institute of Chemical Engineers, Denver, 28–31 Aug 1983
27. Gishler PE (1949) Can J Res 27:104
28. Pratts M (1986) Thermal recovery. Society of Petroleum Engineers, New York
29. Gregoli AA, Rimmer DP, Graue DJ (2000) Upgrading and recovery of heavy crude oils and natural bitumen by in situ hydrovisbreaking. US Patent 6,016,867, 25 Jan 2000
30. Gregoli AA, Rimmer DP (2000) Production of synthetic crude oil from heavy hydrocarbons recovered by in situ hydrovisbreaking. US Patent 6,016,868, 25 Jan 2000
31. Dusseault MB, Geilikman MB, Spanos TJT (1998) J Petrol Technol 50(9):92–94
32. Butler RM, Mokrys IJ (1991) J Can Pet Technol 30(1):97–106
33. Butler RM, Mokrys IJ (1995) Process and apparatus for the recovery of hydrocarbons from a hydrocarbon deposit. US Patent 5,407,009, 18 Apr 1995
34. Butler RM, Mokrys IJ (1995) Process and apparatus for the recovery of hydrocarbons from a hydrocarbon deposit. US Patent 5,607,016, 4 Mar 1995
35. Butler RM, Jiang Q (2000) J Can Pet Technol 39:48–56
36. Yang C, Gu Y (2005a) A novel experimental technique for studying solvent mass transfer and oil swelling effect in a vapor extraction (VAPEX) process. Paper No. 2005–099. In: Proceedings of the 56th annual technical meeting. The Canadian international petroleum conference, Calgary, 7–9 Jun 2005
37. Yang C, Gu Y (2005b) Effects of solvent-heavy oil interfacial tension on gravity drainage in the VAPEX process. Paper No. SPE 97906. In: Society of petroleum engineers international thermal operations and heavy oil symposium, Calgary, 1–3 Nov 2005

38. Banat IM (1995) Biosurfactant production and possible uses in microbial enhanced oil recovery and oil pollution remediation. Bioresour Technol 51:1–12
39. Clark JB, Munnecke DM, Jenneman GE (1981) In situ microbial enhancement of oil production. Dev Ind Microbiol 15:695–701
40. Stosur GJ (1991) Unconventional EOR concepts. Crit Rep Appl Chem 33:341–373
41. Bryant RS, Lindsey RP (1996) World-wide applications of microbial technology for improving oil recovery. In: Proceedings of the SPE symposium on improved oil recovery of the society of petroleum engineers, Richardson, pp 27–134
42. Bryant RS, Donaldson EC, Yen TF, Chilingarian GV (1989) Microbial enhanced oil recovery. In: Donaldson EC, Chilingarian GV, Yen TF (eds) Enhanced oil recovery II: processes and operations. Elsevier, Amsterdam, pp 423–450
43. Khire JM, Khan MI (1994) Microbially enhanced oil recovery (MEOR). Part 1. Importance and mechanisms of microbial enhanced oil recovery. Enzyme Microb Technol 16:170–172
44. Khire JM, Khan MI (1994) Microbially enhanced oil recovery (MEOR). Part 2. Microbes and the subsurface environment for microbial enhanced oil recovery. Enzyme Microb Technol 16:258–259

## *Books and Reviews*

Ancheyta J, Speight JG (2007) Hydroprocessing of heavy oils and residua. CRC Press/Taylor & Francis, Boca Raton

Bower T (2009) Oil: money, politics, and power in the 21st century. Grand Central Publishing, Hachette Book Group, New York

Gudmestad OT, Zolotukhin AB, Jarlsby ET (2010) Petroleum resources, with emphasis on offshore fields. WIT Press, Billerica

Nersesian RL (2010) Energy for the 21st century: a comprehensive guide to conventional and alternative sources, 2nd edn. M.E. Sharpe, Armonk

Speight JG (2008) Synthetic fuels handbook: properties, processes, and performance. McGraw-Hill, New York

Wihbet PM (2009) The rise of the new oil order. Academy & Finance, Geneva

# Chapter 4
# Petroleum Refining and Environmental Control and Environmental Effects

**James G. Speight**

## Glossary

| | |
|---|---|
| Emissions | Gaseous, liquid, or solid by-products introduced into the environment as a result of refining processes. |
| Environmental control | The use of various technologies to control and even prevent refinery emissions from entering the environment. |
| Environmental effects | The effects of refinery emissions on the flora and fauna in the various ecosystems. |
| Refining | The processes by which petroleum is distilled and/or converted by application of physical and chemical processes to form a variety of products. |
| Regulations | The laws by which environmental emissions are controlled. |

## Definition of the Subject

The work summarizes the various process emissions that occur during petroleum refining. There are also general descriptions of the various pollution, health, and environmental problems especially specific to the petroleum industry and places in perspective the government regulations as well as industry efforts to adhere to these regulations. The objective is to indicate the types of emissions and the laws that regulate these emissions.

---

This chapter was originally published as part of the Encyclopedia of Sustainability Science and Technology edited by Robert A. Meyers. DOI:10.1007/978-1-4419-0851-3

J.G. Speight (✉)
CD&W Inc, P.O. Box 1722 Laramie, WY 82073-1722, USA
e-mail: JamesSp8@aol.com

R. Malhotra (ed.), *Fossil Energy: Selected Entries from the Encyclopedia of Sustainability Science and Technology*, DOI 10.1007/978-1-4614-5722-0_4,

## Introduction

Petroleum as an energy source use is a necessary part of the modern world and will be a primary source of energy for the next several decades, hence the need for control over the amounts and types of emissions from the use of petroleum and its products. Furthermore, the capacity of the environment to absorb the effluents and other impacts of process technologies is not unlimited and the environment should be considered to be an extremely limited resource, and discharge of chemicals into it should be subject to severe constraints – as a result, it is necessary to understand the nature and magnitude of the problems involved [1].

Both the production and processing of crude oil involve the use of a variety of substances [2], some toxic, including lubricants in oil wells and catalysts and other chemicals in refining (Fig. 4.1). The amounts used tend to be relatively easy to control, and the spillage of crude oil is more detrimental to the environment.

The purpose of this work is to summarize and generalize the various pollution, health, and environmental problems especially specific to the petroleum industry and to place in perspective government laws and regulations as well as industry efforts to control these problems [3–6]. The objective is to indicate the types of emissions and the laws that regulate these emissions.

## Definitions

Briefly, petroleum production and petroleum refining produce *chemical waste* [6]. If this *chemical waste* is not processed in a timely manner, it can become a *pollutant*. Under some circumstances, chemical waste is reclassified as *hazardous waste*.

*Hazardous waste* is any gaseous, liquid, or solid waste material that, if improperly managed or disposed of, may pose hazards to human health and the environment. In some cases, the term "*chemical waste*" is used interchangeably (often incorrectly) with the term "*hazardous waste*," but chemical waste is always hazardous and the correct use of the terms must be used.

A *pollutant* is a substance present in a particular location (*ecosystem*) – usually it is not indigenous to the location or is present in a concentration greater than the concentration that occurs naturally. The substance is often the product of human activity and has a detrimental effect on the environment, in part or in toto. Pollutants can also be subdivided into two classes: primary and secondary.

$$\text{Source} \rightarrow \text{Primary pollutant} \rightarrow \text{Secondary pollutant}$$

A *primary pollutant* is a pollutant that is emitted directly from the source. In terms of atmospheric pollutants from petroleum, examples are carbon oxides, sulfur dioxide, and nitrogen oxides from fuel combustion operations:

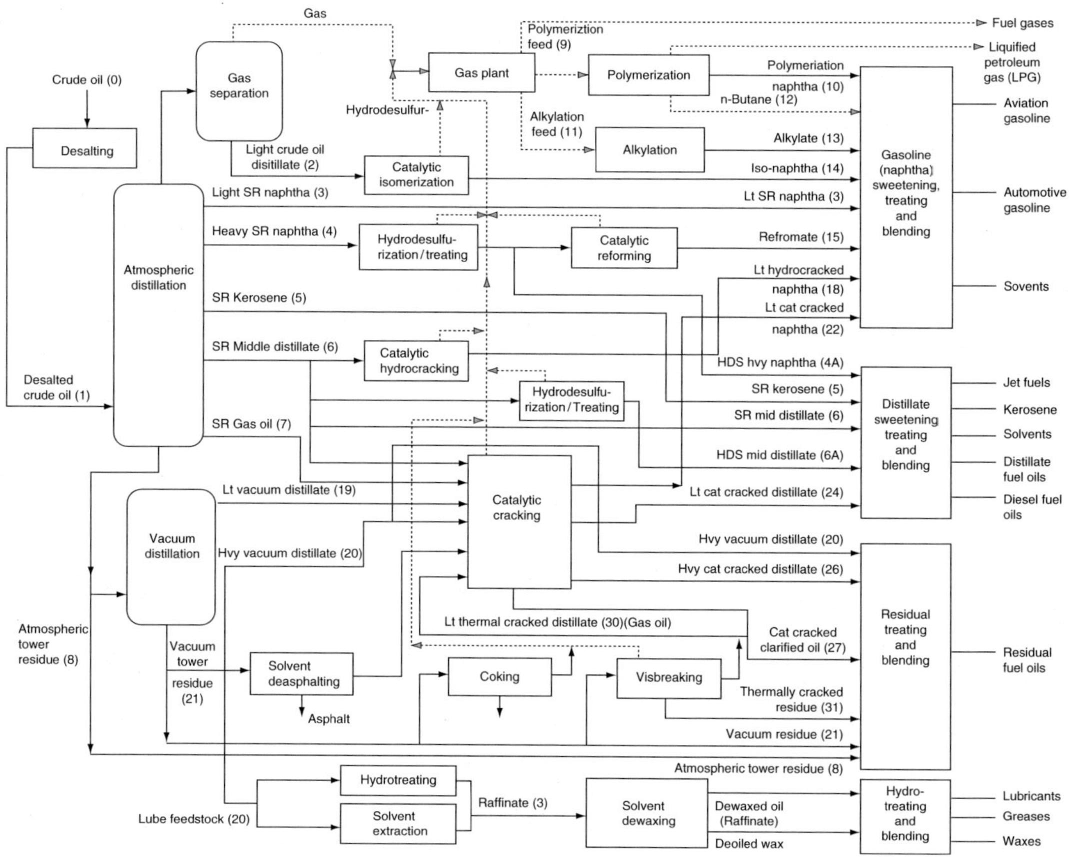


**Fig. 4.1** Schematic overview of a refinery (OSHA technical Manual, Section IV, Chapter 2: Petroleum Refining Processes, http://www.osha.gov/dts/osta/otm/otm_iv/otm_iv_2.html)

$$2[C]_{petroleum} + O_2 \rightarrow 2CO$$

$$[C]_{petroleum} + O_2 \rightarrow CO_2$$

$$2[N]_{petroleum} + O_2 \rightarrow 2NO$$

$$[N]_{petroleum} + O_2 \rightarrow NO_2$$

$$[S]_{petroleum} + O_2 \rightarrow SO_2$$

Hydrogen sulfide and ammonia are produced from processing sulfur-containing and nitrogen-containing feedstocks:

$$[S]_{petroleum} + H_2 \rightarrow H_2S + \text{hydrocarbons}$$

$$2[N]_{petroleum} + 3H_2 \rightarrow 2NH_3 + \text{hydrocarbons}$$

A *secondary pollutant* is a pollutant that is produced by the interaction of a primary pollutant with another chemical. A *secondary pollutant* may also be produced by dissociation of a primary pollutant, or other effects within a particular ecosystem. Again, using the atmosphere as an example, the formation of the constituents of acid rain is an example of the formation of secondary pollutants:

$$SO_2 + H_2O \rightarrow H_2SO_3 \text{ (sulfurous acid)}$$

$$2SO_2 + O_2 \rightarrow 2SO_3$$

$$SO_3 + H_2O \rightarrow H_2SO_4 \text{ (sulfuric acid)}$$

$$NO + H_2O \rightarrow HNO_2 \text{ (nitrous acid)}$$

$$3NO_2 + 2H_2O \rightarrow HNO_3 \text{ (nitric acid)}$$

In many cases, these secondary pollutants can have significant environmental effects, such as participation in the formation of acid rain and smog [5].

An *environmental regulation* is a legal mechanism that determines how the policy directives of an environmental law are to be carried out. An *environmental policy* is a requirement that specifies operating procedures that must be followed. An *environmental guidance* is a document developed by a governmental agency that outlines a position on a topic or which gives instructions on how a procedure must be carried out. It explains how to do something and provides governmental interpretations on a governmental act or policy.

**Table 4.1** Environmental regulations that apply to energy production

| | First enacted | Amended |
|---|---|---|
| Clean Air Act | 1970 | 1977<br>1990 |
| Clean Water Act (Water Pollution Control Act) | 1948 | 1965[a]<br>1972[b]<br>1977<br>1987[c] |
| Comprehensive Environmental Response, Compensation and Liability Act | 1980 | 1986[d] |
| Hazardous Material Transportation Act | 1974 | 1990 |
| Occupational Safety and Health Act | 1970 | 1987[e] |
| Oil Pollution Act | 1924 | 1990[f] |
| Resource Conservation and Recovery Act | 1976 | 1980[g] |
| Safe Drinking Water Act | 1974 | 1986[h] |
| Superfund Amendments and Reauthorization Act (SARA) | 1986 | |
| Toxic Substances Control Act | 1976 | 1984[i] |

[a]Water Quality Act
[b]Water Pollution Control Act
[c]Water Quality Act
[d]SARA Amendments
[e]Several amendments during the 1980s
[f]Interactive with various water pollution acts
[g]Federal cancer policy initiated
[h]Several amendments during the 1970s and 1980s
[i]Import rule enacted

## Environmental Regulations

Environmental issues range from the effects of pollutants on the population at large to effects on the lives of workers in various occupations where sickness or doisability can result from exposure to chemical agents [5, 7, 8].

There are a variety of regulations (Table 4.1) that apply to petroleum refining [6]. The most *popular* is the series of regulations known as the Clean Air Act that first was introduced in 1967 and was subsequently amended in 1970 and most recently in 1990. The most recent amendments provide stricter regulations for the establishment and enforcement of national ambient air quality standards for, as an example, sulfur dioxide. These standards do not stand alone, and there are many national standards for sulfur emissions.

The laws of relevance to the petroleum industry are:

### *The Clean Air Act Amendments*

The first *Clean Air Act* of 1970 and the 1977 Amendments consisted of three titles. *Title I* dealt with stationary air emission sources, *Title II* with mobile air emission sources, and *Title III* with definitions of appropriate terms as well as applicable standards for judicial review.

The *Clean Air Act Amendments of 1990* contain extensive provisions for control of the accidental release of toxic substances from storage or transportation as well as the formation of acid rain (acid deposition). In addition, the requirement that the standards be technology based removes much of the emotional perception that all chemicals are hazardous as well as the guesswork from legal enforcement of the legislation. The requirement also dictates environmental and health protection with an ample margin of safety.

## *The Water Pollution Control Act (The Clean Water Act)*

There are several acts that relate to the protection of the waterways in the United States but of particular interest to the petroleum industry in the present context is the *Water Pollution Control Act* (*Clean Water Act*). The objective of the Act is to restore and maintain the chemical, physical, and biological integrity of water systems.

The original Water Pollution Control Act of 1948 and The Water Quality Act of 1965 were generally limited to control of pollution of interstate waters and the adoption of water-quality standards by the states for interstate water within their borders. The first comprehensive water-quality legislation in the United States came into being in 1972 as the Water Pollution Control Act, which was amended in 1977 and retitled to become the Clean Water Act. Further amendments in 1978 were enacted to deal more effectively with spills of crude oil, with other amendments following in 1987 under the new name of the Water Quality Act.

Section 311 of the Clean Water Act includes elaborate provisions for regulating intentional or accidental discharges of petroleum and of hazardous substances. Included are response actions required for oil spills and the release or discharge of toxic and hazardous substances. As an example, the person in charge of a vessel or an onshore or offshore facility from which any chemical substance is discharged, in quantities equal to or exceeding its reportable quantity, must notify the appropriate federal agency as soon as such knowledge is obtained. The *Exxon Valdez* disaster and the recent spillage of oil into the Gulf of Mexico by BP are well-known examples of such a discharge of chemicals – the chemical being oil.

## *The Safe Drinking Water Act*

The Safe Drinking Water Act, first enacted in 1974, was amended several times in the 1970s and 1980s to set national drinking water standards. The Act calls for regulations that (1) apply to public water systems, (2) specify contaminants that may have any adverse effect on the health of persons, and (3) specify contaminant levels. Statutory provisions are included to cover underground injection control systems. The Act also requires maximum levels at which a contaminant must have no known or anticipated adverse effects on human health, thereby providing an *adequate margin of safety*.

The Superfund Amendments and Reauthorization Act (SARA) set the same standards for groundwater as for drinking water in terms of necessary cleanup and remediation of an inactive site that might be a former petroleum refinery. Under the Act, all underground injection activities must comply with the drinking water standards as well as meet specific permit conditions that are in unison with the provisions of the Clean Water Act.

## *The Resource Conservation and Recovery Act*

Since its initial enactment in 1976, the Resource Conservation and Recovery Act (RCRA) continues to promote safer waste management programs. Besides the regulatory requirements for waste management, the Act specifies the mandatory obligations of generators, transporters, and disposers of waste as well as those of owners and/or operators of waste treatment, storage, or disposal facilities. The waste might be garbage, refuse, and sludge from a treatment plant or from a water supply treatment plant or air pollution control facility and other discarded material, including solid, liquid, semisolid, or contained gaseous material resulting from industrial, commercial, mining, and agricultural operations and from community activities.

The Act also states that solid waste does not include solid, or dissolved, materials in domestic sewage, or solid or dissolved materials in irrigation return flows or industrial discharges. A solid waste becomes a hazardous waste if it exhibits any one of four specific characteristics: (1) ignitability, (2) reactivity, (3) corrosivity, or (4) toxicity. Certain types of solid wastes (e.g., household waste) are not considered to be hazardous, irrespective of their characteristics.

## *The Toxic Substances Control Act*

The Toxic Substances Control Act was first enacted in 1976 and was designed to provide controls for those chemicals that may threaten human health or the environment. Particularly hazardous are the cyclic nitrogen species and that often occur in high-boiling petroleum fractions, distillation residua, and cracked residua.

The Act specifies a *premanufacture notification* requirement by which any manufacturer must notify the Environmental Protection Agency at least 90 days prior to the production of a new chemical substance. Notification is also required even if there is a new use for the chemical that can increase the risk to the environment. No notification is required for chemicals that are manufactured in small quantities solely for scientific research and experimentation.

A *new chemical substance* is a chemical that is not listed in the Environmental Protection Agency Inventory of Chemical Substances or is an unlisted reaction product of two or more chemicals. In addition, the term "*chemical substance*" means any organic or inorganic substance of a particular molecular identity, including any combination of such substances occurring in whole or in part as a result of a chemical reaction or occurring in nature, and any element or uncombined radical. The term "*mixture*" means any combination of two or more chemical substances if the combination does not occur in nature and is not, in whole or in part, the result of a chemical reaction.

## *The Comprehensive Environmental Response, Compensation, and Liability Act*

The Comprehensive Environmental Response, Compensation, and Liability Act (CERCLA), generally known as *Superfund*, was first signed into law in 1980. The purpose of this Act is to provide a response mechanism for cleanup of any hazardous substance released, such as an accidental spill, or of a threatened release of a chemical.

Under this Act, a hazardous substance is any substance requiring (1) special consideration due to its toxic nature under the Clean Air Act, the Clean Water Act, or the Toxic Substances Control Act and (2) any waste that is hazardous waste under RCRA. Additionally, a pollutant or contaminant can be any other substance not necessarily designated by or listed in the Act but that *will or may reasonably* be anticipated to cause any adverse effect in organisms and/or their offspring.

## *The Occupational Safety and Health Act*

The *Occupational Safety and Health Administration* (*OSHA*) came into being in 1970 and is responsible for administering the Occupational Safety and Health Act. Occupational health hazards are those factors arising in or from the occupational environment that adversely impact health.

The goal of the Act is to ensure that employees do not suffer material impairment of health or functional capacity due to a lifetime occupational exposure to chemicals. The Act is also responsible for the means by which chemicals are contained.

## *The Oil Pollution Act*

The Oil Pollution Act of 1990 deals with pollution of waterways by crude oil. The Act specifically deals with petroleum vessels and onshore and offshore facilities and imposes strict liability for oil spills on their owners and operators.

## *The Hazardous Materials Transportation Act*

The Hazardous Materials Transportation Act authorizes the establishment and enforcement of hazardous material regulations for all modes of transportation by highway, water, and rail. The purpose of the Act is to ensure safe transportation of hazardous materials. The Act prevents any person from offering or accepting for transportation a hazardous material (any substance or material, including a hazardous substance and hazardous waste, which is capable of posing an unreasonable risk to health, safety, and property) for transportation anywhere within the United States.

The Act also imposes restrictions on the packaging, handling, and shipping of hazardous materials in which the appropriate documentation, markings, labels, and safety precautions are required.

# Processes and Process Wastes

Enhanced oil recovery (EOR) processes rely upon the use of chemical or thermal energy to recover crude oil that is trapped in pores of reservoir rock after primary and secondary (waterflood) crude oil production has ceased [2, 9].

Chemicals used for enhanced oil recovery include surfactants to reduce the interfacial tension between oil and water, and oil and rock interfaces. Many microorganisms produce biosurfactants and perform this activity by fermentation of inexpensive raw materials such as molasses. Several biosurfactants are being evaluated for use in enhanced oil recovery.

A major issue in enhanced oil recovery processes is the variation of permeability in petroleum reservoirs. When water is injected to displace oil, the water will preferentially flow through areas of highest permeability, and bypass much of the oil. When chemicals are injected, they may also flow preferentially into high-permeability zones with the water, but then will grow and block those zones. When high-permeability zones are blocked, sweep efficiency is improved, and thus oil recovery.

## Transportation

In addition to the conventional meaning of the term *process*, the *recovery* and *transportation* of petroleum also needs to be considered here.

Oil spills during petroleum *transportation* have been the most visible problem. There have also been instances of oil wells at sea "blowing out," or flowing uncontrollably, although the amounts from blowouts tend to be smaller than from tanker accidents.

*Tanker accidents* typically have a severe impact on ecosystems because of the rapid release of hundreds of thousands of barrels of crude oil (or crude oil products) into a small area.

While oil is at least theoretically biodegradable, large-scale spills can overwhelm the ability of the ecosystem to break the oil down. Over time, the lighter portions of crude oil evaporate, leaving the nonvolatile portion. Oil itself breaks down the protective waxes and oils in the feathers and fur of birds and animals. Some crude oils contain toxic metals as well. The impact of any given oil spill is determined by the size of the spill, the degree of dispersal, and the chemistry of the oil. Spills at sea are thought to have a less detrimental effect than spills in shallow waters.

## Refining

Petroleum *refining* is a complex sequence of chemical events that result in the production of a variety of products (Fig. 4.1). In fact, petroleum refining might be considered as a collection of individual, yet related processes that are each capable of producing effluent streams [2, 6].

Petroleum refining, as it is currently known, will continue at least for the next 3 decades. In spite of the various political differences that have caused fluctuations in petroleum imports, it is reality that imports of petroleum and petroleum products into the United States are on the order of 67% of the total requirements [2].

Petroleum, like any other raw material, is capable of producing chemical waste. By 1960, the petroleum refining industry had become well established throughout the world. Effluent water, atmospheric emissions, and combustion products also became a focus of increased technical attention [4, 5, 10–13].

Refineries produce a wide variety of products from petroleum feedstocks and feedstock blends [2, 14]. During petroleum refining, refineries use and generate an enormous amount of chemicals, some of which are present in air emissions, wastewater, or solid wastes (Table 4.2) [2, 6]. Emissions are also created through the combustion of fuels, and as by-products of chemical reactions occurring when petroleum fractions are upgraded. A large source of air emissions is, generally, the process heaters and boilers that produce carbon monoxide, sulfur oxides, and nitrogen oxides, leading to pollution and the formation of acid rain.

**Table 4.2** Emissions and waste from refinery processes

| Process | Air emissions | Residual wastes generated |
|---|---|---|
| Crude oil desalting | Heater stack gas (CO, $SO_x$, $NO_x$, hydrocarbons, and particulates), fugitive emissions (hydrocarbons) | Crude oil/desalter sludge (iron rust, clay, sand, water, emulsified oil and wax, metals) |
| Atmospheric distillation<br>Vacuum distillation | Heater stack gas (CO, $SO_x$, $NO_x$, hydrocarbons and particulates), vents and fugitive emissions (hydrocarbons), steam ejector emissions (hydrocarbons), heater stack gas (CO, $SO_x$, $NO_x$, hydrocarbons, and particulates), vents and fugitive emissions (hydrocarbons) | Typically, little or no residual waste generated |
| Thermal cracking/ visbreaking | Heater stack gas (CO, $SO_x$, $NO_x$, hydrocarbons, and particulates), vents and fugitive emissions (hydrocarbons) | Typically, little or no residual waste generated |
| Coking | Heater stack gas (CO, $SO_x$, $NO_x$, hydrocarbons, and particulates), vents and fugitive emissions (hydrocarbons), and decoking emissions (hydrocarbons and particulates) | Coke dust (carbon particles and hydrocarbons) |
| Catalytic cracking | Heater stack gas (CO, $SO_x$, $NO_x$, hydrocarbons, and particulates), fugitive emissions (hydrocarbons), and catalyst regeneration (CO, $NO_x$, $SO_x$, and particulates) | Spent catalysts (metals from crude oil and hydrocarbons), spent catalyst fines from electrostatic precipitators (aluminum silicate and metals) |
| Catalytic hydrocracking | Heater stack gas (CO, $SO_x$, $NO_x$, hydrocarbons, and particulates), fugitive emissions (hydrocarbons), and catalyst regeneration (CO, $NO_x$, $SO_x$, and catalyst dust) | Spent catalysts fines |
| Hydrotreating/ hydroprocessing | Heater stack gas (CO, $SO_x$, $NO_x$, hydrocarbons, and particulates), vents and fugitive emissions (hydrocarbons), and catalyst regeneration (CO, $NO_x$, $SO_x$) | Spent catalyst fines (aluminum silicate and metals) |
| Alkylation | Heater stack gas (CO, $SO_x$, $NO_x$, hydrocarbons, and particulates), vents and fugitive emissions (hydrocarbons) | Neutralized alkylation sludge (sulfuric acid or calcium fluoride, hydrocarbons) |
| Isomerization | Heater stack gas (CO, $SO_x$, $NO_x$, hydrocarbons, and particulates), HCl (potentially in light ends), vents and fugitive emissions (hydrocarbons) | Calcium chloride sludge from neutralized HCl gas |

(continued)

**Table 4.2** (continued)

| Process | Air emissions | Residual wastes generated |
|---|---|---|
| Polymerization | $H_2S$ from caustic washing | Spent catalyst containing phosphoric acid |
| Catalytic reforming | Heater stack gas (CO, $SO_x$, $NO_x$, hydrocarbons, and particulates), fugitive emissions (hydrocarbons), and catalyst regeneration (CO, $NO_x$, $SO_x$) | Spent catalyst fines from electrostatic precipitators (alumina silicate and metals) |
| Solvent extraction | Fugitive solvents | Little or no residual wastes generated |
| Dewaxing | Fugitive solvents, heaters | Little or no residual wastes generated |
| Propane deasphalting | Heater stack gas (CO, $SO_x$, $NO_x$, hydrocarbons and particulates), fugitive propane | Little or no residual wastes generated |
| Wastewater treatment | Fugitive emissions ($H_2S$, $NH_3$, and hydrocarbons) | API separator sludge (phenols, metals and oil), chemical precipitation sludge (chemical coagulants, oil), DAF floats, biological sludge (metals, oil, suspended solids), spent lime |

$$CO_2 + H_2O \rightarrow H_2CO_3 \text{ (carbonic acid)}$$

$$SO_2 + H_2O \rightarrow H_2SO_3 \text{ (sulfurous acid)}$$

$$2SO_2 + O_2 \rightarrow 2SO_3$$

$$SO_3 + H_2O \rightarrow H_2SO_4 \text{ (sulfuric acid)}$$

$$NO + H_2O \rightarrow HNO_2 \text{(nitrousacid)}$$

$$2NO + O_2 \rightarrow NO_2$$

$$NO_2 + H_2O \rightarrow HNO_3 \text{ (nitric acid)}$$

Hence, there is the need for gas-cleaning operations on a refinery site so that such gases are cleaned from the gas stream prior to entry into the atmosphere.

Fugitive emissions of volatile hydrocarbons arise from leaks in valves, pumps, flanges, and other similar sources where crude and its fractions flow through the system. While individual leaks may be minor, the combination of fugitive emissions from various sources can be substantial. These emissions are controlled primarily through leak detection and repair programs and occasionally through the use of special leak-resistant equipment.

In terms of individual processes, the potential for waste generation and, hence, leakage of emissions is as follows.

## Desalting

Petroleum often contains water, inorganic salts, suspended solids, and water-soluble trace metals. As a first step in the refining process, to reduce corrosion, plugging, and fouling of equipment and to prevent poisoning the catalysts in processing units, these contaminants must be removed by desalting (dehydration) [2].

The two most typical methods of petroleum desalting are (1) chemical separation and (2) electrostatic separation. In chemical desalting, water and chemical surfactant (demulsifiers) are added to the petroleum, heated so that salts and other impurities dissolve into the water or attach to the water, and then held in a tank where they settle out. Electrical desalting is the application of high-voltage electrostatic charges to concentrate suspended water globules in the bottom of the settling tank. Surfactants are added only when the crude has a large amount of suspended solids. A third and less-common process involves filtering heated petroleum using diatomaceous earth.

In the desalting process, the feedstock crude oil is heated to between 65°C and 177°C (150°F and 350°F) to reduce viscosity and surface tension for easier mixing and separation of the water, but the temperature is limited by the vapor pressure of the petroleum constituents. In both methods, other chemicals may be added – ammonia is often used to reduce corrosion and caustic or acid may be added to adjust the pH of the water wash. Wastewater and contaminants are discharged from the bottom of the settling tank to the wastewater treatment facility while the desalted crude is continuously drawn from the top of the settling tanks and sent to the crude distillation tower.

Desalting (Fig. 4.2) creates an oily desalter sludge that may be a hazardous waste and a high temperature salt wastewater stream (treated along with other refinery wastewaters). The primary polluting constituents in desalter wastewater include hydrogen sulfide, ammonia, phenol, high levels of suspended solids, and dissolved solids, with a high biochemical oxygen demand (BOD). In some cases, it is possible to recycle the desalter effluent water back into the desalting process, depending upon the type of crude being processed.

## Distillation

Atmospheric and vacuum distillation units (Figs. 4.3 and 4.4) are closed processes and exposures are expected to be minimal.

Both atmospheric distillation units and vacuum distillation units produce refinery fuel gas streams containing a mixture of light hydrocarbons, hydrogen sulfide, and ammonia. These streams are processed through gas treatment and sulfur

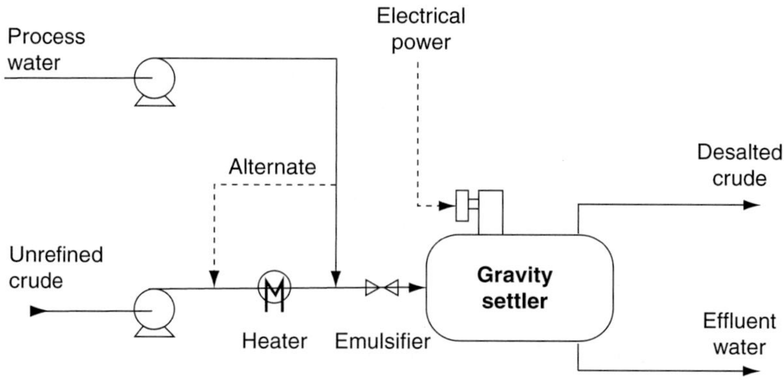


**Fig. 4.2** Schematic of an electrostatic desalting unit (OSHA technical manual, Section IV, Chapter 2: Petroleum Refining Processes, http://www.osha.gov/dts/osta/otm/otm_iv/otm_iv_2.html)

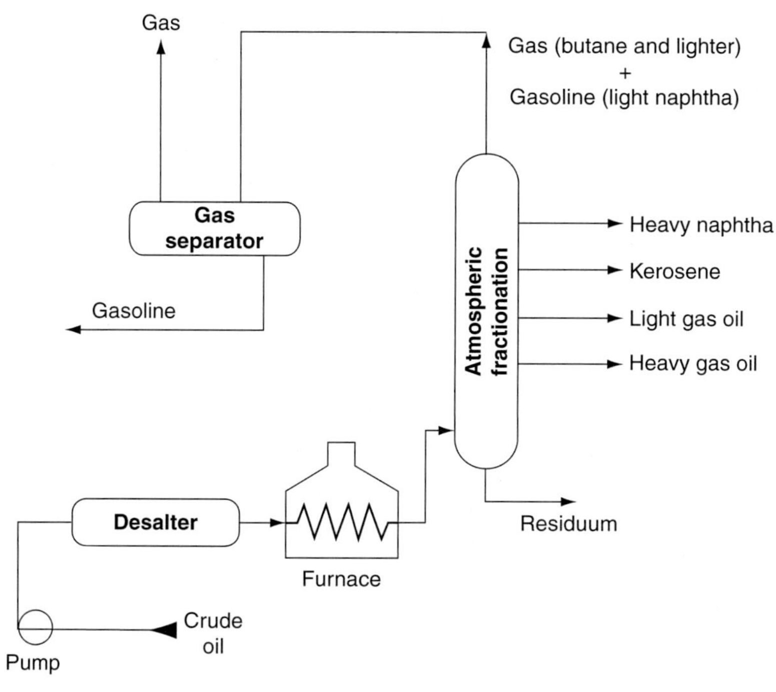


**Fig. 4.3** An atmospheric distillation unit (OSHA technical manual, Section IV, Chapter 2: Petroleum Refining Processes, http://www.osha.gov/dts/osta/otm/otm_iv/otm_iv_2.html)

recovery units to recover fuel gas and sulfur. Sulfur recovery creates emissions of ammonia, hydrogen sulfide, sulfur oxides, and nitrogen oxides.

When sour (high-sulfur) petroleum is processed, there is potential for exposure to hydrogen sulfide in the preheat exchanger and furnace, tower flash zone and

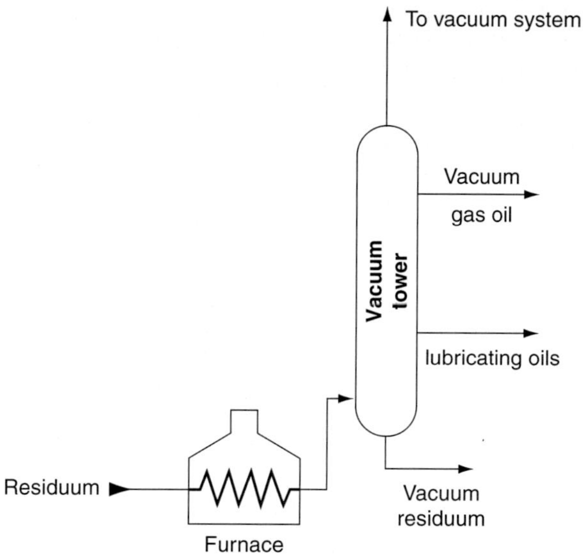


**Fig. 4.4** A vacuum distillation unit (OSHA technical manual, Section IV, Chapter 2: Petroleum Refining Processes, http://www.osha.gov/dts/osta/otm/otm_iv/otm_iv_2.html)

overhead system, vacuum furnace and tower, and bottoms exchanger. Hydrogen chloride may be present in the preheat exchanger, tower top zones, and overheads. Wastewater may contain water-soluble sulfides in high concentrations and other water-soluble compounds such as ammonia, chlorides, phenol, mercaptans, etc., depending upon the crude feedstock and the treatment chemicals. Safe work practices and/or the use of appropriate personal protective equipment may be needed for exposures to chemicals and other hazards such as heat and noise, and during sampling, inspection, maintenance, and turnaround activities.

The primary source of emissions is combustion of fuels in the crude preheat furnace and in boilers that produce steam for process heat and stripping. When operating in an optimum condition and burning cleaner fuels (e.g., natural gas, refinery gas), these heating units create relatively low emissions of sulfur oxides, ($SO_x$), nitrogen oxides ($NO_x$), carbon monoxide (CO), hydrogen sulfide ($H_2S$), particulate matter, and volatile hydrocarbons. If fired with lower grade fuels (e.g., refinery fuel pitch, coke) or operated inefficiently (incomplete combustion), heaters can be a significant source of emissions.

Petroleum distillation units generate considerable wastewater – often an oily sour wastewater and the constituents of sour wastewater streams include hydrogen sulfide, ammonia, suspended solids, chlorides, mercaptans, and phenol, characterized by a high pH.

## Visbreaking and Coking

*Visbreaking* (Fig. 4.5), like many thermal cracking processes, tends to produce a relatively small amount of fugitive emissions and sour wastewater [2, 6]. Usually

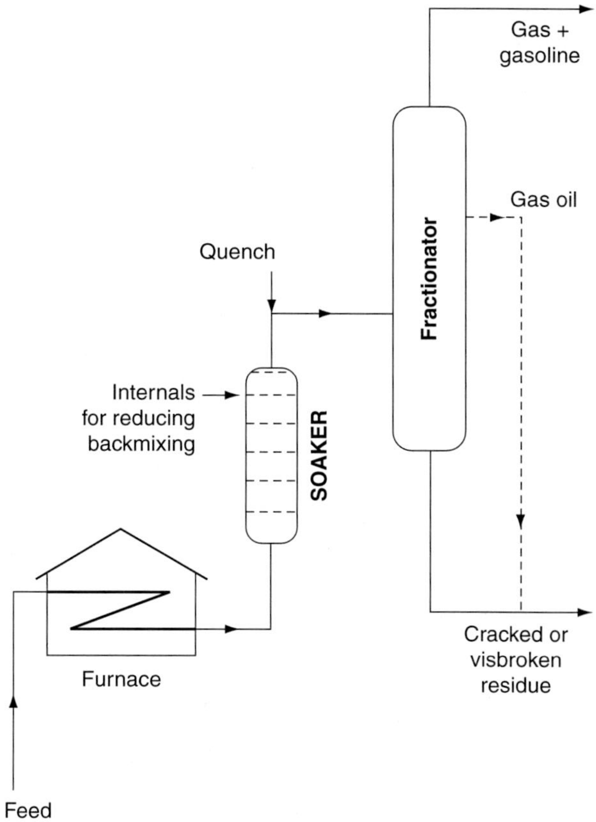


**Fig. 4.5** A soaker visbreaking unit (OSHA technical manual, Section IV, Chapter 2: Petroleum Refining Processes, http://www.osha.gov/dts/osta/otm/otm_iv/otm_iv_2.html)

some wastewater is produced from steam strippers and the fractionator. Wastewater is also generated during unit cleanup and cooling operations and from the steam injection process to remove organic deposits from the soaker or from the coil. Combined wastewater flows from thermal cracking and coking processes are about 3.0 gal per barrel of process feed.

*Delayed coking* is the oldest, most widely used process and has changed very little in the 5 or more decades in which it has been on stream in refineries [2]. *Fluid coking* is a continuous fluidized solids process that cracks feed thermally over heated coke particles in a reactor vessel to gas, liquid products, and coke [2]. Heat for the process is supplied by partial combustion of the coke, with the remaining coke being drawn as product. The new coke is deposited in a thin fresh layer on the outside surface of the circulating coke particle.

*Coking processes* (Figs. 4.6 and 4.7) produce a relatively small amount of sour wastewater from steam strippers and fractionators. Wastewater is generated during coke removal and cooling operations and from the steam injection process to cut coke from the coke drums. Combined wastewater flows from thermal cracking and coking processes are about 3.0 gal per barrel of process feed.

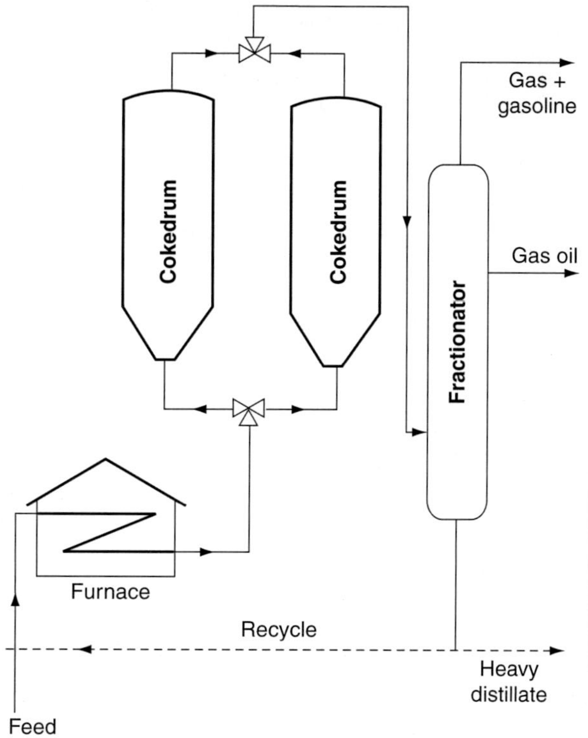


**Fig. 4.6** A delayed coking unit (OSHA technical manual, Section IV, Chapter 2: Petroleum Refining Processes, http://www.osha.gov/dts/osta/otm/otm_iv/otm_iv_2.html)

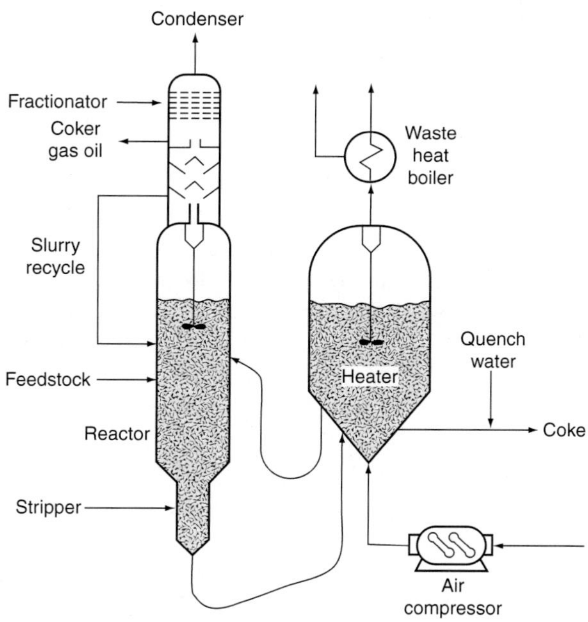


**Fig. 4.7** A fluid coking unit [12]

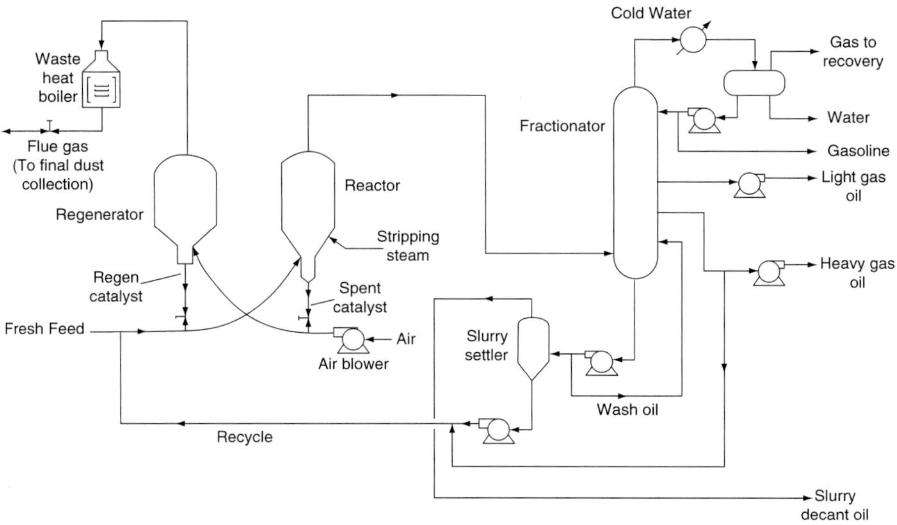


**Fig. 4.8** Schematic of a fluid catalytic cracking unit [12]

Particulate emissions from decoking can also be considerable. Coke-laden water from decoking operations in delayed cokers (hydrogen sulfide, ammonia, suspended solids), coke dust (carbon particles and hydrocarbons) occur.

## Fluid Catalytic Cracking

Fluid catalytic cracking (Fig. 4.8) is one of the largest sources of air emission in refineries [2, 6]. Air emissions are released in process heater flue gas, as fugitive emissions from leaking valves and pipes, and during regeneration of the cracking catalyst. If not controlled, catalytic cracking is one of the most substantial sources of carbon monoxide and particulate emissions in the refinery. In non-attainment areas where carbon monoxide and particulates are above acceptable levels, carbon monoxide waste heat boilers (CO boiler) and particulate controls are employed. Carbon monoxide produced during regeneration of the catalyst is converted to carbon dioxide either in the regenerator or further downstream in a carbon monoxide waste heat boiler (CO boiler). Catalytic crackers are also significant sources of sulfur oxides and nitrogen oxides. The nitrogen oxides produced by catalytic crackers is expected to be a major target of emissions reduction in the future.

Catalytic cracking units, like coking units, usually include some form of fractionation or steam stripping as part of the process configuration. These units all produce sour waters and sour gases containing some hydrogen sulfide and ammonia. Like crude oil distillation, some of the toxic releases reported by the refining industry are generated through sour water and gases, notably ammonia. Gaseous ammonia often leaves fractionating and treating processes in the sour gas along with hydrogen sulfide and fuel gases [6].

Catalytic cracking (primarily fluid catalytic cracking) generates considerable sour wastewater from fractionators used for product separation, from steam strippers used to strip oil from catalysts, and in some cases from scrubber water. The steam stripping process used to purge and regenerate the catalysts can contain metal impurities from the feed in addition to oil and other contaminants. Sour wastewater from the fractionator/gas concentration units and steam strippers contains oil, suspended solids, phenols, cyanides, hydrogen sulfide, ammonia, spent catalysts, metals from crude oil, and hydrocarbons.

Catalytic cracking generates significant quantities of spent process catalysts (containing metals from crude oils and hydrocarbons) that are often sent off-site for disposal or recovery or recycling. Management options can include land filling, treatment, or separation and recovery of the metals. Metals deposited on catalysts are often recovered by third-party recovery facilities. Spent catalyst fines (containing aluminum silicate and metals) from electrostatic precipitators are also sent off-site for disposal and/or recovery options.

Catalytic crackers also produce a significant amount of fine catalyst dust that results from the constant movement of catalyst grains against each other. This dust contains primarily alumina ($Al_2O_3$) and small amounts of nickel (Ni) and vanadium (V), and is generally carried along with the carbon monoxide stream to the carbon monoxide waste heat boiler. The dust is separated from the carbon dioxide stream exiting the boiler through the use of cyclones, flue gas scrubbing, or electrostatic precipitators.

## Hydrocracking and Hydrotreating

*Hydrocracking* (Fig. 4.9) generates air emissions through process heater flue gas, vents, and fugitive emissions [2, 6]. Unlike fluid catalytic cracking catalysts, hydrocracking catalysts are usually regenerated off-site after months or years of operations, and little or no emissions or dust is generated. However, the use of heavy oil as feedstock to the unit can change this balance.

Hydrocracking produces less sour wastewater than catalytic cracking. Hydrocracking, like catalytic cracking, produces sour wastewater at the fractionator. These processes include processing in a separator (API separator, corrugated plate interceptor) that creates sludge [2, 6]. Physical or chemical methods are then used to separate the remaining emulsified oils from the wastewater. Treated wastewater may be discharged to public wastewater treatment, to a refinery secondary treatment plant for ultimate discharge to public wastewater treatment, or may be recycled and used as process water. The separation process permits recovery of usable oil, and also creates a sludge that may be recycled or treated as a hazardous waste.

Like catalytic cracking, hydrocracking processes generate toxic metal compounds, many of which are present in spent catalyst sludge and catalyst fines

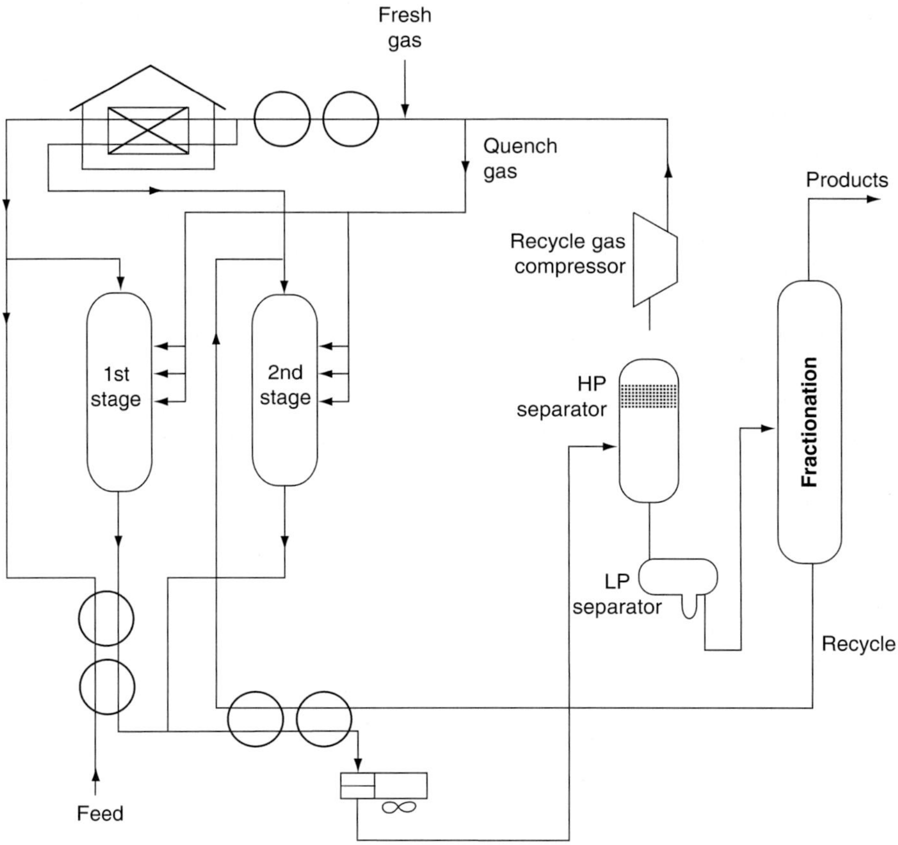


**Fig. 4.9** A two-stage hydrocracking unit (OSHA technical manual, Section IV, Chapter 2: Petroleum Refining Processes, http://www.osha.gov/dts/osta/otm/otm_iv/otm_iv_2.html)

generated from catalytic cracking and hydrocracking. These include metals such as nickel (Ni), cobalt (Co), and molybdenum (Mo).

*Hydrotreating* is the less severe removal of heteroatomic species by treatment of a feedstock or product in the presence of hydrogen [2, 6]. The process (Fig. 4.10) generates air emissions through process heater flue gas, vents, and fugitive emissions [6]. Unlike fluid catalytic cracking catalysts, hydrotreating catalysts are usually regenerated off-site after months or years of operations, and little or no emissions or dust is generated from the catalyst regeneration process at the refinery. Air emissions factors for emissions from process heaters and boilers used throughout the refinery can be calculated (Emissions Factors & AP 42, *Compilation of Air Pollutant Emission Factors*).

Fugitive air emissions of volatile components released during hydrotreating may also be toxic components. These include toluene, benzene, xylenes, and other volatiles that are reported as toxic chemical releases under the EPA Toxics Release Inventory.

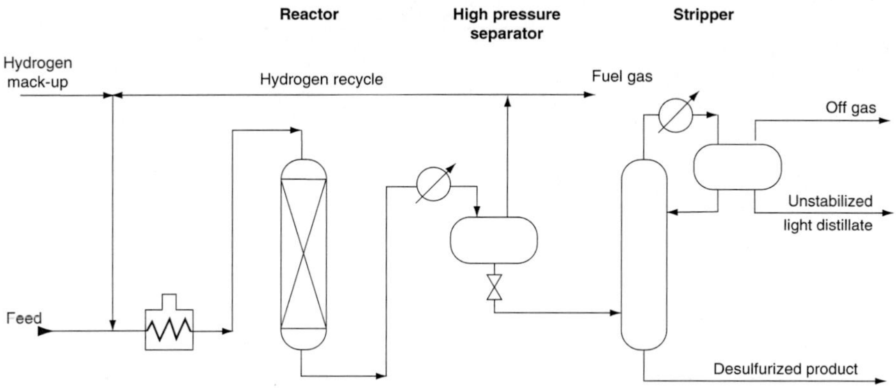


**Fig. 4.10** A distillate hydrotreating unit (OSHA technical manual, Section IV, Chapter 2: Petroleum Refining Processes, http://www.osha.gov/dts/osta/otm/otm_iv/otm_iv_2.html)

Hydrotreating generates sour wastewater from fractionators used for product separation. Like most separation processes in the refinery, the process water used in fractionators often comes in direct contact with oil, and can be highly contaminated. It also contains hydrogen sulfide and ammonia and must be treated along with other refinery sour waters.

Oily sludge from the wastewater treatment facility that result from treating oily and/or sour wastewaters from hydrotreating and other refinery processes may be hazardous wastes, depending on how they are managed. These include API separator sludge, primary treatment sludge, sludge from various gravitational separation units, and float from dissolved air flotation units.

Hydrotreating also produces some residuals in the form of spent catalyst fines, usually consisting of aluminum silicate and some metals (e.g., cobalt, molybdenum, nickel, tungsten). Spent hydrotreating catalyst is now listed as a hazardous waste (K171) (except for most support material). Hazardous constituents of this waste include benzene and arsenia (arsenic oxide, $As_2O_3$). The support material for these catalysts is usually an inert ceramic (e.g., alumina, $Al_2O_3$).

## Alkylation and Polymerization

Alkylation (Fig. 4.11) combines low-molecular-weight olefins (primarily a mixture of propylene and butylene) with isobutene in the presence of a catalyst, either sulfuric acid or hydrofluoric acid [2]. The product is called alkylate and is composed of a mixture of high-octane, branched-chain paraffinic hydrocarbons. Alkylate is a premium blending stock because it has exceptional antiknock properties and is clean burning. The octane number of alkylate depends mainly upon the kind of olefins used and upon operating conditions.

Emissions from alkylation processes (Figs. 4.11 and 4.12) and polymerization processes (Fig. 4.13) include fugitive emissions of volatile constituents in the feed,

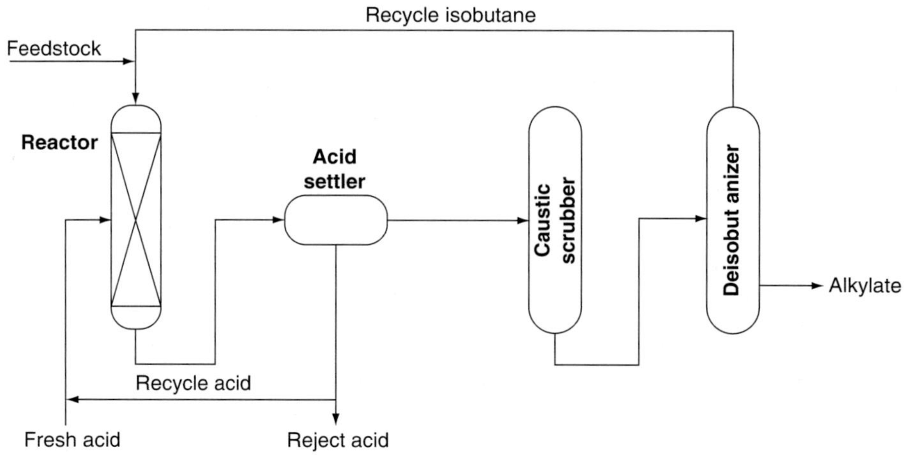


**Fig. 4.11** An alkylation unit (sulfuric acid catalyst) (OSHA technical manual, Section IV, Chapter 2: Petroleum Refining Processes, http://www.osha.gov/dts/osta/otm/otm_iv/otm_iv_2.html)

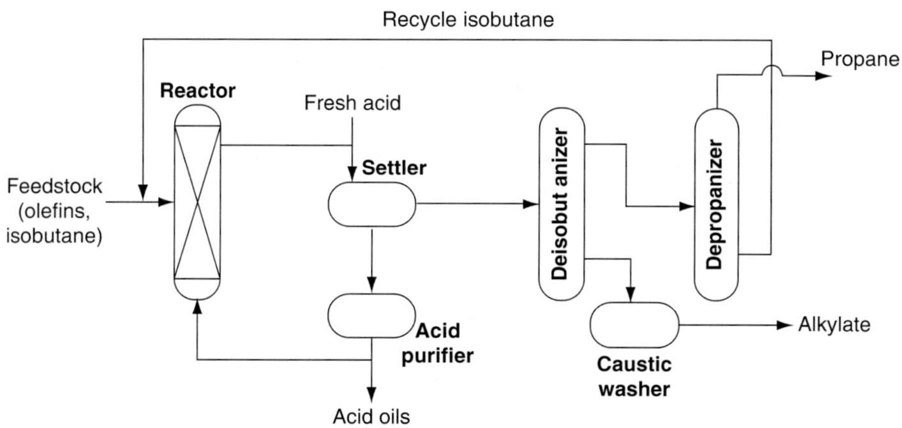


**Fig. 4.12** An alkylation unit (hydrogen fluoride catalyst) (OSHA technical manual, Section IV, Chapter 2: Petroleum Refining Processes, http://www.osha.gov/dts/osta/otm/otm_iv/otm_iv_2.html)

and emissions that arise from process vents during processing. These can take the form of acidic hydrocarbon gases, nonacidic hydrocarbon gases, and fumes that may have a strong odor (from sulfonated organic compounds and organic acids, even at low concentrations). To prevent releases of hydrofluoric acid, refineries install a variety of mitigation and control technologies (e.g., acid inventory reduction, hydrogen fluoride detection systems, isolation valves, rapid acid transfer systems, and water spray systems).

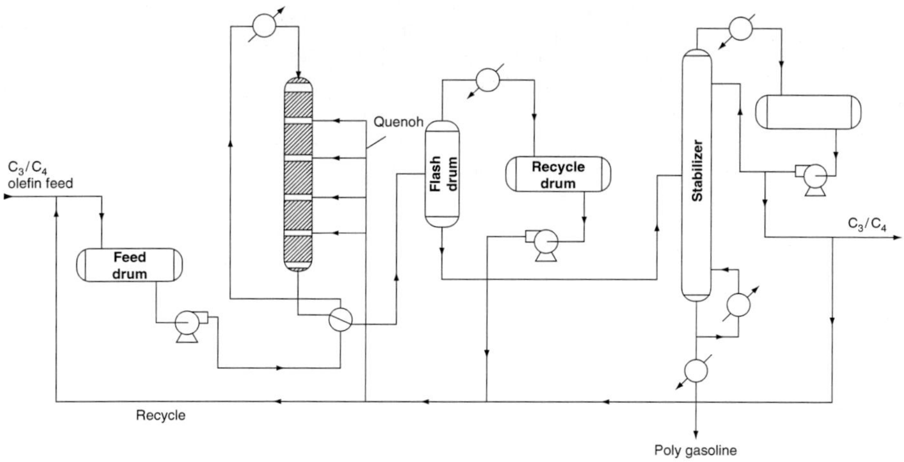


**Fig. 4.13** A polymerization process (OSHA technical manual, Section IV, Chapter 2: Petroleum Refining Processes, http://www.osha.gov/dts/osta/otm/otm_iv/otm_iv_2.html)

In hydrofluoric acid alkylation processes, acidic hydrocarbon gases can originate anywhere hydrogen fluoride is present (e.g., during a unit upset, unit shutdown, or maintenance) [2, 6]. Hydrofluoric acid alkylation units are designed to pipe these gases from acid vents and valves to a separate closed-relief system where the acid is neutralized.

Another source of emissions is combustion of fuels in process boilers to produce steam for strippers. As with all process heaters in the refinery, these boilers produce significant emissions of sulfur oxides, nitrogen oxides, carbon monoxide, particulate matter, and volatile hydrocarbons.

Alkylation generates relatively low volumes of wastewater, primarily from water washing of the liquid reactor products. Wastewater is also generated from steam strippers, depropanizers, and debutanizers, and can be contaminated with oil and other impurities. Liquid process waters (hydrocarbons and acid) originate from minor undesirable side reactions and from feed contaminants, and usually exit as a bottoms stream from the acid regeneration column. The bottom layer is an acid–water mixture that is sent to the neutralizing drum. The acid in this liquid eventually ends up as insoluble calcium fluoride.

Sulfuric acid alkylation generates considerable quantities of spent acid that must be removed and regenerated. Nearly all the spent acid generated at refineries is regenerated and recycled and, although technology for on-site regeneration of spent sulfuric acid is available, the supplier of the acid may perform this task off-site. If sulfuric acid production capacity is limited, acid regeneration is often done on-site. The development of internal acid regeneration for hydrofluoric acid units has virtually eliminated the need for external regeneration, although most operations retain one for start-ups or during periods of high feed contamination.

Both sulfuric acid and hydrofluoric acid alkylation units generate neutralization sludge from treatment of acid-laden streams with caustic solutions in neutralization or

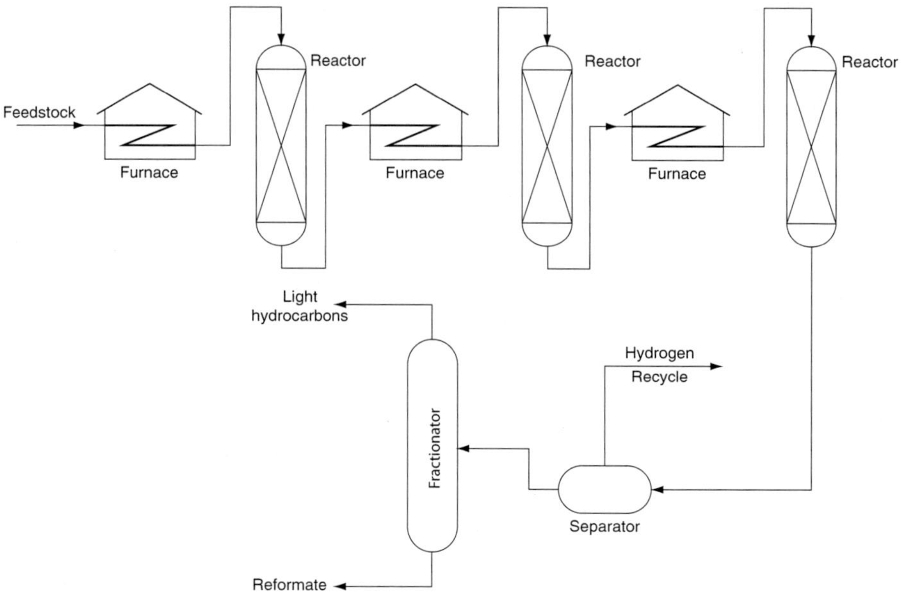


**Fig. 4.14** A catalytic reforming unit (OSHA technical manual, Section IV, Chapter 2: Petroleum Refining Processes, http://www.osha.gov/dts/osta/otm/otm_iv/otm_iv_2.html)

wash systems. Sludge from hydrofluoric acid alkylation neutralization systems consists largely of calcium fluoride and unreacted lime, and is usually disposed of in a landfill. It can also be directed to steel manufacturing facilities, where the calcium fluoride can be used as a neutral flux to lower the slag-melting temperature and improve slag fluidity. Calcium fluoride can also be routed back to a hydrofluoric acid manufacturer.

A basic step in hydrofluoric acid manufacture is the reaction of sulfuric acid with fluorspar (calcium fluoride) to produce hydrogen fluoride and calcium sulfate. Spent alumina is also generated by the defluorination of some hydrofluoric acid alkylation products over alumina. It is disposed of or sent to the alumina supplier for recovery. Other solid residuals from hydrofluoric acid alkylation include any porous materials that may have come in contact with the hydrofluoric acid.

## Catalytic Reforming

Catalytic reforming (Fig. 4.14) converts alkanes to cycloalkanes and to aromatics and emissions from catalytic reforming include fugitive emissions of volatile constituents in the feed, and emissions from process heaters and boilers [2, 6]. As with all process heaters in the refinery, combustion of fossil fuels produces emissions of sulfur oxides, nitrogen oxides, carbon monoxide, particulate matter, and volatile hydrocarbons.

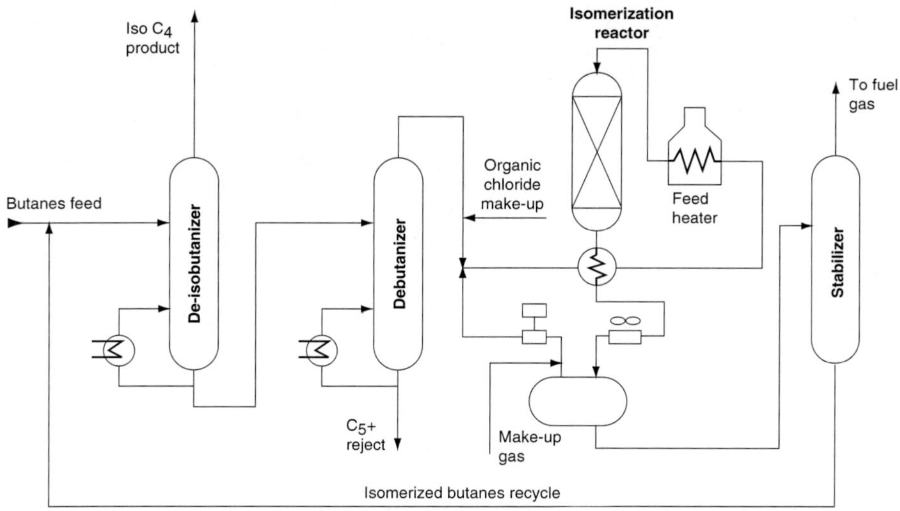


**Fig. 4.15** A butane isomerization unit (OSHA technical manual, Section IV, Chapter 2: Petroleum Refining Processes, http://www.osha.gov/dts/osta/otm/otm_iv/otm_iv_2.html)

Benzene, toluene, and the xylene isomers are toxic aromatic chemicals that are produced during the catalytic reforming process and used as feedstocks in chemical manufacturing. Due to their highly volatile nature, fugitive emissions of these chemicals are a source of their release to the environment during the reforming process. Point air sources may also arise during the process of separating these chemicals.

In a continuous reformer, some particulate and dust matter can be generated as the catalyst moves from reactor to reactor, and is subject to attrition. However, due to catalyst design, little attrition occurs, and the only outlet to the atmosphere is the regeneration vent, which is most often scrubbed with a caustic to prevent emission of hydrochloric acid (this also removes particulate matter). Emissions of carbon monoxide and hydrogen sulfide may occur during regeneration of catalyst.

## Isomerization

Isomerization (Fig. 4.15) converts n-butane, n-pentane, and n-hexane into their respective iso-paraffins of substantially higher octane number [2]. The straight-chain paraffins are converted to their branched-chain counterparts whose component atoms are the same but are arranged in a different geometric structure. Isomerization is important for the conversion of n-butane into iso-butane, to provide additional feedstock for alkylation units, and the conversion of normal pentanes and hexanes into higher branched isomers for gasoline blending.

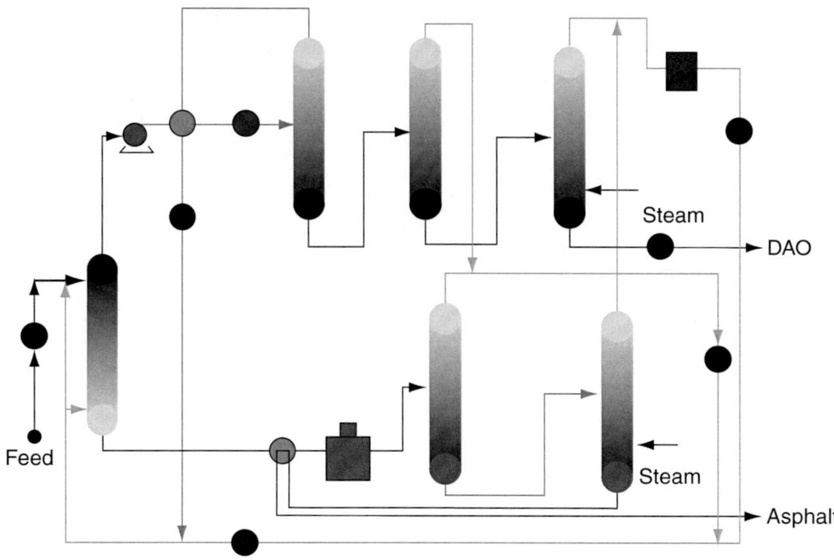


**Fig. 4.16** A deasphalting unit

Isomerization processes produce sour water and caustic wastewater. The ether manufacturing process utilizes a water wash to extract methanol or ethanol from the reactor effluent stream. After the alcohol is separated, this water is recycled back to the system and is not released. In those cases where chloride catalyst activation agents are added, a caustic wash is used to neutralize any entrained hydrogen chloride. This process generates caustic wash water that must be treated before being released.

## Deasphalting and Dewaxing

Propane deasphalting (Fig. 4.16) produces lubricating oil base stocks by extracting asphaltenes and resins from vacuum distillation residua [2]. Propane is the usual solvent of choice due to its unique solvent properties. At lower temperatures (38–60°C, 100–140°F), paraffins are very soluble in propane, and at higher temperatures, (approximately 93°C, 200°F) hydrocarbons are almost insoluble in propane. The propane deasphalting process is similar to solvent extraction in that a packed or baffled extraction tower or rotating disk contactor is used to mix the oil feed stocks with the solvent.

Air emissions may arise from fugitive propane emissions and process vents. These include heater stack gas (carbon monoxide, sulfur oxides, nitrogen oxides, and particulate matter) as well as hydrocarbon emission such as fugitive propane and fugitive solvents. Steam stripping wastewater (oil and solvents) and solvent recovery wastewater (oil and propane) are also produced.

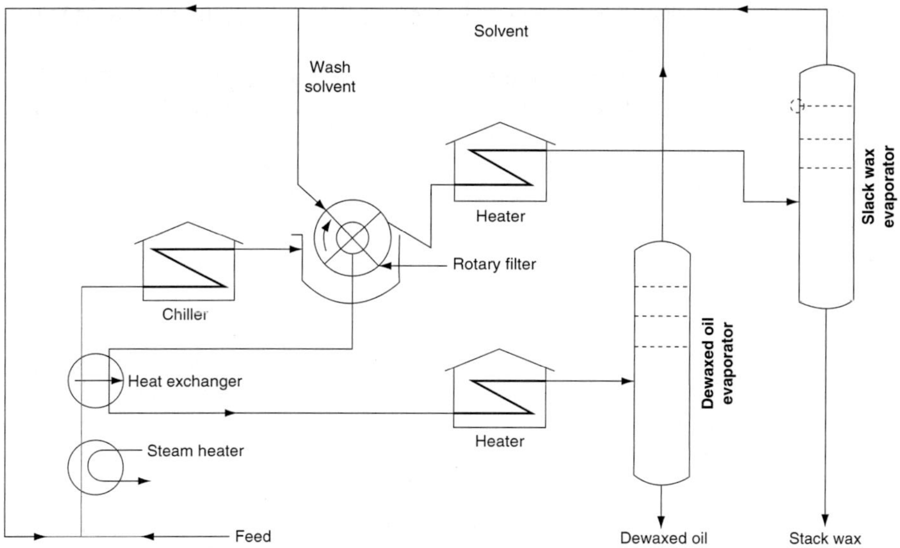


**Fig. 4.17** A solvent dewaxing unit (OSHA technical manual, Section IV, Chapter 2: Petroleum Refining Processes, http://www.osha.gov/dts/osta/otm/otm_iv/otm_iv_2.html)

Dewaxing (Fig. 4.17) processes also produce heater stack gas (carbon monoxide, sulfur oxides, nitrogen oxides, and particulate matter) as well as hydrocarbon emission such as fugitive propane and fugitive solvents [2, 6]. Steam stripping wastewater (oil and solvents) and solvent recovery wastewater (oil and propane) are also produced. The fugitive solvent emissions may be toxic (toluene, methyl ethyl ketone, methyl isobutyl ketone).

## *Gaseous Emissions*

Gaseous emissions from petroleum refining create a number of environmental problems [2, 6]. During combustion, the combination of hydrocarbons, nitrogen oxide, and sunlight results in localized low levels of ozone, or smog. This is particularly evident in large urban areas and especially when air does not circulate well. Petroleum use in automobiles also contributes to the problem in many areas. The primary effects are on the health of those exposed to the ozone, but plant life has been observed to suffer as well.

Refinery and natural gas streams may contain large amounts of acid gases, such as hydrogen sulfide ($H_2S$) and carbon dioxide ($CO_2$) [4, 5]. Hydrogen chloride (HCl), although not usually considered to be a major pollutant in petroleum refineries, can arise during processing from the presence of brine in petroleum that is incompletely dried. It can also be produced from mineral matter and other inorganic contaminants, gaining increasing recognition as a pollutant which needs serious attention.

Acid gases corrode refining equipment, harm catalysts, pollute the atmosphere, and prevent the use of hydrocarbon components in petrochemical manufacture. When the amount of hydrogen sulfide is large, it may be removed from a gas stream and converted to sulfur or sulfuric acid. Some natural gases contain sufficient carbon dioxide to warrant recovery as *dry ice*, i.e., solid carbon dioxide. And there is now a conscientious effort to mitigate the emission of pollutants from hydrotreating process by careful selection of process parameters and catalysts [2, 6, 15].

The terms "*refinery gas*" and "*process gas*" are also often used to include all of the gaseous products and by-products that emanate from a variety of refinery processes [5, 6]. There are also components of the gaseous products that must be removed prior to release of the gases to the atmosphere or prior to use of the gas in another part of the refinery, i.e., as a fuel gas or as a process feedstock.

Petroleum refining produces gas streams that often contain substantial amounts of acid gases such as hydrogen sulfide and carbon dioxide. More particularly hydrogen sulfide arises from the hydrodesulfurization of feedstocks that contain organic sulfur:

$$[S]_{feedstock} + H_2 \rightarrow H_2S + \text{hydrocarbons}$$

Petroleum refining involves, with the exception of some of the more viscous crude oils, a primary distillation of the hydrogen mixture, which results in its separation into fractions differing in carbon number, volatility, specific gravity, and other characteristics [2, 6]. The most volatile fraction, that contains most of the gases which are generally dissolved in the crude, is referred to as *pipe still gas* or *pipe still light ends* and consists essentially of hydrocarbon gases ranging from methane to butane(s), or sometimes pentane(s).

The gas varies in composition and volume, depending on crude origin and on any additions to the crude made at the loading point. It is not uncommon to re-inject light hydrocarbons such as propane and butane into the crude before dispatch by tanker or pipeline. This results in a higher vapor pressure of the crude, but it allows one to increase the quantity of light products obtained at the refinery. Since light ends in most petroleum markets command a premium, while in the oil field itself propane and butane may have to be re-injected or flared, the practice of *spiking* crude oil with liquefied petroleum gas is becoming fairly common.

In addition to the gases obtained by distillation of petroleum, more highly volatile products result from the subsequent processing of naphtha and middle distillate to produce gasoline. Hydrogen sulfide is produced in the desulfurization processes involving hydrogen treatment of naphtha, distillate, and residual fuel, and from the coking or similar thermal treatments of vacuum gas oils and residual fuels. The most common processing step in the production of gasoline is the catalytic reforming of hydrocarbon fractions in the heptane ($C_7$) to decane ($C_{10}$) range.

In a series of processes commercialized under the generic name *reforming*, paraffin and naphthene (cyclic non-aromatic) hydrocarbons are altered structurally in the presence of hydrogen and a catalyst into aromatics, or isomerized to more highly branched hydrocarbons. Catalytic reforming processes thus not only result in the formation of a liquid product of higher octane number, but also produce substantial quantities of gases. The latter are rich in hydrogen, but also contain hydrocarbons from methane to butanes, with a preponderance of propane ($CH_3CH_2CH_3$), *n*-butane ($CH_3CH_2CH_2CH_3$), and *iso*-butane [$(CH_3)_3CH$].

A second group of refining operations that contributes to gas production is that of the *catalytic cracking* processes [2, 6]. These consist of fluid-bed catalytic cracking in which heavy gas oils are converted into gas, liquefied petroleum gas, catalytic naphtha, fuel oil, and coke by contacting the heavy hydrocarbon with the hot catalyst. Both catalytic and thermal cracking processes, the latter being now largely used for the production of chemical raw materials, result in the formation of unsaturated hydrocarbons, particularly ethylene ($CH_2{=}CH_2$), but also propylene (propene, $CH_3.CH{=}CH_2$), *iso*-butylene [*iso*-butene, $(CH_3)_2C{=}CH_2$], and the *n*-butenes ($CH_3CH_2CH{=}CH_2$, and $CH_3CH{=}CHCH_3$) in addition to hydrogen ($H_2$), methane ($CH_4$) and smaller quantities of ethane ($CH_3CH_3$), propane ($CH_3CH_2CH_3$), and butanes [$CH_3CH_2CH_2CH_3$, $(CH_3)_3CH$]. Diolefins such as butadiene ($CH_2{=}CH.CH{=}CH_2$) are also present.

Additional gases are produced in refineries with visbreaking and/or coking facilities that are used to process the heaviest crude fractions. In the visbreaking process, fuel oil is passed through externally fired tubes and undergoes liquid phase cracking reactions, which result in the formation of lighter fuel oil components. Oil viscosity is thereby reduced, and some gases, mainly hydrogen, methane, and ethane, are formed. Substantial quantities of both gas and carbon are also formed in coking (both delayed coking and fluid coking) in addition to the middle distillate and naphtha. When coking a residual fuel oil or heavy gas oil, the feedstock is preheated and contacted with hot carbon (coke) which causes extensive cracking of the feedstock constituents of higher molecular weight to produce lower molecular weight products ranging from methane, via liquefied petroleum gas and naphtha, to gas oil and heating oil. Products from coking processes tend to be unsaturated, and olefin components predominate in the tail gases from coking processes.

A further source of refinery gas is hydrocracking, a catalytic high-pressure pyrolysis process in the presence of fresh and recycled hydrogen. The feedstock is again heavy gas oil or residual fuel oil, and the process is directed mainly at the production of additional middle distillates and gasoline. Since hydrogen is to be recycled, the gases produced in this process again have to be separated into lighter and heavier streams; any surplus recycle gas and the liquefied petroleum gas from the hydrocracking process are both saturated.

Both hydrocracker gases and catalytic reformer gases are commonly used in catalytic desulfurization processes. In the latter, feedstocks ranging from light to vacuum gas oils are passed at pressures of 500–1,000 psi with hydrogen over a hydrofining catalyst. This results mainly in the conversion of organic sulfur compounds to hydrogen sulfide,

$$[S]_{feedstock} + H_2 \rightarrow H_2S + \text{hydrocarbons}$$

The reaction also produces some light hydrocarbons by hydrocracking.

Thus refinery streams, while ostensibly being hydrocarbon in nature, may contain large amounts of acid gases such as hydrogen sulfide and carbon dioxide. Most commercial plants employ hydrogenation to convert organic sulfur compounds into hydrogen sulfide. Hydrogenation is effected by means of recycled hydrogen-containing gases or external hydrogen over a nickel molybdate or cobalt molybdate catalyst.

In summary, refinery process gas, in addition to hydrocarbons, may contain other contaminants, such as carbon oxides ($CO_x$, where $x = 1$ and/or 2), sulfur oxides ($SO_x$, where $x = 2$ and/or 3), as well as ammonia ($NH_3$), mercaptans (R-SH), and carbonyl sulfide (COS).

The presence of these impurities may eliminate some of the sweetening processes, since some processes remove large amounts of acid gas but not to a sufficiently low concentration. On the other hand, there are those processes not designed to remove (or incapable of removing) large amounts of acid gases whereas they are capable of removing the acid gas impurities to very low levels when the acid gases are present only in low-to-medium concentration in the gas.

From an environmental viewpoint, not only are the means by which these gases can be utilized important but also equally important are the effects of these gases on the environment when they are introduced into the atmosphere.

In addition to the corrosion of equipment of acid gases, the escape into the atmosphere of sulfur-containing gases can eventually lead to the formation of the constituents of acid rain, i.e., the oxides of sulfur ($SO_2$ and $SO_3$). Similarly, the nitrogen-containing gases can also lead to nitrous and nitric acids (through the formation of the oxides $NO_x$, where $x = 1$ or 2) which are the other major contributors to acid rain. The release of carbon dioxide and hydrocarbons as constituents of refinery effluents can also influence the behavior and integrity of the ozone layer.

Hydrogen chloride, if produced during refining, quickly picks up moisture in the atmosphere to form droplets of hydrochloric acid and, like sulfur dioxide, is a contributor to acid rain [6]. However, hydrogen chloride may exert severe local effects because, unlike sulfur dioxide, it does not need to participate in any further chemical reaction to become an acid. Under atmospheric conditions that favor a buildup of stack emissions in the area of a large industrial complex or power plant, the amount of hydrochloric acid in rainwater could be quite high.

Natural gas is also capable of producing emissions that are detrimental to the environment. While the major constituent of natural gas is methane, there are components such as carbon dioxide (CO), hydrogen sulfide ($H_2S$), and mercaptans (thiols; R-SH), as well as trace amounts of sundry other emissions. The fact that methane has a foreseen and valuable end-use makes it a desirable product, but in several other situations, it is considered a pollutant, having been identified a greenhouse gas.

A sulfur removal process must be very precise, since natural gas contains only a small quantity of sulfur-containing compounds that must be reduced several orders of magnitude. Most consumers of natural gas require less than 4 ppm in the gas.

A characteristic feature of natural gas that contains hydrogen sulfide is the presence of carbon dioxide (generally in the range of 1–4% v/v). In cases where the natural gas does not contain hydrogen sulfide, there may also be a relative lack of carbon dioxide.

*Acid rain* occurs when the oxides of nitrogen and sulfur that are released to the atmosphere during the combustion of fossil fuels are deposited (as soluble acids) with rainfall, usually at some location remote from the source of the emissions.

It is generally believed (the chemical thermodynamics are favorable) that acidic compounds are formed when sulfur dioxide and nitrogen oxide emissions are released from tall industrial stacks. Gases such as sulfur oxides (usually sulfur dioxide, $SO_2$) as well as the nitrogen oxides ($NO_x$) react with the water in the atmosphere to form acids:

$$SO_2 + H_2O \rightarrow H_2SO_3$$

$$2SO_2 + O_2 \rightarrow 2SO_3$$

$$SO_3 + H_2O \rightarrow H_2SO_4$$

$$2NO + H_2O \rightarrow 2HNO_2$$

$$2NO + O_2 \rightarrow 2NO_2$$

$$NO_2 + H_2O \rightarrow HNO_3$$

Acid rain has a pH less than 5.0 and predominantly consists of sulfuric acid ($H_2SO_4$) and nitric acid ($HNO_3$). As a point of reference, in the absence of anthropogenic pollution sources, the average pH of rain is 6.0 (slightly acidic; neutral pH = 7.0). In summary, the sulfur dioxide that is produced during a variety of processes will react with oxygen and water in the atmosphere to yield environmentally detrimental sulfuric acid. Similarly, nitrogen oxides will also react to produce nitric acid.

Another acid gas, hydrogen chloride (HCl), although not usually considered to be a major emission, is produced from mineral matter and the brines that often accompany petroleum during production and is gaining increasing recognition as a contributor to acid rain. However, hydrogen chloride may exert severe local effects because it does not need to participate in any further chemical reaction to become an acid. Under atmospheric conditions that favor a buildup of stack emissions in the areas where hydrogen chloride is produced, the amount of hydrochloric acid in rainwater could be quite high.

In addition to hydrogen sulfide and carbon dioxide, gas may contain other contaminants, such as mercaptans (R-SH) and carbonyl sulfide (COS). The presence of these impurities may eliminate some of the sweetening processes since some processes remove large amounts of acid gas but not to a sufficiently low concentration. On the other hand, there are those processes that are not designed to remove (or are incapable of removing) large amounts of acid gases. However, these processes are also capable of removing the acid gas impurities to very low levels when the acid gases are there in low-to-medium concentrations in the gas.

On a regional level, the emission of sulfur oxides ($SO_x$) and nitrogen oxides ($NO_x$) can also cause the formation of acid species at high altitudes, which eventually precipitate in the form of *acid rain*, damaging plants, wildlife, and property. Most petroleum products are low in sulfur or are desulfurized, and while natural gas sometimes includes sulfur as a contaminant, it is typically removed at the production site.

At the global level, there is concern that the increased use of hydrocarbon-based fuels will ultimately raise the temperature of the planet (*global warming*), as carbon dioxide reflects the infrared or thermal emissions from the earth, preventing them from escaping into space (*greenhouse effect*). Whether or not the potential for global warming becomes real will depend upon how emissions into the atmosphere are handled. There is considerable discussion about the merits and demerits of the global warming theory [16] and the discussion is likely to continue for some time. Be that as it may, the atmosphere can only tolerate pollutants up to a limiting value. And that value needs to be determined. In the meantime, efforts must be made to curtail the use of noxious and foreign (non-indigenous) materials into the air.

In summary, and from an environmental viewpoint, petroleum and natural gas processing can result in similar, if not the same, gaseous emissions as coal [2, 4, 6]. It is a question of degree insofar as the composition of the gaseous emissions may vary from coal to petroleum but the constituents are, in the majority of cases, the same.

There are a variety of processes which are designed for sulfur dioxide removal from gas streams [2], but scrubbing process utilizing limestone ($CaCO_3$) or lime [$Ca(OH)_2$] slurries have received more attention than other gas scrubbing processes.

The majority of the gas scrubbing processes are designed to remove sulfur dioxide from the gas streams; some processes show the potential for removal of nitrogen oxide(s).

## *Liquid Effluents*

It is convenient to divide the hydrocarbon components of petroleum into the following three classes:

- *Paraffins*: saturated hydrocarbons with straight or branched chains
- *Naphthenes* (*alicyclic hydrocarbons*): saturated hydrocarbons containing one or more rings, each of which may have one or more paraffin side chains

- *Aromatic compounds*: hydrocarbons containing one or more aromatic nuclei (for example, benzene, naphthalene, and phenanthrene) which may be co-joined with (substituted) naphthene rings and/or paraffin side chains

Thermal processing can significantly increase the concentration of polynuclear aromatic hydrocarbons in the product liquid because the low-pressure hydrogen-deficient conditions favor aromatization of naphthene constituents and condensation of aromatics to form larger ring systems. To the extent that more compounds like benzo(a)pyrene are produced, the liquids from thermal processes will be more carcinogenic than asphalt.

The sludge produced on acid treatment of petroleum distillates [2, 6], even gasoline and kerosene, is complex in nature. Esters and alcohols are present from reactions with olefins; sulfonation products from reactions with aromatic compounds, naphthene compounds, and phenols; and salts from reactions with nitrogen bases. To these constituents must be added the various products of oxidation–reduction reactions: coagulated resins, soluble hydrocarbons, water, and free acid.

The disposal of the sludge is a comparatively simple process for the sludge resulting from treating gasoline and kerosene, the so-called light oils. The insoluble oil phase separates out as a mobile tar, which can be mixed and burned without too much difficulty.

In all cases, careful separation of reaction products is important to the recovery of well-refined materials. This may not be easy if the temperature has risen as a consequence of chemical reaction. This will result in a persistent dark color traceable to reaction products that are redistributed as colloids. Separation may also be difficult at low temperature because of high viscosity of the stock, but this problem can be overcome by dilution with light naphtha or with propane.

In addition, delayed coking also requires the use of large volumes of water for hydraulic cleaning of the coke drum. However, the process water can be recycled if the oil is removed by skimming and suspended coke particles are removed by filtration. If this water is used in a closed cycle, the impact of delayed coking on water treatment facilities and the environment is minimized. The flexicoking process offers one alternative to direct combustion of coke for process fuel. The gasification section is used to process excess coke to mixture of carbon monoxide (CO), carbon dioxide ($CO_2$), hydrogen ($H_2$), and hydrogen sulfide ($H_2S$) followed by treatment to remove the hydrogen sulfide. Currently, maximum residue conversion with minimum coke production is favored over gasification of coke.

## *Solid Effluents*

Catalyst disposal is a major concern in all refineries. In many cases, the catalysts are regenerated at the refinery for repeated use. Disposal of spent catalysts is usually part of an agreement with the catalysts manufacturer whereby the spent catalyst is returned for treatment and re-manufacture.

The formation of considerable quantities of coke in the *coking* processes is a cause for concern since it not only reduces the yield of liquid products but also initiates the necessity for disposal of the coke. Stockpiling to coke may be a partial answer unless the coke contains leachable materials that will endanger the ecosystem as a result of rain or snow melt.

In addition, the generation and emission of sulfur oxides (particularly sulfur dioxide) originates from the combustion of sulfur-containing coke as plant fuel. Sulfur dioxide ($SO_2$) has a wide range of effects on health and on the environment. These effects vary from bronchial irritation upon short-term exposure to contributing to the acidification of lakes. Emissions of sulfur dioxide, therefore, are regulated in many countries.

## Future Directions

The petroleum refining industry includes integrated process operations that are engaged in refining crude petroleum into refined petroleum products, especially liquid fuels such as gasoline and diesel as well as processes that produce raw materials for the petrochemical industry.

Over the past 4 decades, the refining industry has experienced significant changes in oil market dynamics, resource availability, and technological advancements. Advancements made in exploration, production, and refining technologies allow utilization of resources such as heavy oil and tar sand bitumen that were considered economically and technically unsuitable in the middle decades of the past century. Along with the many challenges, it is imperative for refiners to raise their operations to new levels of performance. Merely extending today's performance incrementally will fail to meet most company's performance goals.

Petroleum refining in the twenty-first century will continue to be shaped by the factors such as consolidation of oil companies, dramatic changes in market demand, customization of products, and a decrease in the API gravity and sulfur content of the petroleum feedstocks. In fact, in addition to a (hopeful but unlikely) plentiful supply of petroleum, the future of the refining industry will base on the following factors such as (1) increased operating costs or investments due to stringent environmental requirements for facilities and products, (2) accelerating globalization resulting in stronger international petroleum price scenarios. The effect of these factors is likely to reduce refinery profit margins further, and petroleum companies worldwide will need to make significant changes in their operation and structure to be competitive on global basis.

As global petroleum consumption increases and resources are depleted, the production of fuels and petrochemicals from residua, heavy oil, and tar sand bitumen will increase significantly. In fact, over the next decade, refineries will need to adapt to receiving heavier oils as well as a range of bio-feedstocks. It is conceivable that current refineries could not handle such a diverse slate of feedstocks without experiencing shutdowns and related problems.

As feedstocks to refineries change, there must be an accompanying change in refinery technology as petroleum feedstocks are becoming highly variable. At the same time, more stringent anti-pollution regulations are forcing greater restrictions on fuel specifications. There are fundamental limitations on how far current processes can go in achieving proper control over feedstock behavior. This means a movement from conventional means of refining heavy feedstocks by using (typically) a coking process to more development and use of more innovative processes that will produce the maximum yields of liquid fuels (or other desired products) fuels from the feedstock.

Thus, the need for the development of upgrading processes continues in order to fulfill the product market demand as well as to satisfy environmental regulations. One area in particular, the need for residuum conversion, technology has emerged as result of declining residual fuel oil market and the necessity to upgrade crude oil residua beyond the capabilities of the visbreaking, coking, and low-severity hydrodesulfurization processes.

Technological advances are on the horizon for alternate sources of transportation fuels. For example, gas-to-liquids and biomass-to-liquids are just two of the concepts currently under development. However, the state of many of these technologies coupled with the associated infrastructure required to implement them leaves traditional refining of petroleum hydrocarbons for transportation fuels as the *modus operandi* for the foreseeable future, which for the purposes of this text is seen to be 50 years. The near future challenge for refiners will be how to harness new technologies to remain alive in a changing global marketplace.

It is imperative for refiners to raise their operations to new levels of performance. Merely extending current process performance incrementally will fail to meet most future performance goals. To do this, it will be necessary to reshape refining technology to be more adaptive to changing feedstocks and product demand and to explore the means by which the technology and methodology of refinery operations can be translated not only into increased profitability but also into survivability.

Furthermore, environmental regulations could either preclude unconventional production or, more likely, raise the cost significantly. If future US laws limited and/or taxed greenhouse gas emissions, these laws will lead to substantial increase in the costs of production of fuels from unconventional sources. In addition to increases in the volumes of carbon dioxide, restrictions on access to water also could prove costly, especially in the arid or semiarid western States. In addition, environmental restrictions on land use could preclude unconventional oil production in some areas of the United States.

The general prognosis for reduction of emissions or emission cleanup is optimistic. It is considered likely that most of their environmental impact of petroleum refining can be substantially abated. A considerable investment in retrofitting or replacing existing facilities and equipment will be needed although a conscious goal must be to improve the efficiency with which petroleum is transformed and consumed.

# Bibliography

## Primary Literature

1. Ray DL, Guzzo L (1990) Trashing the planet: how science can help us deal with acid rain, depletion of the ozone, and nuclear waste (among other things). Regnery Gateway, Washington, DC
2. Speight JG (2007) The chemistry and technology of petroleum, 4th edn. CRC Press/Taylor & Francis, Boca Raton
3. Majumdar SB (1993) Regulatory requirements for hazardous materials. McGraw-Hill, New York
4. Speight JG (1993) Gas processing: environmental aspects and methods. Butterworth Heinemann, Oxford
5. Speight JG (1996) Environmental technology handbook. Taylor & Francis, Washington, DC
6. Speight JG (2005) Environmental analysis and technology for the refining industry. Wiley, Hoboken
7. Lipton S, Lynch J (1994) Handbook of health hazard control in the chemical process industry. Wiley, New York
8. Boyce A (1997) Introduction to environmental technology. Van Nostrand Reinhold, New York
9. Speight JG (2009) Enhanced recovery methods for heavy oil and tar sands. Gulf, Houston
10. Carson PA, Mumford CJ (1995) The safe handling of chemicals in industry. Wiley, New York
11. Renzoni A, Fossi MC, Lari L, Mattei N (1994) Contaminants in the environment: a multidisciplinary assessment of risks to man and other organisms. CRC Press, Boca Raton
12. Edwards JD (1995) Industrial wastewater treatment: a guidebook. CRC Press, Boca Raton
13. Thibodeaux LJ (1995) Environmental chemodynamics. Wiley, New York
14. Meyers RA (1997) Handbook of petroleum refiing processes, 2nd edn. McGraw-Hill, New York
15. Occelli ML, Chianelli R (1996) Hydrotreating technology for pollution control. Marcel Dekker, New York
16. Hileman B (1996) Environmental hormone disruptors focus of major research intitatives. Chem Eng News 74(34):33

## Books and Reviews

Abraham H (1945) Asphalts and allied substances, vol I. Van Nostrand, New York

Forbes RJ (1958) A history of technology, vol V. Oxford University Press, Oxford

Gary JH, Handwerk GE (2001) Petroleum refining: technology and economics, 4th edn. Marcel Dekker, New York

Hoiberg AJ (1960) Bituminous materials: asphalts, tars and pitches, vol I & II. Interscience, New York

Hsu CS, Robinson PR (2006) Practical advances in petroleum processing, vol 1 and 2. Springer, New York

Khan MR, Patmore DJ (1997) Heavy oil upgrading processes. In: Speight JG (ed) Petroleum chemistry and refining. Taylor & Francis, Washington, DC. Chapter 6

McKetta JJ (ed) (1992) Petroleum processing handbook. Marcel Dekker, New York

Shih SS, Oballa MC (eds) (1991) Tar sand upgrading technology. Symposium series No. 282. American Institute for Chemical Engineers, New York

Speight JG (2000) The desulfurization of heavy oils and residua, 2nd edn. Marcel Dekker, New York

Speight JG, Ozum B (2002) Petroleum refining processes. Marcel Dekker, New York

Speight JG (2008) Handbook of synthetic fuels: properties, processes, and performance. McGraw-Hill, New York

# Chapter 5
# Oil Shale Processing, Chemistry and Technology

**Vahur Oja and Eric M. Suuberg**

## Glossary

| | |
|---|---|
| Bitumen | Orga Processing, Chemistry and Technology solvent-soluble portion of oil shale organic matter. |
| Direct retorting | Retorting in which the process heat requirement is accomplished by combustion of a portion of the oil shale within the retort. |
| Ex situ retorting | Oil shale retorting performed aboveground (requiring mining of the oil shale). |
| Indirect retorting | Retorting in which the process heat requirement is provided by heated gaseous or solid heat carriers produced outside the retort, and then brought into contact with the retort contents. |
| In situ retorting | Heating of oil shale underground in order to release shale oil (also called underground retorting). Confinement of the shale is provided by the surrounding earth, in this case. |
| Kerogen | Macromolecular and organic solvent-insoluble organic matter of oil shale, but more broadly applied to these sorts of organic components of sedimentary rocks. |

This chapter was originally published as part of the Encyclopedia of Sustainability Science and Technology edited by Robert A. Meyers. DOI:10.1007/978-1-4419-0851-3

V. Oja (✉)
Department of Chemical Engineering, Tallinn University of Technology,
Ehitajate tee 5, Tallinn 19086, Estonia
e-mail: vahur.oja@ttu.ee

E.M. Suuberg
Division of Engineering, Brown University, 182 Hope Street,
Providence, RI 02912, USA
e-mail: eric_suuberg@brown.edu

R. Malhotra (ed.), *Fossil Energy: Selected Entries from the Encyclopedia of Sustainability Science and Technology*, DOI 10.1007/978-1-4614-5722-0_5,

| | |
|---|---|
| Oil shale | Oil shale has a wide range of definitions. A general definition that encompasses most is that it consists of a sedimentary rock containing various amounts of solid organic material bound dispersedly in a mineral matrix and yielding principally a shale oil upon pyrolysis. |
| Pyrolysis | A thermochemical decomposition process of organic matter, effected by heating the material in the absence of oxygen. |
| Retort gas | Noncondensable combustible gas, a mixture of gases from an oil shale retort that remains after particulate matter and condensed liquids have been removed. |
| Retort water | Water collected during oil shale retorting from offgas stream together with oil. |
| Retorting | In shale oil recovery technologies, retorting is a pyrolytic process designed to produce oil. Generally, this process is performed by placing oil shale in a sealed vessel heated in order to promote pyrolysis; this reactor is called a retort. |
| Retorting residue or spent shale | Solid waste material remaining after retorting. This material contains most of the mineral matter of the original shale (possibly partially decomposed) and a carbon-rich (char) residue from the organic fraction (which may sometimes be referred to as a semicoke). |
| Shale oil | Room temperature liquid phase organic material, oil product, recoverable from the thermal decomposition of kerogen or separated from the offgas stream of an oil shale retort. |
| Thermobitumen | A bitumen not contained in the original oil shale, but produced by the action of heat on the shale; a thermal degradation (pyrolysis) product of kerogen. |

## Definition of the Subject

Oil shale is a complex mixture of organic and inorganic phases, in which the organic phase is normally that of the greater economic and practical value. In nature there exists a wide range of materials that include these same components, and not all of these materials are candidates for actual use; in some instances, the organic content is simply too low to make processing of the oil shale of any economic interest. Thus, one of the first things that needs to be answered in asking what oil shales it makes sense to study is the question of what sorts of oil shale should be considered as the subject of practical use. This summary begins with a consideration of this fundamental question, and is guided by what has historically been accepted as definitions of oil shales of practical interest. It is recognized that these definitions come from different locations and time periods, with differing demands for the energy or raw materials that the oil shale could provide. This, in turn, means that these definitions can change with time and place, but here they offer a framework for the discussion that follows.

Because in oil shale the organic and inorganic phases form such an intimate mixture, it is critical to consider how this might influence the practical conversion processes to which the oil shale is ultimately subjected. At the same time, it is essential to recognize how the experimental research used to develop these practical processes can be influenced by decisions on how to prepare the oil shale for testing. Isolation of the kerogen material from the oil shale offers advantages in terms of simplifying or enabling certain types of organic fraction analysis, or the evaluation of proposed processing steps in the laboratory. In many instances, this might make good scientific sense, but in other cases, it is important to note that there might be inadvertent changes induced in the material, and in its potential conversion chemistry, by the separation. This summary will only provide a brief overview of some of the aspects of research performed on the different phases present in the oil shale, and offer only this general warning to be cautious regarding the possibility of artifacts induced by separations of the raw material.

The main focus of this entry is on the chemistry of oil shale conversion. Virtually all oil shale conversion processes begin with a thermal degradation of the organic macromolecular structure. The organic fraction of oil shale is largely unrecoverable in any useful form without such a thermal degradation, since the extractable bitumen content in unheated oil shale is normally but a few percent at most. The thermal degradation is critical whether the oil shale is processed to recover oil or burned directly to generate heat, and from that, electricity. In the case of combustion, the thermal degradation is what releases the small organic molecular fragments that can burn in the vapor phase. The less that remains in a solid semicoke phase after the initial heating in a boiler, the more quickly and smoothly the combustion of the shale can take place, and the more efficiently used will be the organic fraction for its heating value content. Most shale oil combustion processes do operate at very high degrees of burnout of the organic matter.

In the case of oil recovery from the oil shale, there is a much-studied set of conditions used to maximize the recovery of oil. There is a balance between the need to achieve sufficient breakdown of the macromolecular structure of the organic phase, which will permit fragments to escape as oil, and the further pyrolysis of the organic phase to release light gases and leave behind a hydrogen-depleted semicoke that has much lower value than oil (and that is embedded in the remaining mineral matrix, making it difficult to utilize). This entry summarizes several aspects of what is known regarding the chemistry of thermal conversion and its kinetics, and offers an overview of some practical technologies in which the thermal conversion of the material takes place.

## Introduction

This entry focuses on the chemical nature of oil shale, and some key aspects of its conversion to oil. One of the first difficulties encountered in this field is deciding what starting materials qualify as "oil shale." A brief review of various practical definitions is first offered. Different practical definitions arise from the different applications of the material. Next, the chemical structure of the shale is considered, including both the inorganic and organic phases contained therein. In this section, there is also consideration of how the organic phase can be analyzed, including how it may be separated from the whole shale.

In section "Definition of the Subject", the chemical processes that result in the practical conversion processes of oil shale are introduced in general terms. In this section it is noted that not all oil shale is necessarily used to produce an oil product. In some instances, the oil shale may be directly burned in devices similar to those used for combustion of coal. In Sect. 3, attention is turned toward retorting to produce oil products. Except for recent decades in Estonia, where direct firing of oil shale for producing electrical power and heat has been practiced, retorting for oil has been the main driver for worldwide interest in oil shale. This section reviews the parameters that influence the results of retorting, as well as providing an overview of some chemical aspects of the process. This review closes in Sect. 4 with a brief consideration of different classes of technology developed for carrying out the retorting process.

It is important to recognize that there is a much more voluminous literature on oil shale chemistry and conversion than has been discussed here. The authors maintain that the present selection does, however, provide a reasonably comprehensive overview of the topic, while emphasizing some of the newer contributions to the literature.

## General Characteristics of Oil Shale

### *Oil Shale Classification*

Oil shale can be defined as a sedimentary rock containing various amounts of solid organic material dispersedly bound in a mineral matrix. Oil shales from different deposits vary in their mineral contents and types, chemical composition of organic matter, geological period of deposition, depositional condition, etc. [22, 31]. Therefore, there are a number of definitions of oil shale based upon application, operational, or scientific point of view [55]. Similarly there are various oil shale classification schemes [32, 54, 55, 76]. Moreover, the term "oil shale" is itself somewhat misleading as the organic matter is not oil-like and the mineral matter is not always classifiable as shale [141].

Following are some selected definitions of oil shale, proposed by several authors in the connection with its energy production or other technological utilization:

1. Gavin [43] defined oil shale as a "compact laminated rock of sedimentary origin, yielding over 33% of ash and containing organic matter that yields oil when distilled, but not appreciably when extracted with ordinary solvents for petroleum."

2. Ozerov and Polozov have described oil shale as "a hard foliated combustible rock formed by joint accumulation of pelagic plants and animals and mineral mass which had been transformed by the action of geographic conditions and chemical, biochemical and hydro-chemical processes" [90].
3. Schlatter has characterized oil shale as "a heterogeneous mixture of organic and mineral matter. It is a fine-grained, tight rock with essentially no permeability or porosity" [114].

It is important to note that in defining oil shale for energy-technological purposes, the following oil shale characteristics have been emphasized:

- A high mineral matter content, which may be about two to five times higher than the organic matter content (generally significantly higher in comparison to coals, which contain mineral matter, by definition, of less than 40%).
- The major portion of oil shale organic matter is insoluble in organic solvents (this in contrast to tar sands).
- The organic matter has hydrogen/carbon atomic ratios typically in the range 1.2–1.6 (which is significantly enriched in hydrogen compared with coals, but hydrogen deficient relative to crude oil).
- The capacity of oil shales to ignite and burn without separation of the organic matter from the mineral matter.
- They undergo thermal decomposition that results in production of a significant amount of liquid organic product or shale oil (higher grade oil shales can yield 100 L or more per ton of dry oil shale).
- They have a moisture content of less than 10–13%.
- They exhibit a low permeability of the rock to gases, vapors, and oils.

The organic matter (often termed kerogen; see below) content of oil shale varies widely, ranging typically from 10% to 40%. As the organic matter of oil shale is the source of energy and/or synthetic fuels value, there have been several attempts to classify oil shale by minimum and maximum organic matter content. The upper limit of organic matter content has been proposed to be about 50% [2, 76], or even higher [141]. The lower organic matter limit is usually defined in relation to the minimum energy requirement of oil shale as an industrial raw material. For example, a lower limit of 5% of organic matter has been proposed to define "commercial deposits." This is because the organic matter content of an oil shale should be at least roughly about 2.5 wt% for the latter to provide the calorific requirements necessary to heat the rock to 500°C [38, 82].

Oil shale grade or energy potential is currently classified or characterized for commercial purposes on either a heating value or an average oil yield basis.

1. The heating value is used to evaluate the quality of an oil shale in the context of direct firing of oil shale in a power plant to produce heat, commonly used to raise steam for electricity generation. The heating value of typical oil shale kerogen is about 40 MJ/kg. For example, the heating value of an Estonian kukersite oil shale kerogen has been reported to be 37.3 MJ/kg [2], while for US Green River formation kerogen the value is around 41.1 MJ/kg and for eastern US Devonian kerogen around 37.5 MJ/kg [99]. This is, of course, not the same as the heating

value of the oil shale as would be determined for the whole rock, using a standard calorimeter (in which the dilution effect of the zero, or negative, heating value mineral matter is important). A combustion grade oil shale is defined as that having a minimum upper caloric value (for the whole oil shale rock) of 3.1 MJ/kg (dry basis) [38], though a heating value limit of not less than about 4.2–6.3 MJ/kg has been suggested for practical purposes [50, 141].

2. Average oil yield, or oil production potential, is determined using a "modified" Fischer assay method (the most widely used standardized technique, ASTM D-3904-80) or an equivalent analytical technique in which the shale is subjected to heating in order to liberate oil from the kerogen. The oil yield depends mostly on kerogen per mass of oil shale, but may also depend upon the fraction of kerogen convertible to oil (there are differences between different kerogens). On the basis of this assay, an "oil shale" is a shale that yields at least 42 L/t of dry shale (or 10 US gal/t, the basis used by the US Geological Survey) [99], though a limit as low as 25 L/t has also been suggested. Similarly, a crude dividing line between lower and higher grade shale has been defined at about 100 L of oil/t of shale (with those yielding below 90 L/t considered "low grade," 90–150 L/t "moderate grade," and those above 150 L/t "high grade" [2, 38]). Commercial grade oil shales range from about 100–200 L/t of shale. However, the Fischer assay does not actually indicate the maximum oil yield that can be actually produced by a given oil shale. Yields greater than Fischer assay have been obtained by hydropyrolysis, donor solvents, and rapid heating, especially for lower oil yield, non-softening oil shales [23, 106].

More than 600 oil shale deposits have been discovered around the world [2]. Oil shales are found in many countries around the world, with nearly 100 major deposits in 27 countries [128]. There exist many surveys/reviews on history of oil shale development throughout the world [98, 128, 129].

Oil shale resources are typically expressed in either of two different ways: (1) as tons of oil shale, and (2) as crude-shale-oil equivalents contained in the oil shale. Global deposits have been estimated to be about 411 metric gigatons or to range from 2.8 to 3.3 trillion barrels ($4.5 \times 10^{11}$ to $5.2 \times 10^{11}$ $m^3$) of recoverable oil equivalent [155].

When talking about known oil shale deposits, there is a need to distinguish between two terms – "resources" and "reserves." Resources typically refer to estimates of all deposits of oil shales, while reserves refer only to those from which oil extraction can be economically profitable with the use of existing technologies. It should be noted that these technologies are constantly developing, and because of that, estimates of reserves will not necessarily remain constant over time, irrespective of use or discovery of new resources.

## *Oil Shale Composition*

Oil shale is composed of mineral and organic matter whose proportions vary with the grade of oil shale.

**Table 5.1** Mineral matter composition in some oil shales presented as oxides in ash, wt%

| Compound | Green River [USA] [99] | Kukersite [Estonia] [88] | El-lajjun [Jordan] [88] | Dictyonema [Estonia] [88] |
|---|---|---|---|---|
| $SiO_2$ | 40–60 | 28.25 | 28.86 | 63.72 |
| $Al_2O_3$ | 10–15 | 9.84 | 7.43 | 16.43 |
| $Fe_2O_3$ | 5–10 | 5.44 | 3.70 | 8.80 |
| CaO | 10–25 | 41.02 | 52.79 | 0.85 |
| MgO | 5–10 | 5.05 | 1.30 | 0.50 |
| $K_2O$ | | 3.60 | | 8.72 |
| $Na_2O$ | | | 0.3 | |
| $SO_3$ | | 6.07 | 1.30 | 0.70 |
| $P_2O_5$ | | | 5.9 | |

## Mineral Matter

Oil shale mineral matter contains several classes of minerals such as carbonates (calcite and dolomite), silicates, sulfides, and others. The inorganic composition of oil shales varies widely. The oil shale mineral composition can change from mostly calcareous to mostly aluminosilicate. For example, a typical mineral composition (in weight percent of the mineral matter alone) of carbonate-rich Kukersite oil shale from Estonia [100] is dolomite 3%, calcite 65%, quartz 8%, illite 15%, orthoclase 5%, chlorite 1%, and pyrite 3%. For the carbonate-rich Mahogany-zone US Green River formation oil shale [99], the composition is somewhat different: dolomite 32%, calcite 16%, quartz 15%, illite 19%, albite 10%, microcline 6%, pyrite 1%, and analcite 1%. In the case of siliceous Eastern US Devonian oil shale [99], a typical mineral composition is quartz 28%, illite 40%, feldspar 12%, pyrite 14%, and other minerals 6%. Quite often, mineral matter composition is represented by the chemical composition of oil shale ash, in which the mineral components have been converted to their oxides during combustion. This is illustrated in Table 5.1. It is, however, very important to keep in mind that the minerals are fundamentally altered by the combustion process, and the ash composition is only an approximate guide to what minerals existed in the raw oil shale.

In addition to the main minerals, oil shales contain a large number of trace elements (also referred to as minor elements). The content of the trace elements is typically in the parts per million (ppm) range. Trace element data for several deposits are available in the literature [92, 117, 141].

## Organic Matter

Oil shale organic matter is composed of solvent-soluble and insoluble organic materials. The organic matter in oil shales is mostly in the form of solvent insoluble cross-linked macromolecular structure or kerogen. However, the term "kerogen" is often used for whole organic matter of oil shale, including a small solvent-extractable fraction of the oil shale (termed bitumen).

Because organic matter is not easily extractable from the whole oil shale, there is an issue of how to quantify it. One simple way is to assume that it is the fraction that can be burned in air. Of course any loss of mineral components during burning (as by dehydration or decarbonylation reactions) can lead to an overestimate of the organic matter.

The organic matter content is thus often expressed on a dry shale basis as "conditional organic matter," $Org_{cond}{}^{d} = 100 - A^{d} - (CO_2)_M{}^{d}$, in which $A^d$ is ash yield upon combustion, $(CO_2)_M{}^{d}$ is carbonate-derived $CO_2$ released by the mineral matter at high temperatures, and a dry basis is indicated by the superscript d [141]. Basically, this formula embodies the concept that what is not left after combustion as ash, or released by carbonate decomposition, must have been organic matter. Both the ash and decomposable carbonate are available from separate analyses. In reality, the actual organic matter content can be somewhat different than that obtainable from the formula above since, for example, the formation of $SO_4{}^{-2}$ ion from combustible sulfur can contribute to the mass of ash, even if the sulfur originated in the organic matter. Also, any other volatilization of mineral components, beyond decomposition of the carbonates, would also lead to errors in estimates of organic matter.

Therefore for specific oil shales more detailed equations have been developed. For example, for Kukersite oil shale, which has a true organic matter content somewhat lower than that typically calculated from the "conditional organic matter," the following formula has been proposed: $Org^{d} = 100 - 1.028A^{d} - 0.964(CO_2)_M{}^{d} + 0.637(SO_4{}^{-2} - 0.07)$, where $SO_4{}^{-2}$ is the sulfate content [104]. For Dictyonema shale, with a true organic matter content somewhat higher than the calculated "conditional organic matter," the following equation has been recommended: $Org^{d} = 100 - A^{d} + 0.125FeO - (H_2O)_M - 0.625S_{pyr} - (CO_2)_M{}^{d}$, where FeO is iron oxide content in ash, $(H_2O)_M$ is the content of mineral water of hydration, and $S_{pyr}$ is the pyritic sulfur content [72]. The difference between true organic matter and "conditional organic matter" is generally not greater than 1–3%.

It is clear that for most technological uses of oil shale, there will be a need to go beyond simple estimates of organic weight fraction in characterizing the material.

## Kerogen

Kerogen is the major fraction of oil shale organic matter, and it is insoluble in organic solvents. It has a solid cross-linked macromolecular structure, and in this respect is somewhat similar to coal. Accordingly, kerogen is swellable, but not soluble, in organic solvents. The term "kerogen" was introduced by Professor Crum Brown to describe the insoluble organic matter in oil shales that gives shale oil on heating and decomposition [158]. Nowadays the term kerogen is used to characterize solid organic matter in all sedimentary rocks [56].

**Table 5.2** Examples of kerogen [141] (or organic matter,* [99]) elemental compositions

| Oil shale | C | H | S | N | O | H/C | O/C |
|---|---|---|---|---|---|---|---|
| Kukersite (Estonia) | 77.3 | 9.8 | 1.7 | 0.4 | 10.8 | 1.52 | 0.105 |
| Green River (USA) | 78.3 | 9.9 | 1.6 | 2.1 | 8.1 | 1.52 | 0.078 |
| Dictyonema (Estonia) | 70.5 | 7.4 | 4.2 | 2.5 | 15.4 | 1.26 | 0.164 |
| Kasphir (Russia) | 67.1 | 8.0 | 10.2 | 1.2 | 13.5 | 1.43 | 0.151 |
| Timahdit (Morocco) | 68.9 | 8.5 | 7.0 | 3.0 | 12.6 | 1.48 | 0.137 |
| Eastern USA Devonian* | 82.0 | 7.4 | 2.3 | 2.0 | 6.3 | 1.08 | 0.058 |

There are several excellent publications on kerogens available in the literature [32, 110, 144, 145]. Oil shale kerogens possess relatively high atomic hydrogen to carbon atomic ratios (typically 1.2–1.6), at least in comparison to coals, for which this ratio is typically less than unity. It is the high hydrogen to carbon ratio (combined with low oxygen to carbon ratio) that makes oil shale kerogens of interest as a potential source of liquid fuels. Heating values of kerogens have been shown to be in the range of 24–40 MJ/kg, though as noted above the typical values are around 35–40 MJ/kg [90, 99].

The macromolecular structure of oil shale kerogens is complex – kerogen is not a "homopolymer." Instead of being built of repeating regular structural units as are synthetic polymers or natural materials such as cellulose, it is composed of highly variable structural units, both in terms of size and chemical character. The exact chemical structure of an oil shale kerogen is difficult to describe, therefore studies of "kerogen structure" can provide data only on average structural parameters. This is the same situation as exists for such natural materials as coals or lignins.

Due to kerogen's insolubility in organic solvents, it was historically somewhat difficult to characterize with respect to its chemical structure. Attempts to liberate it from the oil shale rock would necessarily involve risk of chemical degradation. Even if the approach involves dissolving away the mineral matter (see below), there are concerns regarding whether the kerogen has been altered, and of course, there remains the problem of reliable characterization of the chemical structure of a solid organic macromolecular substance once it is isolated. This latter point is similar to the situation in the field of coal chemistry, where decades of effort have been expended in establishing structures on which there is now reasonably general, though not yet total, agreement. Aspects of kerogen structure have in recent decades also yielded to more sophisticated characterization methods, though again, one can do little more than characterize average structures or structural elements.

As a result of the intractability of the raw kerogen structure, much has been done using its elemental composition as the simplest indicator of properties or predictor of behavior. The composition of kerogens is typically presented based on carbon (C), hydrogen (H), sulfur (S), nitrogen (N), and oxygen (O) content, since these are by far and away the major elements of interest. For more precise compositional characterization of Kukersite oil shale kerogen, chlorine (Cl) is also included [164]. Table 5.2 presents some sample kerogen (or organic matter) elemental compositions of oil shales from major deposits.

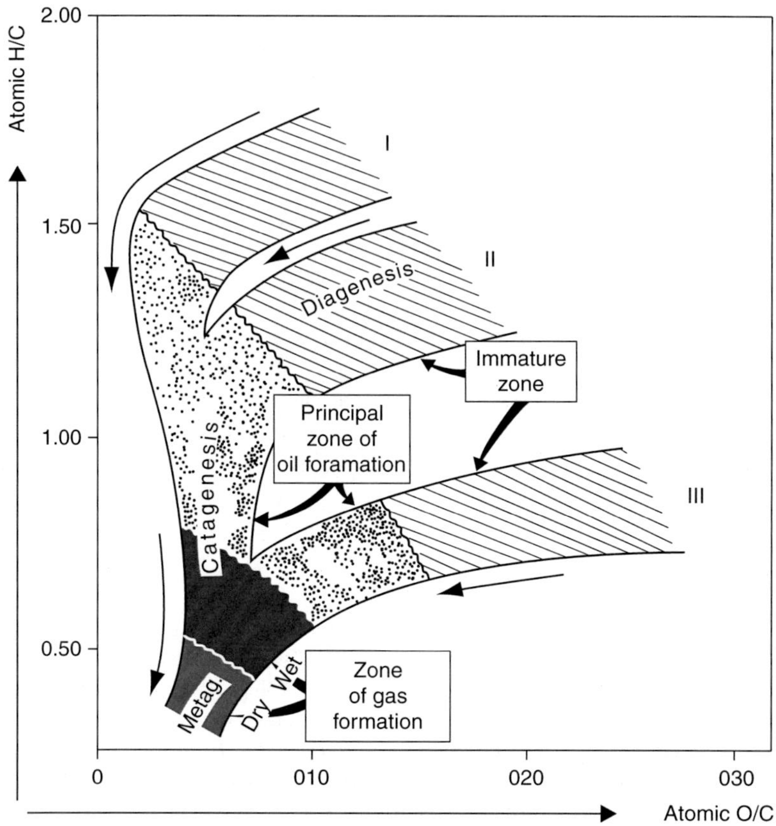


**Fig. 5.1** The van Krevelen diagram, showing the approximate locations of different kerogen types, based upon atomic ratios (from [145], with permission)

Kerogens can be classified in different ways on the basis of elemental composition. One of the most common is based on the Van Krevelen diagram originally developed for classifying coals (Fig. 5.1). Kerogens fall into certain groups on the basis of the atomic ratios of hydrogen to carbon and oxygen to carbon. Kerogens can be grouped as Type I, II, III, or IV with further subgrouping according to sulfur content. Oil shale kerogens are predominantly Type I (Green River Oil Shale from Uinta Basin is a typical kerogen of this group) and II (Torcian Oil Shale from the Paris Basin is an example from this group). The less common coaly shales belong to Type III (Upper Cretaceous sedimentary rock from the Doula Basin, as one example) [33, 145]. The following numerical H/C and O/C ratio ranges have been suggested: for Type I $H/C > 1.25$ and $O/C < 0.15$, for Type II $H/C < 1.25$ and $0.03 < O/C < 0.18$, and for Type III $H/C < 1.0$ and $0.03 < O/C < 0.3$ [145]. Somewhat different H/C and O/C, and also O/S, ratios are found for kerogens at the end of diagenesis, but the above are the values for the materials of practical importance.

**Table 5.3** Yield of bitumen per conditional organic matter. Atomic H/C ratios of some oil shale kerogens and bitumens are shown for comparison [141]

| Oil shale | Yields per conditional organic matter (%) (solvents used) | H/C bitumen | H/C kerogen |
|---|---|---|---|
| Kukersite (Estonia) | 0.71 (benzene:methanol (1:1)) | 1.75 | 1.52 |
| Kultak-Zevardy (Uzbekistan) | 12.01 (benzene:methanol (3:1)) | 1.42 | 1.2 |
| Uktha (Russia, Komyi) | 21.42 (benzene:methanol (3:1)) | 1.58 | 1.21 |
| Green River (USA) | 5.4 (chloroform) | | |
| Dictyonema (Estonia) | 3.36 (benzene:methanol (3:1)) | | |

### *Kerogen Isolation*

An important tool in the field of kerogen study is that of kerogen isolation (or concentration) from the oil shale matrix. This involves the removal of solvent solubles (bitumen) and mineral matter, leaving behind the kerogen. Several reviews are available on isolation or concentration procedures (for information readers are referred to [6, 34, 108, 128, 145]).

Solvent solubles are removable by ordinary solvent extraction. Since all unextracted organics report as kerogen, the efficacy of extraction is important in establishing actual kerogen content. The amount of extractable obtained, and its properties, may vary considerably depending upon the solvent used for extraction. The amount of the extracted matter also increases with an increase in extraction process temperature and with the polarity and chemical activity of the extractant [159]. Typical extraction yields are low, representing only a few percent of organic matter, but varying significantly from shale to shale see (Table 5.3, for examples).

A number of physical separation methods used to separate kerogen from the mineral part of oil shale are described in the literature [42, 107, 135]. These methods involve primarily such physicochemical processes as flotation and centrifugation. Combination techniques, involving such processes as ultrasonication or electrostatic and magnetic separations are also physical methods used in kerogen enrichment [159]. It is worthwhile to note that for certain oil shales, some physical methods are of value, since particle size reduction is accompanied by an increase in particle organic matter concentration.

The most widely used laboratory techniques for kerogen isolation from inorganic matter are based on chemical methods. In this regard, oil shale minerals can be grouped as carbonates (which are soluble in dilute acids), silicates (soluble in HF), and sulfides (which can be removed by oxidizing and reducing agents). Mineral composition determines the reagents (and the efficiency of reagents) to be used for demineralization [108]. For oil shale demineralization with chemical methods readers are referred to [108].

### *Kerogen Degradative Analysis*

Although elemental analysis offers valuable information for classification, other destructive and nondestructive techniques have been used to obtain information on functional groups and on structural functionalities of different oil shales [46, 110]. Earlier quantitative characterizations of kerogen functionalities were based on destructive techniques [146]. The destructive techniques include thermal degradation (pyrolysis, hydrogenolysis, supercritical extraction) and chemical degradation with different chemical reagents based on the specificity of the reagents (for more information concerning this topic see [128, 146]). Destructive analysis techniques are based on obtaining smaller identifiable compounds that have a structural relationship to the kerogen [146].

To describe the pyrolytic behavior of kerogen, detailed knowledge of kerogen structure is critical since the structural properties and functionalities of kerogen strongly influence its pyrolytic behavior. It has been shown in coal pyrolysis studies that devolatilization rates and product compositions can be correlated with coal organic matter chemical structure [123, 127]. So while certain types of functional groups may be directly related to their pyrolytic products, use of this fairly nonspecific method is of limited value.

The analysis of kerogen thermal degradation product molecular weight distributions has provided some insights into the size of the systems being examined. There was a recent study of kukersite (Type I) and dictyonema (Type II) oil shale samples involving Field Ionization Mass Spectrometry (FIMS) pyrolysis under high vacuum. This investigation indicated number-average molecular weight values of 408 and 383 Da with standard deviation values of 141 and 128 for kukersite and dictyonema oil shale pyrolysis products, respectively (estimated using a three-parameter Gaussian distribution function to describe the results) [111].

Earlier studies have, however, pointed to the possibility of higher molecular weights for the decomposition products. Suuberg and Sherman [131] studied pyrolysis of Green River oil shale under vacuum conditions, and obtained a weight-average molecular weight of around 800 Da, using a gel permeation liquid chromatographic technique. This compared favorably with a weight-average molecular weight of around 850 Da obtained for the same Green River shale, using the FIMS pyrolysis method [125]. The latter studies showed the importance of pressure on molecular weight distributions. Oils obtained using a Fischer assay method showed a much narrower molecular weight distribution, shifted to much lower molecular weights. Hence, it is dangerous to compare the molecular weight distributions from different studies unless the mass transport conditions are known to be identical. In any event, all of the above studies point to the typical characteristics of a macromolecular system being thermally degraded to provide fragments exhibiting a distribution of molecular weights (which are all, of course, lower than the original matrix molecular weight).

### *Nondestructive Analysis of Kerogen Structure*

During the last 3 decades there have been a number of qualitative and quantitative studies aimed at characterizing the structure of oil shale kerogens using nondestructive techniques such as solid-state nuclear magnetic resonance (NMR) techniques (cross polarization, magic-angle spinning), Electron Spin Resonance (ESR), Fourier Transform Infrared Spectroscopy, and X-ray photoelectron spectroscopy [66, 82, 115, 120].

One recent comprehensive study on the structure of kerogens was carried out using both X-ray photoelectron spectroscopy and solid-state $^{13}C$ nuclear magnetic resonance techniques [66]. This study determined the average chemical compositions of 18 oil shale kerogens (hydrogen-carbon skeleton, oxygen functionalities, nitrogen functionalities, sulfur functionalities). The data of this study revealed that the average carbon-hydrogen skeletal units of immature Type I and Type II oil shale kerogens (according to H/C and O/C ratios) are generally composed of two to three ring-hydroaromatic/aromatic clusters with a significant number of short attachments and long, roughly linear, alkyl chains. These data also indicated that the number-average molecular weight of the "average repeating structural unit" of oil shale kerogens might be around 400 amu. These values compare favorably with the molecular weight distributions from the actual degradation studies cited above.

Several other studies have been concerned with characterization of kerogen macromolecular structures using solvent swelling in different solvents or solvent mixtures [74, 75, 111, 112]. These studies were performed to evaluate at least semiquantitatively the number-average molecular weight between cross-links (or cross-link density), a fundamental property of a cross-linked macromolecular network. These studies indicate that oil shale kerogen solvent swelling ratios are lower (therefore cross-link density higher and molecular weight between cross-links smaller) than those seen in bituminous coals [76].

Extensive structural data are available for a few oil shale kerogens, especially for oil shales of major deposits. For several kerogens the structural information has been the basis for detailed molecular models of a portion of the structures (e.g., for Rundle Ramsay Crossing [120], Green River [120], and Kukersite [164] (see also [128]). Reviews of the history of development of hypothetical structural models of the Green River Formation kerogen can be found in the literature [128, 158]. For illustrative purposes, a recent two-dimensional model of kukersite oil shale kerogen chemical structure is shown in Fig. 5.2a [164] and another model of Green River Shale in Fig. 5.2b [120]. Again, it needs to be emphasized that what is shown are "averaged" structures that accurately depict elemental ratios and to some extent functional groups, but which cannot be held out as necessarily representing the actual linkages between different groups, and which certainly do not represent the different molecular weight distributions present in the kerogens. It should also be emphasized that kukersite oil shale kerogen is a somewhat unique Type I kerogen due to its high hydroxyl (phenolic) group content [146]. This is apparent in comparing Fig. 5.2a and 5.b. Hence, care should be exercised in using any published model of kerogen structure in trying to understand the chemical behavior of any other type of kerogen. Due to its high oxygen content, kukersite kerogen is also sometimes categorized as straddling Type I and Type II, since the lines between the different kerogen types is not necessarily as hard and fast as might be inferred from Fig. 5.1.

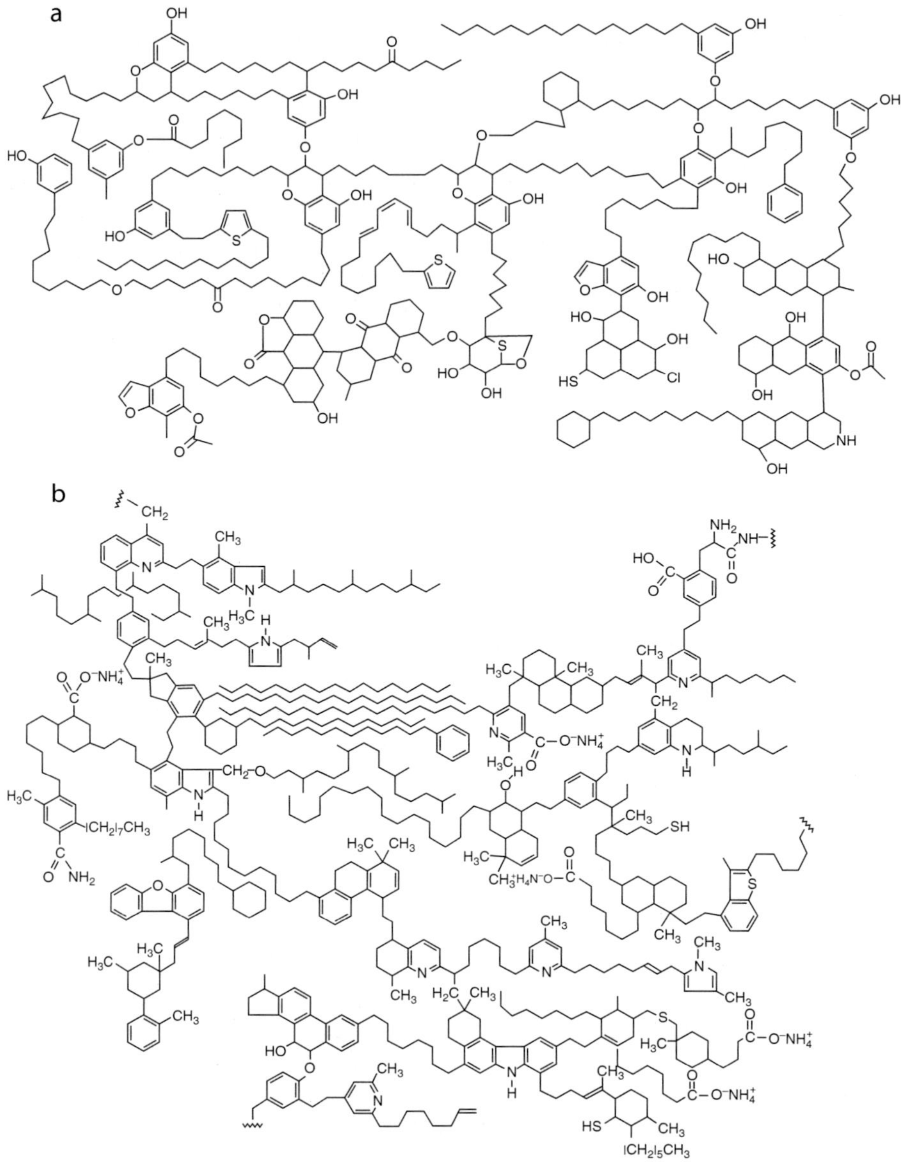


**Fig. 5.2** Two-dimensional models of kerogen chemical structure. (**a**) Model of Estonian kukersite kerogen with molecular mass 6,581 Da and empirical formula $C_{421}H_{638}O_{44}S_4NCl$ (from [164], with permission). (**b**) Model of Colorado Green River kerogen (from [120], with permission)

Bitumen or Bitumoid

Bitumen (bitumoid) is a minor fraction of oil shale organic matter that is soluble in common organic solvents (representing up to 20% of organic material [159]) and is therefore recoverable by solvent extraction. Some authors define the bitumen as a benzene-soluble matter, while others define this material as soluble in petroleum-based solvents, yet others define it more broadly as material soluble in any organic solvents, and finally, some define it as material entirely soluble in $CS_2$. According to [144] there are no standard protocols for extraction procedures, either in terms of the solvent polarity, sample grinding, temperature, duration of extraction, or stirring efficiency. Perhaps the main reason why there has not been greater emphasis on this point is that from a practical point of view, the amounts involved typically represent only a few percent by mass of the organic matter.

These extractables, bitumen, contain biomarkers or organic compounds whose structures are related to the chemical structures of the biological material from which the organic matter was derived [45, 143]. The atomic H/C ratio of bitumen has been shown to be somewhat higher than of corresponding kerogen see (Table 5.3).

## Oil Shale Conversion Processes: An Overview

Oil shale resources are abundant and widespread throughout the world and oil shale has been used for energy-technological purposes for hundreds of years (e.g., there is a shale oil extraction patent dating from 1694, granted by the British Crown, [103]). Despite this, currently only few countries (Estonia, China, Brazil, Germany, and Israel) exploit oil shale on a commercial scale. The main reason for this is that oil shale is a low-grade solid fuel. It cannot compete with coal as a solid fuel, since the latter is available in richer (lower ash content) easily accessible deposits. Oil shale is a potential source of liquid fuels, but suffers there from the fact that the world oil market situation has historically been economically unfavorable in terms of fostering development of upgrading technologies. Generally, there has been need of government subsidies in order to consider development of an industry based upon this resource, and the energy market itself (influenced mainly by crude oil price range and electricity cost) has made oil shale upgrading truly economical only in the few countries mentioned. Moreover, the economic feasibility of oil shale development is certain to be strongly affected by further strengthening environmental regulations (air, water, and land use), socioeconomic issues, and market and investment risks. Taken purely on the basis of greenhouse gas emission per unit of energy, oil shale is unattractive in much the same way as is any other solid fossil fuel, such as coal. Therefore oil shale development in recent decades has been quite chaotic and episodic, and there have been many processes/technologies proposed and many have been abandoned due to economic reasons.

Just as is coal, oil shale is also a multiuse organic-mineral raw material whose economic value has sometimes extended beyond its fuel value. From some oil shales, uranium and other minerals and rare elements can be extracted. As an example, in 1946–1952, a marine type of Dictyonema shale [81] was used for uranium production in Sillamäe, Estonia, and in 1950–1989 alum shale was used in Sweden for the same purpose [36].

Integrated retorting techniques (both ex situ and in situ) have been under consideration to recover shale oil together with nahcolite, alumina, and soda ash from some oil shale deposits [77]. However, the major emphasis in oil shale upgrading has remained on kerogen thermochemical conversion methods (processes using heat to liberate oil). There have also been a few studies (laboratory research) on the use microbial activities to break up kerogen in its mineral matrix [16, 159].

In principle the general thermochemical upgrading processes that have been developed and utilized for coals (combustion, gasification, liquefaction, and pyrolysis) can also be used for oil shale upgrading. The goal is, broadly speaking, to transform the chemical energy locked within oil shale into more concentrated and conveniently useable forms. However, due to oil shale's thermal behavior and character, some types of processes are more favorable than others. It is important to note that the hydrogen/carbon ratio of oil shale organic matter (typically 1.2–1.6) is more favorable to production of liquids than that found in coal (typically 0.7–1), and this reduces the pressure to increase H/C ratio, which has economically been one of the major stumbling blocks of coal to liquid conversion processes. On the other hand, with mineral content of between 70% and 90%, considerably higher than that of commercial coal deposits, oil shale presents the problem of processing a solid with a high mineral rock content, whose handling and heating in any thermal process is economically quite unattractive.

## Direct Combustion

Oil shale can be utilized as a fuel for thermal power plants, where it is burned directly as is coal. Currently there are several countries that utilize oil shale as fuel for power plant fuel: Estonia, Israel, Germany, and China [14, 102]. Some countries have closed their oil shale-fueled power plants (Romania) or turned to other fuels (Russia). Other countries are planning to build such plants (Jordan and Egypt), or burn oil shale together with coal (Canada and Turkey) [14, 51].

There are three commercial technologies in use for oil shale combustion [3]: (1) pulverized combustion or pulverized firing (PF), as is used in the older units in Estonia; (2) fluidized bed combustion (FBC), used by Rohrbach Zement in Dotternhausen, Germany; and (3) circulating fluidized bed combustion (CFBC), used in two new units at the Narva Power Plant in Estonia, Huadian Power Plant in China, and PAMA power plant at Mishor Rotem in Israel. Generally speaking, PF technology is no longer in favor, because it operates at such a high temperature that there can be no sulfur capture by the carbonates present in the mineral matter, and therefore it is environmentally much more problematic. The thermal efficiency of the new Estonian CFBC units has been reported to be in the neighborhood of about 36%, which is typical of this sort of fluidized bed combustion system. The older PF units operated at only about 30% thermal efficiency to electricity [52].

An exceptionally extensive study of the correlation between the heating value and various compositional parameters of Estonian kukersite has been presented by [7]. While the quantitative results of that study are applicable only to this particular oil shale, the methodology may well be generalizable. That study suggested that a good working formula for estimation of the higher heating value of the organic part of oil shale is

$$Q_H = 0.3566C + 1.0623H - 0.1339O + 0.0658S + 0.0106N - 0.0280Cl$$

in which the heating value is in MJ/kg and the element symbols represent the mass percentages of the respective elements in the organic fraction of the oil shale. As part of the same original study by Saar, the lower heating value of the typical dried whole Estonian shale is (see [89])

$$Q_L = 34.45 - 0.302A - 0.4369(CO_2)_c$$

in which A is the dry oil shale ash content and $(CO_2)_c$ represents the content of carbonate mineral $CO_2$ in the dry oil shale.

Since direct combustion of oil shale is not the main focus of this review, the reader is referred to the comprehensive volume recently published by Ots on this topic [89].

## *Low-Temperature Pyrolysis*

Low-temperature pyrolysis up to at about 500°C (also referred as retorting or semicoking or low-temperature carbonization) has been historically the most favored thermochemical conversion process for high-grade oil shales. In this pyrolysis or retorting process, the organic matter is converted to oil (termed retort oil, shale oil, shale crude oil), gas (typically called retort gas), and solid residue (typically termed semicoke in the oil shale industry). The processes can be carried out either by mining the oil shale and then producing oil by heating it in above-ground retorting facilities (termed ex situ processing), or by heating the oil shale underground without mining it and bringing it to the surface (termed in situ processing). Limiting the processing temperature to below 500°C is desirable, in terms of maximizing oil yield.

Oil shales are commonly retorted in commercial ex situ retorts without beneficiation (organic matter is not concentrated). Commercial ex situ shale oil production facilities are in operation in Estonia, Brazil, and China. Commercially produced shale oils are presently used as heating fuels or as feedstocks for producing chemicals [85]. Due to their broad boiling point distributions, high pour points, high viscosities, and high heteroatom contents (nitrogen, sulfur, and oxygen), raw shale oils need to be upgraded to be considered for higher value uses. Retorts could be included within a petroleum refinery. There have been a number of investigations on the use the shale oils as refinery feedstocks for manufacture of motor fuels [87].

## *High-Temperature Pyrolysis and Gasification*

High-temperature pyrolysis (or gasification) under self-generated atmospheres was historically used to produce town gas (e.g., commercial-scale chamber ovens used in Estonia, [30]). Just as in the case of town gas produced from coal feedstocks, gas production from oil shale was abandoned due to economic reasons, largely having to do with the low cost of production of natural gas.

Modern gasification technologies, applied to conversion of coals or biomass, use partial oxidation (steam, oxygen) or hydrogen environments. In these gasification processes, the solid feed material is converted to synthetic gases (composed mainly of hydrogen and carbon monoxide) that can be further used as combustion fuel in, for example, Integrated Gasification Combined Cycle (IGCC) plants designed to produce electrical power, or further catalytically converted (via Fischer–Tropsch reactions) to produce chemicals, including liquid fuels. However, existing commercial gasification technologies that would lend themselves to integration with upgrading plants are not economically/technically suitable for oil shale feed materials, unless the mineral matter is reduced before gasification. There have been no demonstrations of large-scale oil shale beneficiation technologies for concentrating oil shale organic matter (to reduce mineral matter) to levels suitable for proven gasification technologies. Although in principle, oil shale gasification could be more attractive than shale oil extraction via retorting, especially for low oil yield oil shales, this thermochemical conversion process has not been a subject of commercial interest. Over the course of time a number of laboratory-scale to pilot-scale studies have been carried out on this topic [8, 10, 37, 57, 113].

## *Other Thermochemical-Based Conversion Technologies*

Thermal dissolution or heating in presence of solvent has been considered as an alternative to "dry" thermochemical conversion such as classical retorting. These attempts have included those in subcritical and supercritical fluids, reactive and nonreactive fluids, and of special interest have been hydrogen-donor solvents that can provide hydrogen in the conversion process. In the latter case, gaseous molecular $H_2$ can also be introduced. Again, there is strong analogy to methods that have been suggested for direct liquefaction or solvent refining of coals. Examples include processes such as Solvent-Refined Coal processes (SRC-I and the improved version, SRC-II), the Exxon Donor Solvent (EDS) process, and the H-Coal process [8, 94, 99]. A brief review on recent advances in direct coal liquefaction and a list of major direct coal liquefaction processes is given by Shui et al. [118].

The thermal dissolution processes generally involve heating the oil shale to temperatures below those characterized by active devolatilization, as there is a desire to maximize oil yields and minimize gas production. The temperature range of active devolatilization depends somewhat upon gas pressure applied to the system.

The thermal dissolution of oil shales has been mainly of academic or research interest as the H/C ratio of oil shale organic matter is already much more favorable than of coal. For more information on oil shale thermal dissolution, there are recent brief reviews covering the topic [48, 136] and, separately, supercritical extraction [67, 139]. One step toward commercial development is the proposed 1 ton/h so-called Rendall process demonstration plant under construction in Australia by Blue Ensign Technologies Limited. This process is designed to use oil shale from the Julia Creek deposit in Queensland and to hydrotreat it in a two-stage process using a petroleum-based solvent. Again, this proposal is only at a laboratory stage.

A considerable amount of work (up to process-development unit scale) has been published on hydropyrolysis or pyrolysis under high hydrogen pressures, including catalytic hydropyrolysis and hydrogasification [8, 109]. Indeed, this technology was already explored in Estonia in the 1930s [68].

## Shale Oil Production

### *General Considerations Regarding Process Temperatures*

Technologies to recover oil from oil shale are based on thermal breakup of kerogen, the cross-linked macromolecular organic matter of oil shale. All present technological approaches require heat to take kerogen to temperatures at which thermochemical conversion processes are sufficiently fast to be of interest from a practical point of view. As noted above, depending on process conditions the thermochemical conversion results in three general classes of products with different quantities and varying qualities. These products are the oil (more generally non-polymeric organic compounds with a wide molecular weight distribution), a carbon-rich solid residue, and gas (including water vapor).

Kerogen decomposition rates increase with temperature, subject to the Arrhenius rate law. The rates will be quantitatively explored in a later section. Kerogen decomposition is slow at low temperatures and occurs on a time scale of days to years below 350°C, which is why practical processes are generally carried out at higher temperatures. The decomposition becomes fast enough for practical purposes at temperatures of 400°C and above. Cummins and coworkers [25] studied Green River oil shale low-temperature isothermal pyrolysis from 150°C to 350°C. They showed that 2% of kerogen was converted to benzene solubles in 90 days at 150°C and in 360 days 7.7% of kerogen was converted. At 200°C in 90 days 3% was converted, and in 360 days 8.9%. When the temperature was raised to 300°C in 0.5 days 3.3% was converted and in 4 days 12.8% yield was obtained. Finally, at 350°C in 0.5 days 27.6% and in 4 days 65.1% of kerogen was converted into benzene solubles.

The above demonstrates that there is actually no lower temperature limit for kerogen decomposition, consistent with what would be concluded from Arrhenius kinetics; rather, the question lies in establishing the time scale needed at any particular temperature. It is the fact that practical ex situ (i.e., in retort) conversion processes must be carried out on the timescale of minutes or hours, at most, that sets the commonly cited "lower temperature limit" for decomposition. These limits may be extended for in situ techniques, where confinement volume in an aboveground retort is no longer a consideration, but there is still a lower limit at which the process would simply take too long to be practical.

Real processes can be run under conditions varying from slow heating rate and long residence time to fast heating rate and short residence time. For example, in aboveground (ex situ) retorting equipment, heating rates, and residence times can vary from the order of 1°C/min and hours to of order of 1,000°C/min and seconds, respectively. The former is encountered, for example, in moving bed coal gasifier type retorting equipment, which utilizes particles with size on order of 100 mm (e.g., 10–125 mm for the Kiviter process, [101]). The latter high heating rate conditions are encountered in fluidized bed reactors, which use small particles with sizes below a few millimeters. In all these retorting processes (which take place in an inert gas environment or under self-generated gas environment with very low oxygen content, or in a reactive, non-oxygen environment), the oil shale is heated up to the active pyrolytic reaction region from 450°C to 550°C, where oils and gases evaporate and hydrogen-deficient organic matter or semicoke is left in the solid phase.

In situ processes generally involve heating rates below 1°C/min, because of the impracticality of rapid heating in an in situ environment. The relatively low-temperature range of 340–370°C, with time scales of years, is utilized by Shell in its in situ (underground) conversion process (Shell Oil Shale Test project) [77]. The above-cited timescales for kerogen conversion show that it is not the kinetics of conversion alone that determine these long times; rather, they are also influenced by the time needed to heat the large volumes of rock and to allow the released oil to flow out of the shale.

## *Other Processes Taking Place During Retorting*

Destruction of kerogen by heating is accompanied by processes in other oil shale constituents, that is, in bitumen and mineral matter. These processes include:

- Solvent-extractable organic matter or bitumen (non-covalently linked non-polymeric compounds) can vaporize or decompose, depending on their thermal stability. Depending on the oil shale, the bitumen content can constitute up to 20% of organic matter [159], though as noted above it is generally much less than this. If able to quickly escape the hot oil shale and be cooled, these compounds should be captured in a relatively minimally altered form.
- Mineral matter present in oil shale can have catalytic effect on thermochemical conversion processes of kerogen [41, 116]. Also, the minerals themselves can decompose at temperatures of thermochemical conversion of kerogen. Some minerals can partly or fully decompose even under typical retorting conditions (e.g., nahcolite, dawsonite, analcite, smectite, kaolinite, siderite, and pyrite) [76, 89, 92]. These decomposition reactions are generally endothermic. At temperatures around 450°C magnesium carbonate starts to decompose [76]. The interaction of a mineral with organic matter or other minerals present in the oil shale can cause altered thermal behavior, relative to the pure mineral [103]. Rajeshwar and coworkers have summarized thermal characteristics/behavior of minerals (17 minerals) commonly found in oil shale deposits [103]. Thermochemical properties and mineral reactions of oil shale minerals have been reviewed by many authors (e.g., [76, 77, 88, 92, 103]).
- Trace elements of oil shale can distribute between retorting gas, oil, solid residue, and retort water [93]. For example, there may be the presence of trace elements of environmental concern (arsenic, selenium, mercury) in retort oil, gas, or water [8, 92], or trace elements in oil (vanadium, iron, nickel, arsenic) can poison refinery catalysts [92, 93]. Actual concentration values depend on the retorting process and, of course, the trace element composition of specific oil shale. Most of the trace elements are retained in the solid spent shale itself [8].

## *Thermal Effects During Retorting*

Large quantities of heat are required to manufacture a usable fuel (liquid or gaseous) from oil shale. Heat requirements include (1) heating the mineral and organic matter to reaction temperature; (2) providing the heat of reaction related to organic matter decomposition and heat effects covering chemical changes in mineral matter; (3) providing the heat of vaporization of volatiles (oil and water). The thermal management of processes must also include provision for cooling the products of the process back to ambient temperatures. The major amount of process heat is consumed with heating mineral matter. The total heat requirement for heating oil shale to 500°C, including allowing for vaporizing volatiles, is observed to be about 650–750 kJ/kg of oil shale [99]. The heat of the pyrolysis reactions is by itself relatively small [8, 99]. A review of some recent applications of differential scanning calorimetry (DSC), mainly aimed at providing kinetic information, also provides reference to some recent work on thermal effects of pyrolysis [70].

## *Assay Methods: Determining Potential Oil Yield and Its Relationship to Other Properties*

Procedures for determining oil production potentials are known as assay methods. Assay methods generally involve laboratory scale slow heating rate (order of 10°C/min) batch pyrolysis systems or so-called carbonization tests. The procedures were developed to provide a simple and reproducible basis for fossil fuel (coal, lignite, oil shale) characterization and comparison, as opposed to providing scientific insights into the material.

The modified Fischer assay method has been widely used (since 1946) to evaluate the approximate oil potential of oil shale deposits and compare different oil shales. The term "modified" denotes that the original Franz Fischer method [161], used in Germany to characterize coals was taken over to characterize oil shales and has gone through improvement in this process. The technique has been standardized as the American Society for Testing and Materials Method (ASTM method) D-3904-80, D3904-85, and D3904-90 or as the International Organization of Standardization (ISO) standard ISO 647-74 or as the GOST 3168-93 (the GOST is the former Soviet Union Standard).

The standardized modified Fischer assay method (ISO 647-74; GOST 3168-93) consists of heating a 50-g dried and crushed oil shale (90% particles passing through a 1-mm mesh, with not more than 50% particles passing through a 0.2-mm mesh) in a specially designed cast aluminum retort (made of 99.5% aluminum) to 520°C in 70 min with decreasing heating rate, followed by holding the sample at that temperature for 10 min. The process is performed in a self-generated atmosphere. The distilled vapors of oil, gas, and water are passed through a condenser cooled with ice water, and collected into a graduated centrifuge tube. The oil and water are then separated by centrifuging. The quantities reported are the weight percentages of shale oil (and its specific gravity), water, shale residue, and "gas plus loss" by difference.

The above procedures were developed to provide a reproducible basis for oil shale characterization/comparison. Estimated errors are not expected to be greater than 5% (relative) under the standardized conditions used [55]. However, the Fischer assay is a static assay under self-generated "inert" atmosphere and factors such as cracking of volatiles due to long residence time in a hot zone, and char forming reactions [38] can limit the oil yield. Therefore, for example, use of an inert gas sweep [38] provides greater oil yields than from Fischer assay under a self-generated atmosphere. It is not uncommon to encounter laboratory retorting results in which oil yields exceed Fischer Assay results.

**Table 5.4** Fischer assay yields, dry organic matter basis, of some selected Type I and II oil shales. Parent organic matter H/C and O/C ratios are also shown

| | Green River (Utah, USA) [141] | El-lajjun [88] | Kukersite [141] | Eastern USA Devonian (Antrim) [141] | Dictyonema [141] |
|---|---|---|---|---|---|
| H/C ratio of organic matter | 1.52 | 1.42 | 1.51 | | 1.14 |
| O/C ratio of organic matter | 0.08 | 0.05 | 0.11 | | 0.23 |
| Kerogen type | I | I | I/II | II | II |
| Oil yield, wt% | 59.3 | 61.9 | 60.0 | 22.2 | 19.8 |
| Gas yield, wt% | 22.7 | 20.9 | 21.5 | 25.1 | 16.5 |
| Semicoke yield, wt % | 11.8 | 10.8 | 9.1 | 42.5 | 45.4 |
| Retorting water | 6.2 | 6.4 | 9.4 | 10.2 | 18.3 |

Other assays, developed for carbonization tests of fossil fuels, have also been applied to characterizing oil shales, and include the Heinze retort [24], the Gray and King method (developed in England) [152, 161], the original Franz Fischer method [161], the Bureau of Mines oil shale method [161], the Tosco material balance assay (with collection of gaseous products) [8], and the "material balance" Fischer assay method (modification for collection of gaseous products) [47]. Also, results based upon various TGA analyses have been used by many investigators, and are not necessarily directly comparable to the results from Fischer assay methods.

One widely applied oil shale rapid characterization and screening method is Rock-Eval analysis [9, 73]. Rock-Eval pyrolysis has been used to evaluate the maturity of the rock, as well as its richness (organic carbon content) and quality (oil and gas potential) [95]. The technique is based on temperature-programmed heating of a small sample (under 100 mg) in an inert atmosphere (helium or nitrogen) up to 600°C (to 800°C using version Rock-Eval 6). Evolution of hydrocarbon volatiles is monitored by a flame ionization detector, and carbon dioxide ($CO_2$) by other means, such as an infrared analyzer or a thermal conductivity detector, for example. The basic method was designed to measure the amount of free hydrocarbons in the sample ($S_1$), the evolution peak temperature ($T_{max}$) and the amount of hydrocarbons generated through thermal cracking of nonvolatile organic matter ($S_2$), and the amount of $CO_2$ produced during pyrolysis of kerogen ($S_3$). Based on these, the Hydrogen Index (HI) as mg $S_2$ per gram of total organic carbon, Oxygen Index (OI) as mg $S_3$ per gram of total organic carbon, and Productivity Index (PI) as $S_1/(S_1 + S_2)$ can be calculated. Rock-Eval pyrolysis temperature $T_{max}$ and HI are used to evaluate maturity of oil shales. HI and OI have been used to construct a modified van Krevelen diagram.

Thus the approximate grade of oil shale (average oil yield) has been determined historically in laboratory batch retorts such as the "modified" Fischer assay method or the Rock-Eval technique. There are high and low oil yield kerogens based on modified Fischer assay results. Tables 5.4–5.6 are compiled to show how the

**Table 5.5** Fischer assay gas compositions of the selected Type I and II oil shales shown in Table 5.4

| Gas composition, vol% | Green River (Utah, USA) [141] | El-lajjun [88] | Kukersite [141] | Eastern USA Devonian (Antrim) [141] | Dictyonema [141] |
|---|---|---|---|---|---|
| $CO_2$ | 30.3 | 14.5 | 25.0 | 9.7 | 13.6 |
| $H_2S$ | 3.3 | 25.2 | 8.8 | 29.1 | 20.3 |
| $C_mH_n$ | 2.0 | 9.3 | 16.1 | 2.6 | 7.5 |
| CO | 16.8 | 1.0 | 7.9 | 14.3 | 5.7 |
| $H_2$ | 24.2 | 21.6 | 8.2 | 22.4 | 27.1 |
| $C_nH_{2n+2}$ | 21.2 | 28.4 | 34.0 | 19.2 | 25.8 |

**Table 5.6** Fischer assay oil elemental compositions of the some selected Type I and II oil shales shown in Table 5.4 [141]

| | Green River (Utah, USA) | Kukersite | Eastern USA Devonian (Antrim) | Dictyonema |
|---|---|---|---|---|
| C | 84.7 | 83.0 | 83.8 | 83.3 |
| H | 12.0 | 9.9 | 10.6 | 9.3 |
| S | 0.6 | 1.1 | 1.8 | 2.6 |
| N | 2.1 | 0.1 | 0.7 | 1.0 |
| O (by difference) | 0.6 | 5.9 | 3.1 | 3.8 |
| H/C | 1.7 | 1.43 | 1.52 | 1.34 |
| O/C | 0.005 | 0.05 | 0.03 | 0.03 |

composition of kerogen affects the product yield and product composition. Table 5.4 also confirms a general observation that the amount volatiles generated during thermochemical conversion under an inert environment up to about 500°C decreases from kerogen Type I to Type II, and gas to oil ratio and semicoke to oil ratio increase from Type I to Type II [128]. It should be noted that a Van Krevelen diagram-based classification of oil shale behavior (based upon H/C and O/C ratios) provides only a rough prediction of behavior, and better correlations between pyrolysis yields and organic matter types might need additional indicators or different classification approaches [166].

As the modified Fischer assay has been a standardized method for evaluating oil yields of different oil shales, the chemical and physical properties of different oil shale deposits have been correlated with Fischer assay results. Properties that are successfully correlated include, for example, density, thermal conductivity, organic carbon, total hydrogen, aliphatic/aromatic carbon, and unpaired electron concentration [82]. In general, each specific deposit exhibits its own correlations and these correlations need to be derived for the specific deposit, as the relationships are mostly empirical.

Thermochemical conversion of oil shale kerogen to oil produces changes in oil chemical composition relative to the original oil shale organic matter. Analysis of data on hundreds of oil shales (compiled by Urov and Sumberg [141]) indicates that the Fischer assay oil H/C ratio can be either lower or higher than that of the parent organic matter. This is consistent with an observation that shale oil aromaticity can be greater or lower than that of the parent kerogen [82].

There are a number of papers [90, 141] concerned with characterization of Fischer assay oils, examining, for example, elemental composition, molecular weight (varying from 190 to 310 Da), density (or specific gravity varying from 0.8 to 1.04, with only a few exceptions $<1$), heating value (39–44 MJ/kg), viscosity, boiling range (30–70% boiling over up to 350°C), and chemical groups (nonaromatic hydrocarbons, aromatic hydrocarbons, heteroatomic compounds, etc.).

It was already noted that the Fischer assay is a standard method and it does not provide either the maximum or minimum oil yield that can be produced from a given oil shale. Changes in parameters such as reactor configuration, particle size, heating rate, pressure, or pyrolysis environment also can affect product yields and compositions [8, 23, 157]. In real processes these factors are not always independent, as, for example, high heating rates require fine particles or hydrogen environment is used with high pressure.

## *Heating Rate Effects on Retorting Oil Yields*

Fluidized bed retorting, where heating rates reach 1,000°C/min and particle sizes below a few millimeters are used, produce higher oil yields than obtainable from Fischer assay. For example, yields about 110% relative to Fischer assay have been reported in the case of Green River oil shale and about 150% of Fischer assay for eastern US Devonian shale [23]. Fluidized bed pyrolysis shows, in addition to increased oil yield and decreased gas yield, different oil properties such as higher aromaticity, molecular weight, density, and viscosity [23], as compared to Fischer Assay oils. It should also be noted that laboratory-scale fluidized bed retorting experiments usually resulted in higher oil yield than commercial-scale design retorts [23].

In large-scale retorting systems, high heating rates require high heat-transfer rates, and thus, a finely ground feed. Therefore differences in products from low and high heating rate thermochemical conversion processes are influenced also by the effect of particle size. The larger the particle, the slower the mass transfer rate of volatiles to the outer surface of particle, and thus, the more opportunity for the products to undergo secondary reactions that would alter their nature. Hence, it is difficult in most cases to separate the effects of heating rate and particle size.

## *Pressure Effects on Retorting Yields*

The mass transfer of oil out of pyrolyzing particles is affected by the pressure difference between the particle interior, where the products are produced, and the surrounding atmosphere. Delayed escape of the volatiles can result in secondary cracking reactions of higher molecular weight products, resulting in lower oil yields and increase gas and char yields.

With an increase in pressure under an inert environment (or self-generated environment) the oil yield decreases, the gas yield increases, the char yield increases, and oil specific gravity and viscosity decrease. Also, the amount of lighter distillate oil fraction increases [8]. All of these phenomena are consistent with a process of cracking of higher molecular weight oils to lighter oils (with deposition of coke). Different results can be seen when reactive pyrolysis atmospheres (such as hydrogen or steam) are used. On the other hand, these conclusions are not necessarily general, as in high heating rate (1,000 K/s), small particle pyrolysis, oil yields were observed to be quite similar, decreasing only a small amount with increasing pressure above a temperature range of 800–900 K [131]. The oil loss at higher pressures was associated with the occurrence of cracking reactions. The molecular weight distributions of the shale oil were relatively insensitive to pressure, though at higher temperatures, higher pressures tended to slightly favor lower molecular weights. The results of this work also showed clearly that there was an evaporative equilibrium responsible for determining the molecular weight distribution in the oil, as the average molecular weights in the oil were significantly below those of the extractable thermobitumen left in the shale. Hence the effects of pressure are associated with evaporative processes, though these effects are not as significant as they are in determining the yields of tars from coal pyrolysis [127, 132].

## *Effect of Gas Atmosphere on Retorting*

An inert gas sweep of a retort provides greater oil yields than obtained from the Fischer assay procedure under self-generated atmospheres [38]. Pyrolysis in a steam atmosphere has been shown to have favorable effect on products by improving yield and quality of shale oil [38] relative to Fischer assay. In fluidized bed experiments, steam has been studied as one possible fluidization medium [23, 38]. However, the beneficial effect of steam relative to nitrogen is the subject of dispute among different researchers [38].

Pyrolysis under high hydrogen pressure (hydroretorting, hydropyrolysis) is shown to be a potential way to upgrade low-oil grade oil shales to oil. Hydroretorting results in increased oil yield and more aromatic oil with lower heteroatom content and lower molecular weight distribution [106]. Hydrogen pressures of ca 5–10 MPa were sufficient to have an effect [106]. In some studies, higher pressures, 15 MPa, have been used [24]. A hydroretorting assay unit has been developed along with a standardized procedure for its use [106]. In this unit (538°C and hydrogen pressure 6.9 MPa) for low oil yield oil shales (Type II kerogens), oil yields can be increased several times relative to Fischer assay, even 400% over Fischer assay yield [106].

## *Chemical Aspects of Oil Recovery by Retorting*

### Shale Pyrolysis in Comparison to Coal Pyrolysis

The process of recovering shale oil is concerned with decomposition of kerogen, the major portion of oil shale organic matter. Kerogen is a cross-linked macromolecular structure whose thermal behavior depends upon the severity of heating, yielding either mostly hydrogen-rich solvent soluble material or light volatiles (oil, gas) and a carbon-rich solid residue. The overall pyrolysis process is extremely complex with a variety of physical-chemical mechanisms. The composition and yields of pyrolysis products depend on the type and characteristics of oil shale (primarily of kerogen), and as noted above, the particle size, final temperature, heating rate, residence time of vaporizing products in hot zone, pressure, and pyrolysis atmosphere. The possibly important role of secondary reactions has been noted in several places above.

In the period from 1970 to 1990, there were extensive theoretical and experimental studies on coal pyrolysis, and relatively fewer such studies on oil shale. The aim of both research communities was to improve the fundamental understanding of pyrolysis chemistry with an ultimate objective of developing realistic simulation models. The examination of coal pyrolysis tended to place great emphasis on the importance of the interplay between chemical and physical processes encountered in pyrolysis, something that received less attention in the shale pyrolysis work.

Coal pyrolysis is an important phenomenon inasmuch as it is an initial step in all major coal thermochemical conversion processes such as combustion, gasification, carbonization, and liquefaction. The past 4 decades resulted remarkable advances in the modeling of coal pyrolysis. There was a shift away from simple, empirical kinetic models to use of more complex "network" devolatilization models. Those models provide detailed predictions of pyrolytic behavior for a wide variety of coals under variable temperature, pressure, and heating rate conditions. There are many extensive reviews and books published on coal pyrolysis [76, 127], covering the chemistry of pyrolysis, mass transfer effects, comparison of experimental data and models, etc.

In some ways oil shale is similar to coal as the organic matter is mostly a macromolecular three-dimensional structure, however, with different organic compositions and hydrogen-carbon skeletons. Therefore, oil shale pyrolysis is characterized by different product characteristics and by different rates and extents of pyrolytic process events (cross-linking, swelling, softening, resolidification, and devolatilization). The pyrolytic devolatilization processes of oil shales occur at somewhat lower temperatures than do the same sorts of processes in coals. The active devolatilization starts at about 350–400°C with a maximum in the range of 400–500°C and completes at about 500–550°C. The oily pyrolysis products are more volatile (with higher hydrogen/carbon ratio and lower molecular weight distribution) than the coal tarry products. Oja has demonstrated that the kukersite oil shale primary tar (oil) produced during moderate rate pyrolysis (~5°C/s) under a nitrogen sweep in a tubular furnace is easily volatile below the temperatures of active devolatilization [86]. This is not generally the case for coal liquid (tar) pyrolysis products.

It can be assumed that the general principles used to model the mechanisms of coal pyrolysis could also be used to describe oil shale kerogen pyrolysis. For coal, the following pyrolysis event sequence was proposed [127]: disruption of hydrogen bonds; vaporization and transport of non-covalently bonded "guest" molecules; low-temperature cross-linking (in low rank coals) that is accompanied with $CO_2$ and $H_2O$ evolution; bridge breaking and fragmentation of the macromolecular network (depolymerization); hydrogen utilization to stabilize free radicals; vaporization and gas phase transport of light fragments; moderate temperature cross-linking reactions to resolidify the macromolecular network; decomposition of functional groups to produce light species or gases; and high-temperature condensation of the macromolecular network involving hydrogen elimination. In most respects, the same processes must take place in oil shale pyrolysis.

The importance of the extent and timing of cross-linking in the pyrolytic processes of coals has been emphasized in coal pyrolysis models. A number of studies have been performed on swelling of thermally pretreated coals indicating loosening and/or tightening of coal structures in the temperature region prior to tar evolution or active pyrolysis. This phenomenon was found to depend on coal rank: changes in lower rank coals indicate low-temperature cross-linking whereas trends in higher rank coals vary from no observable changes to structural relaxation. It was observed that the extent of low-temperature cross-linking was associated with decomposition of carboxyl groups and possibly hydroxyl groups as formation of new cross-links before devolatilization happened together with $CO_2$ and water release. It has been concluded that low-temperature cross-linking results in low tar yield and can be a reason for the non-softening pyrolytic behavior seen in low rank coals. The importance of such processes in oil shale pyrolysis will be considered below.

## General Observations of Oil Shale Pyrolysis Phenomena

A large number of studies (including reviews) on pyrolysis of oil shales have been conducted and reported over the course of history. The pyrolytic behavior of oil shales have been investigated generally based on individual oil shales or in some cases generalized to oil shale kerogen types (Type I, II, or III in accordance of Van Krevelen diagram) [35, 46, 121]. Although kerogen types (similar to coal ranks) have been shown to be useful for generalizing oil shale kerogen (organic matter) behavior, it should be kept in mind that differences between kerogens make a large difference in behavior, and the trends with kerogen type have not been firmly established.

There have been a number of studies of the changes in kerogen structure (in the solid organic residue) during pyrolysis/heating. These changes involve elemental composition, functional group losses and other structural modifications, as well as changes in the number of free radicals. Techniques employed for tracking the changes include, for example, elemental analysis, Fourier transform infrared spectroscopy, solid-state nuclear magnetic resonance (NMR) spectroscopy, electron microscope or electron spin resonance [28, 119]. There are of course also a great number of studies on volatile product evolution during pyrolysis. This includes evolution of volatile products correlated with kerogen functional group decomposition. Techniques used have included Fourier transform infrared spectroscopy, mass spectrometry, or gas chromatography [105].

For properties of crude shale oils readers are referred to [8, 85, 99, 142].

### Observations Regarding Changes in the Macromolecular Structure of Oil Shale Kerogens During Pyrolysis

In relation to the different oil yields observed in the case of pyrolysis of Type I and Type II kerogens (see Tables 5.4–5.6), solvent swelling experiments on thermally pretreated oil shales have indicated that high oil yield kukersite kerogen (Type I), which exhibits softening pyrolysis behavior during pyrolysis, shows a tendency toward structural relaxation in the low-temperature pre-pyrolysis region [111]. This observation is qualitatively comparable to that seen in high tar-yield softening coals. On the other hand, low oil yield Dictyonema kerogen (Type II), which exhibits non-softening pyrolysis behavior, shows a tendency for pre-pyrolysis cross-linking or structural tightening [163], similar to what happens in low rank non-softening coals [126].

### Mechanisms and Kinetics of Kerogen Decomposition

Mechanisms of kerogen decomposition have been the subject of numerous studies, including many different experimental techniques, and have involved both isothermal and non-isothermal procedures. Techniques used have included thermogravimetry, differential scanning calorimetry, differential thermal analysis, Rock-Eval, and various other types of reactors. These studies tracked changes in at least one pyrolyzing kerogen-related component (parent material or product) as a function of time and temperature. In many such studies, complex materials, such as pyrolysis oil, are considered as single pseudo-components. There are many papers that consider the mechanisms and kinetics of kerogen decomposition, and only a sampling of these is offered here [4, 5, 13, 18, 26, 27, 29, 39, 53, 58, 61, 63, 69, 76, 103, 121, 130, 134, 136, 138, 148, 150, 153, 154, 156]. Despite the existence of a relatively extensive literature on oil shale pyrolytic behavior, very few analyses and reviews are available that offer comparison between the available information or comparison to coal and biomass research results.

Different mechanisms, with various degrees of complexity, have been proposed to describe oil shale pyrolytic behavior. Generally, these are based on tracking intermediate, primary, and secondary pyrolysis products, often without a clear

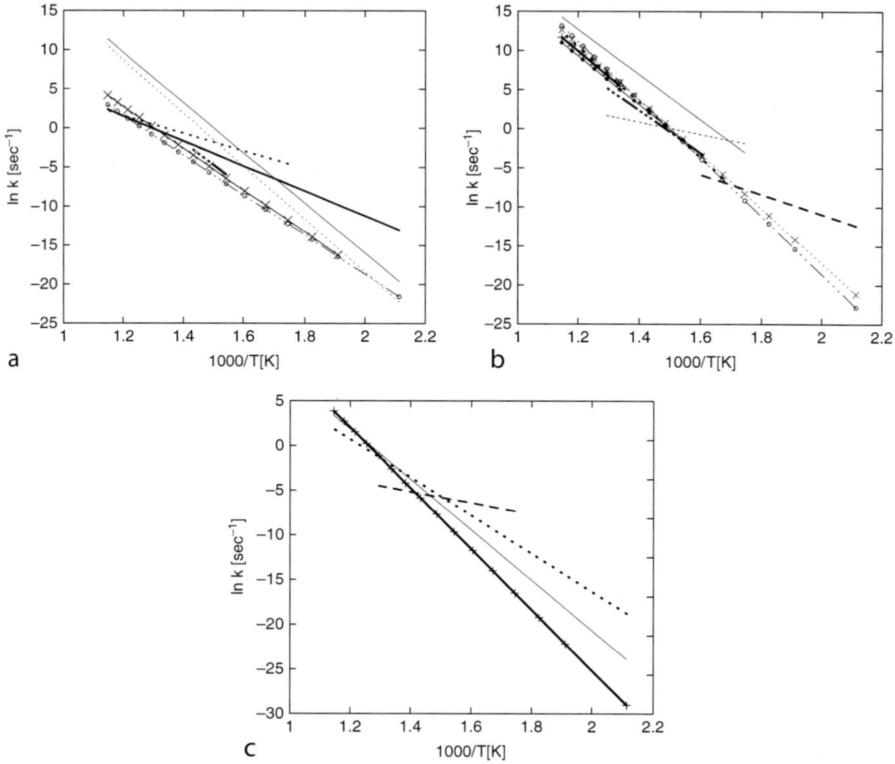


**Fig. 5.3** Comparisons of kinetics of kerogen pyrolysis. (**a**) A comparison of rates on Estonian kukersite kerogens. Thick lines [59, 60, 62]; thin lines from Oja (unpublished results); lines with crosses [96]; line with small circles, [160]. (**b**) A comparison of rates obtained on Colorado Green River Oil Shale. Dark line with crosses [12]; thin solid line [78]; heavy dashed line [25]; heavy dotted line [21]; heavy dashed-dotted line [149]; thin line with small black points [124]; thin dotted line with crosses [96]; thin dashed-dotted line with circles [19]; thin dashed line [84]. (**c**) A comparison of rates obtained on kerogens from Rundle Australian Oil Shale. Heavy line with crosses, [147]; thin line [96]; heavy dashed line [84]; dotted line [91]

identification of all of the products in any of the classes being tracked. Kinetic parameters have been calculated by fitting experimental data to an assumed semi-empirical model of the pyrolysis process. There is, thus, not surprisingly a large variation in reported kinetic parameters (activation energies and frequency factors). Consequently, the results are often of limited generality, applicable to the particular experimental conditions studied, often potentially influenced by heat and mass transfer limitations on observed kinetics.

A sampling of kinetic results is shown in Fig. 5.3. These results have been taken from studies in which both activation energies and pre-exponential factors have been reported. Except as noted, they are based upon a model of pyrolysis that is first order in kerogen, and normally the rate data are those either for total mass loss or the production of oil. Hence, the "model" can be simply represented as:

$$dX/dt = -A\exp(-E/RT)X$$

in which X is a conversion parameter, expressed in terms of remaining unconverted fraction. In many of these studies, a variety of different fitting techniques were employed (e.g., Friedman, Coates-Redfern, and simple Arrhenius). The point here is not to evaluate the different methods or studies one against the other, but to provide a rough sense of how much agreement there is in the literature on pyrolysis kinetics.

It is immediately apparent that despite the clustering of results on the graphs, the actual rates in the temperature regions of most interest can diverge from one another by significant amounts. In the case of the kukersite samples of Fig. 5.3a, the difference near 400°C (1,000/T[K] = 1.49) is about two orders of magnitude. The spread in reported values for Green River oil shale is not as great near 400°C see (Fig. 5.3b, which shows an order of magnitude spread), but there are still significant differences in reported activation energies that lead to differences of even greater magnitude near temperatures of interest. Likewise, there is a significant disparity in the results reported for Rundle Australian oil shale kerogen in Fig. 5.3c. Also, these results point out that there can be significant differences between the kerogens from different samples. The kerogen from kukersite shows typically higher reactivity than is shown by the Green River or Rundle samples. Hence, there is not surprisingly a dependence of kinetics on the origin of the shale.

This comparison of kinetics is not offered to evaluate the relative merits of any one study over another. It is to point out that there remain significant unresolved differences in reported kinetics, and that this is an issue that will seemingly require further work to resolve.

## The Nature and Role of Thermobitumen

At many pyrolysis conditions, thermobitumen (also referred as pyrolytic bitumen or "metaplast" in coal chemistry) is an important intermediate of oil shale pyrolysis. The formation of the thermobitumen has historically caused technological problems in retort design and development due its sticky and plugging character. The thermobitumen can be defined as an intermediate organic substance, which is organic solvent-soluble but nonvolatile at its formation temperatures. Several researchers define the bitumen more specifically as a benzene soluble pyrolysis intermediate [165]. The nonvolatile thermobitumen is used in models as a single pseudo-component that precedes oil generation. In contrast to the thermobitumen, the oil is defined as a volatile organic product of the pyrolysis process. The kinetics of thermobitumen formation have been described in several papers, for Green River oil shale in [18, 165] and for Kukersite oil shale [62, 64, 136].

The observation of thermobitumen formation has been incorporated into a number of mechanisms [8, 76, 83]. Generally pseudo-first-order formation reactions are assumed. For example, a qualitative thermal degradation mechanism

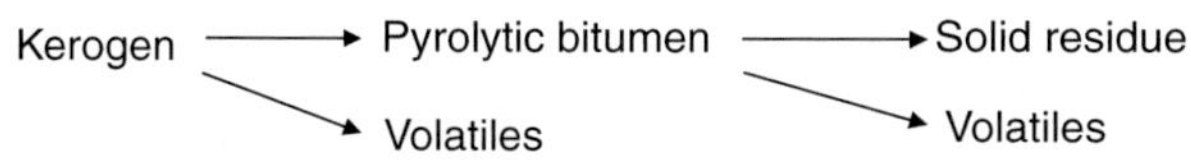


**Fig 5.4** A generally accepted mechanistic scheme for pyrolysis of kerogen

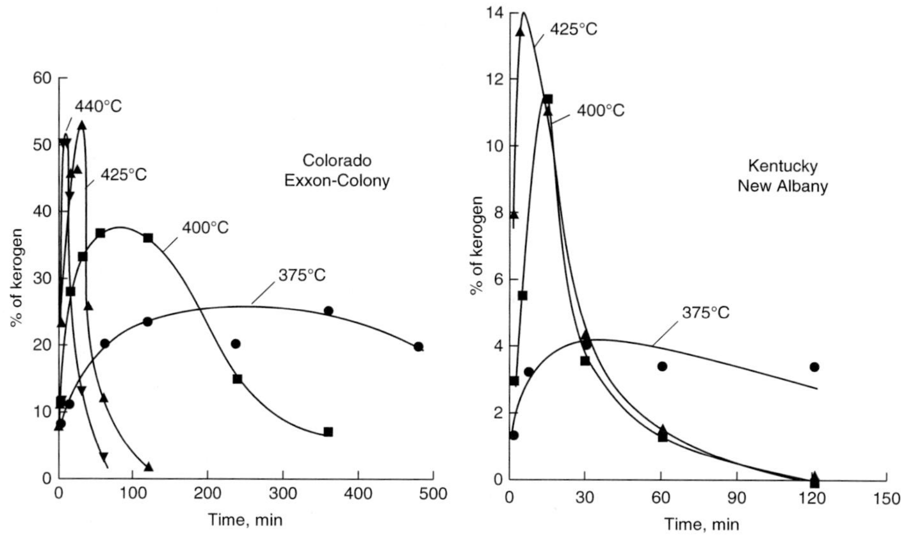


**Fig. 5.5** Time and temperature dependence of pyrolytic bitumen for Colorado and Kentucky oil shales ([165], with permission)

proposed by [1], and now fairly widely accepted in general form by many other researchers, is shown in Fig. 5.4. It embodies the well-established feature that there is a breakdown of kerogen to extractable bitumen that precedes a significant amount of oil release. The thermobitumen formation, its composition, and yield, are related to the time–temperature history of the process and to the oil shale kerogen structure. It is now more generally believed that it is the second stage of volatiles formation (from the thermobitumen) that gives rise to most of the oil product. Light gases may be released in both volatiles' release periods.

Figure 5.5 shows the time and temperature dependence of thermobitumen yields for western US high oil yield Colorado oil shales (Type I kerogens) and eastern US low oil yield Kentucky New Albany oil shale (Type II kerogens). Figure 5.5 indicates that low oil yield non-softening oil shales cannot form considerable amounts of intermediate pyrolytic bitumen. This has been explained by cross-linking behavior of these oil shale kerogens and direct kerogen conversion to residue, but also by autocatalytic reactions between bitumen and kerogen [165].

There are few studies on thermobitumen characterization, but generally its chemical, physical, and thermodynamic properties have been little studied. The properties and composition (elemental composition, molecular weight distribution)

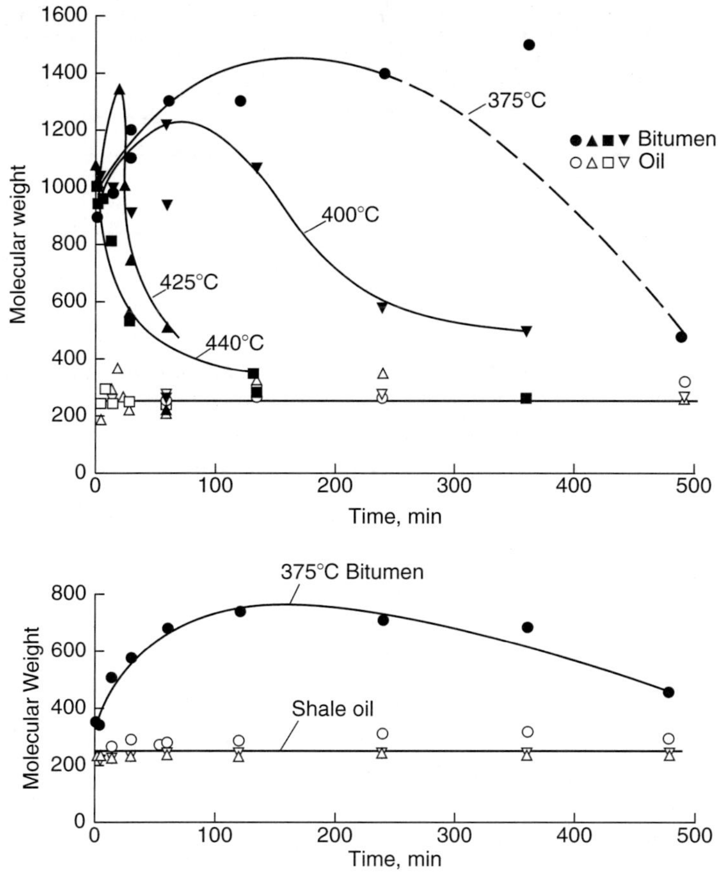


Fig. 5.6 Molecular weights of oil and bitumen versus time and temperature for (**a**) Colorado (Exxon-Colony) and (**b**) Kentucky (New Albany) oil shale ([165], with permission)

of thermobitumen is temperature and time dependent. Figure 5.6 shows the average molecular weights of oil and bitumen as a function of time and temperature for western US Colorado (Exxon-Colony) and eastern US Kentucky (New Albany) oil shales, and Table 5.7 presents thermobitumen characteristics of Estonian kukersite oil shale. The results of Table 5.7 show the expected decrease in bitumen molecular weight as conversion progresses to greater extents at higher temperatures.

These and other available studies [136, 137] indicate that the thermobitumen relative to oil produced at same temperature has lower H/C atomic ratio (and higher aromaticity) and higher number-average molecular weight (depending on oil shale and temperature-time history, the thermobitumen can exceed a value of 1,500 Da). It was noted by Miknis and Turner [165] that atomic H/C and number-average molecular weights of the thermobitumen produced were strongly time and temperature dependent, while oil showed remarkably constant values at all times and

**Table 5.7** Thermobitumen characteristics of Kukersite oil shale (elemental composition of organic matter C 82.1%, H 10.63%, S 0.75%, O + N 6.5% and atomic H/C 1.55) from thermo-chemical conversion in a Fischer retort [65, 162]

| Process conditions (temperature, time) | Yield, % thermobitumen per organic matter | Elemental composition, % | | | | | |
|---|---|---|---|---|---|---|---|
| | | C | H | S | O + N | H/C | Mol. weight |
| 275°C, 456 h | 17.9 | 83.0 | 10.1 | 0.50 | 6.7 | 1.46 | 710 |
| 300°C, 387 h | 69.8 | 85.1 | 8.77 | 0.32 | 5.8 | 1.24 | 1,300 |
| 340°C, 12 h | 59.8 | 83.8 | 9.78 | 0.52 | 5.9 | 1.40 | 1,240 |
| 360°C, 4 h | 63.8 | 83.6 | 9.56 | 0.50 | 6.3 | 1.37 | 790 |
| 380°C, 4 h | 35.6 | 85.3 | 8.85 | 0.53 | 5.3 | 1.25 | 651 |

temperatures. These results on bitumen are in general agreement with the results of [131] on Colorado oil shale, which showed that this fraction had a very high molecular weight (a significant fraction >1,000 Da). In this latter study, the oil itself had a higher number-average molecular weight than observed by Miknis and Turner [165], presumably due to the difference in pyrolysis conditions, which favored evaporation of higher molecular weight fractions.

In addition to direct pyrolysis (retorting) other technological solutions, such as supercritical gas extraction, hydropyrolysis, or thermal dissolution with solvents, take advantage of the chemical processes that occur at thermobitumen formation stage.

## Competitive Reaction Processes and Hydrogen

An important concept in coal thermochemical decomposition is formation of volatile and nonvolatile reactive radical components as a result of bond breaking in the macromolecular structure. Reactions between radical components can lead to char formation, if the radical species involved are located on large fragments of the network structure. Stabilization of radicals by hydrogen can result in volatiles (or potential volatiles), if the hydrogen capping of the structures prevents their incorporation into a new network structure.

On the other hand, the formed volatiles can undergo secondary reactions prior to escaping the reacting particles. This may happen in contact with solid material or in the vapor phase. Such secondary reactions are promoted by long residence times in high-temperature zones within a particle or a reactor. Thermochemical conversion in the presence of hydrogen can result in an increase in non-polymeric volatile hydrocarbons, if a radical secondary reaction pathway can be intercepted by hydrogen capping of a radical. In addition to helping reduce the macromolecular structure to smaller species by the above routes, hydrogen is also consumed in reducing sulfur, nitrogen, and oxygen that exist in the kerogen structure [99]. Hydrogen can be introduced into these roles either directly from the gas phase or from a hydrogen-donor solvent [99]. There are effects of temperature, pressure, and catalysts on these hydrogen reaction pathways [99, 107].

## Technological Aspects of Shale Oil Recovery

The fundamental chemical and heat and mass transfer processes in oil shale retorting have been described to a limited degree in comparison to similar coal and biomass processes. By comparison, there is a rich literature on practical oil shale retorting technologies (including flow sheets, process descriptions, and product characterizations) [8, 30, 99]. There are also reviews/reports covering processes from laboratory scale through process-development unit scale to commercial scale: [15, 20, 85, 97, 101, 122, 140, 142]. The above books and reviews also cover a number of other important topics associated with oil shale processing, such as availability of oil shale reserves [77, 85, 142], oil shale mining [8, 85], research and development needs [77, 85], socioeconomical problems [85], environmental impact [77, 85, 142], oil shale upgrading economics and shale oil cost components [38], shale oil refining [85, 129] and future perspectives [129, 142].

Over hundreds of years of oil shale development, hundreds of retorts (retorting processes) have been invented [85]. Through the long history of oil shale utilization, ex situ (above the ground or surface) retorting technologies have been practically the only commercial technology for oil production from oil shale. In ex situ processing, the oil shale must be mined (using surface mining or underground mining), crushed, and transported to the processing plant for retorting. By comparison, in situ processes cause less land surface disturbance associated with mining. This may involve limited excavation to access oil shale sources or to create permeability by fracturing or drilling heater holes or production wells. Because the oil shale is retorted in the ground, in situ processing requires that the produced oil be extracted through an oil well in the same way as is conventional crude oil [20]. Therefore, commonly cited advantages of in situ process are limited mining and solids' (oil shale and spent shale) handling, and no expense associated with aboveground retorting equipment. The disadvantage of in situ processing involves the complexity in controlling process parameters. There are also concerns raised regarding the ability to capture all of the oil produced underground.

In addition to the issues of technical viability (including process controllability) and capital cost, the choice to use ex situ as opposed to in situ processes depends on factors such as richness of the oil shale, the depth or thickness of the shale bed, the accessibility to deposits, and environmental considerations (associated with escape of oil into groundwater or extraction of materials from spent shale, e.g.,). Again, ex situ processes are often also regarded as environmentally problematic beyond their mining impacts, due to problems of solid residue disposal and potential for groundwater pollution from that source.

Besides the simple ex situ and in situ classifications, there are many other retorting classification schemes used in the literature as technologies can differ in various aspects such as process principle. These include process modes such as batch, semi-batch, or continuous, as well as process flows including cocurrent, counter-current, or crosscurrent. An important classification is based upon heat-transfer mechanism to the shale (direct or indirect, and what the carrier is in direct) as well as heat generation method. In addition, retort style, complexity of technology or by-product processing can be bases of classification.

A vitally important design consideration is always thermal management, both in terms of heat generation and heat transfer to the shale to be retorted. Heat can be transferred to oil shale by different means:

1. *By conduction through a retort wall.* To this group belonged ex situ technologies such as the vertical Pumpherston retort (Bryson retort) [79] or Davidson horizontal rotary retort [80]. In these types of ex situ technologies, heat is transferred through the retort wall. Due to the slow conduction that characterizes such retorts, many such processes are mainly of historical interest as their throughputs per unit of vessel volume are low. Development phase processes of this genre are the Oil-Tech electrically heated vertical retort process (Ambre Energy Limited) [140] and the A.F.S.K. Hom Tov co-pyrolysis process (Mishor Rotem).
   The principle of wall conduction can also be applied to in situ technologies. In that case, it involves placing heating elements or heating pipes into the oil shale deposit. Representative of this approach is the Shell In Situ Conversion Process (ICP) technology (Shell Mahogany Project) [140], which uses electric-resistance heaters combined with a surrounding "freeze wall" to confine the retorting products. Another example is the EGL Oil Shale process [140] in which heat is transferred by superheated steam (in a closed system) through a series of pipes serving as heat exchangers below the oil shale bed. The Geothermic Fuel Cell™ approach (GFL™) developed by Independent Energy Partners, Inc. (IEP) uses a high-temperature fuel cell stack [140].
2. *By gases from shale combustion in the retort* (referred to as the directly heated retorts approach). These technologies utilize heat that originates from burning materials (char and oil shale gas) inside the retort and is therefore also called internal combustion retorting. The heat is transferred through the retort by gases generated in the combustion process. In combustion retorts the retorting gas is diluted and has low calorific value (for syngas classification depending on the heating values, see [77]). A currently used commercial ex situ retorting technology of this type is the Fushun process [101] in China. Other examples of thoroughly investigated ex situ technologies based upon this approach are the Kiviter process (early versions), Nevada-Texas-Utah (NTU), Paraho Direct, Union "A" and Superior Multimineral processes [8, 77]. In situ retorting also utilized this principle in technologies that burned oil shale underground. An example of this was the Occidental Petroleum Corporation Oxy process [85] (modified in situ combustion retorting process).
3. *By gaseous heat carrier heated outside retort* (referred as indirectly heated retorts). In this case, heat is transferred by hot gases heated in an outside furnace and then injected into the retort (or retorting bed). Again, in these systems the retorting gas is diluted and has low calorific value. Currently used commercial ex situ technologies of this group are the current Kiviter process in Estonia and Petrosix process in Brazil [8]. Other main ex situ technology examples are Union

"B," Paraho Indirect, Superior Indirect processes [8], and fluidized bed-based processes. Similarly, in situ technologies can utilize gases that are heated aboveground (such as by combustion), which are then injected into the oil shale formation as heat carriers. Examples of such hot-gas injecting methods are the Chevron CRUSH process and the Petro Probe process and the in situ gas extraction technology (IGE) of Mountain West Energy [140].

4. *By solid heat carrier heated outside the retort* (referred indirectly as heated retort). In this instance, heat is transferred by mixing hot solid particles (heated outside the retort) with the oil shale. Reactive heat carriers, such as oil shale ash, or nonreactive, such as sand or ceramic balls, have been used. In these processes retort gases with high calorific value are produced. These are ex situ processes that typically take place in rotating kiln-type retorts. Currently used commercial ex situ technology of this type includes the Galoter process [44] in Estonia, which utilizes hot shale ash as the heat-carrying medium. Other main technologies of this type include the Alberta Taciuk Process (ATP), TOSCO II (developed by The Oil Shale Corporation, utilizing externally heated ceramic balls 15 mm in diameter as heat carrier) Lurgi-Ruhrgas (retorting is accomplished by mixing with spent shale/sand) [77], the Shell SPHER Process (Shell Pellet Heat Exchange Retorting) [49], and the LLNR HRS (Lawrence Livermore National Laboratory Hot Recycled Solid) process [20].
5. *By other means* such as radio frequency [15], microwave [11], solar [17] or nuclear explosion (e.g., Project Bronco) [8], etc. A number of recent shale oil in situ recovery schemes (e.g., Raytheon/Schumberger; PyroPhase) have been proposed based on volumetric heating by means such as radio frequency or microwave heating [140]. These have thus far been mostly of interest on the research and development level [15].
6. *By combinations of the above*. For example, a combination of solid and gaseous heat carriers is utilized in hot recycle solids fluidized bed retorting systems (e.g., KENTORT II, [23, 133]). Since gas density is about 1,000 times lower than that of oil shale, assuming that roughly same mass heat carrier must be contact the shale, then a large volume of gas is needed to heat the oil shale [18]. The KENTORT II [23, 133] uses recirculating hot spent shale from a gasification section.

The above-mentioned Shell SPHER process and LLNR HRS process could arguably also be included in this class of processes.

An example of an in situ combined heat-transfer mode process could be Electrofrac™ (Exxon Mobil Corporation) [140], which combines wall conduction and volumetric heating through electrically conductive material (used to fill fractures).

There are a large number of different retorting solutions. Generally, the performance of retorts (in situ or ex situ reactors) are typically evaluated or compared based on the percentage of Fischer assay yield achieved and an overall thermal efficiency (as energy efficiency). Typical commercial-scale retorts, with the exception of fluidized bed retorts, have lower oil yields than obtainable from Fischer

**Table 5.8** Some selected ex situ retorting technologies

| Process/retort | References |
|---|---|
| Fushun retort | [101] |
| Brazilian Petrosix | [77, 85] |
| The Nevada-Texas-Utah retort or NTU (also known as the Dundas-Howes retort) | [8, 85] |
| The Paraho direct/indirect process | [85, 99] |
| The Union Oil retort "A" (Union Oil Company) | [8, 76, 77, 159] |
| The Union Oil retort "B" indirectly heated retort (Union Oil Company) | [8, 76, 77, 85, 159] |
| The Union Oil Steam Gas Recirculation (RSG) retorting process (Union Oil Company) | [8] |
| The Superior Oil process | [8, 85] |
| TOSCO II (Tosco corporation) | [77, 85, 99] |
| The Lurgi-Ruhrgas process | [8, 76, 77, 85, 142] |
| Kiviter process | [8, 101, 159] |
| Galoter process | [8, 44, 101, 159] |
| Aostra Taciuk Process (ATP) | [8, 101, 142] |
| Chamber oven (similar to coke batteries) | [159] |
| HYTORT (Institute of Gas Technology, IGT) | [38, 106] |
| Chattanooga process (pressurized fluid bed reactor) | [140] |
| Gas combustion retort, US Bureau of Mines | [8, 76, 77, 142, 159] |
| Oil-Tech vertical surface retort technology (Millenium Synfuels, LLC) | [140] |
| KENTOR II | [133] |
| Shell SPHER processes (Shell Pellet Heat Exchange Retorting) | [49] |

assay. For example, for modified in situ internal combustion retorting, it is around 60% [99] of Fischer assay. For a Fushun generator type of retort, this value is 65% [101] and for the Kiviter type of retort 70% [101]. Higher values are obtained with Petrosix retorting 85–90% [101], Galoter retorting 85–90% [101], and for Taciuk retorting 82% [101].

The thermal efficiency is defined as heat of desired product divided by the total heat of the process [99]. For example, some comparisons may involve using the equation

$$\text{Thermal efficiency} = \frac{\text{Gross Galorific value of(oil + gas)}}{\text{Gross Galorific value of shale + heat of retorting}}$$

which can be found in Probstein and Hicks [99].

More detailed evaluation of processes comes down to desired product yields (oil and gas) and energy consumption per unit of desired product produced, including additional energy requirement such as for gas purification, spent shale disposal, or water treatment.

References offering insights regarding the construction details and operational principles of the retorting equipment referred to above, including product characteristics, are found in Tables 5.8 and 5.9. As there are hundreds of retorting

**Table 5.9** Some selected in situ retorting processes and technologies

| Process/retort | References |
|---|---|
| Occidental Petroleum "Oxy" | [8, 77, 85] |
| Sinclair oil and gas company process | [76, 77] |
| Equity Oil company process | [76, 77] |
| Shell Oil's thermally conductive in situ conversion process | [77, 140, 142] |
| Osborne's in situ process | [77] |
| Dow Chemical Co.'s process | [8, 77] |
| Equity Oil company process | [8] |
| Chevron's Technology for the Recovery and Upgrading of Oil from Shale (CRUSH) process | [140] |
| EGL oil shale process | [140] |
| Electrically enhanced oil production (EEOP) (Electro-Petroleum, Inc.) | [140] |
| PetroProbe process (Petro Probe, Inc.) | [140] |
| Electrofrac™ (Exxon Mobil Incorporation) | [140] |
| Geothermic Fuel Cell™ technology (GFL™) | [140] |
| In situ gas extraction technology (IGE), Mountain West Energy | [140] |
| Radio frequency/critical fluid oil extraction technology (Ryethon and CF technology) | [140] |
| EchoShale in-capsule process (Red Leaf Resources, Inc.) | [140] |
| PyroPhase | [140] |
| American Shale Oil Company (AMSO) process: Conduction, Convection, Reflux (CCR) process (former EGL Resources process) | [140] |

options that have been considered, this is far from being a complete list of possibilities. Nonetheless, it offers some insight into some significant processes. The next sections offer a few more details on selected ex situ and in situ processes.

## *Ex Situ Technologies: An Example and Some Further Considerations*

There are many ex situ retorting technologies that differ by reactor types (fluidized bed, moving packed bed, solid mixers), by operating conditions (temperature, residence time), technical details such as drying of feed, feed distributor, discharge of spent shale, oil and gas separation (oil and gas separation from offgas), heat recovery or by-product utilization. Therefore in addition to the pyrolysis reactor, the heart of retorting process, a particular retorting technology includes various other components (unit operations) in the plant.

Figure 5.7 shows a principal schematic diagram of the Galoter process [44]. The principle of the Galoter process is heating fine-grained oil shale (particles with size below 25 mm) in a horizontal rotary kiln-type retort, with contact of hot oil shale ash obtained from solid retorting residue combustion outside of the retort. The process operates in continuous mode with oil shale and ash in cocurrent flows. The

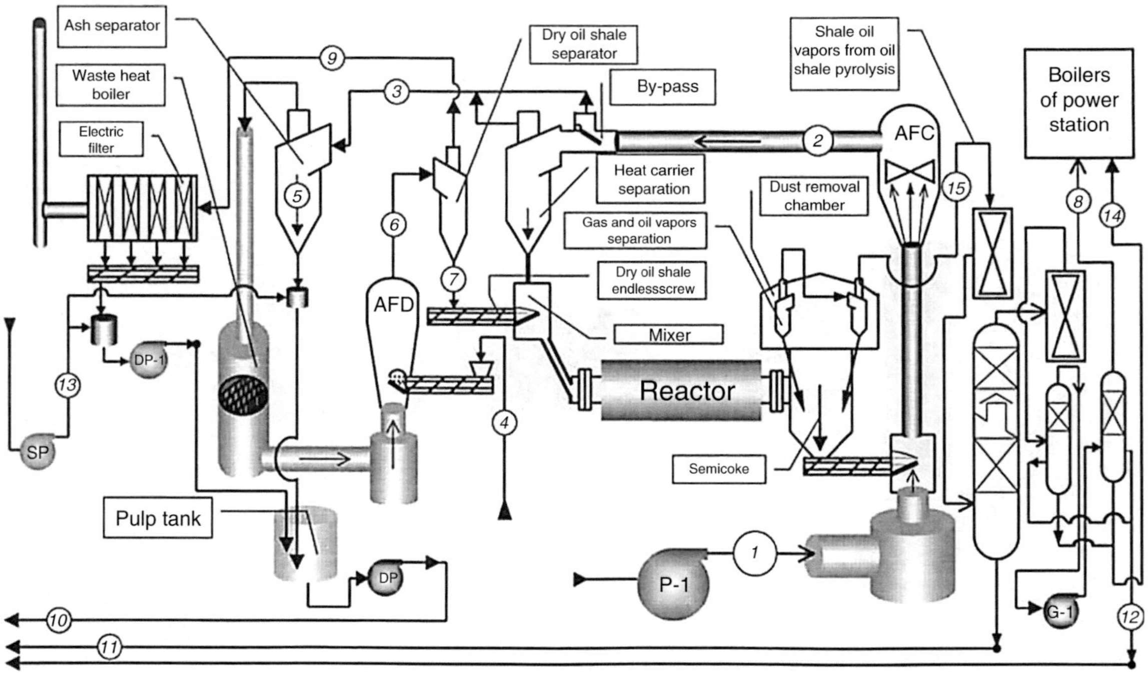


**Fig. 5.7** Galoter Retorting Process ([44], with permission). The key features of this process are the "Reactor" – the drum reactor in which oil shale pyrolysis takes place, after mixing of raw and hot spent shale; "AFC" – aerofountain combustor for spent shale (semicoke) burnout; "AFD" – aerofountain dryer for raw shale; P-1 – centrifugal air blower; DP – dredger pump; DP-1 – small dredger pump of electric filters; SP – pump for settled recycle water; G-1 – centrifugal gas blower; 1 – compressed air given to the retort; 2 – solid heat carrier mixture with stack gas after combustion of semicoke in the AFC; 3 – mixture of ash and stack gas after separation of solid heat carrier required for oil shale pyrolysis process; 4 – raw oil shale feed; 5 – ash separated by first, second, and third stage ash cyclones; 6 – dried oil shale mixture with stack gas after AFD; 7 – dried oil shale separated by first, second, and third stage cyclones prior to entering the system of dry oil shale screw conveyers; 8 – retort gas for firing in power station boilers; 9 – stack gas to ESP for final purification; 10 – ash slurry for storage/disposal; 11 – shale oils to storage; 12 – light oil or gasoline fraction (separated from water) to storage; 13 – recycle water to ash hydroremoval system; 14 – phenolic water to incineration in power station boilers; 15 – oil vapors from retort, after cyclone separators, to condensation system

properties of the shale oil produced (specific gravity, H/C ratio, viscosity, boiling range, and hydrocarbon types) depend upon retorting temperature and residence time, of course, but can also depend on other factors such as condenser construction and temperature of recovery and separation of shale oil and the reactive properties of the hot shale residue heat carrier.

A heat-transfer medium (the heat carrier) can generally be reactive or nonreactive. Examples of reactive heat carriers include the gaseous heat carriers such as hydrogen or solid heat carriers such as oil shale ash. Oil shale ash from carbonate oil shales has chemisorptive properties, due to its alkaline nature, and can reduce the concentration of $H_2S$ in the product gas as well as phenolic compounds in the oil [88]. Also the iron oxide in ash shows potential to reduce $H_2S$ in the gas, but has little effect on phenolic compounds [88]. It should, however, be emphasized that heat carrier ash used in solid heat carrier retorts does not affect significantly the sulfur concentration in oils (i.e., organic sulfur) [40, 88].

It needs to be emphasized again that an important parameter in ex situ retorting design is the heat up time of the retorted shale. This determines overall process time and therefore reactor volume and thus retorting process cost [18], but also impacts retort oil quality. The time to transfer heat to shale depends on the feed oil shale particle size. Often, operational details require specific particle size ranges [8]. It has been indicated that [18] the heat up time of the oil shale particles larger than 1 cm scales roughly as diameter squared and thus a heat up time of 2.5 min can be estimated for 10-mm particles and 4 h for 100-mm particles. Accordingly, the ex situ technologies have been divided in three groups, based upon particle size:

1. Processes with large feed particle sizes (so-called lump oil shale). These processes take place over the longest time scale; the shale residence time in retorts is measured in hours. For example, the industrial Kiviter process [8] uses a feed size of 10–125 mm and the Fushun process of 10–75 mm [2].
2. Processes that use coarsely ground particles. The shale residence time scale is of minutes. For example, the industrial Galoter process [8] uses a feed size of 0–25 mm [2].
3. Processes fed with pulverized oil shale with size below 5 mm. The residence time is shortest, with the time scale of seconds. This is used in fluidized bed retorting units [151]. The fluidized bed retorting systems are less developed than the two former types and no commercial units are yet in use.

## *In Situ Technologies: Further Considerations*

The common principle of in situ processes is heating the oil shale deposit underground. Different versions use different methods of heat transfer, as already described. With in situ technologies there is an opportunity to utilize poorer and deeper oil shale deposits than can be used for conventional aboveground processing [71]. Although a number of field-scale experiments have been conducted over the last decades in several countries [15, 77, 85, 97] no in situ technology-based commercial plant has been placed in operation to date. However, several of the latest innovative technological approaches involve in situ processes [140].

There are different possible ways of classifying in situ technologies (see [77]). The most common is classification into true in situ processes and modified in situ processes. Some define the true in situ processes as involving minimal disturbance of the oil shale ore bed, and the modified in situ as processes as any that use rubblizing either through blasting or partial mining. Therefore the distinction between true and modified in situ processes is a matter of degree of the bed disturbance.

True in situ processes thus do not generally require significant mining. This category of in situ processes uses natural rock permeability or permeability created by artificial fracturing (by explosives or hydrostatic pressure) in place. The true in situ processes follow a general sequence involving dewatering and establishing measures to minimize contact between groundwater and the reaction zone; fracturing or rubblizing to create permeability (if natural permeability is not sufficient); heating the oil shale by ignition, hot fluids, or other means; recovering oil and gas through wells [85].

The modified in situ processes involve partial mining under the oil shale deposit to create void space followed by rubblizing the rest into the created void. To create voids about 10–25% of overall space is mined out [77].

## Final Considerations and Future Directions

This review was intended to capture many aspects of the current state of knowledge and basic research directions in the field of oil shale research. It did so in the context provided by currently and historically explored technologies for oil shale conversion. Generally speaking, there is more new development interest in oil production processes, but the direct combustion processes will remain in use in particular instances (such as in Estonia). While the pyrolytic phenomena that shale undergoes are common to both oil production and combustion processes, there has not been as much emphasis on fundamental exploration of these phenomena from the combustion side, and in that case there are other technologically important issues that have captured more basic research interest (e.g., the behavior of ash in the boilers, the high-temperature corrosion of boilers, the problems of ash handling and disposal, the problems of sulfur capture). For the combustion community, the behavior of the organic portion of the ash has been somewhat secondary, and even the final phases of organic materials' burnout, which has captured the attention of many researchers in the coal combustion community, has been of less interest, because of the inherently higher reactivity of oil shale chars as compared to coal chars.

From the perspective of those who seek to produce oil, the major problems have been those of thermal management and the handling of large amounts of solids, as well as dealing with a range of environmental problems associated with the processing (whether aboveground or in situ). What can be economically done with oil shale in light of the above has dictated the nature of the processes more than consideration of the chemical aspects. Nonetheless, it has been observed that there can be factors whose importance may be sometimes overlooked. Issues of mass transport can materially impact both yields and quality of the product oils, and may be sensitive to factors such as pressure, particle size, heating rate, temperature, and residence time in a hot environment and gaseous environment. All of these factors have been studied, but the general impression is that more work is needed to better establish general trends. Likewise, there have been an enormous number of studies performed on elucidating the kinetics of pyrolysis, but rarely has there been a critical examination of results obtained on the same material examined by different means in different laboratories. As a result, there is a crude understanding of rate processes, but one that for predictive purposes cannot be claimed to be better than an order of magnitude. Despite large numbers of carefully performed studies, there has not yet been closure on the overall nature and rates of the many processes that determine behavior in practical processes.

## Bibliography

### *Primary Literature*

1. Aarna AY (1954) Isothermal destruction of Baltic oil shale. Trans Tallinn Polytech Inst A 57:32–34 (in Russian)
2. EASAC (European Academies Science Advisory Council) (2007) A study on the EU oil shale industry – viewed in the light of the Estonian experience. European Academies Science Advisory Council
3. Alali J, Abdelfattah AS, Suha MY, Al-Omari W (2006) Oil shale. In: Sahawneh J, Madanat M (eds), Natural Resources Authority of Jordan
4. Al-Ayed OS, Matouq M, Anbar Z, Khaleel AM, Abu-Nameh E (2010) Shale pyrolysis kinetics and variable activation energy principle. Appl Energy 87:1269–1272
5. Allred VD (1966) Kinetics of oil shale pyrolysis. Chem Eng Prog 62:55–60
6. Arbiter N (1983) Concentration of oil shale. Miner Process Technol Rev 1:207–248
7. Arro H, Prikk A, Pihu T (2003) Calculation of qualitative and quantitative composition of Estonian oil shale and its combustion products. Fuel 82:2179–2195, 2197–2204
8. Baughman GL (1978) Synthetic fuels data handbook, 2nd edn. Cameron Engineering, Denver
9. Behar F, Beaumont V, Penteado HL (2001) Rock-Eval 6 technology: performances and developments. Oil Gas Sci Technol – Rev IFP 56:111–134
10. Bjerle I, Ecklund H, Svensson O (1980) Gasification of Swedish Black Shale in the fluidized bed. Reactivity in steam and carbon dioxide atmosphere. Ind Eng Chem Process Des Dev 19:345–351
11. Bradhurst DH, Worner HK (1996) Evaluation of oil production from the microwave retorting of Australian shales. Fuel 75:285–288
12. Braun RL, Burnham AK (1986) Kinetics of Colorado oil shale pyrolysis in a fluidized bed reactor. Fuel 65:218–225

13. Braun RL, Rothman AJ (1975) Oil shale pyrolysis: kinetics and mechanism of oil production. Fuel 54:129–131
14. Brendow K (2003) Global oil shale issues and perspectives. Oil Shale 20:81–92
15. Bridges JE (2007) Wind power energy storage for in situ shale oil recovery with minimal $CO_2$ emission. IEEE Trans Energy Convers 22:103–109
16. Bryant RS, Bailey SA, Stepp AK, Evans DB, Parli JA, Kolhatkar AR (1998) Biotechnology for heavy oil recovery. No. 1998.110
17. Burnham AK (1989) On solar thermal processing and retorting of oil shale. Energy 14:667–674
18. Burnham AK (1995) Chemical kinetics and oil shale processing design. In: Snape C (ed) Composition, geochemistry and conversion of oil shales. NATO ASI Series. Kluwer, The Netherland
19. Burnham AK, Braun RL, Coburn TT, Sandvik EI, Curry DJ, Schmidt BJ, Noble RA (1996) An appropriate kinetic model for well-preserved algal kerogens. Energy Fuels 10:49–59
20. Burnham AK, McConaghy JR (2006) Comparison of the acceptability of various oil shale processes. In: Twenty-sixth oil shale symposium, Golden, 16–18 Oct 2006. UCRL-CONF-226717
21. Campbell JH, Koskinas GH, Stout ND (1978) Kinetics of oil generation from Colorado oil shale. Fuel 57:372–376
22. Cane RF (1976) The origin and formation of oil shale. In: Yen TF, Chilingarian GV (eds) Oil shale. Elsevier, Amsterdam
23. Carter SD, Graham UM, Rubel AM, Robl TL (1995) Fluidized bed retorting of oil shale. In: Snape C (ed) Geochemistry and conversion of oil shales. NATO ASI Series. Kluwer, London
24. Citiroglu M, Türkay S, Cepni ZI, Snape CE (1996) Fixed bed pyrolysis and hydropyrolysis of an immature Type I Turkish oil shale. Turk J Chem 20:175–185
25. Cummins JJ, Robinson WE (1972) Thermal degradation of Green River Kerogen at 150°C to 350°C. U.S. Bureau of Mines Report of Investigations 7620
26. Degirmenci L, Durusoy T (2002) Effect of heating rate on pyrolysis kinetics of Gönyük oil shale. Energy Sources 34:931–936
27. Degirmenci L, Durusoy T (2005) Effect of heating rate and particle size on the pyrolysis of Gönyük oil shale. Energy Sources 27:787–795
28. Derenne S, Largeau C, Casadevall E, Damste JSS, Tegelaar EW, Leeuw JW (1989) Characterization of Estonian Kukersite by spectroscopy and pyrolysis: evidence for abundant alkyl phenolic moieties in an Ordovician, marine, type II/I kerogen. Org Chem 16:873–888
29. Dieckmann V, Schenk HJ, Horsfield B, Welte DH (1998) Kinetics of petroleum generation and cracking by programmed-temperature closed-system pyrolysis of Toarcian Shales. Fuel 77:23–31
30. Dinneen GU (1976) Retorting technologies of oil shales. In: Yen TF, Chilingarian GV (eds) Oil shale. Elsevier, Amsterdam
31. Duncan DC (1976) Geologic settings of oil shale deposits and world prospects. In: Yen TF, Chilingarian GV (eds) Oil shale. Elsevier, Amsterdam
32. Durand B (1980) Kerogen. Insoluble organic matter from sedimentary rocks. Technip, Paris
33. Durand B, Monin JC (1980) Elemental analysis of kerogens (C, H, O, N, S, Fe). In: Durand B (ed) Kerogen. Insoluble organic matter from sedimentary rocks. Technip, Paris
34. Durand B, Nicaise C (1980) Procedures of kerogen isolation. In: Durand B (ed) Kerogen. Insoluble organic matter from sedimentary rocks. Technip, Paris
35. Durand-Souron C (1980) Thermogravimetric analysis and associated techniques applied to kerogens. In: Durand B (ed) Kerogen. Insoluble organic matter from sedimentary rocks. Technip, Paris
36. Dyni JR (2006) Geology and resources of some world oil shale deposits. Scientific Investigations Report 2005–5294. United States Department of the Interior, United States Geological Survey

37. Ecklund H, Svensson O (1983) Reactor model for the gasification of Black Shale in the fluidized bed. Comparison with the pilot plant data. Ind Eng Chem Process Des Dev 22:396–401
38. Ekinci E, Yürüm Y (1995) Steam and coprocessing of oil shales. In: Snape C (ed) Composition, geochemistry and conversion of oil shales. NATO ASI Series. Kluwer, The Netherland
39. Ekstrom A, Callaghan G (1987) The pyrolysis kinetics of some Australian oil shales. Fuel 66:331–337
40. Elenurm A, Oja V, Elenurm A, Tali E, Tearo E, Yanchilin A (2008) Thermal processing of dictyonema argillite and kukersite oil shale: transformation and distribution of sulfur compounds in pilot-scale Galoter process. Oil Shale 25:328–334
41. Espitalie J, Madec M, Tissot B (1980) Role of mineral matrix in kerogen pyrolysis: influence on petroleum generation and migration. AAPG Bull 64:59–66
42. Forsman JP (1963) Geochemistry of kerogen. In: Breger IA (ed) Organic geochemistry. Pergamon, Oxford, pp 148–182
43. Gavin JM (1924) Oil shale. U.S. Government Printing Office, Washington, DC
44. Golubev N (2003) Solid oil shale heat carrier technology for oil shale retorting. Oil Shale 20:324–332
45. Gonzalez-Vila FJ (1995) Alkane biomarkers. Geochemical significance and application in oil shale geochemistry. In: Snape C (ed) Geochemistry and conversion of oil shales. NATO ASI Series. Kluwer, London
46. Gonzalez-Vila FJ, Ambles A, del Rio JC, Grasset L (2001) Characterization and differentiation of kerogens by pyrolytic and chemical degradation techniques. J Anal Appl Pyrol 58–59:315–328
47. Goodfellow L, Haberman CE, Atwood MT (1968) Modified Fischer assay equipment, procedures and product balance determinations. In: Joint symposium on oil shale, tar sand, and related materials. American Chemical Society, San Francisco National Meeting, 2–5 April 1968.
48. Gorlov EG (2007) Thermal dissolution of solid fossil fuels. Solid Fuel Chem 41:290–298
49. Gwyn JE, Roberts SC, Hardesty DE, Johnson GL, Hinds GP (1980) Shell pellet heat exchange retorting: the SPHER energy-efficient process for retorting oil shale. American Chemical Society, San Francisco, pp 59–69
50. Hamarneh YM (1984) Oil shale deposits in Jordan. Symposium on characterization and chemistry of oil shales. Am Chem Soc 29(3):41–50
51. Hamarneh Y (2006) Oil shale resources development in Jordan. Natural Resources Authority of Jordan, Amman, 1998. Revised and Updated by Dr. Jamal Alali, Amman
52. Hotta A, Parkkonen R, Hiltunen M, Arro H, Loosaar J, Parve T, Pihu T, Prikk A, Tiikma T (2005) Experience of Estonian oil shale combustion based on CFB technology at Narva Power Plants. Oil Shale 22:381–397
53. Huss EB, Burnham AK (1982) Gas evolution during pyrolysis of various Colorado oil shales. Fuel 66:1188–1196
54. Hutton AC (1987) Petrographic classification of oil shales. Int J Coal Geol 8:203–231
55. Hutton AC (1995) Organic petrography of oil shales. In: Snape C (ed) Geochemistry and conversion of oil shales. NATO ASI Series. Kluwer, The Netherland
56. Hutton A, Bharati S, Robl T (1994) Chemical and petrographic classification of kerogen/macerals. Energy Fuels 8:1478–1488
57. Jaber JO (2000) Gasification potential of Ellajjun oil shale. Energy Convers Manage 41:1615–1624
58. Johannes I, Zaidentsal A (2008) Kinetics of low-temperature retorting of kukersite oil shale. Oil Shale 25:412–425
59. Johannes I, Kruusment K, Veski R (2007) Evaluation of oil potential and pyrolysis kinetics of renewable fuel and shale samples by Rock-Eval analyzer. J Anal Appl Pyrol 79:183–190
60. Johannes I, Kruusment K, Veski R, Bojesen-Koefoed JA (2006) Characterisation of pyrolysis kinetics by Rock-Eval basic data. Oil Shale 23:249–257

61. Johannes I, Tiikma L (2004) Kinetics of oil shale pyrolysis in an autoclave under non-linear increase of temperature. Oil Shale 21:273–288
62. Johannes I, Tiikma L, Zaidenstal A (2010) Comparison of the thermobituminization kinetics of Baltic oil shales in open retorts and autoclaves. Oil Shale 27:17–25
63. Johannes I, Tiikma L, Zaidenstal A, Luik L (2009) Kinetics of kukersite low-temperature pyrolysis in autoclaves. J Anal Appl Pyrol 85:508–513
64. Johannes I, Zaidenstal A (2008) Kinetics of low-temperature retorting of kukersite oil shale. Oil Shale 25:412–425
65. Kask KA (1955) About bituminization of kerogen of oil shale kukersite. Report I. Trans Tallinn Polytech Inst A 63:51–64 (in Russian)
66. Kelemen SR, Afeworki M, Gorbaty ML, Sansone M, Kwiatek PJ, Walters CC, Freund H, Siskin M, Bence AE, Curry DJ, Solum M, Pugmire RJ, Vandenbroucke M, Leblond M, Behar F (2007) Direct characterization of kerogen by x-ray and solid-state $^{13}C$ nuclear magnetic resonance methods. Energy Fuels 21:1548–1561
67. Koel M, Ljovin S, Hollis K, Rubin J (2001) Using neoteric solvents in oil shale studies. Pure Appl Chem 73:153–159
68. Kogerman PN, Kopwillem JJ (1932) Hydrogenation of Estonian oil shale and shale oil. J Inst Petrol Technol 18:833–845
69. Kök MV (2002) Oil shale: pyrolysis, combustion, and environment: a review. Energy Sources 24:135–143
70. Kök MV (2008) Recent developments in the application of thermal analysis techniques in fossil fuels. J Therm Anal Calorim 91:763–773
71. Kök MV, Guner G, Bağci AS (2008) Application of EOR techniques for oil shale fields (in-situ combustion approach). Oil Shale 25:217–225
72. Krym YS (1932) About laboratory methods for determination the tendency of selfignition for coals. Khim Tverd Topl 2–3:7–22 (in Russian)
73. Lafargue E, Marquis F, Pillot D (1998) Rock-Eval 6 applications in hydrocarbon exploration, production and soils contamination studies. Oil Gas Sci Technol 53:421–437
74. Larsen JW, Li S (1994) Solvent swelling studies of Green River Kerogen. Energy Fuels 8:932–936
75. Larsen WJ, Parikh H, Michels R (2002) Changes in the cross-linking density of Paris Basin Toarcian kerogen during maturation. Org Geochem 33:1143–1152
76. Lee S (1991) Oil shale technology. CRC, Boca Raton
77. Lee S, Speight JG, Loyalka SK (2007) Handbook of alternative fuel technology. CRC, Boca Raton
78. Li S, Yue C (2003) Study of pyrolysis kinetics of oil shale. Fuel 82:337–342
79. Louw SJ, Addison J (1985) Studies of the Scottish oil shale industry, vol 1: History of the industry, working conditions, and mineralogy of Scottish and Green River formation shales. Final Report on US Department of Energy. Institute of Occupational Medicine
80. Luts K (1944) Estonian oil shale kukersite, its chemistry, technology and analysis. Revalen Buchverlag, Reval (in German)
81. Maalmann I (2003) Historical survey of nuclear non-proliferation in Estonia 1946–1995. Estonian Radiation Protection Centre
82. Miknis FP (1995) Solid state $^{13}C$ NMR in oil shale research: an introduction with selected applications. In: Snape C (ed) Composition, geochemistry and conversion of oil shales. NATO ASI Series. Kluwer, The Netherland
83. Miknis FP, Turner TF, Berdan GL, Conn PJ (1987) Formation of soluble products from thermal decomposition of Colorado and Kentucky oil shales. Energy Fuels 1:477–483
84. Nutall HE, Guo TM, Schrader S, Thakur DS (1983) Pyrolysis kinetics of several key world oil shales. ACS symposium series, vol 230, pp 269–300
85. Office of Technology Assessment (1980) An assessment of oil shale technology. Congress of the United States, Office of Technology Assessment, Washington, DC

86. Oja V (2005) Characterization of tars from Estonian Kukersite oil shale based on their volatility. J Anal Appl Pyrol 74:55–60
87. Oja V (2006) A brief overview of motor fuels from shale oil of kukersite. Oil Shale 23:160–163
88. Oja V, Elenurm A, Rohtla I, Tali E, Tearo E, Yanchilin A (2007) Comparison of oil shales from different deposits: oil shale pyrolysis and copyrolysis with ash. Oil Shale 24:101–108
89. Ots A (2006) Oil shale fuel combustion. Tallinn Book Printers, Tallinn
90. Ozerov IM, Polozov VI (1968) Principles of oil shale commercial classification. In: United Nations symposium on the development and utilization of oil shale resources, Tallinn
91. Parks TJ, Lynch LJ, Webster DS (1987) Pyrolysis model of rundle oil shale from in-situ $^{1}H$ NMR data. Fuel 66:338–344
92. Patterson HJ (1994) A review of the effects of minerals in processing of Australian oil shales. Fuel 73:321–327
93. Patterson JH, Dale LS, Chapman JF (1988) Partitioning of trace elements during the retorting of Australian oil shales. Fuel 67:1353–1356
94. Perry RH, Green DW (1997) Perry's chemical engineering handbook, 7th edn. McGraw-Hill, USA
95. Peters KE (1986) Guidelines for evaluating petroleum source rock using programmed pyrolysis. AAPG Bull 70:318–329
96. Petersen HI, Bojesen-Koefoed JA, Mathiesen A (2010) Variations in composition, petroleum potential and kinetics of Ordovician-Miocene Type I and Type I–II source rocks (oil shales): implications for hydrocarbon generation characteristics. J Petrol Geol 33:19–42
97. Prien CH (1951) Pyrolysis of coal and oil shale. Ind Eng Chem 43:2006–2015
98. Prien CH (1976) Survey of oil shale research in the last three decades. In: Yen TF, Chilingarian GV (eds) Oil shale. Elsevier, Amsterdam
99. Probstein RF, Hicks RE (2006) Synthetic fuels. Dover, New York
100. Puura E (1999) Technogenic minerals in the waste rock heaps of Estonian oil shale mines and their use to predict the environmental impact of the waste. Oil Shale 16:99–107
101. Qian J, Wang J (2006) World oil shale retorting technologies. In: International conference on oil shale: recent trends in oil shale, 7–9 Nov 2006, Jordan
102. Qian J, Wang J, Li S (2003) Oil shale development in China. Oil Shale 20:356–359
103. Rajeshwar K, Nottenburg R, Dubow J (1979) Thermophysical properties of oil shales. J Mater Sci 14:2025–2052 (Review)
104. Raudsepp H (1953) About the method for determining of organic mass in Baltic oil shales. Proc Tallinn Polytech Inst A 46:3–22 (in Russian)
105. Reynolds JG, Crawford RW, Burnham AK (1991) Analysis of oil shale and petroleum source rock pyrolysis by triple quadrupole mass spectrometry: comparisons of gas evolution at the heating rate of 10°C/min. Energy Fuels 5:507–523
106. Roberts MJ, Snape CE, Mitchell SC (1995) Hydropyrolysis: fundamentals, two-stage processing and PDU operations. In: Snape C (ed) Geochemistry and conversion of oil shales. NATO ASI Series. Kluwer, The Netherland
107. Robinson WE (1969) Isolation procedures for kerogens and associated soluble organic materials. In: Eglinton G, Murphy MTJ (eds) Organic geochemistry. Springer, Berlin, pp 181–195
108. Robl TL, Taulbee DN (1995) Demineralization and kerogen macerals separation and chemistry. In: Snape C (ed) Geochemistry and conversion of oil shales. NATO ASI Series. Kluwer, London
109. Rocha JD, Brown SD, Love GD, Snape CE (1997) Hydropyrolysis: a versatile technique for solid fuel liquefaction, sulphur specification and biomarker release. J Anal Appl Pyrol 40–41:91–103
110. Rullkötter J, Michaelis W (1990) The structure of kerogen and related materials. A review of recent progress and future trends. Org Geochem 16:829–852

111. Savest N, Hruljova J, Oja V (2009) Characterization of thermally pretreated kukersite oil shale using the solvent-swelling technique. Energ Fuel 23:5972–5977
112. Savest N, Oja V, Kaevand T, Lille Ü (2007) Interaction of Estonian kukersite with organic solvents: a volumetric swelling and molecular simulation study. Fuel 86:17–21
113. Schachter Y (1979) Gasification of oil shale. Isr J Technol 17:51–57
114. Schlatter LE (1968) Definition, formation and classification of oil shales. In: United Nations symposium on the development and utilization of oil shale resources, Tallinn
115. Schmitt KD, Sheppard EW (1984) Determination of carbon center types in solid fuel by CP/MAS NMR. Fuel 63:1241–1244
116. Sert M, Ballice L, Yüksel M, Saglam M (2009) Effect of mineral matter on product yield and composition at isothermal pyrolysis of Turkish oil shales. Oil Shale 26:463–474
117. Shpirt MYa, Punanova SA, Strizhakova YuA (2007) Trace elements in black and oil shales. Solid Fuel Chem 41:119–127
118. Shui H, Cai Z, Xu C (2010) Recent advances in direct coal liquefaction. Energies 3:155–170
119. Silbernagel BG, Gebhard LA, Siskin M, Brons G (1987) ESR study of kerogen conversion in shale pyrolysis. Energy Fuels 1:501–506
120. Siskin M, Scouten CG, Rose KD, Aczel T, Colgrove SG, Pabst RE Jr (1995) Detailed structural characterization of the organic material in Rundle Ramsay Crossing and Green River oil shales. In: Snape C (ed) Composition, geochemistry and conversion of oil shales. NATO ASI Series. Kluwer, The Netherland
121. Skala D, Korica S, Vitorovic D, Neumann H-J (1997) Determination of kerogen type by using DSC and TG analysis. J Therm Anal 49:745–753
122. Smith MW, Shadle LJ, Hill DL (2007) Oil shale development from the perspective of NETL's unconventional oil resource repository. DOE/NETL-IR-2007-022
123. Smith LK, Smoot LD, Fletcher TH, Pugmire RJ (1994) The structure and reaction processes of coal. The Plenum chemical engineering series. Plenum, New York
124. Sohn HY, Yang HS (1985) Effect of reduced pressure on oil shale retorting. 1. Kinetics of oil generation. Ind Eng Chem Process Des Dev 24:265–270
125. Solomon PR, Carangelo RM, Horn E (1986) The effects of pyrolysis conditions on Israeli oil shale properties. Fuel 65:650–662
126. Solomon PR, Serio MA, Despande GV, Kroo E (1990) Cross-linking reactions during coal conversion. Energy Fuels 4:42–54
127. Solomon PR, Serio MA, Suuberg EM (1992) Coal pyrolysis: experiments, kinetic rates and mechanism. Prog Energy Combust Sci 18:133–220
128. Speight JG (2007) The chemistry and technology of petroleum, 4th edn. CRC, Boca Raton
129. Speight JG (2008) Synthetic fuels handbook. Properties, processes and performance. McGraw-Hill, USA
130. Strizhakova YuA, Usova TV (2008) Current trends in the pyrolysis of oil shale: a review. Solid Fuel Chem 24:197–201
131. Suuberg EM, Sherman J, Lilly WD (1987) Product evolution during rapid pyrolysis of Green River Formation oil shale. Fuel 66:1176–1184
132. Suuberg EM, Unger PE, Lilly WD (1985) Experimental study on mass transfer from pyrolyzing coal particles. Fuel 64:956–962
133. Taulbee DN, Carter SD (1992) Investigation of product coking by hot recycle solids in the KENTORT II fluidized bed retort. American Chemical Society, San Francisco, pp 800–809
134. Thankur DS, Nutall HE (1987) Kinetics of pyrolysis of Moroccan oil shale by thermogravimetry. Ind Eng Chem Res 26:1351–1356
135. Thomas RD, Lorentz PB (1970) Use of centrifugal separation to investigate how kerogen is bound to the minerals in oil shale. U. S. Bur. Mines, Rep. Invest., 7378
136. Tiikma L, Johannes I, Luik H, Zaidentsal A, Vink N (2009) Thermal dissolution of Estonian oil shale. J Anal Appl Pyrol 85:502–507
137. Tiikma L, Zaidentsal A, Tensorer M (2007) Formation of thermobitumen from oil shale by low-temperature pyrolysis in an autoclave. Oil Shale 24:535–546

138. Torrente MC, Galán MA (2001) Kinetics of the thermal decomposition of oil shale from Puertollano (Spain). Fuel 80:327–334
139. Tucker JD, Masri B, Lee S (2000) A comparison of retorting and supercritical extraction techniques on El-Lajjun oil shale. Energy Sources 22:453–463
140. U.S. Department of Energy (2007) Secure fuels from domestic resources. The continuing evolution of American's oil shale and tar sands industries
141. Urov K, Sumberg A (1999) Characteristics of oil shales and shale like rocks of known deposits and outcrops. Oil Shale 16:1–64
142. US DOE (March 2004) Strategic significance of America's oil shale resources, vol II: Oil shale resources, technology, and economics. U.S. Department of Energy
143. Vandenbroucke M (1980) Structure of kerogens as seen by investigations on soluble extracts. In: Durand B (ed) Kerogen. Insoluble organic matter from sedimentary rocks. Technip, Paris
144. Vandenbroucke M (2003) Kerogen: from types to models of chemical structure. Oil Gas Sci Technol – Rev IFP 58:243–269
145. Vandenbroucke M, Largeau C (2007) Kerogen origin, evolution and structure. Org Geochem 38:719–833
146. Vitorovic D (1980) Structure elucidation of kerogen by chemical methods. In: Durand B (ed) Kerogen. Insoluble organic matter from sedimentary rocks. Technip, Paris
147. Wall GC, Smith SJC (1987) Kinetics of production of individual products from the isothermal pyrolysis of seven Australian oil shales. Fuel 66:345–350
148. Wang Q, Liu H, Sun B, Li S (2009) Study on pyrolysis characteristics of Huadian oil shale with isoconversional method. Oil Shale 26:148–162
149. Wang CC, Noble RD (1983) Composition and kinetics of oil generation from non-isothermal oil shale retorting. Fuel 62:529–533
150. Wang Q, Sun B, Hu A, Bai J, Li S (2007) Pyrolysis characteristics of Huadian oil shales. Oil Shale 24:147–157
151. Wang WD, Zhou CY (2009) Retorting of pulverized oil shale in fluidized-bed pilot plant. Oil Shale 26:108–113
152. Watson NC (1984) A modified Gray-King assay method for small oil shale samples. Fuel 63:1455–1458
153. Williams PT, Ahmad N (2000) Investigation of oil shale pyrolysis processing conditions using thermogravimetric analysis. Appl Energy 66:113–133
154. Wiser WH, Anderson LL (1975) Transformation of solids to liquid fuels. Annu Rev Phys Chem 26:339–357
155. World Energy Council (2007) Survey of energy resources 2007. World Energy Council 2007, United Kingdom
156. Xue HQ, Li SY, Wang HY, Zheng DW, Fang CH (2010) Pyrolysis kinetics of oil shale from Northern Songliao Basin in China. Oil Shale 27:5–16
157. Yang HS, Shon HY (1985) Effect of reduced pressure on oil shale retorting. Ind Eng Chem Proc DD 24:271–273
158. Yen TF (1976) Structural aspects of organic components in oil shales. In: Yen TF, Chilingarian GV (eds) Oil shale. Elsevier, Amsterdam
159. Yen TF, Chilingarian GV (1976) Introduction to oil shales. In: Yen TF, Chilingarian GV (eds) Oil shale. Elsevier, Amsterdam
160. Burnham AK, Braun RL (1999) Global kinetic analysis of complex materials. Energy Fuels 13:1–22
161. Davis JD, Galloway AE (1928) Low-temperature carbonization of lignites and sub-bituminous coals. Ind Eng Chem 20:612–617
162. Kask KA (1956) About bituminizing of kerogen of oil shale-kukersite. Report II. Trans Tallinn Polythec Inst A 73:23–40 (in Russian)
163. Kilk K, Savest N, Yanchilin A, Kellogg DS, Oja V (2010) Solvent swelling of Dictyonema oil shale: low temperature heat-treatment caused changes in swelling extent. J Anal Appl Pyrol 89:261–264

164. Lille Ü, Heinmaa I, Pehk T (2003) Molecular model of Estonian kukersite kerogen evaluated by $^{13}C$ MAS NMR spectra. Fuel 82:799–804
165. Miknis FP, Turner TF (1995) The bitumen intermediate in isothermal and nonisothermal decomposition of oil shales. In: Snape C (ed) Composition, geochemistry and conversion of oil shales. NATO ASI Series. Kluwer, The Netherland
166. Rahman M, Kinghorn RF, Gibson PJ (1994) The organic matter in oil shales from the lowmead basin, Qeensland, Australia. Journ Petrol Geol 17:317–326

## *Books and Reviews*

Ogunsola OI, Hartstein AM, Ogunsola O (2010) Oil shale: a solution to the liquid fuel dilemma. Oxford University Press, USA
Qian J (2010) Oil shale – petroleum alternative. China Petrochemical Press, Beijing
Yen TF, Chilingarian GV (1976) Oil shale. Elsevier Scientific, Amsterdam

# Chapter 6
# Developments in Internal Combustion Engines

**Timothy J. Jacobs**

## Greek Symbols

| | |
|---|---|
| $\eta_c$ | Combustion efficiency |
| $\eta_f$ | Fuel conversion efficiency |
| $\eta_{f,b}$ | Brake fuel conversion efficiency |
| $\eta_{f,i,}$ | Indicated fuel conversion efficiency |
| $\eta_m$ | Mechanical efficiency |
| $\eta_{th}$ | Thermal efficiency |
| $\eta_{th,Carnot}$ | Thermal efficiency of the ideal Carnot cycle |
| $\eta_{th,Otto}$ | Thermal efficiency of the ideal heat engine Otto cycle |
| $\eta_v$ | Volumetric efficiency |
| $\gamma$ | Ratio of specific heats |
| $\gamma_b$ | Ratio of specific heats of the burned mixture |
| $\varphi$ | Fuel–air equivalence ratio. For $\varphi < 1$ mixture is lean. For $\varphi = 1$, mixture is stoichiometric. For $\varphi > 1$, mixture is rich. Note that $\varphi$ is the inverse of the often-used air–fuel equivalence ratio, or $\lambda$. |
| $\rho_{a,i}$ | Inlet air density |
| $\tau$ | Engine torque |

## Symbols

| | |
|---|---|
| BMEP | Brake mean effective pressure |
| $C_{p,b}$ | Constant pressure specific heat of the burned mixture |

This chapter was originally published as part of the Encyclopedia of Sustainability Science and Technology edited by Robert A. Meyers. DOI:10.1007/978-1-4419-0851-3

T.J. Jacobs (✉)
Department of Mechanical Engineering, Texas A&M University,
College Station, TX, USA
e-mail: tjjacobs@tamu.edu

R. Malhotra (ed.), *Fossil Energy: Selected Entries from the Encyclopedia of Sustainability Science and Technology*, DOI 10.1007/978-1-4614-5722-0_6,

| | |
|---|---|
| $c_{v,b}$ | Constant volume specific heat of the burned mixture |
| $f$ | Residual fraction |
| $f_{final}$ | Final calculated residual fraction |
| $f_{final-1}$ | Previous iteration residual fraction to final calculated residual fraction |
| $(F/A)$ | Fuel–air ratio |
| FMEP | Friction mean effective pressure |
| $h_1$ | Specific enthalpy at state 1 |
| $h_2$ | Specific enthalpy at state 2 |
| $h_3$ | Specific enthalpy at state 3 |
| $h_{3a}$ | Specific enthalpy at state 3a |
| $h_5$ | Specific enthalpy at state 5 |
| $h_6$ | Specific enthalpy at state 6 |
| $h_e$ | Specific enthalpy of exhaust mixture |
| $h_i$ | Specific enthalpy of inlet mixture |
| $IMEP_g$ | Gross indicated mean effective pressure |
| $IMEP_n$ | Net indicated mean effective pressure |
| $m$ | Mass |
| $m_1$ | Mass at state 1 |
| $m_2$ | Mass at state 2 |
| $m_3$ | Mass at state 3 |
| $m_4$ | Mass at state 4 |
| $m_6$ | Mass at state 6 |
| $m_a$ | Mass of air |
| $m_f$ | Mass of fuel |
| $m_r$ | Residual mass |
| $m_{total}$ | Total mass |
| $\dot{m}_a$ | Mass flow rate of air |
| $\dot{m}_f$ | Mass flow rate of fuel |
| MEP | Mean effective pressure |
| $M_b$ | Molecular weight of the burned mixture |
| $n_R$ | Number of revolutions per engine cycle |
| $N$ | Engine speed |
| $\overline{R}$ | Universal gas constant |
| $R$ | Gas constant |
| $R_5$ | Gas constant of mixture at state 5 |
| $R_6$ | Gas constant of mixture at state 6 |
| $R_e$ | Gas constant of exhaust mixture |
| $P$ | Cylinder pressure or power |
| $P_1$ | Pressure at state 1 |
| $P_2$ | Pressure at state 2 |
| $P_3$ | Pressure at state 3 |
| $P_{3a}$ | Pressure at state 3a |
| $P_4$ | Pressure at state 4 |
| $P_5$ | Pressure at state 5 |

| | |
|---|---|
| $P_6$ | Pressure at state 6 |
| $P_7$ | Pressure at state 7 |
| $P_b$ | Brake power |
| $P_e$ | Exhaust pressure |
| $P_i$ | Inlet (initial) pressure |
| $P_{in}$ | Net indicated power |
| $P_{limit}$ | Limit pressure |
| PMEP | Pumping mean effective pressure |
| ${}_1Q_2$ | Heat transfer of process 1–2 |
| ${}_6Q_1$ | Heat transfer for process 6–1 |
| $Q_{HV}$ | Heating value of fuel |
| $Q_{HV,f}$ | Heating value of fuel |
| $Q_{HV,i}$ | Heating value of specie i |
| $r_c$ | Compression ratio |
| $s_1$ | Entropy at state 1 |
| $s_2$ | Entropy at state 2 |
| $s_3$ | Entropy at state 3 |
| $s_4$ | Entropy at state 4 |
| $s_5$ | Entropy at state 5 |
| $T_1$ | Temperature at state 1 |
| $T_4$ | Temperature at state 4 |
| $T_5$ | Temperature at state 5 |
| $T_6$ | Temperature at state 6 |
| $T_{cv,adiabatic}$ | Constant volume adiabatic flame temperature |
| $T_e$ | Exhaust temperature |
| $T_H$ | Temperature of a source reservoir |
| $T_i$ | Inlet temperature |
| $T_L$ | Temperature of a sink reservoir |
| $T_r$ | Residual fraction temperature |
| $u_1$ | Specific internal energy at state 1 |
| $u_2$ | Specific internal energy at state 2 |
| $u_3$ | Specific internal energy at state 3 |
| $u_{3a}$ | Specific internal energy at state 3a |
| $u_4$ | Specific internal energy at state 4 |
| $U_1$ | Internal energy at state 1 |
| $U_2$ | Internal energy at state 2 |
| $U_3$ | Internal energy at state 3 |
| $U_4$ | Internal energy at state 4 |
| $v_1$ | Specific volume at state 1 |
| $v_2$ | Specific volume at state 2 |
| $v_3$ | Specific volume at state 3 |
| $v_{3a}$ | Specific volume at state 3a |
| $v_4$ | Specific volume at state 4 |
| $V$ | Cylinder volume |

| | |
|---|---|
| $V_1$ | Volume at state 1 |
| $V_2$ | Volume at state 2 |
| $V_3$ | Volume at state 3 |
| $V_5$ | Volume at state 5 |
| $V_6$ | Volume at state 6 |
| $V_d$ | Displaced volume |
| $V_{max}$ | Maximum cylinder volume |
| $V_{min}$ | Minimum cylinder volume |
| $W$ | Thermodynamic work |
| ${}_1W_2$ | Work for process 1–2 |
| ${}_2W_3$ | Work for process 2–3 |
| ${}_3W_4$ | Work for process 3–4 |
| ${}_4W_5$ | Work for process 4–5 |
| ${}_5W_6$ | Work for process 5–6 |
| ${}_6W_1$ | Work for process 6–1 |
| $W_b$ | Brake work |
| $W_f$ | Friction work |
| $W_{gross}$ | Gross work |
| $W_{ig}$ | Gross indicated work |
| $W_{in}$ | Net indicated work |
| $W_{ip}$ | Pump work |
| $W_{net}$ | Net work |
| $x_i$ | Mole fraction of specie i |
| $y_i$ | Mass fraction of specie i |

## Glossary

| | |
|---|---|
| Combustion | Rapid oxidation of a fuel–air mixture (reactants) converting reactants to products and in the process releasing thermal energy. |
| Lean | Air–fuel mixture is such that there is more air than is chemically necessary to oxidize the available fuel. |
| Products | Species formed as the result of a chemical reaction (in the context of this article species formed as the result of a combustion reaction). |
| Reactants | Species that are to be involved in a chemical reaction (in the context of this article species that are to be involved in a combustion reaction). |
| Rich | Air–fuel mixture is such that there is less air than is chemically necessary to oxidize the available fuel. |
| Stoichiometric | Air–fuel mixture is chemically balanced such that there is the correct amount of air to fully oxidize the available fuel. |

## Definition of the Subject

An internal combustion (IC) engine is a thermodynamic work conversion device that converts chemical energy (typically delivered to the engine in the form of a liquid or gaseous fuel) to work energy (typically in the form of shaft work issuing from a rotating crankshaft). The internal combustion engine markedly distinguishes itself from other types of power-producing equipment – most notably, the heat engine (e.g., steam engine or steam-cycle plant). In the former, chemical energy is released, via combustion (i.e., rapid oxidation) mechanism, internal the same device that converts the released energy to work energy. In the latter, thermal energy is passed into the device via heat transfer; the device thereby converts the thermal energy to work energy. Because of this subtle difference, the thermodynamic limits of maximum efficiency of the internal combustion engine are constrained differently than the thermodynamic limit of maximum efficiency of the heat engine.

There are several types of internal combustion engines; the two most common being the piston/cylinder reciprocating engine and the gas turbine engine. This article constrains its discussion to just piston/cylinder reciprocating engines. As this article is meant to be brief, readers seeking additional and thorough information are referred to the associated cited articles and the authorities listed in "Books and Reviews".

## Introduction

Internal combustion engines are pervasive to our daily activities. They are the primary powerhouse of the transportation industry. They serve as neighborhood and small municipality backup power generators or primary power stations. They provide convenience by powering lawnmowers, leaf blowers, and weed trimmers. They deliver excitement and entertainment in pleasure boats, race cars, and motor bikes. Their scale of usability nearly spans the scales of classical physics, from as small as "micro engines" that fit in the palm of your hand to as large as marine engines that scale three stories and use human-sized doorways for entry into the cylinder block.

The story of the internal combustion engine dates back around 150 years ago, when J.J.E. Lenoir built an engine that combusted coal–air mixtures in a cylinder outfitted with a piston with a two-stroke type fashion (without compression). Soon thereafter, Nicolaus A. Otto and his colleague Eugen Langen expanded on the Lenoir concept, creating an engine that had 11% efficiency compared to Lenoir's 5%. Determined to improve efficiencies of internal combustion engines (which at 11%, were not much better than the steam engine), Otto built the first engine operating on the four-stroke principle which today serves as the primary operating cycle of engines [1]. It is here noted that, although Otto built the first working four-stroke engine, Alphonse Beau de Rochas described in theory the principles of the four-stroke cycle. Further, Beau de Rochas outlined several conditions to achieve maximum efficiency of the internal combustion engine [1].

From this point, and with the aid of several important developers including Rudolf Diesel, Sir Harry Ricardo, Robert Bosch, and Charles Kettering, the internal combustion engine has become one of the most highly efficient, power dense, cost-effective, easily maintained, and versatile power machinery available to consumers. This article will provide some of the important basic information about internal combustion engines, and indicate some of the more recent developments that continue to make internal combustion engines competitive as the preferred power-producing technology.

## The Basics of Internal Combustion Engines

### *Basic Operating Cycle*

As this article concentrates its discussions of IC engines on those of the piston/cylinder reciprocating type, Fig. 6.1 [2] shows the basic geometrical considerations of the piston/cylinder/crankshaft arrangement, which kinematically is described as a crank-slider mechanism. The chemical energy to work energy conversion occurs inside the cylinder, usually bound on the sides by the cylinder walls, on the top by the cylinder head (which typically houses the gas exchange valves, such as the intake and exhaust valves, and other important hardware such as spark plugs and/or fuel injectors), and on the bottom by the piston which reciprocates within the cylinder. A rigid connecting rod fastens the piston to an eccentric location on the rotating crankshaft. The eccentric placement of the connecting rod converts the reciprocating motion of the piston (i.e., boundary motion work) to the rotating motion of the crankshaft (i.e., shaft work). The eccentric placement of the connecting rod to the crankshaft also dictates the important geometrical parameter of the piston engine called the "stroke." The stroke and "bore," or diameter of the cylinder, create the displaced volume, $V_d$, of the piston engine. The maximum volume of the cylinder, $V_{max}$, is attained when the piston is at its bottom-most position; a position referred to as "bottom dead center," or BDC. The minimum volume, or clearance volume, $V_c$, is attained when the piston is at its top-most position; a position referred to as "top dead center," or TDC. The ratio between $V_{max}$ and $V_{min}$ is called the compression ratio, $r_c$, and is given as Eq. 6.1:

$$r_c = \frac{V_{max}}{V_{min}} \tag{6.1}$$

The compression ratio, as will be described in section on "Thermodynamic Analysis of Internal Combustion Engines", is a fundamentally critical parameter

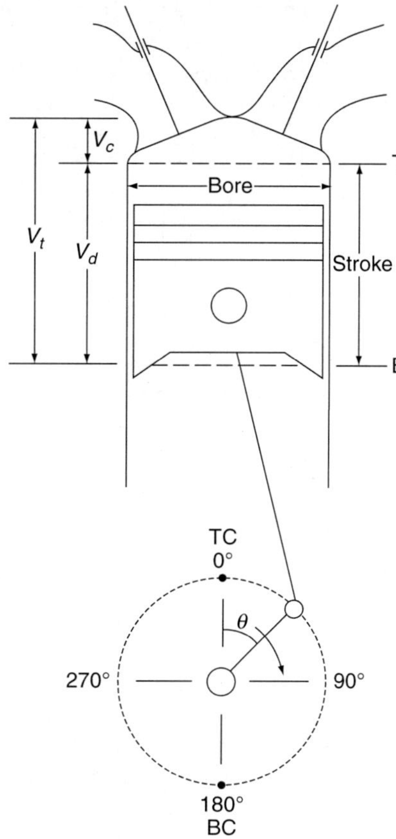


**Fig. 6.1** Illustration of a piston/cylinder arrangement, as often employed in a reciprocating-piston internal combustion engine (Used with permission from [2])

for controlling the efficiency (i.e., the ratio of work energy out to chemical energy in) of an internal combustion piston engine.

There are two major cycles used to exploit the piston engine's conversion of chemical energy to work energy: a "two-stoke" cycle and a "four-stroke" cycle. The earliest prototype engines were of the two-stroke variety (e.g., Lenoir and Otto/Langen engines) [1]. In pursuit of achieving higher efficiency, Otto (for whom the thermodynamic ideal "Otto Cycle" is named) built the four-stroke version of his engine [1]. Today, four-stroke cycle engines are the dominant form; thus, most of the article will center on the details of the four-stroke cycle.

Four-stroke cycle engines require four strokes of the piston to complete one power-producing cycle, as shown in Fig. 6.2 [3]; the reader is also referred to Fig. 6.4a to aid the discussion. Consider first the cycle starting with the piston at TDC and the intake valve open. The piston moves from TDC to BDC, inducting fresh mixture (conventionally, fuel and air in a gasoline engine and only air in a diesel engine) through the open intake valve in what is called the "intake stroke." At some location near BDC, the intake valve closes, and the piston reverses its motion at BDC. Once the valve closes, the piston/cylinder arrangement creates a closed system. As the piston moves from BDC to TDC, the trapped mixture is

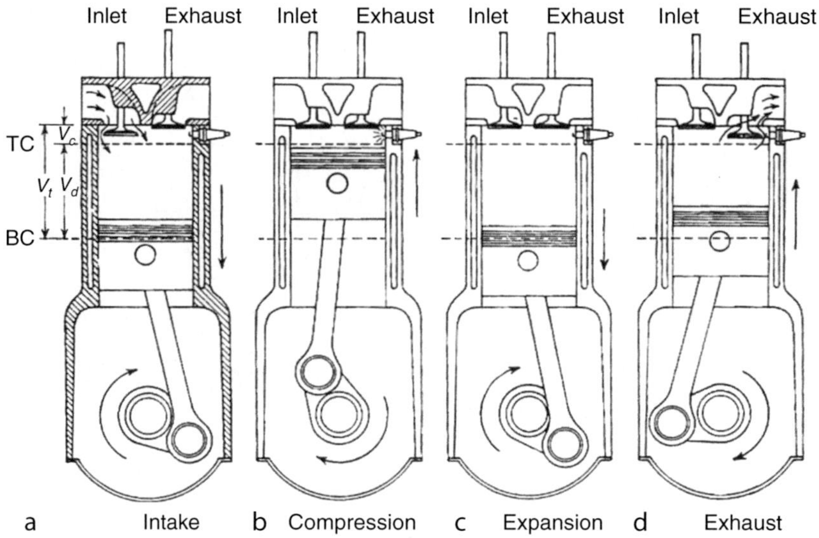


**Fig. 6.2** Illustration of the four-stroke operating principle (Used with permission from [3])

compressed in what is called the "compression stroke," increasing the mixture's temperature, pressure, and decreasing its specific volume (i.e., increasing its density). At a point near TDC combustion is expected to commence. In the case of a spark ignition engine (e.g., a conventional gasoline engine), combustion is initiated by the release of spark at a point near (usually advanced of) TDC. In the case of a compression ignition engine (e.g., a conventional diesel engine), combustion is initiated by injecting liquid fuel directly into the cylinder; the compressed air at elevated temperature and pressure atomizes, vaporizes, and mixes with the injected fuel. After a period of time, the high temperature environment causes chemical reaction and start of combustion. Around start of combustion, the piston reaches TDC, reverses direction, and expands the cylinder volume as combustion converts chemical energy into work energy. This stroke takes on many names, including "power stroke," "combustion stroke," and "expansion stroke." As the piston approaches BDC, the exhaust valve opens, allowing the products of combustion to escape the cylinder. At BDC, the piston reverses direction and motions toward TDC with the exhaust valve open, in what is called the "exhaust stroke." Depending on the engine's crankshaft rotational speed – which can vary from as low as 100 rev/min for large marine-application engines to as high as 15,000 rev/min for race car engines – the four-stroke cycle requires as much as about 1.2 s to as little as 8 ms to complete.

Up to this point, the discussion has centered on the processes occurring in a single cylinder. Most engines, however, are composed of many cylinders and take on various forms (e.g., in-line four-cylinder, "V6", "V8", and "W" engine). In such cases, each cylinder undergoes the same processes but usually out of phase. For example, in an in-line four-cylinder engine (i.e., an engine that has four

**Fig. 6.3** Illustration of the Audi 2.0-L TSI, an example of an in-line four-cylinder engine (Used with permission from [4])

cylinders oriented sequentially in a single-line bank, an example of which is shown in Fig. 6.3), the cylinder processes are typically out of phase by 180° (720°/4). The firing order is usually not linear, however, in order to ensure smooth and continuous operation. For example, a four-cylinder firing order may go 1-3-2-4; that is, as the engine crankshaft rotates cylinder 1 produces a power stroke first, followed by cylinder 3, then cylinder 2, and finally by cylinder 4 (which is then followed again by cylinder 1). Finally, when describing an engine's displacement, it refers to a summation of each cylinder's displacement; thus, each cylinder displacement is on average the total engine displacement divided by the number of cylinders.

## *How an Engine Makes Power*

The reciprocating motion of the piston in a cylinder is the means by which the chemical energy released during combustion is converted into useful work. Work is transferred when a force acts through a displacement; in the case of the piston/cylinder engine, the in-cylinder pressure, $P$, is interpreted as the force and the changing cylinder volume, d$V$, during piston strokes is interpreted as the displacement. Thus, the thermodynamic work of an engine cycle is given by Eq. 6.2:

$$W = \oint P\mathrm{d}V \qquad [\mathrm{kJ/cycle}] \tag{6.2}$$

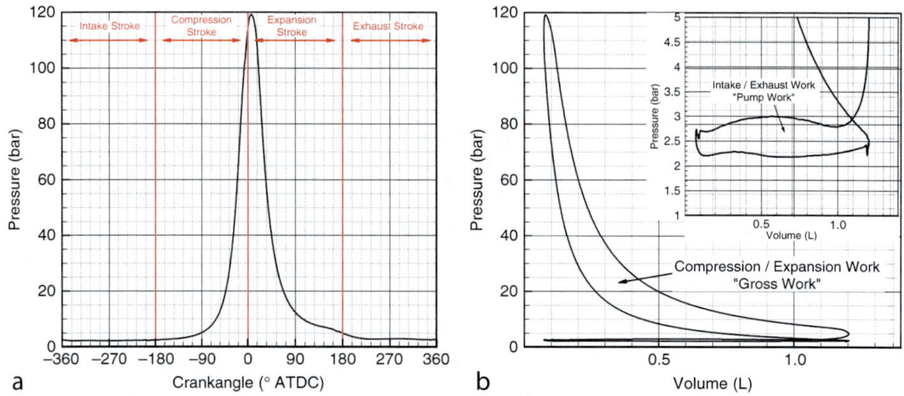


**Fig. 6.4** (**a**) Pressure as a function of crankshaft rotation, or crankangle, in degrees after top dead center (° ATDC), and (**b**) pressure as a function of cylinder volume, also illustrating the areas of the *P–v* plane that render gross work and pump work (Data from author's laboratory, Texas A&M University)

The work of the cycle given by (Eq. 2) is considered *boundary work*, since it results from the changing boundary of the control system (in this context, the control system is that enclosed by the piston/cylinder arrangement, and the moving boundary is manifested by the moving piston). Conventional engines convert this boundary work of the piston to *shaft work* through the piston connecting rod's eccentrically located connection to the crankshaft. The shaft work issuing from the crankshaft is often best interpreted as *torque*, $\tau$; the work given by Eq. 6.2 is related to the torque of the crankshaft via Eq. 6.3:

$$W = 2\pi n_R \tau \qquad [\mathrm{kJ/cycle}] \tag{6.3}$$

where $n_R$ is the number of crankshaft revolutions per power cycle (i.e., $n_R = 2$ for a four-stroke cycle and $n_R = 1$ for a two-stroke cycle). The units for $\tau$ in Eq. 6.3 are kN-m.

The torque of an engine is routinely measured with a dynamometer; in this way, the torque is considered "brake torque," or the amount of resistance torque the dynamometer must apply to "brake" the engine to a certain speed condition. Via application of Eq. 6.3, the "brake work," $W_b$, is determined. Only determining brake work, however, reveals no insight into the in-cylinder work processes. The in-cylinder work processes can be calculated using in-cylinder pressure measurement that is precisely coupled to the in-cylinder volume via a crankshaft encoder and detailed knowledge of the cylinder's and crank-slider's geometries. An example "pressure–crankangle" diagram is shown Fig. 6.4a, where crankangle is reported in degrees after top dead center (° ATDC) relative to "combustion TDC." Also shown in Fig. 6.4a are the four strokes of the four-stroke cycle as described in section "Basic Operating Cycle". In-cylinder pressure is typically measured with a piezo-electric pressure transducer which is able to provide a fast-response indication during the engine cycle [5]. It should be noted that measuring in-cylinder pressure is not a trivial task and great care must be taken to do it properly [6–9]. When using digital equipment (i.e., an analog–digital converter) to electronically record in-cylinder pressure, it becomes necessary to determine the sample rate, which is usually determined by the crankshaft encoder. Varying crankangle resolutions can be used, depending on the level of precision needed of the analysis. For calculating in-cylinder work (described next), a crankangle resolution of up to 10° [6] can be used; for detailed combustion analysis much finer resolution must be used (e.g., about 1° for gasoline engine combustion and 0.25° for diesel engine combustion). The data shown in Fig. 6.4 is recorded every 0.2°. In addition to the crankangle resolution, the engine speed also determines the needed sample rate of the data acquisition system. Finally, it is equally important to know the cylinder volume at each record of pressure when calculating in-cylinder work. This requires precise phasing between the piston's location and the crankshaft encoder; it also requires knowing the precise geometries of the cylinder and crank-slide mechanism. Specifically, the minimum volume (i.e., clearance volume), the piston stroke, and the cylinder bore must be precisely known. It is often best to determine these using precise measuring instruments, rather than rely on manufacturer specifications (which, although have tight tolerances, are nominal values). An example "pressure–volume," or "P–V," diagram is shown in Fig. 6.4b.

Once a precise P–V diagram is determined, the in-cylinder work associated with each process can be determined. The area between the P–V curves, as suggested by Eq. 6.2, represents the in-cylinder work, or "indicated" work (named for the antiquated use of a mechanical stylus-indicator device to record pressure [10]). There are two portions of the typical four-stroke engine cycle, as shown in Fig. 6.4: (1) compression and expansion strokes which in combination result in the "gross work," $W_{ig}$, and (2) intake and exhaust strokes which in combination result in the "pump work," $W_{ip}$. It is important to note that gross work and pump work correspond to the respective strokes of the piston, not necessarily the valve events (i.e., not necessarily the closed portion of the cycle versus open portion of the cycle). In combination, i.e., the sum of gross work and pump work result in the "net work," as given by Eq. 6.4:

$$W_{in} = W_{ig} + W_{ip} \tag{6.4}$$

Note that the subscript "$i$" on the terms in Eq. 6.4 represents "indicated"; since terms "gross," "pump," and "net" only have relevance from indicated data (i.e., in-cylinder pressure data), it is often dropped as a designator on the work terms.

Pump work, which will be nonzero when intake pressure is different from exhaust pressure (nearly all situations), often decreases net work relative to gross work (i.e., the gas exchange process requires the piston to do work on the gas, or, pump work is negative). There are a few situations, with the use of a turbocharger or

**Table 6.1** Summary of various mean effective pressures describing the various work transfers defined for a reciprocating-piston internal combustion engine

| Name | Definition |
|---|---|
| Gross indicated mean effective pressure (gross IMEP) | $\mathrm{IMEP_g} = W_{\mathrm{ig}}/V_{\mathrm{d}}$ |
| Net indicated mean effective pressure (net IMEP) | $\mathrm{IMEP_n} = W_{\mathrm{in}}/V_{\mathrm{d}}$ |
| Pump mean effective pressure (PMEP) | $\mathrm{PMEP} = W_{\mathrm{ip}}/V_{\mathrm{d}}$ |
| Friction mean effective pressure (FMEP) | $\mathrm{FMEP} = W_{\mathrm{f}}/V_{\mathrm{d}}$ |
| Brake mean effective pressure (BMEP) | $\mathrm{BMEP} = W_{\mathrm{b}}/V_{\mathrm{d}}$ |

supercharger for example, when intake pressure is greater than exhaust pressure and pump work is positive. In such situations, net work will be greater than gross work.

The difference between net work and brake work, as shown in Eq. 6.5, is the friction work, $W_{\mathrm{f}}$, of the engine. Friction, of course, always requires work from the system; thus, brake work will always be less than net work. In Eq. 6.5, $W_{\mathrm{f}}$ captures all forms of mechanical friction of the engine, including friction among crank-slider components, in bearings, in valve springs, and in various accessories (e.g., water and oil pumps):

$$W_{\mathrm{f}} = W_{\mathrm{in}} - W_{\mathrm{b}} \tag{6.5}$$

Engine researchers typically quantify all of the above-described work parameters on volume-normalized parameters, which in general represent a "mean effective pressure." The general definition for mean effective pressure, MEP, is given as Eq. 6.6:

$$\mathrm{MEP} = \frac{W}{V_{\mathrm{d}}} \tag{6.6}$$

The various mean effective pressures and their definitions are summarized in Table 6.1.

One of the major benefits of describing the work of an engine in terms of mean effective pressure is that the "size" of the cylinder is removed from consideration. In other words, it is possible to produce more work from an engine that has a larger displaced volume; however, the mean effective pressure may be lower (relative to a lower displaced volume engine) due to a number of other possible influencing parameters that affect and engine's ability to make power (i.e., fuel conversion and volumetric efficiencies, fuel–air ratio, inlet air density, and fuel heating value). To make these types of assessments, refer to Eq. 6.7, which is developed from the basic definition of power, $P$ (i.e., work per unit time), and respective definitions of other involved parameters as shown by Heywood [11]. An example of the use of Eq. 6.7 to assess factors affecting power in a technology comparison is provided in [12].

$$P = \frac{\eta_{\mathrm{f}}\eta_{\mathrm{v}}\rho_{\mathrm{a,i}}(F/A)Q_{\mathrm{HV}}V_{\mathrm{d}}N}{n_{\mathrm{R}}} \tag{6.7}$$

where

$\eta_f$ Fuel conversion efficiency (described in section on "Thermodynamic Analysis of Internal Combustion Engines").

$\eta_v$ Volumetric efficiency, or the engine's effectiveness at "breathing" air. It is defined as $\eta_v = \frac{m_a}{\rho_{a,i} V_d}$, where $m_a$ is the actual mass of air inducted per cycle (kg) and $\rho_{a,i}$ is the density of the intake air at some reference point (kg/m$^3$) (usually the intake manifold, but also may be atmospheric air upstream of the engine air filter). Note that $V_d$ has units of (m$^3$) in Eq. 6.7. It is important to note that volumetric efficiency only quantifies the engine's ability to breath air, i.e., not a fuel–air mixture. Thus, for example, premixing fuel with air prior to induction tends to lower the engine's volumetric efficiency. It is also important to note that volumetric efficiency of the engine depends on the chosen reference point. Thus, if the volumetric efficiency of the whole breathing stream is desired, the reference point is taken as upstream of the intake air filter (for example). If, however, volumetric efficiency of just the intake ports and through the valves is desired, then the reference point is taken as the intake manifold.

$\rho_{a,i}$ As defined above, the density of the intake air at some reference point (kg/m$^3$).

$F/A$ The mass-based fuel–air ratio of the mixture.

$Q_{HV}$ Heating value of the fuel (typically, lower heating value is used since the products leaving the piston/cylinder system with water as a vapor) (kJ/kg).

$N$ Engine speed (rev/s)

To reveal the form of mean effective pressure, Eq. 6.8 is given as a modified form of Eq. 6.7.

$$\text{MEP} = \eta_f \eta_v \rho_{a,i} (F/A) Q_{HV} \tag{6.8}$$

Thus, it is apparent from Eq. 6.8 how an engine with a relatively small displacement might have a higher mean effective pressure than an engine with a large displacement, even though the larger-displaced engine may produce more power. It is also clear from Eqs. 6.7 and 6.8 how the performance (i.e., power) of an engine may be improved beyond the "easy" action of increasing displaced volume. One parameter, for example that benefits the power and mean effective pressure of the engine is the fuel conversion efficiency. This very important parameter is discussed in the next section.

Finally, this section concludes by illustrating a typical power/torque/speed curve. Apparent from Eq. 6.3 and the definition of power, there is a relationship among power, torque, and speed of an engine. This relationship for a typical internal combustion engine is shown in Fig. 6.5. There are a few interesting features to point out in this figure. The first, a practical feature, is the identification of "rated torque" and "rated power". Engines are usually "sized" based on the speed at which

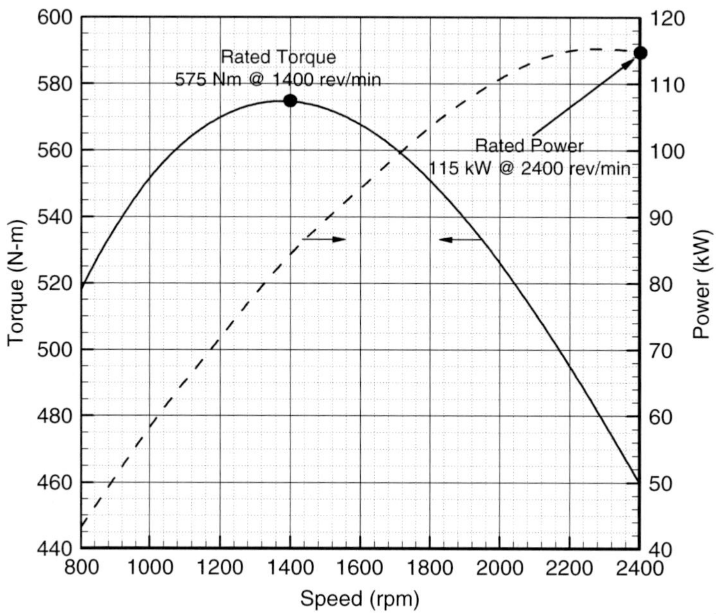


**Fig. 6.5** Torque and power as functions of engine speed for a typical internal combustion engine. Data is collected from 4.5 L medium-duty diesel engine with advanced technology such as turbocharging and exhaust gas recirculation (Data from author's laboratory, Texas A&M University)

they develop maximum torque (also known as rated torque) and the speed at which they develop maximum power (also known as rated power). The second feature, which is apparent from the knowledge that rated torque exists, is the seemingly dependent relationship of torque on engine speed. It is clear from the definition of power that it should have a functional relationship on the speed of the engine, which dictates the amount of work per unit time the engine can deliver. But, based on assessment of Eq. 6.3, there is not a direct relationship between in-cylinder work per cycle and the speed of the engine. In fact, in an ideal sense, the torque of the engine should be constant with engine speed (and, if constraining to an ideal engine, higher than the maximum attained torque of the real engine), similar to how torque of an electric motor is nearly constant with motor speed. Thus, the behavior seen in Fig. 6.5 suggests that there exist factors in a real engine that depend on engine speed and also affect the work produced per cycle (thereby affecting the torque of the engine). These factors predominantly consist of heat transfer and friction; both of which have dependencies on the speed at which the engine operates. Heat transfer of course is a time-dependent phenomenon. At low engine speeds there is more time per cycle for thermal energy to be transferred from the cylinder; thus, torque droops at low engine speeds as heat transfer diminishes the energy available for in-cylinder work production. Friction, too, is a speed dependent function due to the mechanical behavior of the engine's interacting components. At high engine speeds, there is

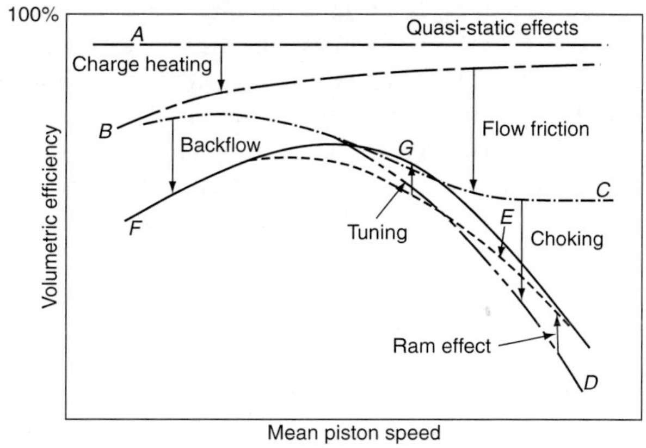


**Fig. 6.6** Example of volumetric efficiency, illustrating the many factors that impact the engine's ability to effectively breathe air (Used with permission from [13]). Improvements to the engine's volumetric efficiency can yield substantial improvements to its ability to make power, evident from Eq. 6.7

increased friction to be overcome on a per cycle basis; thus torque droops at high engine speeds as increased work energy is required to overcome increased friction.

Relating this discussion to Eq. 6.7, both heat transfer and friction effects tend to decrease $\eta_f$. There is, however, another factor of Eq. 6.7 being influenced by heat transfer and friction and thus serving as a major contributor to the behavior of the torque curve shown in Fig. 6.5; this factor is the volumetric efficiency. The maximum amount of work that can be developed per cycle is firstly dependent on the amount of air (or oxidizer) the engine can breathe; the amount of air the engine can breathe ultimately dictates the amount of fuel that can be delivered, which of course serves as the energy carrier to be converted in the cylinder.

A representative volumetric efficiency curve as a function of mean piston speed (which is correlated to engine speed) is shown in Fig. 6.6. There is much important detail in this figure, and the following will describe this in detail; after this discussion, a return to the effect of volumetric efficiency on the torque curve of Fig. 6.5 will be made. First, notice that under ideal situations the volumetric efficiency curve would be 100% and independent of engine speed (like work, trapped mass per cycle ideally has no functional dependency on the number of cycles per unit time, or, speed of the engine). It is noted that, depending on the reference point chosen for quantifying it, volumetric efficiency could be greater than 100% if the engine is, for example, boosted. In such a situation, the chosen reference point is upstream of the boosting device (e.g., turbocharger compressor inlet). Although such a chosen reference point is useful to indicating the quantity of trapped mass in the cylinder, it prevents the quantification of the engine's breathing system outside of the boosting device, and masks opportunities for further improvement to the engine's design.

Returning to the volumetric efficiency curve of Fig. 6.6, and for the specific case of premixed charge engines such as conventional gasoline spark-ignited engines, the premixing of fuel with air prior to induction diminishes the amount of trapped mass of air in the cycle. Such a factor is considered a "quasi-static" effect since it will be present regardless of the speed of the engine. Other quasi-static effects include factors such as, for example, residual fraction due to manifold pressure differences (between exhaust manifold and intake manifold) and exhaust gas recirculation. Thus, notice secondly that the diminishing effect of quasi-static factors is represented as curve “A” in Fig. 6.6. The next phenomenon to be captured in the volumetric efficiency curve is called “charge heating” and is a heat transfer effect. During normal engine operation, the breathing system of the engine reaches a steady state temperature that is higher than the ambient temperature. Thus, the elevated temperature of the intake port results in heat transfer to the intake air, increasing the air temperature, decreasing its density, and decreasing the amount of mass per unit volume. As described above, heat transfer is a time-dependent phenomenon, therefore its presence is most notable at low engine speeds (low mean piston speeds); notice thirdly in Fig. 6.6 the effect of charge heating on the volumetric efficiency curve, which in combination with quasi-static effects lowers it to the curve labeled “B.” Similar to how friction affects work per cycle, friction also affects air flow and has a dependency on mass flow rate (i.e., has a dependency on engine speed). The fourth factor to notice in Fig. 6.6 is the effect of flow friction; its effect in concert with quasi-static and charge heating factors result in the curve labeled “C.” Notice at very high speeds, the flow friction effect seems to level off and play a speed-independent role as engine speed increases; this results from the flow attaining choke conditions where, regardless of the pressure drop across the intake system, the mass flow rate remains constant as its velocity reaches the speed of sound. Thus, the fifth factor to notice in Fig. 6.6 is the effect of choking on the volumetric efficiency curve. The net result of all combined effects including choke is the curved labeled “D.” Interestingly, some phenomena help to increase the engine’s volumetric efficiency; ram effect is one such example. Ram effect occurs at high flow rates where fluid momentum results in continued charging of the cylinder even as the in-cylinder motion of the piston no longer provides the pumping action (i.e., volume increase during intake stroke). The sixth factor to note in Fig. 6.6 is the increase in volumetric efficiency due to ram effect. The combined effects, including ram effect, result in the curve labeled “E.” A similar but diminishing gas exchange phenomenon occurs at low engine speeds, known as backflow. Backflow occurs during the valve overlap period (the period during which both exhaust and intake valves are open as exhaust valves close and intake valves open) where in-cylinder motion becomes quiescent as the piston reaches top dead center and exhaust products of combustion backflow into the intake port. The seventh factor to notice in Fig. 6.6 is the effect of backflow on the engine, which tends to pervade at low engine speeds where exhaust momentum is low. The combined effects, including backflow, on volumetric efficiency result in the curve labeled “F.” Finally, to end on a positive note, the eighth and last factor to observe in Fig. 6.6 is that of tuning. Tuning is the effort to make use of established resonating sound waves in the intake and exhaust systems to either aid in charging the cylinder (intake tuning) or discharging the cylinder (exhaust tuning). Because this charging or discharging benefits relies upon the timing of compression and rarefaction waves being positioned at precise locations in the intake or exhaust systems, fixed geometry intake and exhaust systems can only be tuned at a narrow speed range. The benefit of tuning is shown in Fig. 6.6, where for this particular application tuning is designed for the mid-speed range of the engine. Variable tuning systems are becoming available in production, which through the use of creative flow channeling and blocking, can change the "effective" runner length of the flow passage and allow for tuning across broader speed ranges than allowed by fixed geometry runners. With all factors considered, the final volumetric efficiency curve is the solid curve labeled "G."

Now that the details of the volumetric efficiency curve as a function of engine speed, identified as curve "G" in Fig. 6.6, have been discussed, a return to its effect on engine torque will be made. In fact, it is interesting to note the similarities between the final volumetric efficiency curve (curve "G") of Fig. 6.6 and the torque curve of Fig. 6.5. Such a similarity is not a coincidence, as the maximum amount of work attainable per cycle is a strong function of how well the engine breathes air per cycle. Correspondingly, based on the assessment of Eq. 6.7, significant improvements to an engine's ability to make power can be realized through creative improvements to the engine's volumetric efficiency.

## *Thermodynamic Analysis of Internal Combustion Engines*

At its core, an internal combustion engine is a thermodynamic device. That is, it takes one form of energy (i.e., chemical energy) and converts it into another form of energy (i.e., mechanical "shaft work" energy). There are several considerations that should be given to the thermodynamics of the internal combustion engine, in pursuit of improving its ability to make power and, more importantly, to make power *efficiently*. Efficiency is an important thermodynamic concept which quantifies the ability of a work conversion device to convert one form of energy (e.g., chemical energy) into work energy (i.e., mechanical shaft work). In the context of internal combustion engines, *fuel conversion efficiency* is used, as given by Eq. 6.9:

$$\eta_{\mathrm{f}} = \frac{W}{m_{\mathrm{f}} Q_{\mathrm{HV}}} \tag{6.9}$$

where $m_{\mathrm{f}}$ is the mass of fuel per cycle. Typical maximum fuel conversion efficiencies for internal combustion engines vary from ca. 20% to 45%, depending on engine design and type of combustion system. Fuel conversion efficiency, like power, can be designated as brake or indicated fuel conversion efficiency. The relationship between brake and indicated fuel conversion efficiency is given as Eq. 6.10:

$$\eta_{\mathrm{f,b}} = \eta_{\mathrm{f,i}} \eta_{\mathrm{m}} \tag{6.10}$$

where $\eta_{\mathrm{f,b}}$ is brake fuel conversion efficiency, $\eta_{\mathrm{f,i}}$ is net indicated fuel conversion efficiency, and $\eta_{\mathrm{m}}$ is mechanical efficiency and given by $\eta_{\mathrm{m}} = \frac{W_{\mathrm{b}}}{W_{\mathrm{in}}} = \frac{P_{\mathrm{b}}}{P_{\mathrm{in}}} = \frac{\mathrm{BMEP}}{\mathrm{IMEP_n}}$. In words, fuel conversion efficiency is the ratio of work output to fuel energy delivered. There is another efficiency definition that can be considered, called *thermal efficiency*, which is the ratio of work output to fuel energy released. In other words, since the combustion process converts the fuel's chemical energy to thermal energy, the thermal efficiency indicates the conversion of thermal energy to work energy. It is not possible to precisely measure the energy released by the fuel during the combustion process; it is, however, possible to precisely determine the extent of incomplete combustion, or combustion inefficiency. Thus, *combustion efficiency* is introduced to allow for the determination of the engine's thermal efficiency. Combustion inefficiency is defined as the energy of unreacted fuel in the exhaust to fuel energy delivered to the engine. Combustion efficiency, $\eta_{\mathrm{c}}$, is thusly given as Eq. 6.11 [14]:

$$\eta_{\mathrm{c}} = 1 - \frac{\sum y_{\mathrm{i}} Q_{\mathrm{HV,i}}}{\left[\dot{m}_f \big/ \dot{m}_f + \dot{m}_a\right] Q_{\mathrm{HV,f}}} \tag{6.11}$$

where $y_{\mathrm{i}}$ is an unreacted fuel specie in the exhaust (e.g., carbon monoxide), $Q_{\mathrm{HV,i}}$ is the heating value of specie i, $\dot{m}_f$ and $\dot{m}_a$ are the fuel and air mass flow rates, respectively, and $Q_{\mathrm{HV,f}}$ is the fuel's heating value. Fuel conversion efficiency, combustion efficiency, and thermal efficiency, $\eta_{\mathrm{th}}$, are all related by Eq. 6.12:

$$\eta_{\mathrm{f}} = \eta_{\mathrm{c}} \eta_{\mathrm{th}} \tag{6.12}$$

Note that $\eta_{\mathrm{f}}$ of Eq. 6.12 could be either brake or indicated, in which case the corresponding $\eta_{\mathrm{th}}$ will have the same basis.

At this point, it is instructive to question what the maximum possible fuel conversion or thermal efficiency of an internal combustion engine could be. To begin to answer this question, it should be established that, contrary to what is sometimes simply conveyed, an internal combustion engine is not, by rigid definition, a *heat engine* [15–17]. A heat engine is a device that collects thermal energy, via heat transfer mechanism, from a high temperature reservoir and converts the *available portion* of this thermal energy to useful work. The portion of thermal energy that is not available for conversion to useful work (i.e., the entropy of the thermal energy) is rejected to a low temperature sink via heat transfer mechanism. A practical example of a heat engine is the steam engine. It is the steam engine that caused N.L. Sadi Carnot to establish his two postulates on the maximal conditions of the heat engine [18], thus laying the foundation for what would become the second law of thermodynamics and the mathematical axioms for the property entropy. The fully reversible heat engine cycle, i.e., the Carnot cycle, is the most efficient cycle for converting thermal energy into useful work (which, because of its reversible nature, is the most efficient cycle for converting work energy into

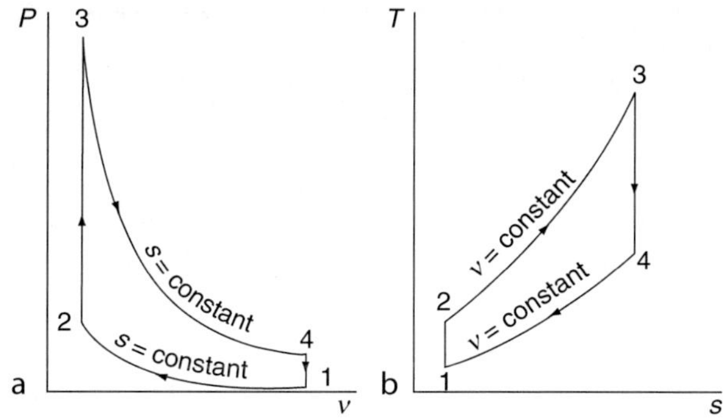


**Fig. 6.7** (**a**) Pressure–volume and (**b**) temperature–entropy diagrams of the air-standard Otto cycle (Used with permission from [19])

thermal energy). These details are provided because a common misconception is that the maximum efficiency of an internal combustion engine is limited by a so-called Carnot efficiency; this really has no technical basis because the internal combustion engine does not operate on the same principle (i.e., the heat engine principle) around which the Carnot cycle is created.

Perhaps much of the misdiagnosis of considering an internal combustion engine a heat engine comes from the use of ideal heat engine cycles to "model" an internal combustion engine; namely the Otto and Diesel cycles. These cycles do serve an important purpose in that certain fundamental parameters affecting the efficiency of the ideal heat engine cycles (i.e., Otto and Diesel cycles) transcend to also affecting the efficiency of the internal combustion engine. An example of such is compression ratio. Thermodynamic analysis of the Otto Cycle, the four processes of which are shown in Fig. 6.7, on (a) pressure–volume and (b) temperature–entropy diagrams reveal that efficiency of the cycle, $\eta_{\mathrm{th,Otto}}$, is given by Eq. 6.13:

$$\eta_{\mathrm{th,Otto}} = 1 - \frac{1}{r_{\mathrm{c}}^{\gamma-1}} \tag{6.13}$$

where $\gamma$ is the ratio of specific heats for the working fluid (e.g., $\gamma_{\mathrm{air}} = 1.4$ at 300 K). This result reveals the asymptotic relationship of increasing thermal efficiency with compression ratio, as shown in Fig. 6.8. Similar behavior is observed in internal combustion engines (i.e., increasing efficiency with increasing compression ratio) as the capability to expand the working fluid (i.e., combustion products), and thus the ability to convert thermal energy to work energy, increases with increasing compression ratio. Another fundamental behavior is apparent Fig. 6.8 related to the changing value of $\gamma$. Notice that thermal efficiency tends to increase with an increase in $\gamma$; similar to the increase in compression ratio, a mixture with a higher $\gamma$ suggests it requires more energy to increase its temperature during a work-transfer (i.e., constant pressure) process compared to a nonwork (i.e., constant volume) process. The corollary to this is more work energy is transferred out of a mixture with higher $\gamma$ as the mixture is expanded. In internal combustion engines, leaner fuel/air mixtures (i.e., those with less than stoichiometric, or chemically complete,

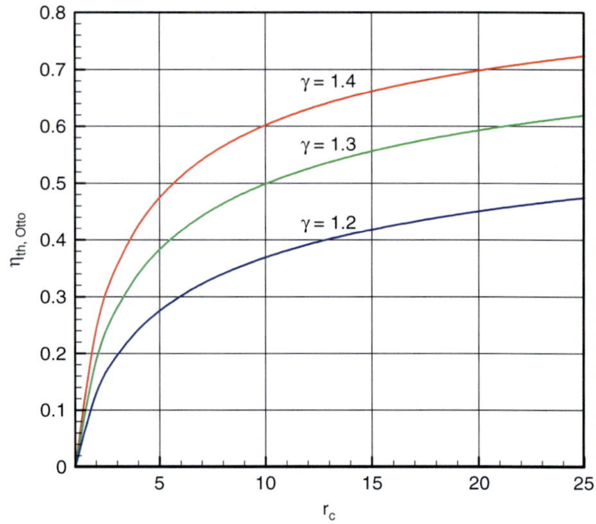


**Fig. 6.8** Otto cycle thermal efficiency as a function of compression ratio for several $\gamma$ values

concentrations of fuel in the mixture) have higher $\gamma$ values, and thus realize an efficiency improvement.

Although the examples of using the Otto cycle to assess the effects of compression ratio and $\gamma$ value on thermal efficiency are helpful, the Otto cycle efficiency cannot be used to predict maximum attainable efficiency limits of an internal combustion engine. There are several inaccuracies in applying the Otto cycle (or any ideal heat engine cycle) to predict internal combustion engine performance and efficiency. Specifically, the assumptions that go into the development of Eq. 6.13 are as follows (refer to Fig. 6.7):

1. Processes 1–2 and 3–4 are *reversible and adiabatic* (i.e., *isentropic*, or constant entropy) processes.
2. Process 2–3 is a constant volume *heat addition* process, where thermal energy of the system increases due to heat transfer with the surroundings at a temperature of no less than $T_3$.
3. Process 4–1 is a constant volume *heat rejection* process, where thermal energy of the system decreases due to heat transfer with the surroundings at a temperature no greater than $T_1$.
4. Properties of the working fluid are constant throughout the cycle.

Of the above four assumptions, all four are impractical (meaning, an actual engine's efficiency will be less than the ideal limit because of real effects, which will be discussed in more detail below). Further, assumptions 2–4 are theoretically improper to impose on an internal combustion engine. Assumptions 2–4 imply a heat engine cycle, which has already been established that an internal combustion engine is not. Instead, an internal combustion engine operates as a collection of processes, all which have ideal limits dictating, in aggregate, the ideal maximum efficiency of the engine. Energy is transferred to the control system (i.e., the

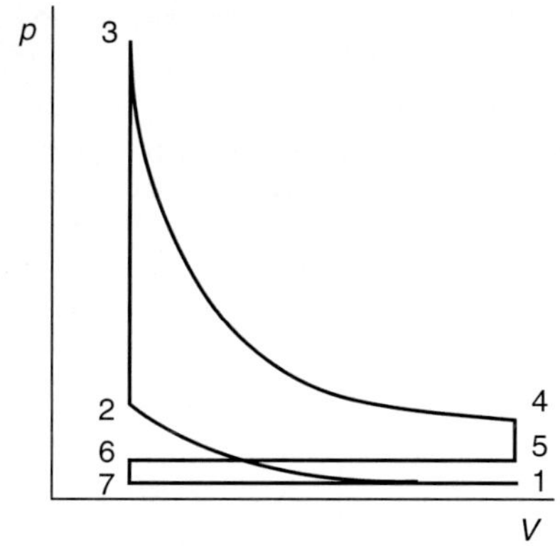


**Fig. 6.9** Pressure–volume diagram illustrating the processes assumed in *Fuel–Air Cycle Analysis* of an internal combustion engine. Note that shown cycle is representative of an engine under "throttled" conditions, where intake pressure is less than atmospheric (Used with permission from [21])

cylinder of the engine) through mass flow rate, as opposed to heat transfer mechanism (as in the heat engine apparatus). This energy, initially in the form of chemical energy, is converted to thermal energy during the combustion process and the expansion of the piston thereby converts the thermal energy to work energy. Additionally, because of the change from reactants to products during the combustion process and the dependence of species' properties on temperature and pressure, it is not appropriate to assume properties of the working substance remain constant during the engine's operation (i.e., assumption no. 4 is not valid).

Instead, maximum ideal efficiency limits of internal combustion engines could be assessed by improving upon the inappropriate assumptions of the Otto heat engine cycle, i.e., allow energy to be transferred into the system via mass flow (i.e., via the intake process) and allow properties of the mixture to change with species, temperature, and pressure. The resulting analysis, referred to as *Fuel–Air Cycle Analysis* [20], employs the following processes (refer to Fig. 6.9) and assumptions in modeling an internal combustion engine cylinder (i.e., the control volume):

Process 6–7–1: Ideal intake process of fuel and air and adiabatic mixing with residual gas from the preceding cycle. Residual gas, or often referred to as "residual fraction," is left-over products of combustion not fully exhausted from the cylinder during the previous cycle's exhaust displacement process.

Process 1–2: Reversible and adiabatic compression of the fuel/air/residual mixture (i.e., the reactants). Chemical species are frozen.

Process 2–3: Complete, adiabatic combustion of reactants to products (the latter of which are assumed to exist in equilibrium concentrations). Combustion, in ideal sense, is modeled as constant volume, constant pressure, or limited pressure. Limited pressure is a combination of constant volume and constant pressure processes. Constant volume combustion occurs until a "limit pressure" is reached, when remainder of combustion occurs at constant pressure.

Process 3–4: Reversible and adiabatic expansion of products, which remain in chemical equilibrium throughout the process.

Process 4 – 5: Reversible and adiabatic "blowdown" of products to exhaust pressure of products remaining in the cylinder. Products remain in fixed composition based on concentrations at State 4.

Process 5–6: Ideal, constant pressure, and adiabatic exhaust displacement of products, remaining in fixed composition.

Ideal gas behavior of the mixture and conserved mass are assumed throughout the analysis. Further, mass transfers only occur during intake and exhaust processes (i.e., no leakage, blow-by, or crevice flow during the cycle). By employing first and second laws and various assumptions to each of the processes, respective states can be thermodynamically fixed, thus allowing determination of work transfers. An item to note about *Fuel–Air Cycle Analysis*, because it accommodates changing species, is that *three* independent properties must be known to fix the states (i.e., two thermodynamic properties, such as temperature and pressure, and species composition). Process 6–7–1 is assumed adiabatic. Further, it is assumed that during the stroke from TDC to BDC, the cylinder pressure is constant and equal to the intake pressure, $P_{\mathrm{i}}$:

$$P_7 = P_1 = P_i$$

$$_6Q_1 = 0$$

With these two statements, and use of first law, the enthalpy at State 1, $h_1$, is given as:

$$h_1 = f\left[h_{\mathrm{e}} + R_{\mathrm{e}}T_{\mathrm{e}}\left(\frac{P_{\mathrm{i}}}{P_{\mathrm{e}}} - 1\right) - h_{\mathrm{i}}\right] + h_{\mathrm{i}}$$

where $f$ is residual gas fraction, $h_{\mathrm{e}}$ is enthalpy of the exhaust (and equal to $h_6 = h_5$), $R_{\mathrm{e}}$ is the gas constant of the exhaust (and equal to $R_6 = R_5$), $T_{\mathrm{e}}$ is temperature of the exhaust (and equal to $T_6 = T_5$), $P_{\mathrm{e}}$ is pressure of exhaust (and equal to $P_6 = P_5$), and $h_{\mathrm{i}}$ is the enthalpy of the fresh intake mixture. Residual fraction, as explained above, is the residual mass, $m_{\mathrm{r}}$ (which is equal to $m_6$, the mass at State 6), divided by the total cylinder mass, $m_{\mathrm{total}}$ (which is equal to $m_1 = m_2 = m_3 = m_4$, the mass of the system at States 1, 2, 3, and 4, respectively), and is related to $r_{\mathrm{c}}$, $P_{\mathrm{e}}$, $T_{\mathrm{e}}$, and the pressure and temperature at State 4, $P_4$ and $T_4$, respectively, by the following:

$$f = \frac{m_{\mathrm{r}}}{m_{\mathrm{total}}} = \frac{m_6}{m_1} = \frac{m_6}{m_4} = \frac{1}{r_{\mathrm{c}}}\frac{P_{\mathrm{e}}T_4}{P_4T_{\mathrm{e}}}$$

It is clear from the equations for $h_1$ and $f$ that a priori knowledge of the cycle is needed in order to fix State 1 (note that $r_{\mathrm{c}}$, $P_{\mathrm{i}}$, $P_{\mathrm{e}}$, $h_{\mathrm{i}}$ are chosen based on desired cycle compression ratio, manifold conditions, inlet composition, and inlet temperature, $T_{\mathrm{i}}$, respectively). Thus, initial values for State 1 are typically assumed for the first iteration; State 4 temperature, pressure, and composition from the initial iteration are then used to estimate $f$ and $h_1$ for the next iteration. This iterative process continues until some convergence criterion is met (e.g., close match between $f_{final}$ and $f_{final-1}$). Rather than assume values of $f$ and $h_1$, it is perhaps more intuitive to assume initial values of $f$ and temperature at State 1, $T_1$; the following expressions [22] are approximations for establishing initial iteration State 1 properties:

$$f = \left\{1 + \frac{T_e}{T_i}\left[r_c\left(\frac{P_i}{P_e}\right) - \left(\frac{P_i}{P_e}\right)^{(\gamma-1)/\gamma}\right]\right\}^{-1}$$

$$T_1 = T_e r_c f\left(\frac{P_i}{P_e}\right)$$

It is also noted that composition of the residual gas must be assumed to fully fix State 1; it is suggested that ideal products of stoichiometric combustion of the fuel–air mixture be used as the residual gas species.

Work analysis of Process 6–7–1 gives the intake work quantity:

$$_6W_1 = P_i(V_1 - V_6)$$

Process 1–2 is assumed reversible and adiabatic (i.e., $_1Q_2 = 0$); from second law, this gives:

$$s_1 = s_2$$

Geometric relationship gives: $v_1 = r_c v_2$

The assumption that chemical composition remains fixed during compression (i.e., Process 1–2) gives chemical composition at State 2; thus, State 2 is fixed. First law analysis of Process 1–2 gives:

$$_1W_2 = U_2 - -U_1 = m(u_2 - u_1)$$

Process 2–3 is the adiabatic combustion process, where reactants become products (products assumed to be in chemical equilibrium) modeled typically in three simple fashions: (1) constant volume combustion, (2) constant pressure combustion, and (3) limited pressure combustion. First law analysis for each of the three cases results in the following, respectively:

1. Adiabatic constant volume combustion

   $$u_3 = u_2$$

   $$v_3 = v_2$$

   $$_2W_3 = 0$$

2. Adiabatic constant pressure combustion

   $$h_3 = h_2$$

$$P_3 = P_2$$

$$_2W_3 = P_2(V_3 - V_2) = P_3(V_3 - V_2)$$

3. Adiabatic limited pressure combustion

Constant volume portion

$$u_{3a} = u_2$$

$$P_{3a} = P_{\text{limit}}$$

$$v_{3a} = v_2$$

Constant pressure portion

$$h_3 = h_{3a}$$

$$P_{3a} = P_3 = P_{\text{limit}}$$

$$_2W_3 = P_3(V_3 - V_2)$$

At this point, it becomes necessary to describe a way in which product species of combustion can be modeled. It is clear from the above discussion that combustion is modeled as adiabatic and (a) constant volume, (b) constant pressure, or (c) limited pressure. In order to capture a representative combustion reaction, it is necessary to assume that several product species may exist and will exist in equilibrium. Also, it is important to clarify that this approach assumes initial and final equilibrium states in predicting combustion; i.e., the thermodynamic process goes from one equilibrium state (reactants) to a different equilibrium state (products). This is in contrast to using a more detailed approach that models the progression of combustion and relies upon reaction kinetics to predict intermediate species and their nonequilibrium concentrations. See, for example, Foster and Myers [16] for overview of detailed engine modeling and, for example, Westbrook and Dryer [23] for overview of kinetic modeling of combustion. While perhaps hundreds of species could be included as products, Olikara and Borman [24] propose a combustion reaction, given as Reaction (6.R1), that might nearly be general for internal combustion engine purposes:

$$\begin{aligned} & x_{13}\left[\mathrm{C}_n\mathrm{H}_m\mathrm{O}_l\mathrm{N}_k + \frac{n+m/4-l/2}{\varphi}(O_2 + 3.7274\,\mathrm{N}_2 \right. \\ & \left. + 0.0444\,\mathrm{Ar})\right] \rightarrow x_1\mathrm{H} + x_2\mathrm{O} + x_3\mathrm{N} + x_4\mathrm{H}_2 \\ & + x_5\mathrm{OH} + x_6\mathrm{CO} + x_7\mathrm{NO} + x_8\mathrm{O}_2 + x_9\mathrm{H}_2\mathrm{O} \\ & + x_{10}\mathrm{CO}_2 + x_{11}\mathrm{N}_2 + x_{12}\mathrm{Ar} \end{aligned} \qquad ((6.R1))$$

where $x_1$ through $x_{12}$ are the mole fractions of the respective product species. $x_{13}$ represents the moles of fuel per mole of products. Solution of the 13 unknowns of course requires 13 independent equations; five of these equations come from the species balance (i.e., balance of C, H, O, N, and Ar species). The remaining eight equations come from the assumption of equilibrium among various product species [24]. As a final note, it is recognized that NO appears as a product species; this represents the equilibrium concentration of NO at the reaction temperature and pressure. Because both the formation and decomposition of NO in the post-flame gas regions are rate limited [25], it is necessary to take a chemical kinetic approach to modeling IC engine exhaust NO concentrations (see, e.g., [26–29]).

Process 3–4 is reversible and adiabatic expansion with the mixture in chemical equilibrium (resulting in different species at State 4 relative to State 3), resulting in the following thermodynamic state (along with geometrical constraint imposed):

$$s_4 = s_3$$

$$v_4 = v_1$$

First law analysis gives the following expression, which is the expansion work of the cycle:

$${}_3W_4 = U_4 - U_3 = m(u_4 - u_3)$$

Process 4–5 is the ideal blowdown process (i.e., mixture that remains in the cylinder expands isentropically, filling the cylinder volume voided by the products irreversibly escaping past the open exhaust valve). Species are fixed based on State 4 concentrations. Thus, State 5 is fixed with the following relationships:

$$s_5 = s_4$$

$$P_5 = P_{\mathrm{e}}$$

Although specific volumes change during the blowdown process (due to mass transfer), total volume is constant; thus, work transfer is 0:

$${}_4W_5 = 0$$

Finally, Process 5–6 is the ideal exhaust displacement process, i.e., cylinder contents are transferred out of the adiabatic control volume via piston displacement at constant pressure.

At this point, it is useful to describe the attainment of various property values indicated above (e.g., $s_1$). Because *Fuel–Air Cycle Analysis* seeks to capture the effects of changing properties on the cycle analysis, the approximations for constant-specific heat ideal gas property equations cannot be used. Instead, it becomes necessary to employ techniques that accommodate changes to mixture properties based on the mixture's temperature, pressure, and species composition. One such technique is to use NASA's Chemical Equilibrium with Applications program [30–34] coupled with the JANAF Thermochemical Tables. By doing so, species at any given temperature and pressure can be predicted (using NASA's Chemical Equilibrium with Applications program) and the corresponding mixture's properties (using JANAF Thermochemical Tables) can be determined. This type of technique, for example, can be programmed into a computer routine.

Based on first law analysis and constant pressure assumption, State 6 is fixed as follows:

$$T_6 = T_5 = T_\mathrm{r}$$

$$P_6 = P_5 = P_\mathrm{e}$$

The exhaust work is given by the following:

$${}_5W_6 = P_\mathrm{e}(V_6 - V_5)$$

Based on the above analysis, the net work and gross work of the cycle are given by:

$$W_\mathrm{net} = {}_6W_1 + {}_1W_2 + {}_2W_3 + {}_3W_4 + {}_5W_6$$

$$W_\mathrm{gross} = {}_1W_2 + {}_2W_3 + {}_3W_4$$

Pump work is as defined by Eq. 6.4.

Further, various efficiencies can be calculated using Eqs. 6.9, 11, and 6.12. Note that although *Fuel–Air Cycle Analysis* is ideal, there may still be incomplete combustion due to the assumption of species existing in chemical equilibrium. Thus, combustion inefficiency will occur at fuel–air mixtures approaching stoichiometric and rich conditions, as dissociation creates concentrations of CO, $H_2$, and other partially oxidized species. Upon execution of the above analysis in a computer routine, for example, the theoretical limits of maximum efficiency of an internal combustion engine can be determined. Examples of such ideal efficiencies for constant volume combustion of isooctane/air mixtures at two different EGR levels are given in Fig. 6.10.

Figure 6.10 reveals interesting behavior of two important internal combustion engine parameters on the various efficiencies, i.e., fuel–air equivalence ratio (often designated as $\varphi$) and EGR level. First, notice that as $\varphi$ approaches stoichiometric (i.e., $\varphi = 1$), $\eta_\mathrm{th}$ decreases, and then begins to increase as $\varphi$ goes rich (i.e., $\varphi > 1$). Further, $\eta_\mathrm{th}$ increases as EGR level increases. To understand this behavior, it is necessary to understand how mixture properties are affected by (a) effect of varying $\varphi$ on species, (b) effect of varying EGR level on species, and (c) effects of $\varphi$ and

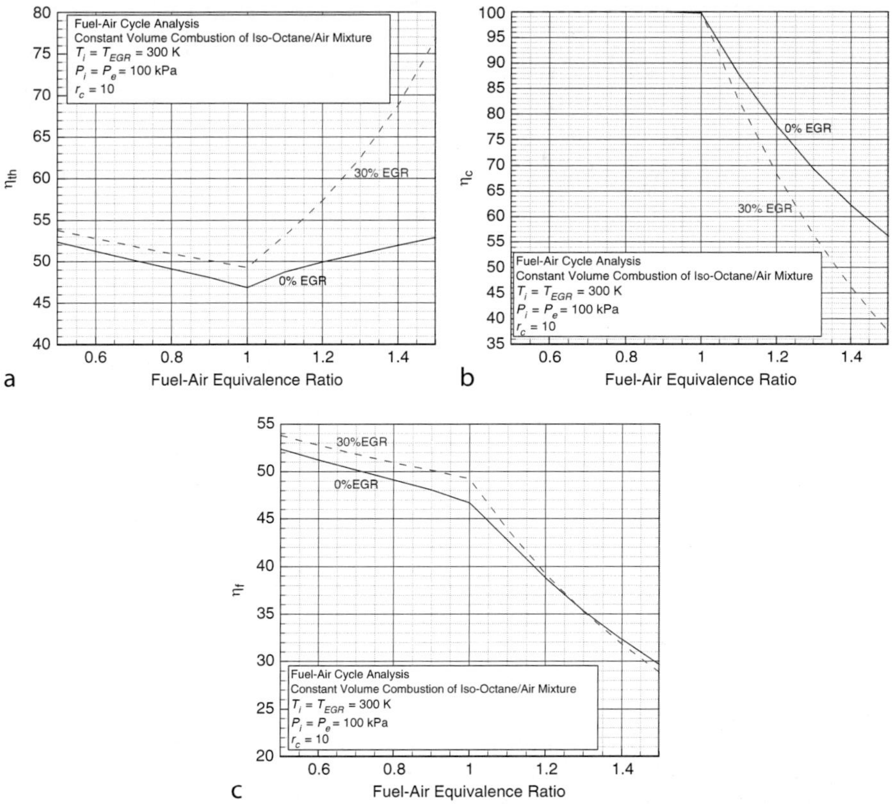


**Fig. 6.10** (**a**) Thermal efficiency, (**b**) combustion efficiency, and (**c**) fuel conversion efficiency of an ideal fuel–air cycle analysis of an internal combustion engine assuming constant volume combustion of isooctane ($C_8H_{18}$) with $T_i = 300$ K, $P_i = P_e = 100$ kPa, $T_{EGR} = 300$ K, and $r_c = 10$

EGR level on temperatures and pressures of the cycle. Ultimately, as fundamentally revealed by Eq. 6.13, a change to mixture properties that results in an increase in $\gamma$ will cause an increase in efficiency. Figure 6.11 [35] summarizes how burned gas mixture properties of an isooctane/fuel mixture are affected by various parameters such as $\varphi$ and temperature. The behaviors of the parameters shown in Fig. 6.11 can be related to $\gamma$ by recognizing that $\gamma$ can be written as given by Eq. 6.14:

$$\gamma_b = \frac{1}{1 - \frac{\bar{R}}{M_b C_{p,b}}} \tag{6.14}$$

where $\gamma_b$ is $\gamma$ for the burned gas mixture, $\bar{R}$ is the universal gas constant (i.e., $\bar{R} \approx$ 8.314 kJ/kg-K), $M_b$ is the molecular weight of the burned gas mixture, and $C_{p,b}$ is the constant pressure specific heat of the burned gas mixture. As an example of the type of analysis that might be done on Fig. 6.11, notice that $M_b$, (Fig. 6.11a)

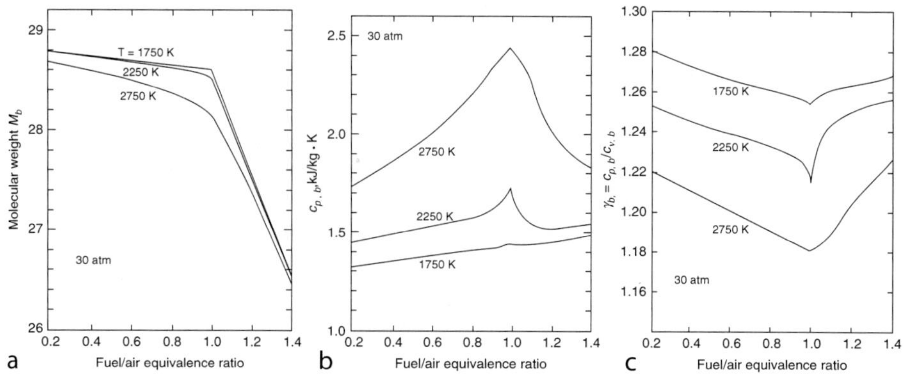


**Fig. 6.11** Burned mass (**a**) molecular weight, (**b**) constant pressure specific heat, and (**c**) ratio of specific heats as functions of equivalence ratio of isooctane/air mixtures at various burned mass temperatures and pressure of 30 atm (Used with permission from [35])

decreases as $\varphi$ increases (for any given temperature of the burned mixture); although concentrations of $CO_2$ ($M_{CO_2} \approx 44$ g/mol) increase as $\varphi$ increases, so too do concentrations of $H_2O$ ($M_{H_2O} \approx 18$ g/mol). Since isooctane, being a paraffin, produces more $H_2O$ than $CO_2$ in the products, the lighter, higher concentration of $H_2O$ dominates the net effect of decreasing $M_b$ with increasing $\varphi$. As the mixture approaches $\varphi = 1$ and becomes rich ($\varphi > 1$), $M_b$ decreases substantially as concentrations of partially oxidized species (e.g., CO and $H_2$) with relatively lighter molecular weights increase. Based on analysis of Eq. 6.14, a decrease in $M_b$ will tend to increase $\gamma_b$, which tends to increase $\eta_{th}$. Also contributing to $\gamma_b$ behavior, however, is $C_{p,b}$, which is shown in Fig. 6.11b. Notice that, up to about $\varphi = 1$, $C_{p,b}$ increases with $\varphi$. A portion of this increase is due to the increasing concentration of species with higher constant pressure specific heats as $\varphi$ increases (e.g., At 1,750 K, $(C_p/R)_{H_2O} \approx 5.94$ and $(C_p/R)_{CO_2} \approx 7.15$ whereas $(C_p/R)_{N_2} \approx 4.28$ and $(C_p/R)_{O_2} \approx 4.46$, as shown in Fig. 6.12b); another portion, however, comes from the increase in burned gas temperature which causes an increase in constant pressure specific heat for most species. Specifically, as $\varphi$ increases, adiabatic flame temperature and thus burned mixture temperature increases up to a peak that occurs slightly rich of stoichiometric (for isooctane; location of peak adiabatic flame temperature will vary depending on the type of fuel), as shown in Fig. 6.12a for isooctane/air–fuel cycle analysis with 0% EGR (note that adiabatic flame temperature, or $T_{CV,adiabatic}$ is equal to State 3 temperature, or $T_3$ of *Fuel–Air Cycle Analysis* and that $T_3$ is partially influenced by the changing State 2 temperature, or $T_2$, as $\varphi$ varies). The effects of increasing burned gas temperature on the various species' constant pressure specific heats are shown in Fig. 6.12b. Thus as $\varphi$ increases $C_{p,b}$ increases as species concentrations change (i.e., increased concentrations of $CO_2$ and $H_2O$) and burned gas temperature increases. An increase in $C_{p,b}$, like $M_b$, will cause $\gamma_b$ to decrease. Thus, for $\varphi$ less than stoichiometric, there are two competing effects on $\gamma_b$; a decreasing $M_b$ and an increasing $C_{p,b}$ as $\varphi$ increases. The net result, as shown

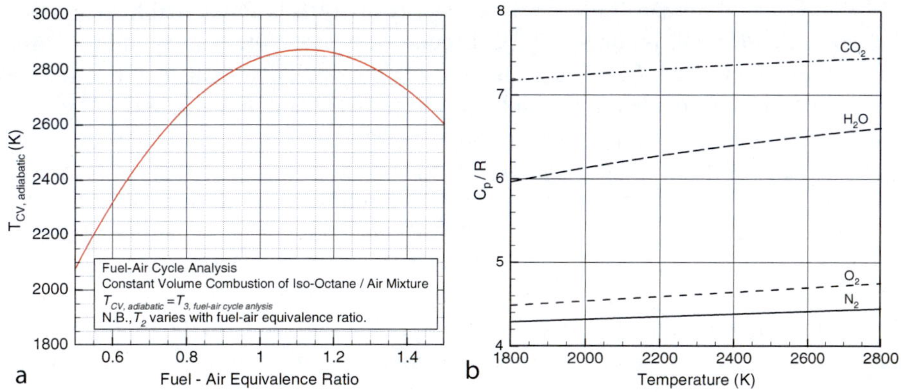


**Fig. 6.12** (**a**) Constant volume adiabatic flame temperature as function of fuel–air equivalence ratio of isooctane with initial temperature and pressure as predicted by *Fuel–Air Cycle Analysis* of isooctane/air mixture and 0% EGR, and (**b**) $C_P/R$ for various species typically found in burned mixtures as function of temperature (Data adapted from JANAF *Thermodynamic Tables* [36])

in Fig. 6.11c, is a decrease in $\gamma_b$ which correspondingly explains the decrease in $\eta_{th}$ as $\varphi$ increases up to stoichiometric (see Fig. 6.10a).

The behaviors of $M_b$ and $C_{p,b}$ change as $\varphi$ increases beyond $\varphi = 1$ (i.e., the mixture becomes rich), causing a corresponding change in $\gamma_b$ and ultimately in $\eta_{th}$. As explained above, $M_b$ decreases more dramatically rich of stoichiometric due to the abundantly increasing concentrations of CO, $H_2$, and other dissociated species which have lower molecular weights than the fully associated (i.e., $CO_2$ and $H_2O$) species. The dramatic decrease in $M_b$ generally dominates any potential increase in $C_{p,b}$, resulting in a net increase in $\gamma_b$. Thus, as $\varphi$ becomes larger than 1, $\eta_{th}$ increases. Further, at certain temperatures (e.g., 2,750 K) $C_{p,b}$ decreases as $\varphi$ becomes greater than 1; the higher temperature of the burned mixture causes higher dissociation, which causes increased concentrations of dissociated species (e.g., CO and $H_2$), which have lower constant pressure specific heats than the fully associated species (e.g., at 2,750 K, $(C_p/R)_{H_2O} \approx 6.6$ whereas $(C_p/R)_{H_2} \approx 4.4$). For such temperatures where $C_{p,b}$ simultaneously decreases along with $M_b$ as $\varphi$ becomes larger than 1, there is a dramatic increase in $\gamma_b$ and a corresponding dramatic increase in $\eta_{th}$.

Although $\eta_{th}$ may tend to increase as $\varphi$ becomes larger than 1, $\eta_c$ tends to decrease (again, as concentrations of dissociated species such as CO and $H_2$ increase as $\varphi$ becomes larger than 1). The result is a general decrease in $\eta_f$. Similar analysis can be conducted to understand the effect of EGR on the various efficiencies. It becomes necessary at this point, however, to assess the extent of the utility of such "first law analysis" – i.e., an analysis centered only on energy transfer – in pursuing future developments of internal combustion engines. For example, an engineer may "dream" of a situation where an engine operates rich at a fuel–air equivalence ratio of 1.4 and 30% EGR, yielding a $\eta_{th}$ of nearly 70%; this certainly outperforms an engine operating lean at a fuel–air equivalence ratio of 0.5

and 30% EGR (see Fig. 6.10a). If only an energy-analysis is conducted to assess this "dream," the idea will immediately be discounted because the corresponding $\eta_f$ for the engine running at 1.4 fuel–air equivalence ratio and 30% EGR is only ca. 32% (compared to the lean operation which is close to 54%). But, the products of the rich combustion still have useful energy, that is, energy that can be converted to useful work. While it is true the cycle cannot convert such *available* energy to useful work, it is more than a "dream" that some other device could convert it (e.g., a fuel cell or a thermoelectric generator). Thus, it might be conceivable to design an engine that allows it to operate at maximal $\eta_{th}$ and couple it to another device that, while not having as high of a $\eta_{th}$, has a high enough $\eta_{th}$ to suitably convert the available chemical energy to useful work, so that the net efficiency of both devices is greater than the $\eta_f$ of the internal combustion engine operating at the lean condition (for example).

Analysis of such "dreams" can only be completed by (a) introducing the concept of *available energy*, or modernly called *exergy*, by employing second law considerations on the analysis and (b) introducing real effects on the engine analysis that include such factors as heat transfer, friction, and other energy "losses" – but not necessarily exergy losses (friction is, of course, an exergy loss). Thus, a brief discussion will follow describing the major benefit of conducting an exergy analysis of an engine system. The idea of available energy – or, that portion of energy which is available to do useful work – is introduced by Gibbs [37] and, in its basic form, suggests that a system with a given amount of energy can transfer that energy out of the system as useful work until the point of equilibrium between the system and its environment (i.e., its surroundings). Using a simplified example, a piston/cylinder arrangement that contains a gas at 2 atm. will, if allowed to interact with its surroundings at 1 atm., expand and transfer work energy until the system pressure is in equilibrium with the surrounding's pressure (i.e., 1 atm.). In a more theoretical sense, it captures the notion that only an orderly flow of energy (e.g., like that established because of a temperature gradient between a system and its surroundings) can be converted into useful work; disorderly energy cannot be converted to useful work. Entropy is the idea of disorder within a system; thus, exergy analysis combines the effect of a system's entropy on its ability to convert its energy into useful work. Because real processes in net increase entropy (i.e., real processes are irreversible and result in entropy generation), systems undergoing real processes destroy exergy; that is, the opportunity to convert energy into useful work is destroyed via some disordering of what had been an orderly flow of energy.

Examples of real effects, or irreversibilities, either in a general thermodynamic system or between a general thermodynamic system and its environment (all of which happen to be present, also, in internal combustion engines) are heat transfer through a finite temperature difference, mixing, unrestrained expansion, combustion, and friction. The general premise of an irreversibility is one which renders a system/environment combination changed as the system undergoes a cycle and returns to its initial state. For example, if a warm body exchanges thermal energy with a cold body (i.e., heat transfer through a finite temperature difference), there is heat transfer, but not work transfer, until the two bodies are in thermal equilibrium.

To return the two bodies to their initial states (i.e., transfer thermal energy from the cold body to the warm body), work energy must be transferred into the system to effect the heat transfer. In the forward process (i.e., the bodies attaining thermal equilibrium), heat transfer occurs without work transfer; in the reverse process (i.e., the bodies returning to their initial states), heat transfer is manifested by a necessary work transfer. A net effect has been made to the surroundings (transfer of work energy) as the system is returned to its initial state; thus, the process is irreversible. If, on the other hand, a device had been positioned between the two bodies of dissimilar temperature that was able to extract the orderly sense of energy motion into useful work (i.e., a heat engine), the irreversibility of the heat transfer through finite temperature difference is diminished. Further, if the heat engine is conceived to be ideal itself, then all the orderly sense of energy motion is converted into useful work; hence, the concept of the Carnot heat engine, or, fully reversible heat engine.

On the one hand, the heat transfer is viewed as a loss; on the other hand, it becomes apparent that there may still be an opportunity to extract useful work from the heat transfer. It is the latter supposition that exergy analysis centers on. To begin a discussion of energy and exergy analysis of an internal combustion engine, a revisit to the assumptions made in the *Fuel–Air Cycle Analysis* is necessary. First, the assumption that the engine is adiabatic is not realistic; all real engines have heat transfer. Thus, it becomes necessary to capture an understanding of the effects of heat transfer on the engine cycle. Second, the assumption of how combustion is modeled is inaccurate. Of course, constant volume combustion in an operating engine is not physically possible due to the finite time required by chemical kinetics to decompose the fuel molecule. Even constant pressure combustion is not realistic because it supposes a constant burn rate with instantaneous starts and ends of combustion. Thus, it becomes necessary to include the effects of "real" combustion on the engine cycle. Third, if it is desired to know the actual torque from the engine (as opposed to what is indicated by the pressure–volume relationship), then a sense of friction must be included in the analysis.

The inclusion of "real effects" in engine cycle analysis is well established. For example, comparisons between actual engine cycles and *Fuel–Air Analysis Cycles* are made by Edson and Taylor [38]. Inclusion of heat transfer, finite combustion rates, and mechanical losses is demonstrated by Strange [39]. The beginnings of detailed combustion modeling, including spatial location of start of combustion and flame propagation, are demonstrated by Patterson and Van Wylen [40]. Such early demonstrations, and the corresponding advancement in digital computers, result in advanced models of heat transfer (e.g., [41–43]), friction [44, 45], and combustion; the latter of which are described in the context of doing a complete engine cycle simulation (e.g., [44, 46–51]). The influences of "real effects" on both energy and exergy perspectives of internal combustion engines are also well established (see, e.g., the reviews by Caton [52] and Rakopoulos and Giakoumis [53]). For example, Fig. 6.13 illustrates both the energy (Fig. 6.13a) and exergy (Fig. 6.13b) distributions of using 0% and 20% adiabatic EGR in a spark ignition engine, as computed by a thermodynamic simulation [54]. Note that when real factors, such as heat transfer and real combustion, are considered, the maximum indicated fuel

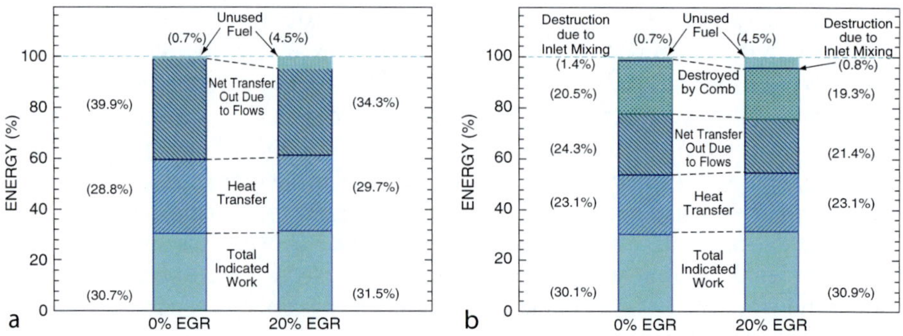


**Fig. 6.13** (**a**) Energy distribution and (**b**) exergy (availability) distribution of various transfer and destruction (in the case of exergy) mechanisms for 0% EGR and 20% EGR in a spark ignition engine modeled using a thermodynamic simulation [54] (Used with permission from Professor J. Caton, Texas A&M University)

conversion efficiency is about 30.7% for non-EGR case (0% EGR) and 31.5% for 20% EGR case. Most of the balance of energy is transferred to the coolant (ca. 29% for non-EGR case) due to nonadiabatic control system and out the exhaust (ca. 40% for non-EGR case) due to underutilized conversion of thermal to work energy. Thus, the importance of considering real effects is clear. It is noted that the data shown in Fig. 6.13 do not include the effect of friction; the brake fuel conversion efficiency for the studied conditions of Fig. 6.13 are available in [54]. It is further noted that friction offers no opportunity for conversion to useful work (i.e., energy lost to friction is all destroyed exergy). Finally, it is noted that EGR seems to have a small to negligible effect on the calculated friction of the simulation [54].

Also clear from Fig. 6.13 is the importance of considering the internal combustion engine from an exergy perspective. If considering only the energy transfers of Fig. 6.13a, it may appear that ca. 70% of the energy is "lost," or unavailable for conversion to useful work. While it is correct to say that a fraction of the fuel's energy is underutilized in the engine control system's (i.e., piston/cylinder chamber) conversion to useful work, it is not correct to say that all this energy is lost. As explained above, exergy provides insight into the fraction of energy that is available to do useful work. For example, Fig. 6.13 illustrates that (for non-EGR cases) 28.8% of the fuel's *energy* is transferred out of the control system via heat transfer, whereas only 23.1% of the fuel's *exergy* is transferred out of the control system via heat transfer. That is, for one unit of fuel, 0.288 unit of energy is transferred via heat transfer, but only ca. 80% (0.231/0.288) of that energy is available to do useful work; thus, 0.231 unit of fuel exergy are a missed opportunity for conversion to useful work. This 0.231 unit of missed opportunity could be exploited by configuring an ideal heat engine between, for example, the cylinder walls and the environment (i.e., representative $T_H$ and $T_L$). Such a conceptualization seems impractical, but exergy is not just passed into the coolant. Exergy is also transferred to the exhaust; ca. 24.3% of the fuel's exergy for the non-EGR case under study in Fig. 6.13b. The high temperature exhaust, with ease of accessibility, offers an opportunity to exploit the fuel's exergy transferred through the exhaust system. There are several practical technologies that offer an opportunity to extract the fuel's exergy transferred out of the engine control system; these are discussed in more detail in the section "Waste Heat Recovery".

One of the important features of Fig. 6.13b is the exergy destroyed due to combustion. Combustion, in any application, is an irreversible phenomenon and thus will render entropy generation (i.e., exergy destruction). Like friction, the ca. 20% of fuel exergy destroyed due to combustion cannot be recovered for conversion to useful work. The major causes for exergy destruction of a typical hydrocarbon-based combustion process include (1) thermal energy exchange among particles within the system, (2) diffusion among fuel/oxidizer particles, and (3) mixing among product species [55]; of these three, the dominant exergy destruction source is the thermal energy exchange among particles. At the onset of combustion, with the system containing reactants, a metastable equilibrium exists as gradients exist between fuel and oxidizer molecules. As combustion proceeds during the process, natural phenomena cause the system to eliminate such gradients and maximize entropy in the pursuit of attaining a more stable equilibrium (i.e., the products state). The maximization of entropy in reducing the gradients is manifested entirely through entropy generation. Using the dominant source of exergy destruction (i.e., entropy generation) during combustion as an example: internal thermal gradients among particles [56] established during the combustion process create microscopic opportunities to do useful work (i.e., thermal gradients can be converted to useful work via a heat engine). Of course, practical implementations of microscopic heat engines do not exist, thus, the thermal gradients are reduced to zero during the combustion process without any conversion to useful work; heat transfer through a finite temperature difference, as described above, generates entropy (destroys exergy).

The general behavior of exergy destruction due to combustion in internal combustion engines is generally well characterized, as summarized by [57]; generally, an increase in combustion temperature decreases exergy destroyed due to combustion. This statement should not be confused with the situation of internal thermal gradients which, as mentioned above, are a major source of combustion-based exergy destruction. While it is tempting to seek ways to minimize (or even eliminate) combustion irreversibilities (and thus minimize or eliminate combustion-based exergy destruction), it should be recognized that such efforts may decrease the conversion of exergy in other ways. Consider, for example, the use of a lean mixture versus the use of a stoichiometric mixture; the former has higher exergy destruction than the later. While a stoichiometric mixture reduces exergy destruction, its mixture composition also causes lower conversion of thermal to work energy during the expansion process (and, coincidentally, increases exergy transfer via heat transfer). Thus, overall efficiency is lower with a stoichiometric mixture. Again, energy- and exergy-based analyses, such as that presented in Fig. 6.13, are necessary to quantitatively make such assessments. In the case of eliminating combustion irreversibilities, a reversible combustion process [58] can be conceptualized [59, 60] where initially separated reactant species are isentropically compressed to their respective partial pressures and a certain temperature; concentrations of individual species, when allowed to interact with each other, will be in equilibrium. After compression, individual species will be collected forming a mixture that is a priori in equilibrium (thus, there are no irreversibilities due to mixing) and allowed to expand isentropically. Because of the increase in moles of the mixture as it expands isentropically maintaining equilibrium along the path, net work is extracted from the collection of processes. Quantitative analysis [60] of such a concept reveals 0% exergy destruction; because of exergy retention in the exhaust products, however, thermal efficiency of the "reversible combustion engine" is around 28% (for a fuel–air equivalence ratio of 1.0, compression pressure and temperature of 10 MPa and 6,000 K, respectively, and an expansion ratio of 18:1). Of course, such concepts are presently impractical; but the notion of reversible combustion is theoretically possible.

In fact, it is briefly noted that, although not explicitly calling his concept one of "reversible combustion," a close inspection of Diesel's original engine design [61] describes a method of nearly attaining reversible combustion (at least, one which eliminates internal thermal gradients). This is described in more detail in section "A Case Study: Diesel Engines Versus Gasoline Engines".

In closing this section on "Thermodynamic Analysis of Internal Combustion Engine", it is noted that assigning a "maximum possible efficiency" of an internal combustion engine is anything but straightforward. *Fuel–Air Cycle Analysis* reveals that the efficiency of an engine cycle depends on several different characteristics of the control system. A more advanced and appropriate analysis of an engine – that which includes real effects of heat transfer, real combustion, friction, and other irreversibilities – gives realistic senses of tangible factors that engineers can strive to improve. Further, exergetic-based analysis offers the important insight of what future opportunities might be exploited in improving the overall system efficiency associated with an internal combustion engine. In returning to the engineer's "dream" engine concept described above, such computed numbers as shown in Fig. 6.13 are more realistic than the ideal *Fuel/Air Cycle Analysis* numbers shown in Fig. 6.10. Further, the exergy-based analysis of Fig. 6.13 could be used to make the necessary assessment of the engineer's dream, to quantify if there are real opportunities of "downstream" exergy conversion when the engine is operated close to maximum theoretical thermal efficiency levels.

## *Spark Ignition Combustion*

As discussed in section on "Thermodynamic Analysis of Internal Combustion Engines", real combustion in an internal combustion engine is not ideal in the sense that it can be modeled as precisely constant volume, constant pressure, or even the combination of the two (limited pressure). Real combustion, instead, involves several complex and interacting features that, on the one hand, make it difficult to predict and model, but on the other hand create opportunities for further development. Conventional engines are either spark ignited or compression ignited. Since the type of ignition results in substantially different combustion features, the two are separated into respective sections. This section describes spark ignition combustion.

Spark ignition combustion is the typical form of combustion found in the commonly called "gasoline" engine (also commonly called either "petrol" engine or "otto" engine); it is more descriptive to refer to such engines as spark ignition engines. The "spark" aspect of the term implies that combustion is initiated by introducing a high voltage electrical arc, typically through the use of a spark plug, into the reactant mixture at a point during the cycle where sustained combustion reaction will proceed. Most spark ignition engines (exceptions are described in section on "Future Directions of Internal Combustion Engines") induct a premixed and homogeneous mixture of fuel and air during the intake process. At a point near the end of compression, the spark plug discharges an arc into the mixture. This arc, in the near-by region of the spark plug, forms a high temperature plasma that evolves into a flame kernel. The flame kernel transitions from the plug region in a laminar sense establishing a flame front with initial velocity close to the laminar flame speed. This initial flame development is referred to as the flame-development period. Because of turbulent motion among particles composing the mixture, the flame then rapidly spreads throughout the mixture in a turbulent fashion, enveloping microscale eddies with flame. This rapid flame propagation is referred to as the turbulent entrainment period. Upon flame envelopment of the turbulent eddies within the mixture, the flame then laminarly burns the microscale eddies in the final stage of the process referred to as rapid burnup. Turbulent entrainment and rapid burn practically occur simultaneously, and together compose what is referred to as rapid-burning period. This described sequence of spark ignition combustion for a typical operating point (i.e., 1,400 rev/min, part-load condition) is shown in Fig. 6.14 [62], and, for the same typical operating point, requires about 10 ms from spark to finish.

One of the important features of Fig. 6.14, along with the shown in-cylinder pressure, is the corresponding mass fraction burned. Using the first law of thermodynamics, it possible to develop an expression that determines the chemical-to-thermal energy conversion rate (commonly called the "heat release rate") of the combustion process. Since the mass of fuel present in the cylinder, along with the fuel's heating value, are known, the mass fraction burned rate of fuel can be calculated from the heat release rate. Figure 6.15 illustrates a typical mass fraction burned curve, as a percent of the total fuel, and defines the above-described periods referred to as flame-development period (shown in Fig. 6.15 as the flame-development angle, $\Delta\theta_d$) and the rapid-burning period (shown in Fig. 6.15 as the rapid-burning angle, $\Delta\theta_b$). These periods are conventionally defined as the angle swept from spark release to 10% mass fraction burned and the angle swept from 10% mass fraction burned to 90% mass fraction burned, respectively.

The mass fraction burned profile of spark ignition combustion can be modified, and the parameters that cause alterations to the mass fraction burned profile are often coupled. Perhaps the most obvious parameter that can affect the mass fraction

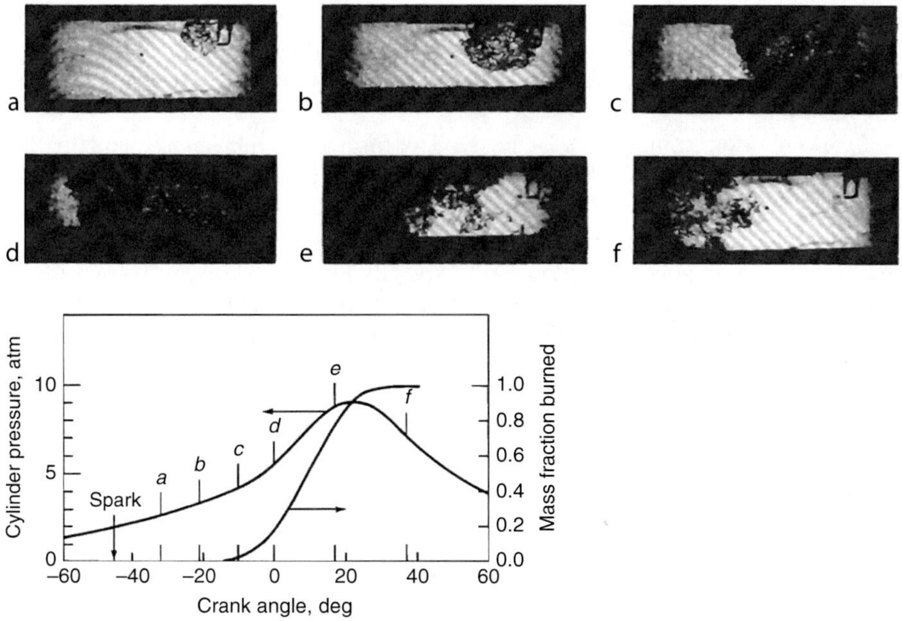


**Fig. 6.14** In-cylinder images capture the sequence of spark ignition combustion with the corresponding result on in-cylinder pressure and mass fraction burned. The various steps of spark ignition combustion shown include the spark release (indicated by "Spark"), flame kernel development (**a–c**), turbulent entrainment and burn up (**d–e**), and flame termination (**f**) (Used with permission from [62])

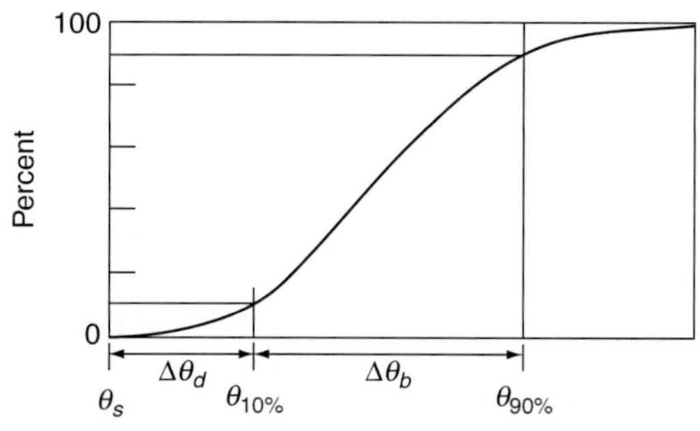


**Fig. 6.15** Mass fraction burn rate (as a percent of total fuel) of spark ignition combustion in an internal combustion engine, illustrating the definitions of the flame-development angle, $\Delta\theta_d$, and the rapid-burning angle, $\Delta\theta_b$ (Used with permission from [63])

burned profile is the spark timing, or sometimes called the spark advance. By adjusting the time in the cycle when combustion is initiated (i.e., the spark advance), the times required for flame development and rapid-burning periods are altered. An advance in spark timing (i.e., spark timing is moved earlier into

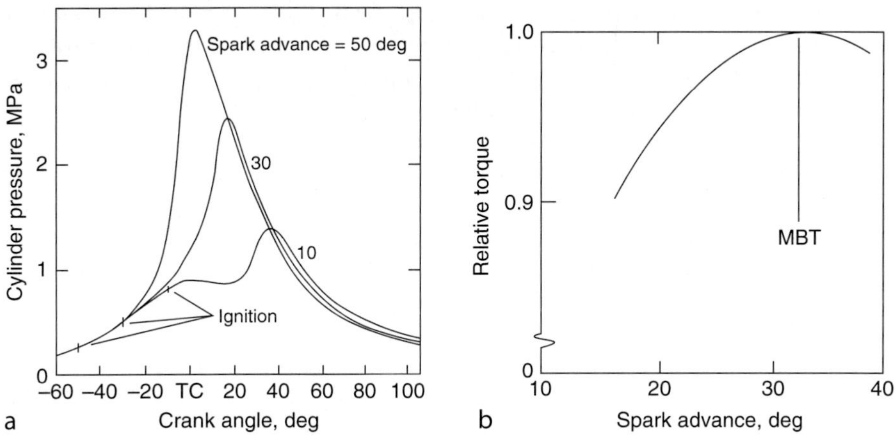


**Fig. 6.16** (**a**) In-cylinder pressure as a function of engine crankangle and (**b**) relative torque to maximum brake torque (MBT) as a function of spark advance of a typical spark ignition engine (Used with permission from [64])

compression stroke) generally results in combustion occurring earlier in the cycle and in-cylinder pressures becoming higher in magnitude. Correspondingly, a retard in spark timing (i.e., spark timing is moved later into the compression stroke, or possibly even into the expansion stroke) generally results in combustion occurring later in the cycle and in-cylinder pressures becoming lower in magnitude. There are several other parameters, however, that can affect the burn profile of spark ignition combustion. Examples of such parameters include the mixture's fuel–air ratio (i.e., stoichiometry), the level of EGR, the level of turbulence in the mixture, the level of heat transfer, initial mixture pressure (i.e., load variation), initial mixture temperature, and engine speed. Analysis of each of these parameters is outside the scope of this article; readers are referred to the "Books and Reviews" for additional reference text on the subject.

One item that is important to mention, however, is the effect of burn profile on engine-scale parameters such as performance, efficiency, and emissions. The rate of combustion, and its corresponding effect on in-cylinder pressure, flame temperature, and mixture gas temperature, alters the performance, efficiency, and emissions of the spark ignition engine. For example, Fig. 6.16 illustrates the effect of spark timing (spark advance) on in-cylinder pressure and relative torque to maximum brake torque (MBT). MBT is the maximum torque attained over a given parametric sweep, most commonly over a spark-timing sweep. Note that there is nearly a single spark timing that yields the maximum torque delivered by the engine. Too early of a timing (advanced) or too late of a timing (retarded) renders a lower torque than MBT. Three spark timings are shown in Fig. 6.16: 50°, 30°, and 10° before TDC (BTDC). The most advanced timing (50° BTDC) renders the earliest burn profile (of the studied timings), the highest magnitude of in-cylinder pressure, and the earliest location of peak pressure (occurring around 0° BTDC). Although it may appear from Fig. 6.16 that the spark advance of 50° BTDC yields the maximum area under the pressure-volume curve (and thus, the maximum work for the cycle), it should be noted that its pressure is lower through most of the expansion stroke than the other two shown spark timings (i.e., 30° and 10° BTDC). This lower pressure through the expansion stroke results from the increased level of heat transfer manifested by the higher gas mixture temperature caused by the faster burn rate-induced higher cylinder pressures. Thus, spark timings advanced of MBT timing result in lower torque due to higher levels of compression work, heat transfer, and friction. Conversely, the most retarded timing (10° BTDC) renders the latest burn profile (of the studied timings), the lowest magnitude of peak pressure, and the latest location of peak pressure (occurring around −40° BTDC). Because of the later-phased combustion, the location of the rise in pressure misses the full opportunity of the expansion stroke. Thus, spark timings retarded of MBT result in lower torque due to expansion losses. It is clear that MBT occurs in the balance of minimized compression work and maximized expansion work, both of which are affected by combustion phasing, heat transfer, and friction. In the example of Fig. 6.16, this balance occurs around 30° BTDC. Notice that peak pressure for this timing is around −20° BTDC; a general "rule of thumb" is that MBT occurs when peak pressure is positioned between −15° and −20° BTDC. As revealed in Fig. 6.16, this rule of thumb extends to the mass fraction burned profile, where maximum brake torque is timed when 50% of the fuel burns by −10° BTDC. In some cases, MBT timing may be "knock limited," meaning that a higher torque could be attained at an earlier timing if fuel knock were not present. Fuel knock is a combustion abnormality of spark ignition engines that, due to the combustion-generated compression of the reactive mixture in the end regions of the cylinder, results from autoignition of the mixture prior to its controlled burn by the propagating flame. Fuel knock can be very damaging because the uncontrolled combustion tends to cause dramatic rises in pressure near critical mechanical components (such as piston rings). The octane rating of a fuel indicates the fuel's resistance to knock in a spark ignition engine; a fuel with a higher octane has a higher resistance to autoignition. Additional discussion about this is provided in the sections. on "A Case Study: Diesel Engines Versus Gasoline Engines" and "Direct Injection, Spark Ignition Engines".

At MBT timing, since fuel flow rate is generally held constant during spark-timing sweeps, efficiency will also be correspondingly maximized. In the case of emissions, however, the trends are not straightforward and a brief discussion of emissions is reserved for the section on "Emissions Formation and Exhaust Pollution".

Finally, it is noted that several parameters affect the spark ignition burn profile. Thus, for each change to a given parameter (e.g., initial pressure, initial temperature, fuel–air equivalence ratio, engine speed, and level of mixture turbulence), a potentially different spark timing will correspond to maximum brake torque. Again, the general trend of spark timing will be such that 50% mass fraction burned occurs at −10° BTDC for maximum brake torque.

## *Compression Ignition Combustion*

In contrast to spark ignition combustion – which uses an electrical arc, or spark, to initiate combustion of a reactive mixture – compression ignition combustion relies on compressive heating – or, the increase in temperature of a gas due to the increase in pressure resulting from the decrease in volume – where combustion initiation is kinetically driven by exceeding the ignition temperature of the fuel–air mixture. Compression ignition combustion is the typical mode of combustion used in the commonly called diesel engine, an engine which more descriptively is called a compression ignition engine.

It is immediately recognized that one challenge of compression ignition engines is the lack of a direct trigger of ignition; the spark acts as the direct trigger of combustion initiation in a spark ignition engine. Conventional applications of compression ignition engines (e.g., the conventional diesel engine) overcome this challenge by using the fuel injection event as the direct trigger. Thus, conventional compression ignition engines typically induct an air and residual mixture (i.e., no premixing of fuel) during the intake stroke. This same unreactive mixture is compressed during the compression stroke until near the point when combustion is desired to begin. At this point fuel is introduced into the mixture, which, due to compression, is a high temperature, high pressure environment. After several complex and coupled processes occur (described below), combustion initiates and chemical to thermal energy conversion takes place.

At this point, it is instructive to briefly describe the role of "glow plugs" often used in compression ignition engines. Glow plugs should not be confused with spark plugs. A glow plug is a resistive heating element inserted into the cylinder of a compression ignition engine to aid in the initial starting of the engine. It acts as a warming device during "cold start"; after engine warm up and stable operation, the glow plugs deactivate. They are not necessary on a cyclic-basis. Spark plugs, on the other hand, are integral components of spark ignition engines; they provide the source of ignition for every combustion cycle of a spark ignition engine.

The method by which fuel is injected into the compressed mixture typically falls into one of two categories: (1) direct injection, or (2) indirect injection. Indirect injection involves injecting fuel into a "prechamber" which is connected to the main chamber. Fuel injected into the prechamber ignites and causes the mixture to issue into the main chamber where main heat release and work extraction occurs. Indirect injection engines are used primarily in applications where motion of the nonreactive mixture in the main chamber is too quiescent for ignition; this typically occurred in the early implementations of small, high speed automotive compression ignition engines. Use of a prechamber, where nonreactive mixture forced through orifices or nozzles connecting the main chamber to the prechamber suitably increased mixture swirl and turbulence, generated sufficient fluid motion to allow ignition to occur. Technological developments in intake port design, combustion chamber design, and fuel injection systems have rendered indirect injection systems nearly obsolete and technically inferior to direct injection techniques.

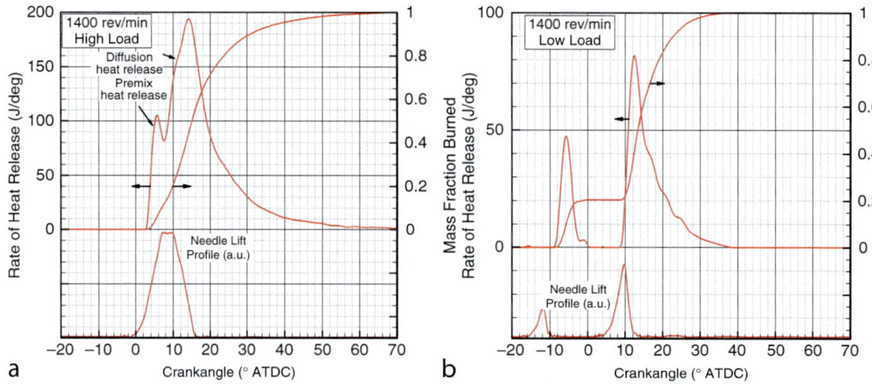


**Fig. 6.17** Fuel injector needle lift profile, rate of heat release, and mass fraction burned for a typical diesel engine operating at 1,400 rev/min (**a**) high-load condition (ca. 75% peak load = 11.3 bar BMEP) and (**b**) low-load condition (ca. 25% peak load = 1.9 bar BMEP) with the use of pilot injection (Data from author's laboratory, Texas A&M University)

Direct injection, as the name implies, indicates that the fuel is injected directly into the cylinder at the point near when combustion is desired to begin. Because of the compressed nature of the nonreactive mixture into which fuel is injected, fuel injection pressures must be very high; i.e., modern fuel injection systems inject fuel at pressures on the order of 100–2,000 bar (ca. 3,000–30,000 psi). The processes of fuel injection, combustion initiation, and subsequent combustion propagation (or burning) are very complex in compression ignition engines. The following discussion will briefly highlight these complexities.

Figure 6.17 illustrates the sequence of events that occur in a typical direct injection compression ignition engine. Two load conditions at 1,400 rev/min are shown, with a high-load condition (ca. 75% peak load which is about 11.3 bar BMEP) shown as Fig. 6.17a and a low-load condition (ca. 25% peak load which is about 1.9 bar BMEP) shown as Fig. 6.17b. Both plots illustrate the fuel injector needle lift profile (which roughly correlates to the fuel delivery rate), the rate of heat release profile, and the mass fraction burned profile. Focusing first on the high-load condition (Fig. 6.17a), notice that fuel injection occurs near 2° BTDC. Fuel in typical compression ignition engines is injected as a liquid. Thus, the fuel must undergo a series of physical processes (e.g., penetration, breakup, atomization, and vaporization) before it undergoes its chemical process of bond fragmentation and eventual ignition. This period of time – i.e., the time between start of fuel injection and ignition (commonly called start of combustion) – is often called the ignition delay period. As will be described below, the ignition delay period of a compression ignition engine is an important parameter that affects the remainder of the burn profile and is affected by several other parameters. The end of the ignition delay period after start of injection will witness start of combustion and noticeable heat release, as indicated in Fig. 6.17a as the positive rate of heat release. During heat release, the mass fraction burned profile steadily increases until end of combustion when all the fuel is most nearly completely burned.

An interesting and important feature of Fig. 6.17 is the rate of heat release profile between start of combustion and end of combustion. Notice in Fig. 6.17a there are two distinct components of the rate of heat release: a premixed component and a diffusion component. These two respective components are nearly an exclusive feature of compression ignition combustion that relies on fuel injection as the ignition trigger. The premixed heat release component results from the complex processes involved in physically preparing the initial "packets" of fuel for chemical decomposition [65]; as described above, the injected liquid fuel must be prepared and sufficiently mixed with the cylinder gases before chemical decomposition can begin. Once ignition occurs, the parcels of fuel that prepared for burning during the ignition delay period react under kinetic control, with relatively high rates of burning for the given mixture temperature. Fuel continues to inject during the premixed portion, and the increase in temperature due to premixed combustion accelerates the physical preparation of the newly introduced parcels of fuel. In spite of this acceleration of fuel preparation, the burn rate becomes limited by the fuel's mixing rate with the air. Since the time scales for mixing are larger than the time scales for chemical kinetic decomposition [66], the burn rate becomes mixing-controlled, rather than kinetically controlled. The mixing-controlled, or diffusion-controlled, component of heat release is observed in Fig. 6.17a as the longer heat release following premixed heat release and is identified as "diffusion heat release." The burn rate of diffusion heat release is generally slower than premixed heat release at any given reaction temperature. Figure 6.17a shows diffusion heat release having a much higher burn rate than premixed heat release; this is manifested, however, by the associated higher mixture temperature effected by the combustion process.

The situation is slightly different for the low-load case, shown as Fig. 6.17b. Conventional compression ignition engines that use fuel injection as the ignition trigger typically also use the amount of fuel delivery as the means to control engine load. Thus, the need to produce less power at a given engine speed is effected by decreasing the amount of fuel delivered during the injection process. This is evident in Fig. 6.17b where the area under the fuel injector needle lift profile is less than that of the high-load case (for the moment, disregard the bimodal fuel injection profile, which will be explained below). Consequently, the distribution of premix versus diffusion heat release is not apparent on the low-load heat release profile. This does not imply that mixing-controlled combustion does not occur during low-load operation; it does imply, however, that combustion is predominantly kinetically controlled at low-load conditions whereas it is predominantly diffusion-controlled at high-load conditions.

The bimodal feature of the fuel injection profile in Fig. 6.17b is the result of the use of a double-injection strategy on this particular engine at this particular condition. Modern advanced engine fuel systems are capable of introducing fuel in sequences, or pulses, in what is commonly called "multi-injection strategies." In the case of the engine highlighted in Fig. 6.17b, the first injection is considered a "pilot" injection, while the second injection is considered a "main" injection. The pilot injection introduces only a small portion (e.g., ca. 10%) of the total fuel to be

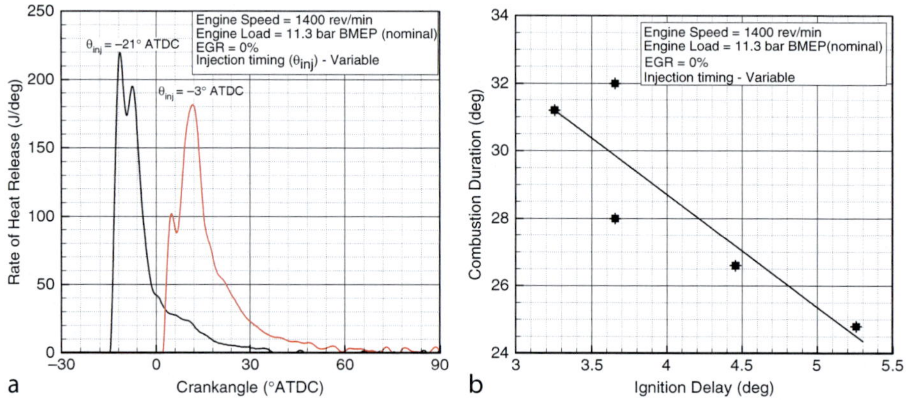


**Fig. 6.18** (**a**) Rate of heat release at an advanced injection timing (21° BTDC) and a retarded injection timing (3° BTDC) and (**b**) the relationship between ignition delay and combustion duration via an injection timing sweep for a typical compression ignition engine operating at 1,400 rev/min, high-load condition (11.3 bar BMEP, nominal) (Data from author's laboratory, Texas A&M University)

delivered in the cycle, creating an opportunity to prepare a smaller portion of fuel for burning during the ignition delay period. This small pilot increases cylinder temperature, so that the ignition delay period of the main injection is much shorter, and there correspondingly is a lower fraction of premixed heat release during the main combustion event. Such a strategy is used, for example, to reduce combustion-generated noise of a diesel engine. Interestingly, diesel engine noise is an "age-old" problem, serving as a primary focus of development for Sir Harry Ricardo (a pioneer in the development of high speed diesel engines) [67].

The above-described use of multiple injections hints at an ever-present relationship within compression ignition combustion systems using fuel injection as the direct trigger; i.e., a relationship among ignition delay, fraction of premix heat release, and fraction of diffusion heat release exists. This relationship is shown in Fig. 6.18a, which illustrates the rates of heat release at two different injection timings of a typical compression ignition engine operating at 1,400 rev/min, high-load condition (11.3 bar BMEP, nominal). Notice for the advanced injection timing there is a substantial level of premix heat release and a correspondingly lower fraction of diffusion heat release relative to the retarded injection timing. The advanced injection timing, due to the injection of fuel into a relatively cool environment when rates of physical and chemical processes are low, results in a relatively longer ignition delay. Because of the longer ignition delay, there is more time for fuel preparation prior to ignition; once ignition occurs, there is a relatively higher concentration of fuel that is prepared for burning and correspondingly combusts in a premix and kinetically controlled fashion. As more fuel is prepared during the ignition delay period for premix burn, there consequently is less fuel available for diffusion heat release; thus, as premix burn fraction increases, diffusion burn fraction decreases. Likewise, at the retarded injection timing, fuel is injected in a relatively hotter environment which enables fuel preparation for burning at a faster rate and combustion commences relatively sooner (i.e., shorter ignition delay). Because of the shorter ignition delay, there is less fuel prepared for burning, thus there is lower fraction of premix heat release and correspondingly larger fraction of diffusion heat release.

Because premix heat release rate is relatively faster (at a given temperature) than diffusion heat release, the "combustion duration" – or, the time taken between start of combustion and end of combustion – for a mostly premix heat release combustion event will be shorter than that of a mostly diffusion heat release combustion event. Thus, there tends to be an inverse relationship between ignition delay and combustion duration, as shown in Fig. 6.18b, where changes to ignition delay are manifested by alterations to injection timing. The shorter ignition delay results in less premix heat release, more diffusion heat release, and correspondingly a longer combustion duration.

Injection timing is just one parameter that can affect ignition delay, and thus the burn profile of a compression ignition combustion event when fuel injection is used as the ignition trigger. Other engine parameters such as EGR level [68], initial temperature and initial pressure [69], injection pressure [70–72], and swirl and turbulence [69, 73] all have certain effects on ignition delay and the resulting relative fractions of premix and diffusion heat release. Like the effect of various parameters of spark ignition engine performance, efficiency, and emissions, the various effects of compression ignition engine parameters on ignition delay and burn profile also have an effect on engine efficiency, performance, and emissions. For example, overly advanced or retarded injection timings will cause decreases in engine torque for a given fuel delivery rate. Additionally, fuel quality is particularly important for conventional diesel combustion operation, where low volatility fuels with high ignitability are generally used to ensure ignition occurs during the cycle. An important fuel parameter – i.e., its cetane number – is used to identify the quality of fuels appropriate for use in a diesel combustion system. A higher cetane corresponds to a fuel with a shorter ignition delay.

Finally, in closing, it is important to recognize the above-described phenomena are largely phenomenological observations that can be made using conventional diagnostics with relatively straightforward analysis. Considerable development (e.g., [49, 74, 75]) has been made to provide substantial insight into the complex fluid, heat transfer, and chemical processes that occur during compression ignition combustion where fuel injection is used as the direct ignition trigger. Description of these details is outside the scope of this work.

## *Emissions Formation and Exhaust Pollution*

A discussion on the basics of internal combustion engines, actually any combustion-based device, is incomplete without a description of the associated harmful species that may exist in the products of combustion. Such harmful species are generally called "exhaust pollution" and many governing agencies around the world place restrictions on the emission of certain pollutants from combustion-based devices. Because the application of the internal combustion engine is so varied, emission regulations tend to be application-oriented. For example, in the USA, emission regulations are placed on passenger vehicles differently than emission regulations placed on heavy trucks or hand-held engine devices (such as lawn mowers). Because of such variability in regulation, the time-oriented nature of the regulations (i.e., regulations have, to this point, been in a state of flux), and variation in regulation among various governing agencies around the world, further description of specific emissions will not be provided. The basic issues at hand, however, can be briefly described.

As shown in Reaction (6.R1), there are several species formed during the combustion reaction of a typical hydrocarbon–air mixture. Some of these species are generally stable and nonreactive in the atmosphere, thus pose little to no harm to the five kingdoms of nature. Other species, however, are either harmful or reactive in ways that lead to harmful consequences. Specifically, there is little scientific debate about the harmful nature of certain combustion products such as CO, NO/ $NO_2$, unburned hydrocarbons (HC), and particulate matter (PM, or the solid/liquid components of exhaust that can be collected on a filter. It is noted that historically sulfur oxides and sulfates have been considered either separately from [76] or in combination with [77, 78] particulate matter). There is, perhaps, continued debate about the potential consequences of other products of combustion; in particular, there is current debate on the consequence of $CO_2$ and the role it may play in the presently observed warming of the planet (i.e., so-called global warming or global climate change). Because of the certainty of the effects of the former species, attention will be given to them and basic information on their formation during combustion. Because of the uncertainty of the effects of the latter species, readers are referred to other literature to uncover the current state-of-debate of $CO_2$ and its potential impact on global trends currently believed to occur (see, e.g. [79, 80]).

The first such species to describe is CO, which due to its fatal effects on human/ animal life is one of the first combustion products to be considered a pollutant [76]. It is well established [76] that all precursor hydrocarbon decomposition reactions firstly form CO. Thus, increased concentrations of CO in engine exhaust result from incomplete reaction of the principal CO oxidation step [26], given as Reaction (6.R2):

$$CO + OH \leftrightarrow CO_2 + H \tag{6.R2}$$

In internal combustion engines, the incomplete oxidation of CO most typically occurs during rich engine operation [76], where available oxidants for final CO oxidation are lacking [81]. Fuel–air mixtures close to stoichiometric or even slightly lean, however, result in nonnegligible concentrations of CO.

The next species to consider is NO. NO emerged as combustion-generated pollutant due to its observed effect of reacting with hydrocarbons in the presence of sunlight to produce tropospheric ozone [82]. Its formation in a combustion system is rather complex, as there are several major pathways through which it can form. These major pathways include thermal (or commonly called Zeldovich), prompt, and fuel-based nitrogen [26]. For reciprocating-type internal combustion engines, the primary NO formation mechanism is the thermal mechanism where atmospheric air serves as the principal source of nitrogen in the mechanism [27]. There are three reactions that compose the mechanism, given as Reactions (6.R3)–(6.R5):

$$O + N_2 \leftrightarrow NO + N \tag{6.R3}$$

$$N + O_2 \leftrightarrow NO + O \tag{6.R4}$$

$$N + OH \leftrightarrow NO + H \tag{6.R5}$$

The forward and reverse reaction rate constants of Reactions (6.R3) and (6.R4) are generally exponentially dependent on temperature. To demonstrate the substantial role temperature plays on NO formation, several simplifying assumptions (see below) are applied to Reactions (6.R3) and (6.R4) to yield Eq. 6.15 [26]:

$$\frac{\mathrm{d}[NO]}{\mathrm{d}t} = \frac{6 \times 10^{16}}{T^{\frac{1}{2}}} \exp\left(\frac{-69090}{T}\right) \left[O_{2,eq}\right]^{\frac{1}{2}} \left[N_{2,eq}\right] \cdot \left(\mathrm{mol/cm^3\text{-}s}\right) \tag{6.15}$$

where [NO] is the concentration (mol/cm$^3$) of NO at time, $t$ (s), and $[O_{2,eq}]$ and $[N_{2,eq}]$ represent the equilibrium concentrations (mol/cm$^3$) at temperature $T$ (K) of oxygen and nitrogen, respectively. The several simplifying assumptions that go into Eq. 6.15 include the following. The first assumption is that the nitrogen chemistry is de-coupled from the combustion reactions. Although combustion reactions generally occur much faster than nitrogen chemistry [25], the presence of O and OH radicals in the thermal mechanism (which are also important species in combustion reactions) may require the chemistries to be coupled for accurate NO prediction [27]. By assuming the chemistries are de-coupled, O, $O_2$, OH, H, and $N_2$ can be approximated by their equilibrium concentrations at equilibrium temperature; assuming equilibrium temperature is the second assumption applied to Eq. 6.15. The third and last assumption applied to Eq. 6.15 is that nitrogen radical (N) is in steady-state concentration (i.e., $\frac{\mathrm{d}[N]}{\mathrm{d}t} = 0$). Finally, it is reinforced that the forward reaction rate constant of Reaction (6.R3) is used from [26] in Eq. 6.15; updated reaction rates are available in Dean and Bozzelli [29]. It is clear from Eq. 6.15 the strong dependency NO formation rate has on temperature.

Because of the dominance of the thermal mechanism on NO formation in internal combustion engines, combustion-based efforts to reduce NO center on reducing the reaction temperature and $O_2$ concentration. Such techniques include altering spark advance [83, 84] or injection timing [85–87] for spark ignition or compression ignition engines, respectively, and introducing EGR into the mixture [87–90].

As described above with NO, unburned hydrocarbons (HC) play a role in the formation of tropospheric ozone. Further, they represent unreacted fuel; if left to discharge into the atmosphere, there is lost opportunity to convert that chemical energy into useful work. Formation of HC species during combustion reaction are complex and varied depending on the type of fuel used [91]. A general cause for HC emissions, however, is insufficient mixing between fuel and air [76], where most of the HC emission species are formed during low temperature ($T < 1{,}000$ K) reactions [26]. Typically, in conventional reciprocating internal combustion engines, most of these species are oxidized as combustion enters high temperature mechanisms [26]. Reciprocating engines, however, contain "sources" for hydrocarbon storage that, if not oxidized upon release in the gas mixture in the later portion of the cycle, emit as HC emissions in the exhaust [92]. Although there are special considerations given to engines operating under cold-start conditions [93], the general storage locations of HC species (and thus, the major source of HC emissions [94]) include cylinder head gasket crevice, spark plug crevices, piston ring pack crevices, and valve seat crevices. In addition to crevice HC storage, other sources of HC emissions [94] include single-wall flame quenching, oil film layers, combustion-chamber deposits, exhaust valve leakage, and liquid fuel (i.e., HC species not vaporized during the process). Because of the relative importance of crevice storage on HC emissions, much of the combustion-based effort to decrease HC emissions has centered on reducing the volume and flow pattern of crevices in the piston/cylinder arrangement.

The final major pollutant to briefly describe is particulate matter, or PM. PM is essentially any exhaust specie that can be collected on a filter; it typically is structured on a solid organic component (which is mostly pyrolized carbon particles, or "soot") upon which organic (e.g., unburned HC) and nonorganic (sulfates) components build. Since soot serves as the building block for PM, much of the research efforts are dedicated to understanding soot formation processes [95–99]. The general path for soot formation begins with the high temperature pyrolysis, or fragmentation to hydrocarbon radical, of the hydrocarbon fuel. Such pyrolysis typically leads to a certain group of hydrocarbon radicals called polyaromatic hydrocarbons. Such polyaromatic hydrocarbons that do not oxidize during the reaction serve as the nucleation site for soot growth. Nuclei then begin to coalesce and substantially increase surface area. Now particles, the high surface area soot particles agglomerate and allow other species (i.e., liquid HC, condensed gaseous HC, and sulfates) to absorb onto the surfaces [99]. Soot formation is strongly dependent on mixture fuel–air ratio and also temperature (both tend to govern the rate of pyrolysis). Formed soot can, however, undergo oxidation as well. Soot oxidation is likewise a function of temperature, but has a stronger dependency on temperature than soot formation [99]. The difference between soot that is formed and soot that oxidizes is ultimately released in the engine exhaust, and is commonly called "net soot release." Compression ignition engines (e.g., diesel engines) are often plagued with high PM emissions, largely due to the heterogeneity induced into the fuel/air mixture by use of fuel injection as a direct trigger for ignition. Spark ignition engines, however, are recently being considered as sources of nanoscale soot particles [100, 101].

It is clear from the above discussion that soot and PM emission processes are complex. From phenomenological considerations, however, Khan et al. [86, 102] and Ahmad [103] provide insight into the general behavior of PM with, in particular, diesel engine combustion. Specifically, it is observed that increases in diffusion heat release generally increase the emissions of PM. A portion of this may be due to the overall lower reaction temperatures, manifested by relatively lower premix heat release, decreasing the rate of soot oxidation through the diffusion flame. This observation is also supported, however, by Dec's [74] observation that precursor soot formation occurs in the standing premix reaction zone within a diffusion flame sheet that exists in diesel combustion. Thus, increased diffusion burn correspondingly results in increased soot emission. Either way, it becomes apparent when relating this discussion to the known effects of injection timing, for example, on diffusion combustion and NO emissions that an attempt to decrease soot by creating higher temperatures with, perhaps, more premix heat release will correspondingly increase NO emissions; hence, the establishment of the conventional "soot-$NO_x$" tradeoff of diesel engines.

As described briefly above with each species, there are in-cylinder and combustion-based methods to reduce the formation of various exhaust pollutants. Advanced development of combustion systems continues to focus on these in-cylinder methods, as described in section on "Future Directions of Internal Combustion Engines". Also in use to eliminate exhaust pollution are exhaust after treatment systems, which conventionally are catalyzed devices (hence, their common name of "catalyst" or "catalytic converter") [104–106]. The basic idea of a catalyst is to promote a reaction that otherwise would not proceed. In aftertreatment of engine exhaust, there are several catalysts in use depending on the species composition of the exhaust. For example, the conventionally named "three-way catalyst" is often used with conventional gasoline spark ignition engines. Such a catalyst is typically an exhaust flow-through device composed of a ceramic monolith (the substrate) with thru-hole passages to allow exhaust gases to flow and a metal-oxide "washcoat" that suspends catalytic particles on the surfaces of the monolithic passages. The monoliths are usually constructed of coerderite and the metal-oxide washcoat is often an alumina washcoat. Catalyzed particles are often from the precious metals group (e.g., platinum, palladium, and rhodium), with platinum being the most commonly used metal. The general operation of the three-way catalyst is to promote the reduction–oxidation reactions among CO and HC (the reductants) and $NO/NO_2$ (the oxidants). In other words, the catalyst promotes the reduction of NO or $NO_2$ to yield stabilized $N_2$ and the oxidation of CO and HC to yield stabilized $CO_2$ and $H_2O$. The efficiency of the three-way catalyst – or, the conversion effectiveness of converting a given species to its more stable species, e.g., CO to $CO_2$ – is strongly dependent on the constituent composition on the inlet mixtures, as shown in Fig. 6.19 [104]. Notice that maximum conversion efficiencies of $NO_x$, CO, and HC occur very near stoichiometric air–fuel ratios; a small departure from stoichiometric conditions – e.g., about 0.5% increase or decrease in $A/F$ ratio – results in a nearly 20% decrease in conversion efficiency. Because of such an

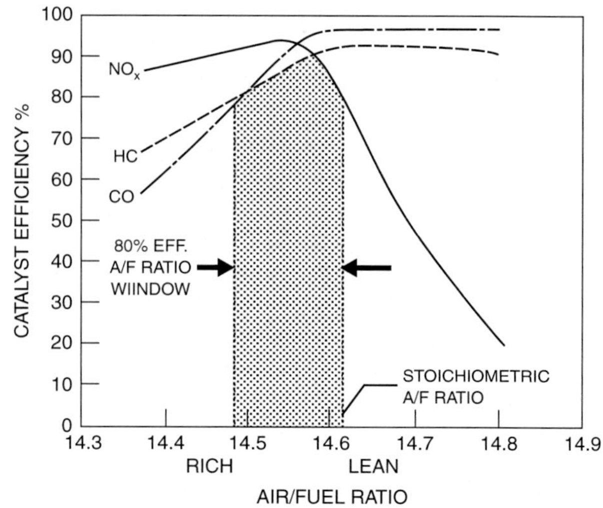


**Fig. 6.19** Catalyst efficiency of a typical three-way catalyst interacting with exhaust from a typical gasoline spark ignition engine as a function of mixture air–fuel ratio (A/F ratio) (Used with permission from [104])

intolerable response of the catalyst, effective control of gasoline spark ignition engines requires precise control of the mixture stoichiometry during operation.

Also clear from Fig. 6.19 is the challenge of outfitting a typical platinum-based catalyst with non-gasoline spark ignition engine technology. Diesel engines which typically operate fuel lean (i.e., oxygen rich), for example, create constituent exhaust species that are difficult to catalyze using conventional techniques. Advanced aftertreatment technologies for diesel engines, or other advanced combustion/engine systems, are under development and beginning to appear in production applications [105, 106].

## A Case Study: Diesel Engines Versus Gasoline Engines

At this point, with the basics of internal combustion engines having been described, it is useful to do a "state-of-the-art" technology comparison between the two dominant conventional internal combustion engines – i.e., the gasoline spark ignition engine and the diesel compression ignition engine – to set the stage for discussions on future directions in internal combustion engines. The elementary comparison is provided in Table 6.2; note that this comparison is neither comprehensive nor general. It is intended to highlight the common state-of-the-art of the two technologies, and create a sense for the trends of future directions in internal combustion engine development.

It is clear from Table 6.2 that the two engine types differ in virtually every way, save for their common use of the kinematic elements of the crank-slider and piston/cylinder components. Consider first the spark ignition engine; its use of spark requires a relatively tight control on the fuel–air mixture equivalence ratio. The

**Table 6.2** Elementary comparison between conventional spark ignition engines and conventional compression ignition engines. The comparison is not comprehensive nor general, rather it is intended to highlight the common state-of-the-art and create a sense for the trends of future directions in internal combustion engine development

| Feature | Conventional spark ignition engine (homogeneous charge spark ignition) | Conventional compression ignition engine (heterogeneous charge compression ignition) |
|---|---|---|
| Also known as | Gasoline, petrol, Otto engine | Diesel engine |
| Method of ignition | Spark | Compression |
| Fuel equivalence ratio | Precisely controlled to stoichiometric | Varies depending on engine load; typically remains lean |
| Fuel/air mixture preparation | Carburetion, throttle-body injection, port injection | Direct injection or indirect injection |
| Degree of fuel/air mixing | Homogeneous | Heterogeneous |
| Fuels used | Gasoline, alcohols, ethanol, hydrogen, high volatility hydrocarbons | Diesel, oils (transesterified), hydrogen, low volatility hydrocarbons |
| Compression ratios | Ca. 8–11 | Ca. 14–21 |
| Method of load control | Throttle restriction of air or fuel/air mixture (depending on mixture preparation) | Fuel injection duration (i.e., fuel quantity) |

flame travel time is minimized when fuel–air equivalence ratio is near stoichiometric, and greater than about 20–40% departure from stoichiometric will typically cause misfire [107]. It is noted that this requirement [107] is only for ignition and flame propagation. The integration of a three-way catalyst with a conventional spark ignition engine requires even more precise control over fuel equivalence ratio, as described in section on “Emissions Formation and Exhaust Pollution” [104]. Thus, conventional spark ignition engines are not able to tolerate substantial departures from stoichiometric conditions. Further, because of this requirement, it is necessary to avoid mixture striations between fuel and air. Thus, conventional spark ignition engines make use of premixing devices (e.g., carburetors or throttle body and port fuel injectors) to prepare the fuel–air mixture prior to induction. The result is a homogeneous mixture of fuel, air, and other residuals such as residual fraction and purposefully introduced constituents such as EGR. In order to assist with the homogenization of the fuel–air mixture, light distillates and aromatics with high volatility are typically used as the fuel for conventional spark ignition engines. The most common, of course, is the mixture of light distillates and aromatics commonly called “gasoline” or “petrol.” Interestingly, largely due to the reactive nature of the homogeneous and near-stoichiometric mixture [108], compression ratios of spark ignition engines are limited to around 11. Compression ratios higher than this promote the undesirable autoignition of certain regions of the fuel–air mixture during the combustion process; a phenomenon known as “fuel knock.” Fuel knock can be very damaging to an engine, mostly due to its occurrence in the end regions of cylinder, where uncontrolled autoignition of the mixture results in excessively high rises in pressure near susceptible components of the piston/cylinder arrangement (e.g., piston rings). As an aside, initially tetra-ethyl lead and eventually other lead alkyls were used in gasoline as anti-knock agents, allowing for compression ratios to increase in spark ignition engines [109]. Concerns over the effect of lead on human and animal life, as well as the implications of lead on catalytic devices, have mostly eliminated lead-based additives from fuels [110]. Finally, and similarly related to the spark's need to ignite a near-stoichiometric and homogeneous mixture, conventional spark ignition engines control their load, or power output, via the use of a throttle. In other words, the power of the engine at any given speed is minimized by decreasing the efficiency by which the engine can induct fresh mixture.

Consider now the conventional compression ignition engine, or commonly called diesel engine. It seems that from its first inception, the diesel engine was to be a compressively ignited machine, as Diesel intended to create an isothermal reaction where air and fuel were (separately) compressed to the combustion temperature and expansion occurred isothermally as chemical energy converted directly into mechanical energy [61]. Of course, Diesel never succeeded in attaining this isothermal process (nor has anyone since); but, the foundation had been laid for a compression ignition engine. By using compression ignition combustion, many of the constraints placed on the spark ignition engine (i.e., near-stoichiometric fuel equivalence ratio, homogeneous mixture, and throttle of intake mixture) no longer need apply to the diesel engine. It also seems that from its initial inception, Diesel intended the fuel to be directly delivered to the compressed air [61]; of course, it is also clear that direct (or indirect) injection of the fuel to compressed air is necessary so as to control start of combustion. Further, to aid in the gradual and isothermal conversion from chemical to mechanical energy, Diesel stipulated that there be a chemical abundance of air (i.e., fuel lean) in the mixture [61]; practical implementations of diesel engines make use of varying fuel-equivalence ratios to control the load of the engine. The direct (or indirect) injection of fuel into the compressed air, contrary to Diesel's conception, establishes a heterogeneous mixture that within itself has widely varying fuel–air ratios and reaction temperatures. Further, since ignition is manifested through compression, fuels with high ignitability (e.g., heavy distillates and aromatics) are required (correspondingly, the fuels also have relatively lower volatility, further establishing the heterogeneous nature of diesel combustion). Also related to the use of compression to ignite the mixture is the need to have high compression ratios (i.e., typically varying between 14 and 21). Finally, as already described, engine load is controlled by the direct and exclusive control of fuel; thus, engine load varies as mixture fuel–air equivalence ratio varies. Due to the heterogeneous nature of diesel combustion, and the dependency of soot formation on mixture stoichiometry under conventional conditions, fuel equivalence ratios rarely exceed 90% of stoichiometric to avoid "smoke limitation."

In the context of the three common attributes of internal combustion engines (i.e., performance, efficiency, and emissions), and the comparison given in Table 6.2, brief comments will be made about the pros and cons of each conventional technology.

**Table 6.3** Comparison of the various parameters that control an engine's ability to make power (Eq. 6.7) between typical applications of conventional spark ignition and conventional compression ignition engines at wide-open throttle or full load condition. It should be noted that these are qualitative assessments of typical technology, and should not be viewed as absolute truths of the respective technologies. Since typical applications of internal combustion engines operate on the four-stroke principle, its effect on the comparison is neutral

| Parameter | Parameter's effect on power[a] | Conventional spark ignition engine | Conventional compression ignition engine |
|---|---|---|---|
| $\eta_f$ | ↑ | | ↑ |
| $\eta_v$ | ↑ | | ↑ |
| $\rho_{a,i}$ | ↑ | | ↑ |
| $(F/A)$ | ↑ | ↑ | |
| $Q_{HV}$ | ↑ | | ↑ |
| $V_d$ | ↑ | | ↑ |
| $N$ | ↑ | ↑ | |
| $n_R$ | ↓ | ↔ | |

[a]In other words, an increase in the parameter will have the listed effect on power

In terms of power, consider Eq. 6.7 and the qualitative relation of each parameter for a given technology at their "wide-open throttle" or full-load operation, as given in Table 6.3. Described in more detail below, diesel engines tend to have higher fuel conversion efficiencies. Because of not premixing the fuel with air and generally lower engine speeds (which allows the avoidance of flow choke), diesel engines tend to have higher volumetric efficiencies (refer to the discussion surrounding Fig. 6.6 and [111] for more detail on factors affecting volumetric efficiency of engines). Typical applications of diesel engines use turbocharging, which increase the inlet mixture density to above atmospheric conditions (typical spark ignition engines operate naturally aspirated). As described above, spark ignition engines generally operate stoichiometric, whereas compression ignition engines generally operate lean even at full power. The heating value of diesel fuel is marginally higher than that of gasoline. Displaced volumes of typical compression ignition engines tend to be larger than those of typical spark ignition engines. Correspondingly, peak power speeds for typical spark ignition engines tend to be higher than compression ignition engines. Mostly due to the larger displaced volumes, but also assisted by higher fuel conversion and volumetric efficiencies, higher inlet density, and higher heating value of the fuel, typical compression ignition engines tend to exhibit higher peak powers than typical spark ignition engines. If the power is normalized by displaced volume to render the specific power, however, spark ignition engines tend to have higher specific power than compression ignition engines (in spite of compression ignition engines having high efficiencies, density, and heating value, the stoichiometric and high speed operation of the spark ignition engine tends to yield higher specific power).

As described in section on "Thermodynamic Analysis of Internal Combustion Engines", engines operating with high compression ratios and lean fuel equivalence ratios will tend to have higher efficiencies than engines operating with relatively lower compression ratios and near-stoichiometric equivalence ratios. As a result, typical compression ignition engines tend to have higher efficiencies than typical spark ignition engines. In some instances, turbocharging that is typically found on compression ignition engines could increase efficiency if intake manifold pressure is boosted to higher than exhaust manifold pressure; the primary function of a turbocharger, however, is to increase inlet mixture density to increase the power capabilities of the engine (as shown in Eq. 6.7). The efficiency improvement of compression ignition engines at part-load conditions becomes amplified as (a) the mixture becomes leaner for the compression ignition engine and (b) the use of throttle to manifest part load in the spark ignition engine introduces a thermodynamic loss parameter in the cycle.

Finally, a brief comparison of emissions between the two engines at wide-open throttle or full load condition is made. Because of their near-stoichiometric operation, engine-out emissions of HC and CO for spark ignition engines tend to be relatively higher for those of compression ignition engines (where the latter uses fuel-lean mixtures, creating an oxygen rich exhaust and high level of oxidation of partially oxidized species such as CO and HC). Engine-out emissions of NO tend to be nearly the same between engine technologies. Engine-out emissions of PM are much higher for compression ignition engines, where mixture heterogeneity creates numerous opportunities for soot formation. Of course, it is important to note that conventional spark ignition engines are typically coupled with an effective catalyst, substantially lowering the catalyst-out emissions of the various species to levels well below the engine-out emissions of compression ignition engines. While the use of aftertreatment systems with conventional compression ignition engines are less straightforward, technology is becoming available to also allow substantial reduction of their engine-out emissions [106].

At this point, it becomes clear both engine technologies have features which are more favorable than the other for power, efficiency, and emissions considerations. For example, the high compression ratio and lean mixture required by compression ignition are attractive from efficiency perspectives. The homogeneous mixture of spark ignition is attractive from uniform combustion and emissions perspectives. The lack of a throttle to control load of a compression ignition engine is attractive, but the direct ignition trigger of a spark ignition engine is also attractive. Thus, it is clear that future engine developments could exploit the favorable features of engine technologies to create the next generation internal combustion engine. The next section will describe such efforts and offer insight into the likely future direction of internal combustion engines.

## Future Directions of Internal Combustion Engines

There are several types of advanced technology that exist for internal combustion engines. Some of this technology is very prevalent on engines (e.g., turbocharging on diesel engines). Some technology is beginning to appear on production models (e.g., variable valve timing). Other technology is still in its development stages, awaiting its potential entry into full-scale production (e.g., homogeneous charge compression ignition combustion). This section will briefly describe such technology.

## *Engine Downsizing*

A general idea that permeates much of the technology under development of internal combustion engines is that of engine downsizing – or, the effort to use advanced technology to enable the use of smaller-sized engines to produce the same power (i.e., increase power density) [112]. The application-oriented benefit is that a smaller engine likely weighs less, thus could improve application efficiency (e.g., better vehicle fuel economy with a lighter engine). The engine itself, however, will likely realize improved efficiency. For example, a smaller engine that uses turbocharging to maintain same power (as a larger-sized engine) will generally operate more often near the location of peak efficiency. Some benefit of such is realized in diesel engines, which typically have best efficiencies near mid-speed, and 75% peak load conditions. More benefit, however, is realized in gasoline engines where throttles are used; a smaller engine with higher power density will require less throttle and thus realize larger gains in efficiency improvement. Overall, the major purpose of engine downsizing is to improve parameters other than $V_d$ in Eq. 6.7. This idea will become more apparent as specific technology is discussed below.

## *Turbocharging/Supercharging/Boosting*

Boosting an engine – i.e., increasing the inlet mixture density to increase the trapped mass per cycle – is a very common means to increase the power density of an engine. As evident from Eq. 6.7, an increase in the inlet air density will directly increase the power of the engine. There are two major types of boosting technology: turbocharging and supercharging. The major difference between the two technologies is the former uses a centrifugal device (i.e., a turbine) to exploit available exhaust energy for conversion to shaft work whereas the latter absorbs shaft work via a direct mechanical connection to the engine. In both cases, the shaft work of the device is coupled to a boosting component – typically, either a centrifugal-based compressor or a positive-displacement compressor – which acts as an "air pump" to increase density of the inlet air. In most applications, a turbocharger uses a centrifugal turbine/compressor configuration while a supercharger uses a positive-displacement compressor.

Diesel engines are typically favorable engines to outfit with boosting devices (further, usually with turbochargers as they assist over the entire engine operating map). Boosting compensates for the diesel engine's typical use of fuel-lean mixtures to improve its power density. Further, since diesel engines do not employ throttles for load control, boosting of a diesel engine provides benefit over the entire operating map. For this latter reason, and due to the general higher efficiency of a turbocharger over a supercharger, turbochargers are the most common boosting device on a diesel engine. Gasoline engines also can be outfitted with boosting devices, with additional complexity to consider. First, boosting devices on gasoline engines usually offer benefit only at wide-open throttle conditions; a point of operation for most applications of engines that is rarely used. Also, for this reason, directly coupled superchargers are often used where improved response to boost is provided. Second, increasing density of the inlet mixture tends to increase the propensity to knock. Thus, while boosting provides additional trapped mass to deliver increased power, a potential retard in timing to avoid fuel knock likely decreases fuel conversion efficiency and introduces a tradeoff in how much additional power can be expected from the boost. Third, because of the typical use of superchargers when boosting is applied to gasoline engines, the system efficiency tends to decrease as shaft work is transferred to provide the boosting action (whereas turbocharging uses available energy of the exhaust).

Often times, boosting devices – in particular, turbochargers – are viewed as devices to increase the overall system efficiency of the engine. In other words, it is thought that because a turbocharger uses exhaust energy that would otherwise be wasted, its conversion to useful work (i.e., boosting) should increase efficiency. This work transfer, however, usually does not leave the control system (i.e., the shaft work of the turbine is directly coupled to the pumping action of the compressor). In some instances, efficiency improvements can be realized if a "negative pumping loop" is created by boosting the intake manifold to a higher pressure than the exhaust manifold. It is also possible to improve the overall engine efficiency if boosting an engine enables the use of a smaller-sized engine for a given application; an effort, as described above, known as downsizing. In most instances, however, overall system efficiency may decrease even with the use of a turbocharger, as the major objective of increasing inlet density to increase power density requires additional energy transfer through the exhaust than out as shaft work. This latter aspect is often realized as an increase in exhaust manifold pressure to "drive" the turbocharger.

An area of technology development for turbocharging is the use of waste-gated and variable geometry turbochargers. In non-waste-gated or non-variable geometry turbochargers, the turbine has fixed geometry and thus fixed flow characteristics. As such, a fixed geometry turbocharger is restrictively designed to provide maximum benefit to the engine at a narrow operating range which usually centers on the peak power condition of the engine. As such, at low speed or low-load conditions, the turbocharger's boosting benefits are diminished. While decreasing the turbocharger's maximum benefit over a broader regime of the engine's operating map, such a constraint also typically creates a dynamic issue during engine accelerations known as "turbo lag." The large turbine designed for maximum engine flow rates requires substantial inertia to rotationally accelerate to maximum boosting benefit. Waste-gated and variable geometry turbochargers offer opportunities to overcome such issues. In the case of a waste-gated turbocharger, usually a smaller turbine with less inertia is used with an exhaust "waste-gate." The smaller turbine provides faster response and better boosting at low-loads and speeds; upon approach of the engine's peak power condition (where either boosting becomes excessive for the intake system or turbocharger speeds exceed maximum limits) a waste-gate opens that allows exhaust energy to bypass the turbine. Thus, the turbocharger is supplied by a fraction of the available exhaust energy providing suitable boost at allowable rotational speeds of the turbocharger at the engine's peak power condition. A variable geometry turbocharger uses a similar concept, except that it is designed to change the flow momentum of the exhaust gases and create multiple pressure ratios for a given exhaust flow rate [113, 114].

Finally, it is noted that outfitting a boosting device to an engine is not a trivial task [115]. There is not an exclusive match between an engine and a given boosting device; instead, the match of a boosting device to its engine application is dictated by the objectives of adding the device, whether it is to exclusively increase power density, efficiency, and/or emissions of the engine system [116–118].

## *Advanced Engine Controls*

Much of the advanced technology that appears on modern engines, or will appear on future engines, is enabled by advanced engine controls. Engine control has always been an integral component of the engine's success at delivering cost-effective and efficient power. Early engine control systems were purely mechanical and only concerned with controlling the level of power (e.g., throttle or rack position) or holding a constant speed with variable load (such as a generator system, using a speed governor). Modern-day engine control systems, however, are virtually all electrical-based, and at the very least sense several aspects of the engine's operation (e.g., cam position, throttle position, manifold temperatures and pressures, and air flow) and typically control most aspects of the engine (e.g., spark timing, injection timing, injection pressure, EGR level, and boost pressure) [119].

Like the engine itself, engine controls are becoming more sophisticated and advanced. Specifically, a general trend to use in-cylinder information as a feedback signal is an example of the type of complexity future engine control systems intend to resolve. Knowledge of in-cylinder pressure, for example, can provide immediate information to the engine controller about load produced by the engine. Additionally, in-cylinder pressure is the major property necessary for assessing the rate of energy release during the combustion process; having such information could allow engine control systems to change parameters based on a desired burn profile in the cylinder [120–122]. Further, along with the continuing advancements in engine model development (e.g., [46–50]), a trend toward model-based engine control [123, 124] intends to decrease engine development time and improve control over the several parameters now present on internal combustion engines.

## Variable Geometry Engine Designs

Along with efforts to effect engine downsizing, and made possible with advancements in engine control systems, is the notion of variable geometry engine designs. In other words, conventional reciprocating internal combustion engines have mechanically fixed geometries, i.e., constant compression ratios and constant displaced volumes. A variable compression ratio is attractive, for example, since it might enable high compression ratio operation for a spark ignition engine at part-load conditions where there is decreased propensity for knock. Similarly, it could be used in diesel engines to avoid excessively high peak pressures at full load conditions. Variable displacement engines are attractive since they enable high peak powers, but use less throttle at part-load conditions (thus diminishing pumping losses).

An early concept of variable compression ratio was proposed by J. Atkinson, for whom the Atkinson Cycle is named. As described by [125], Atkinson's original conception involved mechanical linkages to displace the piston such that the engine's compression ratio is lower than its expansion ratio; the main idea being that more expansion work will yield higher efficiency. Similar in idea, but different in implementation, is the Miller concept [125] which uses either late intake valve closing or late exhaust valve opening to shorten or extend the compression or expansion strokes, respectively. Both concepts are attempted in modern-day applications, using both variable compression ratio techniques [112, 126, 127] and altered valve timing techniques [112, 127, 128]. The latter approach, of using altered valve timing techniques, is fluidly made possible through the use of variable valve timing [112, 127], a concept discussed in more detail in the next section.

While variable compression ratio concepts attempt to increase engine efficiency by way of increasing expansion, variable displacement engines attempt to increase engine efficiency by decreasing the use of throttle (thereby, decreasing the pumping work associated with the gas exchange process of a conventional spark ignition engine). With variable displacement, an engine is able to deliver high power using full displacement; at part-load conditions, rather than use throttle, cylinders can be "deactivated" so that they produce no power and allow the engine to deliver part-load power [129]. The deactivated cylinders typically continue to stroke and exchange gases; the closed portion of the cycle (i.e., compression and expansion) realize some loss due to heat transfer and friction but this is intended to be less than the gain realized through decreased pumping work.

## Variable Valve Timing

Briefly discussed in the above sections is the idea of variable valve timing, or, the ability to change the valve events (i.e., intake and exhaust valve opening and closing) at any given point during the engine's operation [130]. In conventional engine design, the valve events are "fixed" by mechanical positions of the lobes on the cam shaft. The effectiveness of an engine to induct and exhaust mixture depends on many things including, for example, engine speed, engine load, the use of EGR, spark timing, and injection timing. Thus, there are not valve events that will universally yield peak power, efficiency, and/or emission for any given engine design.

The idea of variable valve timing allows the engineer to decouple the valve events from the in-cylinder processes. In other words, flexibility of intake and exhaust is afforded with the use of variable valve timing. This not only allows the valve events to be uniquely tuned for each operating point of the engine for improved performance [131] and efficiency [132], but it also can enable other advanced technology. For example, having variable valve timing allows the implementation of the Miller approach to effecting variable compression ratio at part-load conditions, but enabling conventional operation at peak power conditions [127]. Another example is the use of variable valve timing to control the amount of residual fraction in the cylinder, which can affect emission of certain pollutants [133]; controlling residual fraction is a way to enable advanced modes of combustion, described in more detail below. Finally, variable valve timing can be used to replace a throttle, for example, for load control [134]; in doing so, trapped mass can be controlled without inducing pumping work in the engine. Although not widespread technology, variable valve timing is becoming more prevalent on modern-day engines.

## *Waste Heat Recovery*

Waste heat recovery is the effort to take advantage of temperature gradients created between the engine and its environment. As described in section on "Thermodynamic Analysis of Internal Combustion Engines", thermal energy is transferred out of the system through heat transfer (e.g., through engine coolant) and exhaust flow. Because of the temperature gradient that exists between, for example, the high temperature exhaust and the low temperature surroundings, an orderly flow of thermal energy tends to cause the exhaust system to attain thermal equilibrium with the environment. Conventionally, this orderly flow of thermal energy is wasted (i.e., the heat transfer completely dissipates as generated entropy). It is practically possible to instead intercept the orderly flow of thermal energy and convert it to useful work. In the theoretical limit, this conversion of thermal energy to work energy is given by the completely reversible cycle, often called the Carnot cycle. The corresponding efficiency of useful work converted from thermal energy is thusly the Carnot efficiency, as is given by Eq. 6.16:

$$\eta_{\mathrm{th,Carnot}} = 1 - \frac{T_{\mathrm{L}}}{T_{\mathrm{H}}} \tag{6.16}$$

where $\eta_{th,Carnot}$ is the Carnot efficiency (maximum possible conversion of thermal energy to work energy), $T_L$ is the sink's temperature (e.g., the environment temperature), and $T_H$ is the source's temperature (e.g., the exhaust temperature). Considering that a typical automotive exhaust temperature may be 900 K operating in a 300 K environment, ideal efficiencies could be on the order of 67%. Of course, real process irreversibilities diminish actual device efficiencies from the ideal Carnot efficiency. In spite of the diminished efficiency from ideal, though, waste heat recovery in an automotive application, for example, is reported to decrease vehicle fuel consumption by as much as 7.4% [135].

There are several technologies available to make use of internal combustion engine exhaust waste heat energy; because of its spatial accessibility and concentrated high temperature, exhaust energy has been the focus of much of the waste heat recovery. Example major technologies include: (1) mechanical or electrical turbo-compounding/generating devices, (2) Rankine cycle-type devices, and (3) thermoelectric-type devices. Turbo-compounding or turbo-generating [136] converts exhaust thermal energy to mechanical energy (in the form of pressure and kinetic energy) and couples the mechanical energy either directly to the engine driveshaft (to deliver additional brake power) or to an electric generator. One disadvantage of turbo-compounding (like its turbocharging companion) is the method of converting thermal energy to mechanical energy; the centrifugal device increases an engine's exhaust pressure based on its operation. The increased exhaust pressure affects the engine's operation by altering the pumping work and initial mixture composition (increased exhaust species in the initial mixture); such factors can deteriorate the engine's cycle efficiency.

Rankine cycle-type devices make use of the thermodynamic Rankine cycle – typically with an organic fluid designed to undergo phase change within the temperature ranges of a typical engine exhaust system – to output shaft work for either direct-coupling to the engine driveshaft or electrical generation. In such a device, the engine exhaust stream provides the thermal energy to boil or vaporize the organic working fluid of the cycle. After becoming saturated vapor, the fluid expands through a turbine converting the thermal energy to mechanical energy; the cycle completes with the usual condenser and pump processes. Such a device is reported to increase the combined engine + waste heat recovery efficiency by up to 10% [137]; a potential downside is the added complexity of adding four processes (as opposed to one, for example, in the case of a turbomachine) to waste heat recovery system.

Another example major waste heat recovery device is the thermoelectric device. The thermoelectric effect, first observed by Seebeck in 1821, is the generation of a voltage due to a temperature difference between two junctions of two dissimilar materials [138]; principally, the Seebeck effect describes the operation of a thermocouple measurement of temperature. The practical use of the Seebeck effect to produce electricity as a thermoelectric device has recently emerged with semiconductor materials having favorable properties to transmit electricity with little resistance heating (Joule heating) and thermal conductivity (which would tend to "short-circuit" the thermoelectric device). In fact, the advent of semiconductor materials first made possible practical thermoelectric devices as refrigerators exploiting the Peltier Effect (i.e., the opposite of the Seebeck effect, where an applied voltage creates a temperature difference between two junctions of two dissimilar metals) [138]. Now, however, thermoelectric devices create promise to exploit the available thermal energy in an internal combustion engine's exhaust for conversion to electricity [135, 139].

## *Direct Injection, Spark Ignition Engines*

At the end of the section on "A Case Study: Diesel Engines Versus Gasoline Engines", it is suggested that spark ignition and compression ignition engines each have favorable features for power, efficiency, and emissions, but that each has conventionally designed limitations. Thus, it becomes attractive to design each ignition system's limitations out of the engine, and potentially realize gains in the engine's attributes (i.e., power, efficiency, and emissions). Two now-common technologies exist to do so: (1) direct injection spark ignition combustion and (2) homogeneous charge compression ignition combustion. This and the next section will describe these two technologies.

Direct injection spark ignition combustion attempts to create a stratified fuel–air mixture "charge" around the spark plug so that, at the point of spark release, the spark ignites a near-stoichiometric mixture. It is intended that outside of the stratification zone there is little to no fuel, thus creating an overall lean fuel–air mixture. The use of the word "stratified" to describe the fuel–air mixture is used here, as opposed to heterogeneous, to reinforce the notion that under ideal conditions the fuel–air mixture would be homogeneous (i.e., homogeneously stoichiometric) throughout the fuel–air mixture, but pure air outside the stratification zone (i.e., outside the fuel–air mixture). This is in contrast to the use of "heterogeneous" to describe a mixture (e.g., diesel engine mixture), where it is expected that substantial fuel–air ratio gradients exist throughout the fuel sprays.

In order to manifest the stratified mixture concept (see, e.g., [140–148]), fuel is injected directly into the cylinder. Typically, the combustion chamber of a direct injection spark ignition engine is specially designed to assist the stratification of the fuel–air mixture, and center it on the spark plug. The intake stroke draws air and residual mixture into the cylinder – notably, fuel is not inducted during intake as is done in conventional spark ignition operation. After intake, direct fuel injection into the cylinder usually occurs at some point during the piston's travel from BDC to TDC during the compression stroke. Spark advance is typically timed at around the same point as that in a conventional spark ignition engine (i.e., at a point near the piston reaching TDC-compression). The remaining processes of the direct injection spark ignition engine are basically the same as the conventional spark ignition engine.

There are several potential benefits of direct injection spark ignition combustion. Perhaps the clear benefit is the ability to use lean mixtures in a spark ignition engine. Because the combustion chamber is designed to stratify the mixture and create a stoichiometric mixture near the spark plug, the overall equivalence ratio of the mixture filling the entire chamber can be lean. As described in section on "Thermodynamic Analysis of Internal Combustion Engines", overall lean mixtures possess higher ratios of specific heats ($\gamma$), which translate to higher fuel conversion efficiencies. Another, less obvious, benefit of a stratified mixture is the decreased propensity to fuel knock (or, at least decreased propensity of fuel knock in the regions of the cylinder able to cause harm such as near cylinder walls and piston rings). The lack of a reactive mixture – manifested by charge stratification – in the regions furthest from the spark plug – which are the last to be controllably burned by the propagating flame – decreases the propensity that the mixture will autoignite and burn uncontrollably. Such a feature enables the direct injection spark ignition engine to have increased compression ratios relative to conventional spark ignition engines; again, this is an attribute that promotes an increase in efficiency of the novel engine concept. Finally, the use of direct fuel injection into the cylinder enables the elimination of a throttle to control engine load. In other words, engine load is controlled by the quantity of fuel injected into the cycle, similar to the load control of a diesel engine. Elimination of the throttle, as repeatedly described, will improve part-load efficiency by eliminating pumping losses effected by throttle.

Although the benefits are plentiful, the challenges are also present. From practical perspectives, perfect attainment of a stratified charge is difficult to accomplish. As such, heterogeneities within the stratified mixture emerge and can lead to products of incomplete combustion such as CO, HC, and PM. This, of course, is amplified at full load conditions; thus, peak power attainment through direct injection means alone would likely be limited by smoke limitations (similar to full load limitations of a diesel engine). Further, and like the challenges faced by diesel engines, outfitting direct injection spark ignition engines with after treatment devices is complicated by the use of overall lean mixtures (as described in section on "Emissions Formation and Exhaust Pollution"). In spite of such challenges, direct injection spark ignition engines are in production; continued development of in-cylinder flow modeling tools and engine controls contribute toward the concept's potential success.

## *Homogeneous Charge Compression Ignition Engines*

An attractive feature of the conventional spark ignition engine is its use of a homogeneous mixture. Although this tends to promote knock (as described in section "Direct Injection Spark Ignition Engines"), it provides the benefit of being kinetically rate-limiting as opposed to mixing rate-limiting (as in the case of diesel engines). One issue with using a spark, or single point, to ignite a homogeneous mixture is the establishment of a flame that must propagate the mixture to convert the chemical energy. In order to more quickly react the mixture, in a volumetric sense, multi-point ignition is required; i.e., ignition that occurs in several locations throughout the mixture will result in a faster burn rate. Such a voluminous ignition can be effected through compression of a homogenous mixture.

This is the basic idea of homogeneous charge compression ignition (HCCI); use compression to ignite a homogeneous mixture (similar to a diesel engine, except that diesel engines use compression to ignite an inherently heterogeneous mixture). The homogeneous mixture could, for example, be formed through premixing of fuel and air prior to mixture induction during the intake stroke. A faster burn rate effected by homogeneous charge compression ignition allows heat release to occur at near constant volume conditions; thus establishing the possibility to approach theoretical limits of maximum efficiencies of internal combustion engines. Further, because compression is used to ignite the mixture rather than spark, lean mixtures can be used at part-load conditions which further promotes higher efficiencies of the HCCI concept. Similar to the direct injection spark ignition concept, HCCI engines could eliminate throttles as load is controlled directly by the quantity of fuel mixed with the intake mixture. Finally, the common issue of particulate matter emissions faced by diesel engines (which also use compression ignition) are substantially decreased through the avoidance of locally rich fuel–air mixture zones (due to the use of a homogeneous mixture in an HCCI concept).

Of course, immediately the obvious problem becomes the method of ignition control. HCCI concepts do not have a direct ignition trigger, as do conventional gasoline (i.e., a spark) or diesel (i.e., direct fuel injection) engines. Controlling ignition in the HCCI concept depends on very precise control of the mixture's initial state at start of compression and the compression path followed up to the point of ignition. Several factors which are present – even in tightly controlled research environments – such as heat transfer, turbulence, and the history of preceding combustion events make the practical application of HCCI very challenging. The payoffs, of course, are correspondingly very high.

Practical implementation of HCCI is first reported by Onishi et al. [149] with theoretical developments experimentally provided by Najt and Foster [150]. Several control parameters could be adjusted such as compression ratio (e.g., with the use of variable valve timing), initial temperature, and quantity of residual fraction (e.g., effected either through exhaust gas recirculation or variable valve timing). Identifying the key control parameters, and the optimal way to adjust them during real-time operation of the engine, continue to be on-going research activities [151–156].

## *Advanced Compression Ignition Engines*

A technique to control combustion of an HCCI engine is to use precisely metered amounts of residual fraction, which not only act to alter the kinetics of combustion but also result in substantially lower combustion temperatures. As such, much of

HCCI combustion is characterized by low temperature mechanisms commonly referred to as *low temperature combustion* (LTC). LTC offers a few benefits. First, efficiency improvements in the engine can be realized (in spite of increased exergy destruction due to low reaction temperatures) due to more favorable thermodynamic properties (i.e., higher ratio of specific heats, see Fig. 6.11) of the burned mixture and lower rates of heat transfer. Second, and typically the driver for LTC technology development, lower nitric oxide formation per the discussion in section on "Emissions Formation and Exhaust Pollution".

With this in mind, and reconsidering the prevailing issue of HCCI implementation – i.e., control of start of combustion – it becomes plausible to consider developing an "HCCI-type" mode of combustion in a diesel engine. That is, rather than induct a homogeneous mixture of fuel and air and rely upon indirectly controlled parameters to control ignition, perhaps fuel can be injected directly into the cylinder allowing for better control of ignition. In order to manifest LTC and harvest its benefits (e.g., possibly higher efficiency and substantially emission), high levels of EGR and strategic injection timings are used to extend ignition delay and create a nearly all-premixed combustion event. The long ignition delay, coupled with low temperature mechanisms, establishes the phenomenological observation of two-stage ignition characterized by the presence of cool-flame reactions [150, 157]. Interestingly, because of attainment of LTC, soot precursor formation is substantially abated, and the engine is made to operate with very low emissions of nitric oxide and particulate matter [158–168]. The combustion concept has become known as premixed compression ignition or premixed charge compression ignition combustion. The ability to attain LTC in compression ignition engines is attributed to the advancement of technology now in place on such machines, such as common-rail and electronic fuel pressure systems, variable geometry turbochargers, and exhaust gas recirculation systems.

## Alternative Fuels

The term "alternative fuels" for an internal combustion engine is somewhat baseless, as an internal combustion engine has considerable flexibility in the type of fuel it uses. Of course, conventional fuels are the commonly called "gasoline" and "diesel" fuels, but generally engines have been shown to operate on virtually any gaseous, liquid, and solid dust particle specie that has heating value (i.e., will release thermal energy in a chemical oxidation process). Because of the wide variability of fuels available to internal combustion engines, this topic will not be expanded in this article. There are, however, certain considerations that should be given to the use of a fuel in an engine which was not intently designed for use with such fuel (e.g., use of ethanol in a gasoline engine or use of biodiesel in a diesel engine). First, ignition characteristics of the fuel may not be favorable for the particular engine design. For example, short-chain volatile hydrocarbons do not generally ignite well in conventional unmodified compression ignition engines of typical compression ratios. Likewise, long-chain nonvolatile hydrocarbons do not generally vaporize well in conventional unmodified spark ignition engines. Second, flame temperatures of the combustion of the fuel may exceed material limits of any given engine construction. Third, fuels may react with other support components of the engine system (e.g., solvency of fuels with rubber hoses). Finally, combustion process will likely be altered when using an unconventional fuel in a conventional unmodified engine yielding different emissions, efficiency, and peak power capabilities. Thus, although internal combustion engines have inherent fuel-flexibility, their use with unconventional fuels is not straightforward and requires careful design and engineering considerations.

**Acknowledgments** The author wishes to thank several people who have helped to make this work possible. First, the State of Texas is acknowledged for their financial support of some of the research highlighted here; specifically, their support through the Texas Commission on Environmental Quality and the Houston Advanced Research Center is acknowledged. Second, Professor Jerald A. Caton of Texas A&M University is acknowledged for his contributions to this article and for providing a thorough proof of its contents. Third, Margaret Fisher is acknowledged for her assistance in preparing the copyrighted materials from other sources and securing permissions to use them. Fourth, Wiley is acknowledged for providing permission of copyrighted material *au gratis*. Fifth and lastly, but certainly not least, I acknowledge the contributions and assistance of my graduate and undergraduate students, some of whom have research highlighted in this work. Specifically, these individuals include: Mr. Josh Bittle, Mr. Jason Esquivel, Ms. Sarabeth Fronenberger, Mr. Blake Gettig, Mr. Mark Hammond, Mr. Bryan Knight, Mr. Jeffrey Kurthy, Mr. Jimmy McClean, Ms. Claire Mero, Mr. Yehia Omar, Ms. Gurlovleen Rathore, Mr. Kyle Richter, Mr. Sidharth Sambashivan, Mr. Chris Schneider, Ms. Amy Smith, Mr. Hoseok Song, Mr. Jiafeng Sun, Mr. Brandon Tompkins, Mr. Brad Williams, Mr. R. Kevin Wilson, Mr. Whit Wilson, and Mr. Jesse Younger.

## Abbreviations

**BDC** Bottom dead center
**EGR** Exhaust gas recirculation
**HCCI** Homogeneous charge compression ignition
**IC** Internal combustion
**LTC** Low temperature combustion
**TDC** Top dead center

## Bibliography

### *Primary Literature*

1. Cummins C Jr (1976) Early IC and automotive engines. SAE Trans 85(SAE Paper No. 760604):1960–1971
2. Heywood J (1988) Internal combustion engine fundamentals. McGraw-Hill, New York, p 9

3. Heywood J (1988) Internal combustion engine fundamentals. McGraw-Hill, New York, p 10
4. Automot Eng Int (January 2010), 118(1):47. http://www.sae.org/automag/
5. Brown W (1967) Methods for evaluating requirements and errors in cylinder pressure measurement. SAE Trans 76(SAE Paper No. 670008):50–71
6. Lancaster D, Krieger R, Lienesch J (1975) Measurement and analysis of engine pressure data. SAE Trans 84(SAE Paper No. 750026):155–172
7. Randolph A (1990) Methods of processing cylinder-pressure transducer signals to maximize data accuracy. SAE Trans J Passenger Cars 99(SAE Paper No. 900170):191–200
8. Kuratle R, Marki B (1992) Influencing parameters and error sources during indication on internal combustion engines. SAE Trans – J Engines 101(SAE Paper No. 920233):295–303
9. Davis R, Patterson G (2006) Cylinder pressure data quality checks and procedures to maximize data accuracy. SAE Paper No. 2006-01-1346
10. Amann C (1983) A perspective of reciprocating-engine diagnostics without lasers. Prog Energy Combust Sci 9:239–267
11. Heywood J (1988) Internal combustion engine fundamentals. McGraw-Hill, New York, pp 56–57
12. Tompkins B, Esquivel J, Jacobs T (2009) Performance parameter analysis of a biodiesel-fuelled medium duty diesel engine. SAE Paper No. 2009-01-0481
13. Heywood J (1988) Internal combustion engine fundamentals. McGraw-Hill, New York, p 217
14. Heywood J (1988) Internal combustion engine fundamentals. McGraw-Hill, New York, p 154
15. Lauck F, Uyehara O, Myers P (1963) An engineering evaluation of energy conversion devices. SAE Trans 71(SAE Paper No. 630446):41–50
16. Foster D, Myers P (1984) Can paper engines stand the heat? SAE Trans 93(SAE Paper No. 840911):4.491–4.502
17. The K, Miller S, Edwards C (2008) Thermodynamic requirements for maximum internal combustion engine cycle efficiency, Part 1: optimal combustion strategy. Int J Engine Res 9:449–465
18. Carnot NLS (1824) Reflections on the motive power of heat (Trans and ed: Thurston RH), 2nd edn (1897). Wiley, New York
19. Borgnakke C, Sonntag R (2009) Fundamentals of thermodynamics. Wiley, New York, p 497
20. Heywood J (1988) Internal combustion engine fundamentals. McGraw-Hill, New York, p 177
21. Heywood J (1988) Internal combustion engine fundamentals. McGraw-Hill, New York, p 163
22. Edson M (1964) The influence of compression ratio and dissociation on ideal otto cycle engine thermal efficiency, Digital calculations of engine cycles. SAE, Warrendale, pp 49–64
23. Westbrook C, Dryer F (1984) Chemical kinetic modeling of hydrocarbon combustion. Prog Energy Combust Sci 10:1–57
24. Olikara C, Borman G (1975) A computer program for calculating properties of equilibrium combustion products with some applications IC engines. SAE Paper No. 750468
25. Lavoie G, Heywood J, Keck J (1970) Experimental and theoretical study of nitric oxide formation in internal combustion engines. Combust Sci Technol 1:313–326
26. Bowman C (1975) Kinetics of pollutant formation and destruction in combustion. Prog Energy Combust Sci 1:33–45
27. Miller J, Bowman C (1989) Mechanism and modeling of nitrogen chemistry in combustion. Prog Energy Combust Sci 15:287–338
28. Turns S (1995) Understanding $NO_x$ formation in nonpremixed flames: experiments and modeling. Prog Energy Combust Sci 21:361–385
29. Dean A, Bozzelli J (2000) In: Gardiner WC Jr (ed) Combustion chemistry of nitrogen in gas-phase combustion chemistry. Springer, New York, pp 125–341

30. McBride B, Gordon S (1992) Computer program for calculating and fitting thermodynamic functions. NASA Report No. RP-1271
31. Svehla R (1995) Transport coefficients for the NASA Lewis chemical equilibrium program. NASA Report No. TM-4647
32. Gordon S, McBride B (1999) Thermodynamic data to 20000K for monatomic gases. NASA Report No. TP-1999-208523
33. McBride B, Gordon S, Reno M (2001) Thermodynamic data for fifty reference elements. NASA Report No. TP-3287/Rev 1
34. McBride B, Zehe M, Gordon S (2002) CAP: a computer code for generating tabular thermodynamic functions from NASA Lewis Coefficients. NASA Report No. TP-2001-210959-Rev1
35. Heywood J (1988) Internal combustion engine fundamentals. McGraw-Hill, New York, pp 136–137
36. Stull D, Prophet H (1971) JANAF thermochemical tables, NSRDS-NBS 37. http://www.nist.gov/data/nsrds/NSRDS-NBS37.pdf. Accessed July 5, 2010
37. Keenan J (1951) Availability and irreversibility in thermodynamics. Br J Appl Phys 2:183–192
38. Edson M, Taylor C (1964) The limits of engine performance – comparison of actual and theoretical cycles. In: SAE digital calculations of engine cycles, pp 65–81
39. Strange F (1964) An analysis of the ideal Otto cycle, including the effects of heat transfer, finite combustion rates, chemical dissociation, and mechanical losses. In: SAE digital calculations of engine cycles, pp 92–105
40. Patterson D, Van Wylen G (1964) A digital computer simulation for spark-ignited engine cycles. In: SAE digital calculations of engine cycles, pp 82–91
41. Woschni G (1967) Universally applicable equation for the instantaneous heat transfer coefficient in the internal combustion engine. SAE Trans 76(SAE Paper No. 670931): 3065–3083
42. Hohenberg G (1979) Advanced approaches for heat transfer calculations. SAE Trans 88(SAE Paper No. 790825):2788–2806
43. Borman G, Nishiwaki K (1987) Internal-combustion engine heat transfer. Prog Energy Combust Sci 13:1–46
44. Heywood J, Higgins J, Watts P, Tabaczynski R (1979) Development and use of a cycle simulation to predict SI engine efficiency and $NO_x$ emissions. SAE Paper No. 790291
45. Sandoval D, Heywood J (2003) An improved friction model for spark-ignition engines. SAE Trans J Engines 112(SAE Paper No. 2003-01-0725):1041–1052
46. Blumberg P, Lavoie G, Tabaczynski R (1979) Phenomenological models for reciprocating internal combustion engines. Prog Energy Combust Sci 5:123–167
47. Assanis D, Heywood J (1986) Development and use of a computer simulation of the turbocompounded diesel system for engine performance and component heat transfer studies. SAE Trans 95(SAE Paper No. 860329):2.451–2.476
48. Filipi Z, Assanis D (1991) Quasi-dimensional computer simulation of the turbocharged spark ignition engine and its use for 2 and 4-valve engine matching studies. SAE Trans J Engines 100(SAE Paper No. 910075):52–68
49. Kamimoto T, Kobayashi H (1991) Combustion processes in diesel engines. Prog Energy Combust Sci 17:163–189
50. Reitz R, Rutland C (1995) Development and testing of diesel engine CFD models. Prog Energy Combust Sci 21:173–196
51. Caton J (2003) Effects of burn rate parameters on nitric oxide emissions for a spark ignition engine: results from a three-zone, thermodynamic simulation. SAE Paper No. 2003-01-0720
52. Caton J (2000) A review of investigations using the second law of thermodynamics to study internal-combustion engines. SAE Trans J Engines 109(SAE Paper No. 2000-01-1081): 1252–1266

53. Rakopoulos C, Giakoumis E (2006) Second-law analyses applied to internal combustion engines operation. Prog Energy Combust Sci 32:2–47
54. Shyani R, Caton J (2009) A thermodynamic analysis of the use of exhaust gas recirculation in spark ignition engines including the second law of thermodynamics. Proc Inst Mech Eng Part D: J Automobile Eng 223:131–149
55. Dunbar W, Lior N (1994) Sources of combustion irreversibility. Combust Sci Technol 103:41–61
56. Som S, Datta A (2008) Thermodynamic irreversibilities and exergy balance in combustion processes. Prog Energy Combust Sci 34:351–376
57. Caton J (2000) On the destruction of availability (exergy) due to combustion processes – with specific application to internal-combustion engines. Energy 25:1097–1117
58. Keenan J (1941) Thermodynamics. Wiley, New York, p 269
59. Obert E (1970) Internal combustion engines, 3rd edn. International Textbook Company, Scranton, p 459
60. Patrawala K, Caton J (2008). Potential processes for "reversible" combustion with application to reciprocating internal combustion engines. In: Proceedings of the 2008 technical meeting of the central states section of the combustion institute, Tuscaloosa
61. Diesel R (1897) Diesel's rational heat motor. A lecture delivered at the general meeting of the society at Cassell, June 16, 1897. Original published in Zeitschrift des Vereines Deutscher Ingenieure (Trans: Leupold R). Progressive Age, New York (Reprinted)
62. Heywood J (1988) Internal combustion engine fundamentals. McGraw-Hill, New York, p 391
63. Heywood J (1988) Internal combustion engine fundamentals. McGraw-Hill, New York, p 390
64. Heywood J (1988) Internal combustion engine fundamentals. McGraw-Hill, New York, p 374
65. Lyn W (1963) Study of burning rate and nature of combustion in diesel engines. Proc Combust Inst 9:1069–1082
66. Plee S, Ahmad T (1983) Relative roles of premixed and diffusion burning in diesel combustion. SAE Trans 92(SAE Paper No. 831733):4.892–4.909
67. Ricardo H (1941) The high-speed internal combustion engine (Rev: Glyde HS), 3rd edn. Interscience Publishers, New York
68. Ladommatos N, Abdelhalim S, Zhao H, Hu Z (1998) Effects of EGR on heat release in diesel combustion. SAE Paper No. 980184
69. Meguerdichian M, Watson N (1978) Prediction of mixture formation and heat release in diesel engines. SAE Paper No. 780225
70. Lyn W, Valdmanis E (1968) Effects of physical factors on ignition delay. SAE Paper No. 680102
71. Kamimoto T, Aoyagi Y, Matsui Y, Matsuoka S (1981) The effects of some engine variables on measured rates of air entrainment and heat release in a DI diesel engine. SAE Trans 89(SAE Paper No. 800253):1163–1174
72. Dent J, Mehta P, Swan J (1982) A predictive model for automotive DI diesel engine performance and smoke emissions. Paper presented at the international conference on diesel engines for passenger cars and light duty vehicles. Institution of Mechanical Engineers, London. IMECE Paper No. C126/82
73. Binder K, Hilburger W (1981) Influence of the relative motions of air and fuel vapor on the mixture formation processes of the direct injection diesel engine. SAE Trans 90(SAE Paper No. 810831):2540–2555
74. Dec J (1997) A conceptual model of DI diesel combustion based on laser-sheet imaging. SAE Trans J Engines 106(SAE Paper No. 970873):1319–1348
75. Flynn P, Durrett R, Hunter G, zur Loye A, Akinyemi O, Dec J, Westbrook C (1999) Diesel combustion: an integrated view combining laser diagnostics, chemical kinetics, and empirical validation. SAE Trans J Engines 108(SAE Paper No. 1999-01-0509):587–600

76. Chigier N (1975) Pollution formation and destruction in flames – Introduction. Prog Energy Combust Sci 1:3–15
77. Beltzer M (1976) Non-sulfate particulate emissions from catalyst cars. SAE Trans 85(SAE Paper No. 760038):198–208
78. Khatri N, Johnson J, Leddy D (1978) The characterization of the hydrocarbon and sulfate fractions of diesel particulate matter. SAE Trans 87(SAE Paper No. 780111):469–492
79. Hoffert M, Caldeira K, Benford G, Criswell D, Green C, Herzog H, Jain A, Kheshgi H, Lackner K, Lewis J, Lightfoot H, Manheimer W, Mankins J, Mauel M, Perkins L, Schlesinger M, Volk T, Wigley T (2002) Advanced technology paths to global climate stability: energy for a greenhouse planet. Science 298:981–987
80. Ghoniem A (2011) Needs, resources and climate change: clean and efficient conversion technologies. Prog Energy Combustion Sci 37:15–51
81. Henein N (1976) Analysis of pollutant formation and control and fuel economy in diesel engines. Prog Energy Combust Sci 1:165–207
82. Haagen-Smit A, Fox M (1955) Automobile exhaust and ozone formation. SAE Trans 63(SAE Paper No. 550277):575–580
83. Huls T, Nickol H (1967) Influence of engine variables on exhaust oxides of nitrogen concentrations from a multicylinder engine. SAE Paper No. 670482
84. Starkman E, Stewart H, Zvonow V (1969) Investigation into formation and modification of exhaust emission precursors. SAE Paper No. 690020
85. Hames R, Merrion D, Ford H (1971) Some effects of fuel injection system parameters on diesel exhaust emissions. SAE Paper No. 710671
86. Khan I, Greeves G, Wang C (1973) Factors affecting smoke and gaseous emissions from direct injection engines and a method of calculation. SAE Trans 82(SAE Paper No. 730169):687–709
87. Yu R, Shahed S (1981) Effects of injection timing and exhaust gas recirculation on emissions from a D.I. diesel engine. SAE Trans 90(SAE Paper No. 811234):3873–3883
88. Newhall H (1967) Control of nitrogen oxides by exhaust recirculation, a preliminary theoretical study. SAE Trans 76(SAE Paper No. 670495):1820–1836
89. Benson J, Stebar R (1971) Effects of charge dilution on nitric oxide emission from a single-cylinder engine. SAE Trans 80(SAE Paper No. 710008):7–19
90. Komiyama K, Heywood J (1973) Predicting $NO_x$ emissions and effects of exhaust gas recirculation in spark-ignition engines. SAE Trans 82(SAE Paper No. 730475):1458–1476
91. McEnally C, Pfefferle L, Atakan B, Kohse-Hoinghaus K (2006) Studies of aromatic hydrocarbon formation mechanisms in flames: progress toward closing the fuel gap. Prog Energy Combust Sci 32:247–294
92. Cheng W, Hamrin D, Heywood J, Hochgreb S, Min K, Norris M (1993) An overview of hydrocarbon emissions mechanisms in spark-ignition engines. SAE Trans J Fuels Lubricants 102(SAE Paper No. 932708):1207–1220
93. Henein N, Tagomori M (1999) Cold-start hydrocarbon emissions in port-injected gasoline engines. Prog Energy Combust Sci 25:563–593
94. Alkidas A (1999) Combustion-chamber crevices: the major source of engine-out hydrocarbon emissions under fully warmed conditions. Prog Energy Combust Sci 25:253–273
95. Haynes B, Wagner H (1981) Soot formation. Prog Energy Combust Sci 7:229–273
96. Smith O (1981) Fundamentals of soot formation in flames with application to diesel engine particulate emissions. Prog Energy Combust Sci 7:275–291
97. Kennedy I (1997) Models of soot formation and oxidation. Prog Energy Combust Sci 23:95–132
98. Richter H, Howard J (2000) Formation of polycyclic aromatic hydrocarbons and their growth to soot – a review of chemical reaction pathways. Prog Energy Combust Sci 26:565–608
99. Tree D, Svensson K (2007) Soot processes in compression ignition engines. Prog Energy Combust Sci 33:272–309

100. Hassaneen A, Samuel S, Morrey D, Gonzalez-Oropeza R (2009) Influence of physical and chemical parameters on characteristics of nanoscale particulate in spark ignition engine. SAE Paper No. 2009-01-2651
101. Ericsson P, Samson A (2009) Characterization of particulate emissions propagating in the exhaust line for spark-ignited engines. SAE Paper No. 2009-01-2654
102. Khan I (1969–1970) Formation and combustion of carbon in a diesel engine. Proc Inst Mech Eng 184(3J):36–43
103. Ahmad T, Plee S, Myers J (1982) Diffusion flame temperature – its influence on diesel particulate and hydrocarbon emissions. Paper presented at the international conference on diesel engines for passenger cars and light duty vehicles. Institution of Mechanical Engineers, London. IMECE Paper No. C101/82
104. Kummer J (1980) Catalysts for automobile emission control. Prog Energy Combust Sci 6:177–199
105. Koltsakis G, Stamatelos A (1997) Catalytic automotive exhaust aftertreatment. Prog Energy Combust Sci 23:1–39
106. Johnson T (2010) Diesel emission control in review. SAE Int J Fuels Lubricants 2(SAE Paper No. 2009-01-0121):1–12
107. Taylor C (1985) The internal combustion engine in theory and practice, Vol 2: Combustion, fuels, materials, design (rev. edition). The MIT Press, Cambridge, MA, pp 21–23
108. Taylor C (1985) The internal combustion engine in theory and practice, Vol 2: Combustion, fuels, materials, design (rev. edition). The MIT Press, Cambridge, MA, p 50
109. Heywood J (1988) Internal combustion engine fundamentals. McGraw-Hill, New York, pp 4–5
110. Heywood J (1988) Internal combustion engine fundamentals. McGraw-Hill, New York, p 475
111. Heywood J (1988) Internal combustion engine fundamentals. McGraw-Hill, New York, pp 217–220
112. Clenci A, Descombes G, Podevin P, Hara V (2007) Some aspects concerning the combination of downsizing with turbocharging, variable compression ratio, and variable intake valve lift. Proc Inst Mech Eng D J Automobile Eng 221:1287–1294
113. Van Nieuwstadt M, Kolmanovsky I, Morael P (2000) Coordinated EGR-VGT control for diesel engines: an experimental comparison. SAE Trans – J Engines 109(SAE Paper No. 2000-01-0266):238–249
114. Arnold S, Slupski K, Groskreutz M, Vrbas G, Cadle R, Shahed S (2011) Advanced turbocharging technologies for heavy-duty diesel engines. SAE Trans J Engines 110(SAE Paper No. 2001-01-3260):2048–2055
115. Kessel J, Schaffnit J, Schmidt M (1998) Modeling an real-time simulation of a turbocharger with variable turbine geometry (VGT). SAE Paper No. 980770
116. Hawley J, Wallace F, Pease A, Cox A, Horrocks R, Bird G (1997) Comparison of variable geometry turbocharging (VGT) over conventional wastegated machines to achieve lower emissions. In: IMechE autotech conference, Birmingham, UK, pp 245–259 (IMechE Seminar Publication: Automotive Engines and Powertrains, Paper No. C524/070/97)
117. Hawley J, Wallace F, Cox A, Horrocks R, Bird G (1999) Reduction of steady state $NO_x$ levels from an automotive diesel engine using optimized VGT/EGR schedules. SAE Trans J Engines 108(SAE Paper No. 1999-01-0835):1172–1184
118. Tanin K, Wickman D, Montgomery D, Das S, Reitz R (1999) The influence of boost pressure on emissions and fuel consumption of a heavy-duty single-cylinder DI diesel engine. SAE Trans J Engines 108(SAE Paper No. 1999-01-0840):1198–1219
119. Cook J, Sun J, Buckland J, Kolmanovsky I, Peng H, Grizzle J (2006) Automotive powertrain control – A survey. Asian J Control 8(3):237–260
120. Leithgoeb R, Henzinger F, Fuerhapter A, Gschweitl K, Zrim A (2003) Optimization of new advanced combustion systems using real-time combustion control. SAE Paper No. 2003-01-1053

121. Corti E, Moro D, Solieri L (2007) Real-time evaluation of IMEP and ROHR-related parameters. SAE Paper No. 2007-24-0068
122. Leonhardt S, Muller N, Isermann R (1999) Methods for engine supervision and control based on cylinder pressure information. IEEE/ASME Trans Mechatron 4(3):235–245
123. Yoon M, Chung N, Lee M, Sunwoo M (2009) An engine-control-unit-in-the-loop simulator of a common-rail diesel engine for cylinder-pressure-based control. Proc Inst Mech Eng D J Automobile Eng 223:355–373
124. Turin R, Zhang R, Chang M (2008) Systematic model-based engine control design. SAE Int J Passenger Cars Electron Electr Syst 1(SAE Paper No. 2008-01-0994):413–424
125. Caton J (2008) Results from an engine cycle simulation of compression ratio and expansion ratio effects on engine performance. J Eng Gas Turbines Power 130(5):052809-1–052809-7
126. Wirbeleit F, Binder K, Gwinner D (1990) Development of pistons with variable compression height for increasing efficiency and specific power output of combustion engines. SAE Trans J Engines 99(SAE Paper No. 900229):543–557
127. Sugiyama T, Hiyoshi R, Takemura S, Aoyama S (2007) Technology for improving engine performance using variable mechanisms. SAE Trans J Engines 116(SAE Paper No. 2007-01-1290):803–812
128. Boggs D, Hilbert H, Schechter M (1995) The Otto-Atkinson cycle engine: fuel economy and emissions results and hardware design. SAE Trans J Engines 104(SAE Paper No. 950089):220–232
129. Leone T, Pozar M (2001) Fuel economy benefit of cylinder deactivation – Sensitivity to vehicle application and operating constraints. SAE Trans J Fuels Lubricants 110(SAE Paper No. 2001-01-3591):2039–2044
130. Gray C (1988) A review of variable engine valve timing. SAE Trans J Engines 97(SAE Paper No. 880386):6.631–6.641
131. Payri F, Desantes J, Corberaan J (1988) A study of the performance of an SI engine incorporating a hydraulically controlled variable valve timing system. SAE Trans J Engines 97(SAE Paper No. 880604):6.1133–6.1145
132. Ma T (1988) Effect of variable engine valve timing on fuel economy. SAE Trans J Engines 97(SAE Paper No. 880390):6.665–6.672
133. Meacham G (1970) Variable cam timing as an emission control tool. SAE Trans 79(SAE Paper No. 700673):2127–2144
134. Tuttle J (1980) Controlling engine load by means of late intake-valve closing. SAE Trans 89(SAE Paper No. 800794):2429–2441
135. Stobart R, Wijewardane A, Allen C (2010) The potential for thermo-electric devices in passenger vehicle applications. SAE Paper No. 2010-01-0833
136. Patterson A, Tett R, McGuire J (2009) Exhaust heat recovery using electro-turbogenerators. SAE Paper No. 2009-01-1604
137. Srinivasan K, Mago P, Zdaniuk G, Chamra L, Midkiff K (2008) Improving the efficiciency of the advanced injection low pilot ignited natural gas engine using organic Rankine cycles. ASME J Energy Resour Technol 130:022201-1–022201-7
138. Goldsmid H (1960) Principles of thermoelectric devices. Br J Appl Phys 11:209–217
139. Hussain Q, Brigham D, Maranville C (2010) Thermoelectric exhaust heat recovery for hybrid vehicles. SAE Int J Engines 2:1(SAE Paper No. 2009-01-1327):1132–1142
140. Barber E, Reynolds B, Tierney W (1951) Elimination of combustion knock ~ Texaco combustion process. SAE Trans 59(SAE Paper No. 510173):26–38
141. Davis C, Barber E, Mitchell E (1961) Fuel injection and positive ignition ~ A basis for improved efficiency and economy. SAE Trans 69(SAE Paper No. 610012):120–131
142. Mitchell E, Cobb J, Frost R (1968) Design and evaluation of a stratified charge multifuel military engine. SAE Trans 77(SAE Paper No. 680042):118–131
143. Alperstein M, Schafer G, Villforth F III (1974) Texaco's stratified charge engine – multifuel, efficient, clean, and practical. SAE Paper No. 740563

144. Pischinger F, Schmidt G (1978) Experimental and theoretical investigations of a stratified-charge engine with direct fuel injection. SAE Paper No. 785038
145. Hiraki H, Rife J (1980) Performance and $NO_x$ model of a direct injection stratified charge engine. SAE Trans 89(SAE Paper No. 800050):336–356
146. Ullman T, Hare C, Baines T (1982) Emissions from direct-injected heavy-duty methanol-fueled engines (one dual injection and one spark-ignited) and a comparable diesel engine. SAE Trans 91(SAE Paper No. 820966):3154–3170
147. Giovanetti A, Ekchian J, Heywood J, Fort E (1983) Analysis of hydrocarbon emissions mechanisms in a direct injection spark-ignition engine. SAE Trans 92(SAE Paper No. 830587):2.925–2.947
148. Kato S, Onishi S (1988) New mixture formation technology of direct fuel injection stratified charge SI engine (OSKA) ~ Test result with gasoline fuel. SAE Trans J Engines 97(SAE Paper No. 881241):6.1497–6.1504
149. Onishi S, Jo S, Shoda K, Jo P, Kato S (1979) Active thermo-atmosphere combustion (ATAC) ~ A new combustion process for internal combustion engines. SAE Trans 88(SAE Paper No. 790501):1851–1860
150. Najt P, Foster D (1983) Compression-ignited homogeneous charge combustion. SAE Trans 92(SAE Paper No. 830264):1.964–1.979
151. Martinez-Frias J, Aceves S, Flowers D, Smith J, Dibble R (2000) HCCI engine control by thermal management. SAE Trans J Fuels Lubricants 109(SAE Paper No. 2000-01-2869):2646–2655
152. Law D, Kemp D, Allen J, Kirkpatrick G, Copland T (2001) Controlled combustion in an IC engine with a fully variable valve train. SAE Trans J Engines 110(SAE Paper No. 2001-01-0251):192–198
153. Rausen D, Stefanopoulou A, Kang J, Eng J, Kuo T (2005) A mean-value model for control of homogeneous charge compression ignition (HCCI) engines. J Dyn Syst Meas Contr 127:355–362
154. Shaver G, Gerdes J, Roelle M, Caton P, Edwards C (2005) Dynamic modeling of residual-affected homogeneous charge compression ignition engines with variable valve actuation. J Dyn Syst Meas Contr 127:374–381
155. Bengtsson J, Strandh P, Johansson R, Tunestal P, Johansson B (2006) Multi-output control of a heavy-duty HCCI engine using variable valve actuation and model predictive control. SAE Paper No. 2006-01-0873
156. Chiang C, Stefanopoulou A (2009) Sensitivity analysis of combustion timing of homogeneous charge compression ignition gasoline engines. J Dyn Syst Meas Contr 131:014506-1–014506-5
157. Fish A, Read I, Affleck W, Haskell W (1969) The controlling role of cool flames in two-stage ignition. Combust Flame 13:39–49
158. Takeda Y, Keiichi N, Keiichi N (1996) Emission characteristics of premixed lean diesel combustion with extremely early staged fuel injection. SAE Trans J Fuels Lubricants 105(SAE Paper No. 961163):938–947
159. Akagawa H, Miyamoto T, Harada A, Sasaki S, Shimazaki N, Hashizume T, Tsujimura K (1999) Approaches to solve problems of the premixed lean diesel combustion. SAE Trans J Engines 108(SAE Paper No. 1999-01-0183):120–132
160. Iwabuchi Y, Kawai K, Shoji T, Takeda Y (1999) Trial of new concept diesel combustion system – premixed compression-ignited combustion. SAE Trans J Engines 108(SAE Paper No. 1999-01-0185):142–151
161. Kimura S, Aoki O, Kitahara Y, Airoshizawa E (2001) Ultra-clean combustion technology combining a low-temperature and premixed combustion concept for meeting future emission standards. SAE Trans J Fuels Lubricants 110(SAE Paper No. 2001-01-0200):239–246
162. Kaneko N, Ando H, Ogawa H, Miyamoto N (2002) Expansion of the operating range with in-cylinder water injection in a premixed charge compression ignition engine. SAE Trans J Engines 111(SAE Paper No. 2002-01-1743):2309–2315

163. Shimazaki N, Tsurushima T, Nishimura T (2003) Dual mode combustion concept with premixed diesel combustion by direct injection near top dead center. SAE Trans J Engines 112(SAE Paper No. 2003-01-0742):1060–1069
164. Hasegawa R, Yanagihara H (2003) HCCI combustion in DI diesel engine. SAE Trans J Engines 112(SAE Paper 2003-01-0745):1070–1077
165. Okude K, Mori K, Shiino S, Moriya T (2004) Premixed compression ignition (PCI) combustion for simultaneous reduction of NOx and soot in diesel engines. SAE Trans J Fuels Lubricants 113(SAE Paper No. 2004-01-1907):1002–1013
166. Jacobs T, Bohac S, Assanis D, Szymkowicz P (2005) Lean and rich premixed compression ignition combustion in a light-duty diesel engine. SAE Trans J Engines 114(SAE Paper No. 2005-01-0166):382–393
167. Lechner G, Jacobs T, Chryssakis C, Assanis D, Siewert R (2005) Evaluation of a narrow spray cone angle, advanced injection timing strategy to achieve partially premixed compression ignition combustion in a diesel engine. SAE Trans J Engines 114(SAE Paper No. 2005-01-0167):394–404
168. Jacobs T, Assanis D (2007) The attainment of premixed compression ignition low-temperature combustion in a compression ignition direct injection engine. Proc Combust Inst 31:2913–2920

## *Books and Reviews*

Ferguson CR, Kirkpatrick AT (2001) Internal combustion engines: applied thermosciences, 2nd edn. Wiley, New York

Heywood J (1988) Internal combustion engine fundamentals. McGraw-Hill, New York

Jennings BH, Obert EF (1944) Internal combustion engines: analysis and practice. International Textbook Company, Scranton

Pulkrabek WW (2004) Engineering fundamentals of the internal combustion engine, 2nd edn. Pearson Prentice-Hall, Upper Saddle River

Taylor CF (1985) The internal combustion engine in theory and practice – Vol 1: Thermodynamics, fluid flow, performance (rev), 2nd edn. The MIT Press, Cambridge, MA

Taylor CF (1985) The internal combustion engine in theory and practice – Vol 2: Combustion, fuels, materials, design (rev). The MIT Press, Cambridge, MA

# Chapter 7
# Alaska Gas Hydrate Research and Field Studies

**S.L. Patil, A.Y. Dandekar, and S. Khataniar**

## Glossary

| | |
|---|---|
| In-place | The term used to estimate gas hydrate resources disregarding technical or economical recoverability. Generally these are the largest estimates [1]. |
| Permafrost | A layer of soil or bedrock that has been continuously frozen for at least 2 years. |
| Petrophysical | Physical properties of the reservoir rock. |
| Resources | Resources are those detected quantities of hydrocarbon that cannot be recovered profitably with the current technology but may be recoverable in the future. It also includes those quantities that geologically may be possible but have not been found yet [2]. |

## Definition of the Subject

### *Gas Hydrates*

Natural gas hydrates are solid, crystalline, ice-like materials made up of small molecules of gases, mainly methane confined inside cages of water molecules. These gas hydrates are solid solutions and are formed when water molecules linked

This chapter was originally published as part of the Encyclopedia of Sustainability Science and Technology edited by Robert A. Meyers. DOI:10.1007/978-1-4419-0851-3

S.L. Patil (✉) • A.Y. Dandekar • S. Khataniar
Institute of Northern Engineering, University of Alaska Fairbanks, 425 Duckering Building, Fairbanks, AK755880 USA
e-mail: slpatil@alaska.edu

R. Malhotra (ed.), *Fossil Energy: Selected Entries from the Encyclopedia of Sustainability Science and Technology*, DOI 10.1007/978-1-4614-5722-0_7,

by hydrogen bonding create cavities (host) that enclose a variety of molecules (guest). These molecules can be $CH_4$, $C_2H_6$, and $CO_2$ to name a few. It is interesting to note that there is no chemical bonding between the host water molecule and the caged molecule. Hydrates can form under high pressures or low temperatures, which are generally the conditions that exist in the Arctic or under deep ocean beds. Huge volumes of gas hydrates are expected to exist at various locations around the globe (Fig. 7.3). It is known that 1 $m^3$ of hydrates upon dissociation release about 180 std $m^3$ of gas, making them a huge potential as a future energy resource [3, 4].

## History of Gas Hydrates

The discovery of hydrates dates way back in 1810 when Sir Humphry Davy conducted an experiment and discovered that an aqueous solution of chlorine can be converted into solid form when cooled below 9°C [5]. Faraday confirmed the existence of this solid compound and now it is known that there are more than 100 molecules that can form hydrates when combined with water. Few molecules that are very commonly found to form hydrates are $CH_4$, $C_2H_6$, $C_3H_8$, $CO_2$, and $H_2S$, with $CH_4$ being the most common molecule to form gas hydrates found in existence. It was as early as the 1950s when through x-ray diffraction methods the crystal structure of hydrates was known, and by late 1960 it was realized that the earth's crust is home to vast quantities of natural gas hydrates [5].

It was Powell, in 1948 at the University of Oxford, who was the first to describe clathrate structure and described clathrate compounds as those inclusion compounds *in which two or more components are associated without ordinary chemical union but through complete enclosure of one set of molecules in a suitable structure formed by another* [5]. The clathrate compounds are divided into two categories. One is aqueous clathrates in which water is the host species, commonly known as clathrate hydrates or gas hydrates, and the other category of hydrates where water is not the host molecule are known as nonaqueous clathrates [5]. It is known that water molecules are linked to each other through hydrogen bonding, which results in the formation of cavities having dimensions of 780 and 920 pm [5]. The molecules that are smaller than the dimension of these small cavities and do not react with the hydrogen bonds of the water molecule form gas hydrates under appropriate conditions of temperature and pressure, thereby providing stability to the structure.

## Introduction

Gas hydrates may become an important global source of clean burning natural gas. Scientists have considered the potential of gas hydrate as an energy resource for nearly four decades. However, until 2002, this knowledge has not been systematically applied to determine the economic practicality of gas hydrate resource development. The estimates of in-place unconventional gas hydrate volume are up to 100 Trillion Cubic Feet (TCF) within the existing oil development infrastructure of the Alaska North Slope (ANS) [6]. This region uniquely combines known gas hydrate presence, an undetermined volume of associated free gas accumulations, and existing production infrastructure (Fig. 7.1).

Initial research on Alaskan gas hydrates was focused on the Prudhoe Bay Unit (PBU), Kuparuk River Unit (KRU), and Milne Point Unit (MPU) infrastructure areas of the Alaska North Slope. These areas were being studied to determine gas hydrate reservoir extent, stratigraphy, structure, continuity, quality, variability, and geophysical and petrophysical property distribution. These initial studies characterized reservoirs and fluids, which led to in-place and recoverable resource estimates and commercial potential determination. Studies were conducted to determine best practices and procedures for gas hydrate drilling, data acquisition, completion, and production. This collaborative research provided practical input to reservoir and economic models, determined the technical feasibility of gas hydrate production, and influenced future exploration and field extension of this potential unconventional gas resource.

A successful drilling project to explore, confirm, and evaluate gas hydrates was conducted in February 2007 through a shallow seismic anomaly interpreted as hydrates. In this project, over 400 ft of sediments containing gas- and water-bearing core and well-developed gas hydrate zones (85% recovery) were cored. An extensive suite of electric logs were run to identify the gas hydrate zones, and gas hydrate saturations were computed. Open-hole pressure-transient analyses and fluid sampling tests were conducted using Schlumberger's Modular Formation Dynamics Tester (MDT™).

A second ANS gas hydrate research project was conducted for the Barrow Gas Field (BGF). Barrow, Alaska, is a remote and the northernmost city in the United States. The energy demands of the city and nearby localities are met by natural gas supply from the Barrow Gas Field consisting of three different gas pools: the East Barrow (EB) pool, the South Barrow (SB) pool, and the Walakpa (WAL) pool (Fig. 7.2).

Gas hydrate resource potential of the three BGF pools was analyzed by developing gas hydrate stability models. Material balance studies were performed on East Barrow (EB) gas pool to understand reservoir drive mechanisms and qualitatively estimate the hydrate resource. Field-scale dynamic reservoir simulation models were developed for the EB and WAL gas pools. Production history data were matched, reservoir drive mechanisms were confirmed, free gas and hydrate resources were quantified, hydrate dissociation patterns were examined, optimum locations for drilling infill wells were identified, and future production scenarios were simulated.

The vast amount of unconventional gas in-place combined with evaluation of conventional Alaska North Slope gas commercialization creates industry-government alignment to assess this potential future resource. This region uniquely combines known gas hydrate presence, an undetermined volume of associated free

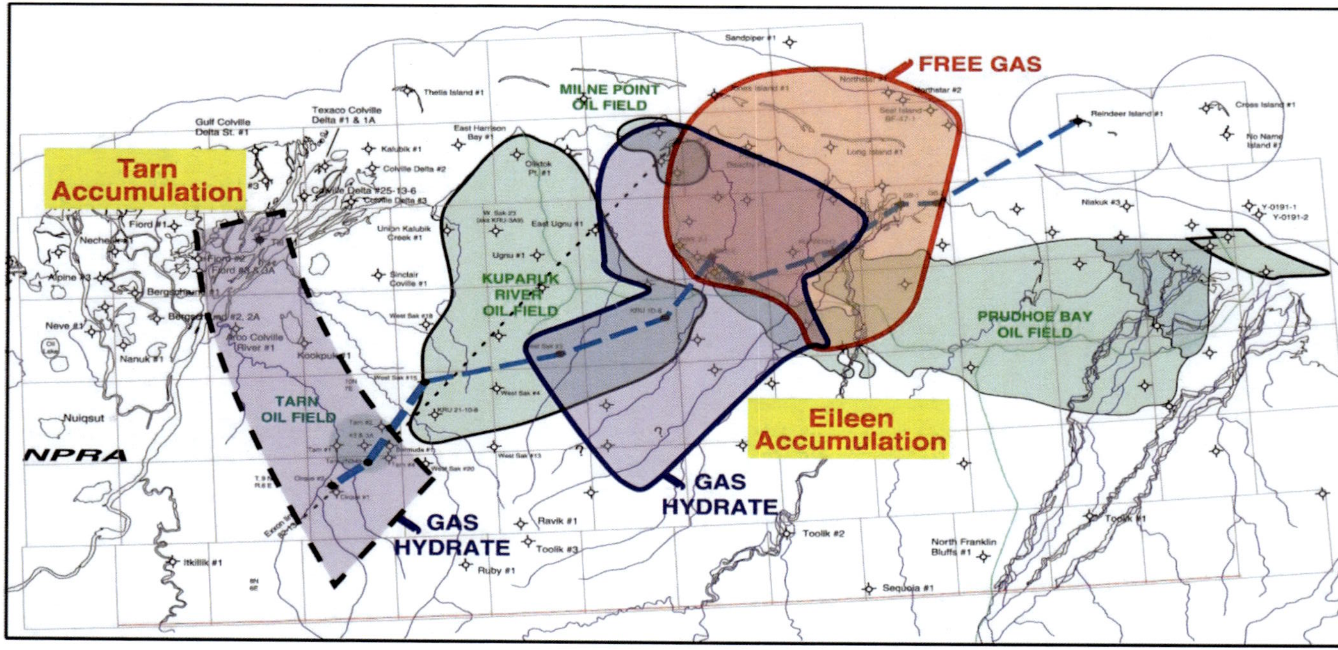


Fig. 7.1 Alaska North Slope oil field developments with Eileen and Tarn gas hydrate trends (Courtesy Collett [32])

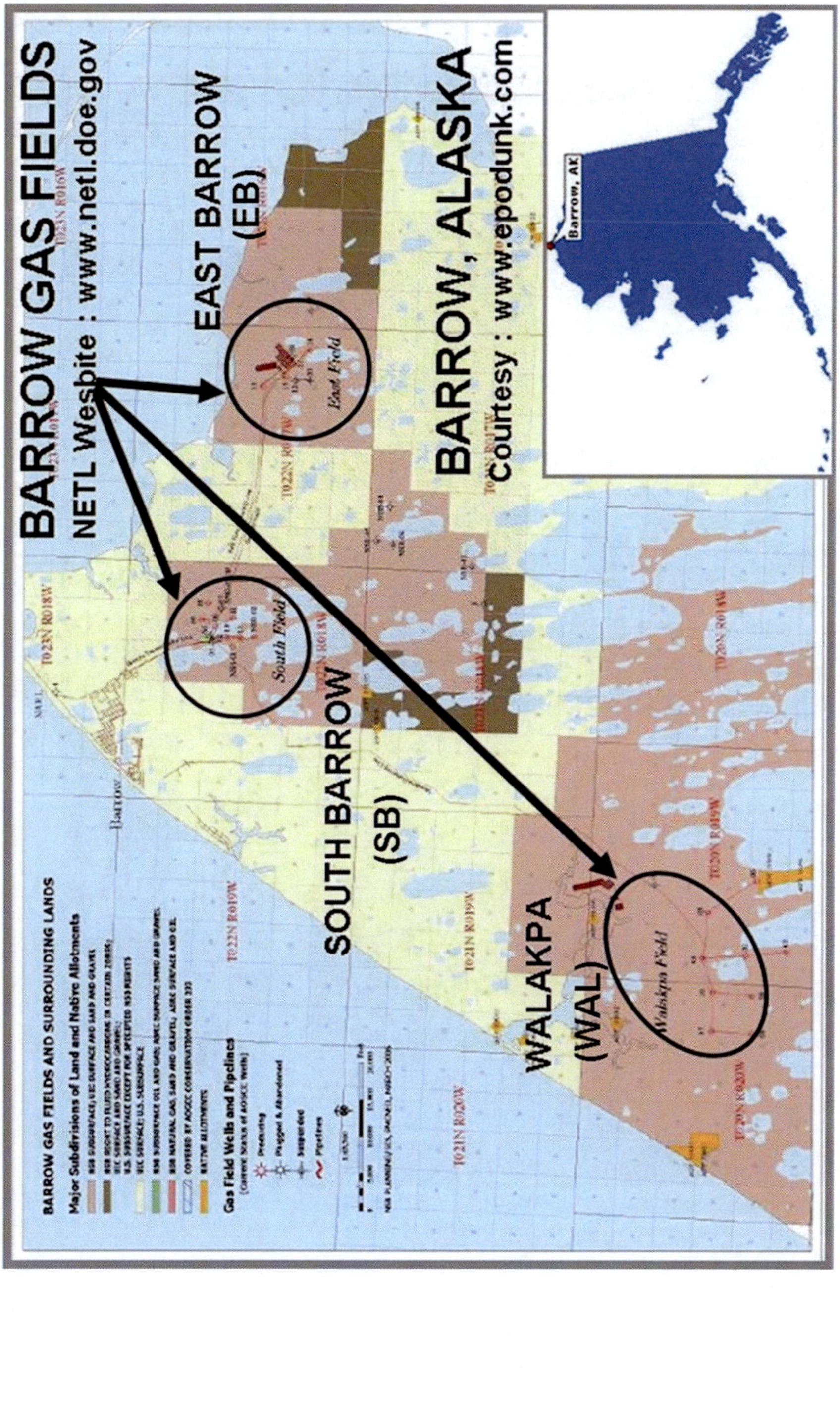


**Fig. 7.2** Barrow gas fields, Alaska (Modified from NETL Website, www.netl.doe.gov)

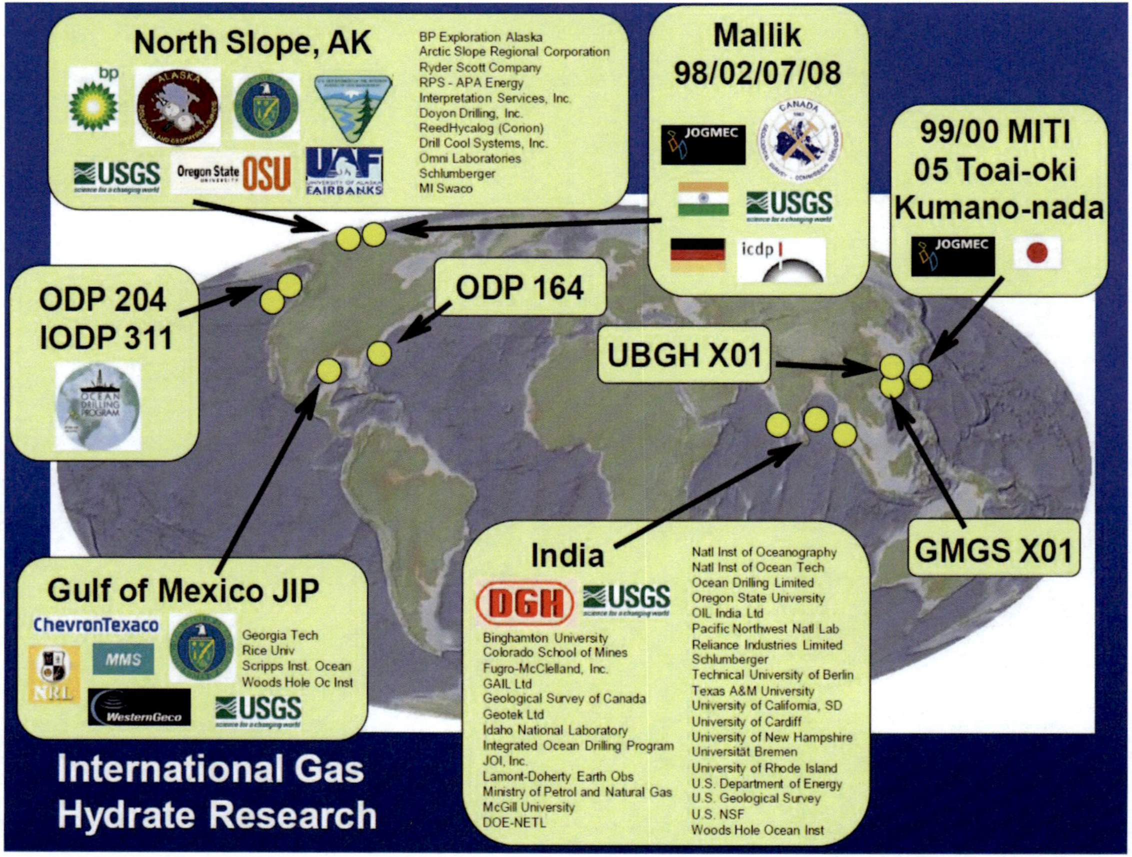


**Fig. 7.3** World distribution of gas hydrates (Adapted from oral presentation at AAPG Annual Convention, San Antonio, Texas, April 20–23, 2008)

gas accumulations, and existing production infrastructure. However, many technical and economic challenges must be resolved before the gas hydrate becomes a viable commercial hydrocarbon resource.

## Global Distribution of Hydrate Resource

Global volumes of gas hydrate estimated till now span several orders of magnitude and there is no certainty in methane hydrate resource estimates across the globe. This uncertainty in estimating the volume of gas hydrates present is majorly because of two factors: (1) The commonly used seismic methods and other geophysical methods that are used in estimating conventional hydrocarbon resources are not sensitive to the huge volumes of hydrates present in the seafloor section. (2) Drilling activities to directly confirm the presence of hydrates have a limited applicability due to limited spatial extent [7]. There is no direct evidence which can be used in estimating the occurrence of in-place gas hydrates. There are 23 locations where the presence of gas hydrates has been confirmed through the recovery of hydrate samples, 20 of them being in the oceanic and the rest 3 in the arctic environments [8, 9]. Since the discovery of gas hydrates, researchers have tried to estimate the amount of gas trapped in hydrates none being certain. Going by even the most conservative approximation of the estimation of the trapped gas, the amount of gas trapped in hydrates is enormous.

HEI (Hydrate Energy International) along with the International Institute for Applied Systems Analysis (IIASA) with funding from the World Bank, United Nations Organizations, and national governments has undertaken the task of assessing the global gas hydrate potential. For this study, data from every continental margin was studied and appropriate depositional models were utilized. The data was reported for 18 regions defined by the United Nations. For the Arctic Ocean, separate resource assessment was done without regard for national boundaries and for the southern ocean (from the coast of Antarctica north to 60° south latitude) [10]. The result of global hydrate resource estimate has been tabulated in Table 7.1 after Johnson [10]. Figure 7.4 depicts the hydrate resource pyramid after Boswell and Collett [11].

The production and commercialization of gas hydrate deposits rely heavily on two important factors: (1) the advancement in technology to prove and produce from these resources overcoming existing obstacles and (2) a natural gas supply market accelerating the need to progress and improve the gas hydrates resource potential [12]. Although various research activities are on way worldwide, gas hydrates are the most unlikely unconventional resource that will be tapped in the near future. But based on the demand for energy in the future, these research and exploration activities need to go on and probably in a much more aggressive way to make these a viable energy source. Figure 7.5 depicts the time line for post 1990 global drilling efforts after Ruppel [12].

**Table 7.1** Gas in-place in hydrate-bearing sands (After Johnson [10])

| Region (United Nations designation) | Gas in-place range (TCF) | Gas in-place median (TCF) |
|---|---|---|
| United States | 1,500–15,434 | 7,013 |
| Canada | 533–8,979 | 2,228 |
| Western Europe | 36–14,858 | 1,425 |
| Central and Eastern Europe | 0–105 | 13 |
| Former Soviet Union | 1,524–10,235 | 3,829 |
| North Africa | 6–1,829 | 218 |
| Eastern Africa | 42–25,695 | 1,827 |
| Western and Central Africa | 79–26,672 | 3,181 |
| Southern Africa | 121–26,369 | 3,139 |
| Middle East | 31–3,848 | 573 |
| China | 10–1,788 | 177 |
| Other East Asia | 14–2,703 | 371 |
| India | 36–6,268 | 933 |
| Other South Asia | 20–3,497 | 557 |
| Japan | 71–471 | 212 |
| Oceania | 38–6,750 | 811 |
| Other Pacific Asia | 64–25,946 | 1,654 |
| Latin America and the Caribbean | 258–31,804 | 4,940 |
| Southern Ocean | 144–45,217 | 3,589 |
| Arctic Ocean | 178–55,524 | 6,621 |
| Total | 4,705–31,3992 | 43,311 |

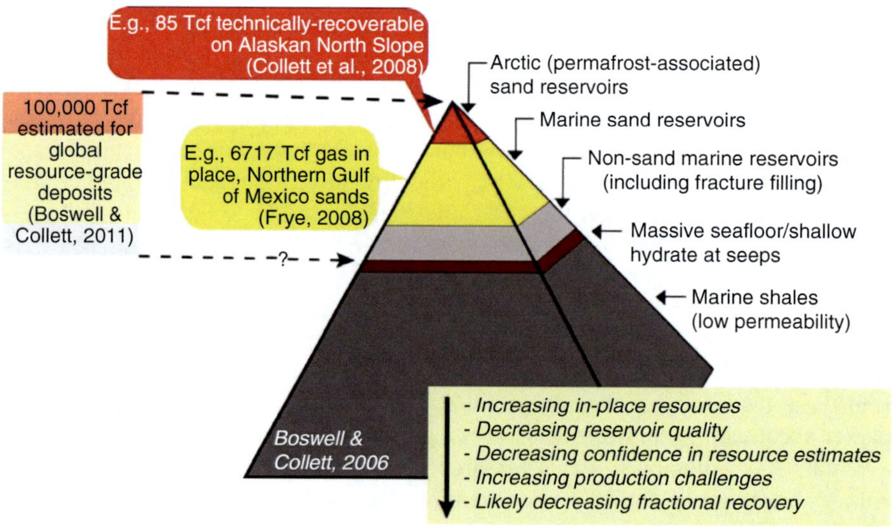


**Fig. 7.4** The hydrate resource pyramid (Ruppel [12])

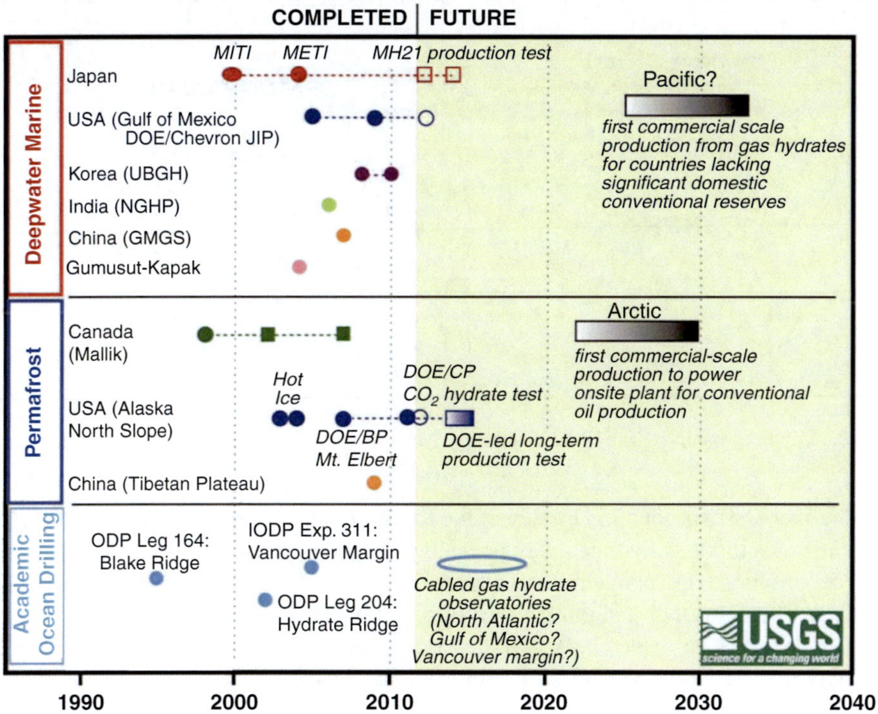


**Fig. 7.5** Timeline of major post-1990 gas hydrate field programs and future activities. *Circles* correspond to logging and/or coring, while *squares/rectangles* denote activities that included/will include production testing. *Solid* symbols denote completed activities, and *open* symbols are potential or planned activities. *Rectangles* filled with *shaded pattern* refer to longer-term production activities that remain prospective (After Ruppel [12])

## Gas Hydrate Resource Potential

Gas hydrates are naturally occurring ice-like substances composed of water and gas, in which a solid water-lattice accommodates gas molecules in a cage-like structure called a clathrate. This clathrate structure can store up to 180 unit volumes of methane gas per 1 unit volume clathrate. This large gas storage capacity makes methane hydrate a very attractive concentrated potential unconventional resource.

Gas hydrates accumulate within reservoirs located in permafrost regions and within subsea sediments of outer continental margins. While methane, propane, and other gases can be included in the hydrate structure, methane hydrates appear to be the most common in nature. The amount of methane sequestered in gas hydrates is probably enormous. However, the in-place and recoverable gas estimates are highly speculative and also dependent upon the conventional petroleum system components of source, charge, reservoir, trap, seal, and a highly uncertain recovery factor, as well as occurrence within the gas hydrate stability field (Fig. 7.6).

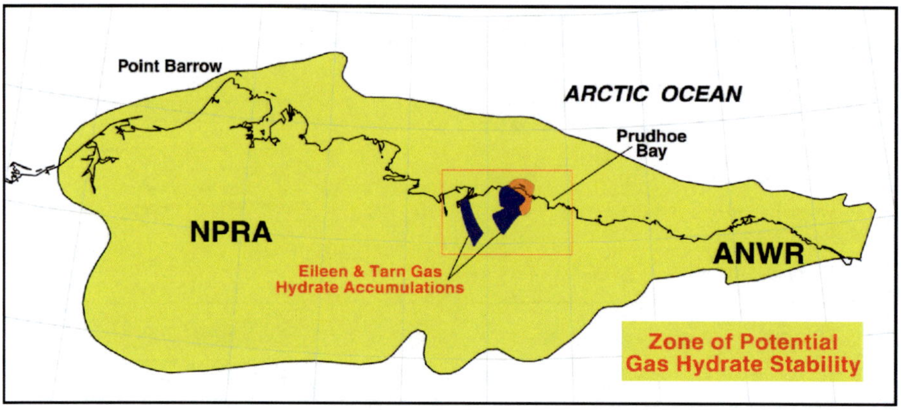


**Fig. 7.6** Zone of potential gas hydrate stability, Alaska, North Slope (Courtesy Collett [32])

The successful production history of the Russian Messoyakha gas hydrate field [6] demonstrates that gas hydrates may provide an immediate source of natural gas that can be produced by conventional depressurization of an adjacent free gas. Gas hydrates also represent a significant drilling and production hazard. Russian, Canadian, and American researchers have described numerous problems associated with gas hydrates, including well control problems and casing failures.

## Review of Global Gas Hydrate Projects

Several countries have active gas hydrate field programs (Fig. 7.3). The following gives an overview/status of these research activities.

### *Japan*

The first remarkable gas hydrate research program in Japan was established in 1995 by the Government of Japan along with the Ministry of Economy, Trade, and Industry (METI). Since then the METI was very proactive in gas hydrate research field. It contributed to Canada's Mallik 2002 Gas Hydrate Production Research Well Program. In 2001, METI established Methane Hydrate R&D Program, which is currently in its Phase 2 (2009–2015). Within the Phase 2 production tests on the eastern Nankai Trough are planned for 2012 and 2014 respectively. Phase 2 also considers environmental impact assessment and development of optimal methane hydrate production approaches as its objectives [13].

## *China*

Guangzhou Center for Gas Hydrate research was founded in 2004 to expand the research and to evaluate gas hydrate resource potential in offshore China. Guangzhou Marine Geological Survey (GMGS) conducted and expedition GMGS-1 in spring 2007 to perform offshore drilling in Shenhu area of the South China Sea. During the expedition eight holes were drilled. Wireline logging was performed on each site. Five wells were cored [14]. Future expeditions to the South China Sea are being considered. Besides, hydrate exploratory drilling takes place in Qinghai-Tibet Plateau.

## *South Korea*

Gas Hydrate Research and Development Organization (GHDO-K), supported by the Ministry of Commerce, Industry, and Energy, is a South Korean gas hydrate program started in 2000. Its main objectives are to confirm gas hydrate resources in the East Sea and to commercially produce gas hydrates by 2015 [14]. It started with seismic data acquisition in the East Sea with shallow coring in the Ulleung Basin in order to define its gas hydrate potential. Ulleung Basin Gas Hydrate Expedition 1 (UBGH1) was the first exploration and drilling expedition launched on behalf of the Korean Institute of Geoscience and Mineral Resources in late 2007. During the expedition five wells were drilled with logging, and while drilling, some of them were cored as well [14]. In July 2010, the UBGH2 expedition was launched. It included vast amount of data acquired and lasted till early October. Subsequently, South Korea is planning field production tests (GHDO-K, 2010).

## *India*

National Gas Hydrate Program (NGHP) of India was initiated in 1997. Its primary goals are performance of ocean drilling, coring, logging, and other activities in order to explore gas hydrate deposits and evaluation of the possibility of their economical production. The NGHP Expedition 1 was successfully conducted during mid-2006. During the expedition, 39 wells were drilled in various sites which are: Kerala–Konkan, Krishna–Godavari, Mahanadi, and Andaman. Special attention was paid to Krishna–Godavari gas hydrate site, where more than 10 wells were drilled [14]. Extensive coring and logging were performed in a number of wells and vast amounts of data were collected. NGHP Expedition 02 was also considered as a future possibility.

## *United States*

In the United States, the Department of Energy (DOE) has been conducting extensive research on gas hydrates for the past several decades. Ocean Drilling Program (ODP) Leg 164 located on Blake Ridge, Atlantic shelf, was the first special attempt to identify gas hydrates [15]. It recovered hydrate cores and documented gas hydrate presence within the sediment [14]. ODP Leg 204 was another deep-sea expedition which discovered gas hydrates in the Hydrate Ridge, located on the Cascadia continental margin. Gulf of Mexico Gas-hydrate JIP consortium was formed in 2001 by industry participants and government agencies in partnership with DOE and led by Chevron. Its goals were studying hydrates both as drilling hazards and also as a potential methane resource. In 2005, it conducted scientific drilling and coring in the gulf [14]. On the North Slope of Alaska, gas hydrates encountered onshore are associated with permafrost. Despite an apparent likelihood of offshore gas hydrate existence in the Arctic Ocean, there is no proof so far. In 2002–2003 Anadarko Petroleum Corporation drilled the Hot Ice gas hydrate research well in Tarn gas hydrate accumulation on the North Slope of Alaska. However, no gas hydrates were encountered [14]. In 2007, the Mount Elbert Gas Hydrate Stratigraphic Test Well was drilled by BP Exploration (Alaska) on behalf of DOE. It was located on the Milne Point, Eileen gas hydrate accumulation. It revealed gas hydrates and even cores were recovered.

## *Canada*

Canada does not have any official gas hydrate program. However, extensive study was performed in Canada of which the remarkable ones are Cascadia and Mallik. Both gas hydrate occurrences are associated with permafrost. ODP Leg 146 and IODP Expedition 311 to the offshore Vancouver Island discovered gas hydrate existence in Cascadia continental margin. Numerous data including seismic, logs, and chloride concentration infer gas hydrate presence. Within IODP Expedition 311, hydrate-bearing cores were retrieved. Mallik-38 onshore well in the Mackenzie River Delta was believed to be drilled through 100 m of hydrate-bearing sandstone. Subsequently in 1998, Geological Survey of Canada on behalf of Japan Petroleum Exploration Company Ltd and Japan National Oil Corporation drilled Mallik 2L-38 gas hydrate research well. Coring and logging both revealed gas hydrates within the sandstone. As a next step, within Mallik 2002 program the partners drilled three more wells: Mallik 3L-38, Mallik 4L-38, and Mallik 5L-38 [14].

## *New Zealand*

In 2010, Gas Hydrate Resources (GHR) program was funded for 2 years by the New Zealand Foundation for Research, Science, and Technology (FRST). It is mostly focused on Hikurangi margin located to the east of the islands. The program plans are: analysis of existing 2D and 3D seismic data acquisition, seafloor sampling in spring 2011, and production modeling [16].

## History of Gas Hydrate Exploration in Alaska

The North Slope of Alaska contains two large gas hydrate accumulations within and near the Prudhoe Bay and the Tarn oil field developments (Fig. 7.1). This confirms the possibility that gas hydrates may represent an important future energy resource. The occurrence of gas hydrates on the North Slope of Alaska was confirmed in 1972 with data from the Northwest Eileen State-2 well, the first dedicated gas hydrate core and test well, located near the Prudhoe Bay Oil Field. Most of the gas hydrates near this Eileen trend discovery well occur in a series of laterally continuous sandstone reservoirs and are geographically restricted to the area overlying the western part of the Prudhoe Bay Oil Field and the eastern part of the Kuparuk River Oil Field. Three-dimensional seismic surveys in the Prudhoe Bay Oil Field indicate the presence of an undetermined amount of free gas trapped down deep below some of the gas-hydrate-bearing units. The volume of gas within the gas hydrates within the Eileen trend is estimated to be about 37–44 trillion cubic feet.

Recently, data from wells along the western margin of the Kuparuk River Field have shown presence of the large Tarn trend gas hydrate accumulation overlying the Tarn oil field. The Cirque-1 well, located about 4 miles southwest of the Tarn oil field, experienced severe well control problems in 1992 after drilling through what appeared to be a free gas interval possibly trapped below the gas hydrate stability zone. The gas-hydrate-bearing stratigraphic interval within the Tarn trend appears to be the up-dip equivalent of the West Sak sands. These sands are estimated to contain more than 20 billion barrels of in-place heavy oil and are the focus of recent development activity. Preliminary analyses of other recently completed wells along the western margin of the Kuparuk River oil field suggest that the volume of in-place gas hydrates within the Tarn trend may exceed that of the Eileen trend.

## Alaska Gas Hydrate Production Potential

The production potential of the Eileen or Tarn gas hydrate accumulations has not been adequately tested. This area is a focus of the gas hydrate research programs by the industry, led by the US Department of Energy. In order to produce gas from gas hydrate, the in situ pressure-temperature regime must change to dissociate gas. Decreasing the reservoir pressure, increasing the reservoir temperature, and/or use of chemicals can allow gas to dissociate from the solid gas hydrate structure. Among the various techniques for production of natural gas from gas hydrates, the most economically promising method is considered to be depressurization of an adjacent free gas.

A growing body of evidence suggests that a huge volume of natural gas is stored as gas hydrates in northern Alaska and that production of natural gas from gas hydrates may be technically feasible. However, numerous technical and economic challenges must be resolved before this potential resource can be developed. Many wells have penetrated gas hydrate during oil production operations on the Alaska North Slope. During this time, gas hydrates were known primarily as a drilling hazard. Industry has considered the resource potential of conventional Alaska North Slope gas during industry and government efforts in working toward an Alaska North Slope gas pipeline. Consideration of the resource potential of this conventional Alaska North Slope gas created the industry–government alignment necessary to also consider the resource potential of the potentially huge (40–100 TCF in-place) unconventional gas hydrate accumulations beneath existing production infrastructure. Two of the wells recently drilled as a part of the industry–government research programs to explore and evaluate gas hydrate production potential on the ANS were the Hot Ice #1 and the Mt. Elbert-01.

## *The Hot Ice No. 1*

The Hot Ice No. 1 well was drilled on the Anadarko Petroleum leases, southwest of the main Kuparuk development, just outside the Kuparuk River Unit boundary. Drilling began in 2003 and ended in 2004. The well was drilled, cored, and logged to a depth of 2,300 ft (700 m), and ultimately abandoned as no hydrates were found [17]. However, the lessons learned from drilling this well were useful in designing future hydrate exploration wells such as the Mt. Elbert.

Hot Ice No.1 well also used water-based polymer drilling mud with KCl (salt) for freezing-point depression and a mud chiller for circulating chilled mud. This was problematic, with freeze-ups and mechanical problems impeding progress. This knowledge influenced decisions with regard to coring fluids and mud chilling on the Mt. Elbert-01 Gas Hydrate Stratigraphic Test Well.

## *The Mt. Elbert-01*

The drilling, coring, and testing of Mt. Elbert-01 Gas Hydrate Stratigraphic Test Well was part of the DOE/NETL-funded "Alaska North Slope Gas Hydrate Reservoir Characterization Project," and was a cooperative effort between industry, academia, and government.

The Mt. Elbert-01 Hydrate Test well was located on a temporary ice pad in the Milne Point Unit on the ANS, south of the B-Pad and east of the E-Pad, with ice road access from the B-Pad. The location of the Mt. Elbert-01 relative to the Milne Point infrastructure is shown in Fig. 7.7. The location of the Mt. Elbert-01 Hydrate Well was chosen so as to drill in an area with "known" hydrate accumulations with

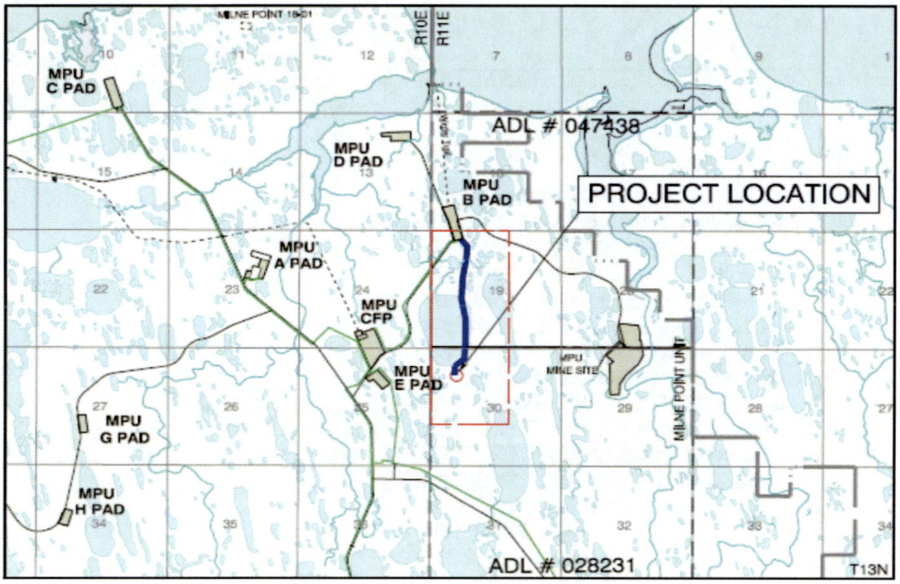


Fig. 7.7 Section Map – Mt. Elbert-01 location relative to Milne Point infrastructure (Courtesy Hanson [18])

isolation (stand-off) from the thermal effects of any adjacent wells. This would allow the best chance to find and test undisturbed hydrates.

Based on well log data from nearby offset wells, it was estimated that the Base of the Ice-Bearing Permafrost (BIBPF) was at approximately 1,850 ft TVDss and the Base of the Gas Hydrate Stability Zone (BGHSZ) was at about 2,950 ft TVDss. Coring operations were planned from three potential gas-hydrate-bearing zones (zones D, C, and B) within this depth range. A total of 23 coring runs were made using the wireline retrievable coring method, which prevented the cored formations from being exposed to pipe tripping effects. This was the first application of wireline retrievable coring on the Alaska North Slope. Figure 7.8 shows a summary of the coring runs. The overall core recovery was 85%. The use of a chilled mineral-oil-based coring fluid and wireline retrievable coring method are believed to be mainly responsible for such high core recovery. The main advantage of a chilled oil-based coring fluid over a chilled water-based coring fluid is that the oil-based fluid would not freeze. Although additives were used, freezing of water-based coring fluids was a major problem in the previous hydrate wells Mallik and Hot Ice #1 [18]. Details of the drilling plan and coring operations for Mt. Elbert-01 are given by Hanson [18].

The logging suite for Mt. Elbert-01 consisted of three electrical resistivity tools and an extended MDT testing. Figure 7.9 shows a sample plot of the well logs, clearly delineating the upper D hydrate zone. Full analysis of Mt. Elbert-01 well logs showed no evidence of significant "contacts" between free gas sands and hydrate-bearing sediments [18]. Mt. Elbert-01 is the world's first successful application of the MDT™ tool in open-hole in gas hydrates.

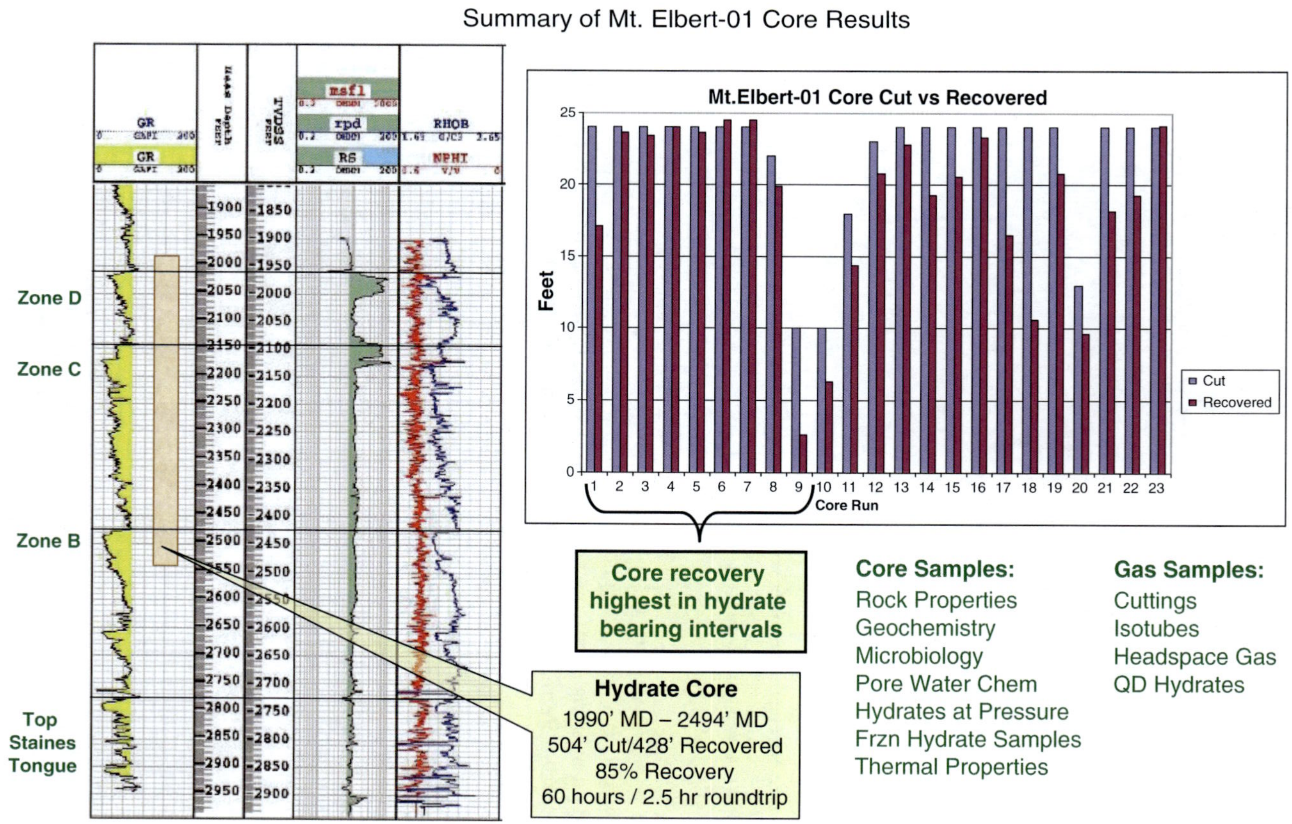


**Fig. 7.8** Summary of core recovery by run relative to the log through the hydrate interval (Courtesy Hanson [18])

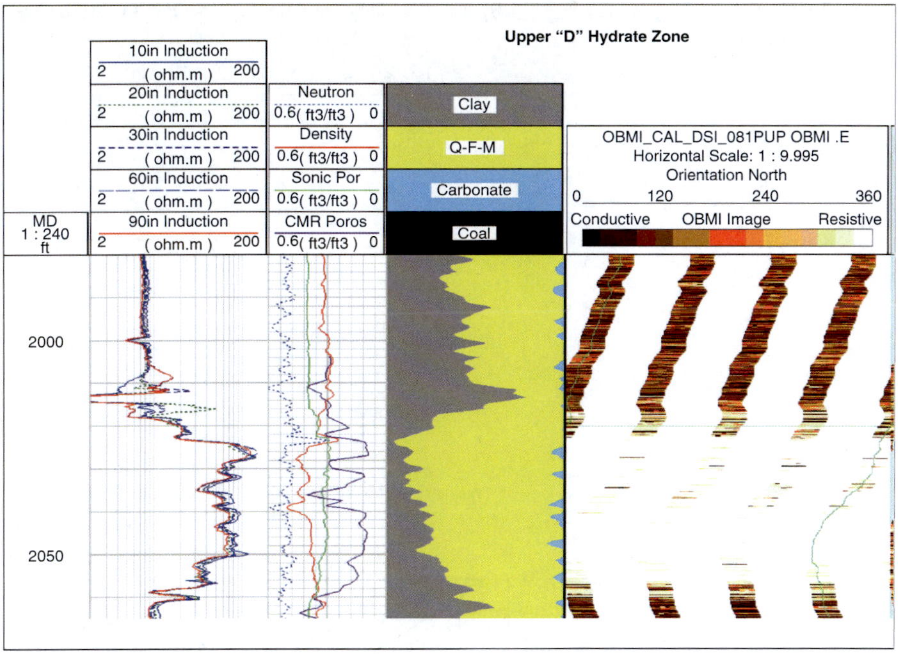


**Fig. 7.9** Composite RT Scanner, Neutron, Density, Sonic, CMR and OBMI Log showing the upper "D" – Hydrate interval (Courtesy Hanson [18])

## North Slope Borough Methane Hydrate Project Review

Barrow, Alaska, is a remote and the northernmost city in the United States. The city serves both as the economic and administrative center of the North Slope Borough (NSB). The energy demands of the city and nearby localities are met by natural gas supply from the Barrow Gas Fields (BGF) consisting of three different gas pools: the East Barrow (EB) pool, the South Barrow (SB) pool, and the Walakpa (WAL) pool (Fig. 7.2). The Barrow Gas Fields have been producing natural gas for the last 25 years. It has been estimated that the proven gas reserves in these fields exceed over 250 billion standard cubic foot (BCF) [19, 20]. Based on the current demands for natural gas in the Barrow region and to enhance natural gas production and expand production facilities in order to provide cost effective and assured source of energy to Barrow and nearby areas, the Barrow gas hydrate project was proposed by the NSB to the US Department of Energy (DOE). In 2006, DOE and NSB jointly funded a research program to characterize and quantify methane hydrate resource potential associated with the BGF, with the ultimate goal of producing gas from the hydrate formations associated with individual gas pools of BGF.

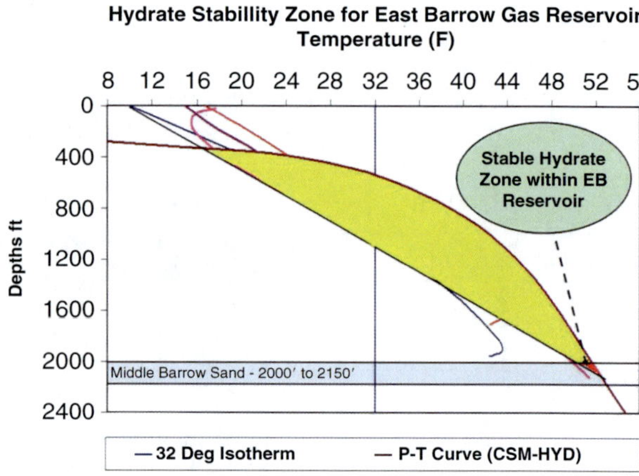


**Fig. 7.10** Hydrate stability model for EB pool

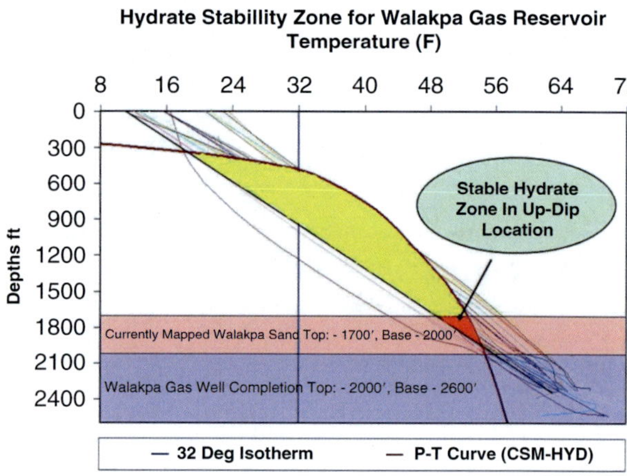


**Fig. 7.11** Hydrate stability model for WAL pool (Courtesy Singh [21])

## *Gas Hydrate Stability Modeling for BGF Pools*

An engineering approach was first adopted to develop methane hydrate stability models for the three gas pools. This was performed on the basis of gas analysis, formation water salinity, and pressure and temperature gradient data available from various sources referenced in [21]. The studied literature indicated a gas composition of practically pure methane (99%) and water salinities in the range of 0–4% w/w. Details of the calculation procedures can be found in [21]. The hydrate stability models for EB and WAL gas pools are provided in Figs. 7.10 and 7.11. Results from the EB pool calculations indicate the existence of hydrate stability zones within the pay zone of 2,000–2,150 ft, which suggests that the in situ hydrates are possibly associated with free gas reservoir. Similar to EB, the hydrate stability model for WAL pool also showed promising results in that stable hydrate zones were not found within deeper zones of the WAL sand where existing gas wells were completed within the 2,000–2,600 ft interval but showed stable

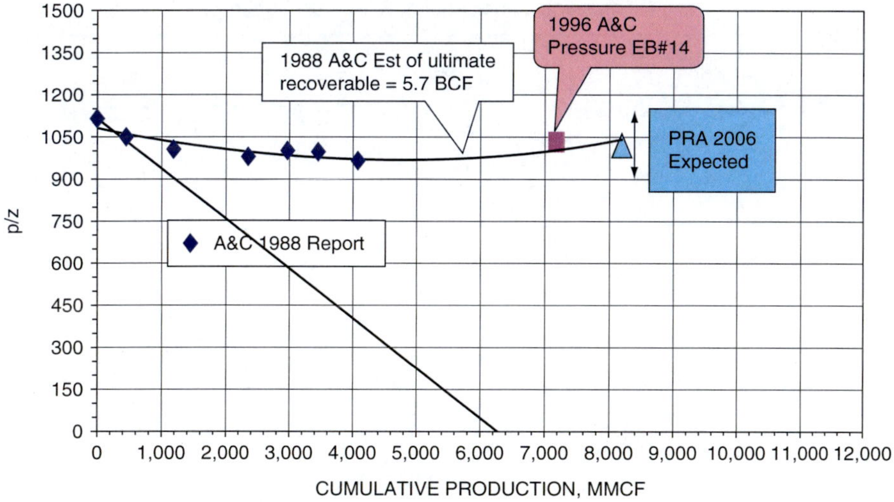


**Fig. 7.12** EB gas pool performance (Stokes and Walsh [23])

envelope existing in up-dip locations of the reservoir between 1,700 and 2,000 ft. To confirm these results, detailed geological and geophysical analysis of unexplored regions of the pools are required to determine the association of gas hydrates with the prospective gas pools. Singh [21] conducted a material balance study to obtain additional confirmation of the presence of hydrates, especially associated with the EB pools.

## *The East Barrow Field Material Balance Study*

The objective of the material balance study was to qualitatively understand the behavior of the EB pool under different drive mechanisms, and to confirm if hydrate dissociation was one of them. Stokes and Walsh [22] performed pressure buildup tests on two EB wells and later [23] compared the static reservoir pressure data with historical pressure data reported by Allen and Crouch [24]. Figure 7.12 represents the historical performance plot of the EB gas pool, which shows an initial drop in the static reservoir pressure with gas production, indicating a conventional response of a volumetric gas reservoir producing by gas expansion. However, long-term production data showed gradual pressure stabilization, which indicated the possibility of water influx from a communicating aquifer. But the possibility of a communicating aquifer was ruled out because (1) there was negligible water production ($\sim$1,600 bbls up to September 2007), (2) cumulative gas production of $\sim$8 BSCF by 2007 was much higher than Allen and Crouch's [24] estimate of $\sim$6 BSCF, and (3) none of the EB wells had watered out. Thus, the EB field performance data of Fig. 7.13 suggested possible contribution from an unknown drive mechanism, which could be gas recharge from dissociating gas hydrates, or water influx from a new aquifer.

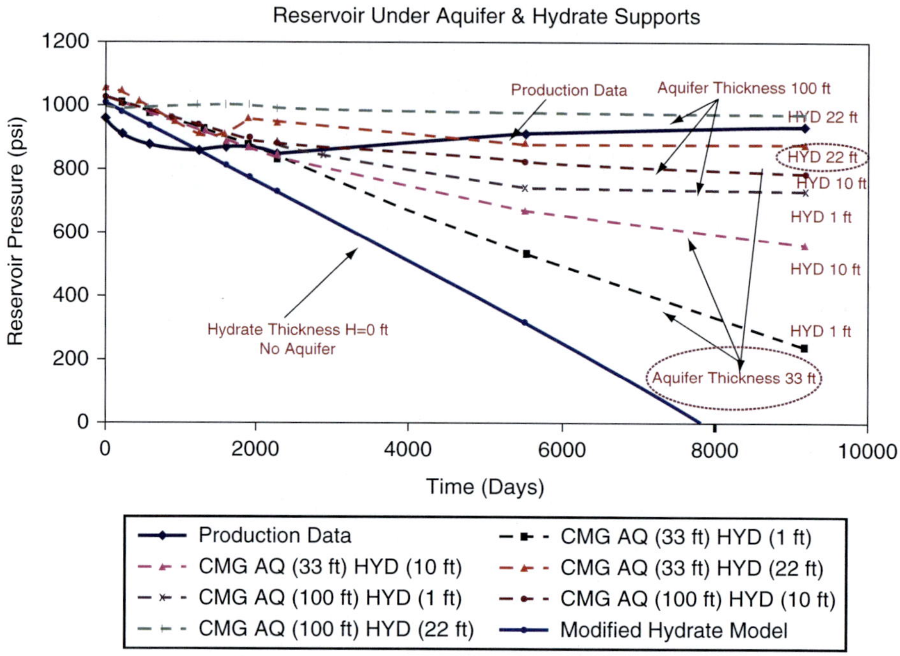


**Fig. 7.13** Hydrate- and aquifer-supported gas reservoir performance prediction for EB pool (Courtesy Singh [21])

In the EB material balance study, the overall performance of a tank type, static reservoir model representing the EB gas pool was compared against the available production data. The following drive mechanisms were studied with the tank model to see which drive mechanism would match actual production data: (1) volumetric gas tank, (2) water drive, (3) dissociating-hydrate drive, and (4) combination of water and hydrate drive. Reservoir and fluid properties and other details are found in Singh [21].

The study showed that the actual production data from the EB field could not be matched by assuming volumetric behavior, water drive, or dissociating-hydrate drive alone. However, a good match of modeled production response and actual production data, shown in Fig. 7.13 by red dotted circles, was obtained in the scenario where a free gas reservoir was in communication with a weak aquifer, but associated with a thick hydrate layer. Thus, a combination of aquifer support and hydrate dissociation is the most likely drive mechanism in the EB pool.

## *Field Scale Reservoir Simulation Study*

Using the reservoir drive mechanism established through the material balance study on the EB pool, Singh [21] conducted numerical reservoir modeling studies using a full-scale, three-dimensional (3D) dynamic reservoir model. A commercial reservoir simulator (CMG-STARS) was used to perform the simulation studies. The static reservoir model of Panda and Morahan [25] was imported and built upon by adding reservoir properties and drive mechanisms. Other details can be found in Singh [21].

The simulation study predicted field-wide performance of the EB pool for a period of 30 years and investigated the impacts of well completion and orientation (vertical wells versus horizontal wells). History matching of the simulation results confirmed the previously estimated combination drive as the primary recovery mechanism for this reservoir. However, gas production by expansion dominates over hydrate dissociation mechanism after approximately 15–20 years of production. The simulation study also indicated that infill drilling using horizontal producing wells would offer the ideal combination of lower water production and higher gas recovery.

## Future Directions

Various methods have been studied and have been proved to be successful in the extraction of gas from gas hydrate deposits. These are depressurization, thermal methods, and the use of inhibitors (e.g., methanol or glycols). Gas hydrate dissociation is an endothermic reaction and as the hydrate deposits dissociate, the surrounding temperatures decrease resulting in further hydrate reformation, thereby reducing the amount of gas that can be produced. Thus any method or a combination of methods used for hydrate dissociation has to not only consider the energy required to dissociate hydrates but also has to take in effect the energy that is required to counter the drop in temperature that takes place due to the endothermic nature of hydrate dissociation [26]. Yutaek and SeongPil [27] have shown how kinetic hydrate inhibitors (KHI) can be used to avoid hydrate reformation, the required concentration of these KHI being 0.5–5 wt%.

Recent experiments conducted by Shiang-Tai and Yen-Tien [28] illustrate how replacement of $CO_2$ with $CH_4$ in methane hydrates can help in the production of methane gas from methane hydrate deposits. When, liquid carbon dioxide is introduced into the methane hydrate system, rapid replacement of solid methane with carbon dioxide molecules takes place. This substitution of methane with carbon dioxide occurs spontaneously and large amounts of carbon dioxide are observed to penetrate methane hydrate molecules and expel methane especially near the hydrate melting point [28]. It has been shown how steam injection can be used effectively in producing gas from gas hydrate deposits [29], and the use of hot-brine injection for the production of gas from natural gas hydrates as well as how various factors like the temperature of the injected solvent, the injection rate, or the injection time impact the production of gas from such deposits has also been shown [30]. It is also proposed that a combination of thermal and depressurization technique can be used to produce from gas hydrates that will help in improving the economic efficiency, paving the way for the development of NGH reservoirs [30].

Weitemeyer and Constable [7] have shown how controlled source electromagnetics can be used to map gas hydrates. Further advancement of this technique has led to the creation of a fixed offset towed sensor that was able to image gas hydrates at MC118 with successful imaging of hydrates in the other three sites in the Gulf of Mexico [7]. Dai [31] has put forward a method to locate and quantify gas hydrate deposit using long-offset reflection seismic information, and since this method is not based on the AVO inversion method it stays away from the uncertainties that are a part of AVO-based seismic studies.

## Conclusions

Significant improvements in the methods for drilling and evaluating potential gas hydrate zones in ANS have taken place in recent years. The drilling and formation evaluation of the Mt. Elbert-01 stratigraphic test well has successfully field-tested the modern geophysical prospecting methods used and has enabled the selection of target zones and field parameters for potential future production testing. The chilled, 100% mineral-oil-based coring fluid provided a significant improvement in wellbore quality as compared to the freezing-point-depressed water-based fluids used in the previous gas hydrate exploration projects. Core recovery, core quality, and electrical well log quality were significantly improved by using oil-based coring fluid, and operational downtime due to freeze-ups was eliminated.

The study of a producing gas field in the ANS has demonstrated the feasibility of producing natural gas from gas hydrates by depressurization to some extent. Diagnostic methodologies were developed to examine the production data from the Barrow gas fields to test the two fields for possible hydrate accumulations and to understand the gas recovery mechanisms. Engineering approach was followed to construct hydrate stability models and material balance procedures for identifying the reservoir drive mechanisms. Based on these results, state-of-the-art dynamic reservoir simulation models were developed using an advanced commercial reservoir simulator to study hydrate reservoir development. Reservoir modeling confirmed the key findings of previous studies and projected the future performance of the reservoir under different scenarios. Finally, hydrate resource potential of EB and WAL pools appears to be promising, and both pools have considerable hydrate potential that can supply large volumes of gas for future years.

## Disclaimer

This entry was prepared as an account of work sponsored by an agency of the US government. Neither the US government nor an agency thereof, nor any of their employees, makes any warranty, expressed or implied, or assumes any legal liability or responsibility for the accuracy, completeness, or usefulness of any information, apparatus, product, or process disclosed, or represents that its use would not infringe privately owned rights. References herein to any specific commercial product, process, or service by trade name, trademark, manufacture, or otherwise do not necessarily constitute or imply its endorsement, recommendation, or favoring by the US government or any agency thereof. The views and opinions of the author expressed herein do not necessarily state or reflect those of the US government or any agency thereof.

This entry is also based on research conducted by two petroleum engineering graduate students (Paul Hanson and Praveen Kumar) from the University of Alaska Fairbanks for their M.S. degree in petroleum engineering. The results of the two US Department of Energy–funded gas hydrate projects in Alaska (BP Alaska Gas Hydrate Project and the North Slope Borough Gas Hydrate Project) are also included. Several other companies, organizations associated, and individuals associated with these two projects have contributed to the work reported here.

**Acknowledgment** The authors are grateful to the US Department of Energy (USDOE), BP Exploration (Alaska) Inc., North Slope Borough (NSB), Petrotechnical Resources of Alaska (PRA), and Petroleum Development Laboratory (PDL) at UAF for their financial support in accomplishing this work. The authors would also like to thank Dr. Tim Collet (USGS), Mr. Robert Hunter (ASRC Energy Services), Mr. Tom Walsh (Petrotechnical Resources Alaska, Inc.), and graduate students Praveen Singh, Andrew Johnson, Paul Hanson, and Vivek Peraser for their contribution, support, and help when needed.

## Bibliography

1. Folger P (2010) Gas hydrates: resource and hazard. CRS report for congress, May 2010, CRS, Washington, DC
2. WEC (2007) Survey of Energy Resources, 2007. World Energy Council, London
3. Islam MR (1991) A new recovery technique for gas production from Alaskan Gas Hydrates. SPE 22924, Dallas, TX, USA
4. Garg R, Ogra K, Choudhary A, Menezes R. Chemical recovery of gas hydrates using fluorine gas and microwave technology. SPE 113556, TX, USA
5. Peter E (1993) Clathrate Hydrates. Ind Eng Chem Res 32:1251–1274
6. Collett TS (1992) Geologic comparison of the Prudhoe Bay-Kuparuk River (U.S.A.) and Messoyakha (U.S.S.R.) gas hydrate accumulations. SPE 24469, Texas, USA. pp 1–20
7. Weitemeyer K, Constable S (2011) Mapping gas hydrates with marine controlled source electromagnetics. In: Proceedings of the 7th international conference on gas hydrates (ICGH 2011), Edinburgh, UK
8. George MJ, Timothy CS, Boswell R, Kurihara M, Matthew RT, Koh C, Dendy SE (2008) Toward production from gas hydrates: current status, assessment of resources, and simulation based evaluation of technology and potential. SPE 114163, Texas, USA
9. Sloan ED, Koh C (2008) Clathrate hydrates of natural gases, 3rd edn. Taylor and Francis, Boca Raton, FL
10. Johnson Arthur H (2011) Global resource potential of gas hydrate – a new calculation. In: Proceedings of the 7th international conference on gas hydrates (ICGH 2011), Edinburgh, UK

11. Boswell R, Collett TS (2006) The gas hydrates resource pyramid. In: Fire in the ice, methane hydrate newsletter. US Department of Energy, Office of Fossil Energy, National Energy Technology Laboratory, Fall Issue, pp 5–7
12. Ruppel C (2011) Methane hydrates and the future of natural gas. In: MITEI Natural gas Report, MIT, MA (Supplementary paper on methane hydrates)
13. Nagakubo S, Arata N, Yabe I, Kobayashi H, Yamamoto K (2011) Environmental impact assessment study on Japan's methane hydrate R&D program. In: NETL, Methane Hydrate Newsletter, January 2011
14. Collett TS, Johnson AH, Knapp CC, Boswell R (2009) Natural gas hydrates. A review. AAPG Memoir 89:146–219
15. Paull CK, Matsumoto R (2000) Leg 164 Overview. In: Proceedings of the ocean drilling program, Scientific Results, vol 164. Ocean Drilling Program, TX, USA
16. Pecher IA, GHR Working Group (2011) Gas hydrates in New Zealand – a large resource for a small country? In: NETL, Methane Hydrate Newsletter, January 2011
17. Kadaster AG, Millheim KK, Thomson TW (2005) The planning and drilling of hot ice #1 – gas hydrate exploration well in the Alaskan Arctic. In: SPE 92764-MS presented at the SPE/IADC drilling conference, 23–25 February 2005, Amsterdam, Netherlands, pp 1–15
18. Hanson P (2007) The Mt. Elbert-01 Stratigraphic test well – lessons from drilling, coring and evaluating gas hydrates on the Alaska North Slope, MS project report, University of Alaska Fairbanks, Fairbanks, December 2007
19. Gruy HJ (1978) Reservoir engineering and geologic study of the East Barrow Field, National Petroleum Reserve in Alaska. Technical Report submitted to Husky Oil NPR Operations, Inc., Anchorage, Alaska, pp 1–62
20. Darkwah SA, Allen WW (1996) Engineering study of the South, East and Walakpa Fields, North Slope Borough, Alaska. Technical Report submitted to the North Slope Borough, Barrow, Alaska, pp 1–32
21. Singh PK (2008) An engineering study to investigate methane hydrate resource potential associated with barrow gas fields, Alaska. MS Thesis, University of Alaska Fairbanks, Fairbanks
22. Stokes PJ, Walsh TP (2006) South and East Barrow gas fields well improvements study. Technical Report submitted to the North Slope Borough, Barrow, Alaska, pp 1–70
23. Stokes PJ and Walsh TP (2007) South and East Barrow gas fields reserves study. Technical Report submitted to the North Slope Borough, Barrow, Alaska, pp 1–36
24. Allen WW, Crouch WJ (1988) Engineering study of South and East Barrow fields, North Slope Alaska, Alaska. Technical report prepared for North Slope Borough Gas Development Project, Barrow, Alaska, pp 1–56
25. Panda MN, Morahan GT (2008) An integrated reservoir model description for East Barrow and Walakpa gas fields. Topical Report prepared under DoE Project No. DE-FC26-06NT42962 submitted to the USDoE, Morgantown, WV, www.netl.doe.gov, pp 1–44
26. Sandilya P, Ganguly S (2011) Conceptual design of an experimental facility for extracting natural gas from natural gas hydrate. In: Proceedings of the 7th international conference on gas hydrates (ICGH 2011), Edinburgh, UK
27. Yutaek S, Seong-Pil K (2011) Dependence of drawdown pressure on the hydrate reformation during methane hydrate production and its inhibition with kinetic hydrate inhibitors. In: Proceedings of the 7th international conference on gas hydrates (ICGH 2011), Edinburgh, UK
28. Shiang-Tai L, Yen-Tien T (2011) Methane hydrate growth and its recovery with carbon dioxide via molecular dynamic simulations. In: Proceedings of the 7th international conference on gas hydrates (ICGH 2011), Edinburgh, UK
29. Lee Y, Shin C, Baek Y, Kim Y, Jang S, Lee J (2011) A study on the dissociation behavior of gas hydrate using the steam stimulation. In: Proceedings of the 7th international conference on gas hydrates (ICGH 2011), Edinburgh, UK

30. Shuxia L, Yueming C, Yongmao H (2011) Experimental study of influence factors of hot-brine stimulation for dissociation of NGH in porous medium. In: Proceedings of the 7th international conference on gas hydrates (ICGH 2011), Edinburgh, UK
31. Dai J (2011) A method to locate and quantify gas hydrate using long-offset seismic. In: Proceedings of the 7th international conference on gas hydrates (ICGH 2011), Edinburgh, UK
32. Collett TS (1993) Natural gas hydrates of the Prudhoe Bay and Kuparuk River area, North Slope, Alaska. AAPG Bull 77(5):793–812

# Chapter 8
# Gas to Liquid Technologies

**Marianna Asaro and Ronald M. Smith**

## Glossary

| | |
|---|---|
| Autothermal reforming (ATR) | The reaction of oxygen and carbon dioxide or steam with methane to form synthesis gas, wherein the exothermic partial oxidation of methane provides energy for the endothermic steam reforming of methane. ATR is also used in reference to the autothermal reformer itself. |
| bbl | Barrels (of oil). |
| BOE | Barrel of oil equivalent. |
| Btu | Also btu, British thermal unit, a measure of energy content. |
| Catalytic membrane reactor (CMR) | A flow-through reactor used to influence an equilibrium-limited reaction to proceed further in the forward direction via selective transport of reactant(s) or product(s) across the membrane. |
| CPOx | Catalytic partial oxidation. |
| CTL | Coal-to-liquids. |
| DME | Dimethylether. |
| DOE | US Department of Energy. |

This chapter was originally published as part of the Encyclopedia of Sustainability Science and Technology edited by Robert A. Meyers. DOI:10.1007/978-1-4419-0851-3

M. Asaro (✉)
SRI International, 333 Ravenswood Avenue, Menlo Park, CA 94025, USA
e-mail: marianna.asaro@sri.com

R.M. Smith
SRI Consulting, 4300 Bohannon Drive, Menlo Park, CA 94025, USA
e-mail: Ronald.Smith@ihs.com

R. Malhotra (ed.), *Fossil Energy: Selected Entries from the Encyclopedia of Sustainability Science and Technology*, DOI 10.1007/978-1-4614-5722-0_8,

| | |
|---|---|
| Fischer–Tropsch reaction | The catalytic conversion of synthesis gas to primarily hydrocarbons, the discovery being credited to Franz Fischer and Hans Tropsch. |
| Gas-Heated reforming (GHR) | Use of heat available by recycling process gas (tail gas) downstream of an ATR for steam methane reforming, in a heat exchanger type reactor. |
| Gas to liquids (GTL) | The conversion of gas to liquid fuels and/or chemicals. |
| Heat exchange reforming (HER) | See GHR. |
| kWh | Kilowatt hours, a measure of energy. |
| Light distillate | A distillation cut of low molecular weight and low boiling range, obtained from refining of hydrocarbonaceous feedstocks, used to produce liquefied petroleum gas (LPG), gasoline, and naphtha. |
| Liquefied natural gas (LNG) | Natural gas that has been converted to liquid form for transport or storage. |
| Middle distillate | A distillation cut of mid-range boiling point, obtained from refining hydrocarbonaceous feedstocks, containing hydrocarbons ranging from $C_5$ through about $C_{20}$ or $C_{22}$. When further distilled, the portion of middle distillates containing $C_5$ through about $C_{15}$ is often referred to as naphtha, and the portion containing $C_{16}$ through up to $C_{22}$ is referred to as diesel. The naphtha is often distilled further to produce gasoline and kerosene/jet fuel, or can be used as feed for a naphtha cracker unit to make light olefins. (Less commonly, the gasoline cut is initially collected along with the light distillates.) |
| Natural gas liquids (NGL) | The purified and condensed portion of natural gas consisting of gaseous hydrocarbons heavier than methane specifically ethane ($C_2H_6$), propane ($C_3H_8$), *n*-butane ($n$-$C_4H_{10}$), and isobutane ($i$-$C_4H_{10}$). |
| Partial oxidation (POx) | The controlled oxidation of natural gas (primarily $CH_4$) with oxygen ($O_2$) such that syngas is formed, rather than forming carbon dioxide ($CO_2$) via complete combustion. |
| Pre-reforming | The use of an adiabatic, preheating zone upstream of an ATR reactor for the purpose of catalytically converting $C_{2+}$ hydrocarbons to a mixture of methane ($CH_4$), hydrogen ($H_2$), carbon monoxide (CO), and carbon dioxide ($CO_2$), thereby allowing oxygen ($O_2$) to be used more efficiently in the reformer. |
| scf | Standard cubic feet, a measure of gas volume. |
| Steam methane reforming (SMR) | The high-temperature catalytic reaction of steam with methane to give synthesis gas. |

| | |
|---|---|
| Synthesis gas (syngas) | A mixture of primarily hydrogen and carbon monoxide produced by gasification or reforming of hydrocarbonaceous materials, used to synthesize fuels or chemicals. |
| Water gas shift reaction (WGSR) | The gas phase reaction of carbon monoxide (CO) with water to form carbon dioxide ($CO_2$) and hydrogen ($H_2$). |

## Definition of the Subject

Like oil and coal, natural gas is not what first comes to mind when considering sustainable fuel sources. Yet as for other fossil sources, conversion of natural gas to transportation fuel is currently more affordable than conversion of renewable resources such as wind and solar, which are technically far away from availability at even a fraction of the scale required to have significant impact on meeting global demand over the coming decades. Given that global proved natural gas reserves are currently estimated as capable of producing more than 1,100 billion equivalent barrels of oil (the energy equivalent of 42 cubic miles of oil) [1], natural gas is a key contributor when considering a sustainable global fuel supply.

The term "gas to liquids" (GTL) is frequently used in reference to the chemical transformation of natural gas to liquid fuels via the Fischer–Tropsch (F-T) technology. In broader usage, the term "GTL" refers to the transformation of natural gas into any liquid, including other fuels, such as liquefied natural gas (LNG), methanol, dimethylether (DME), methyl-*tert*-butyl ether (MTBE), and chemicals such as ammonia (itself an important feedstock, particularly in the fertilizer industry), light olefins, and methanol and dimethylether intended for chemical use.

For clarity, the term "GTL" is used herein to refer to the chemical conversion of natural gas to liquids. However, a practical discussion of GTL should also include LNG, which is produced from natural gas by a phase change rather than a chemical transformation. Strategists in national governments and energy companies alike weigh all of these options when considering how best to distribute and monetize natural gas.

Modern GTL synthesis technology can produce synthetic fuels (synfuels) that burn more cleanly than conventional fuels derived from petroleum. F-T GTL technology can produce a virtually sulfur-free diesel fuel, much cleaner than conventional diesel – reducing smog and acid rain – and can also produce gasoline, jet fuel, or chemicals. GTL performed at or near the gas well also represents an emerging option to transport large quantities of transportation fuels and chemical feedstocks from natural gas, using commercially available tankers.

Production of liquid fuels and chemicals are lower volume applications compared to use of natural gas for power or as a fuel source for heating. Commercial application of GTL might increase in magnitude as the ease of oil recovery decreases and its price increases over the next few decades.

## Introduction

The methane content of natural gas varies greatly within the range of about 50–99%, typically about 95%. The remainder is a multicomponent mixture including the heavier gaseous hydrocarbons ethane, propane, butane, pentane, and some higher molecular weight hydrocarbons; the acid gases carbon dioxide, hydrogen sulfide; mercaptans such as methanethiol and ethanethiol; water, nitrogen gas; helium gas; liquid hydrocarbons; and trace mercury. Thus the preponderance of species in natural gas are high in energy content, with cleanup of contaminants necessary but greatly simplified compared to that required when using coal. Likewise the $CO_2$ emissions from combustion of natural gas [2] are considerably less than from oil [3] or coal [4]: 0.40 lb/kWh compared to 0.61 and 0.79 lb/kWh, respectively.

The bulk of natural gas used today goes directly into power generation (32%), residential and commercial heating (27%), and industrial use (22%). Only a small portion, around 5%, is used as a chemical feedstock, and therefore chemical uses have minimal impact on supply and pricing decisions other than in stranded gas regions where the gas is geographically remote from the main end-user markets. Thus in the Middle East, for example, which has large reserves but relatively small local demand, the proportion of natural gas going into feedstocks is higher (around 27%).

Location and transportation are key factors in defining the need for, and choice of, GTL technology. About 40% of known natural gas reserves are of the stranded type [5], including remote gas (located far from existing pipelines) and deeply buried or otherwise inaccessible gas. Remote gas alone accounts for about 16% of the world's proved natural gas reserves [6]. Related factors determine the type of transportation used, and therefore, the type of GTL processing used, including the size of the reserve, how much time is available to monetize the natural gas, the distance to market, and the gas processing requirements. Environmental influences include a growing resistance to the conventional practice of flaring (or reinjecting) gas that otherwise would be released in association with oil recovery operations (i.e., stranded gas that is also associated gas).

Multiple opportunities exist to achieve better value for natural gas than its local fuel value. First, the gas or its gaseous components can be exported to markets having higher fuel values. Exportation can be performed by pipeline, as pressurized gas, or by ocean tanker as liquefied natural gas (LNG, comprised of 99% methane) or natural gas liquids (NGL, comprised of $C_2$–$C_4$ alkanes). Second, GTL technologies can be used to chemically convert natural gas to higher value liquid fuels such as naphtha or diesel, using F-T technologies; or to fuel grade methanol, DME (directly or via methanol), MTBE (via methanol), or gasoline (via methanol); or to chemicals.

A flowchart of conceptual routes for chemical conversion of the methane in natural gas to liquids is shown in Fig. 8.1. All current gas to liquids chemical production processes require that natural gas is first converted to synthesis gas (syngas), a combination of hydrogen and carbon monoxide. The syngas is subsequently converted in a synthesis section to product. GTL processes of today are therefore of the indirect type, in that they proceed first through the intermediate

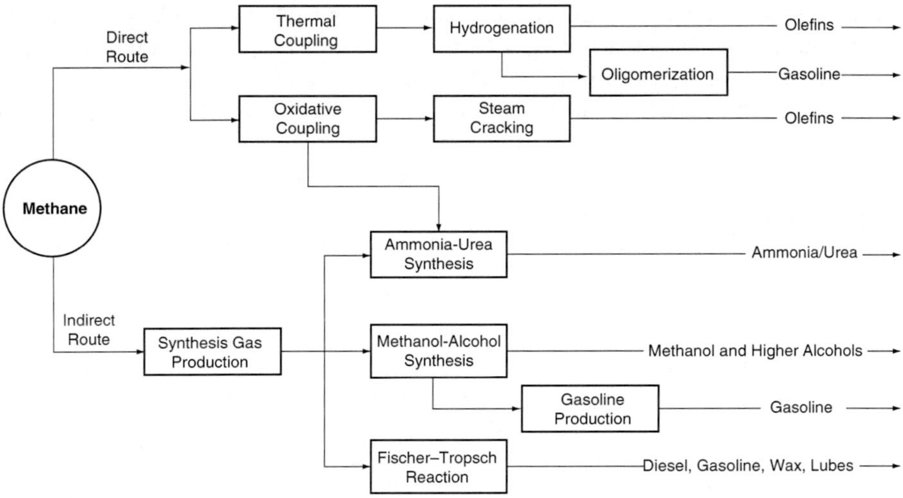


**Fig. 8.1** Conceptual routes for chemical conversion of gas to liquids [7]

syngas instead of converting methane directly to liquid product. Direct syntheses of fuels from natural gas include oxidative coupling of methane and selective oxidation of natural gas to methanol. These have been investigated in the past but poor economics to date have prevented their commercialization.

The intermediacy of synthesis gas is common when converting various hydrocarbonaceous feedstocks such as natural gas, naphtha, residual oil, petroleum coke, coal, or biomass. The lowest cost routes for syngas production are based on natural gas. There are several different approaches to converting natural gas to syngas, the primary methods being catalytic steam methane reforming (SMR), autothermal reforming (ATR), gas-heated reforming (GHR), partial oxidation (POx), heat exchange reforming, and variants thereof. The ratio of $H_2/CO$ produced varies with the process design as well as the H/C ratio in the feedstock. A process will tend to be most economic if the $H_2/CO$ ratio produced is well matched to the ratio needed in the next step. Exceptions include processes configured to produce more $H_2$ than needed, because this extra $H_2$ can be separated and used for power or to augment the syngas ratio of another chemical process. If the $H_2/CO$ ratio is lower than needed, then an additional source of the expensive component $H_2$ must be supplied. With an H/C ratio of 4, methane is well suited for conversion to syngas when the final product slate will contain saturated hydrocarbons or methanol.

The synthesis of hydrocarbons from CO hydrogenation over transition metals was discovered in 1902 when Sabatier and Sanderens produced $CH_4$ from $H_2$ and CO mixtures passed over Ni, Fe, and Co catalysts. In 1923, Fischer and Tropsch reported the use of alkalized Fe catalysts to produce liquid hydrocarbons rich in oxygenated compounds – termed the synthol process. Succeeding these initial discoveries, considerable effort has gone into developing catalysts for the process to produce liquid hydrocarbons. Eventually, a precipitated Co catalyst promoted with $ThO_2$ and MgO supported by Kieselguhr (diatomateous earth) became known as the standard atmospheric process catalyst. In 1936, Fischer and Pilcher developed the medium pressure (10–15 atm) Fischer–Tropsch synthesis process. Following this development, alkalized Fe catalysts were implemented into this medium pressure process. Collectively the process of converting CO and $H_2$ mixtures to liquid hydrocarbons over a transition metal catalyst became known as the Fischer–Tropsch synthesis.

Methanol is a commodity chemical, and one of the top ten chemicals produced globally. The long time interest in methanol is due to its potential use as a fuel and as a feedstock to the chemicals industry. In particular, methanol can be used directly or blended with various petroleum products as a clean burning transportation fuel. Methanol is also an important chemical intermediate used to produce formaldehyde, dimethyl ether, methyl tertiary-butyl ether (MTBE), acetic acid, methylamines, and methyl halides among others.

Production of methanol began in the 1800s, with the isolation of wood alcohol from the dry distillation (pyrolysis) of wood. Research and development efforts at the beginning of the twentieth century involving the reaction of syngas to give liquid fuels and chemicals led to the discovery of a methanol synthesis process (concurrently with the development of the Fischer–Tropsch synthesis). Methanol synthesis is now a well-developed commercial catalytic process with high reaction rate and selectivity (up to 99%). For economic reasons, methanol is produced almost exclusively (over 90%) via the reforming of natural gas.

## Survey of Specific Gas to Liquids Technologies

This survey of GTL processes covers the subtopics of liquefied natural gas, natural gas liquids, methane reforming, Fischer–Tropsch GTL, and methanol synthesis from natural gas.

## Liquefied Natural Gas (LNG)

Liquefaction of natural gas reduces its volume to about 1/600 the volume of the gas as measured under standard conditions. The most compelling justification for the production of LNG is this volume shrinkage that makes LNG, unlike natural gas itself, practical to both store and transport.

Production of LNG has been commercially practiced since 1960. The two major applications of LNG technology are:

(a) Storage of LNG for use in peak shaving plants with seasonal adjustment (mostly in the USA)
(b) Base load LNG plants for international trade with shipping from remote areas to developed countries via dedicated LNG ocean tankers

The capacities of peak shaving plants are more than an order of magnitude smaller than base load LNG plants.

Construction of pipelines depends on the relative values for gas in the supplying and consuming regions. Plans exist for pipeline delivery of gas from isolated eastern Russian gas fields to consuming markets in Asia (Japan, Korea, and China). Nonetheless shipping tends to be preferred over long-range pipeline construction, because pipeline costs are highly capital intensive: a 621 mile (1,000 km) long pipeline would cost some $1 billion, depending on ground conditions. Exporting LNG to Asia and Europe from distant production fields has become economic as a result of improvements to thermodynamic efficiencies of LNG facilities. However, shipping is still expensive at a cost of at least $15 per bbl to transport from the reservoir to the consumer's storage tanks.

Recent industry trends indicate that tankers to hold as much as 135,000 $m^3$ (equivalent to 3.2 billion scf of natural gas) have been specially built, with built capacities expected to soon increase to 165,000 $m^3$. For a shipping distance of 4,000 or 6,600 miles, the shipping requirements for a 1 billion scf/day liquefaction plant can be met by fleets of six or nine tankers, each tanker making 17 or 11 trips per year, respectively. Following transport of the LNG by tanker from the liquefaction site, it is off-loaded at a shore terminal. Vapor generated is compressed for injection into the local distribution pipeline. Independently, a continuous regasification of the LNG from storage tanks is performed, and the amount used for the time period between ships must obviously match the amount unloaded. The LNG pressure is raised by pump to the pressure necessary for it to be vaporized into the distribution pipeline.

## *Plant Considerations*

The liquefaction energy required in a LNG plant typically has been reported as 9–12% of the heat energy in the natural gas, and 9–10% energy shrinkage is a typical number for the modern mega-tonnage capacity plants (without combined cycle systems). Because LNG is stored and delivered at atmospheric pressure and −160°C (−256°F), compression and very deep refrigeration are needed, with associated large consumption of energy. LNG projects have a very high capital cost, in the range of $1.0–1.5 billion for a 3–3.3 million tons per year train on a greenfield site, with 45–60% attributed to off-sites and infrastructure depending particularly on requirements for LNG storage, the marine system, and the heat rejection method. The technology is relatively mature, although economies of scale are being investigated to increase scale size to 4.5–5.5 million tons per year.

The thermal efficiency of LNG plant is determined by two major factors:

(a) Refrigeration cycle efficiency
(b) Power cycle efficiency

Increasing the thermal efficiency of an LNG plant for a given turbine/driver configuration will decrease the cost of production and also minimize both site gas consumption and greenhouse gas emission of $CO_2$ generated by combustion (0.20 t $CO_2$/t of LNG at high thermal efficiency, versus 0.25–0.35 t $CO_2$/t in a typical LNG plant) [8] as well as providing low NOx emission (0.095 kg/t).

The key capital cost elements in LNG facilities are, in descending order:

1. Gas turbines, steam turbines, or motor drivers for refrigeration service
2. Refrigeration compressors, typically over 100,000 kW
3. Steam and power generation including turbines waste heat recovery
4. LNG storage, typically over 250,000 $m^3$ capacity per tank (equivalent to approximately 112,500 t LNG per tank)
5. LNG loading terminal including jetty or causeway
6. Heat rejection system, in most cases, as suggested, seawater cooling
7. Cold box and prechilling for gas liquefaction
8. Natural gas pretreating for $CO_2$ removal gas drying and mercury adsorption
9. LPG and natural gasoline recovery as by-products
10. Fuel gas cold recovery and compression

Although heat transfer for chilling (i.e., the cold box) is less important than the capital investment associated with compression, the correct selection of the optimized refrigeration cycle and associated drivers affects both the compression and heat rejection systems and economics.

The most common refrigeration system in LNG plants involves prechilling with propane followed by use of the mixed refrigerant system. Close to 90% of these plants are licensed by Air Products Corporation, Inc. (APCI). The mixed refrigerant liquefaction systems use a mixture of mostly methane and ethane, in about 1.2–2.0:1 molar ratio. Depending on the feed gas composition, up to 3 mol % nitrogen and 6–12 mol% propane may be added to optimize the LNG boiling curve. The refrigerant cooling curve is adjusted to follow closely the feed gas cooling curve in order to achieve maximum thermodynamic efficiency [9].

One significant exception to the propane prechilling, mixed refrigerant approach is the Phillips Kenai Peninsula plant in Alaska, which commenced operation in 1969. This plant produces about 1.5 million tons per year of LNG by a cascade refrigeration system using pure propane, ethylene, and methane refrigeration cycles. The efficiency of this cascade system has been reported as at the lower end of the scale, about 88% [10]. It is reasonable to assume that the adiabatic efficiency of the refrigeration compressors currently operated by ConocoPhillips in Kenai are on the order of 70%, versus adiabatic efficiencies of 80–85% of more modern centrifugal compressors with three-dimensional blades. Even so, the plant has proved reliable and profitable.

Several patents issued to Phillips (now ConocoPhillips) suggest that improvements in cycle efficiency would result from improved refrigeration load distribution, nitrogen stripping (for nitrogen rich gas), open loop methane refrigeration, and LPG recovery [10–12]. The technology appears driven by the desire to distribute load equally among the propane cycle, ethylene cycle, and methane cycle. This is achieved in part by superheating the methane and ethylene refrigerant vapors fed to the compressors, probably to about −46°C (−50°F). Design and construction considerations then allow the use of six identical gas turbines, such as frame 5D (nominal 30,000 kW), to be used. The open loop methane refrigeration appears also to use the methane refrigeration compressor as the fuel gas compressor.

The loads of the propylene compressor are about 47.3%, the refrigeration ethylene compression loads are about 36.4%, and the methane compressor loads are 16.3% of the total refrigeration load. In practice the combined load on the steam turbines may be about 27% of the total motive power in the facility. The estimated power consumption of the Phillips system, for feed gas at 38°C (100°F) and 650 psig, is 371 kWh/t, assuming very lean gas, such as 96 vol% methane, at a heat rejection temperature speculated to be 38°C (100°F) [10]. This estimate probably includes 15 kWh/t of fuel gas compression [13].

A potential drawback of load equalization is use of up to 5% additional refrigeration power and, in case of fixed speed gas turbines, a reduction in production capacity [13]. Further, this could result in a more complex refrigeration cycle where the propane cycle, ethylene cycle, and methane cycles must be more highly heat integrated [10]. Despite potential drawbacks in power consumption and capacity reduction, the concept of cascade refrigeration may be very sound [14], given the following considerations:

- Even the older, closed loop methane refrigeration as used in Kenai could have an advantage by allowing higher suction pressure to the methane compressor, about 1.7 kg/cm$^2$ (24 psia) instead of an estimated 1.1 kg/cm$^2$ (16 psia) for an assumed open loop compressor.
- Modern design methodology would allow the use of conventional carbon steel metallurgy in the compressors.
- Just one electric motor-driven fuel gas compressor is used for two trains.
- In case of outage of the fuel gas compressor, backup is provided by a draw from the feed gas.
- The ethylene refrigeration cycle, aside from superheating the suction to the first stage to about −46°C (−50°F), apparently comprises only a single side load with a probable goal of obtaining the compression in a single casing – potentially without employing superheating of the suction to the ethylene compressors.
- Use of two side loads reduces refrigeration load by 2% and increases production capacity by 2–3% (although using two casings also increases capital investment, possibly by about $5 million).
- On average about 6.2% of the total power is exported as electric power outside the boundary limits. (During times of high ambient temperature, the start-up steam turbine attached to the gas turbine increases its relative steam consumption and electric power export drops to near zero).

Given the pros and cons of cascade versus conventional mixed refrigerant systems, an objective comparative evaluation could be made only on a case-by-case, site-specific basis for LNG production.

## Methane Reforming in F-T GTL

Reforming is the means by which natural gas is converted to the synthesis gas used as feed for GTL processes. Syngas for use in Fischer–Tropsch GTL is best characterized by the $H_2/CO$ ratio, which should be about 2.0, after reforming, as per the generic F-T stoichiometry of Eq. 8.1.

$$2H_2 + CO \leftrightarrows -CH_2- + H_2O \tag{8.1}$$

There are many process options for reforming technology, used alone or in hybrid reforming approaches, and these are compared in Table 8.1.

The choice of how syngas is produced depends on several factors, including:

- Match between the $H_2/CO$ ratio produced and the $H_2$ demand of the overall process
- Plant and process scale
- The need for, or feasibility of, using an air separation unit (ASU) or oxygen enrichment facilities in the overall process
- Scale and logistics of capital equipment, such as compressors and other gas-handling equipment
- Heat integration and gas recycle options

### *Steam Methane Reforming (SMR)*

Steam methane reforming, or steam reforming for short, is a catalytic conversion of natural gas by reaction with steam. The basic chemistry of steam methane reforming is shown in Eqs. 8.2 and 8.3.

$$CH_4 + H_2O \leftrightarrows 3H_2 + CO \quad \Delta H_{298} = 206\text{kJ/mol CO} \tag{8.2}$$

$$\begin{gathered} CH_4 + 2H_2O \leftrightarrows 4H_2 + CO_2 \\ \Delta H_{298} = 165\text{kJ/mol CO} \end{gathered} \tag{8.3}$$

These reactions are quite endothermic, requiring input of significant transferred heat, and SMR is therefore carried out at high temperature with the catalyst inside tubes within a fired furnace. The amount of steam used is in excess of the reaction stoichiometry requirements, as required to prevent the catalyst from coking.

SMR is the dominant reforming technology used for the production of methanol, ammonia, and other petrochemical products produced from methane at large scale. One reason is that the SMR process is scalable. Another reason is that the high $H_2/CO$

**Table 8.1** Comparison of technologies for syngas generation from natural gas [15]

| Technology | Advantages | Disadvantages |
|---|---|---|
| Steam methane reforming (SMR) | • Most extensive industrial experience<br>• $O_2$ not required<br>• Lowest process temperature<br>• (Best $H_2/CO$ ratio if producing $H_2$) | • $H_2/CO$ ratio higher than typically required for syngas production<br>• Highest air emissions |
| Gas-heated reforming (GHR), heat exchange reforming (HER) | • Compact size and footprint<br>• Application flexibility offers options for incremental capacity | • Limited commercial experience<br>• Usually best coupled with another syngas generation technology, such as ATR |
| Two-step reforming (SMR followed by $O_2$-blown secondary reforming) | • Size of SMR is reduced<br>• Low methane breakthrough favors high purity syngas<br>• Methane content of syngas can be tailored by adjusting secondary reformer outlet temperature | • Increased process complexity<br>• Higher process temperature than SMR<br>• Usually requires $O_2$ |
| Autothermal reforming (ATR) | • Typical $H_2/CO$ ratio produced is close to stoichiometric for F-T and methanol syntheses<br>• Lower process temperature requirement than Pox<br>• Low methane breakthrough<br>• Methane content of syngas can be tailored by adjusting reformer outlet temperature | • Higher process temperature than SMR<br>• Usually requires $O_2$ |
| Partial oxidation (Pox) | • Feedstock desulfurization not required<br>• Absence of catalyst permits carbon formation and, therefore operation without steam, significantly lowering syngas $CO_2$ content<br>• Low methane breakthrough<br>• Low $H_2/CO$ ratio advantageous where ratio $< 2.0$ is required, such as dimethylether synthesis | • Low $H_2/CO$ ratio disadvantageous where ratio $> 2.0$ is required<br>• Very high process operating temperatures<br>• Usually requires $O_2$<br>• High-temperature heat recovery, and soot formation and handling, add process complexity<br>• Low methane content of syngas not easily modified to meet downstream processing requirements |

ratio produced in SMR is stoichiometrically suitable for downstream formation of saturated hydrocarbons and oxygenates. The theoretical ratio for $H_2/CO$ of 3:1 is not reached in practice, the maximum practical value being about 2.8, but in the absence of complex recycle schemes, SMR provides the highest $H_2/CO$ ratio available in reforming of natural gas.

Conventional steam reforming catalysts are 10–33 wt% NiO on a mineral support (alumina, cement, or magnesia). Although natural gas usually contains only small amounts of sulfur compounds, generally in the form of $H_2S$, sulfur compounds are the main poisons of reforming catalysts. Use of uranium oxide or chromium oxide as promoters can impart higher tolerance to sulfur poisoning, but even at a sulfur concentration of 0.1 ppm the catalyst can begin to deactivate and the best practice is to remove sulfur from the raw feed. To maintain a 3-year catalyst lifetime, the sulfur concentration in the reformer feed gas should be less than 0.5 ppm. If the sulfur concentration in the raw feed gas is greater than 1%, the sulfur must be removed by chemical or physical scrubbing. An upfront desulfurization unit is used to absorb $H_2S$ onto a ZnO bed. Any remaining organic sulfur compounds and carbonyl sulfide are partially cracked and absorbed on the zinc oxide bed.

Recent improvements to catalysts and the reforming process have allowed design for operation of side fired reformers in SMR, at conditions not possible with other reforming methods. The most notable developments are noted below.

- Introduction of new generations of catalysts, suitable for pre-reforming and for reforming of heavy feedstocks
  Pre-reforming catalysis converts higher hydrocarbons ($C_{2+}$) in the natural gas feed into a mixture of methane, hydrogen, and carbon oxides.
- Commercial application of advanced reforming at low steam/carbon (S/C) ratios down to 1.5, and high outlet temperatures above 950°C
  Low S/C ratios improve energy efficiency.
- Significant increase of heat flux
  Operation at average heat flux above 150,000 kcal/$m^3$ h has been demonstrated in Haldor Topsøe's Houston, Texas, process development unit (PDU). A high flux unit is now in commercial operation.
- Increase of unit capacity
  Steam reformers can now be constructed for single train reforming units with capacity to provide syngas for a methanol plant of up to 3,000 t/day production.
- Heat exchange reforming
  Heat exchange reforming has been developed and commercialized including with gas-heated option, where the steam reforming reaction is carried out in a unit heated by the product gas from an autothermal reformer.

An obvious advantage of SMR is that it does not need an oxygen plant. However, since steam reformers are more costly than either POx or autothermal reformers, there is a minimal plant size above which the economy of scale of a cryogenic oxygen plant in combination with a POx or autothermal reformer is less expensive than a steam reformer on its own. Use of SMR for Fischer–Tropsch applications does involve recycling of $CO_2$ and removal of excess hydrogen, by means of pressure swing absorbers or membrane separators, to lower the $H_2/CO$ ratio to a level acceptable to the Fischer–Tropsch stoichiometry. Due to the costs involved with these steps, it is most likely that steam reforming will only be considered for F-T GTL when one or more of the following conditions hold:

- A relatively small GTL plant, with a capacity well below 10,000 bbl/day.
- The additional hydrogen can be used for other applications like methanol or ammonia production.
- The natural gas feed has a high $CO_2$ content.
- A suitable water supply can be obtained at low cost.

## *Compact Reforming*

Compact reforming represents a novel mechanical design approach to conventional SMR. Under development by BP, it is currently undergoing extended test runs at their new 300 bbl/day GTL demonstration facility in Alaska. The reformer design resembles a conventional shell and tube heat exchanger, and SMR reactions occur tube side of this reactor, with the tubes filled with conventional Ni-based catalyst. Heat for the endothermic SMR reactions is provided shell side, where the tubes are heated by combustion of a fuel and air mixture. Heat transfer is claimed to occur more efficiently in what is described as a highly countercurrent device.

The shell side combustion zone may also be under elevated pressure, leading to more efficient convective heat transfer to the tubes, and this plus a considerably lower catalyst quantity required considerably reduces the size relative to the conventional SMR furnace design configuration – hence the term "compact" reforming coined by BP. The principal advantages claimed for this approach compared to a conventional SMR are a major reduction of capital cost along with an improvement in thermal efficiency. The reduced space requirements are claimed to make this reforming technology more amenable to installation on barges or oceangoing "plantships" for the conversion of relatively small or temporary sources of natural gas. A drawback of the current state of compact reforming technology is that a large number of parallel units would be required for a world-scale GTL plant.

## *Autothermal Reforming (ATR)*

Autothermal reforming (ATR) has been recognized since the mid-1990s as the technology of choice for production of synthesis gas at large scale. Conceptually ATR is similar to catalytic SMR except that ATR includes the use of oxygen, which renders the reaction less endothermic on a per mole basis than SMR. The stoichiometry for ATR is shown in Eq. 8.4.

$$4CH_4 + O_2 + 2H_2O \leftrightarrows 10H_2 + 4CO \quad \Delta H_{298} = 85.3 kJ/mol\ CO \tag{8.4}$$

In practice the product distribution is strongly affected by the choice of steam/carbon (S/C) ratio. At a S/C ratio of 1.3, the syngas will have a $H_2$/CO ratio of about 2.5, which is higher than the ratio of about 2.0 needed for Fischer–Tropsch synthesis. The $H_2$/CO ratio can be controlled by a combination of lowering the S/C ratio and adding $CO_2$ or $CO_2$-rich gas to the ATR feed, such as by recycling $CO_2$ to the reformer. Decreasing the S/C ratio in the feed gas decreases the amount of $CO_2$ required when adjusting the synthesis gas composition to the desired ratio for F-T synthesis.

A typical ATR process concept for the production of synthesis gas includes the fundamental steps of desulfurization, adiabatic pre-reforming, ATR, and heat recovery. In the desulfurization section, the sulfur present in natural gas feedstock is removed to avoid poisoning the downstream pre-reforming catalyst. Steam, and optionally recycled $CO_2$, is added to the desulfurized natural gas and, after further heating, the resultant mixture is passed to a pre-reformer. In pre-reforming, preheating of the gas is done in a fired heater upstream of the ATR. The use of a pre-reformer upstream of the ATR unit reduces the oxygen consumption per unit of syngas produced during ATR. The exit gas from the ATR is cooled by high-pressure steam production and boiler feed water preheat, and $CO_2$ is removed as needed, using methyldiethanolamine (MDEA, N-methyl-diethanolamine) as an absorbent in a $CO_2$ capture unit. After $CO_2$ removal, the synthesis gas with a $H_2$/CO ratio of 2.0 is available for Fischer–Tropsch synthesis. Various process variants are possible, for example, adjustment of synthesis gas composition by recycling $CO_2$-rich tail gas from the Fischer–Tropsch synthesis.

The result is that ATR offers a reduction in the investment per barrel of product, and the possibility of higher single-line capacity compared to conventional steam reforming. Optimized design in combination with reduction of the S/C ratio can be expected to result in a major increase in the single-line capacity, by more than 25% within the next few years. Currently single train methanol plant units based on ATR can be designed with capacities up to about 7,000 t/day.

The key component of the ATR process is the ATR reactor itself. The ATR reactor has a compact design consisting of a burner, combustion chamber, and a catalyst bed placed in a refractory lined pressure vessel. The pre-reformed natural gas reacts with oxygen and steam in a sub-stoichiometric flame. In the catalyst bed, the gas is equilibrated with respect to the methane steam reforming and shift reactions. Product gas composition is determined by the thermodynamic equilibrium of these reactions at the exit temperature and pressure of the reactor. The exit temperature is determined by the adiabatic heat balance based on the composition and flow of the feed, steam, and oxygen added to the reactor. The product gas is completely free of soot and oxygen.

It is essential that the combined design of burner, catalyst, and reactor ensures that the precursors are destroyed by the catalyst bed to avoid soot formation. The soot limit, that is, the S/C ratio at which soot formation starts, has been investigated by experiment at many combinations of temperature, pressure, so that prediction of acceptable operating conditions for specific feedstocks is now possible.

Additional design parameters of the autothermal syngas generation system that influence cost and thermal efficiency include:

- The preheat temperatures of oxygen and natural gas
  The higher these temperatures are, the less oxygen will be used. The maximum preheat temperatures are determined by safety factors and by the need to prevent soot formation.
- The pressure of the steam generated in the waste heat reboiler
  The higher the steam pressure, the more efficiently can energy be recovered from steam, but the more costly the steam and boiler feed water treatment systems become. The optimum steam pressure will be determined by the relative cost of capital and energy.

The feasibility of soot-free operation of the ATR reactor at very low S/C ratio was demonstrated in Haldor Topsøe's Houston development unit in 1997–1999, with further testing performed since then to explore the limits of the ATR technology with respect to feed gas composition, pressure, and temperatures. Operation at industrial scale at a S/C ratio of 0.6 was demonstrated in a South African plant in 1999 and has been in operation in Europe since the start of 2002 [16]. Two lines have been in operation at Sasolburg since 2004. Haldor Topsøe's ATR technology was also selected for the Oryx plant that Qatar Petroleum and Sasol began operating in 2006 and includes a 6,000 mtd syngas generator; for Chevron Nigeria's Escravos, Nigeria GTL plant scheduled for completion in 2011; and for the methanol portion of Eurochem's gas-to petrochemical complex planned in Lagos State, Nigeria.

Sasol, Syntroleum, and ExxonMobil each have their own proprietary versions of ATR. The most prevalent layout of the syngas production section is a combination of adiabatic pre-reforming and fixed bed ATR. This layout results in high flexibility for variations in natural gas feed and, if needed, in tail gas recycle gas compositions, as well as the ability to reduce oxygen consumption per unit of syngas produced.

ExxonMobil employs a fluidized bed catalytic reformer. Steam-diluted oxygen is fed to the reactor through nozzles, separately from the natural gas. The oxygen reacts exothermically with natural gas in a burning zone near the oxygen inlet. The balance of steam needed in the process is admixed with natural gas and fed at a level below the oxygen inlet to the autothermal reaction zone. A mixture of nickel on alumina catalyst may be combined with a solids diluent (alumina) to form the fluidized bed. The fluidized bed reactor provides superior characteristics of heat and mass transfer for autothermal reforming.

Another ATR approach under consideration is to use slurry bed F-T reactors in series within each train, as the slurry system would facilitate the application of ATR technology at large scale (>10,000 bpd). With this approach, water is removed between stages to enhance the reactant partial pressures in the downstream reactor stage. The removal of other diluents (such as carbon dioxide) between stages may also be considered. This allows lower recycle ratios, increased steam production from the reactor heat removal system, and decreased overall reactor volumes. These advantages will not necessarily compensate for the increased complexity of needing to use reactors in series at larger scale, and detailed studies are required to determine the most cost-effective design approach.

Currently, the most attractive ATR syngas generation technology appears to be oxygen-blown ATR at low S/C ratio. This technology has been chosen for F-T GTL projects that are closest to realization. Extensive testing has proved that technology is ready for application at a large industrial scale with S/C ratios down to 0.6. Application of even lower S/C ratios has been demonstrated at small scale, and the next generation ATR scheme is under development to cut the S/C ratio even lower, to 0.4 or possibly 0.2 [17]. Such low S/C could decrease oxygen requirements and boost the single-line capacity still further.

Two-step reforming is a variant of ATR wherein the exit gas from a fired tubular reformer is further processed in an oxygen-blown, secondary reformer. In both steps, the ATR is carried out in a refractory lined vessel containing a mixer/burner and a catalyst bed. Conditions leading to soot formation are rigorously avoided.

## *Gas-Heated Reforming/Heat Exchange Reforming (GHR/HER)*

Another technology that produces suitable synthesis gas for GTL units is a combination of steam reforming and ATR, performed in separate units, in a process referred to as gas-heated reforming, GHR, or heat exchange reforming, HER. The concept is to use heat available from process gas downstream of the ATR for steam reforming, in a heat exchanger type reactor. The purpose is to reduce oxygen consumption and increase carbon efficiency by optimizing the recycle of tail gas – beyond the range possible with ATR alone – and to achieve this without additional firing that would create an undesired excess of water. The gas-heated reformer is a compact alternative to the fired steam reformer wherein high-grade heat is recycled while the S/C ratio remains below 0.6 [18].

Conventional reformers provide the needed heat by combustion of fuel at atmospheric pressure in a refractory lined duct. Because the convective heat transfer coefficient of a GHR is considerably higher at the elevated pressures of its hot gas side, unit size reductions claimed can be 10–15 times smaller than conventional tubular steam methane reformers. The absence of combustion flue gases results in an inherently higher thermal efficiency for a GHR, since no combustion heat is lost with the warm flue gases discharged to the atmosphere from the stack. Furthermore, the reduced power consumption, which results from eliminating the need for the associated steam methane reformer air and flue gas blowers, contributes to the overall energy saving of a GHR. Much of this thermal efficiency advantage may be offset, however, by the need to generate additional steam externally in an off-site boiler in order to compensate for the lost waste heat recovery potential of a conventional reformer system.

GHR can be combined with ATR to reduce energy consumption and enable the use of smaller air separation plants, while reducing $CO_2$ emissions possibly by as much as 50%. In addition, these GHR/ATR schemes allow shrinking of the required high-pressure steam system and reduction in size of steam turbines. Both series and parallel arrangements of GHR with ATR have been studied and are depicted in Fig. 8.2 [18]. In the series arrangement of HER/GHR, all gas passes through the steam reforming unit and then the ATR. The lower limit for the S/C ratio is typically determined by the steam reforming catalyst [19]. The parallel arrangement allows individual optimization of the S/C ratio by feeding the two reformers independently.

GHR has been employed since the 1980s in some ammonia and methanol plants. In the past, highly efficient GHR was not viewed as an option for high capacity syngas production, mainly because of concerns about metal dusting corrosion. This reformer must operate at higher temperatures in the parallel arrangement than in the series arrangement, to obtain a low methane leakage. Higher operating temperatures may lead to higher cost and more severe conditions, and the main challenge in both series and parallel arrangements is preventing metal dusting corrosion on the shell side of the heat exchange reformer. Mechanical design, including choice of materials of construction, is critical.

A heat exchange reformer designed by Haldor Topsøe for operating in tandem with an ATR at a low S/C ratio was started up at Sasol's facilities in Secunda, South Africa, in early 2003 [16, 20]. The project at Secunda (also referred to as "process gas-heated reforming" or PGR) involved revamping an existing ATR for an increase in syngas production capacity of 30% via GHR.

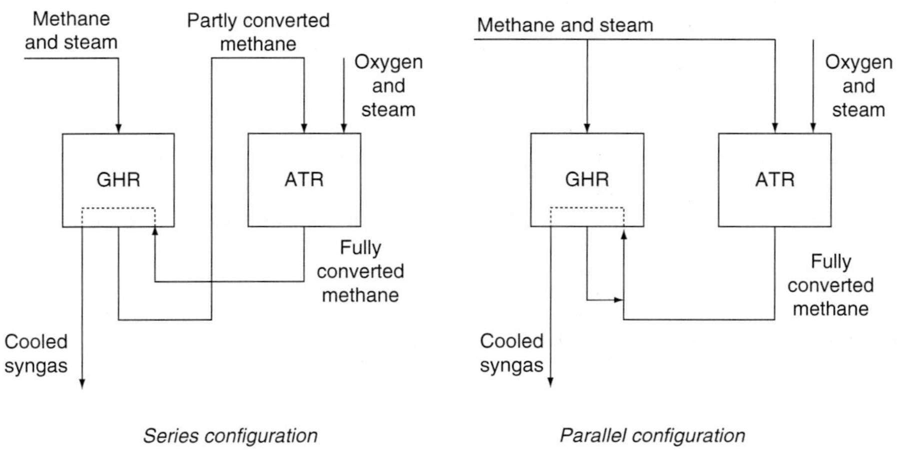


**Fig. 8.2** Series and parallel configuration of the combination of GHR and ATR [18]

Engineers at Johnson Matthey have stated that with mechanical design and configuration improvements, GHR/ATR technology is likely to deliver better efficiency and lower operating costs than conventional ATR for F-T GTL plants of the future. In addition, Air Products and Toyo Engineering Corporation have each studied GHR technology. Although commercial F-T GTL plants using ATR/GHR combinations have yet been built, retrofit projects at large-scale methanol and GTL plants are possible, as well as potential applications of ATR/GHR in refinery hydrogen plants.

## *Air Blown Autothermal Reforming*

It is possible to use air instead of purified oxygen as oxidant in autothermal reforming, and an F-T synthesis process based on air blown ATR was developed by Syntroleum Corporation and demonstrated in their plant in Tulsa, Oklahoma, in 2002–2003. In a very similar operation, air blown secondary reforming is used extensively in the production of synthesis gas in ammonia manufacture. The main advantage for air blown processes is that no air separation unit will be required, which could be a strategic advantage for smaller scale, offshore, barge mounted F-T GTL units. For large-scale land-based units, it may present certain restrictions on the energy system configuration of the plant. The savings of eliminating the capital cost of an air separation cost will be offset by the need for compression of incoming air, for example, which will result perhaps in as much as 50% more power required than that required to produce oxygen.

In addition, the increased volumetric flow of nitrogen-diluted syngas will make all downstream equipment through the F-T synthesis reactors larger. Also, the high content of nitrogen in the synthesis gas excludes the use of internal recycle in the F-T synthesis section and external recycle of tail gas from the F-T reactor back to the syngas section of the plant is excluded. This means that all tail gas must be used as fuel, which because of the high nitrogen content will be of poor quality. At a conversion of hydrogen and CO in the synthesis gas of about 70%, the heating value of the tail gas will be so low that it will not burn unassisted. Either supplemental fuel or catalytic combustion will be required if the energy content in the tail gas is converted to power, and a large amount of power will be generated for export. This may be desirable, however, in remote locations where power may be in short supply.

## *Partial Oxidation (POx) and Catalytic Partial Oxidation (CPOx) Reforming*

When the quantity of oxidant ($O_2$) added to methane is very low and the reaction temperature is raised significantly, uncatalyzed partial combustion of pretreated natural gas occurs. This homogeneous reaction is referred to as gasification or partial oxidation (POx). The resulting syngas is quenched or cooled by steam production, and carbonaceous by-products such as soot are removed by washing. The formation of carbonaceous by-products influences carbon efficiency. For partial oxidation, the syngas $H_2/CO$ ratio produced is slightly below 2.0, which is closest to the optimum needed for the Fischer–Tropsch or methanol syntheses. This low $H_2/CO$ ratio gas results from very little, if any, steam use in the process. Due to the absence of catalyst, the reformer operates at an exit temperature of about 1,300–1,400°C. This high temperature and the absence of catalyst have the following disadvantages as compared to an autothermal reformer.

- Formation of soot and much higher levels of ammonia and HCN appear in the syngas, compared to ATR, which necessitates the use of a scrubber to clean the gas.
- Higher consumption of oxygen.
- Due to the absence of a water gas shift reaction with the POx catalyst, the unconverted methane as well as the methane produced by the Fischer–Tropsch reaction cannot be recycled to the reformer without first removing the $CO_2$ from the Fischer–Tropsch tail gas.

Depending on the energy needs of the plant, the syngas from the POx reformer can either be cooled by means of a water quench or by the production of steam in a heat exchanger. A quench system is the less costly of the two approaches but is also less thermally efficient. In designing a POx-based GTL plant, the choice between a quench or a waste heat boiler will depend on the relative cost of capital and energy. As mentioned above, $CO_2$-rich tail gas separated from the syngas may be recycled to the partial oxidation reactor to improve carbon yield. If recycle is used, the gas production in the POx units must be supplemented by high hydrogen content syngas in an auxiliary steam reformer, as in the Shell process. In Shell's Bintulu GTL plant, each POx unit originally had a capacity rating for production of 3,000 bbl/day of F-T products. Shell has recently scaled up their gasification units to enable production of 10,000 bbl/day of F-T product equivalents.

When a catalyst is employed to assist the partial oxidation, the process is referred to as catalytic partial oxidation (CPOx). The catalyst increases the activity of natural gas conversion and, therefore, provides options for higher throughput, lower operating temperature, or both.

CPOx with no burner (flame) was first practiced on a small scale at low pressures in the 1950s by Haldor Topsøe, and then later with higher pressures by Lurgi. Early versions of CPOx used premixing of hydrocarbon/steam feed and oxygen, with an ignition catalyst placed on top of a fixed bed catalyst, similar to the catalyst used in an ATR-based unit and operating with similar space velocities. Today's CPOx versions feature a combined catalyst system performing both ignition and partial oxidation. The noble metal catalyst operates at very low residence time (very high space velocity). The hydrocarbon feed stream, oxygen, and steam are mixed upstream from the catalytic reactor. All chemical conversion takes place in the reactor and no burner is used. The exit gas composition from CPOx is similar to the exit gas from ATR, given like amount and properties (temperature, pressure, composition) of the inlet streams.

The main potential advantage of catalytic partial oxidation is the smaller size of the reactor. There are also potential disadvantages that will likely outweigh the savings in the cost of the reactor, which is rather modest relative to the total investment in the syngas generation section. The main disadvantages of CPOx relate to the premixing of natural gas hydrocarbon feed and oxygen. The need to premix excludes combinations of CPOx linked in a series arrangement, in particular with pre-reforming or heat exchange reforming, for safety reasons. Premixing hydrocarbon feed with oxygen limits the applicable reactor preheat temperatures to rather low values, kept below the self-ignition temperature of the mixture to maintain control when premixing oxygen and hydrocarbon. A feed gas mixture containing methane and oxygen may spontaneously ignite at temperatures above the autoignition temperature, depending on the actual gas properties, and the industry does not yet have confidence that this issue has been adequately addressed.

When consumption of natural gas and oxygen are compared for CPOx and a combination of pre-reforming and ATR, consumption of both gases are significantly higher for CPOx than for the combination of pre-reforming and ATR. The main reason is that with CPOx operating at a lower inlet temperature than ATR (250–450°C vs. 600°C), a significant part of the hydrocarbon feed is combusted to $CO_2$ and steam (using the larger amount of oxygen) to generate heat to reach the desired syngas exit temperature of 1,050°C.

Considering that a large portion of the syngas process section of a F-T plant is for the air separation unit, CPOx does not appear to be economical under these conditions. CPOx was being investigated by ConocoPhillips but they have recently discontinued this approach.

An alternative for CPOx could be the use of air as an oxidant. This alternative is being pursued by many as a basis for fuel processing systems for fuel cells. For GTL and similar applications operating at high pressures, however, the issue of safety still remains even with air, although to a lesser extent than with oxygen. Oxygen consumption may be conserved (to an amount about 10% less than the amount required for the ATR reactors) if the inlet temperature can be safely raised to a higher level (about 480°C) with a lower oxygen content in the hydrocarbon/oxygen mix. Additional operational issues with CPOx include concerns about catalyst life, carbon dusting, and combustion "flash back" potential in an integrated F-T GTL process plant.

The combination of POx and conventional steam reforming appears to be the nearest competitor to ATR technology. However, catalytic partial oxidation, either oxygen or air blown, does not appear to offer sufficient advantages to outweigh its limitations with respect to operating conditions and inherent safety concerns. CPOx seems to require an oxygen consumption boost, and it is unclear whether this need can be adequately diminished. CPOx is not expected to be important in the GTL industry in general, even if successfully developed for commercialization.

## *Catalytic Membrane Reactors*

Catalytic membrane reactors (CMR) are based on the concept that selective transport of reactant(s) or product(s) can influence an equilibrium-limited reaction to proceed further in the forward direction. The combination of SMR and the WGSR is equilibrium limited, and $H_2$-selective membranes have been used to lower the temperature required for SMR by facilitating the selective removal of $H_2$ as it forms. Both dense (mostly palladium) and porous (such as silica or alumina) $H_2$-perselective membranes have been studied [21]. Throughput limitations, membrane stability, and cost combined with the ready availability of several other reforming technologies have thus far hampered commercial development of $H_2$-permselective membrane reactors for reforming applications.

Another approach to CMR for methane reforming allows a selective passage of reactant oxygen. Nonporous membranes prepared from perovskites or similar ceramic materials allow ionic transport of oxygen from air at low pressure, with up to 100% selectivity for $O_2$. The selectively transported oxygen moves into a reaction mixture at low partial pressure of oxygen, against a large difference in total pressure.

Process concepts based on the use of $O_2$-selective CMR eliminate the need for an air separation unit (ASU) and associated large compressors. Successful elimination of the ASU would impart significant economic potential for this technology; however, difficulties with CMR would need to be overcome first. One problem is in mechanical design, including the mechanical integrity of the membranes and also the membrane-to-metal junctions. Significant consortia efforts formed with US DOE funding led to the development of two novel technologies, known as ITM (ion transport membrane) and OTM (oxygen transport membrane).

Because the production of synthesis gas from natural gas by oxygen-blown ATR accounts for about 60% of the total capital cost of the GTL plant, the ATR and cryogenic ASU are the two most capital-intensive components in the syngas generation process section. ITM/ITM Syngas technology aims to lower capital intensity by combining the ASU and ATR units into a single reactor. In the ITM/ITM Syngas process, $O_2$ in air is selectively reduced and transported across the membrane to the other side, where it partially oxidizes methane to form syngas. The ITM/ITM Syngas concept is particularly attractive for use in locations where space for an ASU would be limited, such as on an offshore platform. The ITM Syngas Team, led by Air Products, has estimated capital cost savings of greater than 30% over a conventional ATR with ASU, as well as an increase in overall fuel efficiency to 61% versus 58% for ATR/ASU, with up to 40% less deck space required [22].

OTM oxygen technology aims to lower capital intensity by replacing cryogenic ASU technology with membrane reactors that allow $O_2$ to selectively pass through the ceramic membrane to the other side, where it partially oxidizes methane to form syngas. The OTM Alliance team, led by Praxair, focused intensively on materials development [23]. This DOE program ended in 2004 and does not appear to be progressing toward commercialization at this time. The OTM approach also has been studied in combination with electrochemical pumping of $O_2$ [24] and with $CO_2$ capture using a solid sorbent [25].

Although membrane materials with satisfactory oxygen permeability at the relevant conditions do exist and progress has been made, the level of oxygen flux in all cases still requires improvement. Concerns persist over mechanical integrity and stability with time. Catalytic membrane reactors are still under development, and it is not possible to predict at this time whether or when CMR will be successfully developed for commercial use. The goal for further development of CMR is to avoid the cost of an oxygen plant.

### *Pre-reforming and Post-reforming*

Additional reforming may be performed upstream or downstream of the main reformer unit, when the main furnace unit has limited heat transfer capability that would otherwise limit overall conversion. The pre-reformer reaction breaks down the heavier hydrocarbons (mostly propane and butane) to methane, upstream of the reformer heater. Post-reforming uses an additional catalyst bed downstream of the main reformer. A down-flow reactor is placed between the outlet transfer line and the steam generator that recovers excess process heat. Adding a post-reformer raises the effective residence time of the reacting gas and thereby boosts capacity. Pre- and, in particular, post-reforming are often considered as retrofits, which may be complicated by having insufficient space or inadequate metallurgy for the higher heat burden.

## Methane Reforming for Methanol/DME Synthesis

Synthesis gas production is best characterized by the generalized stoichiometric ratio M (also referred to as the module), where $M=(H_2–CO_2)/(CO + CO_2)$. Ideally M should be equal to 2.0 for methanol (or for F-T). For kinetic reasons, in order to control by-products formation, a value slightly above 2.0 is normally preferred.

Because about 60% of the capital cost for a methanol plant is in the methane reforming unit, research has aimed at improving reforming processes. Over the years, three process concepts have been considered for preparing syngas targeted for methanol synthesis, described below.

- Single-step tubular reforming
  - Steam/carbon ratio = 2.4.
  - Capacity range up to 2,500 t/day methanol equivalent.
  - Availability of $CO_2$ increases capacity where single-step tubular reforming is attractive.

In the past, the methanol industry was dominated by single-step, steam methane reforming without the use of oxygen. Today, single-step reforming is considered mainly for relatively small plants and for cases where $CO_2$ is present in the natural gas or is available on site or nearby from other sources such as an ammonia plant.

- Two-step reforming
  - Steam/carbon ratio = 1.5–1.8.
  - Capacity range about 1,500–7,000 t/day methanol equivalent.
  - Heavy natural gas makes two-step reforming less attractive compared to single-step tubular reforming.
  - Availability of $CO_2$ is no advantage with two-step reforming.

  Two-step reforming is a combination of a fired, tubular reformer and an oxygen-blown secondary reformer. This concept was applied by Statoil and Conoco in a 2,400 t/day methanol plant in Tjeldbergodden, Norway, commissioned in 1997. The facility has an annual capacity of about 900,000 t of methanol, which corresponds to 25% of Europe's total production capacity [26].
- ATR
  - Steam/carbon ratio = 0.6–0.8.
  - Capacity range about 5,000–10,000 t/day methanol equivalent.
  - Heavy natural gas makes ATR less attractive.
  - Availability of $CO_2$ is no advantage.
  - This scheme is most attractive for the production of fuel grade methanol.

  Using ATR, the composition of synthesis gas is raised to have adequate hydrogen content by taking advantage of the WGSR with removal of $CO_2$.

The main conclusions from early studies were that for relatively small plants (<1,500 t/day methanol equivalent), single-step tubular reforming is advantageous for syntheses gas production. For intermediate scale capacities (corresponding to present world-scale capacities, about 1,500–3,500 t/day methanol), two-step reforming is most attractive. For capacities above about 3,500 t/day, ATR or a combination of ATR and SMR is preferred. For example, the Lurgi-combined reforming process uses a combination of steam and autothermal reforming. Proven single-line capacities for the oxygen plants associated with combined reforming have increased to about 3,000 t/day, corresponding to about 7,000 t/day methanol plant equivalent using two-step reforming, or about 5,000 t/day methanol plant equivalent using ATR alone.

A new process concept for the synthesis loop for methanol has been developed and specially adapted to syngas production by ATR at low steam/carbon ratios. In this scheme, $CO_2$ removal is not applied. Instead, the gas composition is adjusted by recovery and recycle of hydrogen from the purge gas from the synthesis loop.

## Fischer–Tropsch Synthesis

The Fischer–Tropsch (F-T) reaction is a polymerization or oligomerization of carbon monoxide performed catalytically under reducing conditions. As shown in Eq. 8.5, the F-T reaction is thermodynamically favorable by virtue of high exothermicity. Catalysis, temperature control, and downstream processing are used to provide a product slate as close to that of target applications as possible.

$$
\begin{aligned}
&nCO + (2n+1)H_2 \leftrightarrows CH_3 - (C_{n-2}H_{2n-4}) - CH_3 \\
&+ nH_2O \qquad \Delta H_{298} \approx -159\text{kJ/mol CO}
\end{aligned}
\tag{8.5}
$$

Fischer–Tropsch processes were originally developed for use in coal-to-liquid (CTL) processes and are similar when applied to the syngas derived from natural gas. The reader is referred to the CTL entry in this publication for additional discussion of F-T performance variables, operating regimes, and product distributions.

The primary differences between F-T processes applied for GTL and CTL are the range of operating temperatures used, the type of catalyst used, and the type of reactor used. These differences all relate to differing compositions of F-T syngas feed in GTL versus CTL. The stoichiometry of the F-T reaction to make hydrocarbons is about 2.1/1 in $H_2/CO$ as shown in Eq. 8.5. Syngas produced by gasification of coal has a $H_2/CO$ ratio of 0.7–1.0 and is therefore stoichiometrically deficient in hydrogen. Use of an iron-based catalyst for the F-T step of CTL is typical because Fe concurrently catalyzes the water gas shift (WGS) reaction, providing makeup hydrogen as per Eq. 8.6.

$$
CO + H_2O \leftrightarrows CO_2 + H_2 \quad \Delta H_{298} = -41.2\text{kJ/mol CO} \tag{8.6}
$$

In contrast, as shown above in Eqs. 8.2–8.4, the $H_2/CO$ ratio of syngas produced by reforming methane is at or in excess of the value of 2 required by the stoichiometry of the F-T reaction. Thus iron F-T catalysts are less preferred for use in GTL processes than cobalt F-T catalysts, which do not catalyze the WGS reaction. Cobalt is effective in the low-temperature Fischer–Tropsch (LTFT) regime, operated at about 220°C, and not for high-temperature Fischer–Tropsch (HTFT), operated at about 340°C. At the higher temperatures, Co tends to make methane.

Iron F-T catalysts can be used in GTL processes, performed at low temperature to lessen the contribution of iron-catalyzed WGS reaction. Operating in the LTFT regime, a slurry bed reactor (SBR) is typically preferred for several reasons. One reason is that the slurry bed uses a much lower catalyst loading than either the fixed bed or fluidized bed reactor technologies used in HTFT processes. A second reason is that the slurry bed can remove reaction heat more rapidly, to avoid temperature increases that would result in undesired formation of high levels of methane, light hydrocarbons, and, in the extreme, catalyst deactivation due to coking, sintering, and disintegration. Deposition of carbon by the Boudouard disproportionation reaction, Eq. 8.7, is to be avoided, and the heat distribution imparted by the slurry bed is advantageous. A third aspect favoring the slurry bed is that the resultant steam to carbon (S/C) ratio is superior.

$$2CO \leftrightarrows C(s) + CO_2 \quad \Delta H_{298} = -83.4 kJ/mol\ CO \tag{8.7}$$

Even so, Shell does operate a fixed bed, Co-catalyzed F-T process in its GTL plants in Malaysia and Qatar. Shell commissioned the Bintulu plant prior to the widespread consideration of slurry bubble reactor technology for F-T processes and stands by the fixed bed technology. One advantage of fixed beds over slurry beds is the potential for a lesser degree of catalyst attrition that results from impacts and shear forces in the latter.

## *Reaction Pathway*

Fischer–Tropsch chemistry has long been recognized as in effect a polymerization reaction, wherein the monomeric unit is the methylene group, $-CH_2-$ that is generated in situ from carbon and hydrogen atoms of CO and $H_2$. The basic steps of the F-T reaction are as follows:

1. Adsorption of reactant, CO, on the catalyst surface
2. Chain initiation, by CO association followed by hydrogenation
3. Chain growth, by insertion of additional CO molecules followed by hydrogenation
4. Chain termination, by reductive elimination of paraffin from catalyst surface or by β-hydride elimination of olefin from catalyst surface
5. Desorption of product, from the catalyst surface

When tested under relatively comparable conditions, Fe and Co catalysts give sufficiently similar carbon number distributions to suggest a common intermediate for the different F-T metals. The predominance of linearity in F-T products and the formation of terminal, α-olefins suggest that the intermediate responsible for chain growth is a single-carbon unit. For at least the past 50 years, most researchers have agreed that the product distribution observed in F-T is consistent with a single intermediate and chain propagation mechanism for all products (with caveats for $C_1$ and $C_2$, vide infra). Research on the mechanism was most defining during the 1960s and 1970s, and experiments and debate continue on the topic of what that single-carbon unit is, and how it derives from the CO or, for iron, the CO or $CO_2$ present. Candidates include a surface-associated carbon, a metal methylene, a metal hydroxycarbene, a metal alkyl, a metal alkoxide, or a metal carboxylate – a great many choices for a reaction whose mechanism has been under investigation for about a century.

The mechanistic complexity of F-T chemistry is compounded by the use of certain terms, such as carbide or carbene, by different researchers to denote different intermediates. Table 8.2 shows the steps of the F-T reaction according to various mechanisms currently under consideration in the field, presented as composites that capture the essence of each step. The many descriptors in common use for the mechanistic chemistry and intermediates are also provided.

### Formation of $C_1$ Species

The initial reaction of syngas with the F-T metal involves reduction of CO and produces a metal-bound $C_1$ species $CH_2$, CHOH, or $CH_2OH$. Metal-carbides have also been considered for the C1 intermediate, forming by splitting of the C–O bond with formation of water from the O and $H_2$; however, the formation of carbides as discrete species attached with strong, multiple bonds to the surface has been largely discounted by isotopic tracer studies. Many people do still use the term "carbide" to refer to a more loosely associated metal-C1 species, in what has been referred to as the "modern carbide" mechanism. Surface carbiding of iron has been demonstrated under F-T conditions, but Co does not become carbided unless $H_2$ is absent. Given that cobalt is a good hydrogenation catalyst, if formed during F-T the carbide would be hydrogenated very quickly.

The intermediacy of immediate C–O bond cleavage and surface carbides need not be invoked if CO itself is hydrogenated prior to C–O bond scission. Devising experiments that would differentiate exactly to what degree the CO is associated with the metal when scission occurs would be difficult under realistic F-T conditions. In addition, work by Davis has shown that $CO_2$ can initiate chain growth in Fe catalysis (in which case an initial species would be formic acid) [27].

### Chain Initiation and Propagation

Chain initiation occurs when the first C–C bond of the chain is formed, by coupling of two initial $C_1$ fragments or by insertion of carbon monoxide into a metal alkyl bond (or possibly a metal alkoxide bond, in the postulated iron case). Propagation occurs by successive insertion of carbene $C_1$ species into the growing metal-hydrocarbon chain, or by successive cycles of CO insertion and reduction at the metal. The carbene terminology used here is that used by E.O. Fischer in his pioneering discovery of species of this type and refers to any species $M_1C(A_1)(A_2)$ or $M_2(A_1)(A_2)$, where $A_1$ is hydrogen or carbon and $A_2$ is hydrogen, carbon, or oxygen. For monometallic $M_1$, the bonding is formalized as M=C and for bimetallic $M_2$ the bonding is viewed as M-C-M. These $M{=}CH_2$ or $M{-}CH_2{-}M$ carbenes are highly reactive and are well known to insert into metal–carbon bonds, as required by the F-T product distribution, under milder conditions and more quickly than carbon monoxide does.

**Table 8.2** Elements of F-T mechanisms [7]

| *Formation of $C_1$ species* | *Chemistry* | *Common names for $C_1$ species* |
|---|---|---|
| $M_n + CO \rightarrow M_{n-1}{-}C + M{=}O \xrightarrow{2H_2} M{=}CH_2 + (n-1)\ M + H_2O$ | CO dissociation and hydrogenation | Surface-associated carbon (M- – -C), metal carbide (M-C), methylene ($CH_2$), carbene (M=$CH_2$), methylidene (M=$CH_2$), alkylidene (M=$CH_2$) |
| 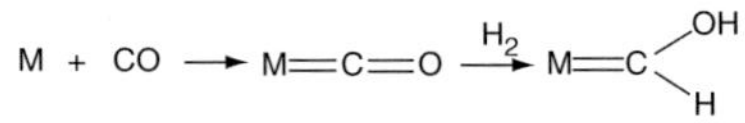 | Direct CO hydrogenation | Enol, hydroxycarbene |
| 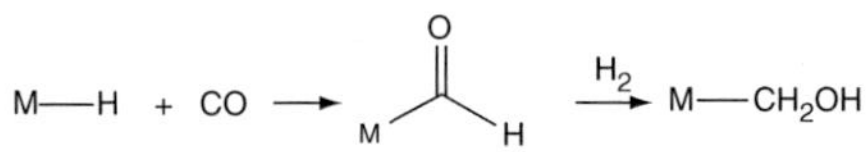 | CO insertion and hydrogenation | Hydroxymethylene, alcohol |
| 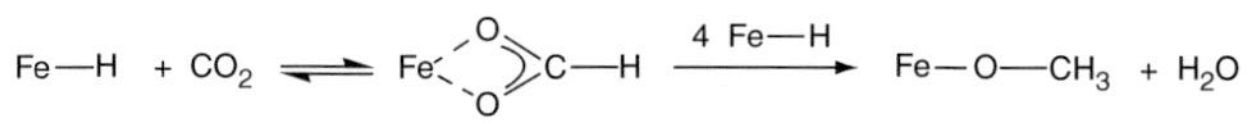 | $CO_2$ insertion and hydrogenation | Oxygenate, metal alkoxide |
| 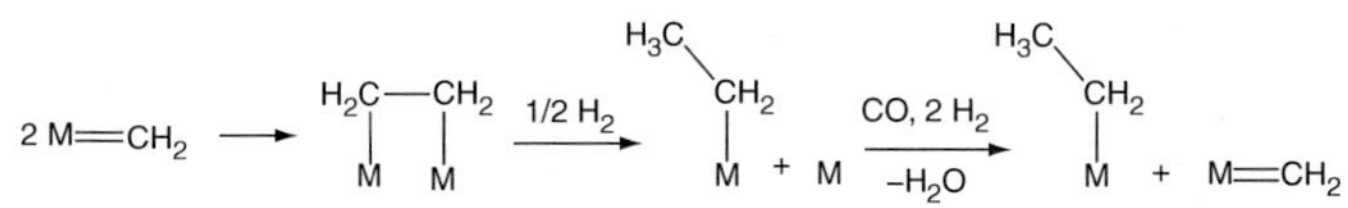 | Carbene coupling | Metal alkyl and methylene ($CH_2$), carbene (M=$CH_2$), methylidene (M=$CH_2$), alkylidene (M=$CH_2$) |
| 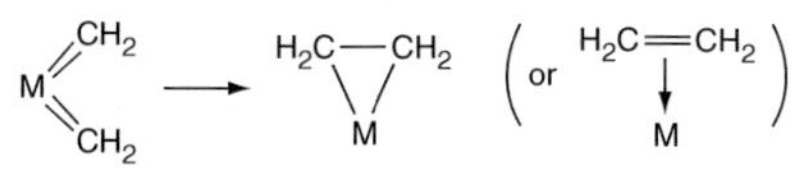 | Carbene coupling (intramolecular) | Metallacycle or olefin complex and methylene ($CH_2$), carbene (M=$CH_2$), methylidene (M=$CH_2$), alkylidene (M=$CH_2$) |
| 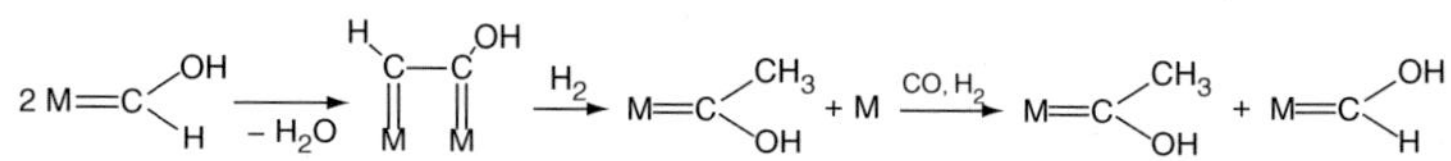 | Carbene coupling | Enol or hydroxycarbene |

(continued)

**Table 8.2** (continued)

| | | |
|---|---|---|
| 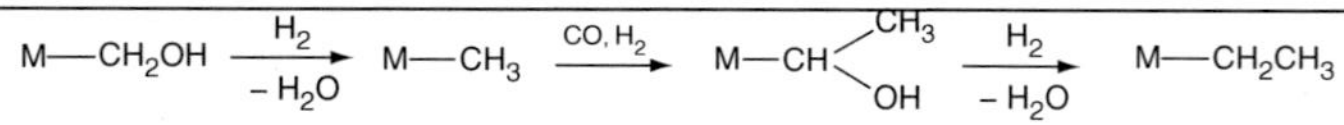 | CO insertion and hydrogenation | Metal hydroxyalkyl or metal alkyl |
| 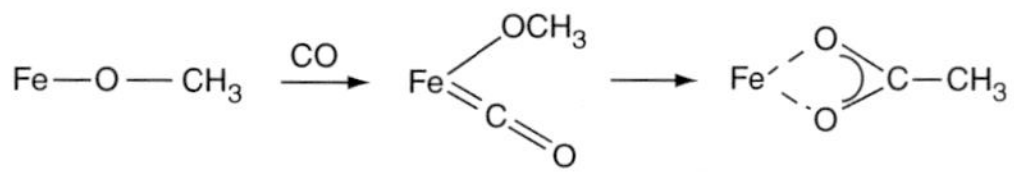 | CO insertion | Oxygenate, metal carboxylate |
| *Chain propagation* | *Chemistry* | *Common names for propagation species* |
| 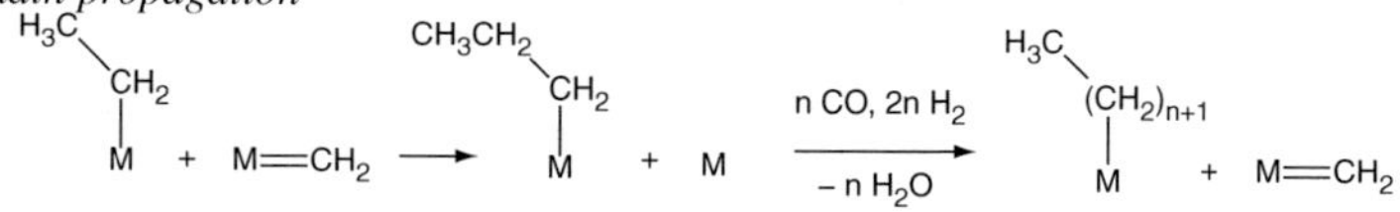 | Carbene insertion into metal–alkyl chain | Metal alkyl and methylene ($CH_2$), carbene ($M{=}CH_2$), methylidene ($M{=}CH_2$), alkylidene ($M{=}CH_2$) |
| 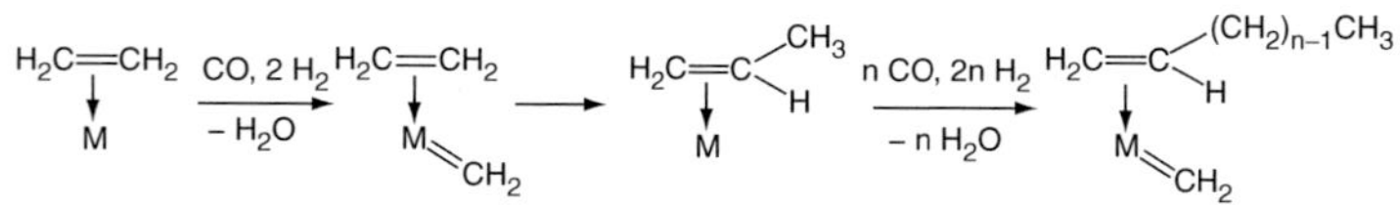 | Carbene insertion into metal–olefin chain | Olefin complex and methylene ($CH_2$), carbene ($M{=}CH_2$), methylidene ($M{=}CH_2$), alkylidene ($M{=}CH_2$) |
| 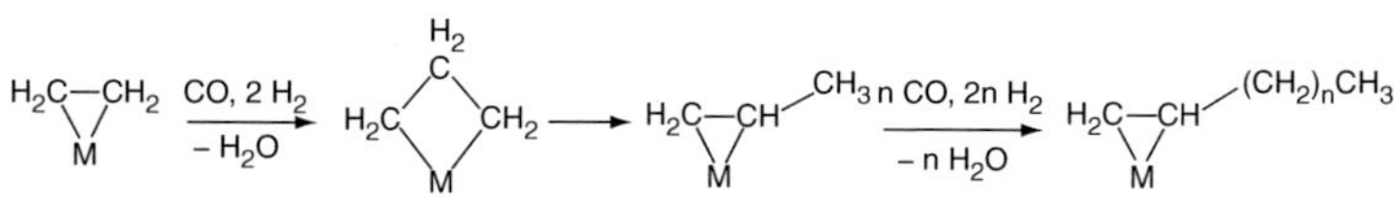 | Carbene insertion into metallacyclic chain | Metallacycle or olefin complex and methylene ($CH_2$), carbene ($M{=}CH_2$), methylidene ($M{=}CH_2$), alkylidene ($M{=}CH_2$) |
| 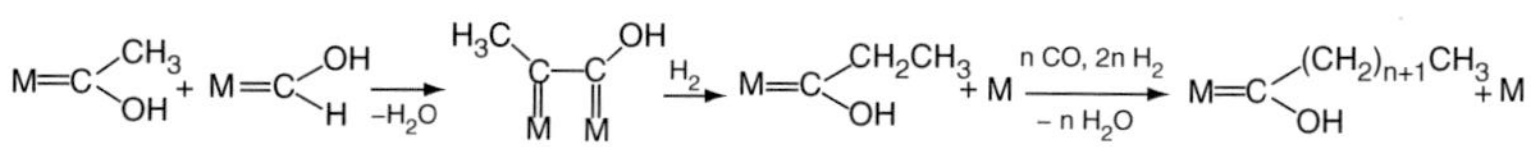 | Carbene coupling, or carbene insertion into metal-carbene chain | Enol, hydroxycarbene |
| | CO insertion and hydrogenation | Metal alkyl |

(continued)

**Table 8.2** (continued)

| | | |
|---|---|---|
| 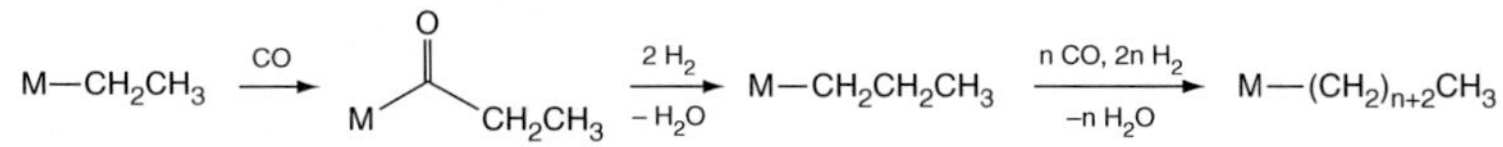 | | |
| 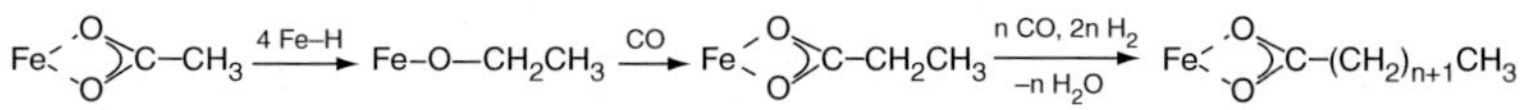 | CO insertion and hydrogenation | Oxygenate, metal carboxylate |
| *Examples of chain termination* | *Chemistry* | *Comments* |
| 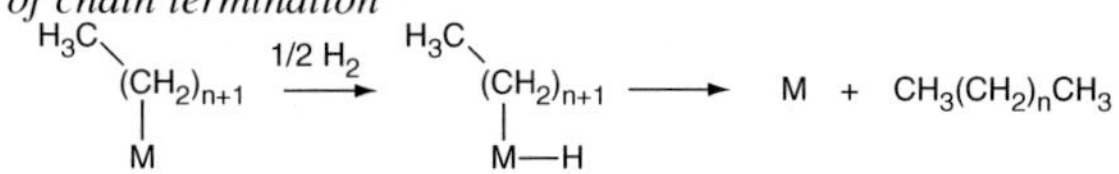 | Hydrogenation, reductive elimination | |
| 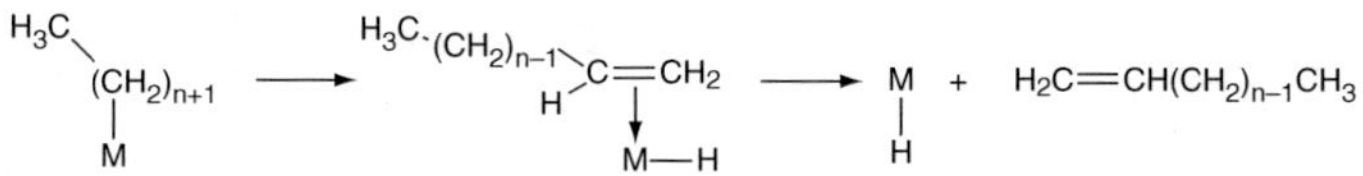 | β-elimination, olefin desorption | (The reverse is olefin re-adsorption) |
| 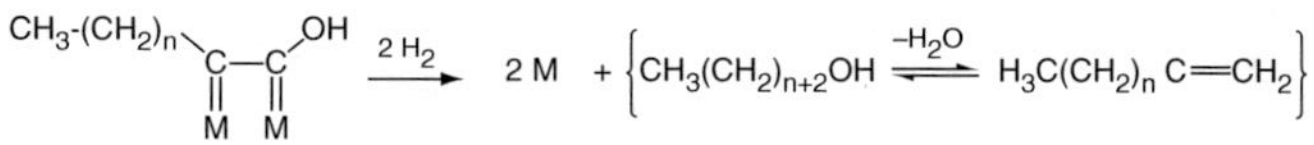 | Hydrogenation, dehydration | |
| 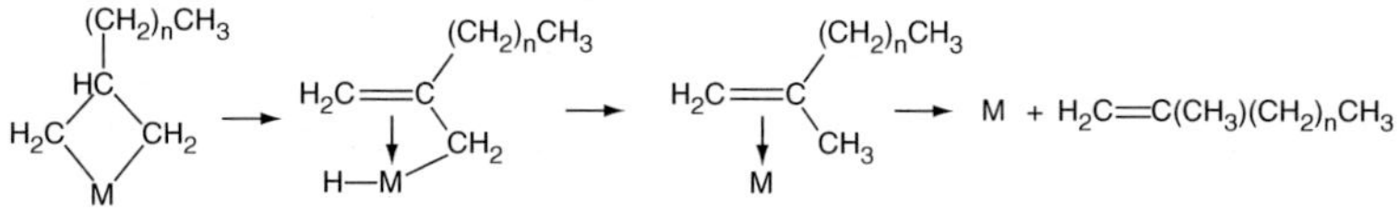 | β-elimination, reductive elimination olefin desorption | |
| 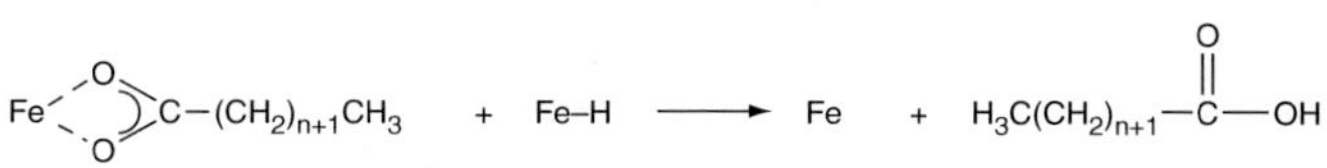 | Hydrogenation | |

Early studies performed by Pettit with $^{13}C$ labeling strongly supported the intermediacy of $M{=}CH_2$ species [28]. Including $^{14}C$ labeled alcohols and olefins in the feed, Emmett found that enols of alkoxides can initiate chain growth. Much of this early mechanistic work, as well as the concept that $CO_2$ can initiate chain growth (Fe catalysis), has been reviewed by Davis [29]. The propagation mechanism also can be considered a carbene mechanism, in that the initial product of CO reaction with a metal alkyl is M–C(=O)R, which has a tautomeric carbene form, $M^+{=}C(O^-)R$. Formation of oxygenates could occur by insertion of CO, to give M–C (=O)alkyl intermediates which are then reduced by $H_2$. Alternatively the alcohols could form by reductive elimination of metal surface-bound OH and the surface-bound chain.

## Chain Termination

The typical chain termination reaction is intramolecular, either reductive elimination or β-hydride elimination. Hydrogenation of a metal on which a chain is actively growing allows for reductive elimination to give the saturated alkane. Loss of an H atom from the beta-position of the chain (the second carbon away from the metal) allows the formation of a metal-hydride and a π-bound olefin, the latter then dissociating easily from the metal center to give an alpha-olefin. The β-elimination is usually quite reversible, and chain growth may resume if an olefin reinserts into the metal-hydride bond.

## ASF Distribution

For F-T reactions, the number of moles of species with carbon number n decreases exponentially with $n$. The weight fraction of product having carbon number, $W_n$, is given by the Anderson–Schultz–Flory (ASF) distribution (Eq. 8.8):

$$W_n = (1 - \alpha)^2 n a^{n-1} \tag{8.8}$$

The parameter $a$ is the Schultz–Flory distribution factor and represents the ratio of the rate of chain propagation to the rate of chain propagation plus the rate of chain termination. The parameter $\alpha$ is the chain growth probability, and $1 - \alpha$ is the probability of chain termination. Both parameter $a$ and parameter $\alpha$ are assumed to be independent of chain length. Either value can be obtained graphically using Eq. 8.8 in log form, with analytical data for $W_n$ as obtained, for example, using gas chromatography. The $\alpha$ values for oxygenated products appear to be about the same as those for hydrocarbons, which suggests that the products all form by the same or similar chain propagation steps and with similar rates of chain termination, albeit with different chain termination chemistries for the oxygenates, olefins, and paraffins.

The parameter $\alpha$ is used most often to characterize product distribution. Values of $\alpha > 0.9$ are representative of processes producing high amounts of wax, $\alpha$ values of about 0.8 maximize diesel cuts, and $\alpha$ values of about 0.7 are characteristic of naphtha production.

## F-T Catalysts

Early studies reported that the level of reaction activities of metal catalysts for the Fischer–Tropsch reaction decline in the order Ru, Ni, Co, and Fe. Later studies indicated that activities declined in the order Ru, Fe, Ni, and Co with alumina as support, and in the order Co, Fe, Ru, and Ni with silica as support. In practice, Ru gives rapid production of high molecular weight wax but this metal is cost-preclusive and supply-limited at plant scale. Nickel gives predominantly methane as product.

Commercially, Co appears to be more active than Fe in that the turnover rates on Co catalysts are about three times higher than on Fe catalysts. In general the observed rates depend on the diffusion rates of reactants and products migrating in and out of the porous catalyst particles, as well as on the intrinsic rate of the F-T reaction at the catalyst surface. The diffusion rates in turn depend on the catalyst's porosity, pore sizes, and particle size, and on whether or not liquid wax is present within the catalyst particles.

The product distribution can be significantly influenced the through use of promoters. The ability of numerous promoters to increase activity with Co catalysts is well documented but their ability to affect selectivity is less clear. In general, any promoter metal that can enhance the even distribution of reduced cobalt on the internal and external surfaces of the catalyst particles could increase the number of active sites available for chain growth and thereby increase the selectivity for wax. Ruthenium appears to do this best, and lanthanum also helps. Rhenium is sometimes used to promote chain growth with cobalt although reports of its efficacy vary.

The traditional industrial support materials for cobalt are alumina, silica, and titania [30]. The choice of support material appears to have little effect on selectivity; however, catalysts using mesoporous supports such as MCM-41 have provided somewhat increased $C_{5+}$ selectivity [31].

The performance of iron catalysts is very dependent on the addition of strong alkali promoters. Oxides of the alkali metals, particularly potassium, promote formation of wax in LTFT catalysis by iron. Use of $K_2O$ also helps keep the iron reduced in carbided form, which may promote the growth of longer chains. Not only is selectivity markedly dependent on the amount of alkali present, but alkali promoters also have a direct effect on the overall rate of reaction. For precipitated iron catalysts, used for LTFT, catalyst activity decreases when the alkali content is raised beyond a threshold level.

Cobalt catalysts produce considerably less $CO_2$ during F-T synthesis than do iron catalysts. The partial pressures of water and $CO_2$ affect selectivity, with the pressure of $CO_2$ being more relevant for iron than cobalt. Cobalt is a better hydrogenation catalyst than iron, and the products of its F-T reaction tend to have a lower olefin/paraffin ratio than those of iron.

## *Catalyst Beds and Reactors for F-T Synthesis*

Axial temperature gradients, poor heat distribution with carbon deposition, and scale-up limitation mean that multi-tubular reactors are usually not suitable for commercial LTFT synthesis. The slurry bed reactor (SBR) is an attractive alternative to fixed bed reactors. When using precipitated catalysts of identical composition, a SBR system is as active as a fixed bed system despite the much lower catalyst loading of the SBR. Slurry bed reactors have become the technology of choice for companies planning new GTL plants, with the exception of Shell as noted above. Both Sasol and Exxon have operated large demonstration slurry phase F-T units with supported cobalt catalysts. A drawback of SBR technology to date is that additional equipment is required to achieve near complete separation of the finely divided catalyst from liquid wax in slurry systems.

Multi-tubular fixed beds do have advantages as research reactors, because they are easy to operate. No equipment is required to separate the heavy wax product from the catalyst, since the liquid wax simply trickles down the bed and is collected in a downstream knock-out pot. Alternatively, laboratory studies may be performed in high velocity, high recycle stirred tank reactors. By varying the fresh feed gas flow to these units, the partial pressures and the kinetics at different conversion levels (tantamount to different bed depths) can be determined.

## *Kinetics*

Various kinetic equations based on research with laboratory reactors have been generated for the F-T reaction. Equations 8.9–8.14 are representative examples of kinetics proposed for Co systems.

$$r = kP_{H_2}{}^2/P_{CO} \tag{8.9}$$

$$r = aP_{CO}P_{H_2}/(1+bP_{CO})^2 \tag{8.10}$$

$$r = aP_{CO}P_{H_2O}{}^{0.5}/\left(1 + bP_{CO} + cP_{H_2O}{}^{0.5}\right)^2 \tag{8.11}$$

$$r = aP_{CO}{P_{H_2}}^2 / \left(1 + bP_{CO}{P_{H_2}}^2\right) \tag{8.12}$$

$$r = c\left({P_{H_2O}}^{0.5} / P_{CO_2}\right) / (1 + 0.93P_{H_2} / P_{H_2O}) \tag{8.13}$$

$$r = d(P_{H_2}P_{CO}) / P_{CO} + 0.27P_{H_2O}) \tag{8.14}$$

Equations 8.15–8.18 are representative of kinetics proposed or Fe systems.

$$r = k{P_{H_2}}^{0.6}{P_{CO}}^{0.4} - \mathrm{fr}^{0.5}{P_{H_2O}}^{0.5} \tag{8.15}$$

$$r = mP_{H_2}P_{CO} / (P_{CO} + nP_{H_2O}) \tag{8.16}$$

$$r = aP_{CO}{P_{H_2}}^2 / (P_{CO}P_{H_2} + bP_{H_2O}) \tag{8.17}$$

$$r = aP_{CO}P_{H_2} / (P_{CO} + bP_{H_2O} + cP_{CO_2}) \tag{8.18}$$

Equation 8.10 for Co responds to an increase in pressure: as the pressure increases, the F-T reaction rate decreases. Cobalt is much more resistant to oxidation than iron, whether by oxygen or by water, which could account for the absence of a $P_{H_2O}$ term in kinetic equations for cobalt catalysts. Even so, Co crystals smaller than a certain size (<5 μm) may be oxidized by water vapor and therefore a more detailed kinetic equation for cobalt could include the partial pressure of water.

The kinetics of Fe catalysis are better studied than those for Co. Sasol performed a large number of experiments and measurements in both fixed and fluidized bed reactors to study iron-based catalyst activity in commercial HTFT and LTFT processes. The key findings are as follows:

- Reaction rate is strongly dependent on hydrogen partial pressure. At low conversion levels, reaction rate is solely dependent on hydrogen partial pressure.
- Reaction rate increases with the partial pressure of CO.
- The partial pressure of $CO_2$ does not appear to have a direct effect on reaction rate. It can, however, affect the partial pressures of other components via the water gas shift reaction and thus may have an indirect effect.
- Reaction rate is markedly depressed by the partial pressure of water.
- The level of hydrocarbon products in the reactor has no apparent effect on F-T reaction rate.

Based on these observations, a satisfactory rate equation for iron-based catalysts is given by Eq. 8.19:

$$r = mP_{H_2}P_{CO} / (P_{CO} + aP_{H_2O}), \quad \text{where } a = k_{H_2O} / k_{CO} \tag{8.19}$$

The above kinetic expression is speculative but has been shown to estimate the overall conversion for both HTFT and LTFT reactors. This kinetic equation also predicts that if the total pressure of the gas feed is increased, the rate is increased by the same factor, that is, the residence time in the catalyst bed remains the same and the degree of conversion remains unchanged. This means that the production rate is increased in proportion to the increase in pressure. Pilot plant tests at Sasol for both the fixed bed LTFT and the fluidized bed HTFT operations confirmed these kinetic predictions.

Given that reforming methane gives as much or more $H_2$ as needed for the F-T reaction, the presence of water gas shift chemistry in the F-T reactor could easily become a liability to efficiency of a GTL process. Using the same reactor under the same operating conditions and feed gas flow, a Co catalyst system will have a much higher syngas conversion than an iron catalyst. With an iron catalyst, after water has been condensed out a larger portion of the tail gas is recycled to the reactor. This means that more reactors are required and there is also a greater need to recycle, both of which increase the total fixed capital and operating costs.

## Commercial Activities in the Fischer–Tropsch Synthesis

In 1993, two large-scale F-T plants were built specifically for operation with natural gas (as opposed to coal). Backed with funding from the South African government, Mossgas (now PetroSA) began operating a 22,000 bbl/day plant using Sasol technology. The gas is produced from a South African offshore field. Also that year, Shell built a 10,000 bbl/day plant at Bintulu, Malaysia. The $600 million plant was expanded to 12,500 bbl/day in May 2000, after a fire destroyed the oxygen plant. Also in 1993, a plant designed by Rentech was built in Colorado Springs, Colorado for conversion of landfill gas to liquids. A lack of methane from the landfill shut down the plant.

A joint venture of Qatar Petroleum and Qatar Shell has recently built a GTL plant in Ras Laffan Industrial City, Qatar, using Shells' cobalt-catalyzed, fixed bed technology. The project is referred to as Pearl and is scheduled for the latter part of 2011. At $18+ billion, the cost of Pearl significantly exceeded budget. This may have resulted at least in part from the incorporation of an additional facility, for producing NGL, into the design. Pearl is expected to generate a total of 140,000 bbl/day in two trains operating at full nameplate capacity, making this plant the world's largest GTL project [32]. In addition Pearl has a capacity for 120,000 barrel of oil equivalents total per day of NGL from two trains. Another new GTL plant has been under construction recently in Qatar, by Sasol, and was operating at about 55% of nameplate capacity in 2010.

## F-T Product Upgrading

Conversion of syngas through the F-T process leads to a distribution of products essentially consisting of *n*-paraffins (>90%) and smaller percentages of alcohols, olefins, and methane. Typical hydrocarbon product cut weight fractions are shown in Table 8.3.

A consequence of aliphatic chain growth by one-carbon units is the theoretical impossibility of synthesizing a F-T product having a narrow range of chain length. This wide molecular weight range of the raw F-T product slate means that downstream separations and refining must be used to create fractions for specific end uses. The crude product from the F-T reaction is most commonly separated into a light, relatively low boiling condensate fraction and a heavy, high boiling wax fraction. A typical boiling range cut temperature would be 372°C. The condensate fraction is a liquid that can be shipped from a remote area to a refinery site for upgrading. Condensate containing, for example, $C_4$–$C_{15}$ products is further separated, into cold separator liquid and hot separator liquid. The cold separator liquid contains $C_{5+}$ in the boiling range of <260°C, retains about 95% of the total oxygenates produced, and has sufficiently low olefin concentration that hydrogenation is unnecessary. The hot separator liquid contains mostly hydrocarbons and has a boiling range of 260–372°C.

### *Hydrocracking Heavy Paraffins (Wax)*

The $C_{16+}$ reactor wax may be upgraded by various hydroconversion reactions including hydrocracking, hydroisomerization, catalytic dewaxing, isodewaxing, or various combinations thereof, to produce middle distillates for use as fuels, lubricants, chemicals, and specialty materials such as nontoxic drilling oils, technical and medicinal grade white oils, chemical raw materials monomers, polymers, emulsions, isoparaffinic solvents, and other specialty products. For example, wax from F-T synthesis is typically first hydrofined, to eliminate alkenes and oxygenated compounds, and then fractionated into different grades of product waxes. The waxes can be sold directly or hydroprocessed by recycle to extinction to yield high quality diesel and kerosene fuels.

**Table 8.3** Hydrocarbon weight fractions in typical F-T product

| Product cut | Carbon No. | wt% |
|---|---|---|
| Gas | $C_1$–$C_4$ | 2.5 |
| Naphtha | $C_5$–$C_9$ | 6.7 |
| Diesel | $C_{10}$–$C_{19}$ | 18.5 |
| Soft wax | $C_{20}$–$C_{34}$ | 27.3 |
| Hard wax | $C_{35+}$ | 45.0 |

Hydrocracking is the most relevant upgrading process, including for maximum production of diesel fuels. Hydrocracking heavy paraffins serves to lower the boiling range of product from that of wax to that of middle distillates, and to improve the cold flow properties of F-T diesel with branched molecules formed during hydrocracking. Because F-T products are predominantly linear, especially those from LTFT, the raw middle distillates have very poor cold flow properties that would hamper their use as transportation fuel unless subjected to hydrocracking.

Another property affected by hydrocracking is the cetane number of diesel. As the pressure of a LTFT reactor increases, the selectivity for wax increases, the degree of branching decreases, and consequently the cetane number of downstream diesel cuts decreases. Subsequent selective hydrocracking of crude wax makes the largest contribution to the final diesel fuel pool. Some chain branching occurs during hydrocracking and thus the cetane number of the diesel fuel produced is somewhat lower than that of straight run F-T diesel. The final diesel pool, however, has a cetane number above 70 – significantly superior to the cetane number for refinery naphtha, which varies from 40 to 50.

Upgrading wax by hydrocracking to middle distillates gives naphtha as the main coproduct. The naphtha fraction may be upgraded to gasoline, via catalytic reforming, or used as a steam cracker feedstock for olefins production. Converting these two naphtha cuts into on-specification gasoline would require a considerable amount of further octane number upgrading. However, this naphtha consists almost entirely of linear alkanes and therefore makes excellent feedstock for production of ethylene by steam cracking, giving much higher selectivity for ethylene than that obtained from steam cracking of naphtha obtained from crude oil.

Feeds derived from LTFT can be hydrocracked under much milder conditions than typical feeds such as vacuum gas oils that are derived from crude oil. In the hydrocracking of crude oil-derived feeds, typically pressures as high as 150 atm are required to prevent coking of the catalyst by aromatic compounds. This is not necessary with paraffinic feeds, wherein LTFT products are hydrocracked at pressures from 35 to 70 atm using commercial hydrocracking catalysts. The relatively low levels of oxygenates, mainly alcohols and lesser amounts of acids or carbonyls, present in F-T waxes are easily and completely hydrodeoxygenated.

The refining of syncrude is often performed in a separate facility from the upstream GTL process. For direct on-site F-T reaction product upgrading, however, the hot separator liquid cut may be combined with F-T wax and processed under much milder conditions than typical crude oil–derived feeds. Conventional hydrocracking of refinery feedstocks is performed in a fixed bed reactor using multiple or single Co/Mo catalyst compositions under operating pressures as high as 2,500 psi, using conventional technologies such as those offered for license by Chevron or UOP. The high operating pressure is required to prevent coking of the catalyst by aromatic compounds contained in the feed. Such high pressures are not required when hydroprocessing paraffinic feeds from LTFT synthesis, which do not contain aromatics, and lower pressures between 500 and 1,000 psi are used to hydrocrack LTFT products with commercial hydrocracking catalysts. During this process, the oxygenates present in the feed are easily and completely hydrodeoxygenated. Typical hydrocracking conditions for F-T products entail operating temperatures of 320–450°C, pressures of 1,000–1,500 psi, and hydrogen treat rates of 500–5,000 scf/bbl.

The low sulfur content of syngas from natural gas enables consideration of hydrocracking catalysts containing a noble metal component, particularly platinum, instead of the less expensive, conventional sulfur-resistant Co/Mo catalysts. Platinum leads to higher hydroisomerization activity and less severe hydroprocessing conditions, with better low-temperature performance of the resultant diesel fuel. For example, ExxonMobil's hydroisomerization process uses a noble metal catalyst to increase the amount of isoparaffins in the distillate fuel and to help the fuel meet pour point and cloud point specifications without the need for additives. Hydroisomerization may be carried out with a single Pt-based catalyst in a fixed bed reactor. Preferred operating conditions use a temperature range of 260–400°C, a pressure range between 500 and 1,000 psi, and a hydrogen treat rate of 1,000–4,000 scf/bbl, and heavy wax products are recycled to extinction through the hydrocracking process.

The processing pressure is dependent on the hydrogenation capacity of the catalyst. When the hydrogen partial pressure is too low, dehydrocyclization of the paraffins starts to occur with the formation of polynuclear aromatics, which eventually would lead to deactivation of the catalyst due to coking. Lower hydrogen/wax ratios lead to a decrease of conversion rate of the $C_{22+}$ fraction and, after adjustment of reaction conditions to achieve the same conversion levels, an increase of both the isoparaffin content and the selectivity to products lighter than diesel.

Isoparaffin content also increases with operating pressure. The hydrocracking process has to meet the following conditions:

- The chain length of the hydrocracked fragments should be predominantly in the desired product's carbon number range.
- The components above said desired range should be hydrocracked in preference to those that are already within or below the desired range.
- The production of less commercially attractive species, for example, light hydrocarbon gases, should be minimized.

Due to the clean nature of the feed, non-sulfided hydrocracking catalysts containing a noble metal component such as platinum can be considered. The use of noble metal catalysts leads to higher hydroisomerization activity and consequently better low-temperature characteristics for the diesel product. Most probably all $C_{14+}$ material would be treated using such catalysts, due to the enhancement of cold flow properties resulting from simultaneous isomerization and hydrocracking. A high blending cetane value is retained, and very little of the diesel range material in the feed is degraded into naphtha and kerosene. Catalysis using a noble metal on amorphous silica-alumina is likely to be the preferred approach for maximum production of diesel blending material.

There are many technology licensors for hydroconversion processes, including ChevronTexaco, UOP, IFP (Axens), Criterion, and Haldor Topsøe. Catalysts can be obtained from these companies as well as from Akzo Nobel and Sud-Chemie, among others.

## *Hydrotreating Light Paraffins*

Hydrogenation is less costly than hydrocracking. For large-scale plants, a condensate hydrotreater should be used to process hydrocarbons that are already lighter than the desired middle distillates and also for straight run middle distillates if cold flow properties are not an issue. A condensate fractionator may be employed if it is desired to produce linear paraffin products. The condensate fractionator can also serve to remove light gases in order to stabilize the condensate for intermediate storage at atmospheric pressure. If a nearby naphtha cracker is available, then the condensate fractionator will provide a naphtha product that need not be hydrogenated to ensure storage stability. Also, a kerosene fraction can be separated using the condensate fractionator.

If both hydrotreating and hydrocracking are used, they may be operated at the same pressure and use a common hydrogen loop. Alternatively, the hydrotreater can be operated in a once through mode, at lower pressure, and the hydrogen tail gas then subsequently pressurized and fed to the hydrocracker loop. For a 30,000 bbl/day equivalent product upgrading system comprised of hydrocracking and hydrotreating, the size of the unit is comparable to those used in oil refineries. There are many process configurations possible depending on site-specific factors like plant capacity, available land, and possible integration with other facilities.

While product upgrading schemes center around three fuel products (diesel, LPG, and naphtha), the upgrading plant is quite amenable to different configurations if the market demand so indicates. Even with cracking and hydrotreating, the ASF distribution means that the F-T process and its associated refinery produce other transportation fuel and products beyond middle distillates. Other saleable products may include solvents, kerosene, jet fuel, illuminating paraffins (sold for use in lighting and appliances in South Africa), and base oils (for lubricant products). Maximizing the recovery of the higher boiling jet fuel while maintaining diesel production will tend to increase economic return.

## *LTFT Naphtha and Blends*

Four LTFT naphthas can be produced using process configurations summarized below in Table 8.4.

**Table 8.4** Process schemes for production of LTFT naphtha

| F-T naphtha | | Production scheme |
|---|---|---|
| SR | Straight run | Fractionation of F-T condensate |
| HT SR | Hydrogenated straight run | Fractionating and hydrotreating of F-T condensate |
| HX | Hydrocracked | Fractionating and hydrocracking of F-T wax |
| SPD | Slurry phase distillate | Blend of HT SR and HX naphthas |

The slurry phase distillate (SPD) naphtha is an excellent feed for the production of lower olefins, particularly ethylene, by steam cracking. Such paraffinic naphtha feed demands less feedstock and lower energy consumption in the cracking process than naphthas from other sources. The optimum olefin yields and selectivities obtainable from any naphtha depend on the cracking severity at commercial conditions. The mass ratio of light olefins, propylene/ethylene (P/E), has an inverse relationship to the cracking severity. High P/E values correspond to low severities and also lower coproduction of methane.

The LTFT naphthas other than SR are essentially entirely paraffinic and therefore have compositions far from those of high-octane naphthas useful in gasoline blends. Even so, the conventional naphthas typically require that the refiner improve cold flow characteristics by changing the boiling point, for compatibility with low-temperature operation. Blends of SPD naphtha with conventional diesels could be used to meet winter grade fuel specifications without decreasing the production volume of diesel and other light products obtained from oil refining. Including F-T naphtha in blends can assist refiners in meeting low sulfur specifications in fuels that have sulfur content near zero (below 1 ppm).

The projected threshold for economic viability of F-T GTL technology, in terms of crude oil price, is now at about $70–75 per BOE. This threshold is based on recycle of reformed NGL (vs. production of NGL for market sales). The investment cost for a modest scale GTL facility producing predominately diesel fuel has been estimated at about $1.1 billion [7]. There are then further opportunities to add value by producing base oils and waxes.

This $1 billion investment value may be taken as a starting point when estimating actual investment costs for a first-of-a-kind plant. There is a substantial gap between such estimate and the final cost of, for example, Pearl. In the Pearl project, the initial configuration was changed to one utilizing NGL for market sales and also the builders experienced escalating capital costs, due to a shortage of major equipment, procurement delays, shortage of skilled engineering labor, and an increased number of reactors from an initially conservative estimate, among other contributing cost factors.

## Methanol

### *General Process Description*

Currently the majority of methanol is synthesized from syngas produced from natural gas, using autothermal reforming, steam reforming, or a combination of steam reforming and autothermal reforming. Most existing commercial plants use steam reforming. Incoming pressurized natural gas feed is preheated and passed through a zinc oxide catalyst bed to remove sulfur compounds that would otherwise poison downstream catalysts. Steam is added to the natural gas stream before it enters the reformer. A typical reformer is a large high-temperature furnace containing a number of vertical tubes that are filled with a nickel catalyst. Once the natural gas is reformed, the resulting synthesis gas is cooled to about 30°C, condensed water is separated from the gas in a process gas separator, and the syngas is compressed using a steam turbine compressor that uses steam generated in an economizer on the syngas outlet from the reformer.

Compressed syngas containing $H_2$, CO, and $CO_2$ is fed to a methanol synthesis reactor that uses a copper catalyst. The methanol reactor outlet gas is cooled to about 40°C and then passed to a pressure letdown tank where unreacted synthesis gas is separated from the product methanol and water. A portion of this unreacted syngas stream is purged and burned as fuel, while the majority is compressed by a recirculating compressor and remixed with feed synthesis gas (from the upstream syngas compressor). The unreacted syngas is thus recirculated back to the methanol reactor, resulting in an overall conversion efficiency of 99%.

Crude methanol collected in the letdown tank usually contains up to 18% water plus impurities such as methyl formate, higher alcohols, ketones, ethers, and esters. This is stored at atmospheric pressure in an intermediate storage tank and later fed to a distillation plant. The distillation plant may be configured in a variety of ways. A typical distillation plant consists of a single distillation column that removes the light ends including dissolved gases, dimethyl ether, methyl formate, and acetone that are burned as fuel. Bottoms from the light ends column are sent to a system that consists of two distillation columns, both of which deliver pure methanol overhead and sequentially remove water and heavy ends such as higher alcohols, ketones, and esters (formed by esterification of lower alcohols with formic, acetic, and propionic acids in column bottoms). Water from the bottoms of the first distillation column is discharged to the process sewer system, while heavier ends separated as bottoms in the second methanol recovery column are sent to the reformer furnace fuel system for combustion. The methanol is at least 99.8% pure after this distillation and is transferred from temporary storage to final product storage tanks where it is directly pumped into ships, railcars, or trucks for delivery to customers and distributors.

## *Reaction Pathway*

Chemical reactions considered in methanol synthesis are shown in Eqs. 8.6, 8.20, and 8.21.

$$H_2O + CO \leftrightarrows H_2 + CO_2 \quad \Delta H = -41.2\text{kJ/mol CO} \tag{8.6}$$

$$\begin{gathered} CO_2 + 3H_2 \leftrightarrows CH_3OH + H_2O \\ \Delta H = -49.0\text{kJ/mol } CO_2 \end{gathered} \tag{8.20}$$

$$CO + 2H_2 \leftrightarrows CH_3OH \quad \Delta H = -90.1\ \text{kJ/mol CO} \tag{8.21}$$

Both CO and $CO_2$ are reactive in the synthesis. Coupling the methanol synthesis with the water gas shift reaction (Eq. 8.6) allows a larger quantity of carbon to become available for most efficient synthesis. Although the direct hydrogenation of either $CO_2$ or CO to methanol is thermodynamically favorable (Eqs. 8.20 and 8.21, respectively), the activation energy for direct hydrogenation is apparently too high and instead the reaction proceeds through a different mechanistic pathway.

The reaction of syngas to produce methanol is about 100 times slower if $CO_2$ is not also present in the mix. Until as recently as the 1990s, the role of $CO_2$ in methanol synthesis was unclear. The water gas shift activity of Cu catalysts is so high that it was difficult to delineate the roles of CO and $CO_2$ in methanol synthesis. Isotopic labeling studies have now unequivocally shown that $CO_2$ is the source of carbon in methanol. The CO undergoes the WGS reaction to make $H_2$ and $CO_2$. The $CO_2$ is thought to keep the catalyst in an intermediate oxidation state, thus preventing the reduction of ZnO followed by the formation of brass.

A long-held mechanism for catalytic methanol synthesis proceeds through a metal-bound formate intermediate, formed by hydrogenation [33, 34]. As depicted in Eqs. 8.20–8.24, $CO_2$ is adsorbed onto the catalyst surface [35], where it reacts with a partially oxidized methane intermediate to form a carbonate, which is then hydrogenated. This intermediate is then further hydrogenated in the rate-limiting step. The copper catalyst sites have high activity for splitting the first C–O bond of $CO_2$ that helps maintain the oxidation state of the active copper sites. At high concentrations, however, $CO_2$ actually decreases catalyst activity and inhibits methanol synthesis. The feed gas composition for methanol synthesis is typically adjusted to contain 4–8% $CO_2$ for maximum activity and selectivity. Although Cu has water gas shift activity, excessive water causes deactivation by blocking active sites. Site blocking lowers catalyst activity but improves selectivity, with a 50% decrease in by-product formation.

Recent modeling efforts using density functional theory (DFT) suggest that the traditional methanol synthesis mechanism proceeding via $CO_2$ accounts for only about two third of the methanol produced [36]. The remaining methanol is said to derive from the CO, mostly via direct reduction by $H_2$. The proposed mechanism is conceptually similar to that for $CO_2$ reduction to methanol and is shown in Scheme 8.1. Overall, methanol synthesis appears rate limited by formation of methoxy ($CH_3O^*$, where * refers to the catalyst surface) at low $CO_2/(CO + CO_2)$ ratios and by $CH_3O^*$ hydrogenation in $CO_2$-rich feeds. Hydrogenation of $CH_3O^*$ as per Eqs. 8.20 and 8.21 is the slow step for both the CO and the $CO_2$ methanol synthesis routes.

$$H_2 + CO + 2^* \rightleftharpoons CHO^* + H^*$$

$$CHO^* + H^* \rightleftharpoons HCHO^* + {}^*$$

$$HCHO^* + H_2 + {}^* \rightleftharpoons CH_3OH^* + H^*$$

$$CH_3OH^* + H^* \rightleftharpoons CH_3OH^*$$

$$CH_3OH^* \rightleftharpoons CH_3OH$$

**Scheme 8.1** Proposed mechanism for reduction of CO to methanol

## *Reactors and Catalysis for Methanol Synthesis*

The two main process features considered when designing a methanol synthesis reactor are (a) controlling and dissipating the large reaction heat, and (b) overcoming the equilibrium constraint to maximize per pass conversion efficiency. Methanol synthesis is a version of the classic trade-off between conversion and selectivity. The maximum per pass conversion efficiency of syngas to methanol is limited to about 25%, and catalyst activities generally decrease as the temperature is lowered. At higher temperatures, catalyst activity increases but so does the chance for competing side reactions to form by-products such as dimethyl ether, methyl formate, higher alcohols, and acetone. Catalyst lifetimes are also reduced by continuous high-temperature operation.

Maintaining reaction temperatures below 300°C is the trade-off strategy employed to minimize catalyst sintering while maximizing conversion of syngas to the target product slate. Although the equilibrium conditions favor low temperatures, methanol converters must be operated at temperatures in the range 200–300°C to ensure the catalysts are active and to use the heat of reaction effectively [37].

Removing methanol as it forms is another strategy used for overcoming the thermodynamic limitations and improving the per pass conversion and process efficiency. Methanol is either physically removed or converted to another product, such as dimethyl ether, methyl formate, or acetic acid.

Numerous methanol reactor designs have been commercialized over the years and these can be roughly separated into two categories, namely, adiabatic and isothermal reactors. Adiabatic reactors often contain multiple catalyst beds separated by gas cooling devices, using either direct heat exchange or injection of cooled syngas, either fresh or recycled. Axial temperature profiles often have a saw tooth pattern: dropping down at the point of heat removal and increasing linearly between the heat exchange sections. The isothermal reactors are designed to continuously remove the heat of reaction so they can operate essentially like a heat exchanger with an isothermal axial temperature profile.

Currently all commercial processes produce methanol from syngas in a vapor phase reactor with a synthesis loop. In a gas phase, tubular or packed bed reactor, reaction heat is dissipated in situ by injecting cool, unreacted feed at various points along the reactor length, or by including internal cooling surfaces for heat exchange. Temperature moderation is also achieved by recycling large quantities of gas, with gas cooling in the recycle loop. Even with these heat transfer measures, the CO content of the feed is limited to about 16% as a means of controlling heat content by limiting the conversion.

The design challenge of removing heat while maintaining precise temperature control is significant, because excessive temperatures seriously diminish the lifetime of the catalyst. A liquid process is of interest, wherein nonvolatile mineral oils make up part of the reactor liquid phase and act as a temperature moderator and heat sink. A new approach investigated by Air Products involves bubbling the syngas through

**Table 8.5** Methanol synthesis catalysts prepared under various aging conditions

| Catalyst | Metals (molar ratio) | Precipitation parameters | Aging parameters |
|---|---|---|---|
| $Cat_{2\text{-const}}$ | Cu/Zn/Al (60/30/10) | pH 7, 70°C | pH $7_{const}$, 70°C |
| $Cat_3$ | Cu/Zn/Al (60/30/10) | pH 8, 70°C | pH $8_{const}$, 70°C |
| $Cat_{2\text{-free}}$ | Cu/Zn/Al (60/30/10) | pH 7, 70°C | Free-aged, 70° |
| $Cat_{Cu/Zn}$ | Cu/Zn (70/30) | pH 7, 70°C | pH $7_{const}$, 70°C |

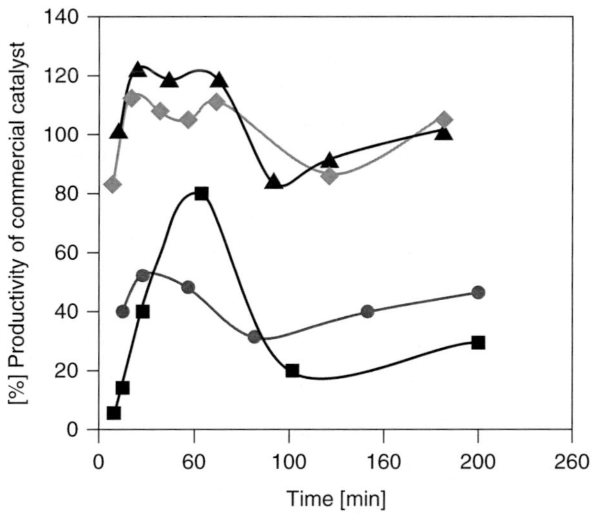


**Fig. 8.3** Methanol productivity as a function of the aging time for selected catalysts: (*triangle*) $Cat_{2\text{-const}}$; (*square*) $Cat_{2\text{-free}}$; (*circle*) $Cat_{Cu/Zn}$; (*diamond*) $Cat_3$ [39]

a slurry consisting of micrometer-sized methanol synthesis catalyst and mineral oil as reaction medium that transfers heat from the catalyst surface, through the slurry, to boiling water in an internal tubular heat exchanger. Concentrations of CO as high as 50% have been used in the slurry system without damaging the catalyst activity.

A typical commercial methanol synthesis catalyst contains of copper and zinc oxide phases with alumina support, that is, $Cu/ZnO/Al_2O_3$. Recent modifications to catalyst preparation at laboratory scale include specialized coprecipitation methods [38], for a greater synergistic effect between copper and zinc, and the maintaining of a strict control of pH and the temperatures of precipitation and calcination [34]. As shown in Table 8.5 and Fig. 8.3, the productivity of commercial catalysts is sensitive to several factors, including conditions used for preconditioning by, for example, aging the catalyst [39, 40].

## *Activities in Methanol Synthesis*

Catalytic methanol synthesis from syngas was classically run as a high temperature, high pressure, exothermic, equilibrium-limited synthesis reaction operated at 320–450°C and 250–350 atm pressure. BASF of Germany introduced the first commercial synthesis of methanol, in 1923, using a zinc/chromium oxide catalyst. New methanol synthesis technologies based on lower reaction pressures (50–100 atm) were introduced in the 1960s. Whereas high-pressure processes require 139,000–146,000 Btu/gal of methanol produced, all modern plants are based on low-pressure technology and are expected to require fewer than 100,000 Btu/gal of methanol produced, on an average yearly basis. Additional advantages of low-pressure processes include lower maintenance costs and reduced downtime.

Johnson Matthey (formerly ICI Chemical Industries) of the United Kingdom and Lurgi of Germany offer the two most widely used technologies for methanol synthesis. The ICI low-pressure process operates at pressures between 50 and 100 atm and temperatures of about 210–290°C. A highly selective copper/zinc alumina catalyst is used in a shot-cooled single or multi-bed reactor. The reaction is quenched on the bed by introducing cold gas through specially designed spargers. Heat recovered from the synthesis loop provides part of the process steam required by the steam reforming furnace.

Lurgi's conventional low-pressure methanol process has a maximum output of 1,830–2,000 t/day. It uses a copper-based catalyst, pressures of 50–100 atm, and temperatures of 230–265°C. Temperature control of the catalyst is achieved by indirect cooling with boiling water. The boiling water in the reactor shell generates 1–1.2 kg of steam/kg of methanol, with a pressure of up to 45 atm. The steam, after moderate superheating, supplies the total energy for driving the recycle gas compressor and is subsequently used for the reboilers in the methanol distillation section. A small quantity of surplus steam is often available for export.

Johnson Matthey, Lurgi, and Haldor Topsøe have been pursuing 10,000 t/day mega-methanol single train plant designs. Lurgi's MegaMethanol® process can accommodate a single train capacity of at least 5,000 t/day and was developed for capacities larger than one million tons of methanol per annum. A schematic of Lurgi's process is shown in Fig. 8.4. Lurgi does not use conventional steam reforming to produce syngas, instead employing autothermal reforming with a newer burner system in what is called the MegaSyn process for converting oxygen and natural gas to syngas.

The first plant to use the MegaMethanol® technology was brought on stream in 2004 at Point Lisas in Trinidad. The plant is owned by Atlas Methanol Company, a joint venture between Canada's Methanex and BP Trinidad and Tobago (bpTT). The 1.7 t/year Atlas methanol plant was built adjacent to the existing Titan Plant, which has a methanol production capacity of 850,000 t/year. Average total methanol production from the plants is 2.5 million tons per year, which makes Trinidad and Tobago the world's leading methanol exporter.

The Point Lisas plant has an on-site air separation plant having a production capacity of 2,800 t of pure oxygen per day. Their mega-methanol production process consists of oxygen-blown autothermal reforming at a pressure of about 40 atm, desulfurization and pre-reforming, a two-step Lurgi methanol synthesis in water- and gas-cooled reactors, and adjustment of syngas composition via hydrogen

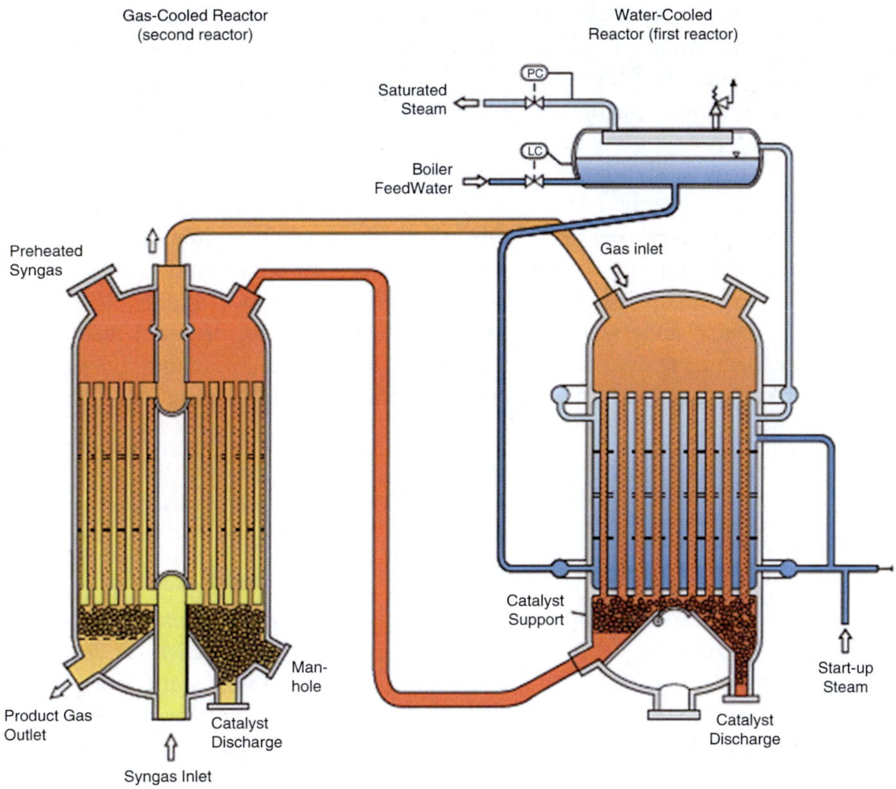


**Fig. 8.4** The Lurgi MegaMethanol® process [41]

recycle. The reformer outlet temperatures range from about 950°C to 1,050°C. The synthesis gas is then compressed using a single casing gas compressor. An integrated recycler in the compressor generates the pressure for the methanol synthesis.

The methanol reactor contains a fixed number of tube sheets containing the copper-based catalyst. The isothermal reactor enables partial conversion of the syngas to methanol, reduces the by-products produced during the process, and increases the yields at low recycle ratios. The Lurgi-combined converter process employs a gas-cooled reactor and a water-cooled reactor. The gas containing methanol from the first reactor is transferred to a downstream water-cooled reactor where cooling takes place. The purge gas is separated and the cooled crude methanol is processed in a three-column distillation unit through which about 40% of the heating steam is saved.

As smaller, high-cost methanol plants continue to be replaced by mega-methanol plants, the industry's aggregate cash cost curve should drive down price to levels that would still provide adequate profitability at prices below $100 per ton. This price environment creates new business opportunities for methanol to compete with other basic fuel and feedstock options. With combustion, a higher heating value of 9,776 Btu/lb, a $100 per ton methanol fuel has an inherent energy cost of $4.6 per million Btu. This cost is comparable to imported LNG at the receiving/regasification terminal. Since the transportation/distribution cost for methanol as a nonviscous liquid is relatively low, methanol fuel cost at the burner tip can be significantly cheaper than natural gas derived from LNG at the burner tip. Not only can methanol be considered under some economic scenarios as a primary fuel for combustion, its very low sulfur content (even when compared to ultralow sulfur diesel) provides an opportunity for methanol to be used as a backup fuel for utility and commercial power plants.

Research is also underway to convert synthesis gases to methanol in a liquid phase rather than using dry, fixed bed reactors. These technologies aim to avoid the problem of low conversion rates and the need for recycling of gases.

## Dimethylether

High oil prices and short supplies of LNG and LPG in Asia have led to the construction of many new DME plants, particularly in China. DME is currently industrially important as the starting material in production of the methylating agent dimethyl sulfate. DME has the potential to be used as a diesel or cooking fuel, a refrigerant, or a chemical feedstock. DME is also used as an aerosol propellant.

### *Reaction Pathway*

DME traditionally is formed in a two-step process where methanol is first synthesized and is then dehydrated, over an acid catalyst such as γ-alumina, at methanol synthesis conditions. The dehydration reaction is shown in Eq. 8.22:

$$\begin{gathered} 2CH_3OH \leftrightarrows CH_3OCH_3 + H_2O \\ \Delta H = -24.5\ \mathrm{kJ/mol\,DME} \end{gathered} \tag{8.22}$$

Alternatively, DME can be synthesized directly from syngas – without the separate process step involving methanol production. Extensive studies have been performed to deduce the stoichiometry of direct DME synthesis from syngas, and the topic is still under debate [42]. Work by Air Products and JFE indicate that the stoichiometry is 1:1 in $H_2/CO$ [43, 44]. The net reaction is shown in Eq. 8.23:

$$\begin{gathered} 3H_2 + 3CO \leftrightarrows CH_3OCH_3 + CO_2 \\ \Delta H = -246\ \mathrm{kJ/mol\ DME} \end{gathered} \tag{8.23}$$

**Table 8.6** Comparative conversions for DME, methanol, and methanol/DME

| Conversion | DME | Methanol | Methanol + DME |
|---|---|---|---|
| Per pass (%) | 50 | 14 | 18 |
| Total (%) | 95 | 77 | 85 |

Net reaction 23 is a combination of Eqs. 8.20–8.22.

Because water produced in the dehydration step is consumed in the WGSR, the conversion of syngas can be higher in DME synthesis than in methanol synthesis. Table 8.6 shows the conversions obtained for direct processes producing DME, methanol, or a mixture of DME and methanol. The observed optimum ratio for DME synthesis is lower than that for methanol synthesis and ideally should be targeted at about 1.

Interestingly, the use of the methanol synthesis reaction 21 instead of 23 would give a net reaction for direct DME synthesis having $H_2/CO$ stoichiometry of 2:1, which does not agree with experimental observation. Thus the direct synthesis of DME apparently proceeds through a different methanol-producing step than occurs in the production of methanol itself.

## *Commercial Activities in DME Synthesis*

Gas phase reaction in a fixed bed was generally used in the past for DME synthesis from methanol. Development of the direct synthesis of DME in the gas phase with improved fixed bed reactors is also of interest [45]. Other improvements to direct DME production include the use of a slurry reactor for liquid phase, low-pressure DME synthesis.

Commercial production of DME originated as the formation of by-products during high-pressure methanol production. Direct, or single-step, synthesis of DME from synthesis gas was developed in the 1980s, prompted by the high prices and uncertain supply of oil prevailing at the time. The objective of direct synthesis was to produce DME as an intermediate step en route to synthetic gasoline. Interest in non-oil-based fuels waned quickly as the oil supply position eased, until the early 1990s when interest was rekindled – this time on a much wider scale – by the increasing demand in Asia for LPG or the LPG-substitute, DME, and by the growing concern to produce cleaner fuels. The companies currently most actively engaged in the development of an economic DME process at large-scale are JFE (Japan), Haldor Topsøe (Denmark), and Air Products & Chemicals (USA).

### Gas Phase DME Production

Direct synthesis of DME in the gas phase with fixed bed reactors has been developed most extensively by Haldor Topsøe [45]. For large-scale processes, the reaction may be conducted in a series of three to four reactors with interstage cooling. Each reactor is operated adiabatically. Alternatively, the adiabatic operation may be carried out at low per pass conversion of CO, to keep the reaction exothermal within limits without interstage cooling (although such an option may not be economically attractive) [46].

Haldor Topsøe's process uses natural gas reforming for production of syngas with the ratio $(n_{H_2} - n_{CO_2})/(n_{CO} + n_{CO_2})$ between 2.04 and 2.1, where n is the number of moles. The greatest CO per pass conversion (65%) and the highest selectivity of DME relative to that of methanol (82/18) were observed using CO-rich syngas. With no $CO_2$, the observed DME selectivity and productivity were substantially higher. Thus in contrast to the synthesis of methanol, described above, the presence of $CO_2$ in the feed syngas was found to have a major negative impact on the DME formation.

The current DME synthesis catalyst is formulated to catalyze the water gas shift reaction at a minimum level with minimal $CO_2$ production. The quality of the residual syngas from the reactors is therefore good enough for recycle without the need for a large, dedicated $CO_2$ removal plant. Ongoing improvements to the gas phase process include the development of bifunctional catalysts to produce DME in a single step in a single reactor [47].

### Slurry Phase DME Production

Air Products did a substantial body of research toward developing a slurry phase process for the direct synthesis of DME [48–52]. The presence of $CO_2$ in the feed gas was observed to have a major impact on DME formation, as expected based on the stoichiometry of Eq. 8.23. In the slurry reactor, syngas is bubbled upward through the solvent, which contains suspended catalyst particles. The catalyst is comprised of alumina particles of about 200 μm average size, with a layer of Zn/Cu oxide methanol synthesis catalyst formed around each alumina particle. The solvent is a light liquid oil (such as Witco-40) or a hydrocarbon such as *n*-hexadecane and acts as both a mass transfer medium for the reaction and a heat transfer medium through which heat from the exothermic reaction is removed by convection [53].

Control of the reaction temperature is even more important in DME synthesis than in methanol synthesis, because the higher equilibrium conversion to DME emits more heat and a hot spot in the reactor could damage the catalyst. In a slurry phase reactor, the heat of reaction is quickly absorbed by the high boiling liquid oil, which has a high heat capacity. Heat is removed from the reactor through immersed internal cooling coils, in which water is vaporized to steam. The temperature within the reactor vessel is well controlled in order to achieve higher conversion with longer catalyst life.

The direct synthesis technology under development by NKK Corporation since 1989 also uses a slurry phase reactor. In this process, natural gas is reformed in an autothermal reformer with oxygen and steam. Carbon dioxide is recycled back from the purification section of the plant to give synthesis gas with a $H_2/CO$ ratio of 1.0. The standard DME synthesis reaction conditions are 260°C and 50 atm pressure. The condensed reactor product is purified in two distillation columns, with $CO_2$ vapor from the first column returned to the reactor, and the bottoms from the first column distilled in the second column. DME is condensed from the second column overheads, and the by-product methanol in the column bottoms returned to the DME synthesis reactor after water removal. The once through CO conversion is 50%. DME selectivity is more than 90%, and the molar ratio of DME to DME + methanol is 0.91. The coproduction of water is very small, with the molar ratio of (DME/DME + methanol + water) = 0.013. With recycle the total CO conversion reaches 95%. A pilot demonstration plant project of 100 t/day DME capacity was begun by NKK at Hokkaido, Japan, in 2002 [54].

Mitsubishi Gas Chemical (MGC) touts higher single pass conversion for its Superconverter [50] slurry DME reactor system, claiming 70–80% conversion efficiency compared with 50–60% for other processes. MGC claims that DME produced at a remote site with low-cost natural gas feedstock in a 5,000–7,000 t/day plant should beat the delivered cost of both LNG and LPG in Japan. Japan DME (MGC, ITOCHU, and Mitsubishi Heavy Industries) have planned a 1 t/year capacity plant in Papua New Guinea.

One key challenge in the economic design of 7,000 t/day plant is the reactor design configuration and catalyst to enable dehydration using a single reactor vessel, as opposed to multiple serial reactor stages or parallel reactor trains. Another important consideration is the ability to achieve sufficiently high methanol conversions to permit reaction product recovery in a single distillation train.

Another process for scales such as 7,000 t/day is referred to as the Jumbo process and is under development by Toyo Engineering using a dehydration reactor based on radial flow design. The process involves integrated coproduction of methanol from natural gas as an intermediate to DME. DME is produced via the vapor phase dehydration of methanol over a fixed bed of an acid catalyst, typically $\gamma$-alumina. Reaction temperatures are typically in the range of 250–400°C, with pressures as high as 250 psig. Gas space velocities have been reported to be in the range of 500–10,000 $h^{-1}$ [56].

Toyo's DME process offers greater flexibility in the methanol/DME product distribution. Since the methanol and DME plants operate independently, the complex could be designed to produce methanol or DME exclusively or in any combination in accordance with market demand. Toyo's design approach would also permit the option of locating their Jumbo DME plant at a different location from that of the methanol plant. The cost of shipping methanol long distances may be less than the cost of shipping an equivalent quantity of DME due to the higher cost of the pressurized refrigerated storage requirements anticipated for oceangoing DME tankers.

## Methanol to Gasoline and/or Diesel

The methanol to gasoline (MTG) process developed by Mobil Oil Corporation involves the conversion of methanol to hydrocarbons over a zeolite catalyst. The methanol conversion process occurs in two steps with chemistry as shown in Eqs. 8.22 and 8.24 (thermochemistry shown for ethylene as product).

$$\begin{gathered}2CH_3OH \leftrightarrows CH_3OCH_3 + H_2O \\ \Delta H = -24.5\ \text{kJ/mol DME}\end{gathered} \tag{8.22}$$

$$\begin{gathered}CH_3OCH_3 \leftrightarrows C_2 - C_5\,\text{olefins} + H_2O \\ \Delta H = -\ 5.5\ \text{kJ/mol DME}\end{gathered} \tag{8.24}$$

Crude methanol (17% water) is superheated to 300°C and partially dehydrated over an alumina catalyst at 27 atm to yield an equilibrium mixture of methanol, diethyl ether, and water (75% conversion of methanol). This effluent is then mixed with heated recycled syngas and introduced into a reactor containing ZSM-5 catalyst at 350–366°C and 19–23 atm to produce hydrocarbons (44%) and water (56%). Because the zeolite has to be regenerated frequently to burn off coke formed during the reaction, the MTG process is rendered continuous by using multiple gasoline conversion reactors operating in parallel with 2–6 week cycles. Mobil's MTG process uses different pressures for syngas production (15–20 atm), methanol synthesis (50–100 atm), and the fixed bed MTG process step (15–25 atm).

The selectivity of Mobil's MTG process to gasoline range hydrocarbons is about 85+%, with the remainder of the product being primarily LPG. Nearly 40% of the gasoline are aromatic hydrocarbons with product distribution of 4% benzene, 26% toluene, 2% ethylbenzene, 43% xylenes, 14% trimethylbenzenes, plus 12% other aromatics. The shape selectivity of the zeolite catalyst results in a relatively high concentration of durene (1,2,4,5-tetramethylbenzene), representing about 3–5% of the gasoline produced. Therefore, MTG gasoline is usually distilled and the heavy fraction is hydroprocessed to reduce the concentration of durene to below 2%. This results in a high quality gasoline with a high-octane number. However, the future of the MTG process remains uncertain because the 1990 Clean Air Act Amendment limits the amount of aromatics in reformulated gasoline.

The first commercial MTG plant came onstream in 1985 in New Zealand (Mobil's Motunui plant), producing 4,500 t/day of methanol and 14,500 bbl/day of high-octane gasoline from natural gas. Gasoline production was later ceased and presently this plant and the nearby methanol plant at Waitara produce 2.43 million tons per year of chemical grade methanol for export [57].

A 100 bbl/day fluidized bed MTG pilot plant was jointly designed and operated near Cologne, Germany, by Mobil (supplying process technology and proprietary catalyst), Union Reinische Brankohlen Kraftstoff AG (acting as the operating agent), and Uhde GmbH (providing engineering services) from 1982 to

1985. Although no commercial plants have been built based on the fluid bed technology, it is considered ready for licensing and/or commercialization [58].

Mobil also developed a methanol to gasoline and diesel process, referred to as the MOGD process. Oligomerization, disproportionation, and aromatization of olefins are the basis for the MOGD process. Selectivity to gasoline and distillate from olefins is greater than 95%. The product slate from MOGD is 3 weight paraffins, 94 wt% olefins, 1 wt% naphthenes, and 2 wt% aromatics [59]. A large-scale test run was performed at a Mobil refinery in 1981. The MOGD process is not currently in commercial practice.

Haldor Topsøe's Integrated Gasoline Synthesis (TIGAS) process was developed to minimize capital and energy costs by integrating methanol synthesis with the MTG step into a single loop (i.e., without isolation of methanol as an intermediate). Whereas Mobil's MTG process uses different pressures for syngas production, methanol synthesis, and the fixed bed MTG process step, the TIGAS process involves catalysts and conditions modified such that the system pressure levels out and separate compression steps are not required. A mixture of methanol and DME is made prior to gasoline synthesis, which results in only one recycle loop from the gasoline synthesis step back to the methanol/DME synthesis step. A 1 t/day demonstration plant was built in Houston, Texas in 1984 and operated for 3 years. The gasoline yield for the TIGAS process (defined as the amount of gasoline produced divided by the amount of natural gas used for feed and as fuel) was 56%. The TIGAS process yields a lower quality gasoline with a lower octane number compared to MTG, because of a reduced selectivity to gasoline range aromatics.

## Methanol to Chemicals

The highest volume chemical processes using methanol as feedstock are formaldehyde, methyl tertiary-butyl ether, and acetic acid. Globally, formaldehyde production is the largest consumer of methanol, followed by methyl tertiary-butyl ether (MTBE) and acetic acid.

Formaldehyde is produced commercially from methanol by three industrial processes:

1. Partial oxidation and dehydrogenation with air in the presence of silver crystals, steam, and excess methanol at a temperature of 680–720°C, otherwise known as the BASF process. The methanol conversion for this process is 97–98%.
2. The same as process 1 except in the presence of crystalline silver or silver gauze at a temperature of 600–650°C. Then, the product is distilled and the unreacted methanol is recycled. The primary methanol conversion for this process is 77–87%.

3. Oxidation with only excess air in the presence of a modified iron/molybdenum/vanadium oxide catalyst at a temperature of 250–400°C, also known as the Formox® process (Perstorp Specialty Chemicals, Sweden). The methanol conversion for this process is 98–99%.

Formaldehyde is used to make resins with phenol, urea, or melamine for the manufacture of various construction board products.

MTBE is produced by reacting isobutene with methanol in the presence of an acidic catalyst. The reaction temperatures and pressures are 30–100°C, and 7–14 atm, respectively, so that the reaction occurs in the liquid phase. Catalysts used are solid acids, including zeolites (H-ZSM-5), and microporous sulfonic acid ion exchange resins such as Amberlyst-15. A molar excess of methanol is used to increase isobutene conversion and inhibit the dimerization and oligomerization of isobutene. At optimum reaction conditions, MTBE yields approaching 90% can be achieved. Currently there are over 140 MTBE plants, with a total installed capacity of about 20 million tons per year, using processes developed by Snamprogetti, Huls (now Oxeno) ARCO, IFP, CDTECH (ABB Lummus Crest and Chemical Research Licensing), DEA (Formerly Duetsche Texaco), Shell (Netherlands), Phillips Petroleum, and Sumitomo.

Greater than 95% of the MTBE produced is used as a fuel additive in the gasoline motor pool. MTBE is also used in the petrochemical industry for the production of isobutene, and it also can be used in a number of chemical reactions including methacrolein, methacrylic acid, or isoprene production.

Acetic acid is produced by methanol carbonylation in the liquid phase using Co, Rh, or Ni catalysts promoted with iodine. The BASF process is initiated by the reaction of methanol with HI to yield methyl iodide. The active catalyst is the metal carbonyl $[Rh_2I_2(CO)_2]$ into which methyl iodide inserts during the rate-limiting step. Acetic acid is formed by the hydrolysis of the eliminated acetyl iodide species $CH_3COI$ that also regenerates HI. The process is run with over 99% selectivity at conditions of 180°C and 30–40 atm.

The Monsanto process is similar to that of BASF but is less severe. All new installed capacity since 1973 is based on the Monsanto process. Even so, the Rh/I catalytic system is very corrosive and requires expensive steels for materials of construction. Complete recovery of the expensive Rh catalyst and recycle of HI are paramount to maintain favorable process economics. The high cost of Rh has lead to the search of other, lower cost metals that could be used as acetic acid process catalysts with similar performance.

Approximately half of the world's production of acetic acid comes from methanol carbonylation and about one third from acetaldehyde oxidation. Methanol carbonylation is the most likely technology of the future. Vinyl acetate, acetic anhydride, and terephthalic acid are all made from acetic acid. Latex emulsion resins for paints, adhesives, paper coatings, and textile finishing agents are made from vinyl acetate. Acetic anhydride is used in making cellulose acetate fibers and cellulosic plastics.

## Ammonia

Ammonia is the basic building block of the nitrogen industry worldwide. Almost all ammonia is produced in the anhydrous form (free of water) by a catalytic reaction of nitrogen (from air) with hydrogen from a hydrocarbon source (usually natural gas). Ammonia is a colorless, pungent, nonflammable gas at normal pressure and temperature and is lighter than air. For storage at atmospheric pressure at sea level, ammonia must be cooled to −33°C and stored as liquid. Lower temperatures are required at higher altitudes. In the usual atmospheric temperature range of 0–40°C, the range of vapor pressure is about 4–15 atm. Approximately 10% of the ammonia produced never reaches the market, because of its volatility and losses, during conversion to other materials and during transportation and storage. Nitrogen fertilizer consumption accounts for more than 85% of the world ammonia market.

### *General Process Description*

The three main steps in producing ammonia are syngas production, syngas purification, and ammonia synthesis. The traditional and current commercial technology for hydrogen production in ammonia production involves steam reforming of natural gas to produce syngas (Eq. 8.2), and the water gas shift reaction (Eq. 8.19), both of which occur in the reformer. Steam reforming is by far the least expensive and most popular method of producing hydrogen for ammonia synthesis. Two-step steam reforming or partial oxidation also can be used to generate hydrogen-rich syngas for ammonia production, but these methods are most useful when the feed contains heavier hydrocarbons in addition to methane.

Raw synthesis gas must be purified before being sent to the ammonia synthesis unit. A shift conversion step is used to remove most of the carbon monoxide from the synthesis gas, because CO acts as a poison to the catalyst used in ammonia synthesis. Shift conversion by the WGSR (Eq. 8.19) also produces more hydrogen. The shift conversion is performed in two stages. In the first stage, called the high-temperature shift (300–450°C), the bulk of the carbon monoxide is oxidized to $CO_2$ over a chromium-promoted iron oxide catalyst. The second stage is a low-temperature shift (180–270°C) in which the remaining carbon monoxide is oxidized to $CO_2$ over a copper/zinc catalyst, leaving a few tenths of a percent in the syngas. Typical catalyst lifetimes for both high-temperature and low-temperature shift catalysts are 3–5 years.

Sulfur tolerant ("dirty shift") catalysts also have been developed. These catalysts can handle larger amounts of sulfur. The company ICI makes dirty shift conversion catalysts that consist of cobalt and molybdenum oxides. They operate over a temperature range of 230–500°C. The controlling factors are the ratio of steam to sulfur in the feed gas and the catalyst temperature.

The synthesis gas is then washed with a solvent – such as monoethanolamine or other amine, potassium carbonate, or sulfinol (sulfolane and an alkanolamine) – to remove most of the carbon dioxide. The solvents typically contain an activator to promote mass transfer. Water scrubbing at high pressures was employed in the past to remove $CO_2$ but is no longer used. Most newer steam reforming plants use methanation to remove residual or trace carbon monoxide and carbon dioxide, by reaction with hydrogen over a nickel catalyst to yield methane and water.

Scrubbing with a liquid nitrogen wash produces a synthesis gas with very low inert gas content. This additional nitrogen is supplied by using air, in excess of the stoichiometric nitrogen requirement, in the secondary reformer or, in the case of plants with no reforming sections such as off-gas plants, by obtaining nitrogen from an associated air separation unit. Older processes involved scrubbing the synthesis gas with copper liquors, such as cuprous ammonium formate, to remove carbon monoxide, and caustic to remove carbon dioxide. Many modern processes include molecular sieve dryers downstream of the methanation step to remove the final traces of water. The synthesis gas is then compressed and finally reaches the ammonia converter, where the contained hydrogen and nitrogen chemically combine over a catalyst into ammonia.

## *Reaction Pathway*

The chemical reaction considered in ammonia synthesis is shown in Eq. 8.25.

$$N_2 + 3H_2 \leftrightarrows 2NH_3 \qquad \Delta H = -92.2\ \text{kJ/mol}\ NH_3 \tag{8.25}$$

Since the reaction is exothermic, maximum conversion at equilibrium occurs at high pressure and low temperature. Ammonia synthesis is achieved in catalytic reactors at pressures ranging from 150 to 345 atm, at a minimum temperature of 430–489°C. A maximum temperature of ~500°C results from the balance between increasing reaction kinetics and decreasing the equilibrium ammonia concentration as the temperature increases. Thermodynamics suggest that a low process temperature is favored. However, the kinetics of the reactions dictates that high temperature is required. Nitrogen is usually fed as air, but in some plants (e.g., partial oxidation plants) the $N_2$ is a purified gas from an associated air separation facility. In terms of the hydrogen source, production methods, product recovery, and overall plant engineering, there are many process variants for ammonia production.

About 50% of the hydrogen in ammonia comes from steam. High temperatures and low pressures favor the highly endothermic methane reforming reaction. Higher pressures tend to lower the methane conversion. In industrial reformers, the reforming and shift reactions result in a product gas composition that closely approaches equilibrium. However, the following side reactions produce carbon in the steam reformer:

Boudouard coking

$$2\,CO \leftrightarrows C(s) + CO_2 \quad \Delta H = -83.3\ kJ/mol\ CO \tag{8.7}$$

CO reduction

$$\begin{aligned} &CO + H_2 \leftrightarrows C(s) + H_2O \\ &\Delta H = -\ 125.5\ kJ/mol\ C \end{aligned} \tag{8.26}$$

Methane cracking

$$CH_4 \leftrightarrows C\,(s) + 2H_2 \quad \Delta H = 80.6\,kJ/mol\ C \tag{8.27}$$

The molar steam to carbon ratio in the reformer is usually between 2 and 6, depending on the feedstock and process operating conditions. Excess steam is used to prevent coking in the reformer tubes. The shift reaction is exothermic and favors low temperatures. Since the shift reaction does not approach completion in the reformer (usually there is 10–15 vol% CO, dry basis, in the reformer effluent), further conversion of CO is performed using external shift conversion catalysts.

## *Reactors and Catalysis for Ammonia Synthesis*

The basic design of an ammonia synthesis reactor is a pressure vessel with sections for catalyst beds and heat exchangers. Over the years, there have been many different designs proposed and constructed for commercial operation such that today, ammonia reactors are classified by flow type (radial, axial, or cross flow) and cooling method (quench or indirect) used.

Axial flow reactors are essentially top-to-bottom flow reactors. The design is comparatively simple; however, a fairly large pressure drop develops across the catalyst bed. A radial flow configuration feeds gas into an annular region between the reactor wall and the outer surface of the catalyst bed. Gas flows through the bed and exits out a central collection tube. This design minimizes pressure drop across a shallow bed with a large surface area. Radial flow reactors tend to be tall vessels with relatively small diameters. The cross-flow reactor configuration has a similar principle in that gas is introduced along one side of the reactor and is collected radially across the reactor by a collector on the other side.

Reactant gas is preheated either by circulating it through heat exchangers or by using water to produce steam. However, removing the heat generated from the exothermic synthesis reaction to maintain control of the reaction temperature is a challenge. Quench converters are designed to introduce cool reactant gas at various points along the length of the catalyst bed. Interbed heat exchangers also can be used to remove heat at specific intervals along the bed, effectively separating the bed into multiple synthesis zones, or continuously along the bed with cooling tubes. These indirectly cooled designs allow for efficient recovery of reaction heat that can be used in other parts of the process.

Early designs based on axial flow were easier to build but not as efficient as the more complex radial flow reactors that have more recently been designed. One example of a radial flow design, by Haldor Topsøe, uses two radial beds with quench gas injection between them. A similar radial flow design uses an interbed heat exchanger in the first catalyst bed. Cold ammonia synthesis gas is introduced from the bottom of the reactor through the second catalyst bed, then through the heat exchanger in the first catalyst bed. The cold gas flow through the second bed also provides indirect heat exchange. An additional heat exchanger is located at the bottom of the reactor to cool the reactor gases.

The Kellogg four-bed axial flow reactor system is a good example of an axial flow reactor design. This quench reactor consists of four catalyst beds held on separate grids. Quench gas is introduced in the spaces between the bed, and a heat exchanger is located at the top of the vessel. Another type of Kellogg reactor is a cross-flow reactor design where gas flows through the catalyst bed perpendicular to the vessel axis. It is available in both quench and indirectly cooled versions.

The ammonia synthesis catalyst is commonly iron, such as magnetite, which may be promoted with aluminum, potassium, and/or calcium. Finely divided iron is pyrophoric, therefore Fe catalysts are reduced to metallic form in situ. Porosity is developed as the catalysts are reduced. During reduction, oxygen is removed from the magnetic crystal lattice without shrinkages. Metallic Fe is formed with essentially the same porous structure as the magnetite precursor. The production and preservation of this highly porous structure during reduction of the ammonia synthesis catalyst precursor leads to highly active catalysts. The addition of structural promoters facilitates the formation of highly porous metallic iron.

Catalyst activity is not directly associated with the Fe surface area but is related to complex interactions of the promoters. Alkali metal promoters in ammonia synthesis catalysts are necessary to attain high activity. Potassium is the most effective alkali promoter, while Li and Na are poor promoters and are not used commercially. Potassium is thought to interact with the Fe crystallites to increase the dissociative sticking probability of $N_2$ on the Fe sites and thereby increase catalytic activity [60].

The most effective catalysts from a process perspective are those that have the highest rate of conversion at the lowest temperatures. Other catalysts are being developed such as promoted Ru on high surface area graphite supports, and these recently commercialized catalysts offer the possibility of greatly decreasing synthesis temperatures and pressures with improved activity at high ammonia concentrations. The Kellogg Advanced Ammonia Process is based on a Ru catalyst that is claimed to be 40% more effective than Fe catalysts, because synthesis efficiency is increased by lowering the process pressure from 150 to 90 atm.

With effective gas cleanup and conditioning, commercial ammonia synthesis catalyst lifetimes of 5–8 years can be achieved in most cases and potentially up to 14 years. Ammonia formation increases as the process pressure is increased. The optimum $H_2/N_2$ ratio is near 2 at high-space velocities and approaches 3 at low-space velocities as equilibrium becomes more dominant. Space velocities in commercial ammonia synthesis vary from 12,000/h at 150 atm to 35,000/h at 790 atm.

Per pass conversion on the order of 10–35% is typically achieved. Ammonia is recovered from the synthesis loop by cooling to condense the synthesis gas at process pressures. The liquid ammonia is separated from the gas effluent, which is recycled back through to the reactor. Earlier plant designs used air or water cooling for ammonia recovery. Modern synthesis plants use refrigeration to condense out the ammonia. The ammonia recovery is not extremely efficient so the recycled gas typically contains 4% ammonia plus any inert gasses (Ar, He, $CH_4$, etc.) that may be in the process stream. Purging some of the gas in the recycle loop before it is recycled minimizes inert gas concentrations, and flashing the liquid ammonia in a pressure letdown step releases any dissolved gases.

Clearly, the ammonia synthesis process consists of many complex unit operations apart from the actual synthesis loop. The way in which these process components are combined with respect to mass and energy flow has a major influence on efficiency and reliability. Many of the differences between various commercial ammonia processes lie in the way in which the above process elements are integrated.

## *Commercial Activities in Ammonia Synthesis*

The entire ammonia industry was originally based on using coal as a feedstock. The industry has now moved toward natural gas as the main hydrocarbon feedstock. By 1990, only 14% of the world ammonia capacity was based on coal or coke. Apart from a few plants operating in India and South Africa, today the majority of coal-based ammonia plants are found in China. More than 20 commercial ammonia synthesis processes have been described [61].

Commercial technology vendors for ammonia include:

- Johnson Matthey
- Linde
- Kellogg Brown and Root
- Haldor Topsøe
- Ammonia Casale
- Uhde

As a result of a surge of major developments in ammonia process technology beginning in the late 1960s, large 1,000–1,500 t/day capacities became the industry standard for new plant constructions. Plants as large as 2,000 t/day are now common. The main ammonia process contractors have each now developed “mega-ammonia technology,” and are offering large plant concepts from 3,000 to 6,000 t/day units. The world’s largest ammonia plant, in Al Jubail, Saudi Arabia, is a collaboration between Saudi Basic Industries Corporation (SABIC) and Saudi Arabian Fertilizer Company. The plant, called SAFCO IV, came onstream in 2007 and has a capacity of 3,300 t/day. This huge plant is the first application of a new ammonia process from Uhde, referred to as the Dual Pressure Process, and utilizes Johnson Matthey's KATALCO™ catalysts [62]. A major factor resulting in much lower average production costs has been the switch from electrically driven compression to the use of steam-driven centrifugal compressors powered by waste heat. During the past 30 years, many of the smaller, older, higher-cost plants in the world have been shut down.

Developments to improve energy efficiency or overcome interruptions of natural gas supply have been examined and some are finding application. The recovery of the hydrogen and ammonia from the synthesis purge gas by a cryogenic unit or membrane is now standard and an additional 5% of ammonia can be produced from the same amount of feedstock. The use of computer controls to improve overall plant performance is increasing.

Substantial improvements have been made over the years in the energy efficiency of carbon dioxide removal systems. The first large-scale ammonia plants in the 1960s typically used monoethanolamine as a solvent, with energy input of over 50,000 kcal/kg-mol of carbon dioxide removed, whereas today's modern plants use improvements such as the dual-step methyl diethanolamine process that can reduce the energy input to about 10,000 kcal/kg-mol or lower of carbon dioxide removed. As described above, improvements to efficiency also have been introduced with new ammonia synthesis catalysts.

The current generation of large ammonia plants is much more fuel-efficient than plants constructed 20 or 30 years ago. A typical world-scale plant constructed in the 1970s consumed about 42 million Btu of natural gas per ton of ammonia produced. Retrofitting such a plant to improve fuel efficiency can reduce gas consumption to about 36 million Btu/t; however, the newest generation of ammonia plants use only about 30 million Btu/t of ammonia, are reported to be easier to operate, and have slightly lower conversion costs. Some newer plants also recover more than 1 million Btu/t by generating electricity, used elsewhere in the complex, from waste heat.

An additional minor trend has been to construct an ammonia plant at a location where a hydrogen off-gas stream is available from a nearby methanol or ethylene operation (e.g., Canadian plants at Kitimat, British Columbia, and Joffre, Alberta). Gas consumption at such production sites ranges from 25 to 27 million Btu/t of ammonia, depending on specific circumstances. Perhaps more important, the capital cost of such a plant is only about 50% of the cost of a conventional plant of similar capacity because only the synthesis portion of the plant is required.

## Direct Conversion of Natural Gas to Liquids

Substantial capital cost could be avoided if methane could be directly converted to methanol, or other high-volume liquid fuels or chemicals, without the intermediate production of syngas. Several direct processes have been investigated over the years, including methane coupling to produce ethane, which can be oligomerized to give gasoline in a reactor similar to that used in the MTG process. Another consideration has been the direct production of aromatics, using an oxidative coupling membrane. Another option is the selective oxidation of natural gas to produce fuels directly. Direct methods would need to have a distinct economic advantage over indirect methods, but to date no direct processes have progressed to a commercial stage. Product yields are generally small while operating in the single pass mode, which makes separations difficult and costly.

### *Oxidative Coupling to Light Hydrocarbons*

In the oxidative coupling reaction, methane and oxygen react over a catalyst at elevated temperatures (700°C) and normal pressures to form ethane as a primary product and ethylene as a secondary product. British Petroleum (now BP) has described experiments at higher temperatures (1,100°C) and short residence times, at which conditions methane reacts with oxygen to produce syngas as well as $C_{2+}$ hydrocarbons in addition to syngas. Several catalysts including zirconia gave $C_{2+}$ selectivities of over 50%.

Work on the lower temperature methane coupling reaction has been done by Phillips Petroleum, Texas A&M University, Akzo Chemie, Amoco (now BP), the University of California at Berkeley, the University of Pittsburg, the University of Tokyo, Idmitsu Kosan, and Union Carbide among others. Unfortunately both the methane and the ethylene may react further to produce $CO_2$, and the single pass combined yield of ethylene and ethane is limited to about 25%. With better catalysts, such as $SrO/La_2O_3$ and $Mn/Na_2WO_4/SiO_2$, a $C_2$ selectivity of about 80% can be achieved, but at a low methane conversion of 20%. About half of the $C_2$ product is ethane and half ethylene. The best $C_2$ selectivities are almost always achieved under oxygen-limiting conditions. Because the reaction is exothermic, a zone within the catalyst bed may be 150–300°C hotter than the external temperature and heat management is a serious problem. This is complicated by the fact that metals normally used for the construction of reactors catalyze the combustion of methane.

The yield of ethylene can be improved considerably by operating in a recycle mode with continuous removal of ethylene, and by applying various techniques for separating products from reactants including cryogenic, membrane, and adsorptive separations. For economic reasons, either molecular sieve or membrane separation systems are most likely to be employed, both of which have severe mass transport limitations not suitable for scale-up. A variation of this recycle system is to use a zeolite catalytic reactor to convert ethylene to benzene and toluene. In principle, aromatic yields in excess of 70% could be achieved; however, to achieve these yields the conversion of ethane to ethylene must be highly selective.

## *Selective Oxidation of Natural Gas to Methanol*

Selective alkane oxidation by transition metal complexes in solution has been the focus of substantial effort since the 1970s. However, the low activity of alkanes and the typically higher activity of the desired products make this process a great challenge. In 1998, Periana et al. of Catalytica Inc. reported developing an effective catalyst for selective oxidation of $CH_4$ to $CH_3OH$ in high yield. The initial catalyst was dichloro-($\eta^2$-(2,2′-bipyrimidyl))platinum(II) complex, (bpym) $PtCl_2$. In well-dried sulfuric acid (102%), $CH_4$ at 3,400 kPa and 220°C was converted to a mixture of $CH_3OSO_3H$ (methyl bisulfate, which presumably would be subsequently hydrolyzed to methanol) and $CH_3OH$ at 72% conversion with selectivity of 81% [63]. Although catalytic in Pt complex, the reaction was very slow and required over half as many moles of catalyst as moles of methane converted. Another catalyst, $(NH_3)_2PtCl_2$, system was more active but less stable. A complete cycle would require the regeneration of concentrated sulfuric acid. The Catalytica process is promising, because it potentially has high yield at relatively low temperatures. On the other hand, it suffers from drawbacks including (1) the catalytic rate is too low; (2) the separation cost is too high; (3) the solvent is too corrosive; (4) reoxidation of $SO_2$ with regeneration of concentrated sulfuric acid is too expensive; and (5) the catalyst is severely inhibited by water.

Cold flame oxidation is another approach to direct conversion of methane to methanol. A pressurized mixture of methane and oxygen is reacted at moderate temperatures of 350–500°C. The reaction mixture is very fuel rich, with methane to oxygen ratios of about 20:1. The main chemical reaction is the oxidation of methane to methanol. However, further oxidation of methanol to formaldehyde often takes place simultaneously. The University of Manitoba reported 90% selectivity for methanol at a single pass conversion of 7.5% in an isothermal process. Although this is among the best results reported to date, the yield of methanol from this process is too low to be of commercial interest.

# Future Directions

The immediate future of F-T and DME gas to liquids would seem to be in their implementation. Acquisitions and ventures are creating "one stop shops" for commercialization. Johnson Matthey has acquired Davy Process technology and ICI, making the parent company a leader in reforming and methanol GTL technologies. The companies Enichem and IFP/Axens have strong capabilities particularly applicable to F-T GTL. Enichem has skills in catalysis, engineering, and the oil and gas business. IFP/Axens has catalysis expertise, plus basic engineering, process development, licensing, and catalyst manufacturing [64]. The future of methanol may be the status quo, because the technology is already mature. Alternatively, GTL methanol plants may expand in number (and possibly size) given the anticipated increased availability of unconventional gas combined with high oil process.

## Bibliography

1. Crane H, Kinderman E, Malhotra R (2010) A cubic mile of oil. Oxford University Press, Oxford
2. United States Environmental Protection Agency (1998) Compilation of emission factors AP-42, v1, 5th edn, Supplement D
3. United States Environmental Protection Agency (2010) Compilation of emission factors AP-42, v1, 5th edn, Supplement E, corrected. Calculation for low-sulfur No. 6 fuel oil
4. United States Environmental Protection Agency (1998) Compilation of emission factors AP-42, v1, 5th ed, Supplement E. Calculation for medium volatile bituminous coal
5. Rajnauth J, Ayeni K, Barrufet M (2008) Gas transportation: present and future. In: CIPC/SPE Gas Technology Symposium 2008 Joint Conference, Calgary, Alberta, 16–19 June 2008
6. Bellussi G, Zennaro R (2007) New developments: energy, transport, sustainability. In: Encyclopaedia of Hydrocarbons, vol III, Chap. 2.6, pp. 161–182, EniChem
7. Smith R, Asaro M (2005) Fuels of the future: technology intelligence for gas to liquids strategies. SRI Consulting, Menlo Park
8. Yost C, DiNapoli R (2003) Benchmarking study compares LNG plant costs. Oil Gas J 101(15):56–59
9. Nielsen R (2001) Fundamentals of mixed refrigerant compared to conventional refrigeration are discussed in "Ethylene Plant Enhancement." PEP Report 29 G, SRI Consulting, Menlo Park
10. Low WR, Andress D, Houser C (1997) Method of load distribution in a cascaded refrigeration process. US 5611216 to Phillips Petroleum Company, 18 Mar 1997
11. Houser C, Yao J, Andress D, Low WR (1997) Efficiency improvement of open-cycle cascaded refrigeration process. US 5669234 to Phillips Petroleum Company, 23 Sept 1997
12. Delong BW (1987) Method for cooling normally gaseous material. US 4680041 to Phillips Petroleum Company, 14 July 1987
13. Netzer D, Nielsen R (2003) Baseload liquefied natural gas by cascade refrigeration. PEP Review 2003–15, SRI Consulting, Menlo Park
14. Smith R, Asaro M (2005) Fuels of the future: technology intelligence for gas to liquids strategies. SRI Consulting, Menlo Park
15. Huffman GP, Feeley III TJ (2000) Fuel science in the Year 2000: where do we stand, where do we go from here? I: Power generation and related environmental concerns – DOE's fine particulate and air toxics research program: responding to the environmental challenges to coal-based power production in the 21st century. Preprints of the Division of Fuel Chemistry of the American Chemical Society, 45(1):108–112, 2000 Spring conference of the American Chemical Society, San Francisco
16. Topsoe_synthesis_g#6D6FFA1.ashx.pdf. Accessed 2 Apr 2011, reprinted from Hydrocarbon Engineering, 2006

17. Christensen TS, Østberg M, Bak Hansen J-H (2001) Process demonstration of autothermal reforming at low steam-to-carbon ratios for production of synthesis gas. In: AIChE Annual Meeting, Reno, 4–9 Nov 2001
18. Wesenberg MH (2006) Gas heated steam reformer modelling. PhD thesis, The Norwegian University of Science and Technology
19. Aasberg-Petersen K, Christensen TS, Charlotte Stub Nielsen CS, Dybkjær I (2003) Recent developments in autothermal reforming and pre-reforming for synthesis gas production in GTL applications. Fuel Process Technol 83(1–3):253–261
20. Loock S, Ernst WS, Thomsen SG, Jensen MF (2005) Improving carbon efficiency in an autothermal methane reforming plant with gas heated heat exchange reforming technology. Paper No. O96-001, 7th World Congress of Chemical Engineering, Glasgow
21. Tsuru T, Yamaguchi K, Yoshioka T, Asaeda M (2004) Methane steam reforming by microporous catalytic membrane reactors. AIChE J 50(11):2794–2805
22. Carolan MF, Chen CM, Rynders SW (2003) Development of the ceramic membrane ITM syngas/ITM hydrogen process. Fuel Chem Div Preprints 48(1):344
23. Robinson ET (S), Sirman J, Apte P, Gui X, Bulicz TR, Corgard D, Hemmings J (2005) Development of OTM syngas process and testing of syngas derived ultra-clean fuels in diesel engines and fuel cells. DE-FC26-01NT41096, Final Report
24. Caro J, Wang H, Noack M, Koelsch P, Kapteijn F, Kannelopolous N, Nolan J (2007) Manufacture of composite membranes and their use for selective partial oxidation reactions of hydrocarbons. EP 1847311 to Universität Hannover, Germany
25. Dupont V, Ross AB, Knight E, Hanky I, Twigg MV (2008) Production of hydrogen by unmixed steam reforming of methane. Chem Eng Sci 63(11):2966–2979
26. http://www.statoil.com/en/OurOperations/TerminalsRefining/ Tjeldbergodden/Pages/default.aspx. Accessed 11 Jan 2011
27. Davis BH (2001) Fischer–Tropsch synthesis: current mechanism and futuristic needs. Fuel Process Technol 71:157–166
28. Brady RC III, Pettit R (1981) The chain propagation step. J Amer Chem Soc 1981:287–1289
29. Davis BH (2001) Fischer–Tropsch synthesis: current mechanism and futuristic needs. Fuel Process Technol 71:157–166
30. Oukaci R, Singleton AH, Goodwin JG Jr (1999) Comparison of patented Co F–T catalysts using fixed-bed and slurry bubble column reactors. Appl Catal A General 186:129–144
31. Manzer L, Schwarz, S (2002) Fischer–Tropsch processes using catalysts on mesoporous supports. US 2002052289 to Conoco, 2 May 2002
32. http://www.zawya.com/projects/project.cfm/pid210307061231?cc. Accessed 11 Jan 2011
33. Spath PL, Dayton DC (2003) Preliminary screening – technical and economic assessment of synthesis gas to fuels and chemicals with emphasis on the potential for biomass-derived syngas. NREL/TP-510-34929
34. Fisher IA, Bell AT (1998) In situ infrared study of methanol synthesis from $H_2$/CO over Cu/$SiO_2$ and Cu/$ZrO_2$/$SiO_2$. J Catal 178(1):153–173
35. Chinchen GC, Denny PJ, Parker DG, Spencer MS, Whan DA (1987) Mechanism of methanol synthesis from $CO_2$/CO/$H_2$ mixtures over copper/zinc oxide/alumina catalysts: use of $^{14}$C-labelled reactants. Appl Catal 30(2):333–338
36. Grabow LC, Mavrikakis M (2011) Mechanism of methanol synthesis on Cu through $CO_2$ and CO hydrogenation. ACS Catal 1(4):365–384
37. Tijm PJA, Waller FJ, Brown DM (2001) Methanol technology developments for the new millennium. Appl Catal A General 221:275–282
38. Liu J, Wei R, Zhang Y, Xu R, Li Z (2009) Preparation of Cu/ZnO/$Al_2O_3$ catalysts for methanol synthesis by improved two-step coprecipitation method. Gongye Cuihua 17(7):22–25. C.A. 2009:1625358
39. Baltes C, Vukojevic S, Schüth F (2008) Correlations between synthesis, precursor, and catalyst structure and activity of a large set of CuO/ZnO/$Al_2O_3$ catalysts for methanol synthesis. J Catal 258(2):334–344

40. Kaluza S, Behrens M, Schiefenhoevel N, Kniep B, Fischer R, Schloegl R, Muhler M (2011) A novel synthesis route for $Cu/ZnO/Al_2O_3$ catalysts used in methanol synthesis: combining continuous consecutive precipitation with continuous aging of the precipitate. ChemCatChem 3(1):189–199
41. Lurgi brochure 0312e_MegaMethanol.pdf
42. Smith R, Naqvi S, Asaro M (2008) Fuels of the future: technology intelligence for coal to liquids strategies. SRI Consulting, Menlo Park
43. Lewnard JJ, Hsuing TH, White JF, Brown DM (1990) Single-step synthesis of dimethyl ether in a slurry reactor. Chem Eng Sci 45(8):2753–2741
44. Ohno Y, Omiya M (2003) Coal conversion into dimethyl ether as an innovative clean fuel. In: 12th ICCS Coal Conversion in DME, 2–6 Nov 2003
45. Haugaard J, Voss B (2001) Process for the synthesis of a methanol/dimethyl ether mixture from synthesis gas. US 6191175 to Haldor Topsøe, 20 Feb 2001
46. Naqvi S (2002) Dimethyl ether as alternative. Fuel PEP Report 245, SRI Consulting, Menlo Park
47. Kang S-H, Bae JW, Kim H-S, Dhar GM, Jun K-W (2010) Enhanced catalytic performance for dimethyl ether synthesis from syngas with the addition of Zr or Ga on a $Cu - ZnO - Al_2O_3/\gamma\text{-}Al_2O_3$ bifunctional catalyst. Energy Fuels 24(2):804–810
48. Bhatt BL, Schaub E, Heydorn E (1993) Recent developments in slurry reactor technology at the LaPorte Alternative Fuels Development Unit. In: International Technical Conference on Coal Utilization & Fuel Systems, LaPorte, pp 197–208, 26–29 Apr 1993
49. Peng X-D, Toseland B, Underwood T (1997) A novel mechanism of catalyst deactivation in liquid phase synthesis gas-to-DME reactions. In: Bartholomew C, Fuentes GH (eds) Catalyst deactivation. Elsevier, Amsterdam
50. Peng X-D (2002) Catalyst activity maintenance for the liquid phase synthesis gas-to-dimethyl ether process. Part II: Development of aluminum phosphate as the dehydration catalyst for the single-step liquid phase syngas-to-DME process. DOE contract DE-FC22-94PC93052, Final Report
51. Peng X-D (2002) Kinetic understanding of the syngas-to-DME reaction system and its implications to process and economics. DOE Contract DE-FC22-94 PC93052, Topical Report
52. Tijm PJ (2003) Development of alternative fuels and chemicals from synthesis gas. DOE Contract number FC22-95PC93052, Final Report
53. Smith R (2009) Dimethyl ether (DME) from coal. PEP Report 245B, SRI Consulting, Menlo Park
54. Ogawa T, Inoue N, Shikada T, Ohno Y (2003) Direct dimethyl ether synthesis. J Nat Gas Chem 12:219–227
55. The Ministry of Industry, Energy and Tourism; Orkustofnun/The National Energy Authority, The Innovation Center Iceland; Mitsubishi Heavy Industries, Ltd.; Mitsubishi Corporation; Hekla hf.; NordicBlueEnergy (2010) A feasibility study report for a DME project in Iceland. IDME Project Feasibility Study – 2009. Accessed 4 Apr 2011
56. Pavone T (2003) Jumbo dimethyl ether production process via Toyo technology. PEP Review 2003–9, SRI Consulting, Menlo Park
57. http://www.methanex.com/ourcompany/locations_newzealand.html. Accessed 4 Apr 2011
58. More detailed process design and economics information can be found in Apanel G (1999) Liquid hydrocarbons from synthesis gas. PEP Report 191A, SRI Consulting, Menlo Park
59. Tabak SA, Yurchak S (1990) Conversion of methanol over ZSM-5 to fuels and chemicals. Catal Today 6(3):307–327
60. Spath P, Dayton D (2003) Biopproducts from syngas, Syngas_products.pdf. Accessed 2 Apr 2011
61. Ullmann's Encyclopedia of Industrial Chemistry (2002) "Ammonia" published online 15 Dec 2006, doi: 10.1002/14356007.a02_143.pub2. Accessed 4 Apr 2011

62. Shah J (2007) SAFCO IV: catalyst start-ups in the world's largest ammonia plant. In: 20th AFA International Annual Technical Conference, Tunisia. 5_03 John_BRIGHTLING_ Johnson Matthey Catalysts_ U.K.pdf. Accessed 4 Apr 2011
63. Xu X, Fu G, Goddard III WA, Periana RA (2004) "Selective oxidation of $CH_4$ to $CH_3OH$ using the Catalytica (bpym)$PtCl_2$ catalyst: a theoretical study" in Studies in surface science and catalysis, Natural gas conversion VII. In: Proceedings of the 7th Natural Gas Conversion Symposium, Dalian, vol 147, pp 499–504
64. Zennaro R, Hugues F, Caprani E (2006) The Eni – IFP/Axens GTL technology: from R&D to a successful scale-up. In: DGMK – SCI conference on synthesis gas chemistry, Dresden

# Chapter 9
# Coal and Peat: Global Resources and Future Supply

**Mikael Höök**

## Glossary

| | |
|---|---|
| Anthracite | Anthracite is the highest rank of coal because it has undergone the greatest degree of metamorphosis away from peat. It features low volatile matter (<10%) and high carbon, giving it the highest energy content of all coals. Semi-anthracite is somewhere in the middle between low volatile bituminous coal and anthracite. |
| Ash | Inorganic residues remaining after combustion. It has less than the initial mineral matter content because of chemical changes during combustion, i.e., the loss of water, carbon dioxide, and sulfurous compounds. |
| Bituminous coal | Bituminous coal lies between subbituminous coal and semi-anthracite in terms of rank. This rank of coal is commonly divided into additional subgroups dependent upon the content of volatile material. |
| Calorific value | Corresponds to the amount of heat per unit mass when combusted. Can be expressed as gross calorific value, which is the amount of heat liberated during combustion under standardized conditions at constant volume so that all of the |

This chapter was originally published as part of the Encyclopedia of Sustainability Science and Technology edited by Robert A. Meyers. DOI:10.1007/978-1-4419-0851-3

M. Höök (✉)
Global Energy Systems, Uppsala University, Uppsala, Sweden
e-mail: Mikael.Hook@fysast.uu.se

R. Malhotra (ed.), *Fossil Energy: Selected Entries from the Encyclopedia of Sustainability Science and Technology*, DOI 10.1007/978-1-4614-5722-0_9,

| | |
|---|---|
| | water in the products remains in liquid form, or as net calorific value, which is the maximum achievable heat release obtainable in a furnace at constant pressure. |
| Carboniferous | A geologic period and system that extends from 360 to 300 million years ago. |
| Coal preparation | Physical and mechanical processes often applied to coal to make it suitable for a particular use. |
| Coalification | The process that transforms vegetation to peat succeeded by the transformation of peat to lignite and later higher coal ranks. |
| Cretaceous | A geologic period and system spanning from 145 to 66 million years ago. |
| Devonian | A geologic period and system spanning from 416 to 360 million years ago. |
| Exploration | The examination of an area by means of surface geological mapping, seismic techniques, geophysical methods, drilling of boreholes and sampling of the underlying sediments and rocks. |
| Jurassic | A geologic period and system extending from 200 to 145 million years ago. |
| Macerals | Microscopic organic constituents of coal that may differ significantly in composition and properties depending on rank and geology. |
| Ordovician | A geologic period and system spanning from 490 to 445 million years ago. |
| Permian | A geologic period and system extending from 300 to 250 million years ago. |
| Reserves | The recoverable volumes that are dependent on certain arbitrary limits in respect of thickness, depth, quality, economics, legal restrictions, and other factors. Reserves are always less than resources. |
| Resources | The amount of coal in place before exploitation; all resources may or may not be economically or technically recoverable; see Reserves. |
| Silurian | A geologic period and system spanning from 445 to 416 million years ago. |
| Stripping ratio | The ratio of the thickness of overburden to that of the total workable coal section. |
| Tertiary | A geologic period and system extending from 66 to 2.6 million years ago. |
| Triassic | A geologic period and system spanning from 250 to 200 million years ago. |
| Volatile matter | Non-moisture component of coal that is liberated at high temperature in the absence of air. |

## Definition of the Subject

Coal is the second most important fuel currently used by mankind, accounting for over 25% of the world's primary energy supply. It provides 41% of global electricity supplies and is a vital fuel or production input for the steel, cement, and chemical industries. However, coal is a fossil fuel formed from organic material by geological processes over millions of years. Hence, coal is a finite resource in terms of human timescales and its continued availability is important to the world economy.

Peat is a related substance, but is classified somewhere between a fossil fuel and biomass. The energy sector uses peat as a fuel to generate electricity and heat. It also has applications in industrial, residential, and other sectors, but global consumption of peat is insignificant in comparison to coal. Peat shares many similarities with coal, and is increasingly often grouped with coal for resource estimates in reports and assessments by public agencies.

Knowing how coal and peat are created is vital to understanding how deposits are formed and what their basic properties are. Geology provides models and methodologies for describing deposits and where to find them. Exploration, drilling, and surveys provide the data necessary to map deposits and assess the resources they contain. Classification schemes are also central to understanding how the terms relate to the underlying data.

Future production of coal and peat is essential for the development of global energy supplies. It is only the produced volumes that can be used in human activities, and a detailed appreciation of the production process is essential in understanding future supply developments. Factors such as economy, technology, legal, and environmental constraints affect the recoverable share of the available resources, i.e., the reserves. Understanding the complexity and the greater whole of the production mechanism and the limitations that are imposed on it require a wide variety of approaches and conceptual infrastructures.

## Introduction

Natural resources are vital in supporting the continued well-being of the world's population. Metals, fossil fuels, minerals, clean water, and many other resources are critical components for many aspects of modern society. Of all available resources, few are as important as energy. In fact, energy has been described as the ultimate resource by several studies [1–4]. In physics, energy is defined as "the ability to do work." Einstein even showed the equivalence between energy and mass [5], implying that everything in the universe is made from energy in various forms. Naturally, it follows that energy influences everything and cannot be substituted for other resources.

Energy resources take many forms, ranging from fossil fuels and uranium to biomass, human labor, or wind power. Currently, over 80% of all primary energy

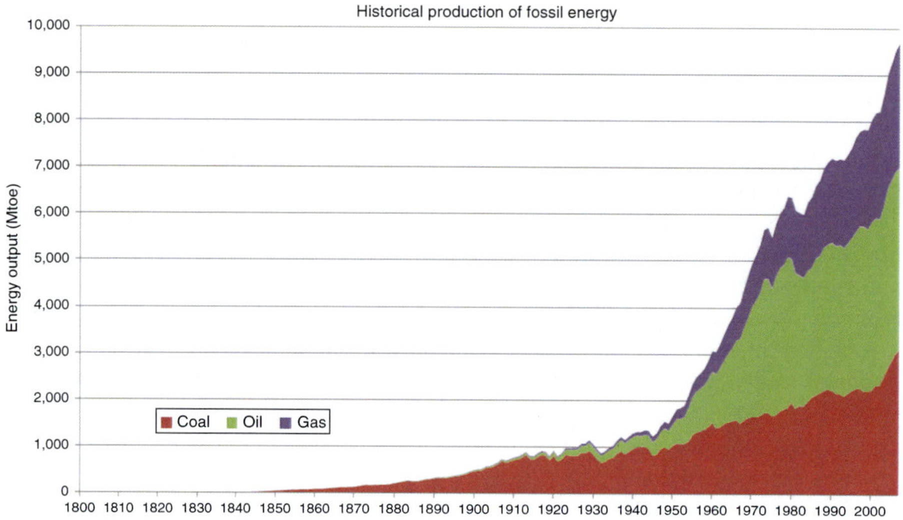


**Fig. 9.1** World production of fossil energy since 1800 based on the author's compilation and interpolation of historical production statistics

used by mankind is derived from fossil fuels [6]. Petroleum dominates, with roughly 35% of world primary energy being derived from this particular hydrocarbon, coal at number two (~26%), followed quickly by natural gas (~21%). The contributions from other energy sources range from 2% to 10% (biomass, nuclear, hydropower) to less than 0.1% (wind, solar).

To understand the conceptual importance of coal, one can begin by studying the exceptional expansion of fossil energy in the last 200 years (Fig. 9.1). The development of modern society, population, and the world economy under this brief period has been remarkable. Nearly all this growth has been powered by fossil energy in various forms, originally coal but later oil and natural gas. The growth soared and mankind both prospered and multiplied in many places, all thanks to cheap and abundant fossil fuels that provided useful energy for a multitude of activities.

The postwar boom also coincides with the green revolution in agriculture and the use of high-yield agricultural techniques based on synthetic fertilizers and high energy use. In fact, fossil energy has been described as the principal raw material of modern agriculture [7–10]. Furthermore, energy is intimately linked to economic growth and development [11, 12]. To summarize, nothing is as important to society as energy and this should never be overlooked when dealing with questions of sustainability.

The rise of the industrial nations in the Western world was chiefly driven by the coal that powered steam engines, transportation, and industry as well as heating homes. Important industrial complexes and cities arose near coal-rich regions, such as the Ruhr or Pennsylvania. In the USA, coal had replaced biomass as the dominant primary energy source by late nineteenth century [13]. Similar events occurred in UK, Germany, and other nations around the early twentieth century. Today, the world is witnessing the rise of China and India as new industrial powers, both countries growth fueled chiefly by coal. However, petroleum eventually surpassed coal as the dominant fuel globally in the 1950s, but coal has remained the second most important energy resource ever since.

Sedimentary layers of coal and peat beds range in age from the Upper Paleozoic to the recent (geologically), and are located all over the world. The layers are a result of the accumulation of vegetable matter in a specialized deposition environment. Local conditions and events can produce different qualities and ranks as well as different degrees of structural complexity.

This chapter provides a fundamental overview of global peat and coal resources, their distribution, and some of the central characteristics in their recovery and utilization. In addition, it presents a brief summary of resource classification schemes and their importance in understanding the amounts that may be available for future supply. Finally, some resource and reserve statistics will be presented with discussion. Estimates of global resources must be dated and can naturally change as more is learnt about their geology. The figures presented here should be seen as a snapshot of the current state of our knowledge. Nevertheless, such estimates indicate where the major deposits and mining activity are currently located and from where future production might originate.

## Coal and Peat: Global Resources

### *Formation of Coal and Peat*

Coal and peat derive from vegetable remains that have been deposited in various environments where it has been transformed by geological processes to potentially useful energy resources. The sedimentation of coal and peat has been extensively studied for over a century. This section describes the general mechanism behind the formation of peat and coal.

Mires are the accepted general term for peat-forming ecosystems of all types. The cycling of matter in most ecosystems is relatively fast and complete. In contrast, mires are characterized by an incomplete cycle, resulting in a positive carbon balance. Peat-forming areas are generally wet places where the annual input of dead organic matter from the photosynthetic activity of the plant cover exceeds annual breakdown. Over time organic material accumulates and becomes preserved. Peat deposits typically occur in waterlogged environments where water prevents oxidation of the organic remains. The presence of water limits the diffusion rate of gasses that leads to low oxygen availability. The large heat capacity of water also induces a low temperature, which inhibits decomposing and decomposition-facilitating organisms and leads to reduced rates of decay.

Peat formation is typically initiated by two mechanisms. One is called terrestrialization, which is when bodies of water are replaced by peat-forming mires. The other mechanism is paludification, which is replacement of dry land by mires due to rising groundwater table or some other hydrological change in the local environment. Peat-producing wetlands are divided into ombrogenous peatlands, which are rain-fed systems, and topogenous peatlands, which owe their origin to a place and its surface/groundwater regime [14]. Moore defined several important subcategories such as bogs, bog forests, marsh, fens, swamps, floating swamps, and swamp forests [15].

The plant debris undergoes several chemical changes due to bacterial decay, compaction, heat, and time, which eventually leads to formation of peat. Pristine plant parts such as roots, barks, or spores are possible components of peat and can be used to determine the type of plants that originally made up the peatland. The peat harvested in the Northern Hemisphere was chiefly formed during the Holocene epoch just after the retreat of the colossal glaciers that covered most of Europe. Consequently, most of the constituent peat-producing plant species are still thriving and continue to produce peat as of today.

The peat must be buried by additional sediments in order to be transformed to coal. Burial leads to compaction of the peat and squeezes out water during the initial stages. A compaction ratio of 9:1 or more is not uncommon [16]. Continued burial and the addition of heat and time will break down the complex hydrocarbon compounds in the peat through a series of possible reactions. Methane and other gaseous alternation products are typically expelled from the deposit, while other elements disperse. The result is that the deposit becomes increasingly carbon-rich. The process begins with plant debris and continues through peat, lignite, subbituminous coal, bituminous coal, anthracite coal, to the pure carbon mineral graphite. This process is referred to as coalification.

Some perspectives on the necessary timescales behind the formation of peat and coal resources are indispensable for a better understanding of the whole picture. The surface part of peat below the living ground layer, being less than 300 years old, amounts on average to 10.2% of the total peat carbon volume [17]. Only the deeper and basal parts of the peat are thousands of years old. In comparison, coal requires much longer time periods to form. Analysis of thick bituminous coal bed layers indicates that optimum peat-forming conditions must have been maintained for 5,000–10,000 years for every meter of clean bituminous coal [18]. Other studies have shown that the global consumption of fossil fuel in 1997 was created from ancient plant matter corresponding to over 400 times the net primary production of the world's current biota [19]. Coal cannot be formed at the same rate society uses it and is therefore a nonrenewable fuel. Peat requires shorter time scales, but is still slowly formed in terms of human time scales.

Studies have shown similarities between the extensive coastal plain peat deposits in Southeast Asia and the North American/European carboniferous coal-bearing sequences in terms of sediment accumulation, mineralogy, and geochemistry [20–23]. Several models provide a good understanding of the cyclic nature of coal sequences, the lateral continuity of coal beds as well as physical and chemical properties. However, no single model can provide a comprehensive explanation of all aspects regarding occurrence, development, and types of coal.

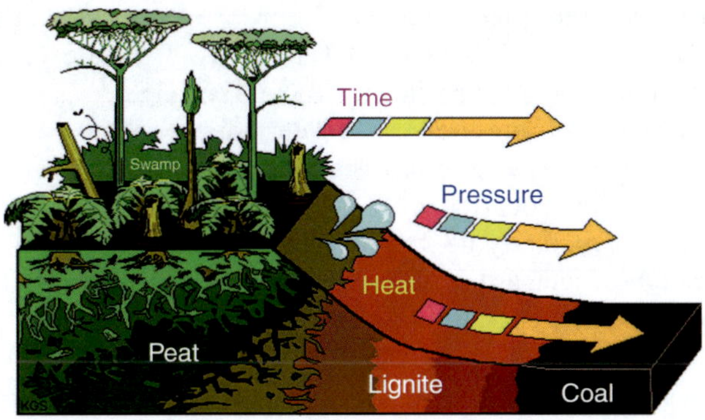


**Fig. 9.2** Principle mechanism of the formation of coal and peat (From Steve Greb and Kentucky Geological Survey. Reproduced with permission)

There are many books and articles on peat and coal geology that can provide more detailed descriptions of sedimentation, structural changes, and other factors and how they influence the formation process. The formation of coal and peat can be described in a simplified manner (Fig. 9.2).

## *Occurrences of Coal and Peat*

The oldest known land plant remains have been dated to the Ordovician/Silurian Period [24], but did not produce any noteworthy accumulations. Certain deposits of Devonian age are the oldest known coal accumulations, but it was not until the Carboniferous/Permian Period that the world's plant cover was sufficiently rich to produce and preserve significant amounts of coal. Although coal and peat are continuously being formed, there are a few major episodes of coal accumulation throughout history. In total, three major coal-formation periods have occurred and these coals form the bulk of the world's coal reserves. They occurred during the Carboniferous/Permian Period, Jurassic/Cretaceous Period, and the Tertiary Period.

Plate tectonics and climate have been described as important factors for answering when and where coal is deposited [25]. Places with favorable conditions have changed over time as continental plates have drifted or when climatic properties have been altered over the long time scales involved in the formation process. Changes in vegetation types can also be traced to certain coal components. Geological age distributions of the world's principal black coal and lignite deposits have been compiled by Walker [26].

During the Carboniferous Period, most of what is today the Northern Hemisphere was located near the equator and consisted of tropical peat mires, which later formed the coal basins of western and eastern Europe, the eastern USA, and former Soviet Union. Coals were also formed in the areas that today make up South America, southern Africa, India, Australia, and Antarctica, but this formation occurred under cooler, more temperate conditions. Coals formed at this time now make up the bulk of the world's black coal reserves. They are generally high in rank, but may have undergone significant structural change.

The climate changed during the Permian Period and the lush vegetation of the northern areas eventually ceased to deposit coal-forming sediments. New, but less abundant coal formation occurred during the Triassic, Jurassic, and Cretaceous Periods. Such coals can chiefly be found in Canada, China, Russia, and the USA. Changes in floral types, such as the onset of angiosperm floras, as well the variety of deposition environments generally lead to the formation of coals that are more complex in their chemical make-up. Analysis of macerals and other constituents in coal provide insight in the specific conditions that were present during their accumulation.

The third major period of coal formation took place during the Tertiary Period, and those coals form the bulk of the world's brown coal reserves as well as important parts of the black coals currently being mined. Tertiary coals are mostly lignite, but in some areas temperature change has resulted in higher-ranking coals.

The age and stratigraphy of all major coal deposits have been studied extensively, in particular those formations that are of economic interest. Meticulous surveys on individual fields are important for development and actual production, but such details for a single field can easily fill an entire bookshelf! More general overviews of the world's most important coal regions have been made by others [26, 27]. Reviews on various pollutants in coal, such as arsenic or metals, can also be found in more specialized literature.

Sometimes it is convenient to express coal deposits in energy terms, especially if they are going to be used as energy sources. Depending on specific properties of the coal in a deposit, the actual energy content may vary significantly. Coal is divided into different ranks depending on the degree of coalification as it matures from peat to anthracite. Coal is broadly divided into these four ranks, namely anthracite, bituminous, subbituminous, and lignite [28, 29], although much more detailed systems can be found.

There are many classification systems used all over the world, each with their own set of definitions. The terms brown coal, black coals, hard coals, stone coals, thermal coal, and steam coal are commonly used, but their meaning may vary from country to country. Lignite and subbituminous coals are commonly seen as low-ranking coals, while bituminous coal and anthracite are considered high-ranking coals. Caution should always be exercised when analyzing global coal data to avoid mixing up different classification schemes.

It should be noted that the ranks tend to overlap in terms of energy and that the calorific value can vary within a rank depending on local geological properties. Relatively high moisture levels and a low carbon content and calorific value characterize low-rank coals, such as lignite and subbituminous coal. Higher-rank coals, such as anthracite and bituminous coal, have a higher carbon and energy content and a lower level of moisture. The following energy contents of coal ranks are common:

Anthracite: 30 MJ/kg
Bituminous coal: 18.8–29.3 MJ/kg
Subbituminous coal: 8.3–25 MJ/kg,
Lignite: 5.5–14.3 MJ/kg

Some of the world's present peat deposits were formed as early as a few hundred thousand years ago, although the main part of the global peat resources was formed during the last 10,000 years [30]. Occurrences of peat are linked to climate, vegetation, local environment, and precipitation that allow peat-forming conditions to arise. There are also less peat occurrences in the Southern Hemisphere, as there is less land there. Present-day peat occurrences can be predicted well by suitable models [31].

Peat is present all over the world and estimates conclude that 2–3% of the global surface is covered by peat-forming areas, equivalent to about 4 million $km^2$. This figure includes some 200,000 $km^2$ of tropical peats in Southeast Asia. A major proportion of the world's peatlands are located in North America and the northern parts of Asia, but significant deposits are also found in northern and central Europe. In total, over 90% of the world peatlands are in the temperate and cold belt in the Northern Hemisphere. The remainder is located in tropical and subtropical latitudes and mostly located under forests. Such accumulations have been identified in the tropical regions of Africa, Latin America, as well as Indonesia and southern/eastern Asia.

Estimates of the total area of the world's peatlands vary considerably from study to study. The basic problem is to separate areas where the peat layer is 30 cm thick at minimum, i.e., peatlands, from areas only covered with wetland vegetation, such as floodplains, coastal lagoons, mangroves, and swamps. Hence, only a few countries have performed accurate peatland mappings. More comprehensive discussions on peat and closer description of specific formations are available in reference literature, scholarly journals, and reports from various geological surveys.

## *Resource and Reserve Classifications*

It is first necessary to look closer at classifications and what the terms actually mean in practice before defining and discussing global resources of coal and peat. Far from all coal and peat being available for utilization by society, some formations are technically unfeasible or just too challenging economically to recover or may even prove to be undesirable to consumers, even though they are geologically available. As a result, it is important to use precise terms when talking about future supply of coal and peat. Especially in outlooks related to sustainable utilization as possible energy sources for society.

Concepts such as "endowment" or "potential" are multifaceted and the meaning of such terms is disputed by different schools of thought. One school argues that "resource endowment" is crustal abundance, while another school maintains that the concept is relatively meaningless for supply assessments and should be avoided. As always, the truth is located somewhere in between the two extremes. Crustal abundance is solely determined by geological processes, which are reasonably well understood by geologists. Furthermore, the physical in situ amounts will also be virtually fixed as a result of the slow underlying formation processes, even though in situ estimates might change in reports as a result of new information.

Available coal accumulations can be divided into reserves and resources. Reserves are defined as being proved and recoverable under present technological and economic conditions. Resources include additional discovered, but uneconomic or unavailable amounts together with undiscovered inferred, assumed, and speculative quantities. This distinction is justified by the idea that over time, and perhaps with new technology and higher prices, production and exploration activities will allow a reclassification of some resources into reserves [32]. The terms resources and reserves are often used synonymously by lay people, but it should be absolutely clear that this is a completely erroneous and misleading practice. Being specific and using the right term is essential to paint a realistic picture in any supply assessment or when planning for a sustainable future. Studies have highlighted this problem and the need to include non-geological factors in future supply estimates [13, 33].

It is necessary to investigate and explore any coal or peat deposits to ascertain whether something can be mined economically and that a product can be obtained that is marketable. Two common points of view are typically used to assess resources and reserves:

1. The degree of geologic assurance
2. The degree of economic feasibility

The method of estimation along with the definition and expression of geological assurance is a key factor for the reliance of coal tonnage estimates. Details surrounding the topography, tectonic variability, structural alteration, and postdepositional erosion provide uncertainties and should be taken into account in resource mappings. For example, heavily disturbed coals with great variation in thickness will need many more data points for an accurate and reliable assessment than relatively undisturbed coals of constant thickness. Data point spacing criteria and how far trends are extrapolated from measurement points are essential for the degree of geologic assurance (Fig. 9.3). Economic feasibility is dependent on the content of sulfur and other unwanted pollutants, thickness, depth, environmental constraints, coal price versus mining costs, and other parameters that affect recovery and consumer desirability.

Sadly there is no international and uniform method for assessment, categorization and designation of coal resources and reserves. Several agencies, countries, and organizations have developed their own sets of definitions and methods to calculate resources and reserves. However, it is possible to roughly convert assessments made within one system to another and form aggregated pictures of the entire world situation. This section will focus on the system used by the United States Geological Survey (USGS), while closer descriptions on other systems can be found in more

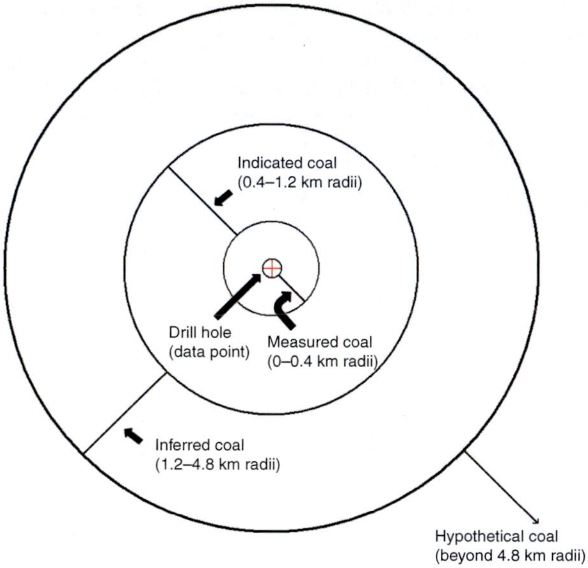


**Fig. 9.3** Reliability categories based solely on distance from drilling measurement points. Measured and indicated coal together form the demonstrated coal category. Based on USGS resource classification system [34]

specialized literature. For comparison, the Australian and Russian systems are conceptually similar but differ in details. In contrast, the UK system is excellent for mining operation, but not very functional for large areas that may contain coal but are not mined. USGS use the following definition for coal resources [34]:

> Coal resources are naturally occurring concentrations or deposits of coal in the Earth's crust, in such forms and amounts that economic extraction is currently or potentially feasible.

This category is divided into a number of subdivisions, such as inferred, indicated, or measured resources depending on geological certainty, depth, and thickness. Technology and proper investment can, at least in theory, recover all available in situ coal if the required economic conditions exist. However, resource numbers are inadequate as they do not take into account economics, the fundamental utilization properties, or even legal issues surrounding extraction in a geographical region. Recoverability is the crucial property for actual production and this leads to a variety of attempts to express the recoverable volumes of coal that can be extracted from the available resources.

Far from all coal being recovered from a coal bed, some amounts will be unavoidably lost in the recovery process for technical reasons. This includes coal that is (1) left to support underground mine roofs, (2) used as barrier pillars adjacent to the mine or property boundaries, or (3) simply left because for a variety of other reasons [34]. Mining and production practices also place constraints based on thickness, depth, rank, and other properties on the recoverable amount of the in situ resource. USGS has chosen the following definition for the reserve base:

> Reserve base: those parts of the identified resources that meet specified minimum physical and chemical criteria related to current mining and production practices, including those for quality, depth, thickness, rank, and distance from points of measurement. The reserve base is the in-place demonstrated (measured plus indicated) resource from which reserves are estimated. The reserve base may encompass those parts of a resource that have a reasonable potential for becoming economically available within planning horizons beyond those that assume proven technology and current economics. The reserve base includes those resources that are currently economic, marginally economic, some of those that are currently sub-economic, and some of the resources that have been or will be lost-in-mining but whose attributes indicate possible future recovery.

Recoverable volumes are also affected by many other factors that make them dynamic. Higher prices, new exploration, or improved technology yield increased amounts of recoverable coal. In a similar way, increased freight rates, extraction costs, or introduction of various restrictions can decrease recoverable volumes. Likewise, political influences can greatly influence demand patterns or even cause direct physical restrictions for mining. USGS [34] finally uses the following terminology to denote reserves:

> Reserves: Virgin and/or accessed parts of a coal reserve base which could be economically extracted or produced at the time of determination considering environmental, legal, and technologic constraints.

Important restrictions are legal prohibitions on coal mining in certain areas and/or the use of a particular mining methodology, such as strip mining. This can prevent the extraction of coal, despite the fact that the coal might be technically and economically recoverable. For example, land-use regulations and federal laws prevent mining near homes, public buildings, or federally funded highways [30]. Accessibility, availability, and restriction factors are linked to the society and its legal and political climate. This can have a major influence on the volumes practically available for recovery. Detailed discussions of how availability of recoverable coal is estimated by the USGS can be found in other works [35–37]. The Energy Information Administration (EIA), the energy authority of the USA, utilizes the following definition to include restrictions and accessibility in their reports and assessments of American coal commonly used by non-geologists, planners, and policymakers:

> Estimated recoverable reserves: cover the coal in the demonstrated reserve base considered recoverable after excluding coal estimated to be unavailable due to land use restrictions or coal currently economically unattractive for mining (and after applying assumed mining recovery rates).

EIA creates this category by applying economic feasibility criteria factoring downward from the reserve base. The "coal reserve" term used by EIA may better be described as "potential coal reserves," as they are generally not proved by any detailed drilling [38]. A full discussion of coal resources and reserve terminology, as used by EIA, USGS, and US Bureau of Mines, can be obtained from the EIA [39]. The estimated recoverable reserves category corresponds to "proved reserves" discussed in the BP Statistical Review of World Energy [40], "proved recoverable reserves" used by World Energy Council (WEC) [41], or "reserves" as used by the German Federal Institute of Geosciences (BGR) [42].

## Coal and Peat Resources in the World

To make accurate coal and peat resource estimates, exploration geologists make detailed studies of relevant accumulations. They take measurements at appropriate intervals and extrapolate using the appropriate classification categories, knowledge of the local geology, and good judgment. In theory, there is no fundamental difference between underground or surface formation maps. However, shallow drilling is cheaper and makes it possible to drill more holes, which makes surface mappings generally more reliable. Because deep boreholes are costly (which means less are drilled, thus making the assessment less reliable) the exploration geologist uses various geophysical methods to supplement the drilling data.

Resources are commonly expressed as in situ tonnage, which is the total amount occurring in place. It can be calculated from the thickness of the deposit, the area of extent, and the relative density. Thickness is commonly determined at each point of measurement, while the area is measured on a map or plan. The relative density is normally taken from the section that will be mined, or alternatively it might be estimated using the known average ash content of a coal or peat seam. The formula for in situ tonnage is:

$$\text{thickness}(\text{m}) \times \text{area}(\text{m}^2) \\ \times \text{relative density}\left(\text{tons}/\text{m}^3\right) = \text{in situ tonnage}$$

At regular intervals, scientists and governmental agencies conduct and publish world coal resource assessments. In particular, since its establishment in 1923, the World Energy Council (WEC) has reviewed all forms of energy resources including coal and peat. Data from the WEC comes from reports by the WEC member committees and constitute a sample that reflects the information available in particular countries. German Federal Institute for Geosciences and Natural Resources (BGR) has also released a number of assessments on global coal resources. These agencies will serve as the main data source for the resource figures presented here.

Peatland resource estimates are commonly quoted on an area basis because initial quantification normally arises through soil survey programs or via remotely sensed data that focuses on geographical extent (Tables 9.1 and 9.2). Estimates of the world's peatland area vary considerably. This is partly due to different wetland classification systems and partly due to insufficient survey data. Peat resources have been estimated to be 5,000–6,000 billion cubic meters, but are hard to quantify in any detail [30]. WEC gives an estimate of 3,500–4,000 billion cubic meters [41]. Uncertainties in water and ash content and many other factors influence the result significantly when converting to tonnage. This can be seen as a reason for circumventing tonnage figures and preferring volumetric units in resource assessments. However, it is clear that the world possesses a huge tonnage of peat overall. The important question is: how much of those resources can be realistically recovered in the future?

**Table 9.1** The distribution of peatlands and other terrestrial wetland ecosystems in the world. Data source: [30]

| Region | Peatlands [km$^2$] | Wetlands [km$^2$] | Total [km$^2$] |
|---|---|---|---|
| Africa | 58,000 | 282,000 | 340,000 |
| Asia | 1,119,000 | 1,149,000 | 2,268,000 |
| Australia and Oceania | 14,000 | 10,000 | 24,000 |
| Central and South America | 102,000 | 330,000 | 432,000 |
| Europe | 957,000 | n. a. | 957,000 |
| North America | 1,735,000 | 657,000 | 2,392,000 |
| *World* | 3,985,000 | 2,428,000 | 6,413,000 |

World coal resource estimates can be traced back to 1936, when the WEC gave a world resource estimate of around 16,000 billion tons (reported as recoverable reserves plus estimated additional in situ amounts). In subsequent assessments, their world coal resources have varied from 10,000 to 14,000 billion tons, with most estimates closer to the lower part of the range. In comparison, BGR estimates of world coal resources have varied from 6,000 to 15,500 billion tons. Using either estimate, it is clear that global coal resources are large and the world production of coal (~6 billion tons) is seemingly insignificant in comparison.

However, there are reasons to doubt these agencies' resource assessments. First, there are large fluctuations in the reported resource amounts as definitions and reporting practices change, coal-bearing areas are added or subtracted and new geological information becomes available.

For instance, total coal resources in the USA decreased by more than 60% between 1913 and 1962. In contrast, between 1968 and 1974 coal estimates were revised upward, in part due to improved accuracy in assessing the amount of lower-rank coal [33] and also by including undiscovered resources in the estimates. Further exploration may show that a substantial part of the undiscovered resources do not exist [33]. Similar modifications, as well as some brief reasoning behind the revisions, can be found in recent world coal resource estimates from the BGR (Table 9.3).

In 2005, BGR reported that they changed coal resource estimates because there was insufficient substantiating information for many countries and many countries had skipped reporting years. In 2006, they incorporated new resources in China and former Soviet Union that lead to a 115% increase in world hard coal resources and a 200% increase in lignite resources. In 2007, BGR made yet another upward revision, increasing world hard coal resources by 68% and lignite resources by 36%. In both cases the majority of the increase came from new resources identified in the USA by a new USGS study [43].

Another important observation and fact of this world is that the peatlands are unevenly distributed with the vast majority concentrated in a few countries (Table 9.2). This trait can also be seen in the assessment literature of world coal resources (Table 9.3). This inequality is a sad fact of this world and a function of

**Table 9.2** Distribution of peat resources for selected countries and regions. Data source: [41]

| Region | Country | Peatlands [$km^2$] |
|---|---|---|
| Africa | Guinea | 5,250 |
| | Nigeria | 7,000 |
| | South Africa | 9,500 |
| | Uganda | 14,200 |
| | Zambia | 11,060 |
| | Total Africa | 58,410 |
| Asia | China | 10,440 |
| | Indonesia | 270,000 |
| | South Korea | 6,300 |
| | Malaysia | 25,360 |
| | Burma | 9,650 |
| | Total Asia | 331,880 |
| Australia and Oceania | Australia | 150 |
| | New Zealand | 2,600 |
| | Papua New Guinea | 6,850 |
| | Total Australia and Oceania | 9,640 |
| Central and South America | Brazil | 15,000 |
| | Chile | 10,470 |
| | Falkland Islands | 11,510 |
| | Guyana | 8,140 |
| | Venezuela | 10,000 |
| | Total South America | 62,400 |
| Europe | Belarus | 23,970 |
| | Estonia | 9,020 |
| | Finland | 89,000 |
| | Germany | 14,200 |
| | Iceland | 10,000 |
| | Ireland | 11,800 |
| | Latvia | 6,400 |
| | Norway | 23,700 |
| | Poland | 12,000 |
| | Russia | 568,000 |
| | Sweden | 64,000 |
| | Ukraine | 10,080 |
| | UK | 19,260 |
| | Total Europe | 876,510 |
| North America | Canada | 1,113,280 |
| | Cuba | 6,580 |
| | Mexico | 10,000 |
| | USA | 214,000 |
| | Total North America | 1,354,220 |

favorable geology in certain parts of the world, bestowing significant deposits of peat and coal on relatively few countries. However, uncertainties are present and there is a problem with a lack of available data for some countries and regions.

**Table 9.3** Estimated total coal resources (recoverable reserves plus other in situ resources), as published by the BGR. All numbers in million metric tons (Mt). Data source: [42]

| Country | 2004 | 2005 | 2006 | 2007 |
|---|---|---|---|---|
| USA | 1,090,404 | 1,057,921 | 1,359,404 | 6,719,682 |
| China | 1,089,787 | 1,089,789 | 4,367,000 | 4,367,000 |
| India | 259,217 | 291,863 | 252,359 | 229,867 |
| FSU | 2,295,275 | 2,326,121 | 2,915,146 | 2,931,293 |
| Australia | 298,400 | 304,801 | 193,700 | 148,200 |
| South Africa | 163,851 | 163,851 | 48,751 | 28,559 |
| Germany | 87,827 | 91,497 | 84,573 | 83,065 |
| Poland | 70,262 | 101,281 | 179,459 | 179,459 |
| UK | 5,220 | 7,220 | 190,005 | 187,132 |
| Indonesia | 10,581 | 64,829 | 26,839 | 23,948 |
| *World* | NA | 6,038,991 | 9,553,840 | 15,510,618 |

To summarize, resource estimates are important from a geological point of view and are vital for understanding where recoverable amounts may be located. However, it is essential to understand that society does not have economic access to all the reported resources. In situ resources include all amounts, even those located deep or of poor quality, making resource estimates largely an academic figure [33]. Resources must be converted into reserves by a detailed evaluation of the degree of certainty of the size of the particular coal resource, and the economic, legal, and environmental constraints placed on its exploitation. Vast resources do not necessarily mean significant future production, as production is dependent on many more factors than just geological availability. This naturally leads to the next section in which reserves are discussed.

## *Coal and Peat Reserves in the World*

Reserves are usually calculated from resources by applying suitable and relevant constraints. Legal, environmental, and geological factors prohibit the recovery of the entire resource so reserve estimates will always be less than resource estimates. Rather than perform a detailed constraint analysis, it is possible to assume a universal recovery factor based on historic recovery factors for similar deposits. The US Bureau of Mines performs this type of calculation regularly [33]. However, the most reliable method is a deposit-by-deposit analysis because it can incorporate local economic conditions as well as the technical recoverability of the minable coal.

Coal resources are primarily of academic interest, and unfortunately provide little information about what quantity of the resource can be practically recovered. For example, let us look at the Gillette coal field in Wyoming and USGS' comprehensive investigation of its coal supply. Luppens et al. [44] assessed the Gillette coalfield and its 11 coal beds and estimated the original coal in place to be 182 Gt, with no restrictions applied. Available coal resources, which are a part of the original resource that is accessible for potential mine development after subtracting all restrictions, are about 148 Gt. Recoverable coal, which is the portion of available coal remaining after subtracting mining and processing losses, was determined for a stripping ratio of 10:1 or less and resulted in a total of 70 Gt. Applying current economic constraints and using a discounted cash flow at 8% rate of return, the coal reserve estimate for the Gillette coalfield is only 9.1 Gt. While economic conditions definitely have the largest impact, reductions caused by technical and availability restrictions are also significant.

Extraction of peat and coal is primarily an economic activity that recovers valuable products from certain deposits with a profit. Prevailing and anticipated price, location of the deposit relative to the market, transportation costs, specific use, and the production cost are all important factors for determining the feasibility of exploitation. Forecasting costs, estimating revenues, and handling inflation over long time periods is an uncertain business. Additional uncertainty comes from the regulatory environment, particularly environmental laws, tax policies, and similar. In essence, profitability is a dynamic factor and can quickly change. To obtain an accurate estimate of the recoverable volume of a deposit requires a meticulous economic study in combination with comprehensive physical surveys.

An often overlooked but fundamental factor is the net energy analysis. Hubbert [45] wrote about net energy in relation to oil but the same principle applies to all energy resources: "there is a different and more fundamental cost that is independent of the monetary price. That is the energy cost of exploration and production. So long as oil is used as a source of energy, when the energy cost of recovering a barrel of oil becomes greater than the energy content of the oil, production will cease no matter what the monetary price may be." In the case of coal and peat, a poor net energy return will prevent deposits that are too deep, too complicated, or too thin from being usable as energy sources. Mining companies leave those resources behind because they require more energy to exploit than they can yield in return. Several researchers have used net energy analysis to determine mining boundaries in an objective manner [46].

Rohrbacher et al. developed a methodology for describing recoverability using geological, technical, and economic factors [47]. They concluded that past coal estimates did not address the amount that might realistically be available for production after environmental and technological restraints are considered. They showed how air quality regulations, mine reclamation legislation and regulations, new safety and health standards, and advancements in mining technology had dramatically changed the economic competitiveness of the coal fields they studied. More importantly, they demonstrated that the generous EIA reserve base would be far less than the reported figure using their methodology. The EIA estimate did not include the effects of mined-out prime reserves, environmental, industrial, economic, and social considerations. Nor did it include the technical aspects of the minability and washability of the individual seams. In contrast, second-generation assessments of available coal amounts in the Pittsburgh coal bed were performed and described by Watson et al. [48]. Their assessment was that half of the original coal resource in place had already been extracted leaving only 15 Gt and those

**Table 9.4** Estimated coal reserves for selected countries. Note how WEC and BP numbers are identical, except for the fact that BP dates them differently. All numbers in million metric tons (Mt). Data sources: [40–42]

| Country or region | WEC Proved at end 2007 | BGR Proved at end 2007 | BP Proved at end 2008 |
|---|---|---|---|
| Australia | 76,200 | 76,900 | 76,200 |
| Brazil | 7,059 | 6,596 | 7,059 |
| Canada | 6,578 | 6,578 | 6,578 |
| China | 114,500 | 192,000 | 114,500 |
| Germany | 6,708 | 40,936 | 6,708 |
| India | 58,600 | 73,510 | 58,600 |
| Indonesia | 4,328 | 6,981 | 4,328 |
| Japan | 355 | 355 | 355 |
| Kazakhstan | 31,300 | 18,947 | 31,300 |
| Poland | 7,502 | 16,329 | 7,502 |
| Russia | 157,010 | 161,553 | 157,010 |
| South Africa | 30,408 | 28,559 | 30,408 |
| Ukraine | 33,873 | 34,375 | 33,873 |
| UK | 155 | 432 | 155 |
| USA | 238,308 | 262,971 | 238,308 |
| *World* | 826,001 | 989,913 | 826,001 |

resources were increasingly being restricted by new environmental laws, encroaching population, and affected by technological factors.

Several organizations regularly compile peat and coal reserve estimates. Many national energy agencies and geological surveys keep reserve records for their own country. The World Energy Council and the BGR collect country level data and publish global coal reserve estimates. The BP Statistical Review of World Energy [40] is another commonly referenced source of reserve estimates but it only reproduces the numbers compiled by World Energy Council without backdating the estimates. Table 9.4 shows the latest reserve figures from these three sources. Several studies have criticized the reserve estimates found in published literature for not adequately accounting for all resource constraints [13, 33, 49–51].

Changes in policy can also cause rapid and significant changes to reserve estimates. For example, when Germany removed their mining subsidy Germany's coal reserves were greatly reduced [52, 53]. In Russia's case, Malyshev [54] shows that according to the best estimates by experienced geologists, mining engineers, and economists around 40% of the Russian coal reserves cannot be regarded as of any practical interest because they are highly unprofitable.

Highly reliable estimates of coal reserves are rare in part because the factors that influence recoverability change frequently. Changes in classification systems, economics, regulations, and other factors can cause significant changes in reserves over time. In the case of the USA, significant downward revisions of the reserves occurred when mining restrictions and other non-geological factors decreased recoverability [13]. Despite the problems with reserve estimates, they are still the best available estimates of how much of a resource that can be economically and technically recovered.

Regulations can directly affect coal recoverability, such as prohibitions due to land-use conflicts, or indirectly, such as taxation. In the USA, the Surface Mining Control and Reclamation Act of 1977 directly prohibits surface mining in national parks, forests, wildlife refuges, trails, wild and scenic rivers, wilderness, and recreation areas. Many national parks have been established over the years, further limiting land available for coal production. Petsch [55] studied Europe and the Ruhr area and concluded that people have become sensitive to intrusions made on their environment, for instance from mining or colliery spoil. Physical limitations from land-use regulations have generally increased over time in many countries. Historically, air quality regulations have increased the demand for low-sulfur coal [13, 56–58]. Expanded environmental awareness in the 1960s and 1970s had lead to heated debates and stricter regulations.

In 2009, President Obama stated that "at a time of such great challenge for America, no single issue is as fundamental to our future as energy." Part of that challenge is reducing the dependence on imported oil and political pressure to limit greenhouse gas emissions. Governments may very well tighten coal regulations and pursue future efforts for addressing anthropogenic climate change and $CO_2$ emissions. Generally, the air-quality regulations in the USA have been an environmental success story although it has done little to mitigate $CO_2$ emissions [59]. Interestingly, tight regulations are not the sole factor in future emissions. Regulatory uncertainty can be significant as businesses delay investments until they can make solid return on investment decisions [60]. Exactly how future $CO_2$ policies will affect coal reserves remains to be seen but it is likely that such policies will reduce the recoverable proportion of coal resources.

Sustainable development, environmental stewardship, and the importance of natural resources for the continued well-being of mankind are driving forces for future policies, at least in any foreseeable outlook. Historical trends point toward increased restrictions and this implies that future energy policies will result in significantly decreased coal availability. However, politics are constantly shifting and sudden change in any direction cannot be ruled out. To summarize, recoverability is a complex issue and future supply of coal will be dependent on many factors ranging from geological availability and quality to socioeconomic acceptability and energy policies.

Because peat is a small energy source, peat reserve estimates are difficult to find and there is no known comprehensive global study. Volumetric data can be converted to tonnage by using the average bulk density of produced peat. USGS Mineral Commodity Survey on peat [61] contains some reserve estimates (Table 9.5). Only a small portion of peatlands are available for extraction due to environmental constraints and other factors.

Only a few countries have a peat industry of significant importance, so the global peat reserves should be regarded as highly uncertain. Economics, demand, environmental regulations, and other factors are still uncertain for many regions. The world possesses large amounts of peat, but that should not be seen as guarantee for future production. If demand for peat increases, reserve estimates will in all likelihood become more accurate.

**Table 9.5** Peat reserves for selected countries. All numbers in million metric tons (Mt). Data source: [61]

| Country | Peat reserves as of 2008 [Mt] |
|---|---|
| Belarus | 400 |
| Canada | 720 |
| Estonia | 60 |
| Finland | 6,000 |
| Ireland | NA |
| Latvia | 76 |
| Lithuania | 190 |
| Moldova | NA |
| Russia | 1,000 |
| Sweden | NA |
| Ukraine | NA |
| USA | 150 |
| Other countries | 10,000 |

As with coal resources, global peat reserves are very unevenly distributed. A small number of nations control the vast majority of the world's coal reserves. The USA, Russia, China, Australia, India, and South Africa control over 80% of the world's coal reserves (Table 9.4) – a remarkable concentration of resources. Peat shows a similar distribution with just a few countries as the world's major reserve holders (Table 9.5).

## *Global Coal and Peat Production*

Production is the activity that actually transforms coal and peat into something useful for consumers and society. Some of the produced volumes are inherently lost in preparation, washing, or transformation to specialized products such as briquettes to meet customer needs. Generally coal and peat are burned to create heat, the heat is transformed into steam, and then the steam into electricity. They are also used in industrial chemical processes, steel-making, and horticulture.

The world produced 6.8 Gt of coal in 2008 [40] and a comparatively tiny amount of peat – only around 25 million tons [61]. Peat's role as an energy source is negligible in the global energy context, while coal is the second most important fuel, second only to oil. As a result, this section will briefly cover peat and coal mining and highlight some important aspects of production.

Currently, just a few countries are the principal producers of peat and all those nations are located in Europe. Finland, Ireland, Belarus, Russia, Sweden, Ukraine, and the Baltic states are the largest producers of peat. Finland is perhaps the most significant peat consumer and derives 6% of its total energy from peat as a way to replace imported fuels [62]. In Finland, over 90% of the total peat production goes to energy production. In contrast, the USA uses most of its peat for horticultural uses [61]. Peat is often used as a practical alternative for smaller heat and/or power plants. There is also significant trade flows of peat among certain countries, but most of the peat is still being consumed by the same country that produced it.

Wetlands that contain peat have to be drained before they can be exploited. Peat in situ contains roughly 90% water and drainage can reduce this significantly. Additional drying by the wind and sun will further reduce moisture content down to 40–50%. The traditional production method was to hand-cut peat into sods or strips, but this has been replaced by mechanical means. Another production technique yields fine peat granules by using a mechanical miller to disturb and granulate the top layer of the peatland. Some of the produced peat goes to the production of peat briquettes, which is a more practical fuel for households. WEC [41] states that the bulk of peat production is obtained from milling, which is more easily used for electricity and heat plants.

Producing coal requires more complex operations and often underground mining methods. Coal mining can be done from the surface by using draglines and other machines to remove the overburden and expose the coal for recovery. Surface mining methods can generally recover a high percentage of the coal and are relatively cost-effective. However, there are limits to the amount of overburden that can be practically removed to obtain access to the coal. Above certain stripping ratios (the amount of overburden that must be removed divided by the amount of recoverable coal), the surface method becomes unworkable and is replaced by deep mining.

In comparison, underground mining is more complicated and expensive and recovers less coal since certain parts must be left to support the roof. Up to 40% of the total coal in a coal seam may be left as supporting pillars, although this sometimes can be recovered at a later stage. Worldwide, about 40% of the production originates from surface mines, while 60% stems from underground operations. Surface mining dominates in some important producing countries [63]. The chosen production method will have a significant impact on the economic as well as technical restrictions that affect the recoverability of the coal.

Since the 1950s, the six largest coal reserve holders have increased their share of world production (Fig. 9.4). China in particular has rapidly increased its coal production and now accounts for over 40% of world coal production. The USA is the second largest coal producer, followed by India, Australia, Russia, and South Africa. Even though coal is mined in over 50 different countries [63], the bulk of production comes from these six nations.

It is impossible for coal and peat production to avoid affecting and disturbing the local environment. Measures that restrict or even prevent mining in environmentally sensitive and other protected areas are affecting legal and economic feasibility of new mines. Mine tailings, acid mine drainage, leakage of associated substances like sodium and heavy metals, land subsidence, impact on local wildlife and landscape changes are problems that are intrinsically associated with production. Avoiding addressing these issues will stir up political or legal controversies [13, 55].

In addition, safety regulations are another complicating factor. Mining companies must agree to follow specific safety laws to receive a mining permit. Some major hazards are explosions (caused by methane gas and dust), roof

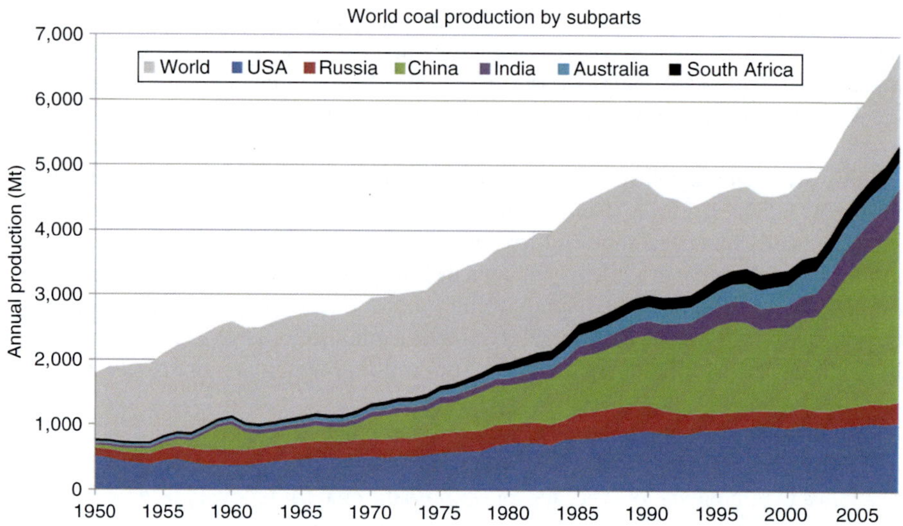


**Fig. 9.4** Total world production 1950–2008. Prior to 1985, the coal production from the Donetsk basin is not included in Russia. In 2008, just six countries accounted for roughly 80% of the total world coal production. These countries will determine the shape of future global coal production. Data source:[40, 64]

collapse, and machine failure. Surface mining is generally safer since the risk of explosions is lower and the risk of roof collapses is almost completely absent. Since many coal deposits are not reachable by surface mining, the company must resort to underground mining. Closer discussions on safety related issues have been made by others [65, 66]. Good safety brings additional costs, while poor safety will hurt the public acceptance of the mining operations as well as potentially risking the demand for the "unsafe" coal. Poor safety may also result in governmental intervention and the closure of mining operations if required safety regulations are not followed.

New technology and recovery methods can change or potentially even revolutionize production. Improved mechanization, use of long-wall techniques that minimize the coal left behind to support the roof, and other measures have improved productivity and economic competitiveness of mining operations [67, 68]. These new mining methods have made previously unreachable deposits feasible for development, especially when they reduce production costs. New inventions will likely continue to improve production even further as research and development continues.

However, there is a danger in expecting too much from new technology. Increasing depletion of the best and most easily accessible coal seams will force future production to be derived from less favorable deposits. This will increase mining costs, which in turn will make energy more expensive across the whole economy. Some studies even show that technological progress will not meaningfully be able to offset depletion [69–71]. The study authors reason that new technologies and mining

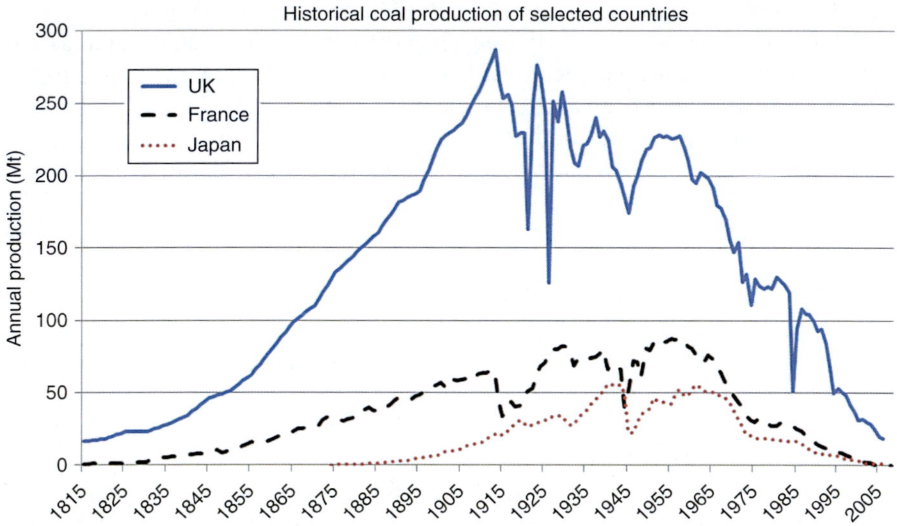


**Fig. 9.5** Historical coal production of UK, France, and Japan. All countries experienced production disturbances around the two world wars and the great depression. Data source: [40, 64]

methods that boost efficiency and decrease labor requirements will be unable to offset the higher production costs of more complicated and labor-intensive deposits.

Coal has been produced in many countries for many decades, which enables us to perform longitudinal studies of production data. Several countries are currently producing only a small share of what they have produced historically (Fig. 9.5). Much of the reduced production can be explained by the move to other energy sources like petroleum or uranium, depletion of commercially feasible deposits, energy policy changes, and similar reasons.

Countries such as the UK still have a significant amount of coal resources left (Table 9.3), but not enough to suggest that coal will be the fuel of choice again. The remaining coal is generally located deeper than the previously mined coal and will be more expensive to mine in terms of labor and capital investment. Given the drawbacks of coal, especially in terms of climate change because of its high carbon content, it is not at all certain that demand for coal will grow again. This naturally leads to the question of future production of coal and peat.

## *Future Production*

Future production of peat and coal is an open and complex question. The resources might be vast, but only a small amount is recoverable once legal, environmental, and economic constraints are accounted for. In particular, it would be a mistake to think that the huge worldwide resource estimates mean that coal and peat will provide a steady energy supply in to the future. In addition to the previous factors, production also depends on social acceptability (i.e., who would want to see mountain-top removal), technical restrictions, the development and competitiveness of alternative energy sources, and expected demand. Although the following overview is unable to discuss all the factors that will affect future production, it will illuminate some of the complexity and leading indicators of future production projections.

There are many methodologies for creating production and supply projections for coal or peat. Some methods are quantitative, and rely on field data, while others are qualitative. Some analysts base their projections on economic theories, others use physical or statistical models derived from natural sciences. The life of the reserves and whether production can match demand are important questions for planning and sustainability.

The economist Adelman [72] has said that the amount of mineral in the earth is an irrelevant, nonbinding constraint to production. He further says that the amount of resource in the ground is probably unknowable but surely is unimportant and of no economic interest. This is correct in the very limited sense that producers will never recover the very last amounts of coal or any other resource from the earth. The ultimate recovery is undefined and governed by non-geological circumstances. However, although the recovery factor might be undefined that does not make it unlimited. It is bound between 0% and 100% of the resource in the earth's crust, which is clearly defined but not known exactly. Thus, the fact that the amount of recoverable coal is undefined and unknown cannot be an argument against the existence of future production limitations. The slow creation of coal and peat in comparison to their rapid extraction from mining imposes an inherent finiteness to the amount of coal and peat available.

Hubbert [73] proposed a model that extrapolates the production curves of finite resources such as petroleum or coal. His model assumes that production begins at zero, just before production has started, and ends at zero, when the resource has been exhausted. In between, the production curve passes through one or more maxima. The actual shape of the production curves may vary, but they are ultimately limited by the amount of resource that was deemed to be recoverable. Hubbert initially proposed a bell-shaped curve for an idealized production behavior, representing various development stages without giving any mathematical description for it. Later, he derived a mathematical framework chiefly based on the logistic function and curve fitting to production and discovery data [74].

Fig. 9.6 shows a simple example of how actual production can be described by mathematical curves. When doing long-term projections, a suitable curve is fitted against historical production and constrained by a reasonable estimate on the recoverable volumes that may be ultimately produced. The total resource in the earth's crust will ultimately place a limit on future production, but the real limit will be lower since it is unrealistic to assume that all geologically existing volumes can be recovered.

Hubbert himself [73] used 2,500 Gt as the ultimate world coal reserves and placed the production peak around 2,150 at an annual output of 6 Gt. This implies

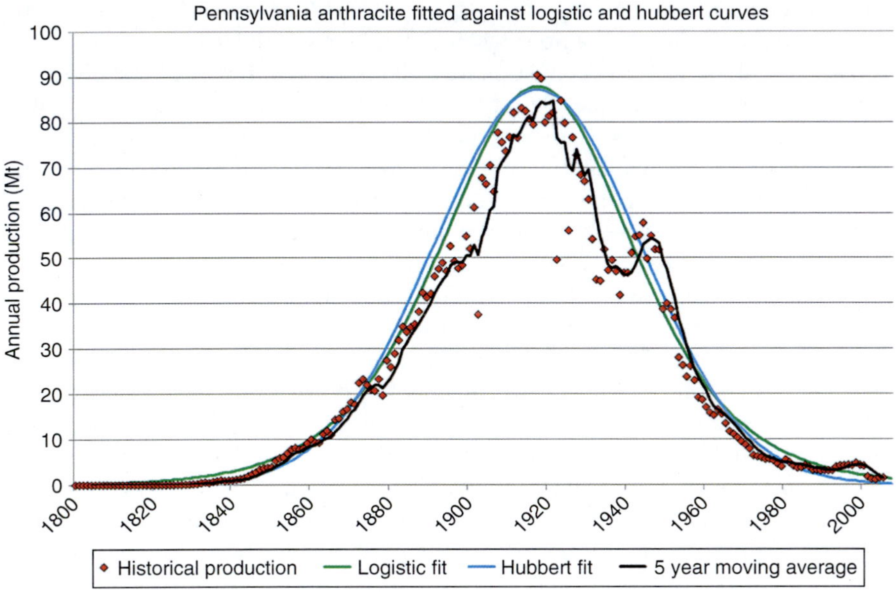


**Fig. 9.6** Historical production of anthracite in Pennsylvania fitted with both logistic and Hubbert curves. Data source: [13]

that there are no near-term problems for coal production. However, his simple approach does not consider all aspects of recoverability. It also agrees poorly with actual production history since current production reached 6 Gt per year in 2008 and production is still increasing.

Hubbert's method is sensitive to the assumed ultimate reserve that constrains the curve so it is essential to have the best possible resource estimate. It has been described as a reasonable model provided that it is acknowledged that reserves are dynamic and that other factors also influence production [75]. Growth curve methods can provide insight into the general long-term flow of finite energy resources, but perturbations caused by sudden and unforeseen near-term economic or political changes cannot be predicted. Therefore, long-term life-cycle projections should not be used as a substitute for meticulous economic studies that forecast perturbations in coal production over the upcoming years or decades [76]. Studies have also used the development of regulations, restrictions, and recoverable volume estimates to create production projections [13].

Other quantitative assessments that have relied on reported reserve and resource assessments found that coal output can increase until 2050 or even 2100 [32, 76]. Thielemann et al. [32] assumed that resources will simply be converted into reserves and then produced due the self-regulating price-driven supply cycle. This model gives no foreseeable production limitations within the next century. Mohr and Evans [77] point out that shrinking coal resource and reserve estimates over the last 25 years seem to indicate that factors other than the self-regulating supply mechanism are playing a role. In any case, future production will still be limited to the countries and regions where the resources are located.

Watson et al. [48] concluded that future coal production from the Pittsburgh coal area is dependent on a complex set of factors, including future transportation costs and environmental policies. Without a holistic view of all factors affecting recoverability and geological availability, no reasonable supply projection can be made. Although predictions of future production will always be uncertain, it is important to remember that production is dependent on much more than just geological existence. Claims that abundant resources will automatically provide future energy supply need to be treated with suspicion – they oversimplify the production situation.

For example, there are significant unexploited coal reserves located in eastern Siberia and far eastern Russia. However, these deposits are far from major population centers and will need major investments in infrastructure before they can be exploited. How fast this can occur is an open question and depends primarily on economic and political factors. Also, Russia might never develop those deposits and may use other energy sources instead. A similar point can be made for the USA and the vast coal resources locked up in Alaska [43]. Underlying assumptions can greatly affect future production projections and one must always remember that production is a complex issue with many factors involved.

Recent concern for the $CO_2$ emissions that come from burning fossil fuels may turn coal and peat into undesired fuels for the future in some countries. On the other hand, countries that care less about the environment might turn to coal and peat as cheap, but polluting, alternatives to imported oil. Clean coal technologies and carbon capture and storage might alleviate much of the emission problems from combustion, which can improve coal's attractiveness. Cheap and reliable wind and solar energy or improved nuclear power might also deliver cheaper energy to consumers than coal and peat if carbon emissions are taxed. Increasing global competition for the remaining petroleum might cause countries to increase their reliance on domestic energy sources. In summary, there are many factors that can affect future production and without a holistic approach it is difficult to make reliable projections.

The current world dependence on fossil fuels also brings about another problem. Since fossil fuels dominate the existing energy system, they will have to power any possible shift to renewable or other energy sources [78]. The effect this has on fossil fuel reserves will depend largely on the size and speed of the shift, since fossil fuel energy used will be additional to that already in use.

In conclusion, there is a lot of coal and peat available in terms of resources. However, far from all the resources are suitable for production, and future production will largely depend on the choices people make. The model and assumptions about recoverability have a significant impact on future production projections. The key point is that huge worldwide resource estimates for coal and peat do not automatically translate into an assured supply. Vast resources are present if society demands coal and peat and is prepared to pay the price for recovering them.

## Future Directions

This section should start with a frank admission that coal will remain one of the most important energy sources for mankind for the foreseeable future. Future development of coal production will be linked to the development of global energy supply, which will affect virtually all aspects of society in some way. However, production is a complex issue that requires a holistic treatment. This chapter has attempted to highlight several of the indicators and parameters that affect future production of coal and peat.

It is also obvious that more work needs to be done in exploration and geology to better understand the occurrence of coal and peat. Better survey data and increased geologic certainty would also help to improve the accuracy of resource estimates – especially in those countries and regions where few assessments have been made or where only old data is available. There is no substitute for a good geological understanding of depositional systems and detailed surveys when making resource estimates.

Estimation of reserves and recoverable volumes involve consideration of numerous economic, technical, legal, and social parameters relative to specific markets and expected demand during a present and period with economic and regulatory uncertainty. Improved understanding and handling of this complex set of factors are required for more accurate and reliable assessments in the future. Recoverability is not a topic that should be taken lightly or oversimplified. Instead it needs holistic treatment to provide reasonable outlooks for future world supply of coal and peat.

## Bibliography

### *Primary Literature*


1. Ehrlich PR, Ehrlich AH, Holdren JP (1970) Ecoscience: population, resources, environment. W.H Freeman and Company, San Francisco
2. Cook E (1977) Energy: the ultimate resource? Resource Papers for College Geography, Issue 77–4, 42 p
3. Simon J (1966) The ultimate resource 2. Princeton University Press, Princeton
4. Bromley DA (2002) Science, technology, and politics. Technol Soc 24:9–26
5. Einstein A (1905) Ist die Trägheit eines Körpers von seinem Energieinhalt abhängig? Ann Phys 323:639–641
6. IEA (2008) Key World Energy Statistics 2008. See also: http://www.iea
7. Pimentel D, Hurd LE, Bellotti AC, Forster MJ, Oka IN, Sholes OD, Whitman RJ (1973) Food production and the energy crisis. Science 182(4111):443–449
8. Green MB (1978) Eating oil: energy use in food production. Westview Press, Boulder
9. Pfieffer DA (2006) Eating fossil fuels: oil, food and the coming crisis in agriculture. New Society Publishers, Gabriola Island
10. Pimentel D (2007) Food, energy, and society. CRC Press, Boca Raton

11. Akinlo AE (2002) Energy consumption and economic growth: evidence from 11 Sub-Sahara African countries. Energy Econ 30:2391–2400
12. Hondroyiannis G, Lolos S, Papapetrou E (2002) Energy consumption and economic growth: assessing the evidence from Greece. Energy Econ 24:319–336
13. Höök M, Aleklett K (2009) Historical trends in American coal production and a possible future outlook. Int J Coal Geol 78(3):201–216
14. Diessel CFK (1992) Coal-bearing depositional systems. Springer, Berlin
15. Moore PD (1987) Ecological and hydrological aspects of peat formation. In: Scott AC (ed) Coal and coal-bearing strata: recent advances, vol 32. Geological Society, London, pp 7–15, Special publications
16. Nadon GC (1998) Magnitude and timing of peat-to-coal compaction. Geology 26(8):727–730. doi:10.1130/0091-7613
17. Mäkilä M (2006) Lot of peat deposits under 300 years old in Finland. Geological Survey of Finland, Peat research report 59/2006
18. Ryer TA, Langer AW (1980) Thickness change involved in the peat-to-coal transformation of bituminous coal of Cretaceous age in central Utah. J Sed Petrol 50:987–992
19. Dukes JS (2003) Burning buried sunshine: human consumption of ancient solar energy. Clim Change 61(1–2):31–44. doi:10.1023/A:1026391317686
20. Neuzil SG, Supardi CCB, Kane JS, Soedjono K (1993) Inorganic geochemistry of domed peat in Indonesia and its implication for the origin of mineral matter in coal. Geol Soc Am Spec Pap 286:23–44
21. Cecil CB, Dulong FT, Cobb JC, Supardi K (1993) Allogenic and autogenic controls of sedimentation in the central Sumatra basin as an analogue for Pennsylvanian Coal-bearing strata in the Appalachinan Basin. Geol Soc Am Spec Pap 286:2–22
22. Gastaldo RA, Allen GP, Huc AY (1993) Detrial peat foundation in the tropical Mahakam River Delta, Kalimantan, Eastern Borneo: sedimentation, plant composition and geochemistry. Geol Soc Am Spec Pap 286:107–118
23. Ruppert LF, Neuzil SG, Cecil CB, Kane JS (1993) Inorganic constituents from samples of a domed and lacustrine peat, Sumatra, Indonesia. Geol Soc Am Spec Pap 286:83–96
24. Wellman CH, Osterloff PL, Mohuiddin U (2003) Fragments of the earliest land plants. Nature 425:282–285. doi:10.1038/nature01884
25. Butler J, Marsh H, Goodarzi F (1988) World coals: genesis of the world's major coal fields in relation to plate tectonics. Fuel 67(2):269–274
26. Walker S (2000) Major coalfields of the world. IEA Coal Research, London
27. Saus T, Schiffer HW (1999) Lignite international. Rheinbraun AG, Cologne
28. American Society for Testing and Materials (2005) Standard classification of coals by rank, ASTM D388-05. ASTM International, West Conshohocken
29. Carpenter AM (1988) Coal classification. IEA Coal Research, London
30. Lappalainen E (1996) General review on world peatland and peat resources. In: Lappalainen E (ed) Global peat resources. International Peat Society, Jyskä
31. Lottes AL, Ziegler AM (1994) World peat occurrence and the seasonality of climate and vegetation. Palaeogeogr Palaeoclimatol Palaeoecol 106(1–4):23–37
32. Thielemann T, Schmidt S, Gerling PJ (2007) Lignite and hard coal: energy suppliers for world needs until the year 2100 – an outlook. Int J Coal Geol 72:1–14
33. van Rensburg WCJ (1982) The relationship between resources and reserves estimates for US coal. Resour Policy 8(1):53–58
34. Wood GH, Kehn TM, Carter MD, Culbertson WC (1983) Coal resource classification system of the U.S. geological survey. US Geological Survey Circular 891. http://pubs.usgs.gov/circ/c891/
35. Eggleston JR, Carter MD, Cobb JC (1990) Coal resources available for development – a methodology and pilot study. US Geological Survey Circular 1055. http://pubs.usgs.gov/circ/c1055/

36. Carter MD, Gardner NK (1989) An assessment of coal resources available for development, central Appalachian region. US Geological Survey open-file report 89–362. http://pubs.usgs.gov/of/1989/of89-362/
37. Luppens JA, Rohrbacher TJ, Haacke, JE, Scott DC, Osmonson LM (2006) Status report: USGS coal assessment of the powder river, Wyoming. U.S. Geological Survey open-file report 2006-1072. http://pubs.usgs.gov/of/2006/1072/
38. American Association of Petroleum Geologists (2007) Unconventional energy resources and geospatial information: 2006 review. Nat Resour Res 16:243–261
39. Energy Information Administration (1996) U.S. coal reserves, appendix A, specialized resource and reserve terminology. http://tonto.eia.doe.gov/ftproot/coal/052995.pdf
40. BP (2009) BP statistical review of world energy 2009. http://www.bp.
41. World Energy Council (1924–2007) Survey of energy resources 2007 and previous reports and statistical yearbooks from previous world power conferences, World Energy Council, London. http://www.worldenergy.org/
42. German Federal Institute of Geology and Natural Resources (1980–2008) Reserves, resources and availability of energy resources. Various editions. http://www.bgr.bund.de/. Accessed 11 Nov 2010
43. Flores RM, Stricker GD, Kinney SA (2004) Alaska coal geology, resources, and coalbed methane potential. USGS report. http://pubs.usgs.gov/dds/dds-077/
44. Luppens JA, Scott DC, Haacke JE, Osmonson LM, Rohrbacher TJ, Ellis MS (2008) Assessment of coal geology, resources, and reserves in the Gillette Coalfield, Powder River Basin, Wyoming. U.S. Geological Survey Open-File Report 2008–1202. http://pubs.usgs.gov/of/2008/1202/
45. Hubbert MK (1982) Response to David Nissens remarks. http://www.hubbertpeak.com/Hubbert/to_nissen.htm
46. Kurleyna MV, Tanaino AS (1997) Open-pit and underground mines – energy analysis of open-pit mining. J Min Sci 33(5):453–462
47. Rohrbacher TJ, Teeters DD, Sullivan GL, Osmonson LM (1993) Coal resource recoverability – a methodology. U.S. Bureau of Mines Circular 9368. http://pubs.usgs.gov/usbmic/ic-9368/
48. Watson WD, Ruppert LF, Tewalt SJ, Bragg LJ (2001) The upper Pennsylvanian Pittsburgh coal bed: resources and mine models. Nat Resour Res 10:21–34
49. Blackmore G, Ehrenreich SB (1987) Reserve data base report of the National Coal Council: advisory report to the U.S. Department of Energy. National Coal Council, Arlington
50. National Petroleum Council (2007) Facing hard truths about energy. http://www.npchardtruthsreport.org/
51. U.S. National Academies (2007) Coal: research and development to support national energy policy. National Academies Press, Washington
52. Storchmann K (2005) The rise and fall of German hard coal subsidies. Energy Policy 33(11):1469–1492
53. Frondel M, Kambeck R, Schmidt CM (2007) Hard coal subsidies: a never-ending story? Energy Policy 35(7):3807–3814
54. Malyshev YN (2000) Strategy for the development of the Russian coal industry. J Min Sci 36(1):57–65
55. Petsch G (1982) Environmental problems of coal production in the federal republic of Germany with particular reference to the Ruhr. Environ Geochem Health 4:75–80
56. Tobin RJ (1984) Air quality and coal – the US experience. Energy Policy 12:342–352
57. Yeager KE, Baruch SB (1987) Environmental issues affecting coal technology: a perspective on US trends. Annu Rev Energy 12:471–502
58. O'Brien B (1997) The effects of title IV of the clean air act amendments of 1990 on electric utilities: an Update. EIA report DOE/EIA-058297 distribution category UC-950. ftp://ftp.eia.doe.gov/pub/electricity/ef_caau1.pdf
59. Ackerman F, Biewald B, White D, Woolf T, Moomaw W (1999) Grandfathering and coal plant emissions: the cost of cleaning up the clean air act. Energy Policy 27:929–940

60. Patiño-Echeverri D, Fischbeck P, Kriegler E (2009) Economic and environmental costs of regulatory uncertainty for coal-fired power plants. Environ Sci Technol 43:578–584
61. U.S. Geological Survey (2009) Mineral Commodity data – peat statistics and Information. http://minerals.usgs.gov/minerals/pubs/commodity/peat/
62. Geological Survey of Finland (2009) Peat resources of Finland. http://en.gtk.fi/Resources/peat_resources.html
63. World Coal Institute (2005) The coal resource – a comprehensive overview of coal. http://www.worldcoal.org/
64. Mitchell B (2003) International historical statistics 1750–2000. Palgrave MacMillan, London
65. Kecojevic V, Nor ZD (2008) Hazard identification for equipment-related fatal incidents in the U.S. underground coal mining. J Coal Sci Eng (China) 15(1):1–6
66. Grayson RL (2008) Improving mine safety technology and training in the U.S. recommendations of the mine safety technology and training commission. J Coal Sci Eng (China) 14(3):425–431
67. Szwilski AB (1988) Significance and measurement of coal mine productivity. Min Sci Technol 6(3):221–231
68. Kulshreshtha M, Parikh JK (2002) Study of efficiency and productivity growth in opencast and underground coal mining in India: a DEA analysis. Energy Econ 24(5):439–453
69. Tilton JE (2003) Assessing the threat of mineral depletion. Miner Energy 18:33–42
70. Rodríguez XA, Arias C (2008) The effects of resource depletion on coal mining productivity. Energy Econ 30:397–408
71. Topp V, Soames L, Parham D, Bloch H (2008) Productivity in the mining industry: measurement and interpretation. Productivity Commission, Melbourne
72. Adelman MA (1990) Mineral depletion, with special reference to petroleum. Rev Econ Stat 72:1–10
73. Hubbert MK (1956) Nuclear energy and the fossil fuels. Shell Development Company, Houston
74. Hubbert MK (1959) Techniques of prediction with application to the petroleum industry. Shell Development Company, Dallas
75. van Rensburg WCJ (1975) 'Reserves' as a leading indicator to future mineral production. Resour Policy 1:343–356
76. Milici RC, Campbell EVM (1997) A predictive production rate life-cycle model for Southwestern Virginia Coalfields. Geological Survey Circular 1147. http://pubs.usgs.gov/circular/c1147/
77. Ion DC (1979) World energy supplies. Proc Geologists' Assoc 90:193–202
78. Mohr SH, Evans GM (2009) Forecasting coal production until 2100. Fuel 88:2059–2067
79. Moriatry P, Honnery D (2009) What energy levels can the Earth sustain? Energy Policy 37:2469–2474

## *Books and Reviews*

Cobb CJ (1993) Modern and ancient coal-forming environments. Geological Society of America, Boulder

George H, Meech J, Workman L (1986) Towards reducing the physical environmental impact of North American surface coal mines; a review of potential selective overburden handling techniques. Min Sci Technol 3(2):81–94

Lappalainen E (1996) Global peat resources. International Peat Society, UNESCO, and Geological Survey of Finland, cop, Jyskä, 359 pp

Schilstra AJ, Gerding MAW (2004) Peat resources. In: Cleveland CJ (ed) Encyclopedia of energy. Elsevier Academic, San Diego

Schissler AP (2004) Coal mining, design and methods of. In: Cleveland CJ (ed) Encyclopedia of energy. Elsevier Academic, San Diego

Seredin VV, Finkelman RB (2008) Metalliferous coals: a review of the main genetic and geochemical types. Int J Coal Geol 76(4):253–289, 1 Dec 2008

Suárez-Ruiz I, Crelling JC (2008) Applied coal petrology: the role of petrology in coal utilization. Academic, New York

Thomas L (2002) Coal geology. Wiley, New York

Yudovich YE, Ketris MP (2005) Arsenic in coal: a review. Int J Coal Geol 61(3–4):141–196, 9 Feb 2005

# Chapter 10
# Coal Preparation

**Gerald H. Luttrell and Rick Q. Honaker**

## Glossary

| | |
|---|---|
| Ash | A measure of coal purity that consists of the noncombustible residue that remains after coal is completely burned. |
| Cut size | The particle size that has an equal probability of reporting to either the oversize or undersize product of a sizing device such as a screen or hydraulic classifier. |
| Cut point | The density corresponding to a particle that has an equal probability of reporting to either the clean coal or reject product of a density-based separator. |
| Dense medium | An aqueous suspension of high-density micron-sized particles (usually magnetite) used to separate coal from rock based on differences in density. |
| Organic efficiency | An indicator of coal cleaning performance calculated by dividing actual clean coal yield by the theoretical maximum yield attainable at the same ash content according to washability analysis. |
| Washability | A laboratory procedure that uses dense liquids (usually organic) to partition particles into various density fractions |

This chapter was originally published as part of the Encyclopedia of Sustainability Science and Technology edited by Robert A. Meyers. DOI:10.1007/978-1-4419-0851-3

G.H. Luttrell (✉)
Department of Mining and Minerals Engineering, Holden Hall Virginia Polytechnic Institute & State University, Blacksburg, VA 24061, USA
e-mail: luttrell@vt.edu

R.Q. Honaker
Department of Mining Engineering, Mining & Mineral Resources Engineering
University of Kentucky, Lexington, KY 40506, USA
e-mail: rhonaker@engr.uky.edu

R. Malhotra (ed.), *Fossil Energy: Selected Entries from the Encyclopedia of Sustainability Science and Technology*, DOI 10.1007/978-1-4614-5722-0_10,

that represent the ideal separation potential of coal (also called float–sink analysis).

Yield — An indicator of separation performance (usually reported as a percentage) that is calculated by dividing the clean coal tonnage by the feed coal tonnage.

## Definition of the Subject

Coal preparation, which may also be called washing, cleaning or processing, is the methodology by which coal feedstocks are upgraded in order to reduce freight costs, improve utilization properties and minimize environmental impacts. The upgrading, which occurs after mining and before transport of the cleaned product to market, is achieved using a variety of low-cost solid–solid and solid–liquid separation processes. Examples of processing technologies used by this industry include screening, classification, dense medium separation, gravity concentration, froth flotation, centrifugation, filtration and thickening. Several of these processes also play an important role in environmental control for the preparation facility.

## Introduction

According to the annual census of coal preparation plants conducted by Coal Age [1], the USA operates 286 coal preparation plants in 12 states. This number is relatively small by comparison to the worldwide fleet which is estimated to be 2,283 plants [2]. The capacity of the plants can range from less than 200 t/h for small operations to 6,000 t/h or more for large industrial plants. The need for such facilities can be attributed to the fact that freshly mined coals contain a heterogeneous mixture of both organic (carbonaceous) and inorganic (mineral) matter. The inorganic matter includes noncombustible materials such as shale, slate and clay. These impurities reduce the coal heating value, leave behind an undesirable ash residue, and increase the cost of transporting coal to market. The presence of unwanted surface moisture also reduces the heating value and can lead to handling and freezing issues for consumers. Therefore, essentially all coal supply agreements with electrical power stations impose strict limitations on the specific energy (heat), ash and moisture contents of purchased coal. Moreover, as the first step in the power cycle, coal preparation plants improve the environmental acceptability of coal by removing impurities that may be transformed into harmful gaseous or particulate pollutants when burned. These pollutants typically include particulates (fly ash) and sulfur dioxide ($SO_2$) as well as toxic trace elements such as mercury. The presence of mineral impurities can also influence the suitability of coal for high-end uses such as the manufacture of metallurgical coke or generation of petrochemicals and synthetic fuels. As such, coal preparation is typically needed to achieve the high levels of coal purity demanded by these secondary markets.

This entry provides a brief overview of some of the technological systems used in coal preparation. In addition, this entry highlights the benefits resulting from coal preparation activities as well as the issues associated with the sustained operation of coal processing facilities. This review suggests that coal preparation will continue to have a significant impact on the cost, recovery and quality of coal supplies. However, future improvements in separation technology and practices are needed to provide further reductions in waste generation and downstream environmental impacts. In many cases, these improvements may also generate incremental revenue due to the recovery of usable coal from waste streams, which could provide a financial incentive for pursuing these activities.

## Assessment of Cleaning Characteristics

The capability of coal preparation to improve coal quality varies widely due to inherent differences in the liberation characteristics of mined coals. The degree of liberation is determined by the relative proportion of composite particles (i.e., particles of intermixed coal and rock that are locked together) present in a particular coal. The composite particles make it impossible to physically separate all of the carbonaceous matter from the host inorganic matter. Consequently, plant operators are often forced to sacrifice coal recovery by discarding some composite particles as waste in order to improve clean coal quality to a level that can meet customer specifications.

The theoretical trade-off between coal recovery and quality can be quantified in the laboratory using washability (float–sink) analysis. An example of experimental data collected from a float–sink analysis is shown in Fig. 10.1. The analysis is performed by sequentially passing a coal sample through containers containing liquids (usually organic) of increasingly higher densities [3]. Pure coal has a relatively low specific gravity ($SG < 1.3$) and is collected as a float product from the first container, whereas pure rock is much denser ($SG > 2.2$) and is collected as a sink product from the last container. Composite particles report as float products in the intermediate containers filled with liquids having densities between that of the first and last containers. After density partitioning, the products from this procedure are dried, weighed and analyzed for quality (ash, sulfur, mercury, etc.). Washability data are very useful for predicting and analyzing the performance of coal preparation plants since most cleaning processes separate coal and rock based on differences in density.

Coal washability has a tremendous impact on how effectively a preparation plant can upgrade a particular coal feedstock. However, the types of processes employed and practices used for operation and maintenance (O&M) can also greatly influence

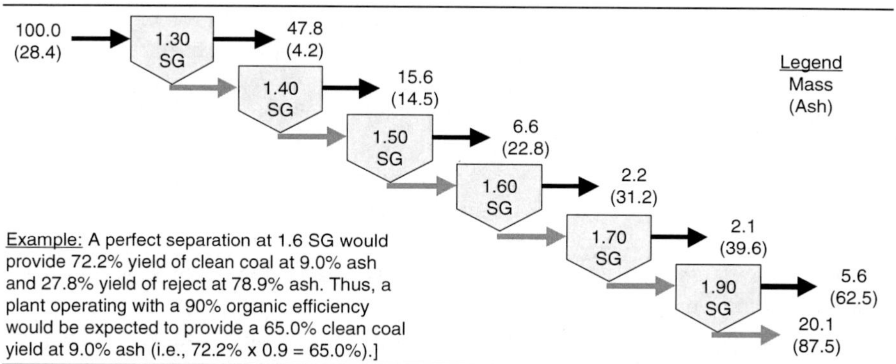


| Specific gravity | | | | Cumulative float | | C Sink | |
|---|---|---|---|---|---|---|---|
| Sink | Float | Mass (%) | Ash (%) | Mass (%) | Ash (%) | Mass (%) | Ash (%) |
| | 1.30 | 47.8 | 4.2 | 47.8 | 4.2 | 100.0 | 28.4 |
| 1.30 | 1.40 | 15.6 | 14.5 | 63.4 | 6.7 | 52.2 | 50.5 |
| 1.40 | 1.50 | 6.6 | 22.8 | 70.0 | 8.3 | 36.6 | 65.9 |
| 1.50 | 1.60 | 2.2 | 31.2 | 72.2 | 9.0 | 30.0 | 75.4 |
| 1.60 | 1.70 | 2.1 | 39.6 | 74.3 | 9.8 | 27.8 | 78.9 |
| 1.70 | 1.90 | 5.6 | 62.5 | 79.9 | 13.5 | 25.7 | 82.1 |
| 1.90 | | 20.1 | 87.5 | 100.0 | 28.4 | 20.1 | 87.5 |
| Totals | | 100.0 | 28.4 | | | | |

**Fig. 10.1** Float–sink (washability) analysis for a 28.4% ash run-of-mine coal

the performance of the preparation facility. This effectiveness is typically reported as an organic efficiency, which is defined as the yield of clean coal actually produced by the separation divided by the theoretical maximum yield of clean coal that could be achieved at the same ash content according to washability analysis. Organic efficiencies are often in the high 90 percentiles for well designed and well run operations, although values well below this desired level are not uncommon for problematic plants. Inefficient processes or practices, which misplace significant amounts of clean coal into waste and/or rock into the clean product, have a large negative impact on profitability. For example, a decrease in organic efficiency of just one percentage point can potentially represent a reduction in plant profitability of 20% or more.

## Coal Processing Operations

During the early part of the last century, coal mines made use of labor intensive sorting methods such as manual hand-picking to remove unwanted rock from mined coal. This inefficient approach was soon replaced by simple mechanical sorting

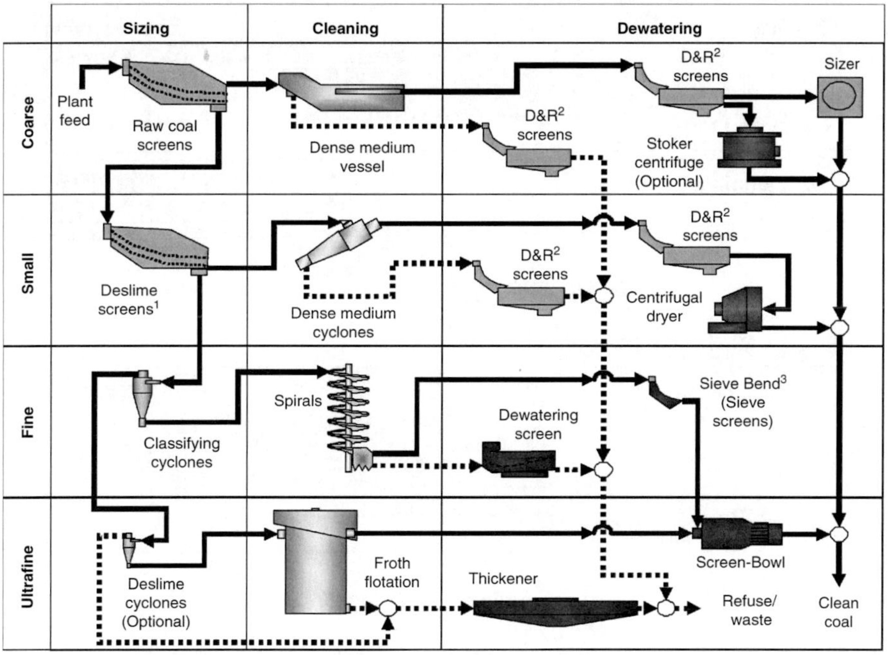


*Notes: 1 -Raw coal and deslimescreen decks may be on the same machine, 2 –D&R screens primarily used to recover dense medium, 3 -Sieve bends may also be combined with clean coal classifying cyclones.*

**Fig. 10.2** Simplified flowsheet for a modern coal preparation plant showing unit operations for sizing, cleaning and dewatering of four different size fractions required for efficiency coal upgrading

processes that reduced misplacement and provided higher levels of productivity. These early plants typically cleaned only the coarser particles (usually larger than >5–10 mm) and either recombined the untreated fine particles back into the washed product or discarded the fines as waste. These inefficient systems dumped large tonnages of coal into waste piles, some of which are being re-mined and reprocessed today. These historic periods were followed by many decades of technology development that ultimately led to the design and operation of relatively efficient plants that are capable of complete or partial upgrading of the entire size range of mined coals. Many modern coal preparation plants operating today are as complex as industrial facilities once employed only by the chemical processing industry.

Figure 10.2 shows a simplified process flowsheet for a modern plant that includes operations for particle sizing, cleaning and dewatering. This sequence of operations, which is called a circuit, may be repeated for different size fractions since the processes used in coal preparation have a limited range of applicability in terms of particle size (Fig. 10.3). Modern plants may include as many as four separate processing circuits for treating the coarse (plus 10 mm), small (10 × 1 mm),

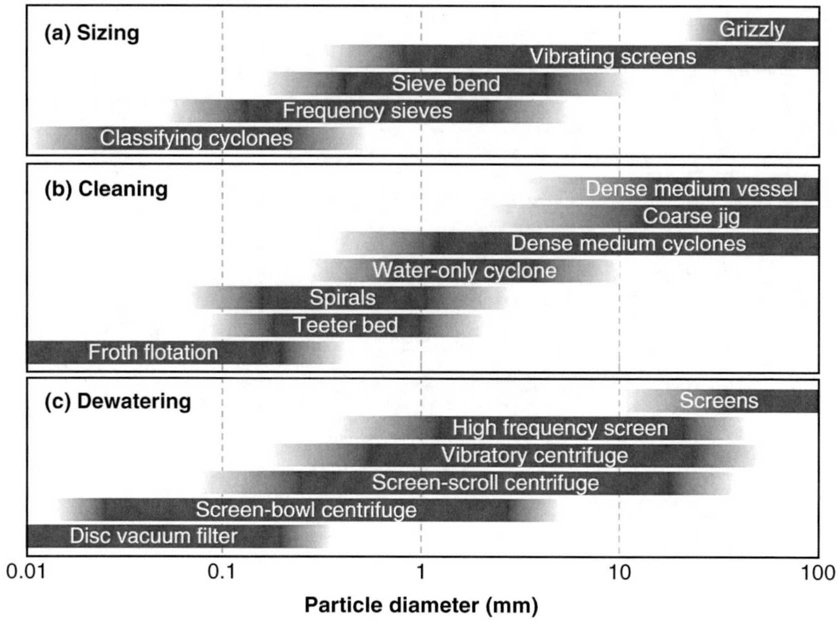


**Fig. 10.3** Effective range of particle sizes treated by various coal preparation processes

fine (1 × 0.15 mm), and ultrafine (minus 0.15 mm) feed material. Although many commonalities exist, the final selection of what number of circuits to use, which types of unit operations to employ, and how they should be configured, is highly subjective and dependent on the characteristic properties of the feed coal in terms of size and composition.

## *Unit Operations for Particle Sizing*

Coal produced by mechanized mining operations contains particles as small as fine powder and as large as several hundred millimeters. Particles too large to pass into the plant are crushed to an appropriate upper size, or rejected where insufficient recoverable coal is present in the coarse size fractions. The crushed material is then segregated into groups having well defined maximum and minimum sizes. Screens are typically employed for sizing coarser particles, while combinations of stationary sieves and classifying cyclones are used for sizing finer particles. Figure 10.3a shows the typical sizes of particles that can be effectively handled by common types of industrial sizing equipment.

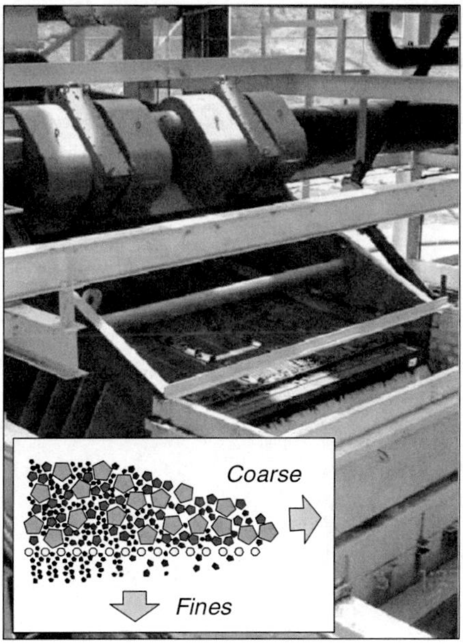


**Fig. 10.4** Vibrating screen used for feed coal sizing

## Screens and Sieves

Screens are mechanical sizing devices that use a mesh or perforated plate to sort particles into fine (particles that pass through the screen openings) and coarse (particles that are retained on the screen surface). Screens are used for a wide range of purposes including the extraction of large rock and trash prior to treatment, the production of optimum size fractions for separation devices and the rinsing/dewatering of products from cleaning operations. The types of screens commonly used in coal preparation include grizzly, vibrating and high-frequency screens. A grizzly is a series of evenly spaced parallel bars that are positioned in the direction of material flow. This simple device is used only for gross sizing of very coarse particles (>50 mm) and is generally applied only for the protection of equipment from large oversize material. Vibrating screens are the most common equipment for sizing particles and are used in almost every aspect of coal preparation. Vibrating screens (Fig. 10.4) utilize an out-of-balance rotating mechanism to create a vibrating motion to sort particles and to move material along the screen surface. High-frequency screens (Fig. 10.5) are normally used for dewatering fine coal or rock by retaining slurry between two opposing inclines. As the name implies, this type of screen operates at a high vibrating frequency, which is beneficial for dewatering fine solids.

The sizing of fine coal is particularly difficult due to the increased likelihood of plugging and the large number of particles that must pass through the screen surface. As such, fine sizing is performed in many cases using sieve bends. A sieve bend (Fig. 10.6) consists of a curved panel constructed from inverted

**Fig. 10.5** High-frequency screen used for dewatering of fine rock

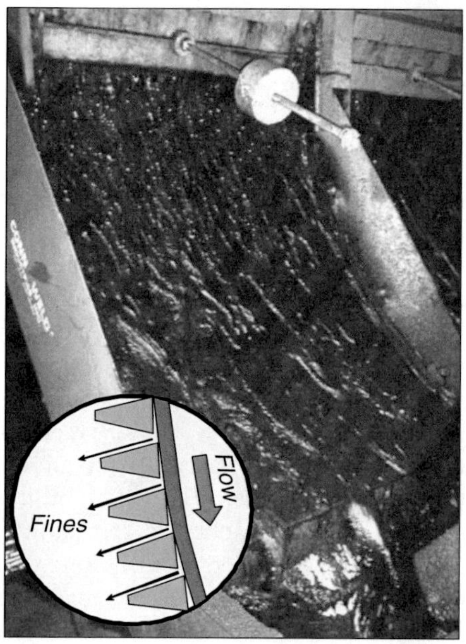


**Fig. 10.6** Sieve bend used to remove fine "slimes" from coal

trapezoidal bars (profile or wedge wire) placed at right angles to fluid flow. The capacity of the sieve is relatively high since the leading edge of the trapezoidal bars "slice" material from the flowing stream. However, due to the unique geometry, the

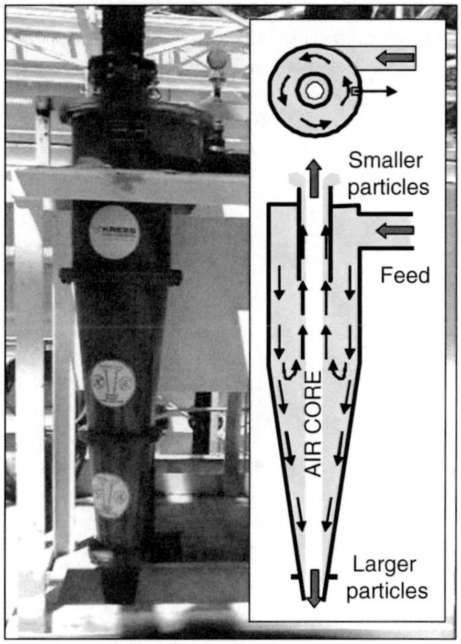


**Fig. 10.7** Classifying cyclone used to hydraulically size fine coal

particle size retained by the sieve is smaller than the spacing between bars. Sieve bends are commonly used for fine particle dewatering and may also be used ahead of vibrating screens to enhance sizing performance.

## Classifying Cyclones

Classification is used in coal preparation for fine sizing where conventional screening and sieving becomes impractical or too expensive. Classification exploits differences in the settling rates of particles of different size (i.e., smaller particles settle slower than larger particles). Classification is usually applied to dilute suspensions of <1 mm diameter particles. This technique is generally less efficient than screening or sieving since the separation can be influenced by particle shape and density as well as particle size. The most common type of classification device used in the coal industry is the classifying cyclone (Fig. 10.7). In this device, dilute slurry is pumped under pressure to multiple parallel units (banks) of cyclones. The rotating flow field within the cyclone creates a centrifugal force that increases the settling rate of the particles. Larger particles are forced to the wall of the cyclone where they are carried to the discharge port (apex) at the bottom of the cyclone, while finer particles exit out the top of the cyclone (vortex finder). Classifying cyclones are commonly applied to obtain cut sizes of 0.10–0.15 mm and represent the only practical option for sizing ultrafine particles (cut size of 50 μm or less).

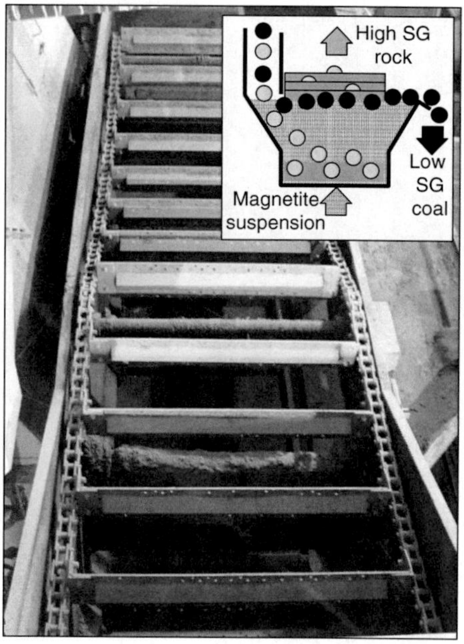


**Fig. 10.8** Chain-and-flight dense medium vessel used to separate coarse coal and rock

Smaller cyclones are typically used to size finer particles since these units provide a higher centrifugal force. Classifying cyclones are very popular in the coal industry since they offer a high unit capacity, operational flexibility and low maintenance.

## *Unit Operations for Coal Cleaning*

The primary purpose of coal preparation is to separate valuable carbonaceous material from contaminants in the surrounding host rock. The separation is typically accomplished using low-cost processes that exploit differences in physical properties that vary according to mineralogical composition of individual particles. Some of the common properties that are used to separate coal and rock include size, density and wettability. As shown in Fig. 10.3b, the effectiveness of different types of separators is limited to a relatively narrow range of particles sizes due to limitations associated with equipment efficiency and geometry.

### Dense Medium Separators

The most common process for upgrading coarse coal (>12.5 mm) is the dense medium vessel. While a wide variety of different designs exist in industry, one of the most popular is the chain-and-flight vessel (Fig. 10.8). This density-based separator consists of a large open tank through which an aqueous suspension of

**Fig. 10.9** Dense medium cyclone used to separate small coal and rock

finely pulverized magnetite (SG = 5.2) is circulated. The magnetite creates a suspension with an artificial density that is between that of coal and rock. Consequently, lighter coal particles introduced into the medium float to the surface of the vessel where they are transported by the overflowing medium to a collection screen. Waste rock, which is much denser, sinks to the bottom of the vessel where it is collected by a series of mechanical scrapers called flights. The flights travel across the bottom of the vessel and eventually drag the waste rock over a discharge lip at one end of the separator. The unit is highly flexible since the quality of the clean coal product can be readily adjusted by varying the density of the medium from 1.3 to 1.7 SG by controlling the amount of magnetite and water introduced into the suspension.

Particles smaller than about 0.5 mm float or sink too slowly in a static vessel to be effectively separated. Therefore, dense medium cyclones (DMCs) are typically used to clean coal and rock in the size range between 0.5 and 12.5 mm. These high-capacity devices (Fig. 10.9) make use of the same basic principle as dense medium vessels, i.e., an artificial magnetite-water medium is used to separate low-density coal from high-density rock. In this case, however, the rate of separation is greatly increased by the gravitational effect created by passing medium and coal through one or more cyclones. Dense particles of rock sink in the rotating medium and are pushed toward the wall of the cyclone where they are eventually rejected out the bottom. Lighter particles of coal rise to the surface of the rotating medium and are carried with the bulk of the medium flow out the top of the cyclone. Recent advances in cyclone technology have allowed the upper particle size limit for DMCs to increase in many current applications to 60–75 mm. This development

**Fig. 10.10** Bank of twin water-only cyclones used to separate fine coal and rock

has made it possible to simplify plant circuitry in some facilities by eliminating the need for a dense medium vessel.

The primary disadvantage of dense medium processes is that they require considerable ancillary equipment to support the operation. A typical circuit requires that the clean coal and rock products pass over drain-and-rinse (D&R) screens to wash the medium from the surfaces of the products and to dewater the particles. The medium recovered from the drain section is circulated back to the separator by a pump, while medium from the rinse section is diluted by spray water and must be concentrated before being fed back. The concentration is achieved using magnetic separations that effectively recover magnetite from nonmagnetic particles that would otherwise contaminate the circulating medium.

## Water-Based Separators

A wide variety of water-based density separators, which avoid the need to utilize a artificial dense medium, are also available for separating coal and rock in the particle size range between 0.15 and 1 mm. The most common methods include water-only cyclones (WOCs) and spirals. A WOC is similar to a classifying cyclone, but typically has a stubby wide-angled conical bottom (Fig. 10.10). Separations occur in a WOC due to differences in the settling rates of coal and rock in the centrifugal field within the cyclone. The separation is also enhanced by the formation of autogenous medium created by the natural fines already in the feed

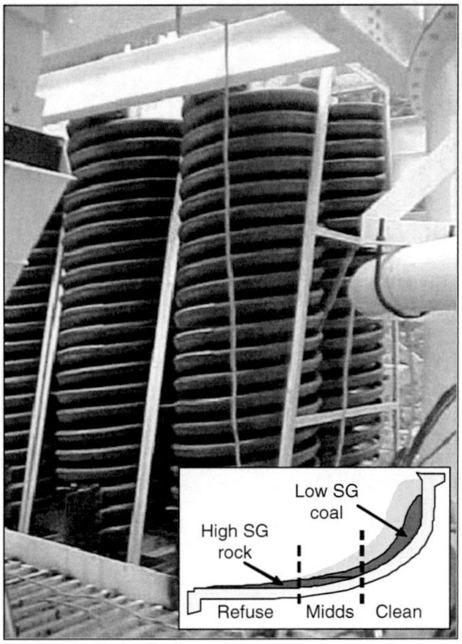


**Fig. 10.11** Bank of spirals used to separate fine coal and rock

slurry. These units are often employed in two stages or in combination with other water-based separators to improve performance.

Another water-based separator for treating fine coal that has grown tremendously in popularity is the spiral concentrator (Fig. 10.11). A spiral consists of a corkscrew shaped conduit with a modified semicircular cross-section. During operation, feed slurry is introduced to the top of the spiral and is permitted to flow by gravity along the helical path to the bottom of the unit. Particles in the flowing film are stratified such that lighter coal particles are forced to the outer wall of the spiral, whereas heavier particles are forced inward to the center of the spiral. The segregated bands of heavy to light materials are collected at the bottom of the spiral. Adjustable diverters (called splitters) are used to control the proportions of particles that report to the various products. A three-product split is usually produced, giving rise to three primary products containing clean coal product, refuse and misplaced "middlings." Because of the low unit capacity (2–4 t/h), spirals are usually arranged in groups fed by an overhead radial distributor. To save space, several spirals (two or three) may be intertwined along a single central axis. Spirals have been successfully utilized in combination with WOCs to improve the efficiency of separating fine coal.

## Froth Flotation Separators

The smallest size fraction of coal (<0.15 mm) cannot be effectively separated using density-based separators. For these particles, a process known as froth flotation must be used. Flotation is a physicochemical process that separates particles based

**Fig. 10.12** Conventional flotation bank used to clean ultrafine coal

on differences in surface wettability. Flotation takes place by passing finely dispersed air bubbles through an aqueous suspension of particles. A chemical reagent, called a frother, is added to promote the formation of small bubbles. Typical addition rates are in the order of 0.1–0.5 lb of reagent per ton of coal feed. Coal particles, which are naturally hydrophobic, selectively attach to air bubbles and are buoyed to the surface of the pulp where they are collected from a coal-laden froth bed that forms atop the cell. Most of the impurities that associate with coal are hydrophilic and remain suspended until they are discharged as a dilute slurry waste. Another chemical additive, called a collector, may be added to improve the adhesion between air bubbles and coal particles. Collectors are commonly hydrocarbon liquids such as diesel fuel or fuel oil. In some plants, clay slimes (<0.03 mm) are removed ahead of flotation using small diameter classifying cyclones to minimize the carryover of the high ash slimes into the froth product with the process water.

Most industrial installations make use of mechanical flotation machines (Fig. 10.12). These machines consist of a series of agitated tanks (4–6 cells) through which fine coal slurry is passed. The agitators are used to ensure that larger particles are kept in suspension and to disperse air that enters down through the rotating shaft assembly. The air is either injected into the cell using a blower or drawn into the cell by the negative pressure created by the rotating impeller. Most commercial units are very similar, although some variations exist in terms of cell geometry and impeller design. A disadvantage of conventional flotation cells is that they allow a small amount of clay slimes to be recovered with the water that reports to the froth product. Therefore, in cases where high clay concentrations are expected, advanced

**Fig. 10.13** Column-type flotation cell used to clean ultrafine coal

flotation processes such as column cells (Fig. 10.13) are often used. Column cells virtually eliminate the recovery of clay slimes by washing the froth with a countercurrent flow of water. This feature allows columns to produce a higher purity coal product without sacrificing coal recovery.

## *Unit Operations for Coal Dewatering*

Solid–liquid separators are used downstream of coal cleaning processes to remove unwanted surface moisture from the sorted products. Excess moisture lowers the heating value of the coal, may lead to handling/freezing problems, and increases the overall costs of transporting the coal to customer sites. As shown in Fig. 10.3c, several different types of mechanical dewatering methods are required depending on the size of particles to be treated. The removal of water from the surfaces of coarser (>5 mm) coal is predominantly carried out using screens, where the prime force involved is gravity. Finer particles have a higher surface area and tend to have correspondingly higher moisture content. There finer particles are typically dewatered using centrifugal methods or filtration systems.

### Centrifugal Dewatering

Centrifugal dryers, which consist of a rotating screen or basket, use centrifugal force to pull water away from the surfaces of coal particles. These devices operate

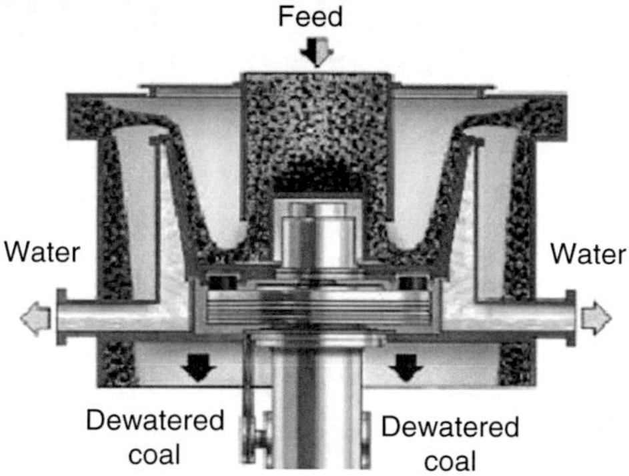


**Fig. 10.14** Vibratory centrifugal dryer used to dewater small/fine coal

in much the same fashion as the spin cycle in a washing machine. For coarser fractions, two popular centrifugal dryers are the vibratory centrifuge and the screen-scroll centrifuge. The vibratory centrifuge (Fig. 10.14) uses a vibrating mechanism to move solids through the unit, while the screen-scroll design uses a screw conveyor to move the solids. For finer particles (<1 mm), both the screen-scroll centrifuge and another popular design, called a screen-bowl centrifuge, may be used. The screen-bowl centrifuge is a horizontal unit that consists of both a solid bowl section and a screen section (Fig. 10.15). Feed slurry is introduced into the bowl section where most of the solids settle out under the influence of the centrifugal field. A rotating scroll transports the settled solids up a beach and across the screen section where additional dewatering occurs before the solids are discharged. Solids that pass through the screen section are typically added back to the machine with the fresh feed. This unit is capable of providing low product moistures, although some ultrafine solids are lost with the main effluent that is discarded from the solid bowl section.

## Filtration Dewatering

Filtration processes may be used to dewater fine coal whenever a high coal recovery is desirable. Filtration involves the entrapment of fine solids as a cake against

**Fig. 10.15** Screen-bowl centrifuge used to dewater fine coal

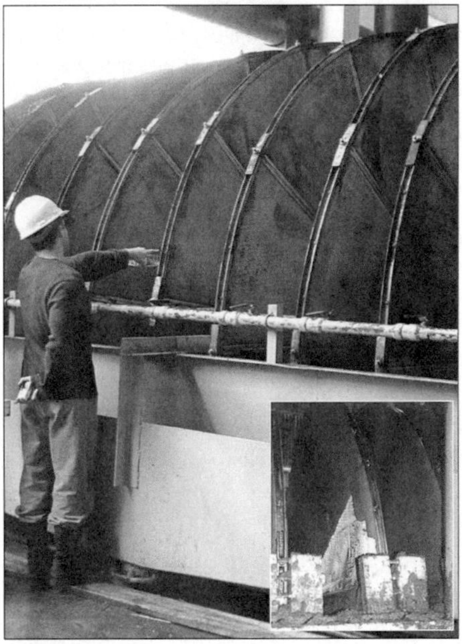

**Fig. 10.16** Disc vacuum filter used to dewater fine coal

a porous filtering media. Traditionally, flotation concentrates have been dewatered using some form of vacuum filtration. These units are capable of high coal recoveries (>97%) while generating product moisture in the 20–30% range. The most popular type of vacuum filter is the disc filter (Fig. 10.16). This device consists

**Fig. 10.17** Thermal dryer used to dry coal to low moisture contents

of a series of circular filtration discs that are constructed from independent wedge shaped segments. At the beginning of a cycle, a segment rotates down into the filter tub filled with fine coal slurry. Vacuum is applied to the segment so that water is drawn through the filter media and a particle cake is formed. The resultant cake continues to dry as the segment rotates out of the slurry and over the top of the machine. The drying cycle ends when the vacuum is removed and the cake is discharged using scrapers and a reverse flow of compressed air.

## Thermal Dewatering

Thermal dryers have been successfully employed in the coal industry to reduce clean coal moisture to single-digit values whenever mechanical dewatering processes are incapable of meeting contract specifications. The most popular design is the fluidized bed dryer (Fig. 10.17), which uses coal, oil or gas as the fuel source to heat the incoming air stream. The amount of fuel required depends on the amount of water that must be evaporated which, in turn, depends on the amount of coal fed to the dryer and the percentage of water in the dewatered product [4]. When operating correctly, thermal dryers can reduce the clean coal moisture to less than 6% by weight at a unit cost proportional to the tonnage of water evaporated [5]. Unfortunately, thermal dryers require high capital costs, often in excess of several million or more. Dryers can also suffer from emission problems associated fugitive dust and poor opacity. Emissions of nitrous oxides, sulfur dioxide, volatile organic

**Fig. 10.18** Conventional thickener used to clarify process water and thicken solids

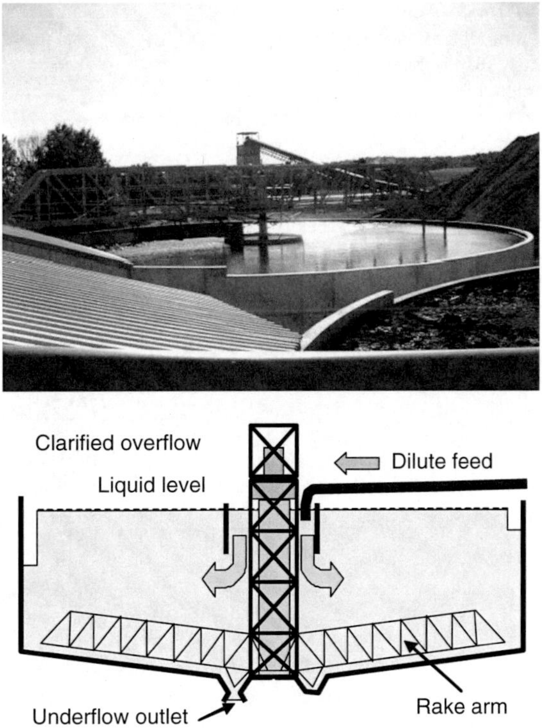


compounds (VOCs) may also present issues for some sites seeking operating permits. Moreover, thermal drying of combustible particles of coal can present safety hazards resulting from accidental fires or dust/gas explosions.

## Thickening and Clarification

Another essential solid–liquid separation process used in coal preparation is thickening. Thickening (or clarification) processes are used to treat the process water so that it can be recycled and reused within the plant. A thickener consists of a large tank in which particles are permitted to settle, thereby producing a clarified overflow and thickened underflow (Fig. 10.18). Thickeners are usually fed dilute slurry (<5–10% solids) that is rejected from the fine coal cleaning circuits. The clarified water that overflows along the circumference of the thickener is recovered and reused as process water. The thickened underflow, which often contains 20–35% solids, is transported to the center of the thickener by a series of rotating rakes. The thickened sludge is then pumped to an appropriate disposal area or is further dewatered prior to disposal. Chemical additives (coagulants and flocculants) are usually introduced ahead of the thickener to promote the aggregation of ultrafine particles so as to enhance water clarity and increase settling

**Fig. 10.19** Active slurry impoundment used for fine waste disposal (before reclamation)

rates. The pH levels must also be carefully monitored and controlled. Conventional thickeners, which are typically 50–200 ft diameter, require constant monitoring to ensure that overflow clarity and underflow density is maintained and the rake mechanism is not overloaded. A torque sensor on the central drive is typically used to monitor rake loading, while a nuclear density gauge is often used to monitor underflow density.

## *Waste Disposal and Storage*

The separation processes used in coal preparation often generate large volumes of waste rock and slurry. As a result, the final step in coal preparation involves the disposal and permanent storage of these waste products into various types of surface and/or underground repositories. Although many variations exist, a common approach is to use separate refuse piles and slurry impoundments. Refuse piles are designed to receive coarse particles of waste rock that can be easily dewatered by screens or sieves. This material is relatively easy to handle and can be transported by truck or belt haulage systems to the disposal area. On the other hand, fine coal wastes are difficult to dewater and are typically discarded in slurry form. The waste slurry contains water, coal fines, silt, clay and other fine mineral particles from the processing plant. In most cases, the slurry is discarded into an impoundment (Fig. 10.19). An impoundment is an engineered structure consisting of a large volume earthen settling basin formed behind a man-made dam or embankment. The dam or embankment is usually constructed from compacted coarse refuse material. The waste slurry is transferred to the impoundment by pumping thickened underflow from the plant thickener through a pipeline. The volume of the impoundment must be sufficiently large to ensure that fine particles settle by gravity before the clarified water at the surface is recycled back to the plant for reuse.

## Beneficial Impacts of Coal Preparation

Coal preparation provides several attractive benefits from both an economic and environmental perspective. The most recognized of these include increased coal reserves, lower transportation costs, improved utilization properties and abatement of pollution.

### *Coal Reserves*

Perhaps no other technology has had a greater impact on expanding the reserve base of economically recoverable coal than coal preparation. Preparation plants employ low-cost physical separation processes to convert run-of-mine coal resources into saleable coal reserves. The percentage of clean coal tonnage generated from each ton of run-of-mine coal is commonly referred to as plant yield. This parameter is one of the most difficult to estimate when evaluating the reserve base for a previously undefined coal property. The yield can be influenced by several factors including the quality of the in-place coal, the washability (separating characteristics) of the run-of-mine coal, the efficiency of the separation processes used by the preparation facility and the strictness of the quality demands imposed by the coal consumer. The quality of the in-place can be affected by small changes to mining practices that directly influence out-of-seam dilution. Variations in washability, which reflect the selective liberation of composite particles of intermixed coal and rock, can also make estimations of coal yield unreliable when coal is subjected to size reduction.

The average yield produced by coal preparation plants has steadily declined over the years due to a depletion of lower ash feeds and increased mining mechanization. For example, Fig. 10.20 shows the yield of clean coal currently obtained from a random survey of several major plants operating in the eastern USA. It is not uncommon for eastern operations to experience yields under 30–35%, thereby producing only 1 t of clean coal from three or more tons of mined product. An estimate complied from production records supplied by coal producers suggests that the average yield is now less than 50% (i.e., 49.8% $\pm$ 3.5%) for the total USA. This situation is expected to worsen as eastern reserves become thinner and more

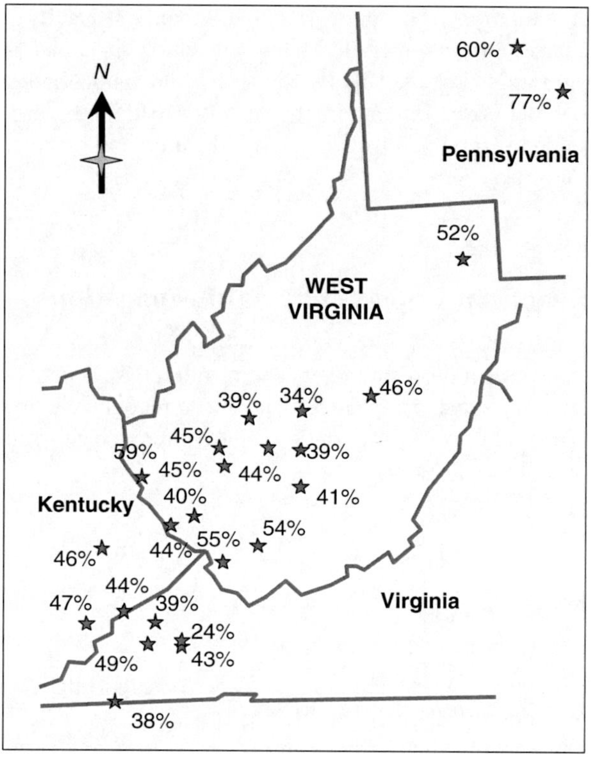


**Fig. 10.20** Examples of clean coal yield values for a selected sampling of eastern US plants

challenging to mine. A study reported by Weisenfluh et al. [6] indicated that nearly 52% of the remaining eastern Kentucky coalfield resources are located in coal seams that are 0.35–0.70 m in thickness, while 31% are in 0.70–1.1 m thick seams. Likewise, a study of the Virginia coalfields found that 30% of the total reserve base (more than 200 million tons) exists in seams with a thickness less than 0.70 m [7]. Consequently, ever increasing amounts of rock from out-of-seam dilution are being mined, loaded and hauled to preparation plants for removal and disposal.

## *Coal Transportation*

Most of the coal consumed in the USA is used for the production of electricity. The cost of transporting the coal to the power station is usually borne by the utility and paid based on the delivered tonnage. The mine operator is also paid by the utility based on tonnage, although the unit price is typically adjusted up or down to account for the actual heat content of the supplied coal fuel. In most cases, this simple pricing structure provides the base economic justification for the operation of coal preparation facilities. The high-ash rock rejected by coal preparation plants contains insufficient heating value to justify the shipment of this material to the utility. The savings in transportation costs are directly proportional to the increase in heating value.

Cost–benefit studies suggest that the economics of coal production are more sensitive to transportation cost than to any other factor. For example, a study by Norton [8] was one of the first to demonstrate how unfavorable changes in the cost of mining, processing and transportation affect profitability. The study concluded that the cost of coal transportation had the greatest overall impact on total revenue. This study also indicated that the clean coal yield was second only to transportation cost in determining revenues. In contrast, increases in the capital and operating costs of the coal preparation plant had only a minor impact on revenue. Therefore, any steps taken to reduce preparation costs (fewer capital improvements, less maintenance, workforce reductions, etc.) usually need to be closely examined to ensure that clean coal yield is not adversely impacted by these cost-cutting measures.

### *Utility Performance*

The thermal efficiency of a power station is very important. A higher efficiency is obviously beneficial from an economic perspective since this provides a proportional improvement in generated revenue for the power station. Moreover, a higher efficiency also reduces the production of greenhouse gases and other pollutants of environmental concern since less coal fuel needs to be burned per unit of electrical power generated.

One method for improving thermal efficiency is to use washed coals of higher quality can significantly improve the thermal efficiency of a boiler [9–12]. Higher quality coals are more reactive and require less excess air for effective combustion, thereby improving efficiency via a reduction in heat lost with the flue gas. Higher quality coals also improve efficiency by avoiding fouling/slagging problems in the boiler, which tend to raise the flue gas temperature and increase heat losses. The extent to which the proper application of coal preparation technology improves thermal efficiency is highly case specific and difficult to predict from purely theoretical considerations. The most reliable data for quantifying efficiency improvements are typically reported based on actual plant studies. One such classic study [13] monitored improvements to boiler performance resulting from switching from an untreated coal (15% ash and 3.5% sulfur) to washed coal (9% ash and 2.8% sulfur) from the same mine. Despite the modest improvement in coal quality, the boiler efficiency increased from by about 1.5 absolute percentage points as a result burning better quality coal. The capacity factor also rose by almost 10% due to less fouling/slagging problems.

The use of coal preparation technologies to improve boiler efficiency has also been a major benefit reported in other nations. The International Energy Agency (IEA) estimates that coal-fired power plants in India can increase thermal efficiency by up to 10 percentage points by switching from unwashed to washed coal [14]. China also expects to make greater use of coal preparation technology to improve thermal efficiencies and environmental performance [15]. The average thermal efficiency in China has been reported to be less than 29% [16], compared to around 38% in OECD countries. As such, coal preparation is expected to continue to have a large impact on the international community.

Another benefit of coal preparation is that it removes impurities that have a significant influence on the O&M costs for a coal-fired boiler. Studies have demonstrated that the removal of abrasive mineral impurities such as pyrite and quartz can substantially reduce wear rates and increase the throughput capacity of utility pulverizers [17, 18]. The impacts associated with the abrasive wear and slagging/fouling of boiler tubes can also be mitigated to a large extent by utilizing washed coals that have been properly cleaned to remove unwanted mineral matter [19–21]. Vaninetti and Busch [22] have provided a detailed description of these problems and have developed empirical formula that can be used to assist in the evaluation of changes to coal quality in specific types of boilers.

The handling characteristics of solid coal are also an important issue. A poor handling coal may hang in railcars, plug chutes and bins and stick to conveyor belts. These problems may result in unscheduled shutdowns, thereby reducing power station availability. Washed coals from preparation plants typically have superior handling characteristics to run-of-mine coals, especially if all or a portion of the ultrafines have been removed. These coals also typically present fewer problems in terms of unwanted dust generation, solids run-off during precipitation events and freezing problems during colder months [23].

## *Pollution Abatement*

Coal preparation plays an important role in reducing the emissions of pollutants that associate with the mineral matter contained in coal. These emissions normally include solid particulate emissions such as fly ash as well as gaseous emissions of precursors associated with acid-rain and air toxics. These emissions are strictly regulated for coal-fired utilities through various legislative acts such as the 1990 Clean Air Act Amendment (CAAA) and 2005 Clean Air Interstate Rule (CAIR). Moreover, coal preparation has a beneficial impact on reducing greenhouse gas emissions by increasing the thermal efficiencies.

Noncombustible impurities present in the feedstocks supplied to coal power stations generate waste steams as either bottom ash/slag or fly ash. Of these, the finest particles of fly ash emitted to the atmosphere are considered to be of greatest environmental concern due to their potential adverse impact on human respiratory health [24]. Power stations make use of several types of effective control technologies to minimize fine particulate emissions. These postcombustion technologies include efficient processes such as electrostatic precipitators (ESPs), fabric filters (FFs), cyclones and wet scrubbers. Modern control systems typically achieve better than 99.5% removal of all particulates and exceed 99.99% in some cases. However, standards for particulate emissions continue to become increasingly stringent as reflected by expanded regulations by the US EPA to include particles finer than 2.5 μm in new ambient air quality standards [25]. As such, there is continued interest in removing greater amounts of particulates upstream of other emission controls using coal preparation technologies.

The separation processes used in coal preparation plants remove noncombustible minerals that ultimately affect the amount and type of particulate matter (PM) that passes downstream to emission control systems. For these systems, proper levels of coal washing can be identified that effectively reduce ash loading and improve removal efficiencies. A recent presentation by American Electric Power to the Asia Pacific Partnership [26] concluded that "*Consistent and proper quality coal is best tool to improve plant operating performance and reduce PM and* $SO_2$ *emissions. Removal of some of the coal ash (includes rocks) at the mine is more economic than in the pulverizer, boiler, precipitator and scrubber.*" Washing also minimizes the total amount of high-surface-area fly ash that is more hazardous to dispose due to its high reactivity. Even in cases where particulate controls are currently deemed adequate, greater use of coal preparation may be required in the future to compensate for deterioration in feedstock quality as higher quality coal reserves become depleted.

Coal preparation has also played an important role in environmental compliance by reducing the sulfur contained in coal feedstocks prior to combustion [27]. Sulfur occurs in coal as three distinct forms (e.g., sulfate, organic and pyritic). Sulfate sulfur is present in very small quantities and is not considered a serious problem. Organic sulfur, which is part of the basic structure of coal, is not removable by conventional cleaning. The precombustion removal of organic sulfur is possible using chemical cleaning methods; however, these elaborate processes are not economically competitive with flue gas scrubbing. Pyritic sulfur is present as discrete inclusions of iron sulfides distributed within the coal matrix. As such, this form of sulfur has the potential to be removed by physical cleaning processes such as coal preparation [28]. Fortunately, many of the high sulfur coals in the USA also contain a high proportion of pyritic sulfur (Fig. 10.21).

Coal preparation plants have been reported to remove up to 90% by weight of pyritic sulfur, although rejections are typically in the 30–70% range due to liberation constraints [28–30]. When compared to the postcombustion control of sulfur, coal preparation offers several distinct advantages including improved market flexibility, lower scrubber loading and concurrent removal of other impurities (ash, trace elements, moisture, etc.). Although coal preparation does not directly impact the nitrogen content of coal, washing has been shown to help reduce $NO_x$ emissions by providing a consistent high quality fuel that makes it easier to control the combustion environment [31].

Coal preparation also has a tremendous impact on the amount of air toxics released to the atmosphere. The 1990 CAAA contained provisions that established new emission standards for a variety of air toxics, which are commonly referred to

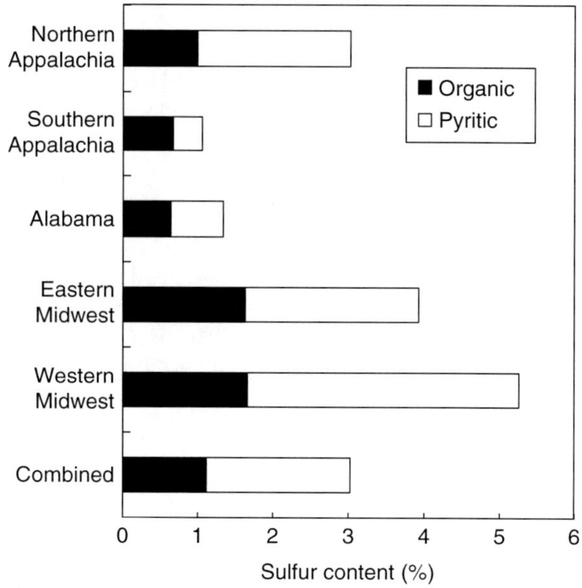


**Fig. 10.21** Distribution of sulfur types typically in US coals (After Cavallaro et al. [29])

as hazardous air pollutants (HAPs). Many of the HAPs identified in the CAAA are present as trace elements in coal [32, 33]. Some of the most noteworthy of these are antimony, arsenic, beryllium, cadmium, chlorine, chromium, cobalt, fluorine, lead, manganese, mercury, nickel, selenium and radionuclides. The concentrations of these elements are known to vary considerably from seam to seam, and in some cases within the same seam [34, 35]. During combustion at electrical utilities, these elements may be released to the atmosphere as solid compounds with the fly-ash and in the vapor phase with the flue-gas. Existing post-combustion control technologies, such as electrostatic precipitators, can be reasonably effective in reducing the concentration of trace elements associated with fly-ash. These commonly include elements such as antimony, beryllium, cadmium, cobalt, lead and manganese. Capture efficiencies of >97% have been reported for electrostatic separators [36]. On the other hand, trace elements such as arsenic, chlorine, mercury and selenium have the potential to volatilize and are less effectively controlled by postcombustion methods.

Studies have shown that many of the HAP precursors identified in the CAAA are associated with the mineral matter commonly rejected by coal preparation plants. This approach to HAPs control is attractive since the waste rock rejected by coal preparation plants is coarser and has a lower reactivity than the high-surface-area ash generated by power stations [37]. In-plant sampling campaigns conducted by various researchers [36, 38] suggest a good correlation between the rejection of mineral matter and the removal of trace elements during physical cleaning. These findings are also supported by laboratory float–sink tests performed using a variety of eastern US coals [39–41]. These data suggest that trace elements are typically rejected at levels of 40–70% by weight using conventional preparation

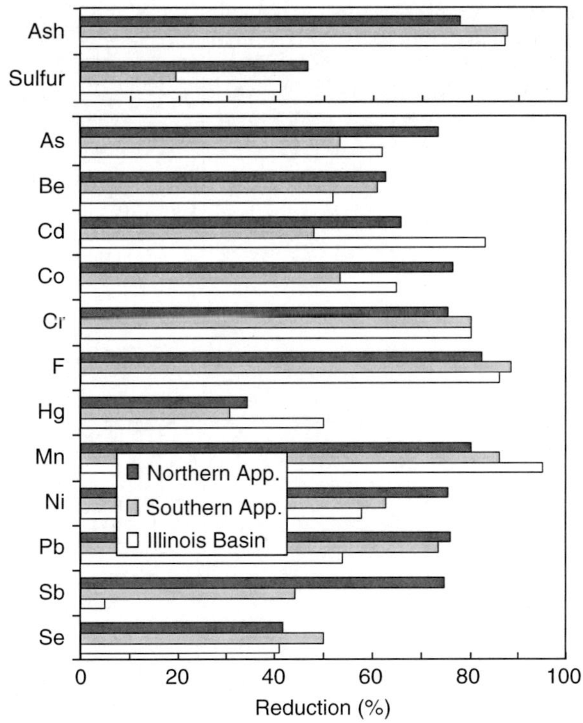


**Fig. 10.22** Reduction in trace element content after coal preparation (After Fonseca et al. [36])

technologies. These values appear to be in good agreement with earlier values reported by Fonseca et al. [36], which showed an average trace element removal by conventional coal preparation of approximately 64% for six different coals (Fig. 10.22). On the other hand, the large degree of variability observed in the data from these and other studies suggest that the rejections of trace elements by coal preparation are very site specific and need to be quantified on a case by case basis.

Of the air toxics reduced by coal preparation, mercury is perhaps the trace element in coal of greatest environmental concern [35]. Mercury can be released during coal combustion and subsequently deposited in the environment. Ecological studies have shown that mercury bio-accumulates in the food chain as higher species consume lower life forms exposed to mercury contamination [42]. Data reported by the US Environmental Protection Agency (EPA) indicate that coal-fired utilities are currently the largest human-generated source of mercury releases in the USA. It is estimated that these plants release approximately 48 t annually [43]. In order to curb these emissions, EPA issued the world's first-ever rule to cap and reduce mercury emissions for coal-fired power plants. Compliance options available to utilities include postcombustion capture of mercury by existing or new flue gas scrubbing technologies as well as precombustion control of mercury by coal preparation and coal switching [44].

A recent study by Quick et al. [45] showed that the mercury content of coal delivered to utilities (based on ICR data) was lower than that of the in-ground coal

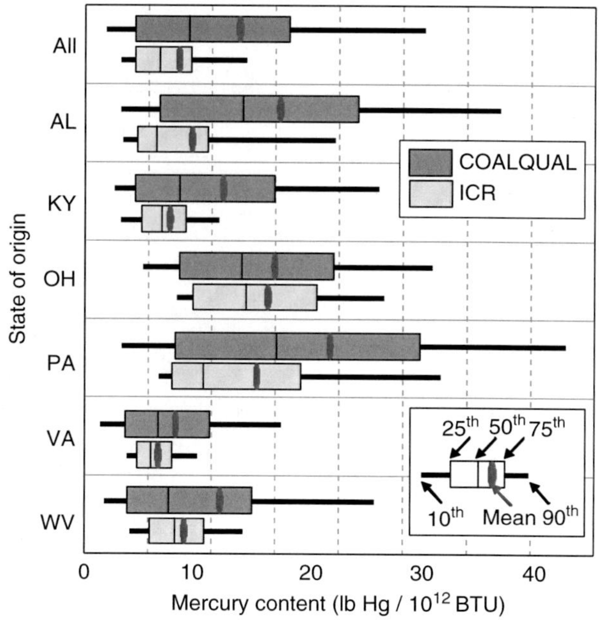


**Fig. 10.23** Comparison of mercury contents for delivered (ICR) and in-ground (COALQUAL) coals (after Quick et al. [45])

resources in the USA (based on COALQUAL data). This comparison is shown in Fig. 10.23 for the primary eastern coal producing states. Based on this study, Quick et al. [45] concluded that "…selective mining and more extensive coal washing may accelerate the current trend toward lower mercury content in coal burned at US electric utilities…" and "…since recent reductions of sulfur emissions from coal-burning electric utilities are largely due to a declining sulfur content of delivered coal, rather than from scrubbing combustion gases, these simple, low-cost approaches to reduce Hg emissions should not be overlooked."

According to Alderman [46], cleaning can reduce mercury by more than 50% in many eastern and western coals and lignites, excluding southern Powder River basin coals. Greater rejections of mercury by coal preparation appear to be limited by inadequate liberation and/or the presence of organically associated mercury. Several studies have suggested that mercury has some degree of association with the iron sulfides present in many run-of-mine coals [47–49]. In fact, the data summarized in Fig. 10.24 suggest that the concentration of many metallic elements found in coal often correlate well with the presence of sulfide minerals [50]. As a result, the rejection level for mercury is often in the same range as the pyritic sulfur rejection for coals subjected to coal preparation. While liberation can be improved by reducing the topsize of the feed coal [51], this approach is difficult to justify in today's marketplace due to the high costs associated with the fine grinding of coal [28]. Moreover, fine particles are also difficult and costly to upgrade, dewater and handle in existing coal preparation facilities.

Another important impact of coal preparation is in the reduction of greenhouse gas emissions resulting from improvements in thermal efficiency, i.e., less $CO_2$ is

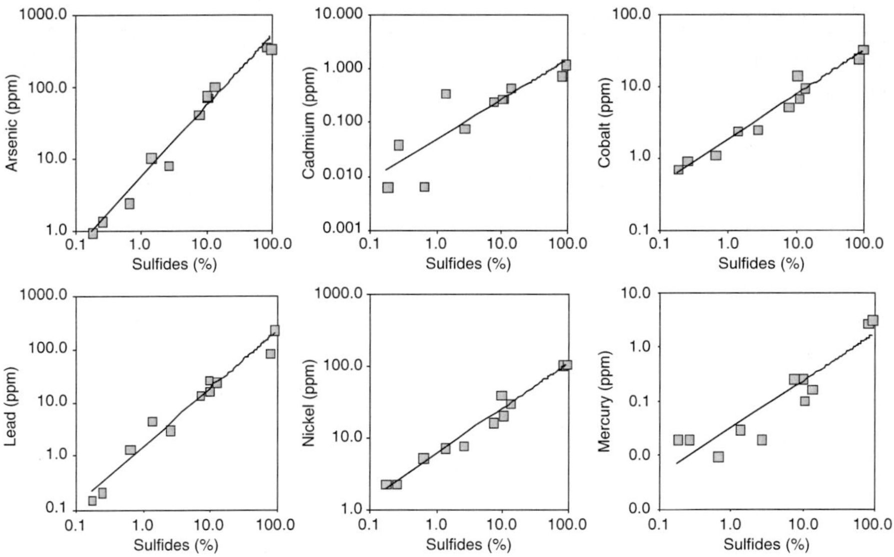


**Fig. 10.24** Effect of sulfide mineral content on the concentration of trace metals in a sample of Pittsburgh seam coal (After Luttrell et al. [50])

produced per unit of electricity generated. Calculations indicate that a one percentage point improvement in thermal efficiency provides a two to three percentage point reduction in $CO_2$ emissions for a typical coal-fired utility. An investigation conducted by Couch [52] indicated that there are more than 4,000 coal-fired boilers (>50 MW capacity) worldwide that could improve thermal efficiencies and reduce $CO_2$ emissions by using coal preparation to improve coal quality. Moreover, von Hippel and Hayes [53] have noted that coal preparation reduces the amount of energy consumed (and $CO_2$ generated) during coal transportation by increasing the specific heating value of the delivered coal. Smith [13] also found that freshly mined coals subjected to coal preparation processes tended to display slower releases of methane, another greenhouse gas.

## Challenges for Coal Preparation

Coal preparation offers may attractive benefits for coal-fired power generation. These typically include lower transportation costs, improved properties for coal utilization and reduced emissions of particulate and gaseous pollutants. However, the industry also faces several challenges that need to be resolved to ensure that preparation plants can continue to operate at a profit without damage to the environment. These concerns can be generally classified as technical factors that

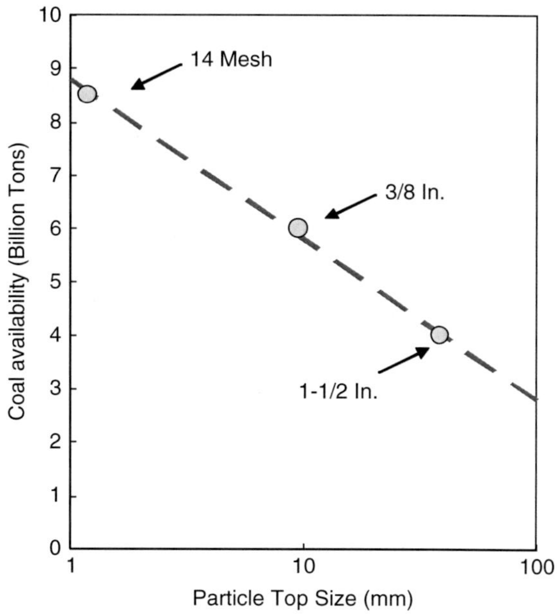


**Fig. 10.25** Effect of decreasing top size on low-sulfur coal availability (After Cavallero et al. [30])

relate to shortcomings in current processing systems, environmental factors that involve improving waste handling and disposal, or health and safety factors that may affect workers or citizens in the surrounding area.

## *Technical Issues*

### Cleaning of Fine Particles

The freshly mined coals that are fed to coal preparation plants are typically crushed to liberate rock before washing and to limit the largest size of particles that enter the plant. Operators prefer to keep the particles as large as possible (usually >50 mm) since fine coal processes are considerably less efficient and substantially more costly. This is unfortunate since the amount of high quality recoverable coal within a given reserve can theoretically be increased by crushing. The size reduction improves the liberation by reducing the population of intermixed composite particles of coal and rock. A study conducted by DOE [30] indicates that the reserves of compliance coal in central Appalachia could be nearly doubled by efficiently cleaning at a particle size of 1 mm (Fig. 10.25). Although a systematic assessment has not been performed to date for trace elements, the size reduction would also be expected to substantially improve the removal of coal-related pollutants other than just ash and sulfur.

Unfortunately, the solid–solid separation processes used industrially to treat fine coal represent the single greatest loss of potentially recoverable coal in a preparation facility. Field studies indicate that the froth flotation process, which is normally used to recover <0.2 mm coal, typically recovers only 60–80% of the organic matter in this size range. This surface-based separation is inherently less effective in removing pyritic sulfur than density-based processes used to treat the coarser sizes of coal [54]. As such, the desulfurization of fine particles is also often poor in many operating preparation facilities. Therefore, the development of effective, low-cost processes for treating fine coal is a major need for the preparation industry. Effective solutions need to be found for improving the recovery, selectivity and capacity of froth flotation processes. This goal may be achieved through fundamental and applied studies that seek to understand and improve flotation chemistry, equipment design and process control. In addition, new types of density separators need to be developed for treating fine coal. Centrifugal separators such as enhanced gravity concentrators, which have been successfully applied in the gold industry, may prove useful for this purpose. These devices have the potential to reject significantly greater amounts of pyritic sulfur (and mercury) that is not efficiently removed using surface-based separators such as froth flotation [55, 56]. However, these machines have several limitations including low throughputs and control difficulties.

## Dewatering of Fine Particles

The solid–solid separation processes employed by modern coal preparation plants require large amounts of process water. After cleaning, the unwanted water must be removed from the surfaces of the particles using mechanical dewatering equipment. Inefficient removal of moisture lowers the heating value, increases transport costs and creates handling/freezing problems for the cleaned coal. Coarse particles can be readily dewatered using simple screening systems, while finer particles require more complicated unit operations such as centrifuges and filters. Unfortunately, the mechanical systems used to dewater fine coal are inefficient and costly [57]. Fines often represent as little as 10% of the total run-of-mine feed; however, this size fraction may contain one third or more of the total moisture in the delivered product.

The availability of low-cost mechanical dewatering equipment that can efficiently remove moisture from fine coal is widely considered to be an important need for the coal preparation industry. Existing technologies for fine coal dewatering tend to produce unacceptably high moistures exceeding 25–35% by weight or intentionally sacrifice half or more of the ultrafines as a waste in an attempt to lower the product moisture. In addition, these processes are typically the most energy intensive processes used in coal preparation, often consuming 6–12 kW/t/h of dry solids processed [58]. Thermal dryer systems can effectively reduce moisture; however, these massive units require very large capital expenditures that are difficult to justify in the coal industry. Also, indirect thermal drying systems (e.g., Holoflite and Torus Disc) typically require 200–400 kW/t/h of product for drying fine coal solids to single-digit moistures [59]. Moreover, stringent air quality standards make it impossible in many cases to obtain new operating permits for thermal dryers. Therefore, the coal preparation industry needs to develop new mechanical solid–liquid separation processes that are substantially more efficient in terms of removing moisture and less expensive to purchase, operate and maintain. Innovative systems are critically needed, which may require fundamental studies to identify controlling mechanisms that can lead to the development "breakthrough" technologies.

## Sizing of Fine Particles

There are many technical challenges for the coal preparation industry that directly relate to size–size separations. Size–size separations are required ahead of solid–solid separations since these units are only effective within a narrow particle size range. Vibrating screens are generally efficient and cost effective for sizing and dewatering coarser particles. On the other hand, screening systems for finer particles, particularly those finer than 0.5 mm, tend to blind easily, wear quickly, and suffer from low throughput and low efficiency. The misplacement of incorrectly sized particles into equipment not designed to handle such sizes can have a large adverse impact on both the separating performance and maintenance requirements for a preparation plant. Another important issue with screening is the desliming of coal products to remove ultrafine mineral sediments that are detrimental to quality and moisture. Many in the industry believe that the ability to screen ultrafine particles at sizes of 0.15 mm and smaller is particularly important.

In addition to new screening systems, breakthrough technologies in ultrafine classification are also needed in the coal preparation industry [60]. Classification is the separation of particles due to differences in settling velocities, which depends not only on particle size, but also on particle density and shape. Firth and O'Brian [61] noted that while existing classification systems were adequate for the coal preparation industry of the past, "…it is apparent that further improvements in yield/ash/moisture relationship achieved by coal preparation plants will require increased efficiency in this size separation step." The ability to classify and better utilize ultrafine particles will increase industry productivity and reduce the generation of wastes. Many in the industry believe that currently there is a lack of efficient ultrafine sizing and desliming technologies for separating at cut sizes of 100 μm and finer.

## On-Line Analysis and Control

Tremendous strides have been made in the automation and control of coal preparation plants during the past several decades [62]. The application of on-line sensors together with programmable logic controllers (PLCs) has allowed modern plants to operate more efficiently and to improve safety by reducing manpower requirements. On the other hand, the industry continues to struggle with the determination in real-time of the quality of their coal products [63]. Analyzers are commercially available for the on-line analysis of many quality parameters for coal including ash, sulfur and moisture, although the measurement accuracy is often poor due to sampling and calibration issues. On-line analyzers do not currently exist that can be used to determine important data such as particle size distributions and washability in real-time. Therefore, improving automation, control and sensor technologies is a key challenge for the industry to overcome. Advances in on-line analysis would enhance the industry's ability to maximize the recovered energy in a marketable product and minimize the generation of unwanted wastes.

## *Environmental Issues*

Environmental compliance in the coal mining industry is strictly controlled by government regulations. History suggests that these regulations were necessary in many cases to ensure uniform environmental stewardship across the industry as a whole. Today, mining companies accept these environmental standards as good business practice and, in some cases, may go beyond simple compliance to promote goodwill and to set an example for others. Nevertheless, potential environmental issues associated with coal processing still exist. These issues have been generically classified by the World Bank to include air emissions, wastewater, hazardous materials, solid wastes and noise [64].

### Coarse Waste Disposal

Of the various challenges facing the coal preparation industry, perhaps none are as significant as those which relate to waste handling and disposal. This importance can be attributed to the fact that coal cleaning operations produce large volumes of waste that must be discarded into refuse piles or impoundments. Refuse piles are designed to receive coarse particles of waste rock that can be easily dewatered. This material is relatively easy to handle and can be safely transported by truck or belt haulage systems to the disposal area with little or no potential for environmental damage. On the other hand, the waste contains solid and liquid components that may present long-term disposal problems depending on the sizes, types and quantities of minerals present and the conditions under which the wastes are stored (dry vs wet, loose vs compacted, etc.). The factors play a key role in establishing the structural integrity (slope stability, surface water run-off, sediment containment, seepage, etc.) and chemical nature (acid generation, metal dissolution, etc.) of the wastes. Solid sediments and dissolved may be transported by rainwater where it would pollute streams or groundwater. Many of these issues can be effectively managed via proper disposal practices and monitoring programs. On the other hand, uncertainties related to the intricate biochemistry and complex hydrology of the waste warrant continued investigation to fully assess the potential for negative impacts associated with long-term disposal of coarse waste. Overall, improved waste characterization, including better methods to define the nature of wastes from coal preparation operations, is considered by many to be a high-priority need for the coal preparation industry.

## Slurry Disposal and Storage

The handling and disposal of fine slurry waste is widely considered to be one of the most difficult challenges facing the coal preparation industry. Fine wastes have historically been discarded into earthen impoundments for permanent disposal. An impoundment is an engineered structure consisting of a large volume settling basin formed behind a man-made dam or embankment. The waste, which is difficult to dewater, is normally pumped from the preparation plant thickener to the impoundment as slurry. The slurry contains water, coal fines, silt, clay and other fine mineral particulates from the processing plant. In most cases, the slurry is retained behind a man-made embankment (earthen dam) constructed from compacted refuse material. The impoundment is designed to have a volume that is sufficiently large to ensure that fine particles settle by gravity before the clear water at the surface is recycled back to the plant for reuse. In some cases, chemical additives may be used to promote settling and to control pH. According to the National Research Council [65], the coal industry discards 70–90 million tons of fine wastes each year into existing impoundments. There are 713 active impoundments and ponds, most of which are located in central Appalachia. In addition, it has been estimated that more than 2.5 billion tons of fine coal waste has been discarded in the past into existing and abandoned impoundments.

Impoundments, like any body of water contained behind a dam, can pose a safety and environmental risk if not properly constructed, monitored and maintained. Potential problems include structural failures, seepage/piping, overtopping, acid drainage and black water discharges. Since the well-known Buffalo Creek dam failure in 1972, strict engineering standards have been mandated by government agencies to regulate the design and operation of impoundments. A detailed report on the design and evaluation of tailings dams for the mining industry has been published by the EPA [66]. No failures of impoundment dams or overspills have occurred since this legislation was enacted. However, several breakthroughs of slurry into old mine workings beneath impoundments have occurred. The most notable was the Martin County incident, which released about 309 million gallons of slurry into streams and rivers in late 2000. A number of small accidental releases of black water have also been reported at various plant sites.

Several alternatives to impoundments have been employed by the coal industry in an attempt to avoid any future potential for environmental damage. These alternatives have been examined extensively by Gardner et al. [67]. For example, some mines use new modes of slurry disposal such as slurry cells and underground injection wells. Slurry cells have been used successfully in some cases (albeit at higher cost), but limitations associated with maintaining less than 20 acre-ft of settling area make this alternative difficult to apply in all cases. Likewise, the use of injection wells has raised public concerns about groundwater contamination and well water quality [68, 69]. To overcome these problems, various types of mechanical solid–liquid separators have recently been investigated as a means of more fully dewatering the fine solids prior to disposal. Notable examples include deep-cone (paste) thickeners and different types of filters (pressure, vacuum, belt press, plate-and-frame, etc.). Deep-cone thickeners have been shown to be capable of producing a paste of 45–55% solids as underflow in waste coal applications [70]. Ideally, the paste can be discarded as a stacked pile, thereby avoiding the need for impoundments to handle waste slurry. This technology has already been demonstrated projects at two mining sites in the eastern USA [71, 72]. Despite these advancements, the Committee on Coal Waste Impoundments [65] that recently examined the issue of slurry disposal concluded that "*. . .although there are alternatives to disposing of coal waste in impoundments, no specific alternative can be recommended in all cases*." Therefore, in the absence of a preferred disposal method, continued development of new and improved processes and practices for slurry disposal is critically needed in the coal preparation industry.

## Process Water Quality

The overwhelming majority of cleaning processes used in coal preparation require large amounts of process water. Nearly all of the process water is supplied by thickening units, which settle out ultrafine suspended solids and recycle clarified water back into the plant. A small amount of fresh make-up water from an external source is usually required to satisfy the balance between moisture contents of solids entering and exiting with the plant. The clarification and recycling of process water provides an effective means of reducing fresh water demands and lowing environmental impacts. On the other hand, plant operators are faced with the difficulty of avoiding the buildup of suspended solids and dissolved substances in the process water. The effects of these components on separating performance and plant maintenance (e.g., rusting and scaling) are not well understood. Deterioration of process water quality is known to reduce sizing efficiency, lower flotation recovery and increase magnetite losses [73]. Evidence also exists suggesting that dissolved ions also adversely impact the performance of dewatering processes [74]. Detailed studies are needed to better understand these problems so that effective solutions can be identified and implemented before they impede preparation plant operations.

There are also growing concerns by the public that chemicals used in the process of coal may be harmful to the environment. The vast majority of coal is upgraded without being contacted with any chemical additives using density-based separation processes. On the other hand, finer coal particles (typically <0.2

mm) are processed using froth flotation circuits, which require small dosages of reagents known as collectors and frothers. Collectors consist of oily hydrocarbons, such as diesel fuel, kerosene and fuel oil, which are insoluble in water and coat fine coal particles. If required, dosages are typically less than 0.3–2 lbs of collector per ton of fine coal processed. Likewise, frothers are added to all flotation systems to promote the formation of small air bubbles and to create a stable froth. Frothers are typically various types of alcohol and/or polyglycol surfactants [75]. Addition rates for frothers are typically less than 10–15 ppm in the process water. In addition, plants add water treatment chemicals into their thickening circuits to enhance settling rates and improve water clarity. These reagents commonly include various types of natural and synthetic coagulants and flocculants. Polymer flocculants, which are added in very small amounts, are diluted in strength to 0.01–0.05% solutions with water to improve performance. These chemical are also widely employed in the purification of drinking water and food as well as in the treatment of sewage, paper-making, oil recovery, storm water run-off and many other types of industrial wastewater [76]. Finally, coal operations may occasionally use small amounts of dust suppressants or freeze inhibitors to address seasonal problems that may arise at the plant site.

Several studies have been carried out in recent years to assess the ecotoxicological impacts of chemicals used by processing plants. For example, several studies have raised concerns that seepage from impoundments containing mine influenced waters (e.g., residual chemicals, particulates, acid drainage) may be impacting the population of fresh water muscles/mollusks in tributaries around coal mining and processing facilities [77]. Many of these issues were addressed in a recent symposium hosted by the Nature Conservancy that addressed the coexistence of coal mining and healthy aquatic ecosystems [78]. Another recent study commissioned by the New Zealand Auckland Regional Council (ARC) examined the effects of residual coagulants and flocculants on natural waters [79]. The study concluded that the negative impacts of these chemicals were "... low level and not likely to be significant in relation to other factors which govern the health of aquatic communities. The benefit of reduced sediment levels in discharges is considered to outweigh the risk of any low level impacts attributable to residual flocculants." However, the study noted that improper application of these chemicals associated with misusage or overdosing may create an environmental risk. Therefore, to avoid these concerns, the development of new technologies that eliminate and/or significantly reduce the additions of process reagents is desirable as a long-term solution to this potential problem. In the near-term, it is likely that chemical manufacturers will continue to develop low-risk "green" reagents for use in coal preparation facilities. Many plants that dispose of waste slurry via underground injection have successfully switched to natural flotation collectors, such as blends of canola oil, vegetable oil, soybean oil, etc., as environmentally friendly replacements for petroleum-based hydrocarbon collectors [80].

#### Air Quality and Dust

The US EPA promulgated standards of performance for air quality for all new and modified coal preparation plants under the 1976 Clean Air Act. These New Source Performance Standards (NSPS) address all types of particulate emissions, including fugitive dust, which may result from producing, handling, transporting and storing coal. Operations impacted by such regulation include crushers and breakers, sizing equipment, cleaning systems, thermal dryers, conveying systems, coal storage areas and coal transfer/loading systems that are directly part of the coal preparation facility. For some processes, such as thermal dryers, opacity less than 10% must be maintained [81]. Dusting problems are not uncommon around material handling transfer points where ultrafine particles have the opportunity to become airborne. These locations may include truck and railcar load-outs as well as conveyor and chute transfer points in and around the preparation plant. Many operations utilize water trucks and water sprays in combination with additions of various chemical and/or crusting agents to lower dusting problems. In addition, some operations make use of enclosures, such as silos and bins, to avoid exposure of coal to the environment. The use of inflatable structures over coal storage piles have also been attempted with limited success [82]. Despite these efforts, fugitive dust still seems to be an issue for many plant sites, often leading to disputes with the public and creating concerns for worker health. Therefore, work is needed to develop better practices and control technologies for reducing dust emissions. The reconstitution of fines via agglomeration and briquetting technologies would be expected to substantially reduce dust emissions, but these processes are currently too expensive to be practical in the coal industry.

## Future Directions

### *Dry Processing*

The low-sulfur coal reserves in the western states have become the most important supply of domestic fossil fuel in the USA during the past few decades. Historically, the majority of coal mined in this region was of sufficient quality to not require any coal preparation except for simple crushing and sizing. More recently, however, increased levels of rock dilution have been noted for coals mined in this region, largely due to the mining of more challenging reserves and the use of larger mining equipment that are less selective [83]. This trend is pushing some coal producers to consider coal washing for the first time. In addition, new federal and state clean air quality requirements are pressuring utilities and coal companies to use precombustion cleaning as a means of reducing $SO_x$ and trace element emissions [84]. One example is the proposed Springerville power plant in New Mexico, which has been tentatively approved contingent upon the use of precombustion cleaning to improve the quality of their coal feedstock.

There are many challenges in using precombustion cleaning to upgrade western coals. The processes traditionally used to wash eastern coals cannot be readily adopted in the west since water resources are lacking and low rank coals often disintegrate in water. As a result, current industry efforts are focusing on the development of dry cleaning processes to upgrade these lower rank reserves. Dry coal separators, such as pneumatic jigs and air tables, are already finding applications at selected mine and utility sites [85–87]. Other developing technologies that may be applicable for this purpose include various types of electrostatic [88, 89] and magnetic separators [90, 91]. The continued development of these low-cost units is believed to be important for upgrading many of the western coal reserves where water is scarce. The development of automated sorters, which use optical, electromagnetic or X-ray detection, to identify and extract rock from coal, also show considerable promise for dry coal concentration [92]. Optical sorters are used for separating particles that are liberated at relatively large sizes (>10 mm). These devices are already being used for separating diamond, gold, uranium and sulfide ores as well as for separating plastic bottles in the recycling industry. Such devices are particularly useful for pre-concentration, which can increase throughput and result in energy savings for other types of solid–solid separators. Continued R&D is necessary to improve separation efficiency (particularly for finer sizes), increase reliability and lower costs for this new generation of coal preparation technology.

## *Conversion Processes*

Coal preparation activities have traditionally been limited to include only those processes that involve physical separations. These processes include unit operations for particle sizing, concentration of organic matter and dewatering/disposal of plant products. These processes are generally considered to be inherently benign since they do not alter the chemical structure of the individual particles contained in the coal. Conversion processes, which include carbonization, gasification and liquefaction operations, are by this definition not considered to be part of the coal preparation industry. On the other hand, a new generation of coal preparation technology is being developed and commercialized that bridges the gap between traditional coal cleaning and coal conversion processes. These gateway technologies have the potential to reduce transportation costs and improve utilization properties for many of the low rank coals located in the Powder River basin. Under this expanded definition, conversion processes that are geared to the production of enhanced solid fuels may also be included under the coal preparation umbrella. Perhaps the best known of these processes include the K-Fuel, Encoal and SynCoal processes [93–95]. Some of these solid fuel production facilities may also produce gaseous or liquid byproducts that are of value in the synthetic fuel market. The upgrading of low rank coals, which are abundant in the USA, is believed to be an attractive means of producing low-emission coals capable of meeting the 2010 CAIR standards. This approach is particularly attractive if the process simultaneously produces syncrude oil as a byproduct [96].

One of the most highly developed and well demonstrated technologies for upgrading low rank coal is the K-Fuel process. This technology is a decarboxylation process that uses heat and pressure to modify the structure of subbituminous coals. By driving off moisture and oxygen, the process has been shown to be capable of reducing emissions of nitrous oxides by 10–20%, carbon dioxide by 8–12% and mercury by as much as 70% [97]. A commercial-size demonstration plant owned by Evergreen Energy is currently being operated in Gillette, Wyoming. Specific concerns associated with this technology include the high cost of thermal processing as well as problems related to disposal of process wastewater and spontaneous combustion of the treated products.

Another widely discussed coal conversion technology is the Encoal Process developed by SMC Mining Company and SGI International. The technology uses a two-step thermal treatment process to produce an enhanced solid coal fuel (char) as well as some derived liquid fuel. In the first processing step, the low rank feed coal is heated until a completely dry solid is produced. The temperature is then increased in a second processing step to promote decomposition and drive off gases via mild gasification. According to published reports [98], the Encoal process generates about one-half ton of solid fuel and one-half barrel of condensed liquids from each ton of feed coal supplied to the thermal reactor. The products, as alternatives to existing fuel sources, are capable of lowering sulfur emissions in coal-fired boilers nationwide [99]. The gaseous products that are not condensed into useful liquid are burned to supply thermal energy for the process. A 1,000 t/day demonstration plant was successfully operated between 1992 and 1998 under DOE sponsorship. However, concerns associated with this process include high treatment costs, excess fines production, dusting problems, wastewater generation and need for coal-char stabilization (to prevent spontaneous combustion).

Another intriguing conversion processes that couples thermal upgrading with physical cleaning to improve the quality of low rank coals is the SynCoal technology. In this process, high moisture coal is processed through vibrating fluidized bed reactors in three sequential stages, two heating stages followed by an inert cooling stage. These reactors remove chemically bound water, carboxyl groups and volatile sulfur compounds. After thermal upgrading, the coal is put through a deep bed stratifier cleaning process to separate the pyrite rich ash from the coal. When fed a typical low rank western coal, the SynCoal process can provide a product with a heating value of up to 12,000 BTU/lb with moisture and ash contents as low as 1% and 0.3%, respectively [100]. A demonstration plant (45 t/h) was successfully operated near Western Energy Company's Rosebud coal mine near Colstrip, Montana. Although the plant closed in 2001, the facility generated more than 2 million tons of products during its lifespan [101]. Similar to the other mild conversion technologies, concerns associated with this process include high treatment costs, dusting problems and product instability.

# Bibliography

## *Primary Literature*


1. Fiscor S (2010) U.S. preparation plant census 2010. Coal Age 9:38–46
2. Kempnich RJ (2003) Coal preparation – a world view. In: Proceedings, 20th international coal preparation exhibition and conference, Lexington, pp 17–39
3. ASTM (1994) Standard test method for determining the washability characteristics of coal, Annual Book of ASTM Standards, 5.05, No. D4371. American Society for Testing and Materials, Philadelphia
4. Miller I (1998) Use of coal mine methane in coal dryers. Coalbed Methane Outreach Program Technical Options Series, U.S. EPA, Air and Radiation Draft 6202J, pp 1–4
5. Meenen G (2005) Implications of new dewatering technologies for the coal industry. In: CAST workshop, 26–28 July, Virginia Tech, Blacksburg
6. Weisenfluh GA, Andrews WA, Hiett JK (1998) Availability of coal resources for the development of coal. Kentucky Geological Survey, Report, U.S. Department of Interior Grant 14–08–0001–A0896, pp 1–32
7. Sites RS, Hostettler KK (1991) Available coal resource study of Appalachia 7.5-minute quadrangle Virginia-Kentucky. Virginia Division of Mineral Resources, Publication 118, pp 1–51
8. Norton G (1979) The economic role of coal preparation in the production and utilization of coal for the market. Norton-Hambelton Consultants, Ann Arbor, pp 1–24
9. Harrison CD, Hervol JD (1988) Reducing electricity generation costs by improving coal quality. EPRI Report CS-5713, Electric Power Research Institute, Palo Alto
10. Davidson PG, Galluzzo NG, Stallard GS, Jennison KD, Brown RE, Jonas TS, Pavlish JH, Mitas DW, Lowe DA, Arroyo JA (1990) Development and application of the coal quality impact model: CQIM™. EPRI Report GS-6393, Electric Power Research Institute, Palo Alto
11. Kehoe DB, Parkinson JW, Evans RJ, Levasseur AA (1990) Environmental and economic benefits of using clean coals in utility boilers. In: Joint ASME/IEEE power generation conference, Boston, 21–25 Oct 1990
12. Harrison CD, Kehoe DB, O'Connor DC, Stallard GS (1995) CQE: integrating fuel decisions. In: Fourth annual clean coal technology conference, Denver, 5–8 Sept 1995
13. Smith SR (1988) Tennessee Valley Authority's experience with switching to improved quality coal. In: Proceedings, conference on reducing electricity generation costs by improving coal quality, 5–6 November, Electric Power Research Institute, Palo Alto
14. Bhaskar U (2007) NTPC looking for sites to set up coal washeries. Wall Street Journal December 3, 2007
15. Glomsrod S, Taoyuan W (2005) Coal cleaning: a viable strategy for reduced carbon emissions and improved environment in China? Energy Policy 33:525–542
16. Blackman A, Wu X (1999) Foreign, direct investment in China's power sector: trends, benefits and barriers. Energy Policy 27:695–711
17. Corder WC (1983) For lower power costs, use high quality coal. In: Proceedings, 3rd European coal utilization conference, COALTECH, Amsterdam, 1:221–240
18. Scott DH (1995) Coal pulverisers—performance and safety. IEACR/79, IEA Coal Research, London, pp 1–83
19. Raask E (1983) Ash related problems in coal-fired boilers. In: Proceedings, engineering foundation conference on fouling of heat exchange surfaces, White Haven, Engineering Foundation, New York, 31 Oct–5 Nov 1982, pp 433–460
20. Couch GR (1994) Understanding slagging and fouling during PF combustion. In: International energy agency, clean coal center, Gemini House, London, IEACR/72, pp 1–118
21. Hatt R (1995) Correlating the slagging of a utility boiler with coal characteristics. In: Engineering foundation conference, Waterville Valley, 16–22 July

22. Vaninetti GE, Busch CF (1982) Mineral analysis of ash data: a utility perspective. J Coal Qual 1(2):22–31
23. Jones C (1998) Fuel management: solve common coalhandling problems. Power 142(6):31–32
24. Smith IM, Sloss LL (1998) PM10/PM2.5 – Emissions and effects. CCC/08, International Energy Agency, Coal Research, London, pp 1–75
25. EPA (1997) The benefits and costs of the Clean Air Act, 1970 to 1990. U.S. Environmental Protection Agency (EPA), pp 1–28
26. Doherty M (2006) Coal quality impact on unit availability and emissions. American Electric Power Site Visit, Asia Pacific Partnership on Clean Development and Climate, Columbus, 30 October–4 November, pp 1–10
27. Couch GR (1995) Power from coal: where to remove impurities? International Energy Agency, Coal Research, London, IEACR/82, pp 1–87
28. Kawatra SK, Eisele TC (2001) Coal desulfurization: high efficiency preparation methods. Taylor & Francis, New York, pp 1–376
29. Cavallaro JA, Deurbrouck AW (1977) An overview of coal preparation. In: Proceedings of the coal desulfurization: chemical and physical methods symposium, New Orleans, 23 Mar 1977, pp 35–57
30. Cavallaro JA, Deurbrouck AW, Killmeyer RP, Fuchs W, Jacobsen PS (1991) Sulfur and ash reduction potential and selected chemical and physical properties of United States coals. U.S. Department of Energy, DOE/PETC/TR-91/6, pp 1–309
31. Couch GR (2003) Coal preparation. International Energy Agency, Clean Coal Center, pp 1–181
32. Obermiller EL, Conrad VB, Lengyel J (1993). Trace element contents of commercial coals. In: Chow W, Connors KK (eds) Managing Hazardous air pollutants: state-of-the-art. Electrical Power Research Institute, Washington, DC. CRC Press, Boca Raton, pp 73–92
33. Meij R, te Winkel H (2007) The emissions of heavy metals and persistent organic pollutants from modern coal-fired power stations. Atmos Environ 41(40):9262–9272
34. Zubovic P (1966) Physicochemical properties of certain minor elements as controlling factors in their distribution in coal. In: Coal science, vol 55, Advances in Chemistry. American Chemical Society, Washington, DC, pp 221–231
35. Swaine DJ (1990) Trace elements in coal. Butterworth, London, pp 1–278
36. Fonseca AG, Tumati RP, DeVito MS, Lancet MS, Meenan GF (1993) Trace element partitioning in coal utilization systems. In: SME-AIME annual meeting, Reno, 15–18 February, Preprint No. 93–261, pp 1–25
37. Jacobsen PS, Blinn MB, Wan EI, Nowak MA (1992) Role of coal preparation in the precombustion control of hazardous air pollutants. In: Proceedings, 8th annual coal preparation, utilization, and environmental control contractors conference, PETC, Pittsburg, pp 412–419
38. Ford C, Price A (1982) Evaluation of the effect of coal cleaning on fugitive elements. Final report, Phase III, BCR Report L-1304
39. Akers DJ (1996) Coal cleaning controls HAP emissions. Power Eng 100(6):33–36
40. Akers DJ, Raleigh CE, Lebowitz HE, Ekechukwu K, Aluko ME, Arnold BJ, Palmer CA, Kolker A, Finkelman RB (1997) HAPs-RxTM: precombustion removal of hazardous air pollutant precursors. Final Report, DOE Contract DE-AC22-95PC95153, U.S. Department of Energy, pp 1–115
41. Palmer CA, Luppens JA, Finkelman RB, Luttrell GH, Bullock JH (2004) The use of washability studies to predict trace-element reductions during coal cleaning. In: Proceedings, 29th international technical conference on coal utilization & fuel systems, Coal Technology Association, Gaithersburg, pp 1365–1376
42. Trasande L, Schechter CB, Haynes KA, Landrigan PJ (2006) Mental retardation and prenatal methylmercury toxicity. Am J Ind Med 49(3):153–158

43. EPA (2001) Results of Part III of information collection request. http://www.epa.gov/ttnatw01/combust/utiltox/utoxpg.html
44. Pavlish JH, Sondreal EA, Mann MD, Olson ES, Galbreath KC, Laudal DL, Benson SA (2003) Status review of mercury control options for coal-fired power plants. Fuel Process Technol 82:89–165
45. Quick JC, Brill TC, Tabet DE (2002) Mercury in U.S. coal: Observations using the COALQUAL and ICR data. Environ Geol 43:247–259
46. Alderman K (2007) Market opportunities for coal preparation. Presentation to the Report Committee of the National Commission on Energy Policy – Coal Study. 17 September, Denver
47. Tkach BI (1975) On the role of organic matter in concentration of mercury. Geochem Int 11:973–975
48. Finkelman RB, Stanton RW, Cecil CB, Minkin JA (1979) Modes of occurrence of selected trace elements in several Appalachian coals. Am Chem Soc Div Fuel Chem 24(1):236–241
49. Minken JA, Finkelman RB, Thompson CL, Chao ET, Ruppert LF, Blank H, Cecil CB (1984) Microcharacterization of arsenic- and selenium-bearing pyrite in Upper Freeport coal. Scan Electron Microsc 4:1515–1524
50. Luttrell GH, Venkatraman P, Yoon R-H (1998) Removal of hazardous air pollutant precursors by advanced coal preparation. Coal Prep 19:243–255
51. Hucko RE (1984) An overview of U.S. Department of Energy coal preparation research. In: SME-AIME annual meeting, Los Angeles, 26 February–1 March, Preprint No. 84–105, pp 1–6
52. Couch GR (2000) Opportunities for coal preparation to lower emissions. International Energy Agency Coal Research, CCC/30, London, pp 1–46
53. von Hippel D, Hayes P (1995) Prospects for energy efficiency improvements in the Democratic People's Republic of Korea: evaluating and exploring the options. Nautilus Institute for Security and Sustainable Development, Berkeley, pp 1–31
54. Adel GT, Wang D (2005) The assessment of fine coal cleanability. Int J Coal Prep Util 25(3):129–140
55. Honaker RQ (1998) High capacity fine coal cleaning using an enhanced gravity concentrator. Miner Eng 11(12):1191–1199
56. Honaker RQ, Wang D, Ho K (1996) Application of the falcon concentrator for fine coal cleaning. Miner Eng 9(11):1143–1156
57. Le Roux M, Campbell QP, Watermeyer MS, de Oliveira S (2005) The optimization of an improved method of fine coal dewatering. Miner Eng 18(9):931–934
58. Yoon R-H, Eraydin M, Keles S, Luttrell GH (2007) Advanced dewatering methods for energy savings in the mineral processing industry. In: SME annual meeting and exhibit, 25–28 February, Denver, Preprint 07–124, pp 1–5
59. van den Broek JJM (1982) From metallurgical coal tailings to thermal feed. AIME Trans 272:49–52
60. Mohanty MK, Palit A, Dube B (2002) A comparative evaluation of new fine particle size separation technologies. Miner Eng 15(10):727–736
61. Firth B, O'Brian M (2003) Hydrocyclones circuits. Coal Prep 23(4):167–183
62. Couch G (1996) Coal preparation: Automation and control. International Energy Agency, London, IEAPER/22, pp 1–64
63. Belbot M, Vaurvopoulos G, Paschal J (2001) A commercial elemental on-line coal analyzer using pulsed neutrons. In: CAARI 2000: 16th international conference on the application of accelerators in research and industry, AIP Conference, Denton, pp 1065–1068. DOI 10.1063/1.1395489
64. Anon (2007) Environmental, health, and safety guidelines for coal processing. International Finance Corporation (IFC), World Bank Group, Good International Industry Practice (GIIP) Technical Reference Document, April 30, pp 1–22
65. NRC (2002) Coal waste impoundments: risks, responses and alternatives. National Academies Press, Washington, DC, pp 1–244

66. EPA (1994) Design and evaluation of tailings dams. Technical Document, EPA 530-R-94-038, 59 pp. http://nepis.epa.gov/EPA/html/Pubs/pubtitleOSWER.htm
67. Gardner JS, Houston KE, Campoli A (2003) Alternatives analysis for coal slurry impoundments. In: SME annual meeting, 24–26 February, Cincinnati, Preprint 03–032, pp 1–5
68. Breen T (2007) Lawmakers want study of coal slurry in drinking water. Charleston Daily Mail, Charleston, 9 January 2007
69. Wilcox L (2007) Coal industry official disputes research about health of people living near mines. The Exponent Telegram, Clarksburg Publishing, Clarksburg, 30 October 2007
70. Parekh BK, Patil D, Honaker RQ, Baczek F (2006) Improving the densification of fine coal refuse slurries to eliminate slurry ponds. In: Proceedings, XV international coal preparation congress, China University of Mining and Technology Press, Beijing, pp 549–554
71. Bethell PJ, Gupta BK, Gross R (2008) Replacement of belt presses with a deep cone thickener at Lone Mountain Processing. In: 25th annual international coal preparation exhibition and conference, Lexington, pp 41–50
72. Henry CD (2008) Fine coal recovery project: Pinnacle mine Smith Branch impoundment. Technical Presentation, Marshall University CBER Summit, pp 1–39. http://www.marshall.edu/cber/summit/S2%20-%201%20Henry%20Beard%20Tech.pdf
73. Osborne DG (1988) Flotation, agglomeration, and selective flocculation. In: Coal preparation technology, vol 1. Graham and Trotman, London, pp 415–477
74. Yoon RH, Luttrell GH (2008) Advanced dewatering systems development. U.S. Department of Energy, Final Report for Contract No. AC26-98FT40153, 31 July, pp 1–225
75. Laskowski JS (2001) Coal flotation and fine coal utilization, vol 14, Developments in mineral processing. Elsevier Science, Amsterdam, pp 1–350
76. Gregory J, Bolto B (2007) Organic polyelectrolytes in water treatment. Water Res 41(11):2301–2324
77. Beaty B (2007) The Nature Conservancy, Clinch Valley Program, Personal Communication
78. Nature Conservancy (2007) Conservation management of the Clinch and Cumberland river systems: a collaborative discussion on coal mining and the aquatic environment. 5–7 September, Abingdon. http://www.cpe.vt.edu/cmrs/agenda.html
79. ARC (2004) Overview of the effects of residual flocculants on aquatic receiving environments. Publication TP226, Auckland Regional Council (ARC), Auckland, pp 1–31
80. Skiles KD (2003) Search for the next generation of coal flotation collectors. In: 20th Annual international coal preparation exhibition and conference, 29 April–1 May, Lexington, pp 187–204
81. EPA (2009) Coal preparation and processing plants new source performance standards (NSPS), US Environmental Protection Agency (EPA), Fact Sheet. http://www.epa.gov/airtoxics/nsps/coal/cpp_nsps_fr_fs_092509.pdf
82. Bowling B (2003) Elk Run may use plant – agreement similar to one made after first dome rupture. Charleston Daily Mail, Charleston, 29 April 2003
83. Bethell PJ (2007) Coal preparation: current status and the way ahead. Presentation to the Report Committee of the National Commission on Energy Policy – Coal Study, Denver, 17 September 2007
84. Honaker RQ, Luttrell GH, Bratton RC, Saracoglu M, Thompson E, Richardson V (2007) Dry coal cleaning using the FGX separator. In: 24th International coal preparation conference, Lexington, pp 61–76
85. Kelley M, Snoby R (2002) Performance and cost of air jigging in the 21st century. In: 19th annual international coal preparation exhibition and conference, 30 April–2 May, Lexington, pp 175–186
86. Lu M, Yang Y, Li G (2003) The application of compound dry separation technology in China. In: 20th Annual international coal preparation exhibition and conference, Lexington, pp 79–95

87. Weinstein R, Snoby R (2007) Advances in dry jigging improves coal quality. Min Eng 59(1):29–34
88. Yoon R-H, Luttrell GH, Adel GT (1995) POC-scale testing of a dry triboelectrostatic separator for fine coal cleaning. In: 11th Annual coal preparation, utilization, and environmental control contractors conference, 12–14 July, U.S. Department of Energy, Pittsburgh, pp 65–72
89. Stencel JM, Schaefer JL, Ban H, Li JB, Neathery JK (2002) Triboelectrostatic coal cleaning: mineral matter rejection in-line between pulverizers and burners at a utility. In: Impact of mineral impurities in solid fuel combustion. Springer, New York, pp 645–651
90. Oder RR (2005) The co-benefits of pre-combustion separation of mercury at coal-fired power plants. In: 98th Annual conference and exhibition, Air & Waste Management Association, Minneapolis, Paper 990
91. Oder RR (2005) The use of magnetic forces to upgrade low rank coals. In: Upgrading low rank coals symposium, coal & aggregate processing exhibition and conference, Lexington, 2–6 May
92. Jong TPR, Mesina MB, Kuilman W (2003) Electromagnetic deshaling of coal. Phys Sep Sci Eng 12(4):223–236
93. Collins KR (2007) EEE: Delivering cleaner coal today. In: 2nd Annual capital one southcoast energy conference, 3–5 December, New Orleans, pp 1–21
94. Federick JP, Knottnerus BA (1997) Role of the liquids from coal process in the world energy picture. In: 5th Annual clean coal technology conference, Tampa
95. Sheldon RW (1997) Rosebud SynCoal partnership: syncoal demonstration technology update. In: 5th Annual clean coal technology conference, Tampa
96. Skov ER, England DC, Henneforth JC, Franklin GR (2007) Syncrude and syncoal production by mild temperature pyrolysis processing of low-rank coals. American Institute of Chemical Engineers, Spring National Meeting, 22–26 April, Houston, Session TB004, pp 1–16
97. Wingfield G (2007) Cleaner coal? Forbes, 17 April 2007. http://www.forbes.com/2007/04/16/clean-coal-congress-biz-wash-cx_bw_0417coal.html
98. DOE (2003) Clean coal technology programs: Completed projects 2003. Clean Coal Technology, U.S. Department of Energy (DOE), National Energy Technology Laboratory (NETL) Clean Coal Technology Compendium, December 2003, 2:1–149
99. DOE (2002) The ENCOAL mild coal gasification project: A DOE assessment. DOE National Energy Technology Laboratory, Technical Report, Clean Energy Document No. 91, March 2002, pp 1–40
100. DOE (1997) Clean coal technology: upgrading of low-rank coals. U.S. Department of Energy, Topical Report No. 10, August 1997, pp 1–28
101. DOE (2006a) Advanced coal conversion process demonstration: Project performance summary – Clean coal demonstration program. Report DOE/FE-0496, March 2006, pp 1–16

## *Books and Reviews*

Arnold BJ, Klima MS, Bethell PJ (2007) Designing the coal preparation plant of the future. Society for Mining Metallurgy and Exploration (SME), Littleton, pp 1–205

Couch GR (2003) Coal preparation. IEA Clean Coal Center, London, pp 1–182

Kawatra KS, Eisele TC (2001) Coal desulfurization: high-efficiency preparation methods. Taylor & Francis, New York, pp 1–360

Kona BB (1997) Coal preparation. Golden Jubilee Monographs. Allied Publishers, Buffalo, pp 1–221

Leonard JW, Hardinge BC (1991) Coal preparation. Society for Mining, Metallurgy and Exploration (SME), Littleton, pp 1–1131

Miller BC (2005) Coal energy systems. Elsevier Academic, Burlington, pp 1–526

Sanders J (2007) Principles of coal preparation, 4th edn. Australian Coal Preparation Society, Dangar, New South Wales, pp 1–560

# Chapter 11
# Coal to Liquids Technologies

**Marianna Asaro and Ronald M. Smith**

## Glossary

| | |
|---|---|
| Coal-to-liquids (CTL) | The conversion of coal to liquid fuels and/or chemicals |
| Coprocessing (of coal) | The simultaneous conversion of coal and waste carbonaceous feedstocks such as petroleum-based residual oil or tar, plastics, or rubbers via once-through direct liquefaction into liquid, solid, and gaseous hydrocarbonaceous materials intended primarily for use as fuel. |
| Direct coal liquefaction (DCL) | The conversion of coal to liquids via dissolution and/or hydroprocessing, without first gasifying the coal. |
| DOE | United States Department of Energy. |
| Fischer–Tropsch reaction | The catalytic conversion of synthesis gas to primarily hydrocarbons, the discovery being credited to Franz Fischer and Hans Tropsch. |
| Gasification (of coal) | The conversion of coal to primarily synthesis gas plus ash at elevated temperature and pressure of oxygen and steam. |

This chapter was originally published as part of the Encyclopedia of Sustainability Science and Technology edited by Robert A. Meyers. DOI:10.1007/978-1-4419-0851-3

M. Asaro (✉)
SRI International, 333 Ravenswood Avenue, 94025 Menlo Park, CA, USA
e-mail: marianna.asaro@sri.com

R.M. Smith
SRI Consulting, 4300 Bohannon Drive, 94025 Menlo Park, CA, USA
e-mail: Ronald.Smith@ihs.com

R. Malhotra (ed.), *Fossil Energy: Selected Entries from the Encyclopedia of Sustainability Science and Technology*, DOI 10.1007/978-1-4614-5722-0_11,

Indirect coal liquefaction (ICL) The conversion of coal to liquids via the intermediacy of synthesis gas.

Light distillate A distillation cut of low molecular weight and low boiling range, obtained from refining of hydrocarbonaceous feedstocks, used to produce liquefied petroleum gas (LPG), gasoline, and naphtha.

Middle distillate A distillation cut of mid-range boiling point, obtained from refining hydrocarbonaceous feedstocks, containing hydrocarbons ranging from $C_5$ through about $C_{20}$ or $C_{22}$. When further distilled, the portion of middle distillates containing $C_5$ through about $C_{15}$ is often referred to as naphtha, and the portion containing $C_{16}$ through up to $C_{22}$ is referred to as diesel. The naphtha is often distilled further to produce gasoline and kerosene/jet fuel, or can be used as feed for a naphtha cracker unit to make light olefins. (Less commonly, the gasoline cut is initially collected along with the light distillates).

Pyrolysis (of coal) A mild gasification process wherein coal is heated in the absence of oxygen at sufficiently mildly elevated temperature as to produce a mixture of hydrocarbonaceous gases, liquids, and solids suitable for use as chemical precursors or fuel.

Synthesis gas (syngas) A mixture of primarily hydrogen and carbon monoxide produced by gasification or reforming of hydrocarbonaceous materials, used to synthesize fuels or chemicals.

Water gas shift reaction The chemically reversible, catalyzed conversion of water and carbon monoxide to hydrogen and carbon dioxide.

## Definition of the Subject

Like oil and natural gas, coal is a fossil resource and as such is not what first comes to mind when considering sustainable fuel sources. Coal contributes to sustainability as a complement to maintaining continuous availability of sufficient fuel to meet growing global demand [1–3].

As for other fossil sources, conversion of coal to transportation fuel is currently more affordable than conversion of renewable resources such as wind and solar, which are technically far away from availability at even a fraction of the scale required for sustainability over the coming decades. Given that proved coal reserves in North America alone are currently estimated as capable of producing more than 840,000,000,000 equivalent barrels of oil (the energy equivalent of 32 cubic miles of oil) [4], the abundance of coal is a boon to a sustained global liquid fuel supply if used wisely.

Just as renewable fuel sources will be benchmarked against fossil sources in terms of economics, so also coal will be benchmarked against renewable fuel sources in terms of environmental sustainability. Today, modern coal-to-liquids (CTL) technology can produce synthetic fuels that burn more cleanly than conventionally produced fuels. CTL technology can produce a virtually sulfur-free diesel fuel, much cleaner than conventional diesel – reducing smog and acid rain – and can also produce gasoline, jet fuel, or chemicals.

Most coal used today is for production of power or for steelmaking. Coke plants use coal to produce high purity carbons for use in steel manufacture. Combustion of coal provides heat and power for industrial process operations. Production of liquid fuels and chemicals from coal are lower volume uses; these applications will tend to increase in importance as the ease of oil recovery decreases and its price increases.

## Introduction

Coal has been mined and used as a source of energy for centuries. The first recorded use of coal was in China, with indications that the Chinese began using coal for heating and smelting about 2,500 years ago, and most historians believe that coal was used in various parts of the world 3,000–4,000 years ago during the Bronze Age [5]. A system for mining coal was present in England by 1180 [6]. Coal provided heat for homes and industry, fuel gas for town lighting, process gas for industry, coke for the iron industry, and fuel for steam locomotives and shipping during the first quarter of the twentieth century. Although electric power has supplanted coal gas for public lighting (town gas), the electric power industry remains significantly coal based today.

Coal resources have been used to produce liquid transportation fuels by several process routes, collectively referred to as coal liquefaction or, more generally stated, as Coal to Liquids (CTL). Early records of coal conversion to liquid fuel date to 1913 and an extraction process developed by Friedrich Bergius in Germany [7]. Bergius treated a mixture of pulverized bituminous coal and coal-derived heavy oil with hydrogen, at high temperature and pressure, to give a liquid product consisting of heavy oils, middle distillates, gasoline, and gases. Various metal compounds have been used to catalyze the process. This type of CTL technology is called Direct Coal Liquefaction (DCL).

DCL technology involves making a partially refined synthetic crude oil directly from coal, which is then further refined into synthetic gasoline and diesel as well as liquefied petroleum gas (LPG). Raw coal oil is highly aromatic and requires further upgrading. Heavy residues generated in the process, referred to as asphaltenes, also include ash from the coal and are either disposed of or gasified to produce part of the hydrogen required for the upgrading. The hydrocarbon fuel products of DCL are similar to hydrocarbon fuels derived from crude oil. A slate of partially refined gasoline-like and diesel-like products, as well as propane and butane, are recovered from the synthetic crude oil mainly by distillation. Liquid yields in excess of 70% by weight of the dry, mineral matter-free coal feed have been demonstrated under favorable circumstances. Overall thermal efficiencies (% calorific value of the input fuel converted to finished products) for modern processes are in the range of 60–70% if allowance is made for generating losses and other non-coal energy imports. These processes generally have been developed to process development unit (PDU) or pilot plant scale. However, no demonstration or commercial-scale plant had yet been built until very recently, primarily because petroleum fuels had been available at lower overall cost.

Development of the DCL approach in the twentieth century proceeded with ambition and then reserve. DCL was first commercialized in the 1920s (the Pott-Broche/IG Farben Process). In the Second World War, Germany produced more than 4 million tons of fuels by DCL. Plants were also built during this time in the UK, France, and Japan. A Japanese-influenced pilot unit at Fushun Coal Mine was operated by the South Manchurian Railway Company from 1938 [8] until 1943 and processed 20,000 t/year of coal. During this same period, the Korean Artificial Petroleum Company operated a 100 t/day plant at its Agochi factory [9].

DCL underwent a new phase of development in the 1970s in response to oil shocks. Major research efforts in the USA during the 1970s and 1980s, in particular, involved work by Hydrocarbon Research Inc. (HRI, now Hydrocarbon Technologies Inc, HTI), cosponsored by the US Department of Energy (DOE) and including substantial R&D at a dedicated facility in Wilsonville, Alabama.

Research on DCL in Japan began in 1974 with the Sunshine project and sponsorship by the New Energy and Industrial Technology Development Organization (NEDO) [10]. Early players included Sumitomo Metals and Hasaki, and Nippon Steel and Kimitsu. The program experienced delays in the next decade, presumably resulting from the drop in oil price. A 150 t/day pilot plant was built at Sumitomo Metal Industries Kashima Steelworks and operated during 1997 and 1998, with a longest run of 1,921 h [11]. Research on DCL in Britain included the two-stage Liquid Solvent Extraction (LSE) and the Supercritical Gas Extraction (SGE) processes from British Coal. Research on DCL in Germany has included the single-stage Kohleöl process, as improved with the Integrated Gross Oil Refining (IGOR+) development by Ruhrkohle AG and Veba Öl AG, as well as the single-stage Imhausen High-Pressure Process [12].

The main focus of DCL research and development continues to be making the process more economically competitive with crude oil refining by decreasing the cost of production, particularly the capital cost of the plant, and the related challenge of finding more efficient ways to deal with tar and solid by-products of the liquefaction.

In addition to the price of oil, increasingly stringent environmental limitations on the content of aromatic compounds and sulfur in motor fuels have presented challenge. Fuels generated by DCL are rich in high-octane aromatics, which is an asset to engine performance but also an environmental issue. Coal-based processes also require desulfurization. The challenge of altering the product slate to meet environmental regulations is compounded by the resultant loss of benefits imparted by aromatics: in facilitating the liquefaction chemistry itself and then in promoting engine performance when using the final fuel product. Desulfurization presents its own performance challenges, because sulfur functions as an engine lubricant.

In the 1920s, Franz Fischer and Hans Tropsch of the Kaiser Wilhelm Institute developed a catalyst to convert hydrogen and carbon monoxide obtained from gasification of coal to hydrocarbon liquids. Because the coal is first gasified before conversion to liquid fuels, the Fischer–Tropsch (F-T) process for CTL is known as a type of Indirect Coal Liquefaction (ICL).

Indirect coal liquefaction (ICL) is a completely different approach from DCL for providing liquid fuels from coal. ICL processes first convert coal to a gaseous intermediate, which is then converted to liquid fuel. Coal gasification technologies rely on incomplete combustion, wherein coal is heated with a quantity of air that is insufficient to allow complete combustion (formation of carbon dioxide and water). The products are condensables and tars, flammable gases, and a solid fuel (or waste). Most coal-derived chemicals are made from the tars, although synthesis gas from coal gasification also can be used for the production of chemicals such as BTX (benzene, toluene, and xylenes), hydrogen, sulfur, methanol, and ammonia. Coal-derived fuels are obtained from the condensables, which are distilled and in some cases hydrotreated to give gasoline, diesel, and fuel oil cuts.

Of particular utility is the gasification of coal to provide the process intermediate synthesis gas, or syngas, a mixture of primarily CO (carbon monoxide) and $H_2$ (hydrogen). The ratio of $H_2/CO$ varies with the source within the range of 0.7–1 for coal. Syngas can be used to generate liquid fuels via the aforementioned Fischer–Tropsch (F-T) technology, which provides hydrocarbon fuels that resemble crude oil derived products. The F-T processes produce synthetic middle distillates that can be used either directly as diesel or in blends with petroleum-derived diesel.

The CTL Fischer–Tropsch process was first taken forward commercially in the 1930s by the German company Ruhrchemie, in conjunction with other partners including Lurgi, to produce gasoline and oil. By 1938 there were nine German plants in operation having a total capacity of 660,000 t/year [13]. Individual reactors yielded about 5,000 L of gasoline/day, with 100 such units used in the first German F-T plants. However, the Bergius process DCL plants were chosen in 1939 when Germany expanded production during World War II, because these were further developed and could be more readily scaled to larger size, eventually processing up to 350 t of coal and yielding 250,000 L of gasoline/day.

By the mid-1940s, natural gas and oil production had become more developed and cost-competitive with coal, and technology for production of synthetic transportation fuels was not considered economic after the Second World War. Since then, research on CTL processes has waxed and waned largely in response to surges and ebbs in the price of oil. However, a perception by some at the time that reserves of crude oil were limited led to continued low-level development of CTL. Sasol purchased rights from the German companies and developed its own CTL process. Construction began in 1950 and an F-T plant came on line in Sasolburg, South Africa during the mid-1950s using Lurgi coal gasifiers and Sasol Synthol reactors to produce gasoline and diesel fuels as well as chemical feedstocks from coal. Even so, in the late 1950s and 1960s the huge oil fields of the Middle East dominated fuel supply chains to the point that further development of CTL technology worldwide essentially disappeared.

Shocks in the price of oil during the 1970s prompted renewed efforts; particularly in South Africa, the USA, Japan, and Britain; to promote an alternative to crude oil: a synthetic crude (syncrude) that would be cost-competitive with oil for refining to transportation fuels. Trade embargoes lasting until the mid-1980s effected large-scale application of F-T ICL in South Africa, at that point up to 60% of the country's transportation fuel requirements being met by coal. With the benchmark price of oil at US $30 per barrel, Sasol brought two F-T ICL plants online, in 1980 and 1983, with the total capacity of the company's three plants at 6,000 t/year [14].

As a result of a drop in the price of crude oil, worldwide developments in indirect coal liquefaction were largely put on hold after the 1980s, with the exception of those in South Africa where Sasol has been using its F-T Advanced Synthol (SAS) reactors since 1989. The ICL approach underwent an R&D resurgence in the 1990s, then in response to an increase in the price of oil [15]. All of these early plants utilized high-temperature Fischer–Tropsch (HTFT) technology with an iron-based catalyst. In 1993, Sasol successfully proved its low temperature Fischer–Tropsch (LTFT) technology with an iron-based catalyst to convert coal-derived syngas to liquid products. The plant is a relatively small scale, 2,500 bpd commercial facility that was recently converted to use natural gas as feed. Sasol's newest generation of LTFT utilizes high performance cobalt based catalysts in the Sasol Slurry Phase Distillate (SSPD) process [14], now used in Qatar to convert natural gas to middle distillates.

Currently, the best option for using coal to produce a high cetane number diesel fuel virtually free of aromatics and of S and N compounds is the F-T process coupled with the applicable downstream work-up of the crude F-T product [16]. Gasification of coal provides syngas, which is converted to a range of hydrocarbons. Alternatively, the syngas can be converted to liquid specialty [17, 18] or commodity chemicals including methanol or dimethylether (DME), either or both of which also can be used as automotive fuels. Methanol can be used alone, or preferably as a bend with conventional gasoline, in a spark ignition internal combustion engine (but not in a diesel engine, because methanol's cetane number is too low). A separate fuel distribution network for methanol also would be required as well as consideration of toxicity (and preventing people from drinking it as a replacement for ethanol). In contrast to methanol, DME can be used as a fuel or fuel additive for automotive transportation in a diesel engine.

DME has an inherent combination of both oxygenate functionality, for antiknock characteristics, and saturated hydrocarbon functionality, for fuel energy content. DME also imparts improved cold-start characteristics as an additive to methanol fuel and has almost twice the fuel content of methanol on a volume basis. DME can be used as a utility fuel or can serve as a blending component or replacement fuel for diesel engines. DME is also an intermediate in production of other fuel, as in ExxonMobil's methane-to-gasoline (MTG) process [19]. DME is a potential feedstock for novel oxygenated fuels [20] and is an alternative source of lower olefins such as propylene [21]. Using low CO content syngas, as derived from coal, a one-step DME process is more efficient and achieves higher syngas conversion than using two or three separate unit operations. DME can be produced with methanol in a dedicated plant, or it can be used as the fuel's production arm of a cogeneration plant to produce power and clean fuels (including even cooking fuel [22]) from coal [23].

ChemSystems Inc., the University of Ohio at Akron and the University of Missouri-Columbia, and Air Products and Chemicals Inc. (APCI) all performed primary early research toward one-step conversion of syngas to methanol in the liquid phase. This development and demonstration work in the USA was performed predominantly in the 1980s and 1990s. Air Products' work with the Alternative Fuels Program of the US DOE additionally focused on dual catalysts for one-step DME production from coal-derived syngas and led to demonstration of processes for conversion of coal-derived syngas to fuels and chemicals, including demonstration at the DOE's Alternative Fuels Development Unit (AFDU) at LaPorte, Texas [24].

In Japan, work on DME was intensified in the 1990s by NKK Corporation [25]. NKK first researched the one-step slurry DME synthesis at bench scale and then teamed with the Center for Coal Utilization, Taiheiyo Coal Mining, and Sumitomo Metal Industries to pilot a 5 t/day DME plant with funding from METI (the Ministry of Economy, Trade, and Industry). The long-range plan at that time was described as producing DME at the site of coal mining and then transporting it to user sites. In terms of fuel value per unit weight, this would be much more efficient than transporting coal itself.

Substantial research and development efforts have been devoted to optimizing product yields and process efficiencies in conversion of syngas to fuels and chemicals. Mechanistic and kinetic studies have been conducted to interpret the fundamentals of specific conversion processes. Methods and hardware to impart temperature control and stability in conversion reactors are under continuous development because of the large excess heat of reaction. In the case of syngas conversions, this research has included discovery of catalytic systems having optimized formulations containing metals with high activity in combination with cocatalysts or additives that improve activity and selectivity for the desired type of product slate. Reactor design and engineering have proceeded in tandem with the catalyst developments in syngas conversions.

## Survey of Specific Coal Liquefaction Technologies

A primary requirement of all coal liquefaction processes is that the fuel value inherent in the coal feedstock must be increased to be more similar to that of conventional liquid fuels. The ratio of elemental hydrogen to carbon, H/C, or even just simply the content of hydrogen itself, is important to determining the fit between feedstock and process options. Coal has a hydrogen content of about 5% whereas that of refined petroleum-based transportation fuel is in the range of 12–15%. Fuel derived from petroleum has an average H/C ratio of about 2 whereas that of coal itself ranges from about 0.7–1.

Table 11.1 shows the full range of process options that have been studied for converting coal to liquids [26]. These approaches include pyrolysis, direct liquefaction, coprocessing, and indirect liquefaction.

## *Pyrolysis*

In pyrolysis processes, the H/C ratio of coal is raised by rejecting carbon. The liquid products have little value, but pyrolysis is used on a massive scale to turn coal into coke for metallurgy, especially steelmaking. The coking process involves heating coal in closed vessels to very high temperatures (up to 2,000°C) in the absence of oxygen. Without oxygen present to cause combustion, at these high temperatures the coal is broken down into lighter molecular weight substances that leave the vessel in the gaseous state. A hard, porous residue containing mostly carbon and inorganic ash is left behind. The amount of volatiles is typically 25–30 wt% of the coal. Pyrolysis is also used to make chemicals.

In coking, the coal beds are deep and the temperatures are kept very high to promote resolidification of the decomposing coal. When making chemicals, the coal beds are thinner and temperatures are lower than for a more complete gasification near the melting or the decomposition temperature of coal to promote production of liquids and gases. In both of these processes, fuel gas containing carbon and hydrogen is formed in significant quantities and high quality as measured by the gas heating value (BTU/scf). Where possible, this gas has been used for process heat but any excess gas also could be sold.

Chemical products can be obtained from high-temperature coking by-products, synthesis gas, and coal liquids. The by-products of the high-temperature coking process are fuel gas, crude tar, and light oils such as benzene, toluene, and xylene (BTX), naphtha, and ammonia. Most coal- derived chemicals today are obtained from the processing of tars. Coke manufacturers burn crude coal tar in coke ovens, refine it to tar-based products, or sell it to coal tar distillers. Tar derived products include ammonium sulfate, sulfuric acid, phenols, cresols, BTX, naphthalene, pyridine, phenanthrene, anthracene, creosote, road tars, roofing pitches, and pipeline enamels. Today, synthesis gas from coal gasification also can be used for the production of petrochemicals such as BTX, sulfur, methanol, and ammonia as well as hydrogen.

The Liquids from Coal (LFC$^{TM}$) process uses mild coal gasification to upgrade low-rank coals to two fuels: a stable, low-sulfur, high-BTU solid fuel similar in composition and handling properties to bituminous coal (referred to as process-derived fuel, PDFTM) and a low-sulfur industrial fuel oil (referred to as coal-derived liquid, or CDL$^{TM}$) [27, 28]. The process was originally

**Table 11.1** Coal to liquids processes

| Mild gasification/ pyrolysis | Single-stage direct liquefaction | Two-stage direct liquefaction | Coprocessing | Indirect liquefaction, F-T | Indirect liquefaction, other |
|---|---|---|---|---|---|
| TOSCOAL – Tosco, USA | H-Coal – HRI, USA | Catalytic Two-Stage Liquefaction, CSTL – HTI/DOE, USA | MITI Mark I and Mark II – Japan | Sasol – South Africa | Methanol-to Gasoline (MTG) – ExxonMobil, USA |
| Char-Oil Energy Development, COED – FMC, USA | Kohleöl/IGOR$^+$ – Ruhrkohle & Veba Öl, Germany (today BP, UK) | H-coal CSTL – Axens, France | Cherry P – Osaka Gas, Japan | Gasel – Axens, France | Direct DME –JFE, Japan |
| Liquids from Coal, LFC – Encoal, USA | Solvent Refined Coal, SRC-I and SRC-II – Gulf Oil, USA | Liquid Solvent Extraction, LSE – British Coal, UK | Solvolysis –Mitsubishi, Japan | GTL.F1 (Statoil, PetroSA, Lurgi), Switzerland | Direct DME –NEDO, Japan |
| EERC/AMAX – Univ. N. Dakota, USA | Exxon Donor Solvent EDS – US | Brown Coal Liquefaction, BCL – NEDO, Japan | Mobil – USA | Syntroleum – USA | LPDME – air products, USA |
| Lurgi-Ruhrgas – Germany | NEDOL – NEDO, Japan | Chevron Coal Liquefaction, CCLP – USA | Pyrosol –Saabergwerke, Germany | Rentech - USA | Topsøe DME –Haldor Topsøe, Denmark |
| Pyrolysis of coal – Inst. of Gas Technol., USA | Imhausen High-Pressure – Germany | Lummus Integrated Two-Stage Liquefaction, ITSL –USA | Chevron – USA | Shell Middle Distillates (SMDS) – Royal Dutch Shell, Netherlands | |
| | | | Lummus Crest – USA | BP, UK | |

(continued)

**Table 11.1** (continued)

| Mild gasification/ pyrolysis | Single-stage direct liquefaction | Two-stage direct liquefaction | Coprocessing | Indirect liquefaction, F-T | Indirect liquefaction, other |
|---|---|---|---|---|---|
| Twin screw – Coal Technol. Corp., USA | Conoco zinc chloride – USA | Kerr-McGee ITSL –USA | | | |
| | Suplex – Sasol, South Africa | Amoco CC-TSL – USA | Alberta Research Council, ARC –Canadian Energy Dev., Canada | ExxonMobil, USA | |
| | | Pyrosol – Saabergwerke, Germany | CANMET AOSTRA – Canada | ConocoPhil-lips, USA | |
| | | Mitsubishi Solvolysis – MHI, Japan | Rheinbraun – Germany | | |
| | | Consol Synthetic Fuel, CSF – Consolidated Coal, USA | Technical University Clausthal, TUC – Germany | | |
| | | Supercritical Gas Extraction, SGE – British Coal, UK | UOP Slurry-Catalyzed – US | | |
| | | | HTI – US | | |

developed by SGI International and then taken further by the ENCOAL Corporation. Development was about 50% financially supported by \$90 million from the US DOE.

A PDU was constructed in 1986 and then a 1,000 t/day demonstration unit was constructed and operated for over 4 years near Gillette, WY in the mid-1990s. Later configurations included a vibrating fluidized bed (VFB) as part of a deactivation loop for solid product, where quenched solids are partially fluidized and exposed to oxidative deactivation to reduce the tendency for spontaneous ignition. The residence time, oxygen content, and temperature of the gas stream are selected to deactivate the coal within the VFB unit. Hydration further stabilizes the solid PDF product, and after treatment in the VFB system the solids are cooled and a controlled amount of water is added to rehydrate the PDF to near its equilibrium moisture content. The CDL portion of the product is condensed. The technology appears to be moving forward, with the transformation of ENCOAL into the Saskatchewan-based company NuCoal, which has memos of understanding with several potential partners including Sinopec.

## *Direct Liquefaction*

In DCL, the H/C ratio of coal is raised by adding hydrogen. The objective of hydrogenation is to maximize the yield of distillate fractions that can be subsequently converted into fuels. Distillate materials comprise the naphtha fraction, typically boiling $C_5$ – 215°C, plus the middle distillate fraction typically boiling between 150°C and 370°C, with the exact cuts depending on product specifications. The liquid products are then refined into diesel, gasoline, and other petroleum products.

The traditional mechanistic view has been that upon heating, inherently weak bonds in coal are broken, and that the resulting and then hydrogen is rapidly transferred from the solvent to cap the resultant free radicals. Rapid hydrogenation is indeed critical to liquid yield; preventing retrogressive reactions leading to formation of solid char. While this picture is largely true, in-depth research showed that it is insufficient. Concurrent with this pathway are solvent-mediated hydrogenolyses of strong C–C bonds – bonds too strong undergo simple thermolysis under reaction conditions – engendered by addition of a free hydrogen atom or a bimolecular H-transfer from solvent species and their radicals [29]. Inclusion of these pathways helps rationalize many otherwise inexplicable phenomena in coal liquefaction, and also helps identify conditions that maximize utilization of hydrogen for liquid production rather than produce the less desirable light gases.

Hydrogen is also critical in the DCL process to reduce oxygen, sulfur, and nitrogen in the coal feedstock. These elements are removed from the crude liquid product as water ($H_2O$), hydrogen sulfide ($H_2S$), and ammonia ($NH_3$), respectively. Oxygenates are reduced to hydrocarbons, because their oxygen content would otherwise lower the fuel value. The nitrogen and sulfur compounds are removed because they would otherwise poison the cracking catalysts in refining operations

**Table 11.2** Typical operating conditions for direct coal liquefaction

| Operating parameter | Typical operating conditions |
|---|---|
| Temperature (°F) | 750–850 |
| Pressure (psia) | 1,500–3,000 |
| Residence time (h) | 0.5 |
| Solvent to coal ratio (lb/lb) | 1.5/2 |

downstream of the DCL unit. DCL products are only partially refined at the DCL plant. They must be further refined into finished liquid fuel products at conventional refineries, where additional $H_2$ is added to bring the H/C ratio up to about 2 for the fuel products.

Many factors affect the rate and extent of coal liquefaction, including temperature, hydrogen partial pressure, residence time, coal type and properties, solvent properties, solvent to coal ratio, ash composition, and catalyst. In general, liquefaction is promoted by increasing the temperature, hydrogen partial pressure, and residence time. However, if the temperature is too high, gas yield is increased and coking can occur. Solvent to coal ratio is important. If the ratio is too low, there will be insufficient hydrogen transfer activity. If the ratio is high, a larger reactor will be necessary to provide the required residence time. Typical operating conditions are listed in Table 11.2.

For the most highly developed processes, coal conversion can be as high as 90% on a mineral ash free basis, with a $C_{5+}$ distillate yield up to 75% and hydrogen consumption of 5–7% wt. When an external catalyst is used, it is typically some combination of cobalt, nickel, and molybdenum on a solid acid support such as silica-alumina [30]. In slurry hydrogenation processes, catalyst lifetime is fairly short because of the large number of potential poisons present in the system.

DCL process concepts have been based on a variety of operating conditions and catalysts and can be performed in a single stage or two stages. In a typical single-stage process, coal is contacted with hydrogen or with hydrogenated solvent at ca. 450°C. The yield of fuel oil is about 3 barrels of oil/t of coal with bituminous coals. The light oil fraction is recycled. A high yield of light hydrocarbons is obtained and the efficiency of hydrogen utilization is low. The product is difficult to refine because of its high aromaticity and nitrogen content. Single stage liquefaction is not as effective for subbituminous coals as for higher rank coal.

Most current DCL processes have settled on a multistage approach. Multistage DCL typically uses dispersed catalysts in the liquefaction stage followed by use of supported catalysts in subsequent upgrading stages, with different pressures and temperatures for the different stages. The technologies were demonstrated in several laboratory and pilot plants and the products were more easily refined than many crude oils. There are many variations of two-stage DCL processes, including use of catalyst in the two stages (both stages being catalyzed, versus thermal and catalyzed stages), distillation between stages, whether solids are separated between the two stages or only after the second stage, recycle of ashy bottoms, and the operating hydrogen balance.

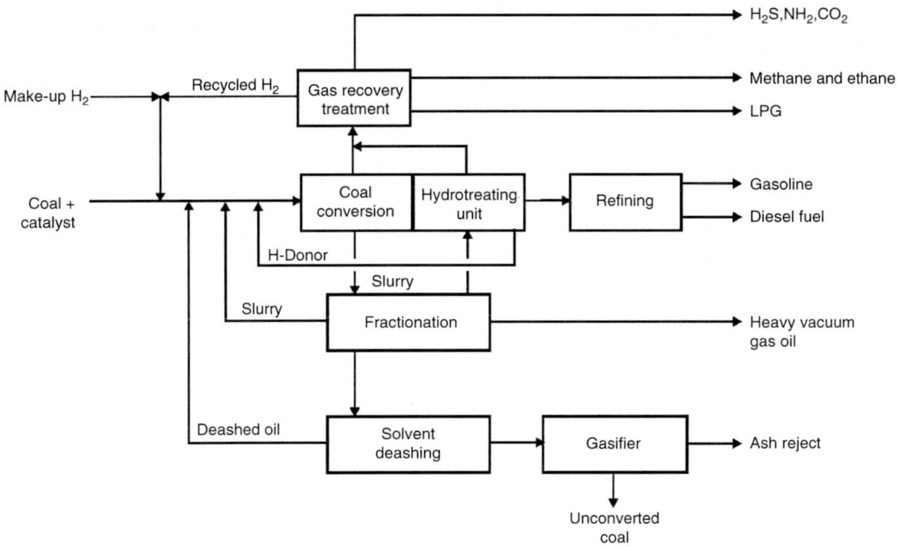


**Fig. 11.1** Block flow diagram of a typical direct coal liquefaction process [31]

A block flow diagram (BFD) for a typical direct coal liquefaction plant using coal slurry hydrogenation is shown below in Fig. 11.1. Coal is ground and slurried with a process-derived solvent, mixed with a hydrogen-rich gas stream, preheated, and sent to a two-stage liquefaction reactor system where the organic fraction of the coal dissolves in the solvent and the dissolved fragments react with hydrogen to form liquid and gaseous products. Sulfur in the coal is converted to hydrogen sulfide, nitrogen to ammonia, and oxygen to water. The reactor products go to vapor/liquid separation. The gas is cleaned, and after removal of a purge stream to prevent buildup of inert materials, the gas is mixed with fresh hydrogen and recycled. The liquid is sent to vacuum fractionation for recovery of distillates. Heavy gasoil is recycled as process solvent and vacuum bottoms are gasified for hydrogen production.

Ash from the gasifier is sent to disposal. Heavy direct liquefaction products contain polynuclear aromatics that are carcinogenic. This problem can be solved, however, by recycling to extinction all material boiling above the desired product end point. A recent development, shown in Fig. 11.1, is to put the ash residue and carbon from recycled solvent directly through the gasifier to produce hydrogen (and power).

The longest experience base with DCL in the USA was a PDU at the R&D facility of Hydrocarbon Technology Inc. (HTI) that consumed 3 metric tons of coal/day. Among the many liquefaction processes developed and tested at pilot plant scale have been Kohleöl (by Ruhrkohle, Germany); Exxon Donor Solvent (by Exxon, now ExxonMobil, USA); Solvent Refined Coal-I (SRC-I) and SRC-II (*by Gulf Oil, USA.*); H-Coal (by HRI, now HTI, and DOE, USA); Catalytic Two-Stage

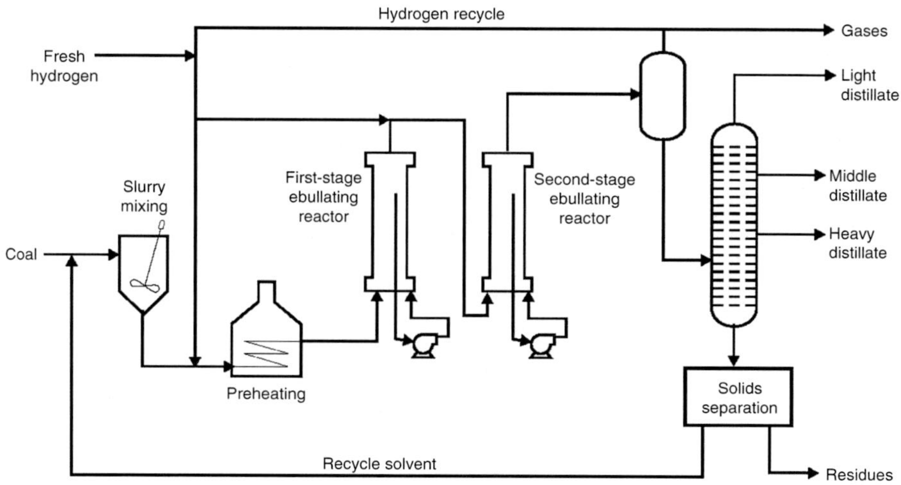


**Fig. 11.2** Process flow schematic of two-stage CTSL direct liquefaction process [33]

Liquefaction (CTSL, by HRI and DOE, USA); Liquid Solvent Extraction (LSE, by British Coal Corporation, UK); Brown Coal Liquefaction (BCL, by NEDO, Japan) and NEDOL (by NEDO, Japan). Only the CTSL, LSE, BCL, and NEDOL processes continued in development at pilot plant scale beyond the late 1980s. Between 1975 and 2000, the USA alone invested $3.6 billion in direct liquefaction technologies; some are now licensed to China [31].

H-Coal was an early single-stage DCL process piloted by HRI and Ashland Synthetic Fuels, Inc. in a 600 t/day pilot plant adjacent to Ashland's refinery in Cattlesburg, Kentucky [32]. The process consists of slurry preparation followed by catalytic hydrogenation/hydrocracking at a single temperature and pressure, 450°C and 2,200 psia respectively, in an ebullating bed reactor. The ebullating bed is a backmixed three-phase isothermal reactor that keeps catalyst in constant motion. It uses the upward flow of liquid to expand the catalyst bed and to distribute liquid, gas, and catalyst evenly across the reactor. The reactor has the characteristics of a uniform temperature similar to continuous stirred tank reactor.

The H-Coal process was a starting basis for subsequent process development sponsored by the US DOE over about 3 decades time. A principal focus of this long-term project was R&D of catalysts that were tolerant of, and would not be rapidly deactivated by, coal-derived mineral matter in the reactor. Typical early catalysts showed rapid deactivation because of coking and metals laydown with loss of surface area. A later process version sought to address this deactivation problem by using two coupled reactor stages, one operating at 400°C and the other operating at 420–425°C. The catalyst was also changed from cobalt/molybdenum/alumina to nickel/molybdenum on bimodal alumina. This Catalytic Two-Stage Liquefaction (CTSL) process is shown in a block flow diagram in Fig. 11.2. The role of catalyst in the first stage of the CTSL process is to promote hydrogenation of the solvent, stabilization of the primary liquefaction products, and hydrogenation of the primary and recycle resid. In the second stage, the catalyst promotes heteroatom removal and thus product quality improvement, conversion of resid to distillate, and secondary conversion to lighter products. The catalyst also aids in avoiding dehydrogenation.

In the first ebullating bed reactor, coal is dissolved with minimal catalysis at about 400°C for bituminous coal (or 435–460°C for subbituminous coal) and 170 bar. The catalyst for this first reaction is typically nickel-molybdenum on alumina and aids in re-hydrogenating the recycled process solvent. The resultant heavy coal liquids are fed to a second ebullating bed reactor and hydrogenated at about 440°C and 170 bar with a catalyst such as that used in the first reactor, or a high-activity iron catalyst. After separation and depressurization, the product of the second reactor is sent to a distillation column at atmospheric pressure, where distillate up to 400°C is removed, with yield from coal of up to 65%. The distillation residue is unconverted coal and asphalt, which are typically gasified for captive fuel value or feed hydrogen.

The CTSL process was further developed over almost 15 years in a pilot plant at the Wilsonville liquefaction test facility. Beginning in 1973, a 6 t/day coal liquefaction plant was built by the Edison Electric Institute (EEI) and Southern Company Services, and the US DOE became the primary sponsor in 1976. Amoco Oil Co. (now BP) joined the project in 1984.

#### Developments at the Wilsonville Facility in the USA

Because of the role of Wilsonville in consistently demonstrating success or failure of a large number of concepts, particular attention has been paid to the results from this facility. The plant began operation in 1974 in a single-stage mode referred to as Solvent Refined Coal (SRC-II, which made liquid fuel as opposed to SRC-I, which made a solid fuel from coal) but evolved to a two-stage operation utilizing two ebullating bed catalytic reactors. Initial efforts in two-stage liquefaction focused on catalytic upgrading of the thermal products, referred to as Nonintegrated Two-Stage Liquefaction (NTSL). This configuration was termed nonintegrated because the coal-derived resid hydrocracking step did not interact with the primary thermal part of the plant. The NTSL concept proved excessively inefficient because of the high hydrogen consumption associated with the thermal part of the operation.

Coal throughput, expressed as space velocity per unit reactor volume, was substantially improved in going from NTSL to Integrated Two-Stage Liquefaction (ITSL) mode. In ITSL, a short contact time thermal reactor is closely coupled to an ebullating bed catalytic reactor where hydrocracking occurs. The thermal resid produced in the ITSL at short contact times is more reactive toward expanded-bed hydrocracking, which permits operation of the hydrocracker at lower temperature and minimizes gas make. Minimizing formation of light hydrocarbon gases in turn would be expected to reduce the rate of catalyst coking, improve hydrogen utilization efficiency, and increase selectivity for higher distillate. A 35% increased yield from $C_{5+}$ distillation was observed.

From 1985 to 1992, work at Wilsonville focused on development of the CTSL, utilizing ebullating bed catalytic reactors in both stages. The ebullating bed allows use of the substantial heat of reaction to preheat coal slurry feeds, and this improves process thermal efficiency. The converted and refined products are separated at high pressure and temperature into lighter oils and gases, and heavy oils and solids. Following pressure letdown and heat exchange, the lower boiling oils are fractionated while the heavier oils are separated from the unconverted coal and ash and recycled as the slurry solvent for the coal. Initial work indicated that distillate yields as high as 78% could be obtained by operating the first stage at low severity (399°C) and by using a large pore bimodal Ni-Mo catalyst. Significant improvement in distillate production can be achieved but again at the cost of increased hydrogen consumption.

Also explored were Reconfigured Integrated Two-Stage Liquefaction (RITSL), where solvent de-ashing was practiced after the hydrocracking stem, and Close-Coupled Integrated Two-Stage Liquefaction (CC-ITSL), where two reactors (thermal/catalytic) were linked directly without any intervening processing steps. Incremental improvements in distillate yield and selectivity were realized by changing the process configuration, but at the expense of increased hydrogen consumption.

The Wilsonville facility was closed in 1992 and was decommissioned shortly thereafter, but development work funded by the DOE continued at the Lawrenceville, N.J. facility [34–36]. The process has now evolved into a generic composite of much of the liquefaction development work funded by the DOE in the 1980s and 1990s. Most recently, a close-coupled configuration has been adopted in which both stages use an active supported catalyst, and HTI has included an in-line hydrotreater after the second-stage reactor to improve product quality. In 1997, with support from the DOE, HTI signed a 2-year agreement with the Chinese Coal Research Institute (CCRI) to carry out a feasibility study for a direct coal liquefaction plant using Shenhua coal [37].

The Kerr-McGee critical solvent de-ashing (CSD) technology (also termed the Residual Oil Solvent Extraction, or ROSE, process) was also tested at Wilsonville, resulting in the development of alternative methods for solids removal from primary liquefaction products. Other DCL process developments in the USA have included the single-stage Exxon Donor Solvent (EDS) process, and two-stage processes have included Lummus's Integrated Two-Stage Liquefaction (ITSL), Kerr-McGee's ITSL, Chevron Coal Liquefaction (CCLP), Consol Synthetic Fuel (CSF), and Amoco's (now BP's) Close-Coupled Two-Stage Liquefaction (CC-TSL) process.

In 1996, Sandia National Labs published results of its study on the efficacy of hydrogenation for removing nitrogen and sulfur from the CTSL crude product [38]. The kerosene fraction from HTI's first proof-of-concept run was used and denitrogenation and desulfurization were achieved under relatively mild conditions. Processing at the highest severity condition (388°C, 103 psig $H_2$, 1.5 g/h/g-cat) decreased nitrogen from 645 ppm in the feed to 7.9 ppm in the hydrotreated product. Sulfur content of 239 ppm in the feed was decreased by hydrotreating to less than 20 ppm.

In the early the 1990s, HTI was engaged in DOE-sponsored research on coprocessing of coal and plastics waste using DCL.

## DCL Developments in Europe

The Kohleöl DCL process [39] is based on technology used at commercial scale in Germany, until after the Second World War, and then improved with the advent of the Integrated Gross Oil Refining ($IGOR^+$) process [40]. Coal is slurried with a process-derived recycle solvent and a "red mud" disposable iron catalyst. The pressurized and preheated slurry is sent to an up-flow tubular reactor operated at about 470°C and 300 bar $H_2$. Products from the top of the reactor pass to a hot separator, the overheads from which are hydrotreated at 350–420°C in a fixed-bed reactor maintained at the same pressure as the main reactor. The hydrotreated products are depressurized and cooled in two stages. The liquid product from the first stage is recycled for use as solvent in the slurry step. The liquid product from the second stage is routed to an atmospheric distillation column to give light oil ($C_5$ – 200°C bp) and a medium oil (200–325°C bp) product. The bottoms from the original hot separator pass to a vacuum distillation column to recover distillable liquids. These are added to the hydrotreating reactor feed and are subsequently largely recycled as solvent. The vacuum column bottoms consist of pitch, mineral matter, unreacted coal, and catalyst, which in commercial operation would be used as a gasifier feedstock for hydrogen production. Greater than 90% conversion can be obtained with bituminous coals, with liquid yields in the range 50–60% on a dry ash-free coal basis. The product distribution is about 19% $C_1$–$C_4$ hydrocarbons, 25% light oil ($C_5$ through 200°C bp), 33% medium oil (200–325°C bp) and 22% unreacted coal and pitch.

The LSE process was developed more fully than other European DCL processes, by British Coal Corporation between 1973 and 1995 [12]. In the LSE process, coal is slurried with a process-derived recycle solvent, preheated, and passed to a noncatalytic digestion step, which consists of two or more continuous-stirred tank reactors. These reactors operate in series at a temperature of 410–440°C and 10–20 bar. The solvent acts as a hydrogen donor, transferring up to 2% by weight of hydrogen to the coal. The digester product is partially cooled, filtered to remove unreacted coal and ash, and dried. The filtered coal extract is distilled to recover the light oil wash solvent and is then preheated, mixed with $H_2$, and sent to one or more ebullating bed reactors in series. There is no interstage separation and the reactors operate at 400–440°C and about 200 bar. The cooled, depressurized reactor products are distilled with cut-point adjusted to maintain solvent balance, resulting in a product that typically boils below 300°C. The column bottoms are routed partially to a vacuum distillation column, used to control the level of pitch in the recycle solvent. The overheads from this column are recombined with the main atmospheric column bottoms stream and recycled as the solvent to the slurrying step. The product distribution is about 15% $C_1$–$C_4$ hydrocarbons, 50% light oil ($C_5$ through 300°C bp), 50% solvent surplus (300–450°C bp), 24% filter cake organics, and less than 1% pitch. A 2.5 t/day pilot plant was built and operated for 4 years at Point of Ayr, North Wales but has since been decommissioned.

## DCL Developments in Japan

The Sunshine project pursued in Japan examined three approaches to direct liquefaction of bituminous coal: solvolysis, solvent extraction, and direct hydrogenation of coal. From these studies, NEDO found that increasing catalyst activity correlated with increased liquid yield; that hydrogenation was most effective for liquefaction at pressures in the 150–250 atm range and temperatures in the 400–450°C range; and that increased solvent viscosity aided production of light oil as fuel. The approaches were consolidated over a time span of 18 years, with demonstration of the resultant NEDOL process completed in 2000. The process uses 170 atm $H_2$ in the liquefaction reactor at 450°C. A process flow diagram of the NEDOL process is shown in Fig. 11.3. [11].

Like H-Coal, the NEDOL process is single-stage, meaning that all liquefaction reactors are operated at a single temperature and pressure. The NEDOL process has been further developed, however, and is said to have low severity conditions and high distillate yield comparable to those of the CC-ITSL and IGOR$^+$ processes.

NEDO also developed the BCL process, in the early 1980s, to use brown coal (lower rank, wetter coal), to the scale of 50 t/day in a pilot-plant constructed at Morwell in Victoria, Australia [41]. It was operated during 1985–1990, processing a total of about 60,000 t of coal. Operations ceased in October 1990. In the 1990s, NEDO worked with Indonesia on developing technology for use with that country's brown coal. Also in Japan, Mitsubishi Heavy Industries (MHI) researched a two-stage DCL process referred to as Mitsubishi Solvolysis.

The development of DCL is depicted from a historical process economic perspective in Fig. 11.4, which shows the cost that the oil product would need to sell for to make specific DCL processes economically competitive. As the process technology advanced the liquids yield from bituminous coals increased from 2.7 to 4.3 bbl/t. The increase in the yield of liquids are partly due to the fact that the later process designs used steam reforming of natural gas to provide hydrogen instead of gasifying a portion of the coal. The costs decreased concomitantly, but today direct coal liquefaction still remains highly capital-intensive.

Overcoming the techno-economic difficulties of DCL could enable a lower cost overall CTL process than ICL, given that DCL does not require the gasification step. Although largely dormant for over a decade in the USA, Europe, and Japan; direct liquefaction is undergoing renewed research effort, most actively pursued in the People's Republic of China (PRC).

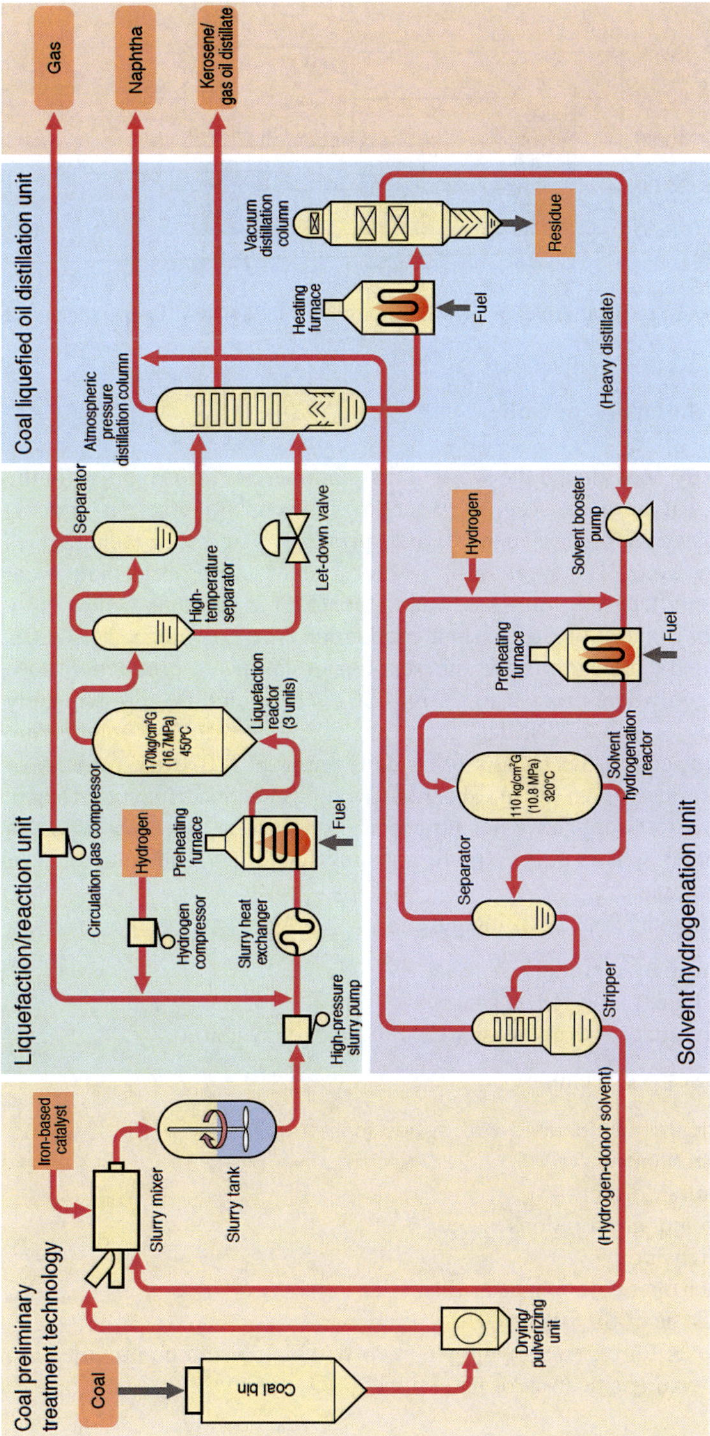


**Fig. 11.3** Flowchart for the NEDOL process for direct liquefaction of bituminous coal

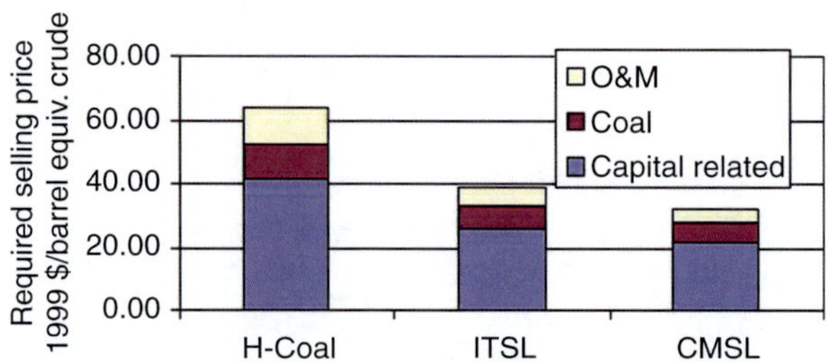


**Fig. 11.4** Relative economics for representative DCL processes [36] (O&M is operations and maintenance)

## Recent DCL Developments in China

Shenhua has recently constructed the world's first commercial DCL facility, in the Inner Mongolia Autonomous Region. Shenhua's direct liquefaction process combines elements of German, Japanese, and, primarily American technologies with its own innovations. The single-train reactor system and process plant were designed by HTI and licensed from Headwaters Inc. (HTI is currently a subsidiary of Headwaters), and operations of its first production line commenced in 2008. Shenhua has planned expansion in two phases, for a total of ten production lines with projected oil equivalent products of 10 million t/year, including approximately 150,000 barrels/day of middle distillate fuels (naphtha + diesel). The engineering design of the plant, called SH-I, calls for a daily input of 7,000 metric tons of Shenhua's subbituminous coal. Also, Shenhua Group, Shanghai Huayuan Group, and Shanghai Electric Group have set up a coal liquefaction research center in Shanghai. The center will mainly explore and develop direct and indirect coal liquefaction technologies.

In summary, advantages of DCL include:

- High-octane gasoline is produced
- More energy efficient than ICL
- Products have a higher energy density than indirect conversion

Disadvantages of DCL include:

- Processing difficulties associated with solids separation
- High aromatic content
- Low cetane number diesel (45)
- Potential water and air emissions issues
- High capital intensity
- May have higher operation expenses than ICL
- Lower feedstock flexibility than ICL
- High requirements for external hydrogen supply (which has to be provided by, e.g., additional coal gasification or natural gas)

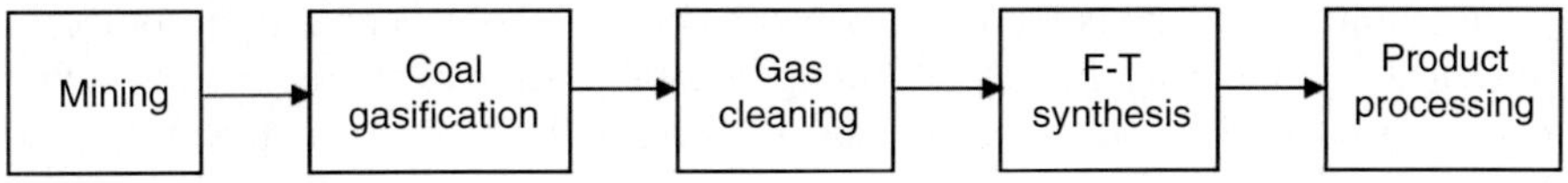


**Fig. 11.5** Key process steps for indirect coal liquefaction [45]

## *Coprocessing*

Coprocessing combines waste materials having high hydrogen content (relative to coal) with coal as feed for conversion to liquid fuel products. Over 44 billion pounds of plastic waste materials, 280 million automotive tires, and about 1.2 billion barrels of waste oil are generated in the USA each year [42]. Coprocessing with coal would allow capture of energy content in these waste materials instead of simply disposing of them. It has frequently been stated that synergistic interactions between coal and heavy oil can result in distillate yields and coal conversion levels greater than expected from individually processing the coal and heavy oil. Several companies researched coprocessing, including Chevron's 6 t/day coprocessing plant run in 1983 at the company's Richmond, California site. The research included synergistic incorporation of pretreated petroleum vacuum resids with coal as feedstocks, and demetallation of resids was also performed [43]. The process was said to have good operability.

The poorer solvent properties of heavy oils detracted from the conceptual simplicity and led to use of forcing conditions and increased hydrogen consumption that have adversely affected economics of coprocessing [44]. Although developments in coprocessing have been essentially on hiatus for over a decade, the topics of energy and reuse of materials remain prominent. Improvements in technologies for gasification, solvation, and $CO_2$ capture might lead to a resurgence of this CTL approach.

## *Indirect Liquefaction*

Indirect coal liquefaction typically refers to processes that first generate synthesis gas by gasification of coal followed by production of solid, liquid, and gaseous hydrocarbons via catalytic reduction of CO in subsequent stages of the process. Frequently, the term ICL refers to the synthesis of liquid transportation fuels by means of Fischer–Tropsch synthesis catalyst. Other ICL technologies include methanol and/or dimethylether from syngas, or methanol-to-gasoline (MTG). MTG was commercialized in 1986 and uses ZMS-5 zeolite catalyst to convert methanol to gasoline.

A basic block flow diagram of key process steps in F-T ICL is shown below in Fig. 11.5.

After mining and coal preparation, beneficiated coal (coal that has been subjected to physical grinding and chemical cleaning methods during coal preparation) is first gasified with oxygen and steam to produce syngas. This syngas is cleaned of all impurities and sent to the F-T reactors where most is converted to liquid hydrocarbon fuels. The liquid products from F-T ICL contain zero sulfur and essentially zero aromatic hydrocarbons, and therefore ultra clean diesel or jet fuels can be produced after additional refining. Carbon dioxide is produced mainly during water gas shift (vide infra) and the F-T synthesis reactions, and can be captured for subsequent storage or use. The unconverted synthesis gas can be either recycled back to the F-T reactors, to maximize liquid fuels production, or combusted in a gas turbine combined cycle power plant to generate internal and external power. Thus ICL plants can be configured to produce liquid fuels only, on a freestanding basis, or to produce a combination of liquid fuels and power for export.

The other major ICL approach produces methanol from the syngas, via a different set of chemistries from F-T, and uses the methanol as a chemical or as a fuel, typically as a blend with petroleum derived gasoline. The methanol serves as oxygenate and fuel extender. Dimethylether (DME) also can be produced from syngas, either directly or via dehydration of methanol.

Synthesis of fuels and chemicals from syngas does not occur in the absence of appropriate catalysts. The basic concept of a catalytic reaction is that reactants absorb onto the catalyst surface, rearrange atoms, and combine into products that desorb from the surface. One fundamental functional difference between different syngas conversion catalysts is whether or not the adsorbed CO molecule is cleaved during reaction at the catalyst surface. In Fischer–Tropsch synthesis, C–O bond cleavage is a necessary reaction step. In methanol synthesis, the CO bond remains intact. Hydrogen plays two roles in catalytic syngas conversions, as a reactant needed for hydrogenation of CO to give the hydrocarbon moieties, and as a reductant to activate and regenerate the metal surface.

Much of the research and process development on coal liquefaction after the 1990s was aimed at matching the fuels synthesis conditions with modern, efficient coal gasifiers such as those developed by Uhde, Texaco (now GE), Shell, and ConocoPhillips. A block flow diagram of a typical F-T CTL process to produce naphtha and diesel middle distillate cuts is shown below in Fig. 11.6, with the liquefaction end of the process indicated within the dash-lined boundary. Coal feed is prepared in the milling and drying section and then sent to the gasifier where it is fully converted to raw synthesis gas (or a coal slurry gasifier may also be employed instead).

In the F-T synthesis step, the syngas is converted to raw product. The lighter ends of the raw F-T product are partially recycled internally and the remainder is sent to the heavy ends recovery unit. Oxygen for coal gasification is supplied by an air separation unit (ASU, not shown). A part of the raw synthesis gas is treated in the water gas CO-shift unit to convert steam and CO to hydrogen and $CO_2$, after which this treated gas is remixed with the untreated portion. The split between the portions is controlled such that the required $H_2/CO$ ratio of the F-T synthesis is always met. The shifted syngas is then submitted to an acid gas removal (AGR)

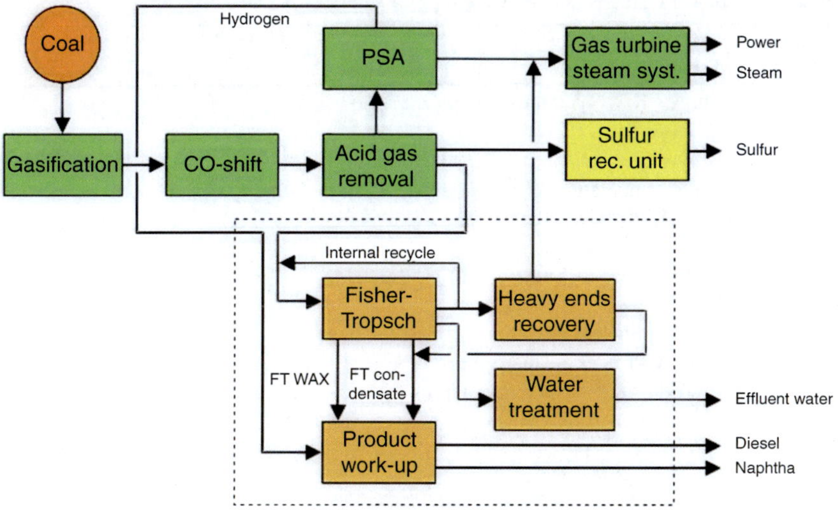


**Fig. 11.6** Fischer–Tropsch coal to liquids process [15] (Figure courtesy Uhde Corporation of America)

step, where COS, $H_2S$, and $CO_2$ are removed to produce clean syngas feed for the F-T synthesis unit. The Rectisol adsorption process is commercially proven in CTL applications and is typically deployed for AGR, applying methanol at low temperatures as a solvent. However, a more energy efficient system, such as polyethylene glycol (PEG)- or glycol diether-based processes (e.g., Selexol or Genosorb) may be used instead of Rectisol, although in these processes the COS has to be removed in a separate, COS hydrolysis step. A small fraction of the clean syngas is sent to a pressure swing adsorption (PSA) unit for production of pure hydrogen, as $H_2$ is required in the product workup section, and the major portion of the syngas is routed to the F-T plant.

The tail gas from the heavy ends recovery unit is sent to a combined cycle plant to produce sufficient power for the entire CTL complex. The F-T waxes and FT condensate products are sent to the product upgrade section, where they are hydroprocessed with hydrogen (from the PSA unit) to produce primarily diesel product. A significant amount of water is coproduced with the hydrocarbons during F-T synthesis hydrocarbons. This water fraction contains impurities such as oxygenates, which have to be removed in a water treatment step. Sour gas ($H_2S$, COS) removed in the AGR unit is sent to a sulfur recovery unit, a Claus plant, where it is converted into elementary sulfur. The tail gas from the Claus plant is hydrogenated and recycled to the acid gas removal unit, or it can be treated in a SCOT (Shell Claus Off-gas Treating) treatment unit. Using the SCOT process, the sulfur components in the Claus tailgas are catalytically converted to hydrogen sulfide, which is removed by solvent extraction and recycled to the Claus plant, and the off gas from the absorber is incinerated.

**Table 11.3** Types of coal gasification technology

| Gasification type | Technology |
|---|---|
| Moving bed | Lurgi – Dry Gas (Germany) |
| | FBDB – Sasol (South Africa) |
| | British Gas and Lurgi – BGL (UK) |
| Fluidized bed | High Temperature Winkler (HTW) – RWE, formerly Rheinbraun – marketing byUhde (Germany) |
| | Kvaerner (Finland) |
| | Foster-Wheeler (Switzerland) |
| | GTI/SES – U-gas (USA) |
| Entrained | SCGP – Shell (USA) |
| | Koppers-Totzek – Uhde (Germany) |
| | PRENFLO$^{TM}$ – Uhde (Germany) |
| | SFG – Siemens (Germany) |
| | GE (USA) |
| | ConocoPhillips (USA) |

As an alternative to the recovery of elemental sulfur, sulfuric acid can be produced from the sour gas components.

The yield of products in a fully self-sufficient CTL plant based on hard coal gasification is approximately 1,875 barrels per ton of coal. The amount of sulfur produced depends on the concentration of sulfur in the coal.

## Gasification

Gasification breaks down coal into molecules of much lower molecular weight, CO and $H_2$, usually using high temperature in the presence of oxygen at high pressure. Gasification technologies introduce air and/or oxygen, often with steam to moderate the temperature in the combustion zone of the gasifier. The product is syngas, a mixture of hydrogen and carbon monoxide. The syngas is subsequently converted to liquid fuels and chemicals in ICL processes. The three main types of gasifiers are the moving bed, fluidized bed, and entrained flow reactors. Gasification technology alternatives are grouped under the three technology types in Table 11.3. All three types work well and are in commercial use.

### Moving Bed Gasifiers

The moving bed may use co-current or counter-current flows and is sometimes referred to as a fixed bed because the coal bed is kept at a fixed, constant height. The moving bed gasifier first provided a convenient means to handle solid residuals (ash) from the gasification. Moving bed gasifiers can accommodate coal with moisture content as high as 35%, as long as the ash content is less than 10%.

Depending on the temperature at the base of the bed of coal, the ash is either dry solid or a molten slag. If excess steam is added, the temperature can be kept below the ash fusion point. The coal bed typically rests on a rotating grate through which dry ash residue falls for removal. To reduce steam usage, British Gas and Lurgi developed a slagging bottom gasifier (BGL gasifier) in which the ash is allowed to melt and drain off through a slag tap. This gasifier has over twice the capacity per unit of cross-sectional area compared to the dry bottom gasifier. Allied Syngas, Advantica, and Envirotherm offer the BGL technology commercially.

### Fluidized Bed Gasifiers

In the fluidized bed gasifier, the feed is fluidized in air, or sometimes oxygen, and steam with upward, turbulent flow for efficient mixing. The temperature is normally below ash melting temperature. The ash is removed as dry particles, which is advantageous if the ash is highly corrosive and would damage the reactor if removed as a molten slag (a problem when gasifying high-sulfur coal or biomass). Throughput of the fluidized bed gasifier is higher than for the moving bed. The low operating temperature of the circulating fluidized bed reactor technologies of Kvaerner and Foster Wheeler are particularly appropriate for low-rank coal (or other reactive feed such as biomass). The higher pressure of the HTW gasifier technology is suitable for coal (and biomass) [46].GTI developed an agglomerating ash fluidized bed gasifier for its U-Gas process in which crushed limestone for sulfur capture is injected with the coal, and adding the limestone aids in processing ashes having very high fusion temperatures. Char and ash that exit the gasifier with the product gas are recycled to the hot agglomerating and jetting zone, where temperatures are high enough to pyrolyze freshly introduced coal, gasify the char, and soften the ash particles. The ash particles stick together and fall to the base of the gasifier where they are cooled and removed. The agglomerating fluid bed gasifier can be blown by either air or oxygen. The U-Gas technology is licensed by Synthesis Energy Systems (SES).

Alkaline metal compounds, especially chlorides and carbonates, can be used as catalysts for coal gasification in a fluidized bed including such compounds as potassium carbonate, zinc chloride, and sodium carbonate.

### Entrained Flow Gasifiers

Entrained flow technologies particularly target issues of solids handling in coal gasification. When bituminous coals are heated to 300–350°C, the particles tend to swell and agglomerate in producing a consolidated cake. The handling of this caking char and the heavy tars that accompany it has been critical to the development of gasification processes. The agglomerate that forms can disrupt gas flow patterns and lower thermal efficiency.

In the entrained flow reactor technology, coal is fed as a dry, pulverized solid or as a fine slurry with liquid and is most commonly gasified with oxygen (rather than air) in co-current flow. It is operated above the ash-melting point, which means that the molten ash forms a liquid slag inside the gasifier that runs down the walls of the gasifier and is removed through a bottom hole. The use of oxygen at high pressure and temperature is a source of both high throughput and versatility. Typically a dry-feed gasifier such as SCGP or PRENFLO can process a wide range of coals, from low-grade soft brown coal to top-grade hard anthracite. These dry-fed entrained-flow gasifiers have single-train capacities of about 3,500–4,000 t/day. Most of the ash is removed as slag.

Uhde's PRENFLO entrained gasifier technology has operated successfully for the last 12 years in the world's largest dry-fed IGCC plant, located in the Elcogas plant in Puertollano, Spain. PRENFLO employs dry-feeding, multiple side-fired burners, and a cooled membrane wall for robustness under the slagging conditions inside the gasifier. The raw gas leaving the PRENFLO™ gasifier is first cooled down to approximately 900°C in a gas quench system with a recycle gas stream. The sensible heat of the raw gas is used to produce saturated high pressure steam, and saturated medium pressure steam, both of which are ultimately used in the steam turbine of the power block to produce power. The raw gas is next sent to gas conditioning operations; consisting of filtration to remove particulates, scrubbing with water to remove chlorine and other soluble components, removal of COS and HCN in a hydrolysis reactor, and removal of $H_2S$ using a methyl-diethanolamine (MDEA) solution. The $H_2S$ fraction from the desulfurization unit is treated in the Claus unit with tail-gas recycle, where liquid elemental sulfur is produced for sale on the local market. Electrical power is generated by a single combustion turbine and a single steam turbine.

A slurry feed may be used with low rank coals as these can exhibit serious particle decrepitation, dusting, and even spontaneous combustion when exposed to drying conditions upon storage, crushing, and handling. A typical coal water slurry contains about 70% mass of coal, pulverized to minus 200 mesh or finer, and water with about 1% mass of an additive to alter the rheological properties of the liquid mixture and thus allow a larger coal loading [47]. Slurry-fed entrained flow gasifiers include the GE gasifier and the E-Gas gasifier (now owned by ConocoPhillips). The GE gasifier coal slurry is fed through single-burner at the top of the gasifier, which is a refractory lined pressure vessel. Refractory stability under slagging conditions depends largely on the ash content and ash properties. Furthermore the burner lifetime of slurry-fed burners is considerably lower than that of dry-fed burners. Reactor exit gas is cooled either by direct water injection or by a radiative cooler located directly below the reactor. The E-Gas gasifier differs from other systems in that it uses a two-stage reactor, also involving refractory lined pressure vessels. The bulk of the feed slurry and all the oxygen are sent to the first (horizontal) stage, where the coal is gasified. Hot gas flows into the second (vertical) stage, where the remainder of the coal slurry is injected. Hot fuel gas is cooled in a fire tube boiler gas cooler and then sent to the F-T unit. The process produces no ash and recycles by-products for uses such as road construction materials.

Today, both solid (dry) and slurry-fed gasifiers are likely to be entrained flow gasification systems. The newer entrained flow gasifiers are considerably more efficient, albeit at the price of making product gas having a higher content of syngas, total $H_2$ and CO, but a lower $H_2$/CO ratio [48]. A critical aspect of long-term sustainability of any gasification technology is that it be versatile enough to accommodate various ranks of coal as well as other feeds such as biomass.

The first commercial scale and currently longest running coal gasification plant in the USA was put into service in 1983 by Eastman at its Kingsport, Tennessee facility. A schematic of the gasification plant is shown in Fig. 11.7. The process uses two high-pressure (1,000 psig) GE gasifiers, one being a spare, to convert about 1,250 t/day of bituminous coal into acetyl chemicals via methanol and CO. Over 99.9% of the sulfur (as $H_2S$) is removed from the syngas via a Rectisol unit designed by Linde AG and is recovered as elemental S by a combination of Claus sulfur recovery units (SRUs) and Shell Claus off-gas treating (SCOT) processes. $CO_2$ is also removed by the Rectisol process and is currently vented, but could easily be captured for sequestration. A cold box, also designed by Linde AG, is used for cryogenic separation of CO from a portion of the syngas, and some of the syngas itself is used in downstream methanol plants based on technology from Lurgi and Air Products. Mercury contaminant originally present in the coal is removed using a packed bed of sulfur-impregnated activated carbon. Originally built for production of chemicals from coal, the Eastman gasification process and associated learning curve with a variety of coals are representative of what would be used for future production of power and fuels from coal in processes such as integrated combined cycle gasification (IGCC) and industrial gasification.

Disadvantages of entrained-flow gasification include a rather high oxygen requirement and a high waste heat recovery duty. However, the ability to feed coal that is dry reduces the oxygen requirement and increases efficiency compared to a single-stage entrained gasifier using slurry feed.

## Fischer–Tropsch

The Fischer–Tropsch (F-T) reaction is a polymerization or oligomerization of carbon monoxide performed catalytically under reducing conditions. Chemically F-T is single reaction, with approximate stoichiometry for linear hydrocarbon production as shown in Eq. 11.1. The greater hydrogen content of methyl groups at the chain termini increase the stoichiometric ratio of $H_2$:CO slightly beyond 2:1, with the actual F-T stoichiometry for paraffinic hydrocarbon production being about 2.1:1.

$$\begin{aligned} &n\mathrm{CO} + (2n+1)\mathrm{H_2} \leftrightarrow \mathrm{CH_3}-(\mathrm{C_{n-2}H_{2n-4}})-\mathrm{CH_3} \\ &\quad + n\mathrm{H_2O} \\ &\Delta\mathrm{H}_{298} \approx -38\ \mathrm{kcal/molCO} \end{aligned} \tag{11.1}$$

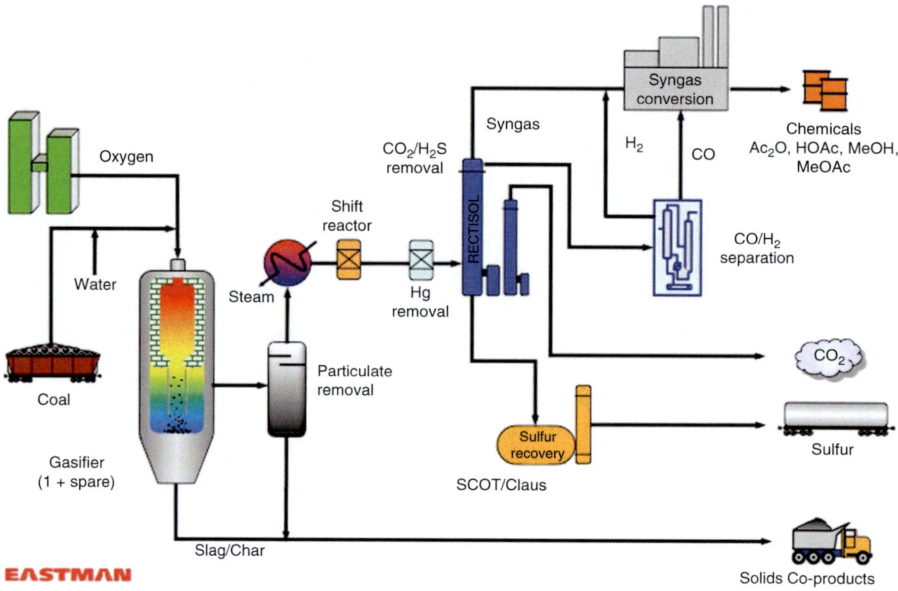


**Fig. 11.7** Flow diagram of Eastman's coal gasification plant (Figure courtesy of Eastman Chemical Company)

Mechanistically, the F-T reaction consists of four steps: (1) formation of metal-bound $C_1$ species from feed CO (or possibly $CO_2$ if present); (2) chain initiation wherein metal-bound $C_2$ species form from intramolecular or intermolecular coupling of the metal-bound $C_1$ species; (3) chain propagation wherein the initial metal-bound $C_2$ species grows as it couples with newly formed metal-bound $C_1$ species; and (4) chain termination wherein metal-bound $C_n$ species undergo hydrogen transfer reactions that disable further propagation [49].Mechanistic specifics remain in debate but it is well accepted that hydrogen transfer reactions play key mechanistic roles throughout the reaction, including metal-catalyzed hydrogenation of CO, β-hydride elimination from the growing chain to give terminal olefins, and intramolecular reductive elimination of hydrocarbon (R) and hydrogen (H) fragments from the catalytic center to give products during chain termination. Hydrogenation is not always complete and oxygenation products such as lower alcohols and carbonyl compounds also form. Olefins form throughout the process via β-hydride elimination, and these either re-hydrogenate during propagation or exit from the catalytic cycle during termination.

## Performance Variables and Parameters, and Product Distribution

The reaction temperature and catalytic metal strongly influence the chemistry and product distribution. The reaction is run in either a high temperature (HTFT) or low temperature (LTFT) regime. The HTFT reaction is always performed with iron as catalyst, at ~340°C, and the LTFT reaction is run at ~220°C when using a Co-based catalyst. The Fe-catalyzed LTFT reaction can be run a little hotter, at 230–270°C. The HTFT reaction is traditionally used to make light syncrude for production of gasoline and diesel. The LTFT reaction produces primarily high molecular weight hydrocarbons, referred to collectively as F-T wax, that are typically hydrocracked to make diesel [50].

Much novel work on stable cobalt catalysts for slurry bed gas to liquids (GTL) applications has been patented in the past 15 years. Although some companies, notably Syntroleum, apparently plan to use cobalt in CTL F-T operations, reports dedicated specifically to the use of Co in F-T CTL are essentially unavailable [51]. This is mainly because unlike iron, cobalt does not undergo the water gas shift reaction (WGSR) at the conditions needed to produce the supplemental hydrogen needed to make saturated hydrocarbons from coal. It is also the case that, while essentially all examples of Co patents use GTL conditions, many of these Co patents state that various feed sources including syngas from coal may be used instead of natural gas [52]. Major research domains of emphasis in F-T CTL catalysis include understanding and fixing problems of catalyst instability and attrition at slurry LTFT conditions [53, 54].

The parameters used to assess and report performance of F-T catalysts are those typically used for catalytic reactions: conversion, selectivity, product distribution, yield, reactor productivity, and catalyst fate or lifetime. The selectivity of a catalyst with a given feed composition is the main determinant of measured product distribution (although product distribution is also influenced by process design). The yield of a given cut is the product of overall conversion and selectivity for that cut. Yield is usually expressed as a function of time and catalyst quantity in these continuous processes. Catalyst fate relates to the mechanism and time frame of catalyst deactivation and therefore determines its useful lifetime.

The conversion for a F-T process is generally reported as %CO converted to any product, although conversion of total feed ($CO + H_2$) is sometimes used. Coproduction of the undesired by-product methane typically rises directly with conversion. F-T synthesis with Co is typically run at about 60% conversion, and conversions in LTFT are typically in the range of 50–70%.

The reactor productivity is the quantity of product produced per quantity of catalyst per unit time, expressed by weights of product and catalyst (e.g., $kg_{prod}/kg_{cat}/h$) or by volumes ($L_{prod}/L_{cat}/h$). Productivity may be reported for bulk product or for specific cuts, or even as the quantity of CO converted per quantity of catalyst per unit time. Productivity of the LTFT reaction performed in a slurry bed reactor is higher than for a fixed bed, primarily as a result of improved heat management in the slurry system. The continuous motion of the liquid-catalyst matrix allows sufficient heat transfer to achieve commercial slurry reactor productivities of about 200 volumes of CO converted/vol catalyst/h.

The product distribution reflects the selectivity of the catalytic system at the reaction conditions, typically reported as a percentage of the total product or as a percentage of the CO converted (carbon basis). Selectivity data also may be

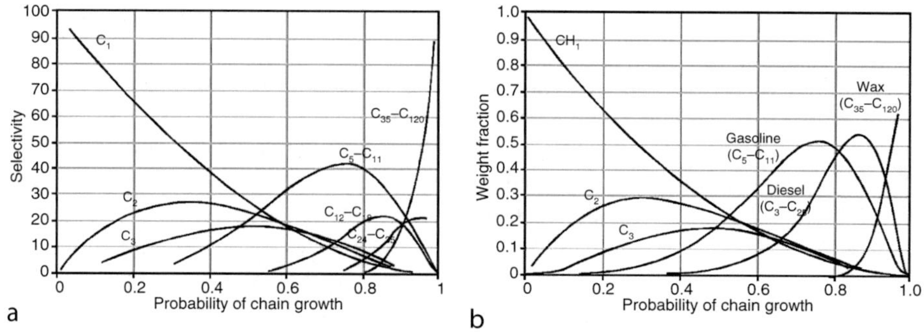


**Fig. 11.8** Product distribution plots for the Fischer–Tropsch reaction [55]. (**a**) General distribution, and (**b**) additional breakdown of the $C_{12+}$ cuts

supplied in absolute quantities, as reactor productivities for specific cuts. The weight fractions of various product cuts obtained within the F-T kinetic regime are shown in Fig. 11.8. The product slate is influenced by many reaction variables; particularly the feed composition, the reaction temperature, the catalytic metal and reaction promoters, and the support material if present.

F-T product slates include gaseous and liquid fractions. The gaseous reaction product includes hydrocarbons boiling below about 650°F (tail gases through middle distillates). The liquid reaction product (the condensate fraction) includes hydrocarbons boiling above about 650°F (vacuum gas oil through heavy paraffins). Commercially, the product boiling below 650°F is typically separated into a tail gas fraction and a condensate fraction of about $C_5$–$C_{20}$ normal paraffins plus higher boiling, branched hydrocarbons. Branching may be advantageous to a number of end-uses, particularly when increased octane values and/or decreased pour points are desired for lube oil additives. The degree of isomerization is preferably greater than 1, and more preferably, greater than 3 mol of isoparaffin per mole of *n*-paraffin. If used in a diesel fuel composition, the products preferably have a cetane number of at least 60.

The F-T reaction produces a wide range of saturated and olefinic hydrocarbons. Figure 11.9 shows the percentage proportions of olefins, alkanes, oxygenates, and aromatics in upgrades syncrude obtained from the F-T reaction at varied conditions of temperature regime and catalytic metal. The HTFT reaction gives aromatics, while the LTFT reaction does not give aromatics with any catalytic metal. Cobalt is a better hydrogenation catalyst than iron, and therefore products of the Co-catalyzed F-T reaction tend to have a lower olefin/paraffin ratio than those of iron. (Not shown in Fig. 11.9, the high temperature reaction also gives more branched hydrocarbons, and with a higher degree of branching, than the low temperature reaction using either Fe or Co.) The LTFT and HTFT both give oxygenates, typically 10–15% for the HTFT reaction and about half of that for the LTFT reaction with Fe (and much less with Co). Most oxygenates produced in the LTFT reaction are alcohols. The HTFT reaction gives alcohols and ketones,

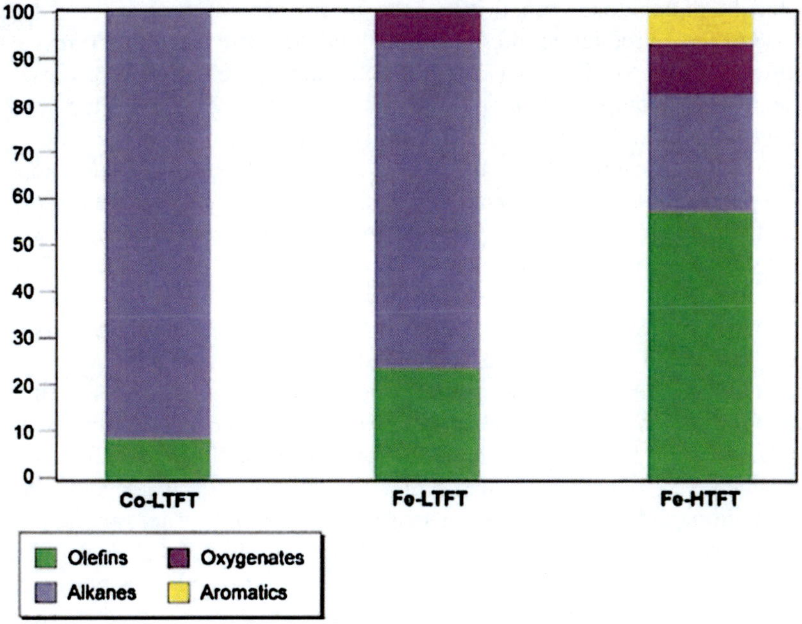


**Fig. 11.9** F-T product slates dependence on temperature and catalytic metal (Figure courtesy of SRI Consulting) [57]

plus a lesser amount of carboxylic acids and aldehydes. The LTFT reaction gives more methanol than the HTFT reaction, and the two temperature regimes give comparable amounts of ethanol.

Product slates for F-T catalysis are controlled by polymerization kinetics, with fairly constant chain growth probabilities that fix the ranges of possible product distributions [56]. The moles of species with carbon number n decrease exponentially with n. The weight fraction of product having carbon number $n$, $W_n$, is given by the Anderson-Schultz-Flory (ASF) distribution (Eq. 11.2):

$$W_n = (1 - \alpha)^2 n a^{n-1} \tag{11.2}$$

The parameter “a” is the Schultz-Flory distribution factor and represents the ratio of the rate of chain propagation to the rate of chain propagation plus the rate of chain termination. The parameter “$\alpha$” is the chain growth probability, and $1-\alpha$ is the probability of chain termination. In using this equation, a and a are considered as independent of chain length. Either value can be obtained graphically using this equation in log form, with analytical data for $W_n$ as obtained, for example, by gas chromatography. In practice, the parameter $\alpha$ is used most often to characterize product distribution [45].

Values of $\alpha > 0.9$ are representative of processes producing high amounts of wax, $\alpha$ values of about 0.8 maximize diesel cuts, and a values of about 0.7 are characteristic of naphtha production. For the intermediate molecular weight cuts of relevance to fuels production, diesel and naphtha, the ASF distribution reveals the limitations on selectivity shown in Fig. 11.8. Because selectivity for diesel cuts does not exceed about 22% but wax selectivity can reach 90 + %, much industrial research has focused on wax production with downstream hydroprocessing of wax to diesel (or to lube oil base stocks). The $\alpha$ values for oxygenated products appear to be about the same as those for hydrocarbons, which suggests that the products all form by the same or similar chain propagation steps and with similar rates of chain termination, albeit with different chain termination chemistries for the oxygenates, olefins, and paraffins.

In practice, there are several deviations from the ASF product distribution inherent to the F-T process. The amount of methane produced is often higher than would be predicted from this equation, because methane is formed by a chain termination mechanism that is separate from those forming the higher molecular weight products (methane is formed by addition of 2H atoms to a metal-bound $C_1$). Catalysts that preferentially catalyze hydrogenation over chain initiation, particularly nickel, increase the probability of termination to give methane. A second deviation from the ASF distribution is that the amount of $C_2$ species formed is often lower than expected based on the ASF equation, although the reason for this is not clear. It has been suggested that an adsorbed ethylene can react with a surface methylene from either side to propagate the chain; whereas higher alkyl adsorbed species would show a lesser rate of chain propagation due to steric constraints.

Some approaches to catalyst design have succeeded in controlling the F-T product distribution by altering the value of alpha. One such approach is the inclusion of alkali metal elements, particularly potassium. Oxides of the alkali metals increase $\alpha$ and promote formation of wax in LTFT catalysis by iron. Use of $K_2O$ helps keep the iron reduced, in the carbide form, which promotes growth of longer chains.

With iron catalysts, alpha can exhibit two regimes of relatively constant value, sometimes referred to as the "two alpha" phenomenon. The value of a can increase with chain length rather than remain constant, and typically changes from about 0.60–0.65 to about 0.90–0.93 at roughly $C_{20}$ [58]. A double $\alpha$ is most evident for low temperature catalyst systems with high wax productivity and is not evident for high temperature systems that make diesel or functionalized hydrocarbons.

## HTFT

The HTFT reaction has been run in a fixed bed reactor (FBR) or fluidized bed reactor. Traditionally, Sasol used a circulating fluidized bed reactor (CFBR), in its Synthol process. The Synthol CFBR was replaced in 1989 with the Sasol Advanced Synthol (SAS) reactor, based on a conventional fluidized bed and providing superior RAMO (Reliability, cost-effective Availability, Maintainability, and efficient

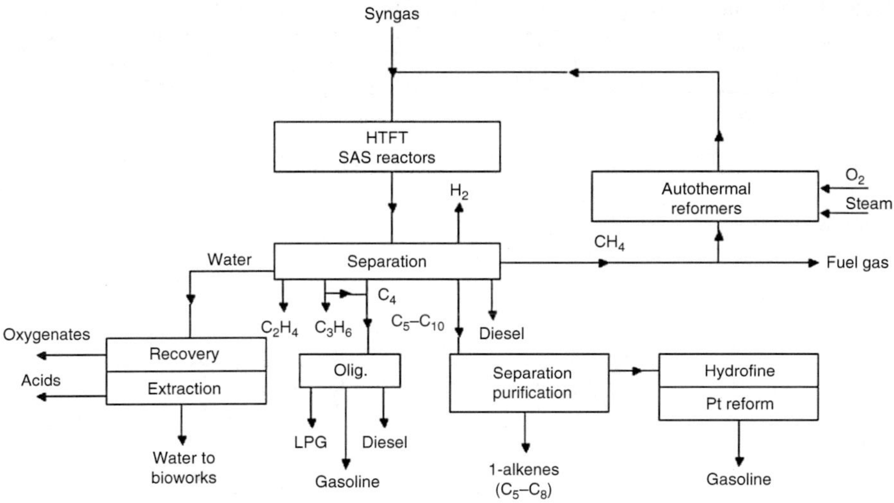


**Fig. 11.10** Secunda plant block flow diagram [59]

Operability). A block flow diagram of the F-T and downstream portion of Sasol's Secunda plant using the SAS technology is shown in Fig. 11.10.

The process is employed to produce a range of fuels and also olefins. The fluidized, iron-based catalyst gives primarily $C_1$–$C_{20}$ raw product. The $C_2$-rich stream is split into ethylene and ethane. Ethane is cracked in a high-temperature furnace to ethylene, which is then purified. Propylene from the light hydrocarbon gases is purified and used in the production of polypropylene. Large quantities of olefins in the $C_5$–$C_{11}$ range are produced at these HTFT conditions and the alpha-olefins (straight chain olefins with an ene group in the terminal position) are recovered for use as chemicals ($C_5$, $C_6$, and $C_8$) are recovered or sent into the fuel pool ($C_7$–$C_{11}$). Oxygenates in the aqueous stream are separated and purified in the chemical work-up plant to produce alcohols, acetic acid, and ketones including acetone, methyl ethyl ketone (MEK), and methyl isobutyl ketone (MIBK).

## LTFT

The early German LTFT plants used fixed bed reactor (FBR) technology with Fe catalysts, through World War II. In 1955, Sasol began using a tubular fixed bed reactor (TFBR), consisting of a shell containing 2,050 tubes of 12 m length and 5 cm diameter packed with extruded precipitated iron-based catalyst. A typical catalyst system would be potassium-promoted precipitated iron (such as a Sasol-type catalyst consisting of K = 1.9 wt%, Cu = 2.4 wt%, Si = 5.9 wt%, Fe = 47.3 wt %, $O_2$ = ~42.5 wt.) This LTFT is referred to as the Arge process and is operated with a shell side temperature of about 220°C and reactor pressure of 25 bar (early

**Table 11.4** Carbon selectivity of selected LTFT and HTFT processes (%) [60]

| Product | TFBR (LTFT) | SBR (LTFT) |
|---|---|---|
| $CH_4$ (methane) | 4 | 7 |
| $C_2$–$C_4$ olefins | 4 | 24 |
| $C_2$–$C_4$ paraffins | 4 | 6 |
| Gasoline | 18 | 36 |
| Middle distillate | 19 | 12 |
| Heavy oil, waxes | 48 | 9 |
| Polar oxygenates | 3 | 6 |

| | TFBR (LTFT) | | SBR (LTFT) | | SASR FBR (HTFT) | |
|---|---|---|---|---|---|---|
| | $C_5$–$C_{12}$ | $C_{13}$–$C_{18}$ | $C_5$–$C_{12}$ | $C_{13}$–$C_{18}$ | $C_5$–$C_{12}$ | $C_{13}$–$C_{18}$ |
| *n*-Paraffins | 95 | 93 | 96 | 95 | 55 | 60 |
| Paraffins | 53 | 65 | 29 | 44 | 13 | 15 |
| Olefins | 40 | 28 | 64 | 50 | 70 | 60 |
| Aromatics | 0 | 0 | 0 | 0 | 5 | 15 |
| Oxygenates | 7 | 7 | 7 | 6 | 12 | 10 |

Arge) or 45 bar (Arge commissioned in 1987), wherein the heat of reaction is dissipated by generation of steam on the shell side. Further scalability could be achieved by more than doubling the number of tubes, to, for example, 5,000, but the reactor size eventually becomes unwieldy. Sasol found that a slurry phase system is more readily scaled, in part due to engineering design and in part because the Co-based catalysts used are more efficient than the Fe-based catalysts used for the LTFT reaction. Sasol commissioned its first slurry bed reactor (SBR) in 1993.

Table 11.4 shows typical product slate selectivities for the LTFT, TFBR; the LTFT, SBR; and the HTFT FBR. Both fixed- and slurry-bed LTFT processes give much more paraffin, less olefin, no aromatics, and less oxygenates compared to the HTFT process. LTFT is particularly suited for a modern coal to diesel (CRD) plant.

As indicated in Eq. 11.1, the F-T reaction is highly exothermic. The reaction heat must be removed rapidly, to avoid temperature increases that would result in undesired formation of high levels of methane, light hydrocarbons, and, in the extreme, catalyst deactivation due to coking, sintering, and disintegration. Deposition of carbon by the Boudouard disproportionation reaction, Eq. 11.3, is to be avoided, and the heat distribution imparted by the slurry bed is advantageous.

$$\begin{aligned} &2CO \leftrightarrow C + CO_2 \\ &\Delta H_{298} = -20.6\ \text{kcal/molCO} \end{aligned} \tag{11.3}$$

Attrition and breakage of catalyst particles is a more difficult issue to overcome than coking, particularly for slurry bed catalyst systems [61]. Attrition refers to abrasive or erosive removal of surface material that is less than about 50 μm in particle diameter. Breakage refers to a decrease in median particle size caused by fracture of catalyst particles. As the concentration of smaller particles builds up, catalytic activity eventually decreases and also the downstream separation of wax from solid catalyst becomes inefficient. Catalyst particles are subjected to hydrodynamically induced mechanical stress at slurry bubble column conditions that has severely limited the lifetime of LTFT iron catalysts.

Precipitated, unsupported iron catalysts are particularly prone to disintegration into smaller particles. Supporting the iron imparts strength to the particle to decrease fracture, but at the expense of decreased volumetric productivity due to dilution of the active catalyst by the support material. Making spherical particles by spray drying [62] can help with fracture or with attrition, but not always with both [63, 64].

The pursuit of mechanically robust iron F-T catalysts remains an area of active research. Traditionally, Fe catalysts have been disposed of rather than regenerated, because iron is inexpensive. However, a few reports of potential use and even regeneration [65] of Fe slurry catalysts have recently appeared, notably from Rentech [66], presumably because management of even low hazard waste volumes is of increasing interest. It is also the case that using a fluidized slurry bed catalyst system allows longer runs, since deactivated catalyst can be removed and fresh catalyst added without taking the reactor off-line as is required for fixed bed reactors. On the other hand, an advantage of fixed bed reactors is that since the wax produced simply trickles down the bed, there is no problem separating the wax from the catalyst. In the slurry system, special internal and/or external devices are required to continuously remove wax from the finely divided catalyst in the slurry.

### Water–Gas Shift Reaction

The water-gas shift reaction (WGSR), Eq. 11.4, is important for CTL because the $H_2/CO$ ratio produced by gasification of coal is lower than that produced by steam reforming of methane in GTL (gas to liquids), by about a factor of 3. Without the WGSR, other sources of hydrogen must be provided in CTL, if hydrocarbon fuels are the intended product slate as per Eq. 11.1, and additional sources of hydrogen in turn typically require additional use of power that itself generates $CO_2$.

$$\begin{aligned} &CO + H_2O \leftrightarrow CO_2 + H_2 \\ &\Delta H_{298} = -9.8\ \mathrm{kcal/molCO} \end{aligned} \tag{11.4}$$

The in situ WGSR occurs concurrently during the iron-catalyzed F-T reaction. Cobalt does not catalyze the WGSR under F-T conditions and therefore iron catalysts produce more $CO_2$ than do cobalt catalysts. In practice the WGSR is one of several factors that affect the overall efficiency of an F-T CTL process when using Fe versus Co as catalyst, another being the greater efficiency of the smaller reactor used for Co compared to Fe, and each must be taken into account by process engineers when designing CTL plants.

Another key aspect of the WGSR is that it can run in reverse. A potential use for some of the by-product heat typically available in hot outlet gases from high temperature gasifiers is to generate steam heat for an external water gas shift reaction, to be conducted downstream of the gasifier, wherein some $H_2$ (from recycle or tailgas) consumes the undesirable by-product $CO_2$.

## Downstream Product Upgrading

Upgrading technologies for F-T syncrude are modifications of those developed for crude oil refining. Refinery process options for producing fuels from F-T middle distillates include wax hydrocracking, distillate hydrotreating, catalytic reforming, naphtha hydrotreating, alkylation, and isomerization. The many technology licensors for hydroconversion processes include ChevronTexaco, UOP, IFP (Axens), Süd-Chemie Sasol Catalysts, and Haldor-Topsøe. Catalysts can be obtained from these companies as well as from AkzoNobel, Süd-Chemie, and Axens among others.

Reforming and alkylation of gaseous products to produce liquid fuel-range products are appropriate when upgrading synthetic crude from the Fe-catalyzed HTFT reaction. Skeletal modification technologies most important to upgrading synthetic crude from HTFT include isomerization, mild hydrocracking, and even fluidized catalytic cracking. Catalytic oligomerization of olefins is especially applicable to the product from an iron-catalyzed, HTFT process because the bulk of the crude product consists of these olefins, particularly $C_3$–$C_5$ olefins (gases). The Sasol Synfuels CTL refinery in Secunda uses an alkene oligomerization process based on solid phosphoric acid (SPA) catalysis to convert most of the light alkene HTFT product into liquid fuels [67].

Currently, the production of a high quality diesel fuel is preferred to the production of gasoline in F-T fuels production. This is because the same factors that count against F-T gasoline, such as product linearity and low aromatic content, serve as very positive factors in favoring high quality, high cetane number in F-T diesel fuel. Straight run F-T diesel makes up about 20% of the total F-T product and has a cetane number of about 75 because it is predominantly linear. For maximum production of high quality diesel, the slurry bed reaction operating in the high wax selectivity mode with cobalt-based catalyst, or the high diesel selectivity mode with iron-based catalyst, is the most efficient approach. Lighter and heavier fractions are usually less desirable due to limited markets and/or lower prices.

The Co-catalyzed process is typically chosen for maximum wax production in LTFT gas to liquid (GTL) processes; however, the use of iron-based catalysts is generally preferred for CTL because coal-derived syngas is deficient in $H_2$ and cobalt does not catalyze the WGSR that supplements $H_2$ production during the iron-catalyzed F-T reaction. When operating in the LTFT mode, iron catalyst can produce up to 50–75% linear wax products. LTFT-derived feeds can be hydrocracked under much milder conditions than typical crude oil derived feeds, such as vacuum gas oils, which require pressures as high as 150 bar to prevent coking of the catalyst by aromatic compounds. This is not an issue with paraffinic feeds, and pressures between 35 and 70 bar are used to hydrocrack LTFT products using commercial hydrocracking catalysts. The relatively low levels of oxygenates present in F-T waxes, mainly alcohols and lesser amounts of acids or carbonyls, are easily and completely hydrodeoxygenated.

After hydrotreatment, the straight run diesel fuel cut has a cetane number ranging from 70 to 75. If both hydrotreating and hydrocracking are used, they may be operated at the same pressure using a common hydrogen loop. Alternatively, the hydrotreater can be operated in a once-through mode, at lower pressure, and the hydrogen tail gas then subsequently pressurized and fed to the hydrocracker loop. For a 30,000 bpd equivalent product upgrading system involving both hydrocracking and hydrotreating, the size of the unit is comparable to those used in oil refineries. There are many process configurations possible depending on site-specific factors including plant capacity, available land, and possible integration with other facilities.

When using the slurry phase LTFT reactor, water is removed from overhead gases in a three-phase separator and then the remaining condensate is sent to a low temperature distillation column, in the heavy ends recovery section, to recover condensate from tail gas in the reactor overheads. The overheads from the low temperature fractionation column are used as fuel gas. Recovered condensates from the column are sent to the product work up section. The waxy reactor product is withdrawn from the side of the F-T reactor and continuously separated from the catalyst. The waxy product is then sent to the first low temperature column, which sends a heavy cut to the hydrocracker and a light cut to the hydrotreater. The wax hydrocracker produces the diesel product, while olefins in the condensate are hydrogenated to paraffins in the hydrotreater. The hydrocracking step typically uses a silica-alumina supported Ni-Mo-based catalyst and can be adjusted to maximize the production of ultraclean diesel product. The products from the hydrotreater and hydrocracker are then combined and fractionated into diesel and naphtha products. Up to approximately 80% of the product is diesel and the remaining portion is naphtha.

Product upgrading schemes center around three fuel products (diesel, LPG, and naphtha), but the upgrading plant is quite amenable to different configurations if market demand so indicates. The ASF distribution means that the F-T process and its associated refinery also produce transportation fuels and products other than diesel and gasoline. Other saleable products may include solvents, kerosene, jet fuel, illuminating paraffins (sold for use in lighting and appliances in South Africa), base oils (for lubricant products), and chemicals. Maximizing recovery of the higher boiling jet fuel while maintaining diesel production will tend to increase economic return.

Major projects planned in F-T ICL include a US $3.3 billion plant in Ningdong with Shenhua Ningxia Coal Group and Sasol Ltd, scheduled to begin construction in 2010 [68].

### Methanol and Dimethylether

Supplementation of liquid fuel supplies with methanol made from abundant coal reserves has become increasingly attractive in countries that are oil and natural gas resource poor. Methanol has a broad range of applications including:

- Fuel substitute for gasoline
- Gasoline fuel blend to extend conventional fuel supplies
- Energy storage medium in coproduction of power and fuel
- Rapid start combustion turbines to meet peak electricity demand
- Advanced methanol powered and plug-powered vehicles
- Direct methanol fuel cells

Methanol can be synthesized from syngas according to Eq. 11.5, where the thermodynamics assume gas phase components. The formation of methanol from syngas is highly exothermic.

$$\begin{aligned} &CO + 2H_2 \leftrightarrow CH_3OH \\ &\Delta H_{298} = -21.6\ \text{kcal}/\text{molCO} \end{aligned} \tag{11.5}$$

The feasible operating temperature for methanol synthesis is strongly affected by the temperature sensitivity of the copper-based catalyst, which is prone to deactivation via sintering of copper or coking [69]. Methanol synthesis occurs at significant rate in the temperature range of 210–260°C, optimizing at about 237°C [70]. Modern commercial methanol synthesis catalysts make use of the high reactivity of copper oxides combined with the dispersant function of zinc oxide, supported on alumina. This combination was pioneered by ICI [71] (Imperial Chemical Industries PLC, acquired by AkzoNobel in 2008) and allows methanol synthesis to be performed at low pressure with recycle (35–55 atm) and relatively low temperature (200–300°C, typically 250°C). The component contents are in the range of about 40–80% CuO and 10–40% ZnO, with 5–10% $Al_2O_3$. For example, the unreduced material RPJ-19, from United Catalysts Inc., has been described as 55 wt% CuO, 36% ZnO, and 8% $Al_2O_3$, with 1% $SiO_2$ [70]. Magnesia may be added to control sintering and increase the available surface area of copper.

The methanol synthesis catalysts are also sensitive to catalyst poisons. The contaminants sulfur and halide are typically removed upstream to $<100$ ppm. Because catalyst lifetime is significantly diminished by excessive temperature, reactor temperature control is very important. Currently, all commercial processes produce methanol from synthesis gas in a vapor phase reactor with a synthesis loop, using packed bed or tubular reactors, and temperature moderation is achieved by recycling large quantities of hydrogen-rich gas. Typically, a gas phase process is limited to about 16% CO in the reactor inlet, as a means of avoiding excess heating by constraining the activity per pass. Cool, unreacted gas is injected at different stages in the catalyst bed or internal cooling surfaces are used to provide further temperature control.

As a means to improve throughput per pass with temperature control, ChemSystems and Air Products developed a fundamentally different, liquid phase reactor system. In this system, referred to as the LPMeOH™ (Liquid Phase Methanol) process, synthesis gas is bubbled through a slurry, consisting of micrometer-sized catalyst particles and methanol in a suspension of mineral oil. The nonvolatile mineral oil acts as a temperature moderator, transferring the heat of reaction from the catalyst surface via the liquid slurry to boiling water in an internal tubular heat exchanger. As a result of maintaining a constant, highly uniform temperature through the entire length of the reactor, the slurry reactor can achieve a much higher per pass syngas conversion than its gas-phase counterparts and also CO concentrations in excess of 50% have been maintained without adverse effect on catalyst activity. As with slurry F-T systems, the liquid phase methanol process benefits from the ability to withdraw spent catalyst slurry and add fresh catalyst on line periodically, without having to shut the system down.

Typical commercial Cu/Zn/alumina methanol synthesis catalysts are also active for the low temperature water gas shift (LTWGS), which enables use of syngas from lower quality, low-hydrogen feedstocks such as coal to be used in making methanol. Reduction of CO or $CO_2$ to methanol with $H_2$ is exothermic, and heat dissipation in fixed bed reactors is inadequate to protect the catalyst or to allow optimal product recovery. While less exothermic than the methanol synthesis reaction, the WGSR also adds heat to the system.

ChemSystems performed pioneering development of the LPMeOH process in the 1970s [72]. The LPMeOH technology was based on early design and demonstration work funded by EPRI (Electric Power Research Institute) in the mid-1970s, wherein methanol is produced in one step using the high-CO content syngas from coal [73]. Air Products then undertook a significant research effort together with ChemSystems and DOE co-funding, to further develop the process in the 1980s and 1990s, as part of the DOE's cogeneration program. While ChemSystems originally used an ebullating bed design with an aromatic (1,2,4-trimethylbenzene) or mineral oil liquid, Air Products used the more stable mineral oil in a slurry bubble reactor to further augment heat and mass transfer.

The LPMeOH process was demonstrated at large pilot scale during 1997–2002, in a collaborative effort between Air Products (technology supplier), Eastman Chemical (syngas supplier, plant operator, and end user of methanol), and Arcadis (fuel and power testing). This facility in Kingsport, TN operated for nearly 6 years, with continuous runs of up to 3 months. The nameplate capacity of 80,000 gpd was exceeded by 15%. Eastman syngas, derived from high-sulfur (3–5%) bituminous coal, was used. Of note is the ability to produce sulfur-free methanol containing less than 1% water, which makes this methanol directly suitable for transportation as well as chemical use. Used as a fuel additive, methanol also showed promise for reducing NOx emissions from gas turbines and diesel engines. Air Products also applied the understanding gained from developing the LPMeOH process in developing the Liquid Phase Dimethyl Ether Process (LPDME™) process, which was demonstrated in the early 1990s.

Sunggyu Lee and colleagues, at the University of Akron and later at the University of Missouri-Columbia, pioneered the mechanistic understanding of the one-step slurry methanol process and also worked on the MeOH/DME slurry process [74]. The LPMeOH process has three distinctive features and benefits compared to prior art using fixed beds. First, the commercial Cu/Zn/alumina methanol synthesis catalyst is slurried instead of fixed, which greatly improves heat transfer via the inert, liquid medium. Second, the slurried catalyst is a fine powder, which avoids mass transfer limitations involving pore diffusion. Third, the improved mass transfer in the well-agitated slurry allows much higher single-pass conversion of the syngas.

The initial method devised to combine steps in liquid-phase DME production involved combining the methanol dehydration catalyst with the slurry phase methanol synthesis catalyst [75]. For DME, typically, the methanol synthesis catalyst is copper on alumina promoted by zinc oxide. The ratio of $CuO/ZnO/Al_2O_3$ is typically about 2:1:5 with a $H_2$:CO ratio of 2:1 but will vary with syngas composition. Alumina is the most common dehydration catalyst. Both CO conversion and selectivity for DME are improved by including 10%wt sulfate in the alumina. Many researchers have reported using zeolites, and recent research with zeolites has been quite active in Asia [76]. Likewise, silicoaluminophosphate (SAPO) molecular sieves have been reported recently for use as dehydration catalysts in direct synthesis of DME. A third catalyst functionality may be included to enhance the water gas shift reaction, although the methanol synthesis catalyst is itself also an effective water gas shift catalyst.

Long-term stability of one-step DME synthesis catalyst systems was noted as a primary difficulty by Mobil in the early 1980s and continued as an issue in Air Products' work. The temperature needs and catalyst stability are considered for three simultaneous reactions in direct synthesis of DME: water gas shift (Eq. 11.4), methanol synthesis (Eq. 11.5), and methanol dehydration (Eq. 11.6) [77]. The methanol synthesis via Eq. 11.5 can be considered in principle as the combination of water gas shift and the reduction of $CO_2$ by hydrogen (Eq. 11.7).

$$\begin{aligned} &2CH_3OH \leftrightarrow CH_3OCH_3 + H_2O \\ &\Delta H_{298} = -5.9\ \text{kcal/molDME} \end{aligned} \tag{11.6}$$

$$\begin{aligned} &CO_2 + 3H_2 \leftrightarrow CH_3OH + H_2O \\ &\Delta H_{298} = -11.7\ \text{kcal/molMeOH} \end{aligned} \tag{11.7}$$

Even so, the mechanism actually in effect during direct synthesis of DME is in some debate and might not involve the explicit intermediacy of methanol as per Eq. 11.6. Figure 11.11 shows the results of modeling done by JFE Holdings (formed in 2002 by the merger of NKK and Kawasaki Steel) based on their pilot plant results, which give best conversion at a ratio of 1:1 CO:$H_2$. Specifically, DME may form directly from coal-derived syngas as per Eq. 11.8. The stoichiometry of Eq. 11.8 being noted for coal-derived syngas, it is also the case that catalysts for

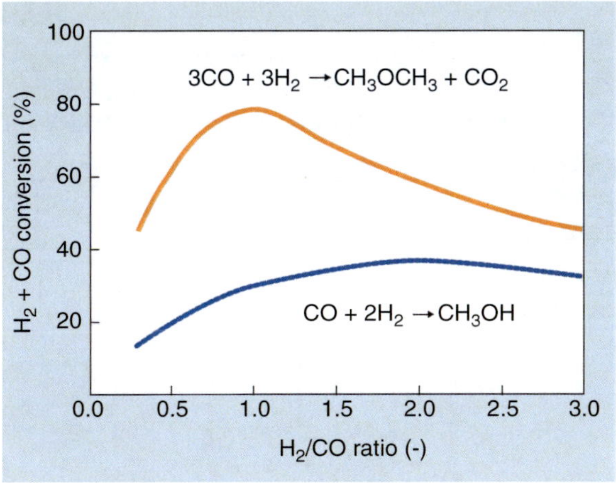


**Fig. 11.11** Equilibrium conversion of CO and $H_2$ to DME (280°C, 5 MPa) [81]

direct DME synthesis have been based on methanol synthesis catalysis and most processes give some combination of both DME and methanol.

$$3CO + 3H_2CH_3OCH_3 + CO_2$$
$$\Delta H_{298} = -19.6\ \text{kcal/molCO} \qquad (11.8)$$

The WGS reaction used for one-step DME/MeOH production is of the low temperature type (LTWGS) and operates best at about 180–240°C, which overlaps substantially with the methanol synthesis temperature range. The overall chemistry is more easily controlled if the WGS catalyst has little or no F-T activity for hydrocarbon production, which means that the LTWGS catalyst needs to be copper-based instead of iron-based. The typical catalyst used for LTWGS in DME or methanol synthesis has the general formula $xCu/yZnO/Al_2O_3$ and has no hydrocarbon-producing F-T activity. A typical low temperature shift catalyst contains 30–70% CuO (prior to reduction), 20–50% ZnO, and up to about 40% alumina [78].

The optimal temperature for the methanol dehydration reaction is set not by catalyst stability but by the equilibrium dehydration reaction thermodynamics. The dehydration proceeds best at about 250°C and yield declines as the temperature is raised further. The catalyst function for dehydration of alcohol to ether is a solid acid. Various options are possible, including acidic aluminas, silica-aluminas, zeolites, and various solid or supported inorganic acids. In practice, γ-alumina is preferred to date. The dehydration rate is directly related to the acid strength of the zeolite and the overall rate of DME synthesis can be controlled by the proportion of ion exchange in the dehydration catalyst [79]. Dehydration is relatively fast above about 240°C and only a few percent of dehydration catalyst is needed in the total catalyst formulation [70]. The LPDME process was demonstrated on a 10 t/day scale for 25 days at Air Products' LaPorte facility using commercially available methanol synthesis and γ-alumina dehydration catalyst in a ratio of 95:5 wt%.

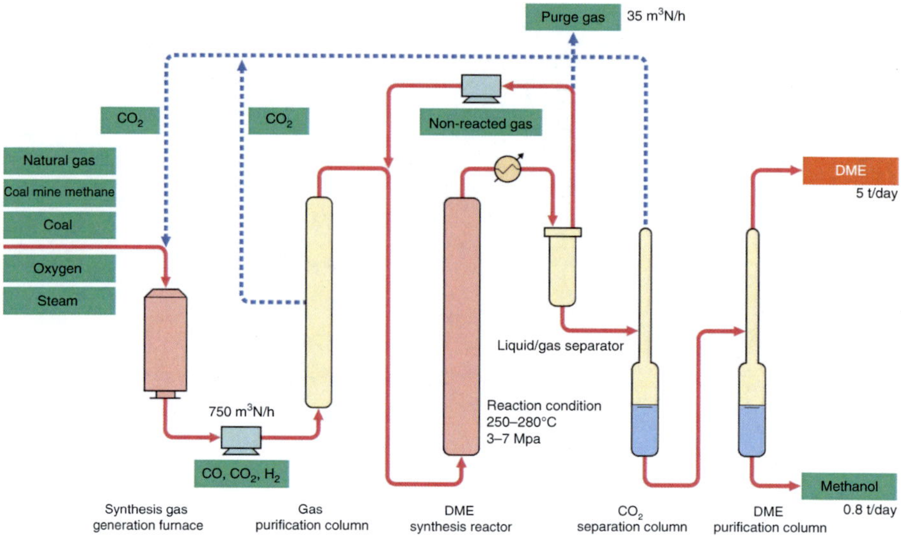


**Fig. 11.12** Process flow diagram of DME direct synthesis from coal (or methane)

The temperature and pressure were approximately 250°C and 50 bar, respectively. Deactivation occurred at a rate of 0.7%/day for both catalyst components [80].

The concern that LPG used heavily in Asia as a fuel and a power source will become increasingly unavailable has sparked significant research to develop DME as a replacement [81]. Properties of DME and LPG are similar in terms of fuel characteristics, low toxicity, and especially handling, distribution, and storage. Some use of DME as an LPG substitute has recently begun in China. To demonstrate the potential replacement of automotive fuel, fleet tests with DME have been performed (the fuel injection system requiring modification to accommodate the lower viscosity of DME compared to diesel, but without modification of the engine itself) [82]. Turbine, boiler, and fuel cell tests with DME have also been performed [83].

JFE in Japan has been a major driver for development and commercialization of one-step slurry DME production. NKK, Taiheiyo Coal Mining, and their partners constructed a 5 t/day direct syngas to DME pilot plant in 1999 at Taiheiyo's site in Hokkaido and operated it intermittently for about 5 years [83]. In 2002, JFE began a 5-year development plan for a 100 t/day pilot plant demonstration in Kushiro, with METI and many collaborators that together formed the DME Development Corp. and the DME International Corp. Interestingly, the feed for syngas production actually used is natural gas; however, JFE opts to bring the 2:1 $H_2$:CO ratio (produced by reforming natural gas) down to 1:1 (more like the ratio produced from coal) by adding $CO_2$ during reforming, and the work therefore also relates to advancement of the state of the art in producing DME and methanol by ICL of coal.

JFE reported an impressive purity from their overall process: 99.5 + % DME and <100 ppm combined water and methanol [84]. In this process, shown in Fig. 11.12, coal is fed to the gasifier and then raw syngas is brought up to a $H_2$:CO ratio of 1:1 via the WGSR using steam from the gasifier. Sulfur and $CO_2$ are removed prior to DME synthesis. Unreacted CO and $H_2$ are separated from liquid products and sent back to the DME synthesis reactor. The DME synthesis makes $CO_2$ as a by-product, plus some methanol and a slight amount of water. The separation of $CO_2$, DME, methanol, and water involves a low temperature extraction (−39 °C) [85]. After downstream distillation, methanol is cycled back to the DME reactor.

A different type of ICL process involving methanol is the gasification of coal followed by conversion to methanol and subsequent conversion of methanol to a marketable gasoline product. A $3 billion plant with a production capacity of approximately 700,000 t/year to be built by TransGas is planned in Mingo County, West Virginia. The plant will use Uhde's PRENFLO direct quench gasifier technology to make syngas, which will be converted to methanol and subsequently to gasoline using ExxonMobil's methanol-to-gasoline technology. Methanol conversion is nearly complete and the total liquids yield is higher than in F-T synthesis, including significantly less tail-gas production. The product of the MTG reaction does not require much further upgrading and refining compared to other coal liquefaction routes. A 100,000 t/year MTG process plant was started in China's Jincheng, Shanxi province in 2009, as part of a CTL demonstration complex.

Like DCL, ICL is complex and characterized by heavy investment, but CTL processes, in general, have been recently described as economically competitive at an oil price as low as $25/barrel [86] or, more typically, at an oil price of about $55/barrel [45], with the cost of coal being that incurred at the mine (i.e., without the cost of shipping and handling).

## Environmental Factors

The overall conversion of coal to liquids not only produces fuel- or chemical-grade products but also consumes water and produces carbon dioxide, as shown in Eq. 11.9 [87].

### Carbon Dioxide

The high carbon content of coal combined with the energy use in ICL processes involving gasification means that carbon dioxide emissions on a well-to-wheel basis are about 1.8 times more for coal than for oil. In terms of fuel preparation alone, it has been estimated that an F-T CTL plant without capture would release about 0.78 t of carbon dioxide per barrel of product, in comparison to a comparable oil refinery emitting about 0.1 t of carbon dioxide per barrel of product. For a typical bituminous coal, approximately 670 lb of carbon (or 2,450 lb of carbon dioxide) would be emitted for every barrel of F-T liquids produced. This compares to about 250 lb of carbon (or 900 lb of carbon dioxide) emitted for every barrel of petroleum fuels produced by refineries. When carbon sequestration is employed for both facilities (92% captured and stored at the F-T CTL plant and 40% at the refinery), the carbon dioxide emissions become approximately equal.

$$(4 + y)CH + 2(y - x)H_2O \rightarrow (4 + x)CH_y + (y - x)CO_2 \tag{11.9}$$

The large quantity of $CO_2$ produced currently poses a significant challenge to development of the industry. At the time of this writing, no requirements exist in the USA to manage carbon emissions from fossil fuel sources, but should carbon management be required, carbon dioxide produced during the conversion process could be captured for subsequent storage in deep saline aquifers or sold for use in enhanced oil recovery (EOR) operations. A range-finding study done in 2004 by Mitretek Systems (now Noblis) for the US DOE considered production of substitute natural gas (SNG) from coal and assumed that the value of $CO_2$ for EOR was \$12/t, which would significantly improve the economics of a CTL plant [88]. At present, there are over 2,500 km of dedicated $CO_2$ pipelines within the US delivering to commercial EOR projects in areas such as the Permian Basin of West Texas, southeastern New Mexico, the Rocky Mountain Region of Utah, Wyoming and Colorado, Louisiana, and the Weyburn Fields in Saskatchewan. With the exception of $CO_2$ from the Dakota Gasification Plant, all of this $CO_2$ is naturally occurring.

Total production related emissions from a 2 million barrels/day F-T CTL industry could be as much as a 0.5 Gt/year. There is already a significant knowledge base regarding how to handle $CO_2$ and inject it into deep geological structures, because injection of $CO_2$ is already used commercially in the USA to augment natural gas production. The transport of large volumes of $CO_2$ via pipeline is also a well-established practice, within segments of the oil and gas industry. Given adequately scaled infrastructure for carbon capture and storage in place, emissions from CTL plants and emissions from use of CTL fuels would be comparable to those associated with the production and consumption of petroleum-based fuels. If sequestration of carbon dioxide becomes mandatory, an additional \$4–\$8/barrel for the price of low-sulfur, light crude oil would be required for profitable operation, depending on specific site logistics [89].

The high sulfur content of most coals is addressed at multiple junctures in the F-T ICL process. Sulfur is removed when the raw synthesis gas is scrubbed to remove $CO_2$ and sulfur compounds (predominantly $H_2S$) and again during purification of F-T product streams. The $H_2S$ containing $CO_2$ streams are treated in a Claus or similar process, where the $H_2S$ is oxidized to elemental sulfur.

### Fischer–Tropsch Reaction Product Water

Water is a net reactant in CTL processes (Eq. 11.9) but is a product of the F-T portion of CTL processes (Eq. 11.1). On a mass basis, approximately 1.3 t of water is produced per ton of hydrocarbon product. After separation of the hydrocarbon product, the F-T reaction water typically contains numerous dissolved oxygenated compounds such as alcohols, aldehydes, acids, ketones, esters, and traces of other hydrocarbons. Purification of F-T reaction water is required in commercial processes, because all water is reused. Distillation removes most of the nonacid oxygenates to leave a bottom stream known as F-T acid water, rich in carboxylic acids. The recovered oxygenated compounds can be purified to chemical products or used as a fuel source. After extraction of the oxygenates, only low concentrations of organic acids remain and these are easily biodegraded by either aerobic or anaerobic digestion to produce water of very high quality, for use as coolant or even for drinking, from the original F-T acid water stream. High purity water suitable for use as boiler feed water can also be produced but requires a third set of treatments, after distillation and digestion, such as adsorption of trace suspended solids, dissolved salts, and organic material on activated carbon.

In addition to distillation-based purification of F-T reaction water, membrane-based, liquid-liquid extraction, and biological operations have been developed to remove and recover carboxylic acids from F-T acid water [90]. The resultant water stream from these processes is generally of a good quality and can be reused or disposed of to the environment.

After physical separation of oil and water streams, plus steam stripping and oil extraction to remove volatiles and aromatics, respectively, sulfur recovery is performed by conversion to elemental sulfur. Tail gas streams are also cleaned for use in recycle or as fuel.

The stream of water from the cooling system (referred to as the blowdown stream) must be cleaned also. This water contains inorganic salts and is treated with coal ash that chemically binds the salts, before the remaining water is collected by evaporation and condensation.

In addition to fuels, chemicals, and purified water, F-T ICL processes generate mineral waste derived from the coal. This non-leachable slag is suitable for sale as aggregate. Other, low-volume solid wastes include spent catalysts, organic sludge from biological wastewater facilities, inorganic sludge from flue gas desulfurization or evaporation ponds, and by-product sulfur if not recovered. These wastes may require specialized treatment to avoid transmission of soluble components if present, such as traces of the elements arsenic, chromium, mercury, molybdenum, and selenium, to local water supplies. However, F-T CTL plants incorporating modern, state-of-the-art gasification plants should accomplish removal of these and other potential pollutants within the normal plant process operations.

#### Utility Water

The CTL process chemistry uses water as a source of hydrogen and oxygen, during gasification of coal, and also produces water during the F-T reaction. The overall chemistry consumes water, but the chemical transformations are not the only use of water during coal liquefaction. The other major uses of water are for cooling, boiler feed water, and in solids handling, including processing of fine coal or dust. Most of the water is recovered after purification, but in a DCL plant, for example, 3–5% of the water being circulated in a cooling tower is lost to evaporation, leaks, and blowdown [91]. Thus, for every 1,000 gal circulated, 30–50 gal of makeup water is needed.

The amount of water needed varies with the process design, choice of gasifier, ambient temperature and humidity, and type of coal. According to the DOE Idaho National Lab, approximately 12–14 barrels of water are used for every barrel of F-T liquid fuel produced, [92] meaning that 40 million gallons of water/day are needed for an 80,000 barrel/day CTL facility. That is enough water to meet the domestic needs of more than 200,000 people (or 1/5 the population of the state of Montana).

The usage level of water in F-T CTL plants is considered a significant logistic issue in geographical areas of low rainfall or otherwise limited water resources. For example, competing use of water for agriculture would be a problem in the Western USA, and competing use of water for thermal-electric plants would be an issue in the Eastern USA. The availability of water is already an issue in Montana and Wyoming, where most of the inexpensive supplies of coal in the USA are located. Where water is in short supply, this logistic requirement becomes an environmental and/or economic issue and therefore has precluded construction of some planned CTL plants. Use of air, instead of water, as coolant may substantially reduce water usage requirements of the exothermic F-T reaction.

## Future Directions

Current commercialization of CTL technology serves the global interest in sustainability of fuel supply. Any additional production of fuels from alternative resources reduces overall world demand for petroleum and benefits all consumers by easing pressure on crude oil prices. As the price of oil increases and its easy availability decreases, countries with large coal reserves and insufficient domestic petroleum to meet increasing transportation fuel demands – particularly the PRC and India – are potential candidates for using CTL technologies to supplement conventional petroleum-based fuel products.

In the case of DCL, contribution to sustained fuel supply requires that the Shenhua project be successful and lead to commissioning of additional units, given estimates of proven coal Shenhua coal reserves in the Shanxi Province/Inner Mongolia Autonomous Region of about 200 billion tons [93]. In this context, it is important to note that parallel efforts are also underway in the PRC and elsewhere to erect commercial scale indirect coal liquefaction plants. Presumably, these plants have lower technical risks because the designs are derived from ICL technology that has already been well proven at full commercial scale, by Sasol. On the other hand, the direct liquefaction project, while supported by a 6 t/day pilot plant constructed in Shanghai, has no precedent on a larger scale. The more than

1,000-fold scale up from pilot plant to commercial unit is an ambitious objective that carries more than the typical amount of technical risk. These plants will also incorporate technologies specific to regional coals and environmental regulations.

If DCL technology is successful in the PRC, it will garner increased global interest, perhaps in combination with indirect liquefaction in a hybrid systems configuration. In this scenario, the low value residue from the direct liquefaction process would be gasified to provide a low cost source for synthesis gas production, rather than incurring a cost penalty for disposal. A recent US National Coal Council (NCC) report suggests that this might be a good time and justification to revisit the direct liquefaction pathway, particularly if the scale-up in the PRC is successful. The NCC report also suggests the speculative possibility of hybrid direct and indirect plants.

In the case of ICL, the F-T technology is already well commercialized but currently has a very limited worldwide aggregate production volume. Computational science is a rapidly expanding area of research that offers the potential to shortcut development time by providing a theoretical basis for subsequent R&D and technology design activities. It provides the opportunity to quickly explore novel processing strategies to guide and accelerate experimental research. The computational work focuses on the complex and critical chemical and physical aspects of converting coal to premium fuels: the fundamentals of catalyst activity/selectivity; hydrodynamics; separation of small catalyst particles from the liquid product; optimum system integration to achieve high process efficiency, minimal pollutant emissions and $CO_2$ capture; and computational frameworks to enable virtual demonstration of the entire fuel life cycle.

Laboratory research and additional modeling, followed by larger-scale bench and pilot-scale testing, could experimentally verify the computational findings. However, systems engineering is an important component activity needed to evaluate the technical and economic feasibility of: novel processes resulting from computational research; advanced integrated gasification and F-T processing concepts and cleanup technology; advanced reactor types needed to improve the process efficiency when utilizing specific regional coals; novel in situ reactions/processing; and modeling of the gas-solid–liquid physics of the F-T reactor to help achieve the highest throughput and liquid product quality.

Life cycle analyses performed on a “mine-to-wheels” basis likely will be used to identify those parts of the overall system impacting health and the environment now. Estimates of the impact of $CO_2$ on global warming, and estimates of the impacts of global warming itself, may also be attempted but these projections appear at present as far too broad in range to aid in resolving issues of energy sustainability. The greatest current push for the future use of coal appears to be in developing sustainable concepts for capturing and then storing or using the $CO_2$, because doing so obviates a significant portion of the large uncertainty as to potential effects of that $CO_2$ were it not captured. A related type of research involves combining coal use with chemistries or processes that generate less (or consume some amount of) $CO_2$, to decrease the overall output of $CO_2$ per barrel of oil equivalent produced.

In addition to the freestanding F-T CTL process configuration, the basic plant can be reconfigured to capture about 90% of the carbon that otherwise would be released into the atmosphere within the plant boundary. Also, another option, not involving $CO_2$ recycle, would be to design a polygeneration plant that produces both fuels and power.

More production from existing mines and the opening of new coalmines will be required to accommodate growth in the CTL industries as well as growing global demand for electric power. These facilities will require land for construction and support operations such as roads, railroads, and storage facilities that create changes in land use, alter topography, and impact ecological systems. Mitigation and reclamation strategies would need to be implemented to offset some of these changes to land and ecological resources. Currently, there are very strict reclamation regulations in place and they are being further enhanced to encourage a reforestation initiative to better restore the land, while concurrently improving water quality and providing a source for carbon sequestration. Even if the environmental risks are addressed, there is a very good possibility of public reluctance to accept the need for new facilities, particularly these coal-based plants. Measures would need to be taken to involve the general public and other state, local, and nongovernmental entities to assure them that these plants could effectively protect human health and safety and the environment.

However, F-T CTL plants would be expected to have very low environmental criteria pollutant emissions. They could also be designed to accommodate carbon capture and pressurization for subsequent sequestration in saline aquifers or oil reservoirs for enhanced oil recovery (EOR). Therefore, these plants could remove a good deal of uncertainty associated with possible new environmental regulations. Early plants that would sell or demonstrate $CO_2$ use for EOR would be encouraged.

Regarding methanol and DME, natural gas is currently the preferred feedstock for methanol production but companies such as Eastman Chemical in particular have rendered the coal-based process commercial, and cost-effective technology for $CO_2$ mitigation could pave the way for significant increase in commercial production of methanol and DME from coal in Asia and elsewhere.

Future CTL plants will use advanced clean coal gasification technology to produce transportation fuels and optionally electric power. Pollutant emissions will be minimal because coal-derived sulfur will be removed and converted into elemental sulfur. Nitrogen oxides will be minimized using low-NOx burners and selective catalytic reduction (SCR) in the flue gas stream and mercury will be removed, perhaps by some combination of pre- and post-combustion processes. Using air coolers where possible will minimize water use, and solids emissions will consist of non-leachable slag from the gasification process. Because of the sensitivity of the F-T catalyst to poisons, all contaminants must be removed to near-zero levels (ppb levels), thus ensuring that overall plant emissions would be close to zero.

Certainly coal conversion to hydrocarbon liquids and electricity is now economically attractive from a cash cost viewpoint. However, large capital investments are required and small to medium coal conversion plants will not become a reality unless there are effective forms of state assistance with the initial capital investment. This would be motivated by the desire for independence from imported energy and the more efficient and cleaner utilization of the local energy resource. The Chinese government has accepted this approach, and China is rapidly becoming the pioneer for developing commercial CTL technology options. However, it is the duty of technology providers to ensure that technology is made available to allow coal use to be as efficient as possible.

Site-specific early design studies would provide the ability to obtain information on environmental baselines for CTL plants. These plants would be ready for $CO_2$ separation and capture and the information obtained would define resource requirements. Site-specific information would also address where the resources, such as coal and water, are coming from, how they are delivered, and how waste products are to be reused or disposed. Additionally, current R&D activities cosponsored by the DOE and industry are being pursued to improve $CO_2$ separation and capture and define $CO_2$ storage sinks.

Making fuels from coal via either DCL or ICL processes has the advantage that fuel infrastructures already in place for petroleum crude oil products can be used largely unchanged when a shift is made to coal-derived fuels. The advantage of fuel infrastructure compatibility offered by synthetic hydrocarbon fuels is not so great in the PCR at present, where a liquid hydrocarbon fuel infrastructure for transportation fuels is in a highly developmental state. The challenges are primarily economic, as prospective air pollutant regulatory constraints worldwide give high value to clean synthetic fuels having emission characteristics superior to crude oil derived fuels, whereas coal-based synfuels processes incur significant cleanup costs. Much, although not all, of the contaminants needing to be cleaned up result from the biological origins of coal itself. In that regard, clean coal research might also provide insight on more efficient processing and cleanup in emerging biomass-to-liquids processes.

## Bibliography

### *Primary Literature*

1. Appert O (2008) CTL could provide long-term, transport-fuel supply. Syngas Refiner, IV 9:14–18
2. Gray D, White C, Tomlinson G, Ackiewicz M, Schmetz E, Winslow, J (2007) Increasing security and reducing carbon emissions of the US transportation sector: a transformational role for coal with biomass. DOE/NETL-2007/1298, p 61
3. De Klerk A (2009) Overview of coal-to liquids technology. In: De Klerk (ed) Beyond Fischer–Tropsch. Elsevier, Amsterdam, p 1. Arno de Klerk comments, "Coal-to-liquids technology started as an ironic twist of fate seen against the present day drive to move towards renewable sources of energy, since coal replaced whales as feed material for the production of lamp oil (kerosene). Coal is not a renewable energy source, except when viewed on a geological time scale, but whales are, yet, the whaling industry came close to making whales a non-renewable resource by converting all whales into oil." The anecdote illustrates a growing understanding among energy experts, that even with conservation global demand for energy is

growing and can best be met by careful use of both renewable and nonrenewable resources. On this point, see also Crane et al
4. Crane H, Kinderman E, Malhotra R (2010) A cubic mile of oil. Oxford University Press, New York
5. Thomson E (2002) The Chinese coal industry: an economic history. Routledge, London
6. Lister G (1944) Chronological records of coal mining, transport etc. In: Northumberland and Durham from A.D. 1180 to 1839. Ramsden Williams, Consett
7. Perry H (1974) Coal conversion technology. Chem Eng 81(15):88–102
8. Murakushi N (1981) The transfer of coal-mining technology from Japan to Manchuria and manpower problems: focusing on the development of the Fushun coal mines. Japanese Experience of the UNU Human and Social Development Programme series, p 47
9. NEDO (2006) Clean coal technology in Japan. 4A1. Coal liquefaction technology development in Japan, pp 57–58. www.nedo.go.jp/kankobutsu/pamphlets/sekitan/cct2006e.pdf. Accessed 10 April 2010
10. Sugawara A, Kurosawa S, Hatori H, Saito K, Yamada, Y, Sugihara M, Wasaka S, Yoshida H, Seo T, Susuki T, Inoguchi M, Sohnai M (1998) Coal conversion technologies on the new sunshine program in Japan. Preprints of symposia – American Chemical Society, Division of Fuel Chemistry 43(2): 330–334
11. NEDO (2006) Clean coal technology in Japan. 4A2. Bituminous coal liquefaction technology (NEDOL), pp 59–60. www.nedo.go.jp/kankobutsu/pamphlets/sekitan/cct2006e.pdf. Accessed 10 April 2010
12. Ghosh TK, Prelas MA (2009) Energy resources and systems: volume 1: Fundamentals and non-renewable resources. Springer Science + Business Media, Berlin
13. Dry ME (2002) The Fischer–Tropsch process: 1950–2000. Catal Today 71:227–241
14. Jager B, Espinoza R (1995) Advances in low temperature Fischer–Tropsch. Catal Today 23(1):17–28
15. Radtke K, Heinritz-Adrian M, Marsico C (2006) New wave of coal-to-liquids. An opportunity to decrease dependency on oil and gas imports and an appropriate approach to a partial revival of domestic coal industries. VGB PowerTech 86(5):78–84
16. Williams RH, Larson ED (2003) A comparison of direct and indirect liquefaction technologies for making fluid fuels from coal. Energy Sustain Dev VII(4):103–129
17. Grant-Huyser M, Maharaj S, Matheson L, Rowe L, Sones E (2004) Ethoxylations of detergent-range oxo alcohols derived from Fischer–Tropsch alpha-olefins. J Surfactants Deterg 7(4):397–407
18. Price JG (2004) Chemicals from synthesis gas. US 6740683 to Sasol Technology, 25 May 2004
19. Degnan TF Jr, Chen NY, Somorjai GA (2009) Heinz Heinemann's legacy at ExxonMobil: an illustrious career in industrial catalysis. Catal Lett 133(1–2):227–230
20. Brown D, Bhatt B, Hsiung T, Lewnard J, Waller F (1991) Novel technology for the synthesis of dimethyl ether from syngas. Catal Today 8:279–304
21. Sardesai A, Lee S (2005) Alternative source of propylene. Energy Sources 27(6):489–500
22. Larson ED, Yang H (2004) Dimethyl ether (DME) from coal as a household cooking fuel in China. Energy Sustain Dev, VII 3:115–126
23. Jones G, Jr Holm-Larsen H, Romani D, Sillis R (2001) DME for power generation fuel: supplying India's southern region. PETROTECH-2001 January 08–12, New Delhi, India
24. Bhatt B, Schaub E, Heydorn E (1993) Recent developments in slurry reactor technology at the LaPorte alternative fuels development unit. In: International technical conference on coal utilization & fuel systems, Clearwater, April 26–29, pp 197–208
25. Shikada T, Ohno Y, Ogawa T, Ono M, Mizuguchi M, Tomura K, Fujimoto K (1998) Direct synthesis of dimethyl ether from synthesis gas. In: Parmaliana A, Sanfilippo D, Frusteri F, Vaccari A, Arena F (eds) Studies in surface science and catalysis (Natural Gas Conversion V) 119: 515–520

26. Miller CL, Cicero D, Ackiewicz M, Anderson J, Schmetz E, Winslow J (2006) Coal conversion – a rising star? In: 23rd international Pittsburgh coal conference, Pittsburgh, September 25–28
27. White LC, Frederick JP (1995) ENCOAL mild coal gasification project. In: Proceedings of the 12th annual international Pittsburgh coal conference, Pittsburgh, pp 151–156
28. US Department of Energy, National Energy Technology Laboratory (2002) The ENCOAL® mild coal gasification project. A DOE assessment. DOE/NETL-2002/1171
29. McMillen DF, Malhotra M (1989) The role of hydrogen transfer in bond-cleavage and bond-forming processes during coal conversion. In: Schindler HD (ed) Coal liquefaction. A research needs assessment DE-AC01-87ER30110 final report, v2, ch 4, US Department of Energy, pp 30–49
30. Haenel MW (2008) Catalysis in direct coal liquefaction. In: Gerhard E (ed) Handbook of heterogeneous catalysis, vol 6, 2nd edn. Wiley-VCH Verlag, Weinheim, Germany, pp 3023–3036
31. Miller L (2008) Coal conversion technology. Congressional noontime briefing. Rayburn House Office Building, Washington DC, April 24
32. Eccles RM, DeVaux GR (1982) H-coal commercialization: current status. Energy Progress 2(2):111–115
33. Cleaner Coal Technology Programme (1999) (PDF). Technology status report 010: coal liquefaction. Department of Trade and Industry. http://www.dti.gov.uk/files/file18326.pdf. Accessed 10 April 2010
34. Styles GA (1982) The Wilsonville advanced coal liquefaction R&D facility – accomplishments and process evolution. Energy Progress 2(3):160–162
35. Valente AM, Cronauer DC (2003) Progress in coal liquefaction including a discussion of Wilsonville. Preprints of symposia – American Chemical Society, Division of Fuel Chemistry 48(1): 147–148
36. Burke FP, Brandes SD, McCoy DC, Winschel RA, Gray D, Tomlinson G (2001) Summary report of the DOE direct liquefaction process development campaign of the late twentieth century: topical report. DE2002-794281/XAB; DOE/PC-93054-94 Department of Energy, Washington DC
37. Nolan P, Shipman A, Rui H (2004) Coal liquefaction, Shenhua group, and China's energy security. Eur Manage J 22(2):150–164
38. Stohl FV, Lott SE, Diegert KV, Goodnow DC (1996) Results of hydrotreating the kerosene fraction of HTI's first proof of concept run. US DOE, SAND–96-0990C; CONF-960807–2
39. Friedrick F, Strobel BO (1983) The KOHLEOEL experimental plant at Bergbau-Forschung. In: Proceedings of the 3rd European coal utilisation conference, Amsterdam, pp 227–235
40. Strobel BO, Loering R (1992) IGOR – taking the short cut in coal hydrogenation. Preprints of papers – American Chemical Society, Division of Fuel Chemistry 37(1):448–455
41. NEDO (2006) Clean coal technology in Japan. 4A3. Brown coal liquefaction technology (BCL), pp 61–62. www.nedo.go.jp/kankobutsu/pamphlets/sekitan/cct2006e.pdf. Accessed 10 April 2010
42. Ramdoss PK, Tarrer AR (1997) Modeling of two-stage coal coprocessing process. Energy Fuels 11(1):194–201
43. Wender I, Tierney JW (1995) Effect of pretreating of host oil on coprocessing. Final report. DOE/PC/91054–T15, Contract Number AC22-91PC91054, Oct. 1 1995 Oct 01
44. Miller RL, Giacomelli GF, McHugh KJ, Baldwin RM (1989) Coprocessing of coal and residuum under low-severity reaction conditions: effect of basic nitrogen promoters. Energy Fuels 3(2):127–131
45. Smith R, Asaro M, Naqvi S (2008) Fuels of the future: technology intelligence for coal to liquids strategies. SRI Consulting, Menlo Park
46. Koornneef J, Junginger M, Faaij A (2007) Development of fluidized bed combustion – an overview of trends, performance and cost. Prog Energy Combust Sci 33(1):19–55

47. Senapati PK, Das D, Nayak A, Mishra PM (2008) Studies on preparation of coal water slurry using a natural additive. Energy Sources A 30(19):1788–1796
48. SRI Consulting (2006) Coal gasification. Process Economics Program report 154A, ch 4, SRI Consulting, Menlo Park, pp 4–46
49. Davis BH (2001) Fischer–Tropsch synthesis: current mechanism and futuristic needs. Fuel Process Technol 71(1–3):157–166
50. Leckel D (2005) Hydrocracking of iron-catalyzed Fischer–Tropsch waxes. Energy Fuels 19:1795–180
51. Davis BH (2007) Fischer–Tropsch synthesis: comparison of performances of iron and cobalt catalysts. Ind Eng Chem Res 46(26):8938–8945
52. Smith R, Asaro M (2005) Fuels of the future: technology intelligence for gas to liquids strategies. SRI Consulting, Menlo Park
53. Bukur DB (2005) Attrition studies with catalysts and supports for slurry phase Fischer–Tropsch synthesis. Catal Today 106(1–4):275–281
54. Bukur DB, Carreto-Vazquez VD, Pham HN, Datye AK (2004) Attrition properties of precipitated iron Fischer–Tropsch catalysts. Appl Catal, A 266(1):41–48
55. Dry M (2004) Chemical concepts used for engineering purposes. In: Steynberg A, Dry M (eds) Studies in surface science and catalysis 152: Fischer–Tropsch technology. Elsevier, Amsterdam, p 215
56. Donnelly TJ, Yates IC, Satterfield CN (1988) Analysis and prediction of product distributions of the Fischer–Tropsch synthesis. Energy Fuels 2:734–739
57. Smith R (2009) SRI Consulting Coal to gasoline. Process Economics Program report 271, SRI Consulting, Menlo Park. Data for the figure were obtained from reference 58
58. Donnelly TJ, Satterfield CN (1989) Product distributions of the Fischer–Tropsch synthesis on precipitated iron catalysts. Appl Catal 52(1):93–114
59. Dry M (2004) Chemical concepts used for engineering purposes. In: Steynberg A, Dry M (eds) Studies in surface science and catalysis 152: Fischer–Tropsch technology. Elsevier, Amsterdam, pp 196–257
60. Jager B, Kelfkens RC, Steynberg AP (1994) A slurry bed reactor for low temperature Fischer–Tropsch. In: Curry-Hyde HE, Howe RF (eds) Natural gas conversion II. Elsevier, Amsterdam, pp 419–425
61. Adeyiga AA, Bukur DB, Carreto-Vazquez V, Ma W, Nowicki L (2004) Attrition resistance and catalytic performance of spray-dried iron Fischer–Tropsch catalysts in a stirred-tank slurry reactor. Ind Eng Chem Res 43(6):1359–1365
62. Bai L, Chang J, Hao Q, Li Y, Liu F, Wang H, Xiang H, Xu B, Yi F, Zhang C (2007) Effect of reduction temperature on a spray-dried iron-based catalyst for slurry Fischer–Tropsch synthesis. J Mol Catal A: Chem 261(1):104–111
63. Bartholomew C, Bukur DB, Datye AK, Nowicki L, Pham HN, Xu J (2003) Attrition resistance of supports for iron Fischer–Tropsch catalysts. Ind Eng Chem Res 42(17):4001–4008
64. Zhao R, Goodwin JG, Jothimurugesan K, Gangwal K, Spivey JJ (2001) Spray-dried iron Fischer–Tropsch catalysts.1. Effect of structure on the attrition resistance of the catalysts in the calcined state. Ind Eng Chem Res 40(4):1065–1075
65. Demirel B, Bohn MS, Benham CB, Siebarth JE, Ibsen MD (2005) Method and apparatus for regenerating an iron-based Fischer–Tropsch catalyst. US 6838487 to Rentech, 4 January 2005
66. Benham CB, Bohn MS, Yakobson DL (1996) Process for the production of hydrocarbons. US 5504118 to Rentech, 2 April 1996
67. van der Merwe W (2010) Conversion of spent solid phosphoric acid catalyst to environmentally friendly fertilizer. Environ Sci Technol 44(5):1806–1812
68. AsiaPulse News (2009) SHENHUA NINGXIA COAL, SASOL'S COAL LIQUEFACTION PJT BEGINS 2010. Yinchuan, 22 June 2009
69. Peng X, Toseland B, Underwood T (1997) A novel mechanism of catalyst deactivation in liquid phase synthesis gas-to-DME reactions. In: Bartholomew C, Fuentes GH (eds) Catalyst deactivation. Elsevier, Amsterdam

70. Gogate M, Lee S (1991) A single- stage, liquid-phase dimethyl ether synthesis process from syngas. I. Dual catalytic activity and process feasibility. Fuel Sci Technol Int 9(6):653–679
71. Cornthwaite D (1972) British patent 1296212 to Imperial Chemicals Inc. (ICI)
72. Espino RL, Plezke TS (1975) Methanol production. US 3888896 to ChemSystems, 10 June 1975
73. Sherwin M, Blum D (1979) Liquid-phase methanol. Final report. Electric Power Research Institute, EPRI-AF-1291
74. Lee S, Sardesai A (2005) Liquid phase methanol and dimethyl ether synthesis from syngas. Top Catal 32(3–4):197–207
75. Air Products Liquid Phase Conversion Company (2004) Commercial-scale demonstration of the liquid-phase methanol (LPMEOH) process. DOE/FE-0470, US Department of Energy
76. Joo O, Jung K, Han S (2002) Modification of H-ZSM-5 and gamma-alumina with formaldehyde and its application to the synthesis of dimethyl ether from syn-gas. Bull Korean Chem Soc 23:1103–1105
77. Lee S, Parameswaran VR, Wender I, Kulik CJ (1989) Roles of carbon dioxide in methanol synthesis. Fuel Sci Technol Int 7(8):1021–1058
78. Cai Y, Davies S, Wagner J (2003) Water gas shift catalyst. US 6627572 to Süd-Chemie, 30 September 2003
79. Kim J, Park MJ, Kim SJ, Joo O, Jung K (2004) DME synthesis from synthesis gas on the admixed catalysts of $Cu/ZnO/Al_2O_3$ and ZSM-5. Appl Catal, A 264(1):37–41
80. Tijm PJ (2003) Development of alternative fuels and chemicals from synthesis gas. DOE contract number FC22-95PC93052, final report
81. NEDO (2006) Clean coal technology in Japan. 4A4. Dimethyl ether production technology (DME), pp 63–64. www.nedo.go.jp/kankobutsu/pamphlets/sekitan/cct2006e.pdf. Accessed 10 April 2010
82. Hayashi H, Todoroki A, Yasuto A, Ohno Y (2002) NKK DME diesel vehicle development and fleet test in Japan. In: International symposium on alcohol fuels, Sao-Paulo, November 12–15
83. Ohno Y, Tanishima S, Aoki S (2005) Coal conversion into dimethyl ether as an innovative clean fuel. In: International conference on coal science and technology, ICCST, Okinawa, Japan, October 09–14
84. Mogi Y, Ohno Y, Ogawa T, Inoue N, Shikada T (2000) Development of slurry phase dimethyl ether synthesis technology. In: Pittsburgh coal conference, Pittsburgh, September 11–14, pp 398–408
85. Ohno Y, Yagi H, Inoue N, Okuyama K, Aoki S (2007) Slurry phase DME direct synthesis technology – 100 tons/day demonstration plant operation and scale up study. In: Noronha FB, Schmal M, Sousa-Aguiar EF (eds) Studies in surface science and catalysis: natural gas conversion VII. Elsevier, Amsterdam
86. Liu Z (2005) Clean coal technology: direct and indirect coal-to-liquid technologies. InterAcademy Council. http://www.interacademycouncil.net/Object.File/Master/10/335/CleanCoal technology_ coal liquefaction.pdf. Accessed 24 April 2010
87. Probstein RF, Hicks RE (2006) Synthetic fuels. Dover, Mineola
88. Gray D, Salerno S, Tomlinson G, Marano JJ (2004) Polygeneration of SNG, hydrogen, power, and carbon dioxide from Texas lignite. Mitretek technical report for the DOE, MTR-04-2004-18
89. Camm F, Bartis JT, Bushman CJ (2008) Federal financial incentives to induce early experience producing unconventional liquid fuels. Report prepared for the United States Air Force and the National Energy Technology Laboratory of the DOE. http://www.rand.org/pubs/technical_reports/TR586/, RAND corporation. Accessed 10 April 2010
90. Kohler LPFD, Du Plessis GH, Du Toit FJ, Koper EL, Phillips TD, Van Der Walt J (2006) Method of purifying Fischer–Tropsch derived water. US 7153432 to Sasol Technology, 26 December 2006
91. US Department of Energy (2006) Emerging issues for fossil energy and water. DOE/NETL-2006/1233

92. Boardman R (2007) Gasification and water nexus. Presented March 14, 2007 at the GTC workshop on gasification technologies, Denver
93. Wei W (2003) Current issues of China's coal industry: the case of Shanxi. In: Coate B, Brooks R, Fraser I, Xu L (eds) Proceedings of the 15th annual conference of the association for Chinese economics studies Australia (ACESA), RMIT (Royal Melbourne Institute of Technology) Business Research Development Unit, Melbourne

## *Books and Reviews*

Dry ME (2001) High quality diesel via the Fischer–Tropsch process – a review. J Chem Technol Biotechnol 77(1):43–50

Egan CJ (1976) Method of power generation via coal gasification and liquid hydrocarbon synthesis. US 3986349 to Chevron Research Company, 19 October 1976

Furstner AE (1995) Active metals: preparation characterization applications. In: Furstner A (ed) Supported metals. Wiley, New York

Iglesia E, Reyes S, Madon RS (1993) Selectivity control and catalyst design in the Fischer–Tropsch synthesis: sites, pellets, and reactors. Adv Catal 39:221–302

Iglesia E, Reyes S, Soled S (1993) Reaction-transport selectivity models and the design of Fischer–Tropsch catalysts. In: Becker ER, Pereira CJ (eds) Computer-aided design of catalysts and reactors. Marcel Dekker, New York, p 640

Jin Y, Wang J, Wang T (2007) Slurry reactors for gas-to-liquid processes: a review. Ind Eng Chem Res 46(18):5824–5847

O'Brien R, Xu L, Bao S, Raje A, Davis B (2000) Activity, selectivity and attrition characteristics of supported iron Fischer–Tropsch catalysts. Appl Catal, A 196(2):173–178

Rahmin II (2008) GTL, CTL finding roles in global energy supply. Oil Gas J 24:22–31

Sloane D (2008) Clean coal technologies accelerating commercial and policy drivers for deployment. International Energy Agency/Coal Industry Advisory Board, p 70

Tijm PA, Waller F, Brown D (2001) Methanol technology developments for the new millennium. Appl Catal, A 22(1–2):275–282

Williams A, Pourkashanian M, Jones JM (2001) Combustion of pulverized coal and biomass. Prog Energy Combust Sci 27:587–610

# Chapter 12
# Mining Industries and Their Sustainable Management

**Sandip Chattopadhyay and Devamita Chattopadhyay**

## Glossary

| | |
|---|---|
| Abandoned mines | Mines for which the owner cannot be found, or for which the owner is financially unable or unwilling to carry out cleanup. They may pose environmental, health, safety, and economic problems to communities, the mining industry, and governments in many countries. |
| Acid (rock or mine) drainage | Many metal ore bodies and coal deposits contain significant quantities of sulfide minerals – often including the ore minerals themselves. When such minerals are brought to the surface, they react chemically with air and water producing sulfuric acid, which may dissolve other minerals containing potentially toxic elements. This *acid drainage* from coal and metal mining around the world can pollute water and the surrounding land, affecting plant and animal life. Acid drainage is known as *acid mine drainage* when it is closely associated with mining activities, and *acid rock drainage* when this phenomenon occurs naturally, without human intervention. Both phrases are in |

This chapter was originally published as part of the Encyclopedia of Sustainability Science and Technology edited by Robert A. Meyers. DOI:10.1007/978-1-4419-0851-3

S. Chattopadhyay (✉)
Tetra Tech Inc, 250 West Court Street, Suite 200 W, Cincinnati, OH 45202, USA
e-mail: sandip.chattopadhyay@tetratech.com

D. Chattopadhyay
CH2M Hill, 1 South Main Street, Suite 1100, Dayton, OH 45402, USA

R. Malhotra (ed.), *Fossil Energy: Selected Entries from the Encyclopedia of Sustainability Science and Technology*, DOI 10.1007/978-1-4614-5722-0_12,

common use, although particular stakeholder groups have particular preferences related to the controversial nature of this issue.

**Acidophile** An organism that thrives in a relatively acidic environment.

**Ammonification** The biochemical process whereby ammoniacal nitrogen is released from nitrogen-containing organic compounds.

**Amorphous** Irregular, having no discernible order or shape. Rocks or minerals that possess no definite crystal structure or form, such as amorphous carbon.

**Bioleaching** Extraction of metal from solid minerals into a solution is facilitated by the metabolism of certain microorganisms.

**Biomining** Extraction of specific metals from their ores through biological means, usually bacteria. It is an actual economical alternative for treating specific mineral ores, involving percolation and agitation techniques.

**Community** The people living around the mine who are directly affected (both positively and negatively) by the mine's activities.

**Contaminated land/water** Land/water containing concentrations of potentially toxic materials (organic or inorganic) elevated above the natural background concentrations in a particular area. In relation to mining, land or water contamination may occur through fuel spills, run-off from waste rock dumps, leaks from tailings impoundments, wind-blown dust from tailings and waste rock, smelter emissions, and drainage from mine workings. Contaminated groundwater is caused by the seepage of contaminated waters into aquifers.

**Crystalline** A substance in which the constituent atoms, molecules, or ions are packed in a regularly ordered, repeating three-dimensional pattern.

**Denitrification** A microbially facilitated process of nitrate reduction that may ultimately produce molecular nitrogen through a series of intermediate gaseous nitrogen oxide products.

**Dissimilatory reduction** Sulfate-reducing bacteria reduce sulfate in large amounts to obtain energy and expel the resulting sulfides as waste; this is known as dissimilatory sulfate reduction. They are anaerobes, which use sulfate as the terminal electron acceptor.

| | |
|---|---|
| Electrowinning | The recovery of metal by electrolysis. An electric current is passed through a solution containing dissolved metals, and this causes the metals to deposit on a cathode. |
| Extractant | An immiscible liquid used to extract a substance from another liquid. |
| Gypsum | A sedimentary rock consisting of hydrated calcium sulfate. |
| Heap leaching | To dissolve minerals or metals out of an ore heap using chemicals. For example, a cyanide solution percolates through crushed ore heaped on an impervious pad or base pads during heap leaching of gold. |
| Macrophyte | An aquatic plant that grows in or near water and is either emergent, submergent, or floating. They provide cover for fish and substrate for aquatic invertebrates, produce oxygen, and act as food for some fish and wildlife. |
| Mesophile | An organism that grows best in moderate temperature (typically between 15°C and 40°C). The term is mainly applied to microorganisms. It is also used to describe mesophilic anaerobic digestion, which takes place optimally around 37–41°C or at ambient temperatures between 20°C and 45°C, where mesophiles are the primary microorganisms present. |
| Methanogenesis | Production of methane by biological processes that are carried out by methanogens. A methanogen is a single-celled microorganism and is a member of the Archaea. Archaea are unique because unlike most life on Earth that rely on oxygen and complex organic compounds for energy, Archaea rely on simple organic compounds (e.g., acetate) and hydrogen for energy. |
| Mining life cycle | The processes of exploration, mining development, extraction, processing, refining, smelting, and mine closure. |
| Ore | Mineral-bearing rock that can be mined and treated profitably under the existing economic conditions, or those conditions which are deemed to be reasonable. |
| Reclamation | Process of converting derelict land (land that requires intervention before beneficial use) to usable land and may include engineering as well as ecological solutions. |

| | |
|---|---|
| Remediation | Environmental cleanup of land and water contaminated by organic, inorganic, or biological substances. |
| Restoration | Seeks to artificially accelerate the processes of natural succession by putting back the original ecosystem's function and form. |
| Tailings | Mineral wastes produced from the processing operations after the valuable minerals have been extracted. |
| Waste rock | The mineral wastes produced during mine development – including overburden and barren rock – and those parts of an ore deposit below the economic cutoff grade. Often, and particularly in some metal deposits, the waste rock may contain sufficient sulfide mineral concentrations to generate long-term acid drainage problems. |

## Definition of the Subject and Its Importance

Mining, minerals, and metals are important to the economic and social development as they are essentials for modern living. However, supplies of minerals, such as coal, are limited, and sustainable management of natural resources requires the maintenance, rational and enhanced use as well as a balanced consideration of ecology, economy, and social justice. Mining industry's recognition and acceptance of its sustainable development is growing. In the mining and metals sector, this means that investments should be:

Leaders in the global mining business community have begun to assign a new strategic significance to the term sustainability. No longer confined to the economic realm, sustainability embraces a broad spectrum of organization characteristics related to social and environmental responsibility. Profitability alone is inadequate as a measure of success, and that many of the nonfinancial concerns associated with sustainability are fundamental drivers of long-term shareholder value. Failure to recognize these strategic issues threatens the very survival of a business enterprise.

Large amounts of material are involved in large-scale mining and minerals extraction. The problems arising from the change in the chemistry of million tons of natural ore during the processing steps (such as, grinding, calcining, and roasting operations) and their resultant bioavailabilities are not well understood. Mining produces large volumes of waste, and decisions regarding waste handling and other aspects of operations are often difficult and expensive to reverse; they need to be made correctly initially through mine closure planning. Another challenge is the environmental legacy left by mining. The environmental issues of current mining operations are daunting enough. But in many ways far more troubling are some of the continuing effects of past mining and smelting. The loss of biodiversity is the other great challenge of mining sustainability. The loss of biodiversity is an irreversible loss. Conservation practices that guarantee a minimum impact on biodiversity must be adopted and implemented.

As important as the methods of mining and beneficiation is how the minerals are used for sustainable development. An integrated approach for the use of minerals must be developed. Current patterns of minerals used raise concerns about efficiency and the need for more equitable access to resources worldwide. The environmental and health impacts of different mineral products in use need to be carefully managed. Where the risks associated with use are deemed unacceptable or are not known, the costs associated with using certain minerals may outweigh the benefits.

The importance of sustainable development in mining industry includes actions at all levels to:

- Support efforts to address the environmental, economic, health, and social impacts of mining throughout the life cycle, including workers' health and safety. A range of partnerships, furthering existing activities at the national and international levels, among interested governments, intergovernmental organizations, mining companies and workers, and other stakeholders to promote transparency and accountability for sustainable mining and minerals development.
- Enhance the participation of stakeholders, including local and indigenous communities, to play an active role in minerals, metals, and mining development throughout the life cycles of mining operations. This includes efforts after mine closure for rehabilitation purposes, in accordance with national regulations and taking into account significant transboundary impacts.
- Foster sustainable mining practices through the provision of financial, technical, and capacity-building support to developing countries and countries with economies in transition for the mining and processing of minerals. This includes small-scale mining, and, where possible and appropriate, improve value-added processing, upgrade scientific and technological information, and reclaim and rehabilitate degraded sites.

## Introduction

Mining takes a significant toll on the environment. The intensity of the environmental impact depends on what is being mined, where and how. Unless it is meticulously planned and carefully executed, mining can devastate lands, pollute and deplete water resources, denude forests, wipe out wildlife, and defile the air.

The history of mining has been one of boom and bust periods that were a balance between available natural resources, production costs, and the ability to sustain profitability. In recent years, the mining industry leadership has emphasized that the modern-day mining business has become more complex with the advent of stricter regulations, better-informed stakeholders, and closer in-depth scrutiny of current and/or proposed operations. Increased emphasis is being placed on the importance of sustainable operations that are focused on effectively integrating environment, social, and economic impacts.

The quantity of mining waste produced fluctuates yearly, and as the individual mines and quarries manage their wastes according to local conditions, there are no definitive statistics. After being removed, *waste rock*, which often contains acid-generating sulfides, heavy metals, and other contaminants, is usually stored above ground in large free-draining piles. This waste rock and the exposed bedrock walls from which it is excavated are the primary sources of pollution caused due to mining. The US mining industry produces approximately 8,000,000 tons per year (t/year) of process residue that may contain hazardous species as well as valuable by-products. These process residues are generated by smelter off-gas cleaning at approximately 5,500,000 t/year, and baghouse dust and wastewater treatment at approximately 2,100,000 t/year [24]. Comparable statistics was obtained for other countries. It was estimated that 96.4 million tons of mining and quarrying waste was produced in 2004 in the UK [4]. The Canadian mineral industry typically generates one million tons of waste rock and 950,000 t of *tailings* per day, totaling 650 million tons of waste per year [29]. The right technology may be able to recover marketable by-products from process residues to generate revenue and reduce disposal costs for the mining industry. The process residue that cannot be reused can impact the water resources, and the effect may be manifest throughout the life cycle of the mine and even long after mine closure.

## Mining Life Cycle

Minerals are nonrenewable resources, and the mines have finite lives. Mining represents a temporary use of the land, and during this temporary use of the land, the *mining life cycle* can be divided into the following stages: exploration, development, extraction and processing, and mine closure.

Exploration is the work involved in determining the location, size, shape, position, and value of an ore body using prospecting methods, geologic mapping, and field investigations, remote sensing (aerial and satellite-borne sensor systems that detect ore-bearing rocks), drilling, and other methods. Building access roads to a drilling site is one example of an exploration activity that can cause environmental damage.

The development of a mine consists of several principal activities: conducting a feasibility study, including a financial analysis to decide whether to abandon or develop the property; designing the mine; acquiring mining rights; filing an Environmental Impact Statement (EIS); and preparing the site for production. An example of site preparation impacting the environment is removal of the overburden (surface material above the ore deposit that is devoid of ore minerals) by excavation prior to mining.

Extraction is the removal of ore from the ground on a large scale by one or more of three principal methods: surface mining, underground mining, and in situ mining (extraction of ore from a deposit using chemical solutions). After the ore is removed from the ground, it is crushed so that the valuable mineral in the ore can be separated from the waste material and concentrated by flotation (a process that separates finely ground minerals from one another by causing some to float in a froth and others to sink), gravity, magnetism, or other methods, usually at the mine site, to prepare it for further stages of processing. The production of large amounts of waste material (often very acidic) and particulate emission have led to major environmental and health concerns with ore extraction and concentration. Additional processing separates the desired metal from the mineral concentrate.

The closure of a mine refers to the cessation of mining at a site. Planning for closure is often required to be ongoing throughout the life cycle of the mine and not left to be addressed at the end of operations. It involves completing a *reclamation* plan plan and ensuring the safety of areas affected by the operation, for instance, sealing the entrance to an abandoned mine. The Surface Mining and Control Act of 1977 states that reclamation must "restore the land affected to a condition capable of supporting the uses, which it was capable of supporting prior to any mining, or higher or better uses." *Abandoned mines* can cause a variety of health-related hazards and threats to the environment, such as the accumulation of hazardous and explosive gases when air no longer circulates in deserted mines, use of these mines for illegal residential or industrial dumping, and others. Many closed or abandoned mines have been identified by federal and state governments and are being reclaimed by both industry and government.

## Coal Mining

Mining can affect both air and water. A fossil fuel that has been mined intensively is coal. As an example of environmental impact, effects of coal mining are presented below.

Coal is and will continue to be a crucial element in a modern, balanced energy portfolio, providing a bridge to the future as an important low cost and secure energy solution to sustainability challenges. A review of the Annual Energy Outlook 2010 report from the Energy Information Administration (EIA) indicates that world coal consumption will increase by 56%, from 132 quadrillion BTU in 2007 to 206 quadrillion BTU in 2035. Figure 12.1 presents the data for coal consumption [6] versus total energy consumption [17] between 1980 and 2030.

The growth rate for coal consumption is uneven, averaging 1.1% per year between 2007 and 2020 and 2.0% per year between 2020 and 2035. The slower growth rate for the earlier period is largely from a decline in coal consumption in 2009 during the global economic recession. With economic recovery, world coal consumption

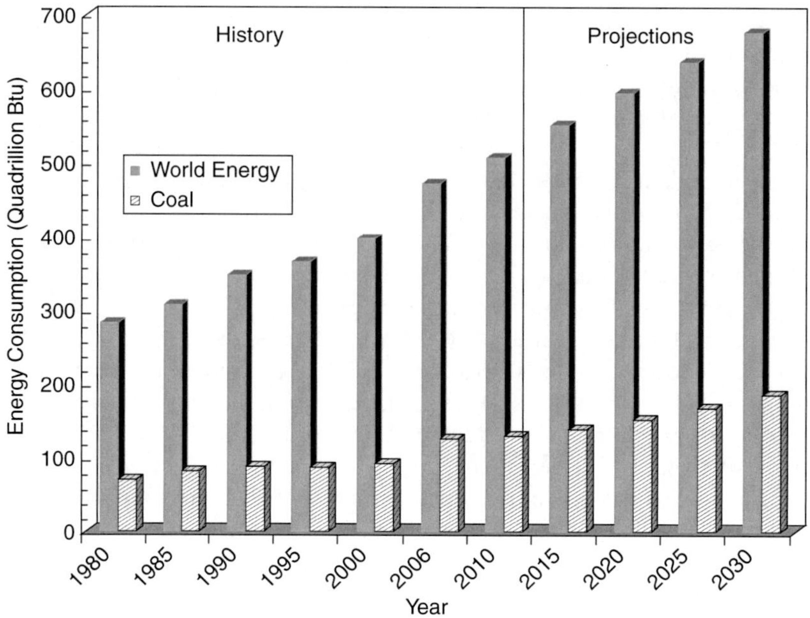


**Fig. 12.1** Global energy consumption 1980–2030

rebounded, and it is expected to return to its 2008 level by 2013. The US energy demand is expected to grow at an average annual rate of 1.1% for the next 25 years. Among the largest consumers of energy are China and India, and together they accounted for about 10% of the world's total energy consumption in 1990 and 20% in 2007. In a separate report by Waddell and Pruitt [31], it is projected that coal utilization will increase worldwide by 44% by 2025.

The process of coal excavation consists of several cycles: the cycle of excavation, disposal, and recultivation. The first step is reconnaissance of the terrain (preliminary survey of ground) that has been identified for mining. Preparation of the terrain may necessitate demolition of constructions, felling of trees, relocation of watercourses, roads, etc. This is usually followed by excavation of the overburden and then transporting and disposing the material using specialized equipment (mining machinery). In present days, bucket wheel excavators, conveyor belts, and stackers are mainly used. When the coal seam appears, such a seam is excavated and transported to the crushing plant.

Some of the important characteristics of coal excavation are listed below:

- Notwithstanding the occupancy of large areas and the change in the purpose of the land, end of mining will result in bringing back large portions of the used areas to the previous state. It means that though certain areas are temporarily occupied by mining activities, for example, for a period of 10 or more years, they can be reused as before following this period.

- Area occupied by the external dump, although it is generally higher than the terrain before mining activities, can be brought back to the state so that it can be reused.
- At the end of mining activities, the mine becomes the owner of large areas of agricultural land and forests. The owner of an agricultural land and forest is obliged to work the soil and control forest husbandry.

## Pollution from Coal Mining

The process of coal excavation causes deterioration of the original morphological and pedological structure of the terrain and soil, and results in the release of harmful substances and/or mineral dust into the air. Such a release is primarily related to the deterioration of the upper seam structure during mining operations. The negative impact may occur as a result of the excavation of the upper seam and its inadequate disposal, and of mixing the upper seam with the lower one, as well as other barren materials. The impact of lignite excavation also represents a potential contamination of the upper seam due to the precipitation of dust from the air.

In addition to the aforementioned impacts, mining in a particular area will result in an increase in the area population not only due to the increased number of mine workers but also due to the development of related industries. Such development will often result in disappearance of the arable upper seam due to the building up of infrastructure facilities (roads, railroad tracks, waterways, industrial areas, etc.) and of the change in the purpose of the soil in the vicinity of the mine.

Mining can cause physical disturbances to the landscape, creating eyesores such as waste-rock piles and open pits. Such disturbances may contribute to the decline of wildlife and plant species in an area. In addition, it is possible that many of the pre-mining surface features cannot be replaced after mining ceases. Mine subsidence (ground movements of the earth's surface due to the collapse of overlying strata into voids created by underground mining) can cause damage to buildings and roads.

Opencast coal mining leads to the disturbance of geological layers where different sedimentary products are being mixed, which means that a completely new, anthropogenic land, i.e., substratum, is created, without any resemblance to the original land, which is called a deposol. This is an example of visual pollution where the characteristics of the landscapes' appearance have changed due to mining. The opencast coal exploitation results in morphological modifications of the terrain, such as the creation of large-scale depressions and the formation of outside overburden dumps.

### *Pollution of Air*

Coal mines emit constituents that cause or contribute significantly to air pollution and that may reasonably be anticipated to endanger public health and welfare. In each stage of mining, from exploration to ore recovery to downstream processing, there is the potential for air quality impacts due to emissions of particulates (dust, diesel, and silica). The monochromatic appearance of the mine areas is due to generation of large quantities of fugitive dusts during mining operations. Coal mining areas are black, bauxite and iron-ore rich regions are red, while limestone gives a chalky white hue. Dust results from blasting, handling, processing, or transporting of soil and rock or can arise from bare or poorly vegetated areas in combination with air movements. It is one of the most visible, invasive, and potentially irritating impacts of mining, and its visibility often raises concerns. Many dusts contain metals which are potentially hazardous, and are known to cause certain diseases. It has the potential to severely affect flora and fauna near the mine and to impact the health of mine workers and local residents. Dust also affects the agricultural productivity of the area. The level of dust generated, its behavior (particle size, density, travel distance), and types of health and environmental risks depend on many factors including mine type, local climate, topography, working methods, types of equipment used, the mineralogy and metallurgical characteristics of some ores, and the land use of the area around the mine.

- EIA reported that estimated recoverable reserves of coal in USA stand at 275 billion tons, an amount that is greater than any other nation in the world. All of the energy growth forecasts have major carbon dioxide ($CO_2$) emission consequences. The world energy outlook [32] indicated that $CO_2$ emissions will increase from 26.6 gigatonnes (Gt) in 2005 to 34.1 Gt in 2015 and 41.9 Gt in 2030 [10]. In the Alternative Policy Scenario [32], $CO_2$ emissions rise to 31.9 Gt in 2015 and to 33.9 Gt in 2030. EIA of the US Department of Energy (DOE) in its International Energy Outlook (IEO) 2007 forecasted that $CO_2$ emissions in the Reference Case will increase from 26.9 Gt in 2004 to 33.9 Gt in 2015 and to 42.9 Gt in 2030. In the past, global $CO_2$ emissions rose from 23.5 Gt in 2000 to 27.1 Gt in 2005 [11], which is an increase of over 15%. Based on this, one can say that the forecasts made by IEA and IEO are optimistic.
- Methane is the second most emitted greenhouse gas after $CO_2$, and is more than 20 times more potent than $CO_2$ in terms of its heat-trapping capabilities. Methane is a by-product of coalification, the process by which organic materials convert into coal. It is stored throughout the surrounding rock strata in varying sized pockets and, due to the greater overburden pressures, often increases in concentration the deeper the coal seam. Because methane can create hazardous working conditions for miners, it must be removed from underground mines. While methane escapes during the processing, transport, and storage of coal, 90% of the emissions come from the actual coal mining process from all three categories of coal mines: surface mines, underground mines, and abandoned mines.

In 2004, methane accounted for 14.3% of the total anthropogenic greenhouse gas emission load. Since pre-industrial times, methane has contributed to 22.9% of the greenhouse gas load in the Earth's atmosphere. Methane is more abundant in the Earth's atmosphere now than it has been at any time during the past 400,000 years, and the average atmospheric concentration of methane has increased 150% since

1750 due to human activities [25]. The US EPA [27] has concluded that recovering methane from coal mines would significantly reduce the amount of greenhouse gases emitted into the atmosphere because every ton of methane recovered is equal to approximately more than 20 t of carbon dioxide emissions. While methane is a danger in coal mines, it can, if captured, be a valuable commodity: natural gas. To date, underground mines have the most potential to generate a profitable amount of methane. Mines that have adopted a "methane to markets" program as promoted by the US EPA have been successful at methane capture and sale because research has been developing for over three decades to create the best technologies.

- Particulate matters (PM), including total suspended particulates, $PM_{10}$ (particle size less than 10 μm), and $PM_{2.5}$ (particle size less than 2.5 μm), are released during coal mining operations. Coal mining activities that can lead to the release of particulate matter are blasting, truck loading, bulldozing, dragline operation, vehicle traffic, grading, and storage piles.
- Volatile organic compounds (VOCs), including non-methane organic compounds, are often vented along with methane, from coal mining operations. VOCs are precursors to ground-level ozone, a criteria air pollutant, and are regulated under the Clean Air Act.
- Nitrogen oxides (NOx) are a group of gases that are known to be ground-level ozone and $PM_{2.5}$ precursors and that include nitrogen dioxide ($NO_2$), a criteria pollutant. Sources of NOx at coal mines include fugitive emissions from overburden and coal-blasting events, tailpipe emissions from mining equipment, point source emissions from stationary engines, coal-fired hot water generators, and natural gas–fired heaters.
- For some pollutants, such as sulfur dioxide, the minerals processing industry is the largest source of emissions.
- Certain kinds of ore, for example uranium, are radioactive, and pose health problems to workers and adjacent communities. Radioactivity follows with whatever radioactive materials escape into the atmosphere or water, and accompanies this material to its final destination. Besides targeted minerals that are radioactive, the host rock may be radioactive, and pose problems for workers. Radioactivity is also present in a gas, radon, which can cause problems for workers in underground mines, if it is present.

## *Impacts of Dust*

The impacts of dust, especially from coal mine area are included below:

*Effects on animals*: Coal dust is a tumorigenic agent in experimental animals. Coal dusts are associated with lymphomas and, at the higher dose, adrenal cortex tumors in rats exposed to 6.6–14.9 mg/m$^3$ for 6 h/day intermittently for 86 weeks.

*Effects on humans*: Coal dust causes pneumoconiosis, bronchitis, and emphysema in exposed *community*. Coal workers' pneumoconiosis (CWP) is characterized by development of coal macules, a focal collection of coal dust particles with a little reticulin and collagen accumulation. These lesions may be visible as small opacities (less than 1 cm in diameter) on X-rays. Complicated CWP is characterized by lesions consisting of a mass of rubbery well-defined black tissue that is often adherent to the chest wall. This is associated with decrements in ventilatory capacity, low diffusing capacity, abnormalities of gas exchange, low arterial oxygen tension, pulmonary hypertension, and premature death. The disease may progress after the cessation of exposure.

*Effects on plant*: Plants exposed to toxic dust exhibit lesser growth, become more prone to disease attack and rodent attack, their stomata and other holes get chocked, and they also gasp for oxygen. Fruits, vegetables, and cereals from these plants contain toxins.

*Effects on infrastructure*: Biochemical reactions occur when constituents present in the coal dust starts reacting with the constituents of the host surface materials, resulting in increase in the rate of weathering/weakening.

## *Pollution of Water*

The potential impacts of mining on the water [28] are:

- Disruption of hydrological pathways
- Seepage of contaminated leachate into aquifers
- Depression of the water table around the dewatered zone
- Disposal of saline mine water into rivers

The impacts of mining arising from the disruption of hydrological pathways, seepage of contaminated leachate into aquifers, and depression of the water table tend to be relatively localized and limited compared to disposal of mine water. Disposal of mine water is a worldwide problem, occurring wherever operating mines, both underground and opencast, are found [19]. The quality of the mine water depends largely on the chemical properties of the geological materials that come in contact with the water.

The mining industry is a major producer of acidic sulfur-rich water that typically poses a risk to the environment. Mining increases the exposed surface area of sulfur-bearing rocks allowing for excess acid generation beyond natural buffering capabilities found in host rock and water resources. Pyrite ($FeS_2$) is the major sulfur mineral in coal. When coal is mined, fresh sulfur-bearing minerals in the coal and rocks are exposed to air and water. Problematic mine drainage is given many names including *acid rock drainage* (ARD), *acid mine drainage* (AMD), and mining influenced water (MIW). AMD forms during metal or coal mining when sulfur-bearing minerals are exposed to water and air, forming sulfuric acid. Heavy metals, leached from rocks, can combine with the acid and dissolve, creating highly toxic runoff. The general distinction between ARD and AMD depends on whether drainage quality has been degraded by mining or is of poor quality due, in part, to natural causes.

MIW is the term used to refer all mining-related water because acidic, neutral, and alkaline water can all transport metals and other contaminants.

The origin of acidic metal-rich mine drainage water is the accelerated oxidation of $FeS_2$ and other sulfidic minerals. The reaction of pyrite with oxygen and water produces a solution of ferrous sulfate and sulfuric acid. Ferrous iron can further be oxidized producing additional acidity. The following reaction shows the reaction of pyrite with oxygen and water to produce hydrogen ions, sulfate ions, and soluble metal ions.

$$2FeS_2(s) + 7O_2 + 2H_2O \rightarrow 2Fe^{+2} + 4SO_4{}^{-2} + 4H^+$$

Further oxidation of $Fe^{+2}$ (ferrous) to $Fe^{+3}$ (ferric) occurs when sufficient oxygen is dissolved in the water or when the water is exposed to sufficient atmospheric oxygen.

$$2Fe^{+2} + {}^1\!/\!{}_2O_2 + 2H^+ \rightarrow 2Fe^{+3} + H_2O$$

$Fe^{+3}$ can either precipitate as $Fe(OH)_3$, a red-orange precipitate seen in waters affected by ARD, or it can react directly with pyrite to produce more $Fe^{+2}$ and acidity as per the following reactions:

$$2Fe^{+3} + 6H_2O \leftrightarrow 2Fe(OH)_3(s) + 6H^+$$

$$14Fe^{+3} + FeS_2(s) + 8H_2O \rightarrow 2SO_4{}^{-2} + 15Fe^{+2} + 16H^+$$

In undisturbed natural systems, this oxidation process occurs at slow rates over geologic timescale. Iron- and sulfur-oxidizing bacteria are also known to catalyze these reactions at low pH, thereby increasing the rate of reaction by several orders of magnitude [18]. Bacteria such as *Thiobacillus ferrooxidans* and *Ferroplasma acidarmanus* are known to specifically accelerate such reactions.

A few management options for saline mine water, as summarized by Annandale et al. [1], are:

1. Pollution prevention at source
2. Reuse and recycling of water to minimize the volume of polluted water
3. Treatment of effluents, if the problem cannot be solved through prevention, reuse, and recycling
4. Discharge of treated effluent
5. Utilization of gypsiferous mine water for irrigation

The term "gypsiferous" refers to rocks and soils containing more than 2% *gypsum*, i.e., calcium sulfate ($CaSO_4 \bullet 2H_2O$). There are concerns regarding potential use of mine water for agricultural crops [5], due to the amount of salt that would leach and potentially contaminate groundwater. The next few sections highlight successes and limitations with respect to environmental impact and sustainable use of such waters for irrigation of agricultural crops.

## Restoration Process

It is the obligation of the mining company to restore the working cavity in the best possible way, through overburden disposal activities and technical reclamation, to the existing natural environment in terms of functionality and aesthetics. The following measures can be applied for the recultivation of deteriorated surfaces:

- Technical measures – to contribute to the improvement of resistant and deformable characteristics of the dump and directly influence the enhancement of the stability of slopes.
- Bio-technical measures – to contribute to faster achievement and maintenance of the dump stability along with the technical measures.
- Biological measures – that imply the implementation of agricultural and forest improvements that contribute to the stability and maintenance of reclaimed areas, but they are much more significant from the aspect of area revitalization and establishing natural biocenoses (a group of interacting organisms that live in a particular habitat and form a self-regulating ecological community)

The process of reverting an industrial land back to an agricultural terrain is a very slow process and good planning steps are required. Depending on the size of the coal deposit, the process of turning an agricultural, forest, or urban environment into an industrial (mining) one, and then, by the recultivation process, back again into the agricultural or forest one, may require long time, even many decades.

There are ongoing arguments in favor of preventing further expansion of the utilization of fossil fuels, such as coal. One of the restorative measures is to perform the recultivation of a dump even though it has a relatively low impact on the environment. Talking about fertile alluvial lands, it is necessary to preserve the fertile solum (a set of related soil horizons that share the same cycle of pedogenic processes) through selective excavation in order to bring the soil back to agricultural production or other use. The forestation of soil and terrains deteriorated by opencast lignite mining will prevent the further deterioration processes, contribute to the maintenance of the ecological balance in nature, and enhance the absorption of $CO_2$ from the air and increases the content of the oxygen therein.

The control of particulate emission is a fundamental part of a mine environmental management plan because of the increasing public awareness of human health issues and expectations of environmental performance, and the duty of care required of mine operators by government and the community. Particulate emission management system can result in cost savings, increased profits, and improved government and community relations, as well as easier access to resources and financial support in the future.

There are a number of systems to address emissions of harmful air pollutants from coal mines. It is technologically feasible to capture or flare methane from coal mines, instead of releasing methane directly into the atmosphere. Many mines have already taken steps toward capturing methane emissions for economic reasons. Twenty-three mines in Alabama, Colorado, Pennsylvania, Virginia, and West Virginia have methane drainage facility and they recover methane within the range of 3–88% efficiency [7]. Of these 23 mines, 12 sell recovered methane for natural gas energy use and two mines use the methane to heat mine ventilation air and to generate onsite power. Flaring is an option if capture is not technologically feasible. Although flaring occurs at a small number of mines in the USA, there is a long and safe history of flaring at working coal mines in the UK and Australia. US EPA estimates that nearly 50% of all of the US coal mine methane emissions, or more than 1.25 million tons of methane, can be reduced at a zero net cost, while nearly 90% can be reduced at a cost of less than $15/t. However, the benefit of reducing methane could be as much as $240/t of methane reduced. Any efforts to address methane will most likely also address VOC emissions due to the fact that VOCs and methane are often released together from coal mines.

Examples of particulate matter emission control measures currently in use include, but are not limited to: (a) storing coal in enclosed coal silos or barns; (b) paving coal mine access roads; (c) watering or treating with dust suppressant any unpaved roads; (d) enclosing conveyor transfer points; (e) use of dust collection baghouses or other controls to reduce emissions from transfer points and crushers within processing plants; (f) fitting out-of-pit conveyors with hoods or otherwise containing emissions; (g) fitting out-of-pit dump hoppers with water sprays, a baghouse, or other controls; (h) treating haul roads with dust control chemicals or water; (i) watering short-term haul roads; and (j) regularly maintaining haul roads to reduce dust re-entrainment.

With regard to $NO_x$, mines have reduced emissions by as much as 75% through the use of borehole liners and changing their blasting agent blends. Other mitigation measures that may be effective at reducing $NO_x$ include reduced blast sizes, changed composition of explosive agents, and changed placement of blasting agents. Reduction of impacts due to radioactivity consists of ventilation for underground mines, dust control, standard containment measures for solid waste, and standard precautions for discharges from processing.

## Sustainability and Mining

The Earth's resources have played a vital role for human communities. Nowadays, the management of our natural resources has become an urgent issue at both national and international level. The extraction and use of natural resources are also causing environmental problems, which require urgent solutions. For mining operations, resource extraction not only has a massive impact on ecosystems, it also releases pollutants contained in the rock, and consumes large amounts of water and energy. The transportation of resources from remote areas requires an ecologically intensive transport infrastructure.

Due to the progressive exploitation of mineral deposits and the availability of new technologies, deposits with lower ore content are now being mined. This means that an increasing amount of ore and other non-usable material has to be extracted to produce the same amount of metal. For example, toward the beginning of the twentieth century, copper ore mined by US mineral industry consisted of about 2.5% of usable metal by weight; today the proportion has dropped to 0.51% [8]. This activity impacts even more severely on ecosystems and water resources and increases the volume of mining waste, resulting in even more radical changes to entire landscapes. The rising global demand for resources accelerates this trend. After extraction, the subsequent stages in the raw materials' life cycle entail further environmental pollution. Comparative analyses of industrial sectors show that the highly resource-intensive industries are associated with above-average levels of emissions of greenhouse gases and other pollutants.

Climate change and the overexploitation of natural resources are two sides of the same coin. Climate change will impact water supplies, exacerbating existing pressures on water resources caused by population and economic growth. Given the combination of these stressors, the sustainability of water resources in future decades is a concern in many parts of the world. The sustainability of water resources is defined as the maintenance of natural water resources in adequate quantity and with suitable quality for human use and for aquatic ecosystems. Human needs for resources, like water, land, continue to grow with increasing population, primarily for direct consumption, but also secondarily for energy production, and agricultural and mining activities.

The privilege of mining can be enjoyed in the best possible way, but there is the responsibility of looking out for the future generations, preserving some of the natural resources by utilizing reserves in compliance with the principles of the sustainable development. There is also a need to reclaim and to revive deteriorated surfaces resulting from mining.

The nine sustainable development challenges that the mining industries face are:

- Ensuring long-term viability of the mining industry
- Controlling, using, and managing the land
- Using minerals to assist with economic development
- Making a positive impact on local communities
- Managing the environmental impacts of mines
- Maximizing the use of minerals so as to reduce waste and inefficiency
- Giving stakeholders access to information to build trust and cooperation
- Managing the relationship between large corporates and small-scale mining companies
- Sector governance: clearly defining the roles, responsibilities, and instruments for change expected of all stakeholders.

Mining, Minerals and Sustainable Development (MMSD) North America has developed an approach to assess how a mining/mineral project or operation contributes to sustainability. The assessment considered tracking the record of mining and minerals in the past and its current contribution to and detraction from economic prosperity, human well-being, ecosystem health, and accountable decision-making. A set of seven questions were developed as a means of assessing whether the net contribution to sustainability over the term of a mining/mineral project or operation will be positive or negative and a way of discovering how current activities can be improved and aligned with the emerging concept of sustainability [12]. The seven-part numbering used by MMSD (see Table 12.1) has been intended as an aid to communicate and not to imply a particular sequencing of steps or prioritization of topics. These questions follow a hierarchy of objectives, indicators, and specific metrics. In this way a single, initial motivating question cascades into progressively more detailed elements, which can be tailored to the project or operation being assessed for application throughout its full life cycle.

The focus on applying the seven questions to sustainability approach is not so much on how mining can be sustainable (mining as a discrete activity cannot continue indefinitely) but on how mining can contribute to sustainability [9].

Sustainability Opportunity and Threat Analysis is another simple operational tool addressing the social, economic, and environmental dimensions of the issues under consideration and it can be applied to evaluate the viability of a mining operation and its ability to contribute to sustainable development objectives. An impact-based model of sustainable mine development is shown in Fig. 12.2. This analysis involves constructing inventory of sources of impacts and following key steps including:

(a) Scoping (addressing the reasons for the mining process and the environment, and agreeing to the scope of the exercise)
(b) Information gathering (to emphasize the importance of collecting and organizing relevant information into a suitable framework)
(c) Identifying risks (systematically reviewing impact areas under consideration, and identifying opportunities or threats)
(d) Analyzing and evaluating the risks (qualitative scales of likelihood and consequence can be assigned to identify opportunity or threat to create overall risk rating and prioritized list)
(e) Treating risks (control measures to address opportunities or threats considered high priority)
(f) Reporting and reviewing to represent a broad, scanning exercise that can be picked up by existing business planning and monitoring processes. It also provides an opportunity for subsequent evaluation of relevant metrics for the issues identified by the process, allowing operations to measure both impacts and progress toward agreed objectives.

Most mine designs are based on traditional mining engineering factors, such as the quality of the commodity being mined, geology, topography, hydrology, land ownership, geography, infrastructure, etc. Environmental compliance and sustainability are to be considered in mine design and operation as a modifying factor to those designs.

**Table 12.1** Seven questions to sustainability – How to assess the contribution of mining and minerals activities

| Seven fundamental questions | Framework for guiding a sustainability assessment |
|---|---|
| 1. *Engagement*<br>Are engagement processes in place and working effectively? | Processes of engagement are committed to, designed, and implemented so that they:<br>• Ensure all affected communities of interest have the opportunity to participate in the decisions that influence their own future<br>• Are understood, agreed upon by implicated communities of interest, and consistent with the legal, institutional, and cultural characteristics of the community and country where the project or operation is located |
| 2. *People*<br>Will people's well-being be maintained or improved? | Project/operation lead directly or indirectly to maintenance of people's well-being:<br>• During the life of the project/operation<br>• Post-closure |
| 3. *Environment*<br>Is the integrity of the environment assured over the long term? | Project or operation lead directly or indirectly to the maintenance or strengthening of the integrity of biophysical systems so that they can continue in post-closure to provide the needed support for the well-being of people and other life forms |
| 4. *Economy*<br>Is the economic viability of the project or operation assured, and will the economy of the community and beyond be better off as a result? | Assurance of the financial health of the project/company and contribution of the project or operation to the long-term viability of the local, regional, and global economy in ways that will help ensure sufficiency for all and provide specific opportunities for the less advantaged |
| 5. *Traditional and nonmarket activities*<br>Are traditional and nonmarket activities in the community and surrounding area accounted for in a way that is acceptable to the local people? | Project or operation contribute to the long-term viability of traditional and nonmarket activities in the implicated community and region |
| 6. *Institutional arrangements and governance*<br>Are rules, incentives, programs, and capacities in place to address project or operational consequences? | Institutional arrangements and systems of governance in place that can provide certainty and confidence that:<br>• Capacity of government, companies, communities, and residents to address project or operation consequences is in place or will be built<br>• Capacity will continue to evolve and exist through the full life cycle including post-closure |
| 7. *Synthesis and continuous learning*<br>Does a full synthesis show that the net result will be positive or negative in the long term, and will there be periodic reassessments? | Placement of an overall evaluation and periodic reevaluation based on:<br>• Consideration of all reasonable alternative configurations at the project level<br>• Consideration of all reasonable alternatives at the overarching strategic level for supplying the commodity and the services it provides for meeting society's needs<br>• Consideration of all reasonable alternatives at the overarching strategic level for supplying the commodity and the services it provides for meeting society's needs |

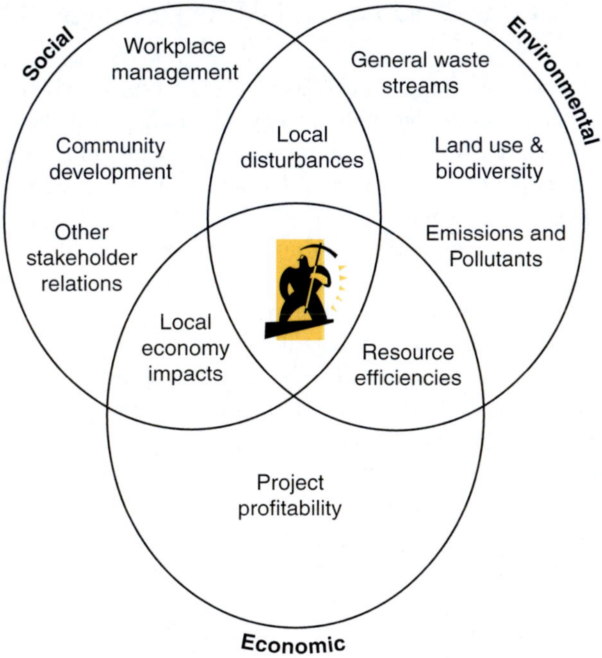


**Fig. 12.2** Sustainable development model for impact-based mining operations

While practices may become more responsive to sustainability, mine design continues to be governed by established mining engineering approaches.

A review of the available literature on engineering optimization does not reveal any focus on mine design, environmental protection associated with mines, and sustainability. Mathematical multi-criteria optimization approaches, however, have been used in resource management. Mine design optimization would need to consider all constraints, system parameters and characteristics, and desired outcomes in order to build a useful and reliable model. Since optimization of mine design, and in particular coal mine design, to address sustainability along with other parameters has not been widely practiced, identifying the appropriate parameters for measurement and the mathematical or logical relationships between these parameters is not a trivial task.

Another important constraint on mining operations is the statutory and regulatory frameworks within which they are required to operate. Many of these legal and policy structures tend to create an adversarial approach by instituting a system where one or more parties must respond to actions, proposals, decisions, etc. of another party. The participation is often late in the design process, or after design has been completed, and thus any change to mine or reclamation design necessary as a result of public or regulatory agency input creates a retrofitted design, which cannot possibly be optimal. To optimize the design of mining and reclamation operations, traditional mining engineering considerations and environmental and sustainability goals must be accounted for simultaneously.

Several technologies are discussed that are considered to promote sustainable development.

## Irrigation with Lime-Treated Acid Mine Drainage (AMD)

Higher crop yields were obtained under sprinkler irrigation with treated mine water compared to dryland production, without any foliar injury to the crop. Possible nutritional problems, for example deficiencies of potassium, magnesium, and nitrate, occurring due to calcium and sulfate dominating the system, can be solved through fertilization. Such soils need to be managed and fertilized differently to those on which crops are produced under normal farming conditions. Sugar beans, wheat, maize, and potatoes were very successfully produced under irrigation with calcium sulfate and magnesium sulfate-rich mine water. In an experimental setting, soil salinity increased with time, but the values of soil saturated electrical conductivity stabilized at relatively low levels, due to gypsum precipitation [1]. Measurements taken between 1997 and 2007 showed that soil salinity increased from a low base and oscillated around 250 mS $m^{-1}$.

Land preparation and fertilization management are, however, critical for successful crop production, especially on rehabilitated soil. During short to medium term (up to 8 years) irrigation with gypsiferous mine water, negligible impact was noticed by Annandale et al. [1] on groundwater quality. They operated the system with flexibility and managed with the multiple objectives like maximum crop production, water use, job creation, economic return or maximum gypsum precipitation, and minimum salt leaching. Gypsum precipitation was also shown to be taking place in the soil. The presence of gypsum did not create any physical and/or chemical property changes that could adversely affect crop production and soil management. Crop production under irrigation with coal-mine water, rich in calcium, magnesium, and sulfate is, therefore, feasible, and sustainable if properly managed.

Pasture production with sodium sulfate-rich mine effluent is also feasible, but requires a well-drained profile and a large leaching fraction to prevent unsustainable build up of salt in the soil. Unfortunately, this type of water does not present much of an opportunity for gypsum precipitation, which is able to drastically reduce the salt load of the receiving water in the case of calcium and sulfate-rich mine water. The application of calcium nitrate as a nitrogen source to the crop adds calcium to the soil and removes some sulfate from the water system by enhancing gypsum precipitation. Measurement of the hydraulic conductivity of the soil is recommended to monitor the effect of the water on the infiltration rate of the soil, as high sodium levels are likely to cause deflocculation or dispersion of clay particles.

## Application of Phytotechnology

Phytotechnology can be applied to address issues related to stabilization of tailings and hydraulic control for drainage so that human and ecological exposures to contaminants associated with mining solid wastes and mine impacted waters are low. Implementation of phytotechnology is a common component of mining reclamation and *restoration* projects by the establishment of a plant cover as a final remedy. In certain cases, application of phytotechnology can be used for removal of metals from *contaminated media*. Establishing phytotechnology requires careful plant species selection and soil amendments that equates to an initial investment; however, these systems, once established can be maintained with minimal effort.

There are six basic phytoremediation mechanisms that can be used to clean up contaminated sites: phytosequestration, rhizodegradation, phytohydraulics, phytoextraction, phytodegradation, and phytovolatilization (see Table 12.2).

The particular phytotechnology mechanism used to address contaminants depends not only on the type of contaminant and the media affected, but also on the cleanup goals. Typical goals include containment through stabilization or sequestration, remediation through assimilation, reduction, detoxification, degradation, metabolization or mineralization, or both. Applying phytotechnology to impacted sites entails selecting, designing, installing, operating, maintaining, and monitoring planted systems that use the various mechanisms mentioned above. The goal of the system can be broadly based on the remedial objectives of containment, remediation, or both. Furthermore, the target media can be soil/sediment, surface water, or groundwater, and these can be either clean or impacted. In some cases, groundwater transitioning to surface water can be addressed as a riparian situation where target media are combined.

One of the main advantages of phytotechnology is that this technology can be applied to a variety of metals in mining sites and to impacted soil/sediment, surface water, and groundwater. In addition, it can be applied to various combinations of chemical types and impacted media simultaneously. Additional advantages are:

(a) Considered a green and sustainable technology
(b) Does not require supplemental energy, although monitoring equipment may use solar power
(c) Can improve the air quality and sequester greenhouse gases
(d) Minimal air emissions, water discharge, and secondary waste generation
(e) Lower maintenance, resilient, and self-repairing
(f) Inherently controls erosion, runoff, infiltration, and dust emissions
(g) Passive and in situ
(h) Favorable public perception
(i) Improves aesthetics, including reduced noise
(j) Applicable to remote locations, potentially without utility access
(k) Provides restoration and land reclamation during cleanup and upon completion
(l) Can be cost-competitive

The benefits of using the phytotechnology-based techniques are the relative lower costs, labor requirements, and safer operations compared to the more intensive and invasive conventional techniques. Phytotechnology generally provides long-term remedial solutions. Plantations may require irrigation, fertilization,

**Table 12.2** Phytotechnology mechanisms to sequester constituents of interest [13]

| Mechanism | Description | Cleanup Goal |
|---|---|---|
| Phytosequestration | Ability of plants to sequester selected contaminants in the rhizosphere zone through exudation of phytochemicals and on the root through transport proteins and cellular processes | Containment |
| Rhizodegradation | Exuded phytochemicals can enhance microbial biodegradation of contaminants in the rhizosphere | *Remediation* by destruction |
| Phytohydraulics | Ability of plants to capture and evaporate water off the plant and take up and transpire water through the plant | Containment by controlling hydrology |
| Phytoextraction | Ability of plants to take up contaminants into the plant with the transpiration stream | Remediation by removal of plants |
| Phytodegradation | Ability of plants to take up and break down contaminants in the transpiration stream through internal enzymatic activity and photosynthetic oxidation/reduction | Remediation by destruction |
| Phytovolatilization | Ability of plants to take up, translocate, and subsequently transpire volatile contaminants in the transpiration stream | Remediation by removal through plants |

weed control (mowing, mulching, or spraying), and pest control. Establishment of phytotechnology systems include various expenditures, such as earthwork, labor, planting stock, planting method, field equipment, heavy machinery (typically farming or forestry equipment), soil amendments, permits, water control infrastructure, utility infrastructure, fencing, security, etc. About 10–15% of the initial capital costs can be added as a contingency for replanting [13].

Phytotechnology are appropriate only under specific conditions. The major limitations are depth, area, and time. The physical constraints of depth and area depend on the plant species suitable to the site (i.e., root penetration) as well as the site layout and soil characteristics. Phytotechnology typically require larger tracts of land than many alternatives. Time can be a constraint since phytotechnology generally take longer than many other alternatives and are susceptible to seasonal and diurnal changes. Additional limitations include a plant's tolerance to specific constituents of concern or site conditions, availability of water as irrigation source, climate (challenging for plant establishment in areas with short growing season or in arid environments), and pests, infestations, or other nuisances.

## Biomining and Bioprocessing

Mining industries are increasingly aware of the potential of microbiological approaches for recovering base and precious metals from low-grade ores, and for remediating acidic, metal-rich wastewaters that drain from both operating and abandoned mine sites. Biological systems offer a number of environmental and economical advantages over conventional approaches, such as pyrometallurgy, though the microbial application is not appropriate in every situation. *Biomining* (metal extraction) and bioprocessing are currently utilized in full-scale operations to process low-grade deposits and reprocessing earlier metal-containing wastes. This usually results in the production of less chemically active tailings, lower energy inputs, and other environmental benefits (zero production of noxious gases). Recently, there have been major advances in the field of microbiology, which will allow greater control of *bioleaching* operations, resulting in greater efficiencies and faster rates of metal extraction. Mineral processing using microorganisms have been exploited for extracting gold, copper, uranium, and cobalt. Engineering systems ranging from crude *heap leaching* systems to temperature-controlled bioreactors have been used, depending on the nature of the ore and the value of the metal product. A typical example of mineral bioprocessing mechanism is discussed below.

Sulfide minerals may be divided into acid soluble (such as zinc sulfide or sphalerite) and acid insoluble (such as pyrite and arsenopyrite). Two routes (the "thiosulfate" and "polythionate" mechanisms) have been proposed for the biological oxidation of these sulfide minerals [21]. Acid-soluble sulfides are readily degraded by sulfur-oxidizing *acidophiles*. The mineral is first subjected to proton-mediated dissolution, forming the free metal and hydrogen sulfide:

$$\mathrm{MS} + 2\mathrm{H}^+ \rightarrow \mathrm{M}^{2+} + \mathrm{H_2S}$$

The hydrogen sulfide so formed is microbially oxidized to sulfuric acid, allowing the process to continue:

$$H_2S + 2O_2 \rightarrow H_2SO_4$$

Acid-soluble sulfides may also be attacked by ferric iron, producing ferrous iron and polysulfide:

$$MS + Fe^{3+} + H^+ \rightarrow M^{2+} + 0.5H_2S_n + Fe^{2+}$$

Polysulfide may be further oxidized by ferric iron to produce elemental sulfur, which, in turn, is oxidized to sulfuric acid, and will further accelerate mineral dissolution via proton attack:

$$0.5H_2S_n + Fe^{3+} \rightarrow 0.125S_8 + Fe^{2+} + H^+$$

$$0.125S_8 + 1.5O_2 + H_2O \rightarrow 2H^+ + {SO_4}^{2-}$$

Sulfides that are resistant to proton attack are oxidized by ferric iron, producing thiosulfate as an initial by-product:

$$FeS_2 + 6Fe^{3+} + 3H_2O \rightarrow 7Fe^{2+} + {S_2O_3}^{2-} + 6H^+$$

Accelerated oxidation can result in low pH, high concentrations of dissolved metals and, in some cases, elevated temperatures. These conditions limit the diversity of life-forms that occurs in commercial bioleaching operations. Single-celled organisms live only in extremely acidic liquors (pH $<$1–4) and are obligate acidophiles, some are thermophilic (to varying degrees) and some can fix carbon dioxide. The primary microorganisms involved in mineral oxidation are those that catalyze the oxidation of ferrous iron and/or reduce sulfur, while others contribute to the process indirectly by, for example, removing materials that accumulate during ore dissolution [14].

Engineering approaches used in biomining are: (a) irrigation-based principles (dump- and heap-leaching, and in situ leaching) and (b) stirred tank processes [20]. The dump leaching involves gathering low-grade copper-containing ore of large rock/boulder size into vast mounds or dumps and irrigating these with dilute sulfuric acid to encourage the growth and activities of mineral-oxidizing acidophiles, primarily iron-oxidizing *mesophiles*. Copper can be precipitated from the metal-rich streams draining from the dumps by displacement with iron. Other developments on the engineering and hydrometallurgical aspects of biomining have involved the use of thin layer heaps of refractory sulfidic ores (mostly copper, but also gold-bearing material) stacked onto water-proof membranes, and recovery of solubilized copper using solvent extraction coupled with *electrowinning*. In situ bioleaching has been developed to scavenge for uranium and copper in otherwise worked out mines.

## Bioremediation

Water draining from active and abandoned mines and mine wastes are acidic and often contain elevated concentrations of metals (iron, aluminum and manganese, and others) and metalloids (arsenic, chromium, selenium, and uranium). The basis of bioremediation of AMD derives from the abilities of some microorganisms to generate alkalinity and immobilize metals, thereby essentially reversing the reactions responsible for AMD.

Microbiological processes that generate net alkalinity are mostly reductive processes, and include *denitrification*, *methanogenesis*, sulfate reduction, and iron and manganese reduction. *Ammonification* (the production of ammonium ion from nitrogen-containing organic compounds) is also an alkali-generating process. Photosynthetic microorganisms, by consuming a weak base (bicarbonate) and producing a strong base (hydroxyl ions), also generate net alkalinity:

$$6HCO_3^-(aq) + 6H_2O \rightarrow C_6H_{12}O_6 + 6O_2 + 6OH^-$$

The reduction of soluble iron (ferric iron) does not decrease solution acidity, however, the reduction of solid phase (*crystalline* and *amorphous*) ferric iron compounds does.

$$Fe(OH)_3 + 3H^+ + e^- \rightarrow Fe^{2+} + 3H_2O$$

where $e^-$ represents an electron donor, which is generally an organic substrate.

Bacteria that catalyze the *dissimilatory* reduction of sulfate to sulfide generate alkalinity by transforming a strong acid (sulfuric) into a relatively weak acid (hydrogen sulfide).

$$SO_4^{2-} + 2CH_2O + 2H^+ \rightarrow H_2S + 2H_2CO_3$$

Besides the ameliorative effect on AMD brought about by the resulting increase in pH, the reduction of sulfate is an important mechanism for removing toxic metals from AMD, since these metals (e.g., zinc, copper, and cadmium) form highly insoluble sulfides.

$$Zn^{2+} + H_2S \rightarrow ZnS + 2H^+$$

The bioremediation technologies, like constructed wetlands and compost bioreactors for AMD, are passive systems. In case of the constructed wetlands, the solid-phase products of water treatment are contained within the wetland sediments. The key major advantages of these passive bioremediation systems are their relatively low maintenance costs. The limitations of these technologies are:

(a) They are often relatively expensive to install.
(b) May require more land area than is available or suitable.
(c) Their performance is less predictable than chemical treatment systems.
(d) The long-term fate and stability (in the case of compost bioreactors) of the deposits that accumulate within them is uncertain [15].

Aerobic wetlands are generally more effective to treat mine waters that are net alkaline. The reaction involves oxidation of ferrous iron, and subsequent hydrolysis of the ferric iron produced, which is a net acid-generating reaction.

$$4Fe^{2+} + O_2 + 4H^+ \rightarrow 4Fe^{3+} + 2H_2O$$

$$4Fe^{3+} + 4H_2O \rightarrow 4Fe(OH)_3 + 12H^+$$

In the event the pH of the mine water decreases significantly as a result of these reactions, additional amendments (such as an anoxic limestone drain) may be required. To maintain oxidizing conditions, aerobic wetlands are relatively shallow systems that operate by surface flow. Though *macrophytes* are planted for aesthetic reasons, they also regulate water flow, accelerate the rate of ferrous iron oxidation, and stabilize the accumulating iron precipitates.

In contrast to aerobic wetlands, the compost bioreactors (installations that are enclosed entirely below ground level and do not support any macrophytes) are anaerobic. The microbially catalyzed reactions in compost bioreactors generate net alkalinity and biogenic sulfide, and thus can treat mine waters that are acidic and metal-rich. The reductive reactions that occur within compost wetlands are driven by electron donors that derive from the organic matrix of the compost itself. The indigenous iron- and sulfate-reducing bacteria are generally considered to have the major roles in AMD remediation in compost bioreactors.

## Metal Recovery: Biotic and Abiotic

Innovative and alternative techniques that facilitate the economic control and recovery of metal values are beneficial not only for protection of human health and the environment, but also for the recovery of these commodities and resource conservation. AMD, rich in metals, should be considered not only as a serious environmental problem, but as an important resource of metals of considerable value to many industrial concerns.

Recently developed metal recovery processes from AMD and other acidic streams enable us to recover and recycle these metals cost effectively. A two-stage process was developed by Tabak et al. [23] to separate the sulfate-reducing bacteria (SRB) from AMD via sulfate reduction and production of biogenic hydrogen sulfide from the four-stage and six-stage metal sequential separation and biorecovery units. The four-stage selective sequential batch-type metal precipitation process was able to separate metal sulfides and hydroxides at reasonably high recovery rate and at high precipitate purities. Copper sulfide and zinc sulfide were precipitated in Stage 1, aluminum hydroxide was precipitated in Stage 2, ferrous sulfide was precipitated in Stage 3, and other metals were precipitated in Stage 4. The effluent water contained only calcium and magnesium in trace concentrations and both sulfate and metal sulfide concentrations were below detectable limits.

von Fahnestock [30] recently developed a breakthrough technology to efficiently remove sulfate ions and metal cations from AMD by using the Acid Mine Drainage Value Extraction Process (AMD VEP), which is a novel adaptation of liquid–liquid extraction process. The process converts AMD to purified water as well as saleable products, such as potassium sulfate and iron sulfate. This technology results in simultaneous cost-effective isolation and concentration of useful metals and sulfate ions from mine pool water. A 30-gpm (113.5-L/m) demonstration plant was built and was operational within 15 months in St. Michael, PA. In this technology, AMD water, laden with sulfate and iron, feeds into the water purification stages where the AMD water is sequentially contacted with an *extractant* solution in a countercurrent flow path. The extractant solution is formulated to efficiently pull the sulfate and iron from the aqueous phase.

The water purification stages are composed as a set of four mixer-settler units, which are two-compartment tanks. The different stages are described below:

1. In the first compartment, the extractant and the AMD are mixed in a chamber. The residence time is between 60 and 90 s.
2. In the second stage, the combined effluent flows from the mixing chamber into the settling chamber where the organic extractant phase disengages from the water phase. The extractant, containing iron and sulfate, overflows an exit weir into the settling chamber, and is separated cleanly from the water phase, which underflows the same weir and exits as a separate stream with proportionately less iron and sulfate.
3. In the recovery stages of the AMD VEP, the extractant, loaded with iron and sulfate, flows sequentially from the water purification stages to the metals recovery and sulfate recovery stages of the process. Iron and sulfate are sequentially recovered as usable products during these stages. The metals-recovery stages are a set of two to three metal cation recovery tanks. The extractant flows counterclockwise with an aqueous sulfuric acid solution to form an iron sulfate concentrate that is harvested for reuse. The sulfuric acid and the extractant are mixed and separated in a similar manner to that in the mixer-settler tanks of the extraction section.
4. The sulfate-recovery stages remove sulfate from the extractant, which again flows through a series of mixer-settler tanks in a countercurrent fashion. The extractant is contacted with potassium carbonate (a basic aqueous solution) to produce a potassium sulfate concentrate. The extractant exits the last mixer settler of the sulfate-recovery section regenerated and ready to contact a new stream of AMD feedwater in the water purification stages. The potassium sulfate ($K_2SO_4$)

concentrate leaves the sulfate-recovery stages and is collected and stored for sale as fertilizer or reuse in a product tank.

An engineering analysis indicated that the maximum cost to treat and recover metals by AMD VEP process is $8.00 dollars/1,000 gal [3] dependant on product values and reagent pricing. AMD VEP, based on abiotic processes, is one of the few recent technologies to purify AMD water sufficiently to enable its discharge to surface waters. In addition, the product water can be made useful for industrial, municipal, agricultural, residential, and other uses. Importantly, these process products enable the system to focus on meeting seasonal and regional demands. For example, potassium sulfate fertilizer can be produced for 6 months a year for the turf grass industry, and the same system can produce sodium sulfate deicer for highways for 6 months a year, by changing operating conditions and reagent feeds.

## Water Quality and Acid Mine Drainage: Pre-mine Predictions and Post-mine Comparisons

Necessity of accurate prediction of acidic drainage from proposed mines is recognized by both industry and government as a critical requirement of mine permitting long-term operation. Substantial emphasis has been placed on prediction of acid drainage associated with coal development in the Eastern USA [2], and metal mining in the Western USA and in Canada [16]. The prediction of acid-generating potential from any geologic formation is dependent on the ability to characterize the presence and quantity of both acid-forming minerals and neutralizing minerals in the materials that are expected to be unearthed during mining operations. Typically, samples are collected by drilling during exploration, analyzed and interpreted with respect to their risk of acid formation. In these analyses, the amount of sulfur present in geologic materials is measured and attributed to being either an acid-forming mineral such as pyrite ($FeS_2$) or non-acid-forming mineral such as gypsum ($CaSO_4 \cdot 2H_2O$). The relative amount of acid-forming minerals is then contrasted to the amount of neutralizing minerals such as calcite ($CaCO_3$) to develop a prediction of the probability of acid generation.

The acid base accounting (ABA) of a material is the balance between total acid-generating potential (AP), which is the total amount of acidity that would be produced if all sulfide in a material is completely oxidized, and total acid-neutralizing potential (NP), which is the amount of acid that could be consumed by neutralizing minerals. AP and NP are converted to $CaCO_3$ equivalents and reported as grams of $CaCO_3$ per kilogram rock. ABA is typically calculated from analysis of sulfide S and carbonate C, assuming a 1:1 molar ratio of sulfide S (AP) and carbonate C (NP).

Converting chemical analysis for sulfide S ($S_{FeS_2}$) and carbonate C ($C_{CaCO_3}$) can be expressed as [26]:

$$AP = S_{FeS_2}(10) \times (3.12)$$

$$NP = C_{CaCO_3} \times (10) \times (8.33)$$

where

$S_{FeS_2}$ = concentration sulfide sulfur in sample (weight% S)
3.12 = molecular weight of $CaCO_3$/molecular weight of sulfur
$C_{CaCO_3}$ = concentration carbonate carbon in sample (wt% C)
8.33 = molecular weight of $CaCO_3$/molecular weight of carbon

The ratio of neutralization potential (NP) to acid potential (AP) is commonly presented in graphical interpretations with the inference that geologic materials with an abundance of NP are unlikely to generate acidic drainage. Skousen et al. [22] reported that NP:AP ratios $<1$ commonly produce acidic drainage, NP:AP ratios between 1 and 2 may produce either acidic or neutral drainage, and NP:AP ratios $>2$ should produce alkaline water.

However, this index does not always accurately predict the resultant acid generation from a mine. Out of 56 mines evaluated by Skousen et al. [22], 11% did not conform to the expected results based on NP:AP ratios, including four sites with ratios $>2$ that eventually produced acidic drainage. Mineralogical variation between each geologic domain causes dissimilar reactivity to weathering conditions and leads to laboratory variability in assessment. Forecasting future water quality impacts from AMD based on laboratory and field data should not be considered routine and robust, rather they should be considered an area of uncertainty and ongoing research.

## Future Directions

The starting point is typically the acceptance of sustainable developmental principles at board-room level as corporate goals, and then informing the workforce, investors, and others of that commitment. Relevant employees need to be engaged as a first step in the practical application of sustainable development principles, followed by the gradual extension of training in sustainable methods of working to the workforce as a whole.

## Bibliography

1. Annandale JG, Beletse YG, Stirzaker RJ, Bristow KL, Aken ME (2009) Is irrigation with coal-mine water sustainable? In: Proceedings of international mine water conference, Pretoria, South Africa
2. Brady KBC, Hornberger RJ, Fleeger G (1998) Influence of geology on postmining water quality: Northern Appalachian Basin. In: Brady, KBC, Smith MW, Schueck J (eds) Coal Mine

Drainage Prediction and Pollution Prevention in Pennsylvania: Harrisburg, Pennsylvania. Department of Environmental Protection, p. 8–1 to 8–92
3. Conkle HN (2008) Engineering analysis report, value recovery from mine or rock drainage water, with water purity enhancement, using innovative low cost technology. Battelle, Columbus
4. DEFRA (2007) Waste strategy for England 2007. Presented to parliament by command of her majesty. Department for Environment, Food and Rural Affairs, PB12596
5. Du Plessis HM (1983) Using lime treated acid mine water for irrigation. Water Sci Technol 15:145–154
6. EIA (2010) Annual energy outlook 2010 with projections to 2035. U.S. Energy Information Administration, Office of Integrated Analysis and Forecasting, U.S. Department of Energy, Washington, DC. DOE/EIA-0383
7. Earth Justice (2010) Petition for rulemaking under the clean air act to list coal mines as a source category and to regulate methane and other harmful air emissions from coal mining facilities under section 111. Denver, CO
8. Gardner G, Sampat P (1998) Mind over matter: recasting the role of materials in our lives. Worldwatch Institute, Washington, DC, p 18
9. Hodge RA (2004) Mining's seven questions to sustainability: from mitigating impacts to encouraging contribution. Episodes 27(3):177–184
10. IEA (2007a) World energy outlook 2007 – China and India insights. OECD/IEA, Paris
11. IEA (2007b) $CO_2$ emissions from fuel combustion, 1971–2005, 2007 edition. OECD/IEA, Paris
12. IISD (2002) Seven questions to sustainability: how to assess the contribution of mining and minerals activities. Task 2 work group, mining, minerals and sustainable development North America (MMSD) North America. International Institute for Sustainable Development, Winnipeg
13. ITRC (2010) Technology overview of phytotechnologies. Prepared by The Interstate Technology & Regulatory Council, Mine Waste Team. Washington DC
14. Johnson DB (2001) Importance of microbial ecology in the development of new mineral technologies. Hydrometallurgy 59:147–158
15. Johnson DB, Hallberg KB (2002) Pitfalls of passive mine drainage. Rev Environ Biotechnol 1:335–343
16. MEND (2001) List of potential information requirements in metal leaching/acid rock drainage assessment and mitigation work. Mining environment neutral drainage program. W. A. Price, CANMET, Canada Centre for Mineral and Energy Technology
17. Mishra PC, Jha S (2010) Dust dispersion modeling in opencast coal mines and control of dispersion in Mahanadi coalfields of Orissa. Bioscan 2:479–500
18. Nordstrom DK, Southam G (1997) Geomicrobiology – interactions between microbes and minerals. Mineral Soc Am 35:261–390
19. Pulles W (2006) Management options for mine water drainage in South Africa. In: WISA, mine water division (ed) Mine water drainage-South African perspective. Water Institute Southern Africa, Johannesburg
20. Rawlings DE (2002) Heavy metal mining using microbes. Annu Rev Microbiol 56:65–91
21. Schippers A, Sand W (1999) Bacterial leaching of metal sulfides proceeds by two indirect mechanisms via thiosulfate or via polysulfides and sulfur. Appl Environ Microbiol 65:319–321
22. Skousen J, Simmons J, McDonald LM, Ziemkiewicz P (2002) Acid-base accounting to predict post-mining drainage quality on surface mines. J Environ Qual 31:2034–2044
23. Tabak HH, Scharp R, Burckle J, Kawahara FK, Govind R (2003) Advances in biotreatment of acid mine drainage and biorecovery of metals: 1. Metal precipitation for recovery and recycle. Biodegradation 14:423–436
24. U.S. EPA (1995) Identification and description of mineral processing sectors and waste streams. Office of Solid Waste, Washington, DC. RCRA Docket No. F-96-PH4A-S0001

25. U.S. EPA (2006a) Global mitigation of non-$CO_2$ greenhouse gases. Office of Atmospheric Programs, Washington, DC. EPA 430-R-06-005
26. U.S. EPA (2006b) Management and treatment of water from hard rock mines. Engineering Issue. Office of Research and Development National Risk Management, Research Laboratory Cincinnati, OH. EPA/625/R-06/014
27. U.S. EPA (2009) Identifying opportunities for methane recovery at U.S. coal mines: profiles of selected gassy underground coal mines 2002–2006. Coalbed Methane Outreach Program, EPA 430-K-04-003
28. Younger PL, Wolkersdorfer C (2004) Mining impacts on the fresh water environments: technical and managerial guidelines for catchment scale management. Mine Water Environ 23:S2–S80
29. SDWF (2011) Mining and water pollution. Available from safe drinking water foundation. http://www.safewater.org/PDFS/resourcesknowthefacts/Mining+and+Water+Pollution.pdf. Accessed 11 May 2001
30. von Fahnestock M (2010) AMD value extraction process. Water Cond Purificat 52(1):38–40
31. Waddell S, Pruitt B (2005) The role of coal in climate change: a dialogic change process analysis. The generative dialogue project launch meeting, New York, NY, 6–8 Oct 2005
32. WEO (2007) World energy outlook 2OO7: China and India insights. International Energy Agency. 61 2007 01 1 P1, Paris

# Chapter 13
# $CO_2$ Reduction and Coal-Based Electricity Generation

**János Beér**

## Glossary

| | |
|---|---|
| °C | Deg. Celsius |
| °F | Deg. Fahrenheit |
| AUSC | Advanced ultra-supercritical steam |
| Btu | British thermal unit |
| CCS | Carbon capture and storage (sequestration) |
| CFBC | Circulating fluidized bed combustion |
| $CO_2$ | Carbon dioxide |
| COE | Cost of electricity |
| DOE | U.S. Department of Energy |
| EIA | Energy Information Administration (USDOE) |
| EPRI | Electric Power Research Institute |
| GW | Gigawatt |
| Hg | Mercury |
| GNP | Gross national product |
| HHV | Higher heating value |
| HRSG | Heat recovery steam generator |
| IEA | International Energy Agency |
| IGCC | Integrated gasification combined cycle |
| MIT | Massachusetts Institute of Technology |
| Mpa | Megapascal |
| MW | Megawatt |
| MWh | Megawatt-hour |

This chapter was originally published as part of the Encyclopedia of Sustainability Science and Technology edited by Robert A. Meyers. DOI:10.1007/978-1-4419-0851-3

J. Beér (✉)
MIT, 02139 Cambridge, MA, USA
e-mail: jmbeer@mit.edu

R. Malhotra (ed.), *Fossil Energy: Selected Entries from the Encyclopedia of Sustainability Science and Technology*, DOI 10.1007/978-1-4614-5722-0_13,

| | |
|---|---|
| NETL | National Energy Technology Laboratory (USDOE) |
| $NO_x$ | Nitrogen oxides |
| NRC | National Research Council (USA) |
| OECD | Organization for Economic Cooperation and Development |
| Oxy | Oxygen-blown combustion |
| Oxy/FGR | Oxygen-blown combustion with flue gas recirculation |
| PC | Pulverized coal |
| PC/SC | Supercritical steam/pulverized coal |
| PC/USC | Ultra-supercritical steam/pulverized coal |
| PM | Particulate matter |
| psi | Pounds per square inch |
| R&D | Research and development |
| RD&D | Research, development, and demonstration |
| $SO_2$ | Sulfur dioxide |
| TCR | Total capital requirement |

## Introduction

The provision of electric power is one of the prerequisites of prosperity; there is strong correlation between electric power-generating capacity and per capita gross domestic product (GDP) (Fig. 13.1) [8]. Across the world, economic indicators signal that there will be continued growth and increased electricity demand; also, at least 3.6 billion people lack adequate access to electricity and 1.6 billion have no electricity at all.

During the next 15 years, 1,200 GW of new capacity is projected to be added to the world's present electric generation capacity of about 4,000 GW. The EIA projects that by 2030 global demand for electricity will grow by more than 75%.

The U.S. Energy Information Administration (EIA) data for 2008 show that 54% of the electricity supply in the USA was generated by coal, 12% by natural gas (increasing to an estimated 32% by 2020), and about 2% by oil, with the rest produced by nuclear power (21%), hydropower (9%), and by renewable solar or wind (2%).

Electricity generation (billion kilowatt-hours) by fuel type from 1970 to 2020 in the USA is illustrated in Fig. 13.2 [20]. The increase in the use of coal and, especially, natural gas for power generation projected for the period from 2000 to 2020 is to meet growing demand for electricity and to offset the projected retirement of nuclear power plants.

Because of the large coal reserves in major developing countries such as China, India, and Indonesia, where most of the new power-generating plants are being installed, it can be expected that coal will remain the dominant source of power generation worldwide during the first half of this century. There is also a strongly growing demand for natural gas, a clean fuel capable of being used in power generation with high efficiency.

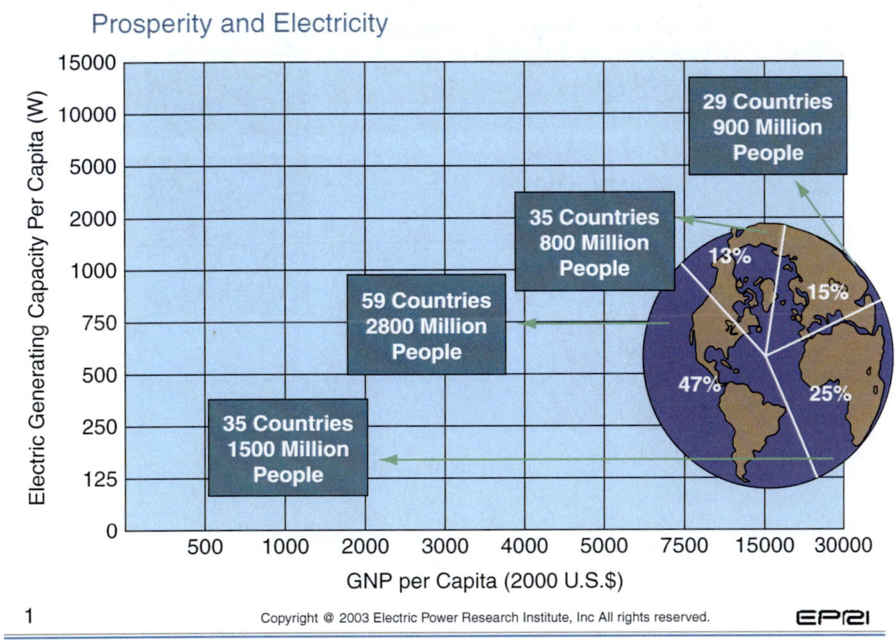


Fig. 13.1 Prosperity and electricity (Source: EPRI [8])

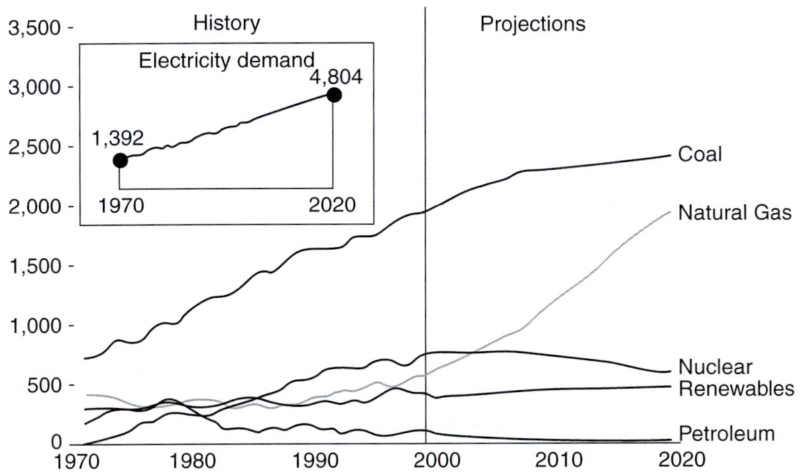


**Fig. 13.2** US Electricity generation by fuel, 1970–2020 (billion kilowatt-hours) (Source: US EIA [5])

Coal as an energy source has the attraction of broad availability (there are large reserves in several countries around the world), low cost, and economic utilization by mature technologies.

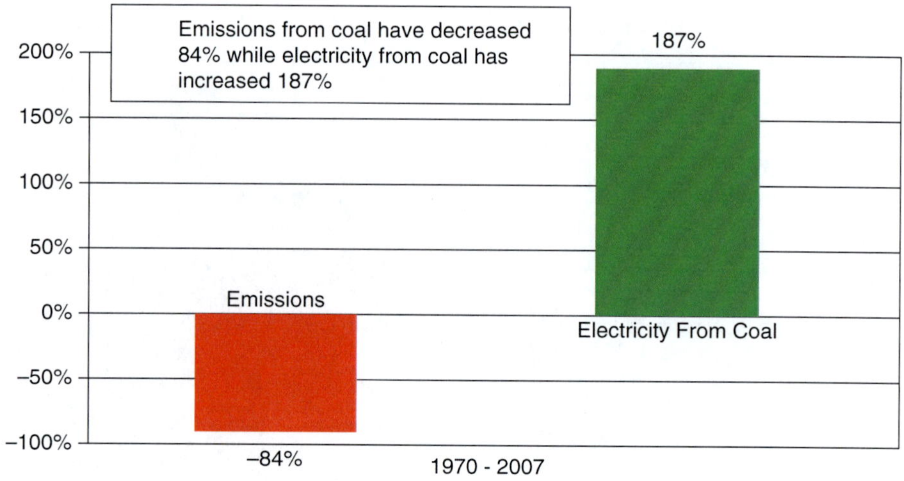


**Fig. 13.3** US emissions of criteria pollutants and electricity from coal during 1970–2007 (Source: NCC Report [14])

The disadvantages are due mainly to environmental and health impacts at the mining phase and in the course of coal utilization. Both the mining and the utilization of coal are subject to increasingly tightening standards by governmental regulations.

Because of no acceptable practical alternative to coal and natural gas, industrial research and development programs since the 1970s have been concentrated on methods of reducing the emissions of so-called criteria pollutants ($SO_2$, $NO_x$, Hg, and fine particulates). New environmental technologies developed and demonstrated under the Clean Coal Technology Program of the U.S. Department of Energy (DOE) during the past 40 years have become commercial and are being applied worldwide.

US emissions reductions have been significant in the face of increasing output of electricity generation coal (Fig. 13.3) [14].

The reduction of $CO_2$ emission remains a challenge. Carbon capture and geological sequestration (CCS) is the key enabling technology for the long-term significant reduction of $CO_2$ emissions from coal-based power generation. It is expected that CCS will become commercial (deployed without significant government subsidy) for base load power generation around 2020–2025 following the construction and operation of several demonstration plants during the present and next decade.

Before the advent of commercial CCS, however, higher efficiency technology options are available for coal-based power generation. Improved efficiency will reduce the amount of fuel used for the generation of a given electric energy output and hence reduce the emissions of all pollutants. Also, when the time comes for CCS deployment, higher efficiency will reduce the energy penalty and make new plant or retrofit of existing plant less expensive [1, 2, 6].

Main higher efficiency technology options discussed below are:

- Pulverized Coal Combustion in Ultra-supercritical Steam Cycle
- Integrated Coal Gasification Combined Gas Turbine-Steam Cycle (IGCC)
- PC or CFB/Oxy/Combustion in Supercritical Steam Cycle

## Pulverized Coal in Ultra-Supercritical Steam Cycle (PC/USC)

The efficiency of PC/SC power plant can be increased in small steps to 43% (HHV) and beyond, as illustrated in Fig. 13.4 [18].

The first two steps in the diagram concern the waste gas heat loss, the largest of a boiler's heat losses, about 6–8%. *The air ratio*>, usually called excess air factor, represents the mass flow rate of the combustion air as a multiple of the theoretically required air for complete combustion. Excess air increases the boiler exit gas mass flow and, hence, the waste gas heat loss. Improved combustion technology, e.g., finer coal grinding and improved burner design, permit lowering the excess air without sacrificing completeness of combustion. Some of these remedies require additional expenditure in energy, e.g., for finer coal grinding, and for increasing the momentum flux of the combustion air through the burners, but this increase in parasitic energy is usually small compared to the efficiency gain due to the reduced excess air.

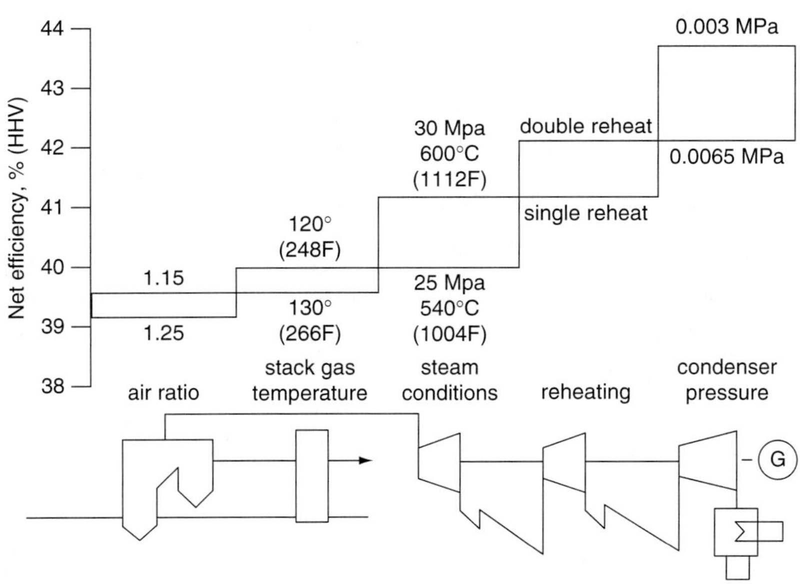


**Fig. 13.4** Steps to improve the efficiency in PC/Supercritical Coal Plant (Schilling [18])

*The boiler exit gas temperature* can be reduced by appropriate boiler design limited only by the dew point of the flue gas. There is a close relationship between the excess air of combustion and the low limit of exit gas temperature from a boiler fired by a sulfur-bearing fuel. Higher excess air leads to an increase in the oxidation of $SO_2$ to $SO_3$, with $SO_3$ promoting sulfuric acid formation in the combustion products. Sulfuric acid vapor increases the dew point of the flue gas and hence raises the permissible minimum exit gas temperature. At an exit gas temperature of 130°C (266°F) a reduction of every 10°C (18°F) in boiler exit temperature increases the plant efficiency by about 0.3%.

*The Rankine cycle efficiency* is proportional to the pressure and temperature of heat addition to the cycle, and is inversely proportional to the condenser pressure, and therefore to the temperature of the cooling medium. The usual design basis for condenser pressure in the USA is 2.0" Hg abs. (67 mbar). Power plants in Northern Europe with access to lower-temperature cooling water use condenser pressure of 1.0" Hg abs. (30 mbar) pressure. This difference can produce an efficiency gain of more than 2 percentage points [19].

As steam pressure and superheat temperature are increased above 221 bar (3,208 psi) and 374°C (706°F) the steam becomes supercritical (SC); it does not produce a two-phase mixture of water and steam as in subcritical steam, but instead undergoes a gradual transition from water to vapor with corresponding changes in physical properties.

Ultra-supercritical (USC) steam generally refers to supercritical steam at more than 1,100°F temperature. EPRI's terminology for 1,300°F and 1,400°F plants is Advanced Ultra-supercritical (AUSC).

The thermodynamic efficiency of the Rankine steam cycle increases with increasing temperature and pressure of the superheated steam entering the turbine. It is possible to increase further the mean temperature of heat addition by taking back partially expanded and reduced temperature steam from the turbine to the boiler, reheating it, and reintroducing it to the turbine. In the usual designation of steam parameters the second and third temperature refers to single and double reheat, e.g., 309 bar/594°/594°/594°C, respectively.

The average efficiency of the existing US coal-fueled electricity generating fleet is 33%, based on the higher heating value (HHV) of the coal.

Pulverized coal plants with USC parameters of 300 bar and 600/600°C (4,350 psi, 1,112/1,112°F) can be realized today, resulting in efficiencies of 43% (HHV). These plants are 21% more efficient than the mean of today's US fleet of coal-fired plants; i.e., they would use 21% less coal for the same power generation and emit 21% less $CO_2$. There are several years of experience with these 600°C (1,112°F) plants in service, with excellent availability [4, 11].

Further improvement in efficiency achievable by higher ultra-supercritical steam parameters is dependent on the availability of new, nickel alloys for boilers and steam turbines (Fig. 13.5) [10, 17].

Two major development programs in progress, the Thermie Project of the European Commission and a US program managed by EPRI for the USDOE's NETL and the Ohio Coal Development Office (OCDO), aim at steam parameters

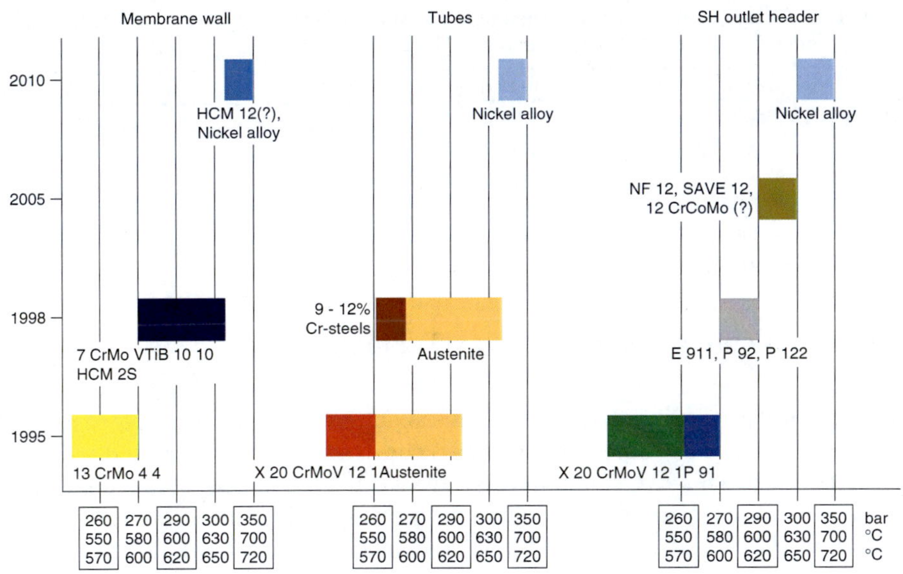


**Fig. 13.5** Stages in materials development and related advanced steam parameters (Henry et al. [10])

of 375 bar, 700°C/720°C (5,439 psi, 1,292°F/1,328°F), and 379 bar, 760°C/760°C (5,500 psi, 1,346°F/1,400°F), respectively. The plant efficiency increases by about 1 percentage point for every 20°C rise in superheat and reheat temperature. An advanced 700°C (1,293°F) USC plant will likely be constructed during the next 7–10 years constituting a benchmark for 46% efficiency (HHV) coal-fired power plant.

It is estimated that before 2020, the time when CCS technologies may begin to become commercially available, about 13 Gigawatts (GW) new coal-based capacity will be constructed in the USA (EIA 2010). If more efficient presently available (1,112°F) USC technology is utilized instead of subcritical steam plants, $CO_2$ emissions would be about 570 MMt less during the 30 years' lifetime of those plants, even without installing a $CO_2$ capture system.

High efficiency coal-based power generation is also important to long-term solutions of reducing $CO_2$ emissions by using CCS, as it mitigates the significant energy cost of CCS application. Because of the reduced coal use for a given electricity output, the plant has a smaller footprint with respect to size of coal handling and emission control systems. These savings and the use of modern analytical techniques that enable optimal use of Ni alloys can minimize the cost of USPC technology.

Coal consumption and $CO_2$ emission comparisons for presently available 500 MW PC/SubC, PC/SC and PC/USC plants without CCS are presented below in Table 13.1 [13].

**Table 13.1** Comparative coal consumptions and emissions of air-blown pulverized coal combustion technologies without CCS (MIT Coal Study [13])

| Performance | Subcritical | PC/SC | PC/USC |
|---|---|---|---|
| Heat rate (Btu/kWh) | 9,950 | 8,870 | 7,880 |
| Gen. efficiency (HHV) | 34.3% | 38.5% | 43.3% |
| Coal use (MMt/year) | 1.54 | 1.37 | 1.22 |
| $CO_2$ emitted (MMt/year) | 3.47 | 3.09 | 2.74 |
| $CO_2$ emitted (g/kWh) | 931 | 830 | 738 |

Assumptions: 500 MW net plant output; Illinois #6 coal; 85% capacity factor

## Coal Gasification Combined Gas Turbine-Steam Cycle (IGCC)

Gasification-based technologies use partial oxidation of coal with oxygen as the oxidant to produce a synthesis gas (syngas) consisting mainly of CO and $H_2$. The gas is cleaned to remove contaminants before it is used as fuel in a combustion turbine. The exhaust gas of the gas turbine raises steam in a heat recovery steam generator (HRSG) for a steam turbine-electric generator set. The combined cycle efficiency improves through the reduced effect of the steam condenser's heat loss. As with combustion technologies, higher efficiency results in lower emissions per MWh.

While the IGCC concept is being successfully demonstrated in two plants in the USA and two plants in Europe, utility power generation demands introduce new challenges that will require further RD&D to overcome. The gasification process operates best under steady-state conditions. The load change conditions associated with utility electricity generation will burden the technology. The many chemical syngas cleanup processes will have to respond to these changes on a real-time basis. In addition, the gasifier and associated gas cleanup systems will be exposed to a much larger range of fuel quality than experience has demonstrated. Again, this variation introduces conditions that require more RD&D to commercialize.

In the prevalent designs the gasifier operates at high pressure. Bituminous coal is pulverized and fed as coal-water slurry [7]. The temperature in the gasifier is above 1,300°C, so that the coal ash can be removed as liquid slag. The product syngas undergoes rigorous cleanup of particulates, sulfur and nitrogen oxides, and mercury, prior to entering the gas turbine (GT) combustor. Precombustion cleanup of the syngas is economically favorable compared to postcombustion flue gas cleanup because of the lower volume flow rate, undiluted by nitrogen, and elevated pressure. The syngas can be cooled prior to cleanup by radiant and convective heat exchangers which raise steam, and improve plant efficiency. However, because of the additional cost and operational problems due to fouling, the convective heat exchanger is often omitted in recent designs at the detriment of the plant efficiency.

Another element of efficient design is the integration of subsystems with the main generating plant. For example, air can be fed from the main compressor of the GT to the Air Separation Unit that produces oxygen for the gasification process, and the nitrogen can be returned to the gas turbine combustor as a diluent to reduce $NO_x$

**Table 13.2** Evolutionary improvements due to DOE advanced gasification research

| | Change in cost and performance of IGCC system due to R&D | | | | | |
|---|---|---|---|---|---|---|
| | Increase in efficiency (%) | Capital cost reduction ($/kW) | O&M reduction ($/year) | Reduction in COE ($/MWh) | Availability impro vement (%) | Emissions |
| DOE gasification project areas | | | | | | |
| Warm gas cleanup | 1–2 | 70–100 | Minimal | 1.8 | 0 | 500 ppb sulfur[a] |
| Instrumentation (temperature measure ments)[b] | 0.5–1 | 0 | Minimal | Minimal | 1–2 | – |
| Materials (refractory)[c] | 0 | 0 | 2 million | 0.5 | 4–6 | – |
| Non-DOE or non-gasification technical areas | | | | | | |
| Heat recovery | 3 | | | | | |
| Industry learning and evolution | 1–2 | | | | | |

*Source*: National Research Council

*COE* cost of electricity, *O&M* operation and maintenance, *kW* kilowatt, *MWh* megawatt-hour

[a]Air Products & Chemicals, Inc. analysis

[b]Internal report by Parsons

[c]The DOE turbine program is considered a complementary program to advanced gasification program

formation. The gain in improved efficiency, however, is weighed against operating a more complicated plant of somewhat reduced availability.

The current IGCC units have, and next generation IGCC units are, expected to have electricity generating efficiencies that are less than or comparable to those of supercritical PC generating units [9].

Current units typically gasify high-heating value, high-carbon fuels. Polk IGCC with a Texaco-GE coal-water-slurry gasifier and radiant syngas cooling operates at 35.4% (HHV) generating efficiency. The Wabash River IGCC with a coal-water-slurry fed E-Gas gasifier, radiant and convective syngas cooling, and no integration operates at about 40% generating efficiency. The IGCC in Puertollano Spain with a dry-feed shell-type gasifier, radiant and convective heat exchangers, and combustion turbine-air separation unit integration has a generating efficiency of about 40.5% (HHV). Supercritical PC units operate in the 38–40% efficiency range, and ultra-supercritical PC units in Europe and Japan are achieving 42–46% (HHV) generating efficiency.

A National Research Council Panel reviewing the DOE's Advanced Gasification Research Program estimated the improvements upon successful completion of the research. The research areas and anticipated benefits are listed in Tables 13.2 and 13.3 [15].

**Table 13.3** Evolutionary improvements due to DOE Advanced Gasification Research

| | Change in cost and performance of IGCC system due to R&D | | | | | |
|---|---|---|---|---|---|---|
| | Increase in efficiency (%) | Capital cost reduction ($/kW) | O&M reduction ($/kWh) | Reduction in COE ($/MWh) | Availability improvement (%) | Emissions |
| DOE gasification project areas | | | | | | |
| Ion transport membrane air separation[a] | 1 | 75 | Minimal | 1.4 | 0 | – |
| Stamet pump (from backup material)[b] | 0.5 | 40–100 | Minimal | 1.4–1.8 | 0 | – |
| Non-DOE or non-gasification technical areas | | | | | | |
| DOE turbine program[c] | 2–3 (for combined cycle power island) | 60–100 | | | | |

*Source*: Gary Stiegel [22], U.S. Department of Energy National Energy Technology Laboratory (for data on DOE gasification project area cost and performance)

*COE* cost of electricity, *O&M* operation and maintenance, *kW* kilowatt, *kWh* kilowatt-hour, *MWh* megawatt-hour

[a]Air Products & Chemicals, Inc. analysis

[b]Internal report by Parsons

[c]The DOE turbine program is considered a complementary program to advanced gasification program

## $CO_2$ Capture and Compression in PC and IGCC Plant

$CO_2$ capture from pulverized coal combustion (PC) involves postcombustion cleanup, the separation and recovery of $CO_2$ that is at low concentration and low partial pressure in the exhaust gas. Chemical absorption with amines is presently the only commercially available technology. The $CO_2$ is first captured from the exhaust gas stream in an absorption tower. The absorbed $CO_2$ must then be stripped from the amine solution using a large amount of steam, regenerating the solution for recycle to the absorption tower. The recovered $CO_2$ is cooled, dried, and compressed to a supercritical fluid. It is then ready to be piped to sequestration.

The use of steam for $CO_2$ removal reduces the steam available for power generation.

To maintain constant net power generation the coal input, boiler, steam turbine/generator, and emission control equipment must all be increased in size. The thermal energy required to recover $CO_2$ from the amine solution reduces the efficiency by 5 percentage points. The energy required to compress the $CO_2$ to a supercritical fluid is the next largest factor, reducing the efficiency by 3.5

**Table 13.4** $CO_2$ emissions, efficiency, and cost estimates for advanced generation technologies without and with CCS

| | PC/SC | | PC/USC | | IGCC | | SCPC/Oxy |
|---|---|---|---|---|---|---|---|
| | Without | With | Without | With | Without | With | With CCS |
| $CO_2$ emitted g/kWh | 830 | 109 | 738 | 94 | 824 | 101 | 104 |
| Efficiency, % HHV | 38.5 | 29.3 | 43.4 | 34.1 | 38.4 | 31.2 | 30.6 |
| TCR $/kW | 2,800 | 4,524 | 2,865 | 4,408 | 3,016 | 3,996 | 4,016 |
| COE c/kWh | 6.00 | 9.64 | 5.93 | 9.25 | 6.44 | 8.24 | 8.82 |

*Source*: $CO_2$ and efficiency data from MIT 2007. TCR and COE data modified after Booras [21] to include the effect of the recent increase in construction costs

percentage points. All other energy requirements amount to less than 1 percentage point.

R&D is in progress pursuing the use of alternative sorbents, such as chilled ammonia to reduce the energy intensity of the $CO_2$ capture process.

The stakes are high because a successful solution would be applicable to both new plants and to the retrofit of existing plants with CCS [6].

*IGCC* lends itself favorably for efficient $CO_2$ capture and sequestration because $CO_2$ can be separated from a relatively small volume of fuel gas (syngas) at high pressure. Without CCS, IGCC is more expensive, and has lower efficiency and availability than PC/SC and PC/USC technology but, if CCS were available today, equipped with CCS, IGCC would be cheaper. By recirculating part of the liquid phase supercritical $CO_2$ the water in the slurry feed can be replaced by $CO_2$. This could reduce the oxygen required for the vaporization of the water in the slurry and for the attainment of the high temperature necessary for trouble-free liquid slag removal from the gasifier.

For coals of lower heating value such as subbituminous coals or lignite, the COE gap is, however, substantially narrowed or is even reversed.

It is noteworthy that there is significant cost and performance loss attached to the capture and compression of $CO_2$ from both combustion and gasification plants in preparation of its sequestration [5].

Results of studies presented in Table 13.4 provide information on $CO_2$ emitted, efficiency, estimates of total plant cost and cost of electricity for IGCC, PC/USC technologies without and with $CO_2$ capture and compression, and for SC/PC/Oxy.

## PC/Oxy/USC

When oxygen, instead of air is used as oxidant for combustion, the mass flow rate of combustion products is significantly reduced and the flue gas $CO_2$ concentration is greatly increased. In order to avoid unacceptably high temperatures in the boiler, combustion products, mainly $CO_2$, are recirculated from the end of the boiler to the combustion chamber. This restores the furnace gas temperature to air combustion levels resulting in an $O_2$ volume concentration of about 30%, compared to 21% for air-blown combustion. This difference is due to the higher specific heat of $CO_2$ than that of the replaced nitrogen, and also, to $CO_2$'s high radiative emissivity. Flue gas recirculation (FGR) increases the $CO_2$ concentration in the flue gas to beyond 90%, the complement being $N_2$, due to air leakage and about 3% O2 required for complete burn out of coal. This makes the flue gas ready for sequestration after the removal of condensables but without energy-intensive gas separation. If avoidance of corrosion in the compressor and pipeline requires further exhaust gas polishing, the fivefold reduced flue gas volume leads to strongly reduced capital and treatment costs relative to those for an air-blown combustion plant [12].

The presently available Cryogenic Air Separation process consumes a significant fraction of the generating plant's output and reduces its efficiency by 6.4 percentage points. The development of membrane-type oxygen processes with greatly reduced energy requirements are urgent R&D targets

## CFBC with Oxy/FGR

CFBC with an external heat exchanger lends itself favorably to oxy/fuel application because the solids circulation provides an effective means, additional to flue gas recycle, for controlling the combustion temperature. Solids, consisting of sorbent, coal ash, and coal char particles, are precipitated from the gas stream exiting the combustor (riser) section of the CFB boiler and are split into two solids streams: one that is recirculated to the riser without cooling, and the other that is cooled in an external heat exchanger before recirculation. The cooling of the combustor by the cold solids permits the reduction of the rate of flue gas recirculation, with the result that the $O_2$ concentration in the feed stream can rise to above 40% without exceeding a limiting combustor temperature level of about 850°C required by the thermodynamic stability of $CaSO_4$ and by smooth fluidization. The corresponding lower gas mass flow leads to reduced size and cost of the boiler and of the post combustion plan. The advantages also include increased fuel flexibility including the capability of firing biomass, which may allow additional $CO_2$ credits.

The oxygen-fired CFB concept has been validated in Alstom's 3 $MW_{th}$ pilot scale test facility [16]. The results show that the technology can be developed from that of conventional CFB boilers, so that a 100–350 $MW_e$ oxy-fired demonstration plant could be developed in the short term followed by a commercial size plant in the midterm.

## Concluding Remarks

Coal will continue to play a large and indispensable role in electricity generation, in a carbon-constrained world, under any scenario. The key enabling technology for $CO_2$ emissions mitigation in coal combustion and gasification plants is $CO_2$ capture and sequestration (CCS).

CCS has to be demonstrated at scale, integrated with power generation, and the legal framework of sequestration has to be developed before CCS becomes commercial, probably by the 2020–2025 period.

Before the advent of CCS there will be about 45 GW new coal-based electricity generating capacity constructed in the USA (about 1,000 GW worldwide) and the question arises of what the technology options are for these new plants.

Increased efficiency of power generation is the most predictable and cost-effective method for $CO_2$ emissions reduction. In coal plant without CCS it is also the only practical method for mitigating $CO_2$ emissions now, and it remains important for future plants equipped with CCS to reduce the energy cost of $CO_2$ capture.

Pulverized coal combustion in Rankine cycle steam plant is the prevailing utilization technology. Compared to average efficiency of the existing coal-based fleet, up to 24% reductions in $CO_2$ and pollutant emissions can be achieved today in commercial ultra-supercritical new plants.

Coal gasification combined cycle (IGCC) plants are successfully demonstrated in the USA and in Europe. Today, without CCS, IGCC is more expensive, and has lower availability than high efficiency PC/SC and PC/USC plants but, if CCS were available today, equipped with CCS IGCC would be cheaper. Also, IGCC has the capability of less expensively lowering further criteria pollutant emissions.

Use of oxygen instead of air in combustion significantly reduces the mass flow rate of combustion products with the concomitant reduction of flue gas capital and treatment costs. Because of the high concentration of $CO_2$ in the flue gas it is expected that following the removal of pollutants and agents that could be corrosive in the compressor and the pipeline, the flue gas can be sequestered without the separation of $CO_2$.

The RD&D challenge is in the development of the membrane oxygen system that is less energy intensive than the presently available cryogenic air separation.

There are important technology developments in progress that can change the performance and economics of advanced technology options by the time CCS will be commercial. A broad portfolio of advanced clean coal technology RD&D should be aggressively pursued to meet the CCS challenge.

Demonstration of $CO_2$ sequestration at scale and integrated with power generation will give the public more confidence that a practical carbon emission control option exists and maintains opportunities for the lowest cost, widely available energy source to be used to meet the world's pressing energy needs in an environmentally responsible manner.

## Bibliography

1. Amor AF, Viswanathan R (2004) EPRI, supercritical fossil steam plants: operational issues and design needs for advanced plants. In: Fourth international conference on advances in materials technology for fossil power plants, Hilton Head Island

2. Beér JM (2007) High efficiency electric power generation, the environmental role. Progr Energy Combust Sci 33:107–134
3. Beér JM (2004) Electric power generation; fossil fuel. In: Cleveland CJ, Ayres RU et al (eds) Encyclopedia of energy, vol 2. Elsevier, Amsterdam, pp 217–227
4. Blum R, Hald J (2002) ELSAM Skaerbaek
5. Booras G, Holt N (2004) Pulverized coal and IGCC plant cost and performance estimates. In: Gasification technologies conference, Washington, DC
6. Dalton St (2007) Efficiency of generation-the role of very efficient/low emission coal. In: International conference on coal utilization and fuel systems, Clearwater
7. Dooher JP (2009) Physio-chemical properties of low rank coal/liquid $CO_2$ slurries as gasifer feed-stocks. In: The 34th international technical conference of coal utilization and fuel systems, Clearwater
8. EPRI (2003) Electricity technology roadmap. Electric Power Research Institute, PaloAlto, CA
9. EPRI (2006) CoalFleet RD&D augmentation plan for advanced combustion based power plants. EPRI Report 1013221
10. Henry JF, Fishburn JD, Perrin IJ, Scarlin B, Stamatelopoulos GN, Vanstone R (2004) In: Twenty-ninth international conference on coal utilization & fuel systems, US DOE. ASME, Coal and Slurry Technology Association, Washington, DC, pp 1028–1042
11. Kjaer S, Klauke F, Vanstone R, Zeijseink A, Weissinger G, Kristensen P, Meier J, Blum R, Wieghardt K (2001) The thermie project of the EC. Powergen Europe, Brussels
12. Lavasseur AA, Chapman PJ, Nsakala NY, Kluger F (2009) Alstom's oxy-firing technology development and demonstration-near term $CO_2$ solution. In: The 34th international technical conference on clean coal and fuel systems, Clearwater
13. Deutch J, Moniz J (eds) (2007) MIT the future of coal. MIT Press, Cambridge
14. National Coal Council reports 2008 and 2009. National Coal Council, Washington, DC
15. National Research Council (2005) Report of the panel on DOE's IGCC R&D program. National Academies Press, Washington, DC
16. Nsakala L et al (2003) Greenhouse gas emissions control in circulating fluidized bed boilers. US DOE/NETL [48] Cooperative Agreement, May 2003
17. Palkes M (2003) Boiler materials for ultra supercritical coal power plants. Conceptual Design ALSTOM Approach NETL-DOE, USC T-1
18. Schilling HD (1993) VGB Kraftwerkstechnik 73(8):564–576 (English Edition)
19. Termuehlen H, Empsperger W (2003) Clean and efficient coal fired power plants. ASME Press, NY
20. US Energy Information Administration (2000). Annual energy OUTLOOK
21. Booras G (2008) Economic assessment of advanced coal based power plants with $CO_2$ capture MIT carbon sequestration forum IX. Cambridge, MA
22. Stiegel G (2005) US DOE Gasification program overview in National Research Council Report. Washington, DC

# Chapter 14
# Pulverized Coal-Fired Boilers and Pollution Control

**David K. Moyeda**

## Glossary

| | |
|---|---|
| Anthracite | Coal which typically contains 86–97% carbon. Anthracite is considered the highest rank of coal as it has the highest energy content of all coals. |
| Ash | Inorganic residues remaining after combustion. |
| Baghouse | See fabric filter. |
| Bituminous coal | Coal which typically contains 45–86% carbon. Bituminous coal lies between subbituminous coal and anthracite in terms of rank, and is commonly divided into additional subgroups dependent upon the content of volatile material. |
| Calorific value | Corresponds to the amount of heat per unit mass when combusted. Can be expressed as gross calorific value, which is the amount of heat liberated during combustion under standardized conditions at constant volume so that all of the water in the products remains in liquid form, or as net calorific value, which is the maximum |

This chapter was originally published as part of the Encyclopedia of Sustainability Science and Technology edited by Robert A. Meyers. DOI:10.1007/978-1-4419-0851-3

D.K. Moyeda (✉)
GE Energy, 18A Mason, Irvine, CA 92618, USA
e-mail: david.moyeda@ge.com

R. Malhotra (ed.), *Fossil Energy: Selected Entries from the Encyclopedia of Sustainability Science and Technology*, DOI 10.1007/978-1-4614-5722-0_14,

achievable heat release obtainable in a furnace at constant pressure.

| | |
|---|---|
| Carbon dioxide ($CO_2$) | A heavy, colorless gas that results from the combustion of fossil fuels and from natural sources. |
| Carbon monoxide (CO) | A colorless, odorless gas produced by incomplete combustion of fossil fuels. |
| Coal | A solid fossil fuel consisting primarily of carbon, hydrogen, nitrogen, oxygen, sulfur, and nitrogen. Coal also contains ash, minerals which do not burn, and moisture. Coal is typically classified or ranked by its volatile matter, fixed carbon content, and calorific value. |
| Dry FGD | A process that removes sulfur oxides from the flue gas and results in the formation of a dry product or waste. |
| Electrostatic precipitator (ESP) | A device for removing particulate from a gas stream based upon using an electric field to charge the particles in the gas and move them to a collecting surface. |
| Fabric filter | A device for removing particulate from a gas stream based upon filtering the gas through a filer media. |
| Flue gas desulfurization (FGD) | Technologies that are used to remove sulfur oxides from the flue gas. |
| Lignite | Coal which contains 25–35% carbon and which has a lower calorific value than subbituminous and bituminous coals and typically higher moisture and volatile content. Lignite is the lowest range of coal. |
| Low-$NO_X$ burners (LNB) | Technology for reducing $NO_X$ emissions by controlling fuel and air mixing in the flame. |
| Nitric oxide (NO) | A colorless gas resulting from the combustion of fossil fuels. |
| Nitrogen dioxide ($NO_2$) | A reddish-brown gas that can be emitted from the combustion of fossil fuels or is formed by atmospheric reaction of nitric oxide (NO) and oxygen ($O_2$). |
| $NO_X$ | Refers to the total nitric oxide (NO) and nitrogen dioxide $NO_2$ concentration. |

| | |
|---|---|
| Overfire air (OFA) | Technology that reduces $NO_X$ emissions based upon air staging. |
| Reburning | Technology that reduces $NO_X$ emissions based upon staging fuel in a fashion that permits fuel fragments to reduce (or reburn) nitric oxide (NO) in the flue gas. |
| Selective catalytic reduction (SCR) | Technology that reduces $NO_X$ emissions by mixing ammonia into the flue gas and reacting the ammonia with $NO_X$ over a catalyst. |
| Selective noncatalytic reduction (SNCR) | Technology that reduces $NO_X$ emissions by mixing an amine-based reagent into the flue gas at a temperature which selectively promotes the reaction of amine ($NH_2$) with nitric oxide to form molecular nitrogen ($N_2$) |
| Subbituminous coal | Coal which typically contains 35–45% carbon and which typically has a lower calorific value than bituminous coal and higher moisture and volatile content. |
| Sulfur dioxide ($SO_2$) | A colorless, irritating gas resulting from the combustion of sulfur contained in fossil fuels, particularly coal. |
| Sulfur oxides | Refers to sulfur dioxide ($SO_2$) and sulfur trioxide ($SO_3$). |
| Volatile matter | Non-moisture component of coal that is liberated at high temperature in the absence of air. |
| Wet FGD | A process that removes sulfur oxides from the flue gas and results in the formation of a product or waste that is a solution or slurry. |

## Definition of the Subject

Fossil fuels, such as coal, natural gas, and fuel oil, are used to generate electric power for industrial, commercial, and residential use. Due to its relatively low cost and abundance throughout the world, coal has historically played a significant role in energy production and approximately 41% of the world power generation was supplied by coal-fired power plants in 2008 [1]. While increased discoveries of natural gas and fuel oil resources, and growth in renewable energy sources, such as wind, solar, and geothermal energy is projected to reduce the use of coal for power generation, energy from coal will continue to be used to satisfy the world's energy demands.

One drawback in the use of coal for power production is that it produces high levels of air pollutants, such as particulate, sulfur oxides, and nitrogen oxides. Coal also produces higher carbon dioxide emissions than other fossils such as natural gas and fuel oil. In addition, due to the large quantities of coal that are used for power production, emissions of toxic metals, such as mercury, which are contained in very small quantities in the coal, can also be a concern.

Over the past several decades, a number of technologies have been developed to reduce the air pollutant emissions formed from coal combustion. Modern power plants can be equipped with advanced technologies that reduce particulate, sulfur oxide, and nitrogen oxide emissions to the most stringent levels. These technologies are the focus of this entry.

## Introduction

The abundance of coal throughout the world led to its use in China as early as 1000 B.C. and by the Romans in Britain before 400 A.D. [2]. While the use of coal in Briton largely disappeared when the Romans left in the fifth century, coal use in England increased in the thirteenth century, and by the beginning of the seventeenth century, coal was the dominate source of energy [3]. During the industrial revolution of the eighteenth and nineteenth century, the invention and development of the steam engine led to an increase in coal production and use for industrial processes and transportation [4–6]. In the late nineteenth century, advances in electricity and the invention of a reliable incandescent lamp set the stage for the use of coal to generate electrical power with the first coal-fired central generating plant in the USA established in 1882 [7, 8]. The growing demand for electrical power led to further developments in steam-generating boiler technology, such as pulverizing the coal prior to introducing it to the boiler furnace. Pulverized coal firing enabled the construction of larger boilers and power plants and became the predominant firing method for large steam-generating power plants beginning in the late 1920s [9, 10].

The air pollution associated with coal combustion was recognized in England as early as the thirteenth century, where the burning of coal in urban areas created a smoky environment and led to bans on coal use [11]. While improvements in combustion methods and the use of chimneys or stacks to disperse the smoke overcame some of the objections to coal burning, fundamentally, the growing need for low-cost energy offset public concerns over the unhealthy aspects of coal combustion, and such bans were largely ignored [3]. As the world entered the twentieth century, widespread industrialization and the increased use of coal resulted in degraded air quality in London and other English cities [12]. By the middle of the century, severely degraded air quality in Los Angeles, California [13], and air quality disasters in London, England [14, 15], and Donora, Pennsylvania [16], heightened public awareness of the dangers of air pollution and government recognition of the need for research to understand the formation of and problems associated with air pollutants and for the development of regulations to limit their emissions [17].

For coal combustion, the primary air pollutants of concern are particulate matter (PM), sulfur dioxide ($SO_2$), and oxides of nitrogen (NO and $NO_2$, which are referred to as $NO_X$) [18]. These pollutants originate from the ash, sulfur, and nitrogen species present in the coal. Particulate matter reduces visibility and contributes to regional haze. In addition, "coarse" particles (from 2.5 to 10 μm in diameter, $PM_{10}$) and "fine" particles (smaller than 2.5 μm in diameter, $PM_{2.5}$) can accumulate within different areas of the respiratory system and aggravate health problems such as asthma or lead to increased respiratory symptoms and disease [19–21]. Exposure to sulfur and nitrogen oxides in sufficient concentration can lead to respiratory symptoms, particularly for those susceptible to these problems [22, 23]. In addition, sulfur and nitrogen oxide emissions can react with water in the atmosphere to form acids that deposit in lakes and soils leading to acidification [24–26], which is referred to as "acid rain." Nitrogen oxides also react with volatile organic compounds to form photochemical oxidants, such as ozone ($O_3$), which present a hazard to human health and plants in high concentrations and which cause visible "smog" in urban areas [27, 28].

This article will introduce how PM, $SO_2$, and $NO_X$ emissions are formed during coal combustion and will discuss the technologies that have been developed to control these emissions from pulverized coal-fired power plants. The technology discussion will focus on the primary technologies that have been applied and are available on a commercial scale.

## Air Pollutant Emissions from Coal Combustion

The primary air pollutants regulated from coal-fired power plants worldwide are carbon monoxide (CO), sulfur dioxide ($SO_2$), oxides of nitrogen (NO and $NO_2$, which are referred to as $NO_X$), and particulate matter (PM) [29]. In general, CO emissions from pulverized coal-fired power boilers are low (<50–200 ppmv) as the combustion system tends to be operated with sufficient excess air to maximize combustion efficiency [30]. Local standards, rather than national standards, are generally established to limit CO emissions and can vary widely. $SO_2$ and $NO_X$ emissions are regulated as both can be a health hazard in high concentrations [22, 23] and are contributors to dry and wet acid deposition [24–26]. In addition,

NOx emissions contribute to ground-level ozone formation through the reaction with volatile organic compounds [27, 28]. Particulate emissions can lead to reduced visibility and degraded air quality. Small particulates emitted from combustion (<10 μm) or formed by reactions of $SO_2$ and $NO_X$ in the environment (<2.5 μm) are associated with increased respiratory symptoms and disease [19–21].

In addition to these primary pollutants, since the early 1990s, there has been significant interest in the emissions of hazardous air pollutants (HAPs), such as toxic organic compounds, acid gases, and metals, from coal combustion. Based upon comprehensive studies performed to characterize HAPs from fossil fuel–fired power plants [31, 32], the US Environmental Protection Agency (EPA) identified mercury as the primary HAP of concern due to the contribution of coal-fired power plants to the total anthropogenic emissions of this compound and to the risk of exposure to methylmercury through the consumption of contaminated fish [33]. Methylmercury is a highly toxic neurotoxin that bioaccumulates up the food chain [34]. The concern over mercury emissions from coal combustion is shared in other developed areas of the world [35–37]. In addition to mercury, other HAPs of concern from coal-fired power plants due to their levels of emissions and potential health risks include carcinogenic metals, such as chromium, nickel, and arsenic, and acid gases, including hydrogen chloride (HCl) and hydrogen fluoride (HF), which are soft-tissue irritants and can cause respiratory disease [38].

A by-product of the combustion of fossil fuels is the generation of carbon dioxide ($CO_2$). In the atmosphere, $CO_2$ and other gases, such as water vapor, absorb and reemit infrared radiation reflected from the earth's surface, resulting in a "greenhouse" effect that raises surface temperature of the earth [39]. Since coal has a higher carbon-to-hydrogen ratio than other fossil fuels, such as natural gas and fuel oil, $CO_2$ emissions from coal combustion are higher than those from the combustion of other fossil fuels [40, 41]. The high carbon footprint of coal-fired power plants has led to an increased demand for the generation of power from alternative sources, such as cleaner fuels (e.g., natural gas) and renewable sources (e.g., biomass, wind, and solar), to the development of high-efficiency coal-fired power plants (e.g., integrated gasification combined cycle (IGCC)), and to the research and development of methods for carbon capture and sequestration from large coal-fired power plants [42, 43].

This article focuses on the primary pollutants from coal. The levels of $SO_2$, $NO_X$, and PM from coal-fired power plants vary widely and are determined by the coal type and combustion system design. Table 14.1 compares the major properties of several coals, low, medium, and high-sulfur bituminous coals; subbituminous coal; and lignitic coal. As can be seen in the table, the sulfur, nitrogen, and ash content contained in coals can vary significantly. Bituminous coals are characterized by a higher heating value in terms of energy release per mass of fuel consumed than lower rank subbituminous coal and lignite, which have higher moisture contents in comparison to bituminous coal. Bituminous coals also tend to have higher sulfur and nitrogen content in comparison to the low-rank coals. In the remainder of this entry, the impacts of coal ash, sulfur, and nitrogen content and boiler design and operation on emissions from coal-fired power plants will be discussed.

**Table 14.1** Typical coal properties

| Property | Units | Bituminous low sulfur | Bituminous medium sulfur | Bituminous high sulfur | Subbituminous | Lignite |
|---|---|---|---|---|---|---|
| *Ultimate analysis* | | | | | | |
| Carbon | % dry | 69.36 | 70.06 | 62.08 | 65.54 | 66.15 |
| Hydrogen | % dry | 5.32 | 5.00 | 4.44 | 4.15 | 4.20 |
| Nitrogen | % dry | 1.50 | 1.66 | 1.07 | 0.95 | 0.96 |
| Sulfur | % dry | 1.04 | 3.08 | 7.40 | 0.79 | 0.37 |
| Oxygen | % dry | 12.73 | 7.54 | 6.15 | 14.00 | 20.72 |
| Ash | % dry | 10.05 | 12.66 | 18.86 | 14.57 | 7.60 |
| *Proximate analysis* | | | | | | |
| Volatile matter | % a.r. | 42.97 | 38.15 | 34.87 | 31.16 | 27.02 |
| Fixed carbon | % a.r. | 43.21 | 43.10 | 42.79 | 34.88 | 33.38 |
| Ash | % a.r. | 9.62 | 11.77 | 18.05 | 11.26 | 4.97 |
| Moisture | % a.r. | 4.20 | 6.98 | 4.29 | 22.70 | 34.63 |
| *Dry, ash-free basis* | | | | | | |
| Nitrogen | % d.a.f. | 1.67 | 1.90 | 1.32 | 1.11 | 1.04 |
| Sulfur | | 1.16 | 3.53 | 9.12 | 0.92 | 0.40 |
| Volatile matter | | 49.86 | 46.95 | 44.90 | 47.18 | 44.74 |
| Fixed carbon | | 50.14 | 53.05 | 55.10 | 52.82 | 55.26 |
| *Gross heating value* | | | | | | |
| | kJ/kg | 27,244 | 27,386 | 25,663 | 20,002 | 16,866 |
| | Btu/lb | 11,713 | 11,774 | 11,033 | 8,599 | 7,251 |

*a.r.* as received
*d.a.f.* dry, ash-free

**Table 14.2** Major constituents in ash for typical US coals

| State coal type | | Pennsylvania bituminous | Utah bituminous | Wyoming subbituminous | North Dakota lignite |
|---|---|---|---|---|---|
| *Ash composition* | | | | | |
| $SiO_2$ | wt.%, S-free | 45.3 | 60.0 | 43.3 | 22.0 |
| $Al_2O_3$ | | 24.2 | 22.7 | 17.2 | 20.4 |
| $Fe_2O_3$ | | 20.3 | 4.1 | 6.3 | 11.8 |
| $TiO_2$ | | 1.2 | 1.2 | 1.4 | 0.5 |
| $O_2O_5$ | | 0.6 | 1.5 | 2.5 | 0.1 |
| CaO | | 4.8 | 4.6 | 22.7 | 30.3 |
| MgO | | 1.1 | 1.9 | 4.0 | 8.0 |
| $Na_2O$ | | 1.4 | 1.1 | 1.7 | 5.1 |
| $K_2O$ | | 1.3 | 1.8 | 0.5 | 1.4 |

## *Particulate Emissions*

Particulate emissions from pulverized coal combustion result primarily from the mineral matter included in the coal. The mineral matter can be classified as being inherent (chemically bound into the coal matrix), included (present in the coal matrix), or extraneous (soil and rock mixed into the coal during mining) [44]. The composition of the mineral matter varies according to the geographic location, type of coal, and how much extraneous material is entrained into the coal during mining. The primary minerals in coal are quartz, aluminosilicates, iron sulfides, and carbonates [45]. These minerals can be broken down into the major ash constituents shown in Table 14.2 [46]. The primary constituents are silica and alumina oxides ($SiO_2$ and $Al_2O_3$). The iron (Fe) and alkali metal (calcium (Ca), magnesium (Mg), and sodium (Na)) content vary widely but are important as these metals have a major influence on the slagging and fouling characteristics of the resulting ash [47]. Slagging refers to the buildup of molten or partially fused ash on the furnace walls or radiant heat transfer surfaces of the boiler. Fouling refers to the deposit of ash on the convective heat transfer surfaces such as the superheater and reheater.

During combustion, the mineral matter in coal undergoes several transformations to become particulate matter that is entrained in the flue gas. The size and composition of the particulate matter depends upon the mineral composition, how it is included in the coal, the presence of other species, and the time-temperature history of the coal particles as they are being burned [48]. There are two primary mechanisms for ash formation during combustion. First, ash may remain with the burning coal particle, melt at high temperature and coalesce to form liquid droplets, and then agglomerate with other ash particles [49]. Fragmentation of the coal char leads to the formation of additional ash particles smaller than the parent particle [50]. Second, at high temperatures, the more volatile metals will vaporize [51]. These vapors can then undergo homogeneous nucleation to form very fine particulate that grow larger by the condensation of additional volatile species onto the particle surface or can form larger particles by coagulation and chain agglomerate formation. Ash particles formed by the first mechanism represent the bulk of the fly ash mass and are greater than 1 μm in diameter, with a typical size range of 3–50 μm, and those by the second mechanism are less than 0.5 μm in diameter, with a peak around 0.1 μm.

The total particulate emissions in the flue gas from a coal-fired boiler depend upon the design of the combustion system and the amount of unburned carbon resulting from incomplete combustion. For pulverized coal-fired boilers, 70–90% of the ash will typically end up in the flue gas (fly ash), and the remaining 10–30% of the ash will collect on the walls of the boiler and end up as ash removed from the bottom of the boiler (bottom ash) [9]. Cyclone-fired boilers and other boilers with wet bottom designs collect more bottom ash and generate approximately 15–30% fly ash. Boiler load can also impact particulate emissions with lower boiler loads resulting in lower emission rates depending upon boiler design and fuel characteristics [52]. Periodic cleaning of heat transfer surfaces, such as the superheater, reheater, economizer, and air preheater, via soot blowing can lead to short-term increases in particulate loading in the flue gas.

## *Sulfur Oxides Formation*

Sulfur is present in coal in inorganic and organic forms. The primary inorganic form is as iron sulfide ($FeS_2$), which is typically in pyrite (cubic) form. A small amount of sulfur is also present as inorganic sulfates, typically as salts of minerals such as calcium and iron ($CaSO_4 \cdot 2H_2O$ and $FeSO_4 \cdot 7\ H_2O$). The organic compounds containing sulfur are larger chain hydrocarbons and are believed to consist of thiols, sulfides, and thiophenes [53]. The proportions of inorganic and organic sulfur vary depending upon the coal; however, organic sulfur typically comprises between 30% and 50% of the total sulfur [54]. The remainder is primarily pyrite, as inorganic sulfates are typically less than 0.1% of the total sulfur.

During the combustion process, the inorganic and organic sulfur is released and converted to sulfur dioxide, $SO_2$. For pyrite, this process involves decomposition to FeS and subsequent oxidation [55]. For the organic sulfur, this process involves decomposition and oxidation of the parent compound to release the sulfur-bearing species. The gas-phase sulfur chemistry is complex [56], but in fuel-lean flames, the released sulfur species are readily oxidized to $SO_2$. Under fuel-rich conditions, hydrogen sulfide ($H_2S$) and carbonyl sulfide (COS) are also present in low concentrations [57, 58], but these species are readily oxidized to $SO_2$ in the high-oxygen environment of practical combustion systems.

For bituminous and subbituminous coals, nearly all of the sulfur in the coal (90–95%) is released as sulfur oxides, both $SO_2$ and sulfur trioxide, $SO_3$, with the remainder ending up as sulfates in the ash. For pulverized coal-fired boilers, emissions of $SO_2$ are independent of the boiler design and operating conditions [59]. For lignitic and subbituminous coals containing ash with a high alkali (calcium and sodium) content, a higher fraction of the sulfur is retained in the ash. Sulfur oxide emissions from lignites can be correlated with the sulfur, sodium oxide ($Na_2O$), and silica ($SiO_2$) content of the ash [60].

The fraction of sulfur that is emitted as $SO_3$ is small and is typically 0.5–1.5% of the total sulfur in the coal [61]. While equilibrium favors the formation of $SO_3$ at low temperatures, the reaction kinetics are too slow to permit appreciable $SO_3$ to form prior to the flue gas exiting the stack [49]. The $SO_3$ that does form is a concern as, at low temperatures, it reacts with moisture to form sulfuric acid, which can condense out of the flue gas and cause corrosion. The coal sulfur content and fly ash composition and boiler design and operating variables, such as excess oxygen and flue gas residence time-temperature profile, are all factors that can influence the concentration of $SO_3$ that is emitted [62].

## *Nitrogen Oxides Formation*

During combustion, $NO_X$ emissions can form through three main mechanisms. The first process is fixation of nitrogen in the air (atmospheric nitrogen) via the overall reaction:

$$N_2 + O_2 \rightarrow 2NO \tag{14.1}$$

The detailed reaction is believed to be initiated by a free oxygen atom (O) attacking the very stable nitrogen molecule ($N_2$) to form NO and a free nitrogen atom (N), which can then attack $O_2$ [63]. As this chain reaction can only be initiated at high temperatures, NO formed from this process is referred to as "thermal $NO_X$." The second mechanism is through the reaction between hydrocarbons in the flame front with molecular nitrogen [64]. As this reaction can produce NO levels higher than those expected from thermal NO formation alone within the initial stages of combustion, it is referred to as "prompt $NO_X$." The third mechanism is through oxidation of fixed nitrogen species present in the fuel. NO formed from this process is referred to as "fuel $NO_X$." For pulverized coal combustion, the majority of the $NO_X$ emissions result from fuel nitrogen rather than thermal $NO_X$ formation [65, 66]. Prompt $NO_X$ is estimated to be less than 5% [67].

The nitrogen content in coal typically varies between 1% and 2%, with some coals having nitrogen contents higher and lower than this range. Analytical techniques suggest that the majority of the nitrogen is contained in five-member (pyrrolic) or six-member (pyridinic) aromatic ring structures [68, 69], with pyrrolic nitrogen being the dominate form for all coal ranks. A smaller fraction of nitrogen is present in other species which may include quaternary nitrogen (i.e., a nitrogen molecule with four bonds) or aromatic amines [70]. The form and distribution of nitrogen within the coal matrix is expected to impact how the nitrogen is released during the combustion process and the potential for its conversion to NO.

As a coal particle devolatilizes during combustion, a fraction of the nitrogen is released with the volatiles, while a fraction of the nitrogen stays in the coal char [71]. The volatile nitrogen may be contained in tars or in the form of cyanic and amine species, which are believed to be the products of rapid secondary pyrolysis of the hydrocarbon forms. As the tars undergo further breakdown, a fraction of the nitrogen is released primarily in the form of HCN [72], and a fraction is incorporated into soot compounds [73]. The nitrogen species released during devolatilization can undergo oxidation to form NO via the overall reactions:

$$HCN + O_2 \rightarrow NO + CO + H \tag{14.2}$$

$$NH_3 + O_2 \rightarrow NO + \frac{1}{2}H_2O + H_2 \tag{14.3}$$

In the absence of oxygen, the fixed nitrogen species can be reduced to $N_2$. After the volatiles are released, further nitrogen species, primarily as HCN, can be released from the coal char by thermal dissociation of the remaining organic solids. As the coal char burns, NO can be formed by heterogeneous oxidation of nitrogen remaining in the char [74].

Pilot-scale studies to evaluate the impacts of coal type and combustion conditions on the formation of $NO_X$ emissions from pulverized coal combustion have shown that as the total fuel nitrogen content increases, $NO_X$ emissions increase, but that the fraction of fuel nitrogen evolved with the volatiles also impacts $NO_X$ emissions [75]. Large differences in $NO_X$ emissions can result from coals having similar nitrogen contents and burned under the same conditions when fuel nitrogen is evolved at different rates. Fuel and air contacting also has a strong influence on $NO_X$ emissions with rapid and intimate contact of air with the fuel during devolatilization leading to higher $NO_X$ emissions than those which result from the slow mixing processes typical of a diffusion flame [76]. The staging inherent in a diffusion flame lowers $NO_X$ emissions by allowing fuel nitrogen to evolve in an oxygen-free environment where it can be reduced to $N_2$ rather than being oxidized to NO [77]. In a practical pulverized coal flame, coals which evolve fuel nitrogen early in the combustion process tend to produce lower $NO_X$ emissions than those that retain more nitrogen in the char.

The coal type and the boiler design and combustion system all impact the $NO_X$ emissions produced from pulverized coal-fired power plants [75, 78]. Low-rank coals (i.e., subbituminous and lignite) produce lower $NO_X$ emissions than high-rank coals (i.e., bituminous and anthracite). This is primarily due to the lower nitrogen and higher volatile content of the low-rank fuels. For power generation, the majority of pulverized coal fired boilers use either an array of circular burners located on one or two walls of the furnace (i.e., wall or opposed wall firing) or columns of burners arranged on the furnace corners firing in a tangential pattern (i.e., tangential firing). Tangential firing results in longer, slower mixing flames and produces lower $NO_X$ emissions than wall-fired boiler designs, which tend to have shorter, faster mixing flames. Other firing system designs, such as cyclone and arched-fired boilers, that are designed to remove the coal ash in a molten state tend to produce the highest $NO_X$ emissions due to intense fuel and air mixing and higher temperatures. For a given boiler design, firing configuration, and coal, $NO_X$ emissions can be approximately correlated with the total heat liberation per unit of cooled surface area [79]. Older boiler designs tend to have a high ratio of heat release per surface area resulting in high $NO_X$ emissions, while more modern designs use a lower ratio to reduce $NO_X$ emissions.

## Particulate Matter Control Technologies

To remove dust from a gas stream, such as fly ash contained in boiler flue gas, a force must be applied that causes the particles to divert from the flow direction of the gas long enough for the particles to contact a collecting surface. The collecting force can be gravitational, centrifugal, inertial, direct interception, diffusional, or electrostatic [80]. Some collection devices may use a combination of these forces. The effectiveness of the collecting force depends upon the characteristics of the particulate matter to be collected; hence, the particulate matter size distribution and chemical and physical characteristics must be known to select the most appropriate collecting device [81].

Modern coal-fired power plants use two primary devices to remove and collect fly ash: electrostatic precipitators and fabric filters. Older boilers and industrial boilers may also use mechanical collectors, such as cyclones, to remove larger fly ash particles prior to the remaining particulate being collected in an electrostatic precipitator or fabric filters. Venturi scrubbers have also been used for large coal-fired boilers; however, these are less common and would not be used for compliance with current emissions regulations. The remainder of this section will provide an introduction to electrostatic precipitators and fabric filters.

### *Electrostatic Precipitator*

If a dust-laden gas is passed through a strong electrical field, the dust will become charged and will begin to flow in the direction of the ion flow, enabling them to be removed from the gas. To apply this phenomenon in practice requires four steps: (1) the particles need to be charged, (2) the particles need to be collected on a surface, (3) the particles need to be removed from the surface in a fashion that minimizes their reentrainment into the gas flow, and (4) the particles need to be removed from the device [82]. The collecting device that accomplishes these steps is typically referred to as an electrostatic precipitator (ESP). As shown in Fig. 14.1, in an ESP, an electric potential is set up between a discharge electrode and a collection electrode, and the dust laden gas is passed between the electrodes at a relatively low velocity. When the potential is high enough, corona discharge is

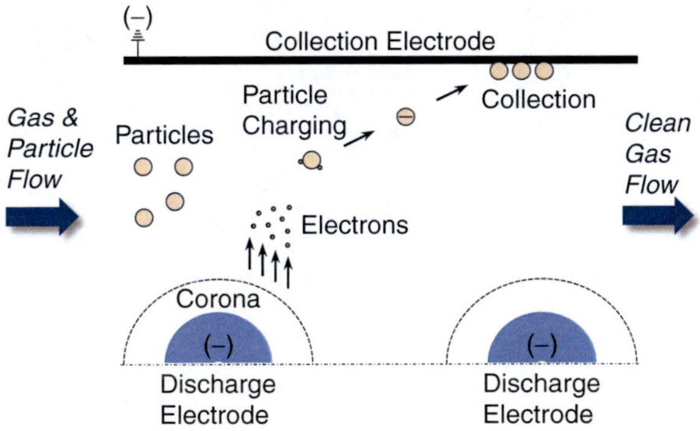


**Fig. 14.1** Principle of electrostatic precipitator operation

generated near the discharge electrode that begins to ionize the gas molecules; these molecules then migrate to the lower potential collecting electrode. In the process, the molecules attached themselves to the dust particles which then also migrate to the electrode. Upon reaching the electrode, the particles lose their charge and stick to the electrode. As the dust layer on the collection electrode builds up, the electrical current is reduced, and it becomes necessary to vibrate the plate to remove the particles. This is accomplished through a rapper system that strikes the top of the collection plate periodically to remove the dust. The dust then falls into the collection hopper where it can be removed.

ESPs are generally classified by the method in which the particulate is removed from the collection electrode, e.g., wet versus dry, and by the geometry of the electrodes, e.g., wire in tube or wire and plates. The most common design for coal-fired power plants is the dry, wire, and plate type as illustrated in Fig. 14.2. The main components of an ESP are a gas-tight casing or outer shell, discharge electrodes, collection electrodes, a high-voltage transformer rectifier or high-frequency power supply for application of electrical power, rappers to remove particulate from the collection electrode, and pyramidal-shaped hoppers to remove particulate from the system [83]. The collection electrodes or plates are arranged in multiple mechanical fields, where each field consists of a series of equally spaced plates perpendicular to the gas flow. The discharge electrodes or wires are suspended in the gas passage between each pair of plates. An ESP will have one or more mechanical fields. A high-voltage pulsed direct current or direct current is applied to the discharge electrode. When the voltage is high enough, a corona discharge will be generated that causes particles to be moved to the collection electrode.

The performance of an ESP is impacted by the design of the ESP, the strength of the electric field, and the particle characteristics. The theoretical collection efficiency of an electrostatic precipitator can be expressed by the Deutsch-Anderson equation [84]:

$$\eta = 1 - e^{-\left(\frac{A}{V}\right)\omega} \qquad (14.4)$$

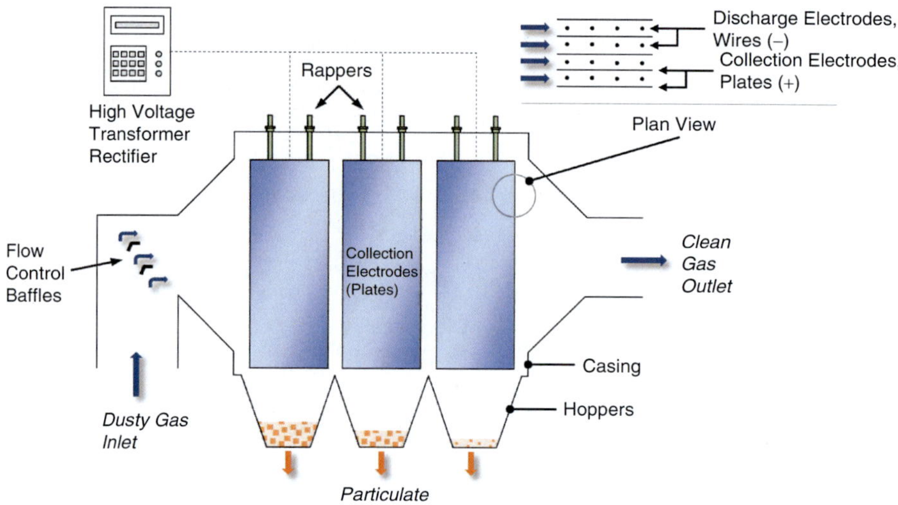


**Fig. 14.2** Overview of electrostatic precipitator

where $\eta$ is the collection efficiency for a given particle size, $A$ is the surface area of the collecting electrode, $V$ is the volumetric gas flow rate, and $\omega$ is the migration velocity of the particle. The term ($A/V$) is referred to as specific collection area (SCA) of the ESP. Equation 14.4 shows that the collection efficiency of an ESP can be improved by either increasing the surface area of the collecting electrodes or decreasing the volume of gas to be treated. The theoretical migration velocity of the particle can be calculated by comparing the electrostatic forces on the particle to the drag force and, with simplifying assumptions, can be expressed as [85]:

$$\omega = \frac{E_c E_p d}{4\pi\mu_g} \tag{14.5}$$

where $E_c$ is the charging field strength, $E_p$ is the precipitating (collecting) field strength, $d$ is the particle diameter, and $\mu_g$ is the gas viscosity. Equation 14.5 shows that the particle migration velocity is proportional to both the electric potential and the particle size. As particle size is reduced, voltage must be increased to maintain migration velocity and collection efficiency. Equation 14.5 suggests that continuing to increase the electric field will continue to increase the particle migration velocity and, hence, the ESP collection efficiency. While this is true, a practical limit is reached where the field breaks down and sparking between the electrodes occurs [80]. Sparking destroys the electrical field and lowers the collection efficiency.

The electrical conductivity of the particulate matter has an impact on the electrical field between the discharge and collection electrodes. The resistance to electrical conductivity is called the "resistivity" [86]. As the dust layer builds up on the collection electrode, the voltage at which sparking occurs is lowered due to the increased electrical field at the dust layer. If the resistivity of the ash is too high, charge will build up on the collected particles and can become high enough to cause an electrical breakdown which causes ions of the opposite polarity to be injected back into the gap which reduces the charge on the particles in the gas flow and which can cause sparking. This breakdown is referred to as "back corona" [87]. The resistivity of fly ash is a function of the particle composition, gas temperature, and concentrations of water vapor and $SO_3$ in the flue gas [88].

Properly designed and operating ESPs can remove between 99% and 99.9% of the total particulate matter. Older equipment is generally less efficient. As noted above, the primary factors that influence the ESP performance are the particle resistivity, the particle size distribution, and the flue gas temperature and flow rate [83]. As these parameters vary from the conditions assumed in designing the ESP, the ESP collection efficiency can be impacted. Other factors that can impact the ESP performance are the electrical conditions in the ESP, the gas flow distribution, and the particle reentrainment during rapping [89]. The applied voltage in each field of the ESP needs to be optimized to maximize the acceptable spark rate with the specific ash characteristics. The gas flow distribution entering the ESP should be optimized to ensure a uniform flow distribution between the collecting plates, and the ESP should be designed to minimize the potential for gas flow to bypass the plates (sneakage). The rapping system design and frequency should be optimized to minimize the reentrainment of particles into the gas stream.

### *Fabric Filter Baghouse*

If a dust-laden gas is passed through a fiber bed or filter, the dust will be removed from the gas and will collect on the filter. The principle mechanisms of filtration include impaction, interception, and diffusion, which are related to the relative particle and fiber size and velocity through the filter [90]. These mechanisms are illustrated in Fig. 14.3. Particles will impact on a fiber when they are too large to follow the gas streamlines around the filter. Interception will occur when a particle follows a streamline close to a fiber and is attracted to the surface by van der Waals forces. Particles that are very small are influenced by Brownian diffusion and can come into contact with the fiber and be removed from the gas. Other forces, such as electrostatic and gravity, can also play a role in removing particulate from the gas.

The filtration media can be flat and supported in a frame or in bags which are supported on cages. As bags are most common for large systems, the containing device is commonly called a baghouse. In a baghouse or fabric filter, the dust-laden gas passes through the suspended filtration media. Particles impact on the filter and are held. As collection proceeds, a deposit begins to build up that also serves as a means of collecting particulate. Eventually, the deposit must be removed or the

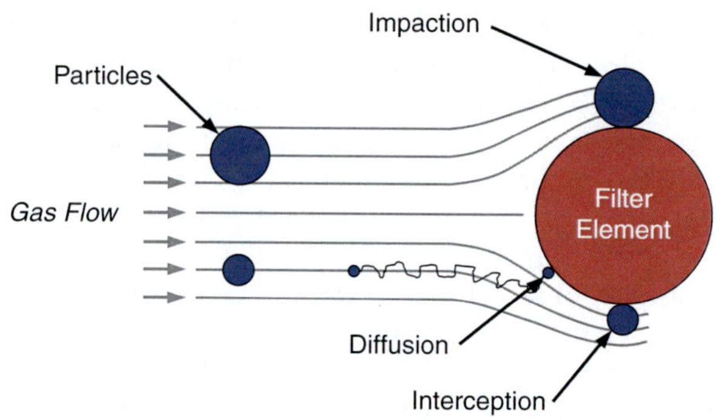


**Fig. 14.3** Principal mechanisms involved in fabric filtration

pressure drop will be too high. As this can be accomplished by several means, baghouses are most commonly classified by the cleaning method [91]. Typical cleaning methods include mechanical, reverse airflow, and pulse-jet cleaning. Mechanical cleaning typically involves flowing the gas on the inside of the bag, stopping the gas flow, and shaking the bag to remove particulate. Reverse airflow cleaning consists of periodically flowing gas in the opposite direction to the normal gas flow to remove the particulate buildup from the filter. In a pulse-jet system, the gas flows from the outside to the inside of the bag. Periodically, a pulse of compressed air is injected down the center of the bag to flex the bag and remove particulate.

Reverse airflow and pulse-jet baghouses have been applied to coal-fired boilers throughout the world [92, 93]. Simplified overviews of the cleaning methodology and main components for reverse airflow and pulse-jet fabric filters are illustrated in Figs. 14.4 and 14.5. The similar components for each type of fabric filter include a gas-tight casing, fabric filters, supporting frames, and pyramidal-shaped hoppers to remove particulate from the unit. Large-scale systems will be designed with multiple compartments to facilitate cleaning and maintenance. Reverse airflow baghouses include a fan and ductwork and valves for flowing cleaned gas back through the unit. Pulse-jet baghouses include a compressed air header and distribution plenum to periodically introduce compressed air into the throat of the filter bags. While significant experience exists with the use of reverse airflow baghouses on large utility boilers, interest in pulse-jet baghouses is increasing since this design uses higher air-to-cloth ratios than the reverse airflow design, resulting in smaller equipment size and lower capital costs [94].

The filtration media used in fabric filters can be woven or felted. Woven fabrics are typically used in shaker-type baghouses as they have good mechanical strength. Felted fabrics tend to be used for reverse air and pulse-jet fabric filters [95]. For proper operation, the type of fabric must be matched to the specific application. The gas stream temperature and chemical composition are critical factors that influence the fabric selection. For coal-fired power plants, the baghouse is installed after the air preheater, where normal gas temperatures are in the range of 120–175°C (250–350°F) and where the presence of sulfur oxides and moisture increases the potential for corrosion. In this application, the utility industry has primarily used

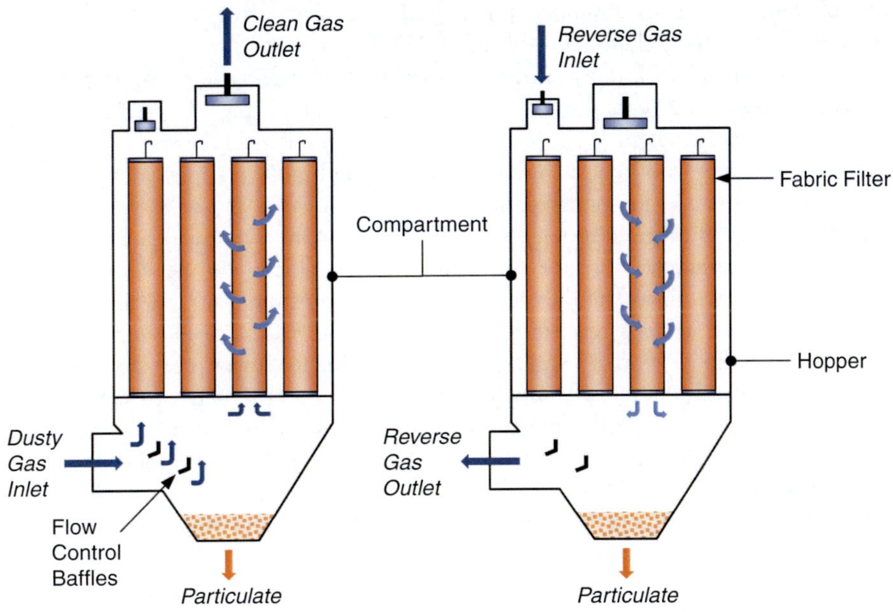


Fig. 14.4 Reverse air flow baghouse features

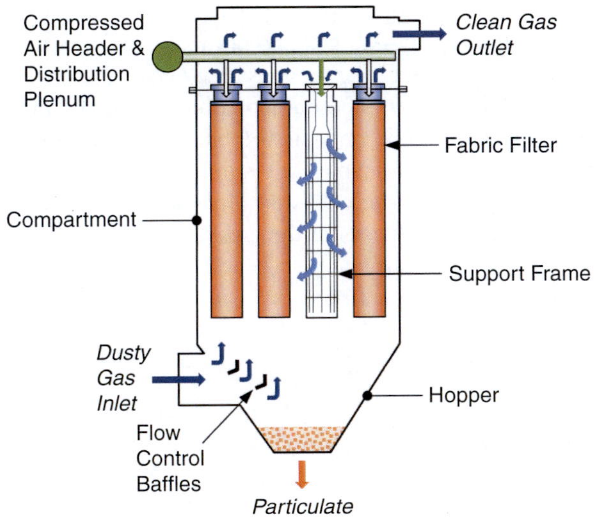


Fig. 14.5 Pulse jet baghouse features

coated fiberglass fabrics for reverse airflow baghouse designs as these fabrics are resistant to chemical attack and can withstand temperatures up to 260°C [96]. The primary coating in use today is polytetrafluoroethylene (PTFE), which improves bag lifetime and provides resistance to acidic attack and thermal excursions.

For pulse-jet baghouses equipped units fired with low-sulfur coal, the bags are typically made of polyphenylene sulfide (PPS) [97].

The performance of a fabric filter is impacted by the particle properties, fabric properties, operating characteristics, cleaning method, and interrelationships between these parameters [98]. While the fundamental mechanisms involved in removing particles from the gas are understood, the highly complex nature of the collection process and the participation of the dust layer in the collection process make the design of fabric filters more empirical in nature. The two primary design and operating parameters are the air-to-cloth ratio, which sets the velocity ratio passing through the filter and, hence, the relative size of the unit, and the pressure drop, which sets the energy consumption requirements [99]. These two parameters are related as baghouses with a small A/C ratio will have a lower pressure drop than baghouses with a higher A/C ratio, resulting in a trade-off between capital and operating costs. The bag cleaning method influences the A/C ratio and pressure drop as more energetic cleaning methods can reduce the A/C ratio (and unit size) for a given pressure drop [100].

Fabric filters are highly efficient collection devices for both coarse and fine particulate, with typical efficiencies of 99–99.9%. Due to the collection mechanisms involved in fabric filtration, particle removal performance is not as sensitive to particle size for these devices as with ESPs. In comparing the costs of fabric filters to ESPs, fabric filters become more economic where highly efficient removal of submicron particles are required [101]. As noted above, the design A/C ratio and operating pressure drop impact fabric filter performance. Fabric filter performance is also impacted by flue gas and particle characteristics. The flue gas temperature impacts the volume of the flue gas and, hence, the operating A/C ratio, while the flue gas moisture content and composition can impact the characteristics of the dust cake. The particle size impacts the buildup (and removal) of the dust cake on the filter and, along with the particle composition, impacts the adhesion of particles to the filter surface and the cohesion of particles to each other. Reactions between the particles and constituents of the gas can also impact the cohesiveness of the dust cake. Finally, the cleaning cycle interval and intensity will also impact the overall performance of a fabric filter.

## $SO_2$ Emissions Control Technologies

The primary mechanisms for removal of a gaseous pollutant from a gas stream involve (1) absorption into a liquid, (2) gas/solid reaction, and (3) adsorption onto a solid. The first two mechanisms are the primary methods for removal of sulfur dioxide from coal-fired boiler flue gas. Commercial processes based upon these mechanisms can be categorized into whether the process is regenerable, where the sulfur compound is separated from the absorbent, or nonregenerable, where the sulfur compounds are thrown away with the absorbent [102]. The product from regenerable processes can be concentrated $SO_2$, hydrogen sulfide, elemental sulfur, or sulfuric acid. $SO_2$ removal processes can be further characterized into wet processes, where the product is a solution or slurry, and dry processes, where a dry product is produced [103]. The majority of non-regenerable processes produce a throwaway waste; however, some wet processes can be modified to produce a gypsum by-product that can be sold if site-specific conditions permit.

The primary technologies for removing sulfur dioxide from coal-fired power plants are wet limestone ($CaCO_3$) and dry lime ($Ca(OH)_2$) scrubbing [104], typically referred to as wet and dry flue gas desulfurization (FGD). The preference for these technologies over regenerable processes is largely driven by the low cost and high availability of limestone and lime, by the relative simplicity of these processes in comparison to other tail end processes, and by the fact that power plants are not chemical companies. For wet scrubbing, limestone tends to be preferred over lime due to lower cost, particularly if a ready supply is located close to the power plant. In comparison to lime, limestone requires finer grinding, has higher transportation costs, requires larger equipment, and is less responsive with respect to pH control [105]. Thus, absorbent selection is done by performing a comparison based upon site-specific technical and economic factors. For wet FGD systems that produce gypsum, which will be sold for wallboard manufacturing, the use of hydrated lime rather than limestone can lead to improvements in the product quality in some cases [106]. In dry scrubbing, lime is required due to the need for rapid reaction times as lime has a higher reactivity than limestone.

A wide variety of wet and dry FGD processes exist, but the basic processes are similar [107–109]. In addition to or instead of lime and limestone, several processes introduce sodium compounds, such as sodium carbonate (e.g., trona), where these materials are available locally to improve process efficiency [110]. Seawater scrubbing, which uses seawater to remove the $SO_2$ and discharge in into the ocean, has also been applied to coal-fired power plants located in a coastal environment [111]. While wet and dry scrubbing are highly efficient technologies, they have high capital and operating costs. This has led to interest in lower capital cost $SO_2$ removal technologies such as furnace sorbent injection [112, 113]. The remainder of this section will discuss the basic wet limestone and dry lime scrubbing processes.

### *Wet Flue Gas Desulfurization*

In wet FGD, flue gas containing $SO_2$ is input into a scrubbing tower or absorber where it is brought into contact with an alkaline solution or slurry containing partially dissolved limestone ($CaCO_3$). The $SO_2$ is absorbed into the solution where it reacts to form calcium sulfite ($CaSO_3$) and calcium sulfate ($CaSO_4$). The detailed chemistry of the process is complex and depends upon the composition of the slurry and extent of dissolution of the limestone [105]. The primary steps

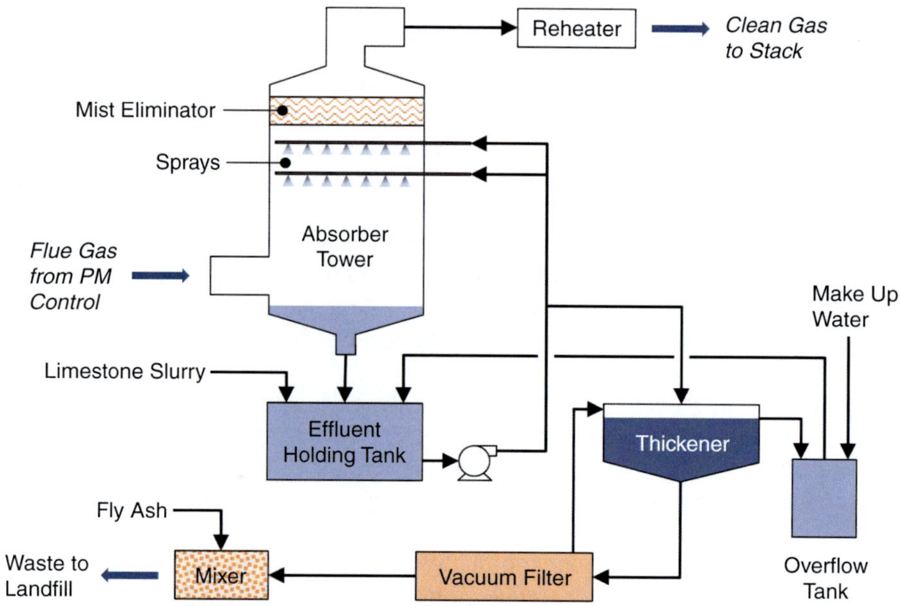


**Fig. 14.6** Basic limestone wet scrubbing process

involved in the process are gas/liquid mass transfer of the $SO_2$, dissolution of the limestone, oxidation of $SO_3$ to $SO_4$ in solution, and crystallization of $CaSO_3$ and $CaSO_4$. By controlling the solution pH and concentration of $CaCO_3$ in the solution, high levels of $SO_2$ removal can be achieved. As $SO_2$ absorbs into the solution as bisulfate ($HSO_3^-$), the primary product from the process is calcium sulfite. However, due to the high oxygen content of flue gas (3–10% by volume), a fraction of the calcium sulfite will oxidize to calcium sulfate. Typically, the product from this process is a waste sludge. It is possible to force oxidation of calcium sulfite into calcium sulfate by bubbling air through the solution which improves the ability to remove water from the waste and permits the material to be upgraded for use for gypsum ($CaSO_4$ •2HO).

The basic components of the limestone wet scrubbing process are illustrated in Fig. 14.6. Flue gas from the particulate control device (ESP or baghouse) enters the bottom of an absorber tower (typically a spray tower) and flows upward. The absorbing solution is sprayed into the gas stream. The $SO_2$ is absorbed into the solution where it reacts with the limestone to form solids. The resulting slurry is then captured in an effluent holding tank where ground limestone slurry and recycled solution are added. This solution is then recycled to the absorber. A portion of the solution is sent to the thickener to concentrate the waste solids as a sludge. The sludge from the thickener is then dewatered in a vacuum filter to produce a filter cake, which is mixed with fly ash to stabilize the waste prior to sending it to a landfill. Enhancements to the basic process include the use of a pre-scrubber upstream of the main absorber to quench the flue gas and protect the materials used in the absorber, incorporation of a second scrubbing loop into the pre-scrubber, and blowing air into the effluent holding tank to force oxidation of calcium sulfite to calcium sulfate [114]. This latter improvement minimizes scale formation, improves the characteristics of the sludge, and eliminates the need for adding fly ash for stabilization. Forced oxidation is also used to produce solids that can be used to produce gypsum for wallboard manufacturing. If lime rather than limestone is used, forced oxidation is not required for scale control. The cleaned flue gas exiting the absorber after passing through a mist eliminator designed to remove entrained droplets from the gas, passes through a reheater, and is then sent to the stack. A reheater is used to protect equipment downstream of the scrubber from condensation and corrosion, to reduce the visible plume, and to improve rise and dispersion of the stack gas.

The performance of a wet FGD system is impacted by a number of design and operational factors, most of which are related [114]. The design of the absorber tower must provide for effective contacting of the flue gas with the absorber solution and for separation of the cleaned gases from the liquid. From an economic standpoint, the size of the tower is linked closely to the process chemistry and the relative flows of flue gas and absorber solution, which is typically expressed as the liquid-to-gas ratio. Operational factors that impact performance include the $SO_2$ concentration in the flue gas, pH of the slurry, solids concentration in the slurry, concentration of other components in the slurry such as magnesium and chloride ions, residence time of the slurry in the reaction tank, and degree of slurry oxidation [105, 115]. High utilization of the limestone (or lime) is important and most systems operate with a calcium-to-sulfur molar ratio of close to 1:1. Liquid pH is monitored and used to control the limestone addition rate to maximum limestone utilization, to minimize the potential for scale in the system, and to reach a specific emissions target. The limestone grind fed to the FGD system can also be important, with finer grinds leading to a smaller-sized reaction tank, higher $SO_2$ removals, and an increase in gypsum purity [116].

Wet FGD systems are capable of achieving over 95% $SO_2$ removal, with some systems being designed for 96–98%, and can be applied to both low and high sulfur coals. As discussed above, a number of parameters can impact system performance and must be optimized to ensure effective $SO_2$ control as well as maintain system reliability and availability. Overall, wet FGD systems represent a significant capital, operating, and maintenance expense for a coal-fired power plant. The main operating and maintenance (O&M) costs differ for various designs and should be considered in selecting an FGD process [117]. Experience with existing plants has led to a number of equipment improvements, such as large, single-tower, absorbers with higher flue gas velocities and improved mist eliminator designs, and process improvements, such as the use of fine limestone and solution buffers [103]. These improvements can lead to higher $SO_2$ removal efficiencies, lower capital costs, and reduced O&M costs.

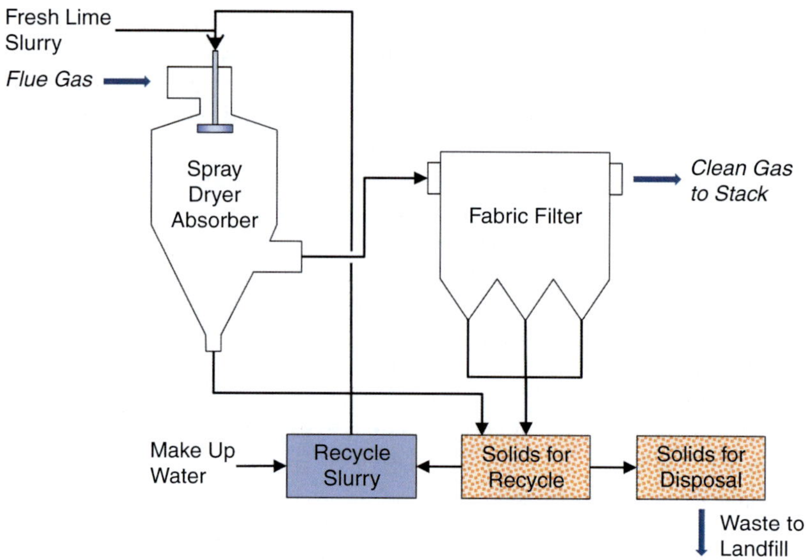


**Fig. 14.7** Basic lime dry scrubbing process

## *Dry Flue Gas Desulfurization*

In dry FGD, flue gas containing $SO_2$ is input into a spray dryer absorber where it is brought into contact with an alkaline solution containing hydrated lime ($Ca(OH)_2$). The $SO_2$ is absorbed into the solution where it reacts with the alkali material to form solid calcium sulfite ($CaSO_3$). Absorption and reaction occur while the thermal energy of the flue gas vaporizes the water in the droplets to produce a fine powder. The primary processes involved in spray drying consist of atomization of the alkaline slurry into a spray of droplets, contacting of the spray with the flue gas, absorption of $SO_2$ into the droplets, reaction with the suspended alkali material, drying of the spray by evaporation of the moisture in the droplets, and, finally, separation of the solids from the gas [118]. Due to the high oxygen concentrations of the flue gas, a fraction of the $CaSO_3$ will be oxidized into $CaSO_4$.

An overview of the lime dry scrubbing process is shown in Fig. 14.7. Hot flue gas from the air preheater outlet enters the top of the cylindrical spray dryer where an atomizer sprays the alkaline slurry from the feed preparation system into the flue gas. $SO_2$ is absorbed into the spray droplets where it reacts with the lime to form a solid. The water in the droplet evaporates leaving a fine powder that can be collected from the spray dryer and in a subsequent particulate collection device. Twin fluid and rotary or spinning disk atomizers can be used; however, most large utility applications use a rotary atomizer. Typically, fabric filters are used for removal of the fly ash and spent absorbent. Collection of the solids on the fabric filter permits additional $SO_2$ removal to occur resulting in increased sorbent utilization.

The spray dryer absorber needs to be designed to maximize $SO_2$ removal, yet produce a dry product. Since absorption and reaction of $SO_2$ is fastest when surface liquid is present, it is necessary to operate as close to the saturation point of the flue gas as possible and to maximize the residence time available for reaction while still yielding dry solids at the absorber walls and outlet [119]. The quality of water that can be used for the slurry is limited by the boiler outlet temperature, which is set by the boiler thermal efficiency, and the desired approach to saturation. The concentration of solids in the slurry is limited by the viscosity of the material which impacts the practical aspects of handling and atomizing the material. These two factors constrain the $SO_2$ removal that can be achieved for a given inlet $SO_2$.

Dry FGD systems are capable of achieving over 95% $SO_2$ removal and are typically applicable to low and medium sulfur coals. There are technical constraints on the application of this technology to high sulfur coals as an increase in sulfur content requires an increase in the solids' content of the slurry and a point can be reached where the viscosity of the material is too high for proper pumping and atomization [103].

## $NO_x$ Emissions Control Technologies

Technologies to reduce $NO_X$ emissions from combustion systems can be placed into two broad categories: (1) combustion modification technologies which reduce the formation of $NO_X$ during combustion and (2) postcombustion technologies which eliminate $NO_X$ from the flue gas following combustion [28]. Combustion modification techniques will generally seek to lower flame temperature or to control the flame stoichiometric ratio depending upon whether thermal $NO_X$ or fuel $NO_X$ emissions are being targeted. For pulverized coal combustion, as discussed previously, the majority of the $NO_X$ emissions result from the nitrogen in the fuel; hence, the most effective combustion modifications are those which control the stoichiometric ratio during coal devolatilization. Postcombustion techniques tend to rely on the use of additives to reduce the $NO_X$ to molecular nitrogen, $N_2$. Removal of $NO_X$ by wet scrubbing is difficult due to the insolubility of NO and poor solubility of $NO_2$. Scrubbing technologies that can be applied to industrial plants have not been widely applied to large pulverized coal-fired power plants due to high capital and operating costs and the low initial $NO_X$ concentrations [120].

For coal-fired power plants, the primary combustion modification techniques are low-$NO_X$ burners, overfire air, and reburning technology [121]. Each of these technologies relies on controlling the combustion process in the furnace. The primary postcombustion technologies are selective noncatalytic reduction (SNCR) and selective catalytic reduction (SCR) [122]. These technologies involve the use of a reagent that reduced $NO_X$ to $N_2$. Combustion modification technologies are generally the most cost-effective when evaluated against postcombustion technologies on a cost per mass of pollutant removed [123]. SCR technology is the most expensive technology but also provides the highest level of $NO_X$ emission reduction. SNCR technology provides a modest level of $NO_X$ reduction, with cost-effectiveness typically between that of combustion modifications and SCR.

In retrofitting existing units with $NO_X$ emissions control technologies, selection of the most cost-effective technology or set of technologies generally requires a boiler-specific analysis that includes assessment of the $NO_X$ emission goals or limits and the expected performance and costs of the specific technologies. Retrofit costs for $NO_X$ control technologies are very site specific and depend upon the ease of the retrofit and the boiler design and operating characteristics, such as the furnace size, firing configuration, and the condition of existing equipment [124]. The remainder of this section will provide an introduction to $NO_X$ control technologies.

## *Low-Nox Burners*

Low-$NO_X$ burners reduce $NO_X$ formation in the combustion process by delaying the mixing of fuel and air in the flame [125]. One means of aerodynamically staging fuel and air mixing in a coal flame for a wall-fired boiler is illustrated in Fig. 14.8. In this low-$NO_X$ burner design, the mixing of fuel and air is controlled by dividing the combustion air into multiple streams, with separate control over each stream. Coal entering the flame is initially burned in a fuel-rich core which gradually becomes fuel lean as air is progressively mixed into the flame. Staging of the air into the flame permits volatile nitrogen compounds released in the early portion of the flame to be processed in a reducing environment, resulting in lower $NO_X$ emissions. Staging of the flame also reduces peak flame temperatures and, therefore, can reduce thermal $NO_X$ formation as well.

For wall-fired boilers, low-$NO_X$ burner designs vary from manufacturer to manufacturer [126–128], but all tend to have similar features, including separate control over the combustion air split into two or more passages, control over the degree of swirl imparted to the air streams, and a flame stabilizing device. The inner and outer air split and swirl settings are adjusted to generate a stable low-$NO_X$

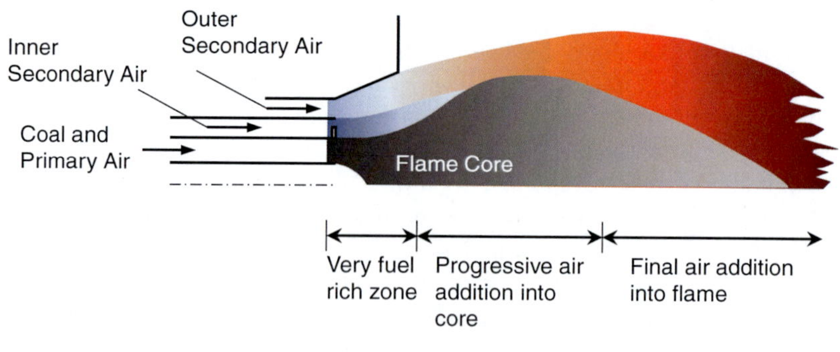


**Fig. 14.8** Typical low-$NO_X$ burner design approach

emissions flame for a specific coal. These settings can also be adjusted to ensure that the flames do not impinge on the side or rear walls of the boiler.

For tangentially fired boilers, low-$NO_X$ burner designs typically consist of alternating passages of air and fuel [129]. The fuel nozzles direct the fuel into the center of the furnace, while the air nozzles direct the air closer to the furnace walls. Offsetting the air from the fuel generates a fuel-rich flame in the center of the furnace which leads to low-$NO_X$ emissions. The offset air also provides a measure of protection along the wall from rich conditions that can lead to wall corrosion. Additional air staging is achieved by adding a fraction of the combustion air through air ports located on the top of the burner stack. The air added to the top of the burner is typically called "close-coupled overfire air" or CCOFA.

The performance of low-$NO_X$ burners in a particular application is very dependent upon the boiler and coal characteristics. Low-$NO_X$ burners generally produce longer flames than the original equipment; hence, the trade-off between optimal burner performance and acceptable carbon-in-ash levels is strongly tied to the furnace geometry. $NO_X$ control levels for the application of low-$NO_X$ burners to wall-fired utility boilers typically range from 40% to 50%.

## *Overfire Air*

Overfire air is a combustion modification technology that, like low-$NO_X$ burners, stages the combustion process to reduce NO emissions [130]. To apply overfire air to a coal-fired boiler, combustion air is diverted from the main combustion zone and is injected through ports located on the walls above the burners. In this process, the primary zone is operated slightly less fuel lean than usual, and fuel and air mixing is delayed. This delay reduces the formation of fuel NO and also reduces peak flame temperatures, resulting in a reduction in thermal NO formation as well. Overfire air is added to complete combustion of unburned fuel.

For pulverized coal-fired boilers, the performance of overfire air is dependent upon the overfire air system design and the burner/boiler characteristics. Optimal $NO_X$ control for a particular system can be achieved provided that the overfire air jets mix effectively, that good control over the main flame can be maintained as combustion air is diverted to the overfire air system, and that there is sufficient residence time in the furnace for carbon burnout to occur. As with low-$NO_X$ burners, the $NO_X$ control performance of overfire air is limited by acceptable carbon-in-ash. For pulverized coal-fired industrial boilers, $NO_X$ reductions between 15% and 30% are generally expected with the application of overfire air.

## *Reburning*

Reburning is a combustion modification technology that removes $NO_X$ from combustion products by using fuel as the reducing agent [131]. As illustrated in

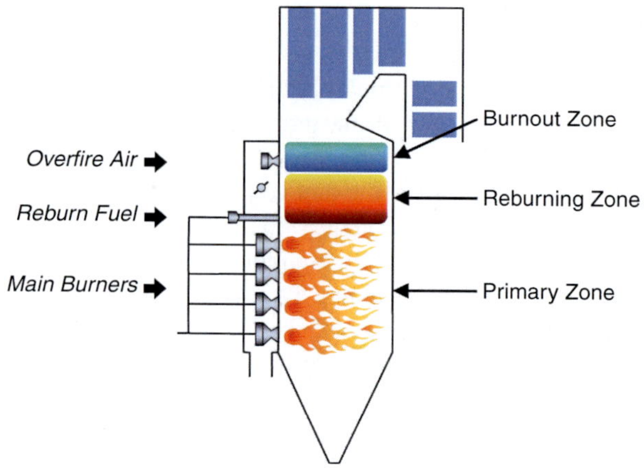


**Fig. 14.9** Application of reburning to a utility boiler

Fig. 14.9, application of reburning to a coal-fired boiler conceptually divides the furnace into three zones (primary zone, reburning zone, and burnout zone) which are typically defined as follows:

*Primary Zone*: The primary zone consists of the normal firing system and constitutes the bulk (typically 80–90%) of the total heat release. This zone is operated fuel lean at an excess air level which is close or slightly below normal operation. $NO_X$ formed in this zone enters the reburning zone.

*Reburning Zone*: The reburning fuel is injected downstream of the primary zone to create a slightly fuel-rich, $NO_X$ reduction zone. The reburning fuel provides the remainder, normally 10–20%, of the total heat input. In this zone, hydrocarbon radicals generated during breakdown of the reburning fuel react with NO molecules to form other reactive nitrogenous species, such as HCN, which react with other NO molecules to form molecular nitrogen, $N_2$.

*Burnout Zone*: In this final zone, combustion air is added to complete the oxidation of the reburning fuel and to bring the boiler back to a normal operating excess air level.

Studies of the reburning process have shown that natural gas, fuel oil, and coal can be used as reburning fuels [132]. As shown in Fig. 14.10, hydrocarbon fuels with low fuel-bound nitrogen, such as propane or natural gas, provide the lowest $NO_X$ emissions since nitrogen in the reburning fuel tends to contribute to the final $NO_X$ exiting the process. In Fig. 14.10, the reburning zone stoichiometric ratio is defined as the ratio of the total air supplied to the primary and reburning zones to the total stoichiometric air requirements of the primary and reburning fuels. The parameters which control the performance of the reburning process have been defined through an extensive series of experimental studies [133]. For boiler applications, the most critical parameters are the reburning zone stoichiometric ratio and the initial $NO_X$ level. As shown in Fig. 14.10, most fuels exhibit a maximum $NO_X$ reduction (i.e., minimum $NO_X$ emissions) at a reburning stoichiometric ratio of about 0.9, which corresponds to a reburning fuel input of approximately 18–19% of the total heat

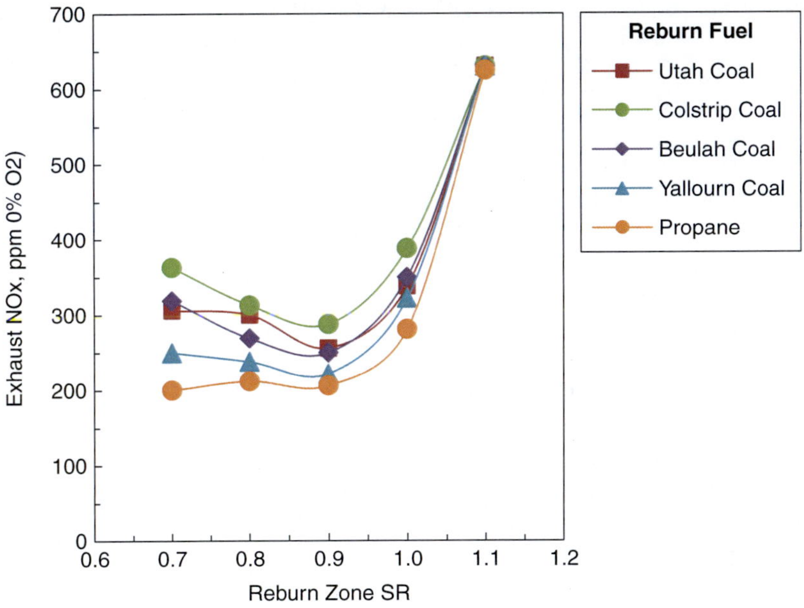


**Fig. 14.10** Impacts of reburning zone stoichiometry and fuel on reburning performance

input at a primary zone excess air level of 10%. $NO_X$ reduction with reburning increases as the $NO_X$ level from the primary zone increases.

Similar to other combustion modification techniques, the most significant factors which limit the performance of reburning in full-scale applications are the design of the reburning system and the boiler characteristics. The boiler geometry must permit adequate mixing of the reburning fuel and overfire air to be achieved and must provide adequate residence time for the process to be implemented above the primary combustion zone. Reliable methodologies for scaling the reburning process to boilers with varying characteristics have been developed [134]. The use of these methodologies has led to $NO_X$ control levels between 50% and 70% for application of reburning to full-scale utility boilers, independent of the firing configuration.

## *Selective Noncatalytic Reduction*

Selective noncatalytic reduction treatment process in which an amine-containing agent, such as ammonia ($NH_3$) or urea ($CO(NH_2)_2$), is injected into combustion gases to react with and reduce NO formed during the combustion process [135, 136]. At the proper temperature, $NH_2$, generated from decomposition of the injected reagent, reacts with NO via a complex set of reactions to form $N_2$. The chemistry of the SNCR process is illustrated in the following global reactions that are based upon the use of ammonia as the reagent:

$$NH_3 + OH \rightarrow NH_2 + H_2O \tag{14.6}$$

$$NH_2 + NO \rightarrow H + OH + N_2 \tag{14.7}$$

$$H + OH \rightarrow H_2O \tag{14.8}$$

Initially, the injected ammonia is activated by OH to form a reactive amine, $NH_2$, radical. This radical then selectivity reacts with NO to form molecular nitrogen, $N_2$, and H and OH radicals. The OH radical can help to activate additional ammonia or can react with H radicals to form water, $H_2O$.

Figure 14.11 shows the impacts of gas temperature on the $NO_X$ reduction achieved with urea or ammonia in tests performed in a pilot-scale combustion facility. As shown in this figure, at the proper flue gas temperature (approximately 950–1,000°C), the injected reagent selectively reduces NO to molecular nitrogen, $N_2$. As temperatures increase, $NO_X$ reduction performance is reduced and a portion of the reagent can be oxidized to NO. As temperatures decrease, a portion of the reagent exits the process unconverted. The unconverted reagent exiting the process is referred to as ammonia slip.

The performance of SNCR on utility boilers depends upon the reagent used, the quantity of reagent injected, and the boiler design and operating characteristics [137]. For maximum performance, the injection system must be designed to provide effective mixing of the reagent within the optimal temperature window. In addition, the boiler design should provide adequate residence time at the proper temperatures. For pulverized coal-fired boilers, the optimal temperature window generally resides within the convective heat transfer sections, where access is limited and quench rates are high [138]. Wide variations in temperature at the point of injection due to stratification or to changes in load can also limit performance. Multiple injection locations can help to reduce the negative impacts of temperature variations. Due to concerns over the handling and storage of anhydrous ammonia, aqueous ammonia or urea is generally preferred for application of SNCR to utility boilers.

Application of SNCR to coal-fired boilers is expected to reduce $NO_X$ emissions by 30–60% from uncontrolled levels. The higher levels of reduction are generally achieved in boilers with slower quench rates (temperature decay versus residence time) in the upper furnace and with good access to the optimal temperature window.

## *Selective Catalytic Reduction*

Selective catalytic reduction is a high-efficiency flue gas treatment process in which ammonia is added to the flue gas to react with NO over a catalyst [139]. The primary overall reactions for this process can be expressed as:

$$4NO + 4NH_3 + O_2 \xrightarrow{\text{Catalyst}} 4N_2 + 6H_2O \tag{14.9}$$

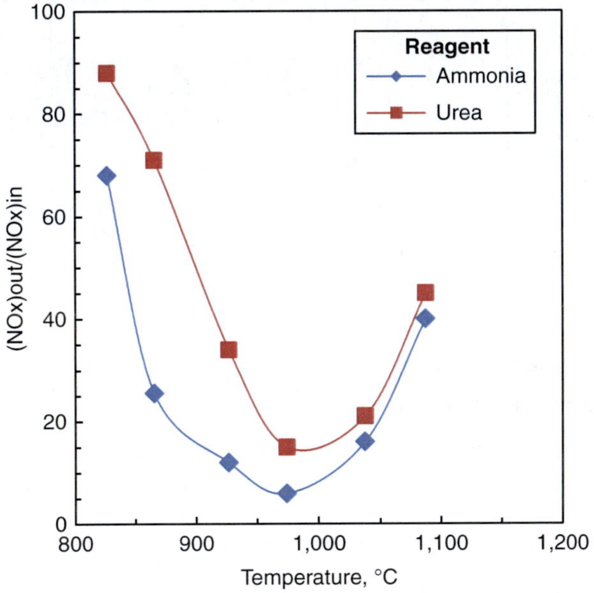


**Fig. 14.11** Impacts of gas temperature in SNCR performance

$$2NO_2 + 4NH_3 \xrightarrow{\text{Catalyst}} 3N_2 + 6H_2O \tag{14.10}$$

SCR catalysts are made of a solid ceramic material that contains active catalytic components. The ceramic may be in the form of a honeycomb monolith or it may be wash coated onto a ceramic, metal, or fiber substrate [140]. The typical operating temperature range for available catalysts is shown in Fig. 14.12. Noble metal catalysts, such as platinum-palladium formulations, function at lower temperatures (150–290°C; 300–550°F) than base metal and zeolite catalysts but are susceptible to deactivation by sulfur oxides. Base metal catalysts, such as vanadium-titanium formulations, function at temperatures in the range of 260–490°C (500–900°F) and are more resistant to sulfur deactivation. Zeolite catalysts operate at a temperature range above that of the other catalysts (340–590°C, 650–1,100°F) and have shown a good resistance to poisoning from trace elements in coal such as arsenic. The majority of SCR technology applications to pulverized coal-fired boilers use vanadia-titania catalysts as they are widely available and have good resistance to deactivation by sulfur oxides in the flue gas. Catalyst formulations are specific to the catalyst manufacturer and may include the addition of other base metal oxides to increase the catalyst activity, to increase poison resistance, to minimize ammonia slip, and to reduce $SO_2$ oxidation to $SO_3$.

To meet the temperature requirements for vanadia-titania catalysts, the catalyst needs to be installed in the flue gas path between the boiler economizer and air preheater. The arrangement for both new and retrofit SCR systems depends upon

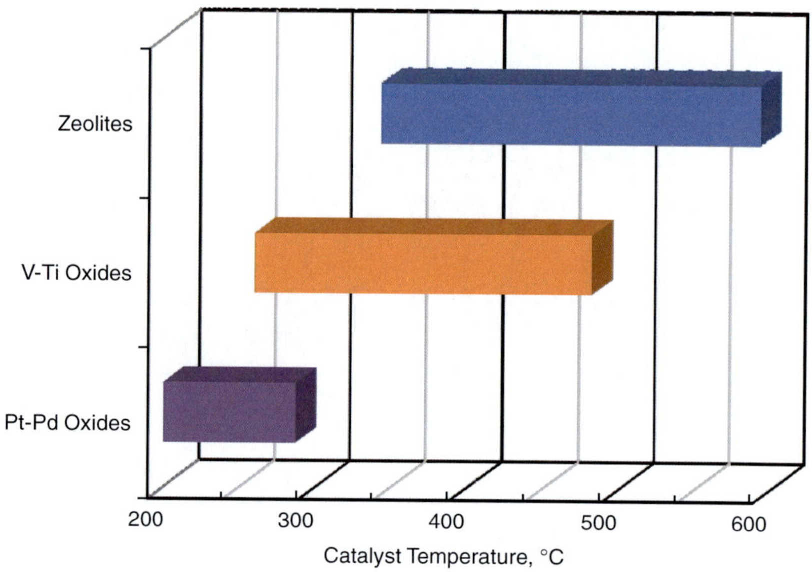


**Fig. 14.12** Operating range for various SCR catalyst formulations

the boiler and balance of plant equipment arrangement [141, 142]. Typically, one or two layers of catalyst are incorporated into a catalyst housing or reactor. Flue gas from the economizer exit is either routed in a horizontal duct to the top of the reactor where it then flows downward over the catalyst. Alternatively, the ductwork arrangement may need a vertical upward duct followed by a 180° transition into the top of the reactor. Ammonia is injected into the flue gas upstream of the SCR reactor using an ammonia injection grid that typically consists of a series of injection lances spaced across the duct and designed to provide good mixing of the ammonia with the flue gas. Static mixers and flow conditioning devices such as turning vanes are used to ensure uniform distribution of the ammonia and flue gas within the ductwork.

A number of factors need to be considered in designing an SCR system, including reactor location and design, catalyst type and configuration, and the location and design of the ammonia injection grid. The catalyst space velocity and uniformity of ammonia and flue gas distribution all influence system performance. Coal properties are an important consideration for catalyst selection [143]. In general, since the SCR system is separate from the combustion process and since the extensive retrofit requirements permit the design to be optimized, the overall control performance is less influenced by boiler characteristics than the other $NO_X$ control technologies described previously. On the other hand, the cost of installing an SCR system is very sensitive to site-specific factors due to the extensive retrofit requirements [144]. $NO_X$ reduction in the range of 70–90% is expected for application of SCR to coal-fired boilers.

## Future Directions

Significant steps have been taken to reduce air pollutant emissions from coal-fired power plants throughout the developed countries in the world. Promulgation of more stringent emission limits in various counties will require further application of these technologies to smaller power plants and place constraints on new power plants. The cost of the air pollution control technologies is a challenge with both retrofit and new technologies and, hence, research and technology development is focused on reducing both the capital and operating costs of these technologies, on improving performance to meet stricter standards, and on improving equipment reliability and availability. In addition, due to concerns over emissions of air toxic compounds from coal-fired power plants, the potential positive and negative impacts of existing air pollution control equipment on these emissions is an area of active investigation.

## Bibliography


1. International Energy Agency (2010) World energy outlook 2010. International Energy Agency, Paris
2. World Coal Insitute (2009) The coal resource, a comprehensive overview of coal. World Coal Insitute, London
3. Freese B (2003) Coal: a human history. Perseus, Cambridge, MA
4. Stuart R (1824) A descriptive history of the steam engine. John Knight and Henry Lacey, London
5. Burn RS (1854) The steam engine. H. Ingram, London
6. Thurston RH (1903) A history of the growth of the steam engine, 4th edn. D. Appelton, New York, Revised
7. Tagliaferro L (2003) Thomas Edison: inventor of the age of electricity. Lerner, Minneapolis
8. Woodside M (2007) Thomas A. Edison: the man who lit up the world. Sterling, New York
9. Babcock & Wilcox Company (1992) Steam: its generation and use, Fortieth edition. Babcock & Wilcox, Barberton
10. Combustion Engineering, Inc (1981) Combustion: fossil power systems, 3rd edn. Combustion Engineering, Windsor
11. Heidorn KC (1978) A chronology of important events in the history of air pollution meteorology to 1970. Bulletin of the American Meteorological Society, vol 59, issue 12. American Meteorological Society, Toronto, pp 1589–1597
12. Thorshiem P (2006) Inventing pollution: coal, smoke, and culture in Britain since 1800. Ohio University Press, Athens
13. Jacobs C, Kelly WJ (2008) Smogtown: the lung-burning history of pollution in Los Angeles. The Overlook Press, New York
14. Brimblecombe P (1987) The big smoke: a history of air pollution in London since medieval times. Methuen, London
15. Greater London Authority (2002) 50 years on: the struggle for air quality in London since the great smog of December 1952. Greater London Authority, London
16. Davis DL (2002) When smoke ran like water: tales of environmental deception and the battle against pollution. Basic Books, New York

17. Reitze AW (2005) Stationary source air pollution law. Environmental Law Institute, Washington, DC
18. Martineau RJ, Novello DP (2004) The clean air act handbook. American Bar Association, Chicago
19. U.S. Environmental Protection Agency (2005) Review of the National Ambient Air Quality Standards for particulate matter. Policy assessment of scientific and technical information. OAQPS staff paper. EPA-425/R-05-005a. Office of Air Quality Planning and Standards, Research Triangle Park
20. U.S. Environmental Protection Agency (2009) Integrated science assessment for particulate matter. Office of Research and Development, Research Triangle Park. EPA/600/R-08/139F
21. U.S. Environmental Protection Agency (2004) Air quality criteria for particulate matter. EPA/600/P-99/002aF. Office of Research and Development, Research Triangle Park
22. U.S. Environmental Protection Agency (2008) Integrated science assessment for sulfur oxides – health criteria. EPA/600/R-08/047F. Office of Research and Development, Research Triangle Park
23. U.S. Environmental Protection Agency (2008) Integrated science assessment for oxides of nitrogen – health criteria. EPA/600/R-08/071. Office of Research and Development, Research Triangle Park
24. Glass NR (1979) Environmental effects of increased coal utilization: ecological effects of gaseous emissions from coal combustion, vol 33. U.S. National Institute of Environmental Health Sciences, Research Triangle Park, pp 249–272
25. Bennett DA, Goble RL, Linhurst RA (1985) The acidic deposition phenomenon and its effects: critical assessment document. EPA/600/8-85/001. U.S. Environmental Protection Agency, Office of Research and Development, Research Triangle Park
26. National Acid Precipitation Assessment Program (2005) National Acid Precipitation Assessment Program report to Congress: an integrated assessment. National Acid Precipitation Assessment Program
27. U.S. Environmental Protection Agency (1997) Nitrogen oxides: impacts on Public Health and the Environment. EPA-452-R-97-002. Office of Air and Radiation, Research Triangle Park
28. U.S. Environmental Protection Agency (1999) Nitrogen oxides (NOx), why and how they are controlled. EPA-456/F-99-006R. Office of Air Quality Planning and Standards, Research Triangle Park
29. McConville A (1997) Emissions standards handbook. IEACR/96. IEA Coal Research, London
30. U.S. Environmental Protection Agency (2000) Air quality criteria for carbon monoxide. EPA 600/P-99/001F. Office of Research and Development, Washington, DC
31. U.S. Environmental Protection Agency (1998) Study of hazardous air pollutant emissions from Electric Utility Steam Generating Units. Final report to congress, vols 1 and 2. EPA-453/R-98-004a, b. Office of Air Quality Planning and Standards, Research Triangle Park
32. U.S. Environmental Protection Agency (1997) Mercury study report to congress, vols I–VII. EPA-452/R-97-003, 010. Office of Air Quality Planning and Standards, Research Triangle Park
33. U.S. Environmental Protection Agency (2000) Regulatory finding on the emissions of hazardous air pollutants from Electric Utility Steam Generating Units, vol 65(245). Federal Register, Research Triangle Park, pp 79825–79831
34. National Research Council (2000) Toxicological effects of methylmercury. National Academy Press, Washington, DC
35. European Commission (2001) Ambient air pollution by mercury (Hg). Position paper. Office for Official Publications of the European Communities, Luxembourg
36. Environment Canada (2010) Risk management strategy for mercury
37. UNEP Chemicals Branch, DTIE (2010) The global atmospheric mercury assessment: sources, emissions and transport. United Nations Environment Programme, Geneva

38. US Environmental Protection Agency (2011) National Emission Standards for Hazardous Air Pollutants from Coal and Oil-Fired Electric Utility Steam Generating Units and Standards of Performance for Fossil Fuel-Fired Electric Utility, Industrial-Commercial- Institutional, and Small Industrial-Commercial-Institutional Steam Generating Units, vol 76(85). Federal Register, Research Triangle Park, pp 24976–25147
39. Mitchell JFB (1989) The "Greenhouse" effect and climate change, vol 27, No 1. American Geophysical Union, Washington, DC, pp 115–139
40. U.S. Environmental Protection Agency (1995) Compliation of air pollutant emissions factors, 5th edn. AP-42. Office of Air Quality Planning and Standards, Research Triangle Park
41. Eggleston HS, Buendia L, Miwa K, Ngara T, Tanabe K (eds) (2006) 2006 IPCG guidelines for National Greenhouse Gas Inventories. National Greenhouse Gas Inventories Programme. Intergovernmental Panel on Climate Change, Hayama, Kanagawa
42. Ciferno JD, Fout TE, Jones AP, Murphy JT (2009) Capture carbon from existing coal-fired power plants, Chemical Engineering Progress. American Instutite of Chemical Engineers, New York, pp 33–41
43. IRA Greenhouse R&D Programme (2007) $CO_2$ capture ready plants. Report No. 2007/4. International Energy Agency, Cheltenham
44. Newton GH, Schieber C, Socha RG, Kramlich JC (1990) Mechanisms governing fine particulate emissions from coal flames. DOE/PC/73743-T8. U.S. Department of Energy, Pittsburgh
45. Gluskoter HJ (1978) An introduction to the occurrence of mineral matter in Coal. [ed.] Richard W. Bryers. Ash deposits and corrosion due to impurities in combustion gases. Hemisphere Publishing Department, Washington, DC, pp 3–19
46. Chen Y, Shah N, Huggins FE, Huffman GP, Linak WP, Andrew Miller C (2004) Investigation of primary fine particulate matter from coal combustion by computer-controlled scanning electron microscopy, vol 85. Fuel Processing Technology, Elsevier, Amsterdam, pp 743–761
47. Folsom BA, Heap MP, Pohl JH (1986) State-of-the-art review summary and program plan. Effects of coal quality on power plant performance and costs, vol 1. Report No. CS-4283. Electric Power Research Institute, Palo Alto
48. Miller CA, Linak WP (2002) Primary particles generated by the combustion of heavy fuel oil and coal. Review of research results from EPA's National Risk Management Research Laboratory. EPA-600/R-02-093. U.S. Environmental Protection Agency, Office of Research and Development, Research Triangle Park
49. Flagan RC, Seinfeld JH (1988) Fundamentals of air pollution engineering. Prentice Hall, Englewood Cliffs
50. Flagan RC (1978) Submicron particles from coal combustion. In: Proceedings of the seventeenth symposium (international) on combustion. The Combustion Institute, Pittsburgh, pp 97–104
51. Kramlich JC, Chenevert B, Park J, Hoffman DA, Butcher EK (1996) Suppression of fine ash formation in pulverized coal flames. Final Technical Report prepared for DOE Grant No. DE-FG22-92PC92548. U.S. Department of Energy, Pittsburgh Energy Technology Center, Pittsburgh
52. Acurex Environmental Corporation (1993) Emission factor documentation for AP-42 Section 1.1 bituminous and subbituminous coal combustion. U.S. Environmental Protection Agency, Office of Air Quality Planning and Standards, Research Triangle Park
53. Attar A, Dupuis F (1979) Data on the distribution of organic sulfur functional groups in coal. Preprints of the Papers of the Spring National Meeting of the Division of Fuel Chemistry, vol 24(1). American Chemical Society, Honolulu, pp 166–177
54. Pershing DW, Silcox GD (1988) $SO_x$ fundamentals. Combustion of solid fuels – a one week intensive course. International Flame Research Foundation, Noordwijkerhout

55. Davis K, Dissel A, Valentine J (2001) The evolution of pyritic and organic sulfur from pulverized coal particles during combustion. The 2nd joint meeting of the US Sections of The Combustion Institute. The Combustion Institute, Oakland
56. Muller III CH, Schofield K, Steinberg M, Broida HP (1978) Sulfur chemistry in flames. In: Proceedings of the seventeenth symposium (international) on combustion. The Combustion Institute, Pittsburgh, pp 867–879
57. Kramlich JC, Malte PC, Grosshandler WL (1981) The reaction of fuel-sulfur in hydrocarbon combustion. In: Proceedings of the eighteenth symposium (international) on combustion. The Combustion Institute, Pittsburgh, pp 151–161
58. Shao D, Hutchinson EJ, Heidbrink J, Pan W-P, Chou C-L (1994) Behavior of sulfur during coal pyrolysis. J Anal Appl Pyrolysis 30:91–100. Elsevier, Amsterdam
59. Castaldini C, Angwin M (1977) Boiler design and operating variables affecting uncontrolled sulfur emissions from pulverized-coal-fired steam generators. EPA-450/3-77-047. U.S. Environmental Protection Agency, Research Triangle Park
60. Folkedahl BC, Zygarlicke CJ (2004) Sulfur retention in North Dakota lignite coal ash. Preprints of the papers of the spring national meeting of the Division of Fuel Chemistry, vol 49(1). American Chemical Society, Anaheim, pp 167–168
61. Blythe G, Dombrowski K (2004) SO3 mitigation guide update. Report No. 1004168. Electric Power Research Institute, Palo Alto
62. Srivastava RK, Miller CA, Erickson C, Jambhekar R (2004) Emissions of sulfur trioxide from coal-fired power plants. J Air Waste Manag Assoc 54:750–762. Air and Waste Management Association, Pittsburgh
63. Glassman I (1987) Combustion, 2nd edn. Academic, Orlando
64. Fenimore CP (1971) Formation of nitric oxide in premixed hydrocarbon flames. In: Proceedings of the thirteenth symposium (international) on combustion. The Combustion Institute, Pittsburgh, pp 373–389
65. Pershing DW, Wendt JOL (1977) The influence of flame temperature and coal composition on thermal and fuel $NO_x$. In: Proceedings of the sixteenth symposium (international) on combustion. The Combustion Institute, Pittsburgh, pp 389–399
66. Sarofim AF, Beér JM (1990) The fate of fuel nitrogen and ash during combustion of pulverized coal, Chapter 4. In: Lemieux PM, Air and Energy Engineering Research Laboratory (eds) Pulverized coal combustion: pollutant formation and control, 1970–1980. U.S. Environmental Protection Agency, Air and Energy Engineering Research Laboratory, Research Triangle Park
67. Hayhurst AN, Vince IM (1980) Nitric oxide formation from N2 in flames. Prog Energy Combust Sci 6:35–51. Elsevier, Amsterdam
68. Kelemen SR, Gorbaty ML, Vaughn SN, Kwiatek PJ (1993) Quantification of nitrogen forms in argonne premium coals. Preprints of the papers of the spring national meeting of the Division of Fuel Chemistry, vol 38(2). American Chemical Society, Denver, pp 384–392
69. Mitra-Kirtley S,. Mullins OC, Branthaver J, Van Elp J, Cramer SP (1993) Nitrogen XANES studies of fossil fuels. Preprints of the papers of the fall national meeting of the Division of Fuel Chemistry, vol 38(3). American Chemical Society, Chicago, pp 762–768
70. Solomon PR, Fletcher TH (1994) Impact of coal pyrolysis on combustion. In: Proceedings of the twenty-fifth symposium (international) on combustion. The Combustion Institute, Pittsburgh, pp 463–474
71. Pohl JH, Sarofim AF (1977) Devolatilization and oxidization of coal nitrogen. In: Proceedings of the sixteenth symposium (international) on combustion. The Combustion Institute, Pittsburgh, pp 491–501
72. Rees DP, Smoot LD, Hedman PO (1981) Nitrogen oxide formation inside a laboratory pulverized coal combustor. In: Proceedings of the eighteenth symposium (international) on combustion. The Combustion Institute, Pittsburgh, pp 1305–1311
73. Chen JC, Niksa S (1992) Suppressed nitrogen evolution from coal-derived soot and low-volatility coal chars. In: Proceedings of the twenty-fourth symposium (international) on combustion. The Combustion Institute, Pittsburgh, pp 1269–1276

74. Zhang H (2001) Nitrogen evolution and soot formation during secondary coal pyrolysis. PhD dissertation, Brigham Young University
75. Pohl JH, Chen SL, Heap MP, Pershing DW (1983) Correlation of NOx emissions with basic physical and chemical characteristics of coal. In: Proceedings of the 1982 joint symposium on NOx control, vol 2. Electric Power Research Institute, Palo Alto
76. Pershing DW, Heap MP, Chen SL (1990) Bench-scale experiments on the formation and control of NOx emissions from pulverized coal combustion, Chapter 9. In: Pulverized coal combustion: pollutant formation and control, 1970–1980. U.S. Environmental Protection Agency, Air and Energy Engineering Research Laboratory, Research Triangle Park
77. Chen SL, Heap MP, Pershing DW, Martin GB (1982) Influence of coal composition on the fate of volatile and char nitrogen during combustion. In: Proceedings of the nineteenth symposium (international) on combustion. The Combustion Institute, Pittsburgh, pp 1271–1280
78. Bartok W, Crawford AR, Piegari GJ (1971) Systematic field study of NOx emission control methods for utility boilers. APTD-1163. U.S. Environmental Protection Agency, Office of Air Programs, Research Triangle Park
79. Payne R, Heap MP, Pershing DW (1981) NOx formation and control in pulverized-coal flames. In: Proceedings of the low rank coal technology development workshop, 17–18 June 1981. San Antonio
80. Licht W (1980) Air pollution control engineering, basic calculations for particulate collection. Marcel Dekker, New York
81. Strauss W (1975) Industrial gas cleaning, 2nd edn. Pergamon Press, New York
82. Kinsey JS, Schliesser S, Englehart PJ (1985) Control technology for sources of PM10. Draft report prepared for EPA Contract No. 68-02-03891, work assignment 4. U.S. Environmental Protection Agency, Office of Air Quality Planning and Standards, Research Triangle Park
83. U.S. Environmental Protection Agency (1982) Control techniques for particulate emissions from stationary sources, vol 1. EPA-450/3-81-005a. Emissions Standards and Engineering Division, Research Triangle Park
84. White HJ (1977) Electrostatic precipitation of fly ash. J Air Pollut Control Assoc 27(1):15–21. Air Pollution Control Association, Pittsburgh
85. U.S Environmental Protection Agency (2011) Electrostatic precipitator plan review. APTI Course No. SI:412B. U.S. Environmental Protection Agency, Research Triangle Park
86. White HJ (1977) Electrostatic precipitation of fly ash. Fly ash and furnace gas characteristics. J Air Pollut Control Assoc 27(2):114–120. Air Pollution Control Association, Pittsburgh
87. Turner JH, Lawless PA, Yamamoto T, Coy DW, Mckenna JD, Mycock JC, Nunn AB, Greiner GP, Vatavuk WM (2002) Electrostatic precipitators. In: Mussatti DC (ed) EPA Air pollution control cost manual. U.S. Environmental Protection Agency, Office of Air Quality Planning and Standards, Research Triangle Park. Chapter 3 in Section 6, Particulate matter controls
88. Szabo MF, Shah YM (1979) Inspection manual for evaluation of electrostatic precipitator performance. EPA-340/1-79-007. U.S. Environmental Protection Agency, Division of Stationary Source Enforcement, Washington, DC
89. White HJ (1977) Electrostatic precipitation of fly ash. Precipitator design. J Air Pollut Control Assoc 27(3):206–217. Air Pollution Control Association, Pittsburgh
90. National Air Pollution Control Administration (1969) Control techniques for particulate air pollutants. NAPCA Publication No. AP-51. National Air Pollution Control Administration, Washington, DC
91. Turner JH, Lawless PA, Yamamoto T, Coy DW, Mckenna JD, Mycock JC, Nunn AB, Greiner GP, Vatavuk WM (2002) Baghouses and filters. In: Mussatti DC (ed) EPA air pollution control cost manual. U.S. Environmental Protection Agency, Office of Air Quality Planning and Standards, Research Triangle Park. Chapter 1 in Section 6, Particulate matter controls

92. Cushing KM, Merritt RL, Chang RL (1990) Operating history and current status of fabric filters in the utility industry. J Air Waste Manage Assoc 40(7):1051–1058. Air and Waste Management Association, Pittsburgh
93. Belba VH, Theron Grubb W, Chang R (1992) The potential of pulse-jet baghouses for utility boilers. Part 1: a worldwide survey of users. J Air Waste Manage Assoc 42(2):209–217. Air and Waste Management Association, Pittsburgh
94. Sloat DG, Gaikwad RP, Chang RL (1993) The potential of pulse-jet baghouses for utility boilers. Part 3: comparative economics of pulse-jet baghouse precipitators and reverse-gas baghouses. J Air Waste Manage Assoc 43:120–128. Air and Waste Management Association, Pittsburgh
95. EC/R Incorporated (1998) Stationary source control techniques document for fine particulate matter. Report prepared for EPA Contract No. 68-D-98-026, Work Assignment No. 0-08. U.S. Environmental Protection Agency, Research Triangle Park
96. Felix LG, Cushing KM, Grubb WT, Giovanni DV (1988) Fabric filters for the electric utility industry. In: Guildelines for fabrics and bags, vol 3, CS-5161. Electric Power Research Instutute, Palo Alto
97. Electric Power Research Instutute (2005) Utility boiler baghouse update. Report No. 1010367. Electric Power Research Instutute, Palo Alto
98. Dennis R (1974) Collection efficiency as a function of particle size, shape and density: theory and experience. J Air Pollut Control Assoc 23(12):1156–1161. Air Pollution Control Association, Pittsburgh
99. Carr RC, Smith WB (1974) Fabric filter technology for utility coal-fired power plants. J Air Pollut Control Assoc 34(1):79–89. Air Pollution Control Association, Pittsburgh
100. Carr RC, Cushing KM, Gallaer CA, Smith WB (1992) Fabric filters for the electric utility industry. In: Guidelines for fabric filter design, vol 5, CS-5161. Electric Power Research Institute, Palo Alto
101. Bustard CJ, Cushing KM, Pontius DH, Smith WB (1988) Fabric filters for the electric utility indistry. In: General concepts, vol 1, CS-5161. Electric Power Research Institute, Palo Alto
102. Devitt T, Gestle R, Gibbs L, Hartman S, Klier R (1978) Flue gas desulfurization system capabilities for coal-fired steam generators, vol II. Technical report. U.S. Environmental Protection Agency, Office of Research and Development, Washington, DC, EPA-600/7-78-032b
103. Wojciech J, Singer C, Srivastava RK, Tsirigotis PE (1999) Status of $SO_2$ scrubbing technologies. EPRI-DOE-EPA combined utility air pollution control symposium. TR-113187-V1. Electric Power Research Institute, Palo Alto
104. Srivastava RK (2000) Controlling $SO_2$ emissions: a review of technologies. EPA/600/R-00/093. U.S. Environmental Protection Agency, Office of Research and Development, Washington, DC
105. Kohl A, Riesenfeld F (1985) Gas purification, 4th edn. Gulf, Houston
106. Weiler H, Ellison W (1997) Wet gypsum-yielding FGD experience using quicklime reagent. EPRI-DOE-EPA combined utility air pollutant control symposium. The mega symposium. TR-108683-V2. Electric Power Research Institute, Palo Alto
107. United Engineers and Constructors, Inc (1991) Economic evaluation of flue gas desulfurization systems. GS-7193, vol 1. Electric Power Research Institute, Palo Alto
108. United Engineers and Constructors, Inc (1992) Economic evaluation of flue gas desulfurization systems. GS-7193, vol 2. Electric Power Research Institute, Palo Alto
109. United Engineers and Constructors, Inc (1995) Economic evaluation of flue gas desulfurization systems. GS-7193-V3, vol 3. Electric Power Research Institute, Palo Alto
110. Fox MR, Hunt TG (1990) Flue gas desulfurization using dry sodium injection. Presented at the EPA/EPRI 1990 $SO_2$ control symposium. s.n., New Orleans
111. Radojevic M (1991) Scrubbing of flue gases with sea-water. Presented at the AFRC/JFRC international conference on environmental control of combustion processes, 7–10 October 1991. s.n., Honolulu

112. Zhou W, Maly P, Brooks J, Nareddy S, Swanson L, Moyeda D (2010) Design and test furnace sorbent injection for $SO_2$ removal in a tangentially fired boiler. Environmental Engineering Science, vol 27, Number 4, Mary Ann Liebert, New Rochelle, pp 337–345
113. Nolan PS (1996) Emission control technologies for coal-fired power plants. Presented at the People's Republic of China Ministry of Electric Power seminar, 22–25 April 1996, s.n., Beijing
114. Henzel DS, Laseke BA, Smith EO, Swenson DO (1982) Handbook for flue gas desulfurization scrubbing with limestone. Noyes Data Corporation, Park Ridge
115. Miller SF, Miller BG, Scaroni AW (1997) Limestone performance in a pilot-scale forced oxidation scrubber. EPRI-DOE-EPA combined utility air pollutant control symposium. The mega symposium. TR-108683-V2, Electric Power Research Institute, Palo Alto
116. Brogen C, Klingspor JS (1997) Impact of limestone grind size on WFGD performance, August. EPRI-DOE-EPA combined utility air pollutant control symposium. The mega symposium. TR-108683-V2. Electric Power Research Institute, Palo Alto
117. Blythe G, Horton B, Rhudy R (1999) EPRI FGD operating and maintenance cost survey. EPRI-DOE-EPA combined utility air pollution control symposium. TR-113187-V1, Electric Power Research Institute, Palo Alto
118. Masters K (1985) Spray drying handbook, 4th edn. Wiley, New York
119. Burnett TA, Anderson KD (1981) Technical review of dry FGD systems and economic evaluation of spray dryer FGD systems. EPA-600/7-81-014. U.S. Environmental Protection Agency, Office of Research and Development, Washington, DC
120. Lachapelle DG, Brown JS, Stern RD (1974) Overview of environmental protection agency's NOx control technology for stationary combustion sources. Presented at the 67th annual meeting American Insitute Of Chemical Engineers
121. Srivastava RK, Hall RE, Khan S, Culligan K, Lani BW (2005) Nitrogen oxides emission control options for coal-fired electric utility boilers. J Air Waste Manage Assoc 55:1367–1388. Air and Waste Management Association, Pittsburgh
122. Hjalmarsson A-K (1990) NOx control technologies for coal combustion. IEACR/24. IEA Coal Research, London
123. U.S. Environmental Protection Agency (1992) Evaluation and costing of NOx controls for existing utility boilers in the NESCAUM region. EPA-453/R-92-010. Office of Air Quality Planning and Standards, Research Triangle Park
124. U.S. Environmental Protection Agency (1994) Alternative control technologies document: NOx emissions from utility boilers. EPA-453/R-94-023. Office of Air Quality Planning and Standards, Research Triangle Park
125. Heap MP, Folsom BA (1990) Optimization of burner/combustion chamber design to minimize NOx formation during pulverized coal combustion, Chapter 10. In: Pulverized coal combustion: pollutant formation and control, 1970–1980. U.S. Environmental Protection Agency, Air and Energy Engineering Research Laboratory, Research Triangle Park
126. Sommer TM, Jensen AD, Melick TA, Orban PC, Christensen MS (1993) Applying European low NOx burner technology to U.S. installations. Presented at the 1993 EPRI/EPA joint symposium on stationary combustion NOx control, 24–27 May 1993. EPRI/EPA, Miami Beach
127. LaRue AD (1989) The XCL burner – latest developments and operating experience. Presented at the 1989 EPRI/EPA joint symposium on stationary combustion NOx control, 6–9 Mar 1989. EPRI/EPA, San Francisco
128. Vatsky J, Sweeney TW (1991) Development of an ultra-low NOx pulverizer coal burner. Presented at the 1991 EPRI/EPA joint symposium on stationary combustion NOx control, 25–28 Mar 1991. EPRI/EPA, Washington, DC
129. U.S. Department of Energy (1996) Reducing emissions of nitrogen oxides via low-NOx burner technologies. Topical report number 5. U.S. Department of Energy, Pittsburgh
130. Campbell LM, Stone DK, Shareef GS (1991) Sourcebook: NOx control technology data. U.S. Environmental Protection Agency, Research Triangle Park

131. Wendt JOL, Sternling CV, Matovich MA (1973) Reduction of sulfur trioxide and nitrogen oxides by secondary fuel injection. In: Proceedings of the fourteenth symposium (international) on combustion. The Combustion Institute, Pittsburgh, pp 897–904
132. U.S. Environmental Protection Agency (1996) Control of NOx emissions by reburning. EPA/625/R-96/001. U.S. Environmental Protection Agency, Cincinnati
133. Chen SL, McCarthy JM, Clark WC, Heap MP, Seeker WR, Pershing DW (1986) Bench and pilot scale process evaluation of reburning for in-furnace NOx reduction. In: Proceedings of the twenty-first symposium (international) on combustion. The Combustion Institute, Pittsburgh, pp 1159–1169
134. Payne R, Moyeda DK (1994) Scale up and modeling of gas reburning. In: Moussa A, Presser C, Rini MJ, Weber R, Woodward G, Gupta AK (eds) Combustion modeling, scaling and air toxins. ASME FACT, vol 18. American Society of Mechanical Engineers, New York, pp 115–122
135. Lyon RK (1987) Thermal DeNOx, controlling nitrogen oxides emissions by a noncatalytic process. Environ Sci Technol 21(3):231–236. American Chemical Society, Washington, DC
136. Jødal M, Nielsen C, Hulgaard T, Dam-Johansen K (1990) Pilot-scale experiments with ammonia and urea as reductants in selective non-catalytic reduction of nitric oxide. In: Proceedings of the twenty-third symposium (international) on combustion. The Combustion Institute, Pittsburgh, pp 237–243
137. Muzio L, Quartucy G (1993) State-of-the-art assessment of SNCR technology. Topical Report No. TR-102414. Electric Power Research Institute, Palto Alto
138. Berg M, Bering H, Payne R (1993) NOx reduction by urea injection in a coal fired utility boiler. Presented at the 1993 EPRI/EPA joint symposium on stationary combustion NOx control, 24–27 May 1993. EPRI/EPA, Miami Beach
139. U.S. Department of Energy (2005) Selective catalytic reduction (SCR) technology for the control of nitrogen oxide emissions from coal-fired boilers. Topical Report Number 23. U.S. Department of Energy, Pittsburgh
140. Pereira CJ, Amiridis MD (1995) NOx control from stationary sources: overview of regulations, technology, and research frontiers, Chapter 1. In: Umit S, Agarwal SK, Marcelin Ozkan G (eds) Reduction of nitrogen oxide emissions. American Chemical Society, Washington, DC, pp 1–13
141. Nischt W, Woolridge B, Bigalbal J (1999) Recent SCR retrofit experience in coal-fired boilers. Presented at POWER-GEN international, 30 Nov–2 Dec 1999. New Orleans
142. Tonn DP, Uysal TA (1998) 220 MW SCR installation on new coal-fired project. Presented at the Institute of Clean Air Companies ICAC forum '98, 18–29 Mar 1998, Durham
143. Khan S, Shroff G, Tarpara J, Srivastava R (1997) SCR applications: addressing coal chacteristics concerns. EPRI-DOE-EPA combined utility air pollutant control symposium – the mega sumposium. Technical report TR-108683-V1. Electric Power Research Institute, Palo Alto
144. Foerter D, Jozewicz W (2001) Cost of selective catalytic reduction (SCR) application for nox control on coal-fired boilers. EPA/600/R-01/087. U.S. Environmental Protection Agency, Office of Research and Development, Washington, DC

# Chapter 15
# Natural Gas Power

**Raub W. Smith and S. Can Gülen**

## Glossary

| | |
|---|---|
| Brayton cycle | The thermodynamic cycle describing the operation of a gas turbine. In a combined cycle, it is the *topping* cycle due to its relative position vis-à-vis Rankine cycle on a temperature–entropy surface. |
| Carnot cycle | Also known as the Carnot engine, it is the embodiment of the second law of thermodynamics in the form of a theoretical cycle comprising two isentropic and two isothermal processes. No heat engine operating in a thermodynamic cycle can be more efficient than the corresponding Carnot engine defined by the constant mean-effective heat addition and heat rejection temperatures. |
| Cogeneration | See combined heat and power (CHP). |
| Combined cycle power plant | A fossil-fired power plant that combines two types of prime movers, usually one or more gas turbines and one or more steam turbines (STs), whose operation are governed by their respective |

This chapter was originally published as part of the Encyclopedia of Sustainability Science and Technology edited by Robert A. Meyers. DOI:10.1007/978-1-4419-0851-3

R.W. Smith (✉) • S.C. Gülen
GE Infrastructure-Energy, 1 River Road, Bldg 40, 4th Floor, Rm 412, Schenectady, NY 12345, USA
e-mail: raub.smith@ge.com; can.gulen@ge.com

R. Malhotra (ed.), *Fossil Energy: Selected Entries from the Encyclopedia of Sustainability Science and Technology*, DOI 10.1007/978-1-4614-5722-0_15,

| | |
|---|---|
| | thermodynamic cycles, i.e., Brayton and Rankine. |
| Combined heat and power (CHP) | The term used for fossil-fired power plants, which, in addition to their primary product, electric power delivered to the grid, also supply a secondary product in terms of useful thermal energy. |
| Combustor | A mechanical device to facilitate controlled mixing and reaction of an oxidizer (in almost all cases air) and a fuel (in almost all cases a pure hydrocarbon or a mixture thereof in gaseous or liquid phase) to generate high-temperature gaseous product for expansion in a turbine and useful shaft work generation. |
| Compressor | A mechanical device that increases the pressure of a gas by reducing its volume. There are different types of compressors, e.g., axial, radial, and reciprocating, which are suitable to different types of operating regimes. |
| Efficiency | Unless specified otherwise, the thermal efficiency of a power-generating system, which is the dimensionless ratio of generated kWh of electricity to the amount of energy required to generate it. It is the inverse of the heat rate with a suitable conversion factor. |
| Emissions | Gases and solid particles (usually undesirable) released into the air as by-products of a combustion process (e.g., in the boiler of a fossil-fired power plant, gas turbine combustor, or other internal combustion engine) to create electric power or propel a vehicle. |
| Firing temperature | The temperature of the gas turbine combustor exhaust gas at the inlet to the first stage rotor, which is the starting point of useful shaft work generation. |
| Gas turbine | A prime mover or internal combustion engine comprising a compressor, combustor, and an expander connected via |

a common shaft, through which air is compressed, burned, and expanded to generate useful shaft work for electric power generation (or thrust in an aircraft jet engine).

Generator — A device that converts the mechanical shaft power generated by a prime mover into electrical power.

Global warming — The apparent increase in the average temperature of the earth's near-surface air and oceans since the mid-twentieth century and its projected continuation (per *Wikipedia*).

Greenhouse effect — The containment of heat from solar radiation striking the earth's surface due to the earth's atmospheric "greenhouse" gases such as carbon dioxide and methane. These gases absorb and emit radiation within the thermal infrared range and are believed to be a primary cause of global warming.

Heat rate — Amount of energy required to generate 1 kWh of electricity. It is the inverse of the thermal efficiency with a suitable conversion factor.

Heating value — The thermal energy produced by completely burning a unit mass of fuel in a combustor to produce carbon dioxide and water. If the water is in a gaseous phase, the heating value is referred to as net or lower heating value (LHV). If the water is in a liquid phase, the heating value is referred to as gross or higher heating value (HHV).

Heat recovery steam generator (HRSG) — Also known as the heat recovery boiler (HRB), HRSG is a cross-flow tubular heat exchanger that recovers the exhaust heat from a prime mover (e.g., a gas turbine) and produces steam at high pressure and temperature that is used in a steam turbine (ST) for additional power generation. HRSG is the key equipment that "combines" gas and

| | |
|---|---|
| | steam turbines in a combined cycle power plant. |
| Rankine cycle | The thermodynamic cycle describing the operation of a steam turbine. In a combined cycle, it is the *bottoming* cycle due to its relative position vis-à-vis Brayton cycle on a temperature-entropy surface. |
| Steam turbine | A prime mover or the power-generating part of an external combustion engine comprising one or more sections connected via a common shaft, through which steam flows, expands, and discharges to a condenser to generate useful shaft work for electric power generation or propulsion. |

## Defining the Subject

Natural gas is an important fossil fuel that has played an increasingly significant role in worldwide electric power generation since the 1980s. The key driver underlying the importance of natural gas as a vital enabler of modern living has been its relative advantage vis-à-vis other fossil fuels in terms of emissions and pollutants. In comparison to coal, the primary fossil fuel used for electric power generation in the world on a constant consumption basis, natural gas emits nearly 45% less $CO_2$, 80% less nitrogen oxides ($NO_x$) with negligible amounts of sulfur oxides, particulates, and mercury.

The environmental advantages of natural gas are further amplified by the significantly higher thermal efficiency of the power plants that burn it for electric power generation in comparison to other variants of oil or coal. Natural gas burning modern gas turbines can readily reach efficiencies of 56–57% in combined cycle configurations. This is well above the average efficiency of existing coal burning steam turbine plants (e.g., low 30s) as well as the most advanced ultra-supercritical designs thereof (e.g., low 40s).

With the beginning of the twenty-first century, economic projections pointed to natural gas as the fastest growing fuel for worldwide electric power generation, with an expected average annual rate of growth of about 3.7% between 2005 and 2030. During this period, natural gas will account for almost one-fourth of the world's total net electricity generation [1]. While such long-term forecasts are highly susceptible to uncertainties such as the impact of potential governmental policies or legislation limiting the use of fossil fuels in favor of nuclear and renewable sources of energy (largely driven by concerns over global warming), it is almost a foregone conclusion that natural gas will continue to be an important part of the world's energy and electric power generation portfolio in the foreseeable future.

Given this impact of natural gas on future electrical production, it is important to establish a basic understanding of the history, current state of the art and future possibilities of power generation using natural gas combustion in order to have a complete picture of future energy sustainability.

## Introduction

Natural gas is a fossil fuel comprising hydrocarbons, primarily methane ($CH_4$) up to 90% on a volumetric basis. It differs from the other fossil fuels in that it is naturally available as a gas (hence the name) as opposed to a solid (i.e., coal) or liquid (i.e., oil). Like other fossil fuels, the basic mechanism (thermogenic) of natural gas formation involves the extremely prolonged high-pressure compression of organic matter remains (e.g., plants, animals that lived millions of years ago on earth). The gaseous phase is the result of extremely high temperatures associated with the depth of the organic matter, typically 1–2 miles below the earth's crust. At even deeper levels, natural gas consists almost entirely of pure methane.

Another (biogenic) mechanism for forming natural gas is the chemical transformation of organic matter by microorganisms. The resulting gas contains methane and carbon dioxide. A well-known example of this mechanism is the landfill gas that results from the decomposition of waste materials deposited into large landfills. Other biogas formation mechanisms are sludge digestion in the tanks of sewage treatment plants (sewage gas) and anaerobic fermentation of agricultural waste. Biogas formation takes place near the earth's surface in the absence of oxygen and the resulting gas usually leaks to atmosphere. Controlled capture of landfill gas and other biogases for utilization in electric power generation is a promising technology. (Uncontrolled energy release (explosion) from the trapped landfill gas leading to death and destruction in the poverty-stricken areas of the world is an unfortunate event that appears in the news from time to time.) Biogas utilization is also important because the main constituent, methane, itself is a principal greenhouse gas (GHG). In fact, methane is nearly 20 times more effective than $CO_2$ in trapping heat emanating from the earth's surface. Therefore, combustion of methane (with resultant $CO_2$ emissions) that otherwise would leak into the atmosphere is a preferable trade-off from GHG reduction point of view.

Coal mine gas is another type of natural gas that is released during pit coal mining. Depending on the time of capture and method, there are different types. The most common types are coal bed methane (CBM) and coal mine methane (CMM). The former is largely pure methane and its production is independent of coal mining. The latter is released during active mining and presents a significant

**Table 15.1** Composition and heating values of typical natural gases

| Component (%) | Coal bed methane | Coal mine methane | Landfill gas | Biogas | Average NG |
|---|---|---|---|---|---|
| $O_2$ | 2.1 | 12.6 | 0–1 | 0–2 | |
| $N_2$ | 8.2 | 46.8 | 2–5 | 0–10 | 1.6 |
| $CH_4$ | 85.9 | 40.0 | 45–60 | 50–75 | 93.1 |
| $CO_2$ | | | 40–65 | 25–50 | 1.0 |
| $C_nH_{2n}$ | 3.8 | | | | 4.3 |
| LHV, kcal/nm$^3$ | 7,762 | 3,200 | 3,350–4,775 | 3,580–5,975 | 8,275 |
| (Btu/scf) | 872 | 360 | 375–535 | 400–670 | 930 |

danger of violent explosion to miners. Proper ventilation of mine shafts to remove and vent CMM into atmosphere is essential for the safety of mining operation. The methane content of CMM is variable (25–60%) and can change suddenly, which might be problematic for its use as a fuel. A third type of coal mine gas is abandoned mine methane (AMM), which can seep from mines that are no longer active. Annual coal mine methane emissions account for about 8% of total anthropogenic methane emissions or about 28 billion m$^3$ of carbon dioxide equivalent [3]. Utilization of CMM and AMM for power generation is critical to the prevention of release of a GHG into atmosphere.

Energy content of the natural gas is measured in terms of Btu (kJ or kcal) per lb (kg) or per standard cubic feet (cubic meters). In many practical calculations, natural gas is assumed to be 100% pure methane, which has an energy content of 21,515 Btu/lb (50,044 kJ/kg or 11,953 kcal/kg) on a lower heating value (LHV) basis or 914 Btu/ft$^3$ (34,050 kJ/m$^3$ or 8,133 kcal/m$^3$) at normal (standard) conditions (herein defined as 77°F (25°C) and 1 atm). The actual composition may vary depending on the source and treatment. Based on the mean of over 6,800 samples of pipeline quality natural gas taken in 26 major metropolitan areas of the USA [4], the methane content is 93.1% with a heating value of 930 Btu/ft$^3$ (8,275 kcal/m$^3$). A sampling of composition and heating value of most common natural gas variants is given in Table 15.1.

Production and distribution companies measure natural gas in volumetric terms, i.e., in multiples of cubic feet such as millions of cubic feet, trillions of cubic feet, etc. In power generation calculations, where natural gas is the fuel burned in the combustor of the gas turbine, the more practical approach is to use the mass flow rate (lb/s or kg/s) along with the LHV value or the product of the two, which is referred to as heat consumption (MMBtu/h or kW).

The purpose of a power plant is to generate electric power. Therefore, the key performance metric of an electric power generator is the net thermal efficiency, which is the ratio of the electric power measured at the generator terminals (after subtracting all the power that is spent to keep the plant running) to the plant's fuel energy consumption. The most efficient way to burn natural gas for electric power generation is in a gas turbine combined cycle (GTCC) power plant. Today's state-of-the-art CC power plant has a rated net efficiency of 58% on an LHV basis.

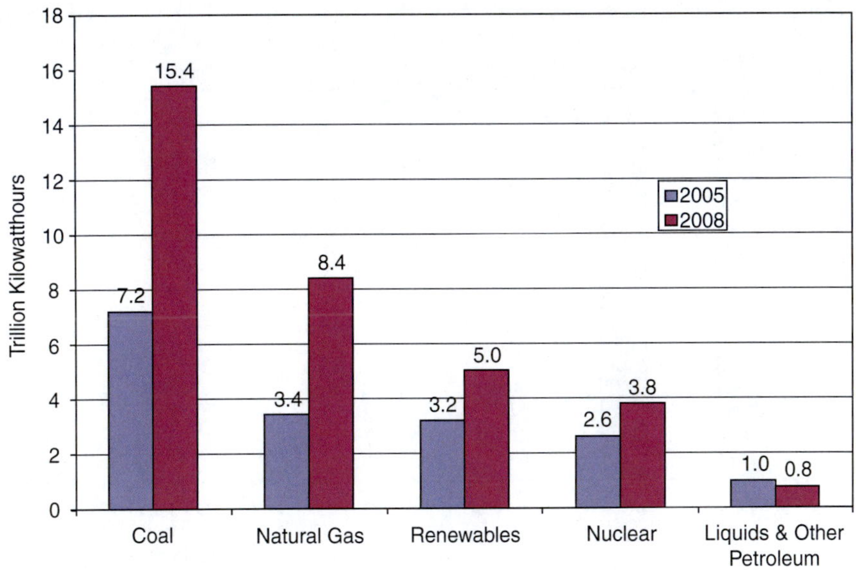


**Fig. 15.1** World Net Electricity Generation. Renewables include hydroelectric (Source: Energy Information Administration (EIA) [1])

In other words, for each 1 lb/s of natural gas (about 23.5 $ft^3/s$ as 100% methane) that enters the combustor of the gas turbine, (using the formula 58% × 21,515 Btu/lb × 1 lb/s × 1.05506 kWs/Btu = 13,166 kW) about 13.2 MW of electric power (net) is delivered to the grid. This is enough power to sustain nearly 13,000 average homes for 1 h at an average power consumption of 1 kW.

In 2005, natural gas accounted for 3.4 trillion kWh of net electricity production for about 20% of the world total (see Fig. 15.1). This contribution is projected to reach 8.4 trillion kWh in 2030 or about 25% of the world total corresponding to a total natural gas consumption of 55.5 trillion $ft^3$ (i.e., 35% of total world production of 158.6 trillion $ft^3$). Using the calculation above as a rough guide, this translates into an average natural gas–fired power plant thermal efficiency of 56.4%. The natural gas reserves of the world are estimated at about 6,186 trillion $ft^3$ with top 20 countries contributing nearly 90% of it (see Fig. 15.2). When one considers the uncertainty in the cited projections, it appears safe to say that there are ample deposits of natural gas to sustain electric power production well into the twenty-first century.

As mentioned in the beginning, natural gas is a superior fossil fuel in terms of emissions and pollutants vis-à-vis other fossil fuels, i.e., coal and oil. This is not to say that natural gas is totally harmless, especially when one considers the emissions of $CO_2$ – the most abundant anthropogenic greenhouse gas in the atmosphere and, as such, the main culprit suspected in driving global warming. A modern gas turbine combustor generates 2.75 lbs of $CO_2$ per lb of natural gas burned. In other useful relative terms, this is approximately 129,000 lb (58,500 kg) of $CO_2$

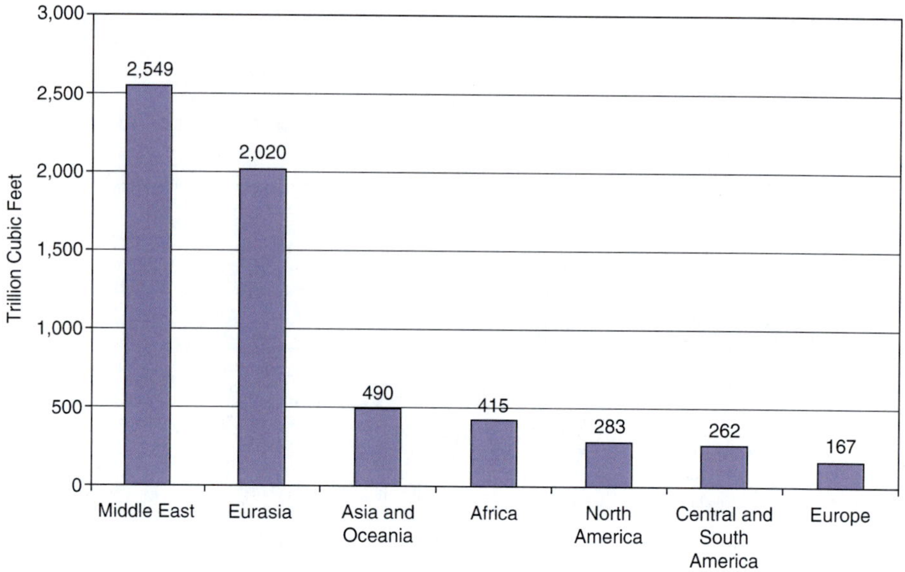


**Fig. 15.2** World Natural Gas Reserves, January 2008. Asia, Oceania and Europe exclude countries that were part of the former USSR whereas Eurasia includes only those. Source: Energy Information Administration (EIA) [2]

per billion Btu (1.06 million MJ) of energy input or 1,125 lb/MWh generated by a gas turbine (GT) in simple cycle configuration (cf., 750 lb in combined cycle configuration). Since not all power generation equipment performs at today's higher efficiency level, real averages would be higher, e.g., 1,300 and 800 for simple and combined cycles, respectively. For the projected natural gas–based electric power generation in 2030 cited above, at a rate of 1,300 lb/MWh, this translates into 5 billion metric tons of $CO_2$ released into earth's atmosphere. (In order to get an idea about the magnitude of this number, consider that the estimated weight of carbon dioxide in the atmosphere was about 2,163 billion metric tons before the industrial revolution.) This is approximately half of the total 2030 projection of $CO_2$ emission from natural gas combustion, which itself is about 20% of the 42.3 billion metric tons for all fossil fuels.

While an important contributor to the greenhouse effect, natural gas is extremely advantageous from the perspective of two other significant combustion-related pollution phenomena, i.e., smog and acid rain. Smog (whose primary constituent is ozone) is formed by a chemical reaction of carbon monoxide, nitrogen oxides, volatile organic compounds, and heat from sunlight [5]. Acid rain is formed when sulfur dioxide and nitrogen oxides react with water vapor and other chemicals in the presence of sunlight to form various acidic compounds in the air [5]. Natural gas combustion products are virtually free of the two main culprits contributing to these severe pollution problems, namely, sulfur oxides ($SO_x$) and particulate matter, and they have 80% less $NO_x$ than products of coal combustion.

In general (notwithstanding some future fuel cell designs discussed later), the most efficient means to utilize natural gas as a fossil fuel in electric power generation is to burn it in internal combustion (IC) engines. The two most widely used IC engines in electric power generation are gas turbine and reciprocating internal combustion or diesel engine. (Note that the term *diesel* in this context refers to the thermodynamic cycle of the compression-ignition (CI) IC engine and not the fuel.) While possible in theory, natural gas is not used with external combustion technology (e.g., boiler and steam turbine plants) due to the pronounced superiority of the IC engines in efficiency, emissions, power density (i.e., favorable cost vs. size trade-off) and operational flexibility.

Modern gas turbines are large air-breathing turbomachines with extremely large power output. For example, consider a 50-Hz (3,000 rpm) 300+ MW net power output unit as listed in a 2008 trade publication [6]. This machine ingests air at ISO conditions (15°C and 1 atm at 60% relative humidity) at a rate of nearly 1,550 lb/s (705 kg/s), compresses it to a pressure that is 18 times that of the ambient and combusts it with about 35 lb/s (16 kg/s) of natural gas (100% $CH_4$) generating 312 MW net electric power for a net thermal efficiency of 39.3%. At the inlet to the expander section of the gas turbine, where they produce useful shaft work, the combustion products are at nearly 1,500°C (2,732°F). This is well above the melting point of the most advanced superalloy materials that are used in the manufacture of turbine expander components. In order to ensure the survival of the turbine parts under those extreme conditions for thousands of hours of continuous operation, nickel-based superalloy components are protected by thermal barrier coatings (TBC) and internally cooled by using the "cold" air extracted from the turbine compressor. (Consider that without cooling and TBC, the first stage vanes of a modern GT would survive barely ten seconds before melting away.) Utilizing those and other technologies, some of which are adopted from the advanced military and civilian aircraft engine–related research and development, land-based heavy-duty industrial gas turbines are true marvels of human engineering.

Diesel or compression-ignition IC engines are no slouches themselves. In fact, the efficiency of a tri-fuel (one that can burn natural gas, light and heavy fuel oils) unit rated at nearly 17 MW net electric output is 47.3% on an LHV basis when operating with natural gas fuel [7]. They are eminently suitable to landfill, biogas, and coal mine gas applications. They are at a disadvantage because of their relatively low power density and cost. The sample multifuel diesel engine cited above has a power density of 0.05 kW/kg of engine weight or 215 kW/$m^2$ of engine footprint. These numbers compare very unfavorably to the power density of a heavy-duty industrial gas turbine. For a typical advanced machine with the performance similar to the sample cited above, the power density is about 0.8 kW/kg of engine weight or about 3,000 kW or more per $m^2$ of engine footprint. In order to appreciate the power density of a GT vis-à-vis renewable energy–based systems, consider that the power density of typical wind and solar radiation are 0.4–0.8 (depending on altitude and speed) and 1.4 kW/$m^2$, respectively [8]. One would need hundreds of acres of wind turbine farms or solar collector fields to replace a single advanced F-Class gas turbine. This gross disparity in the power density of the two IC engine technologies is reflected in the annual unit

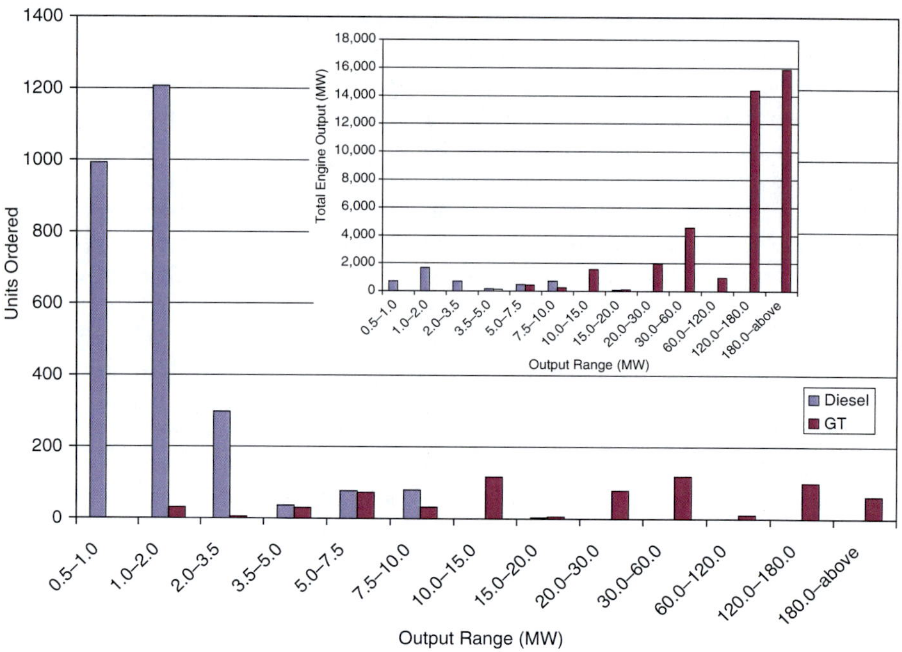


**Fig. 15.3** Dual-fuel and natural gas diesel engine and gas turbine orders, June 2005 through May 2006 [9]

order numbers displayed in Fig. 15.3. In terms of the numbers of units ordered in 2005–2006 [9], reciprocating IC engines dominate (nearly 2,700 units vs. 666 for gas turbines) but the significant bulk of the orders are for units rated at 3.5 MW or less. In terms of the total MW rating of the orders, all dual-fuel and natural gas diesel engines add up to 4.4 GW as opposed to more than 40 GW for all gas turbines.

At this point, it should be fairly obvious that a treatise on the subject of electric power generation using natural gas is essentially a treatise on modern land-based or heavy-duty industrial gas turbines. This is especially true for the past and the present of the technology. The future of electric power generation using natural gas as a fuel source might have a place for the fuel cell technology (with or without gas turbines). Ongoing research activities include solid oxide (SOFC), molten carbonate (MCFC), and polymer electrolyte (PEFC) technologies. From where one stands today, the most promising use for these technologies seems to be found in the fields of transportation and portable power. With the exception of distributed power generation, in which small-scale (1–500 kW) units power individual residences, office buildings, etc., the prospect of fuel cell–based technologies to replace internal combustion engines and specifically gas turbines as large-scale base load power-generating units in the near term, especially in a cost-effective manner, is practically nil.

## *A Brief History of Natural Gas*

Natural gas has been around for quite a while. In China, nearly 2,500 years ago, early innovators used natural gas escaping naturally from the ground to light their temples and heat brine for distillation. Bamboo sections, split lengthwise, were glued together, and bound with twine, allowing the gas to be transported. In the early seventeenth century, French explorers near Lake Erie reported seeing natives burning gas naturally seeping from the ground. The first industrial use of natural gas can be traced back to England in the late eighteenth century when gas produced from coal was used to light houses and streets. In the USA, the first natural gas well was dug in 1821 in Fredonia, NY, four decades before Colonel Drake hit upon oil and gas at a depth of 69 feet in Lake Erie, PA. The well, dug by William Hart, was 27 ft deep in a creek bed. The extracted gas was transported through hollow log pipes to Fredonia for use in street lamps. Throughout most of the nineteenth century, natural gas was used almost exclusively for the same purpose, i.e., lighting. Following the invention of the Bunsen burner in 1885, natural gas also presented households with a more convenient means of cooking and heating. The main impediment to a wider utilization of natural gas was the lack of a transportation mechanism. The primary method, still in use today, is through an underground pipeline. While the first significantly long pipeline was constructed in 1891 (120 miles long between central Indiana and Chicago), it was only after World War II when technology advances made construction of longer and more reliable pipelines feasible. Until that time, natural gas was either allowed to vent into the atmosphere (intentionally or unintentionally) or burned in flares (as a by-product of coal mining and oil exploration) or just left in the ground when found alone. In fact, when the first astronauts looked at earth from the orbit, the most prominent sights on the earth's surface were the flares from the oil fields in Middle East and Africa. Even now about 150 million $m^3$ of natural gas, roughly equivalent to annual US residential consumption, is flared to the atmosphere. The contribution of the flared natural gas (NG) to global GHG emissions is about 400 million metric tons, equivalent to nearly one third of the annual US car exhaust emissions.

Today natural gas is used for much more than street lighting or household cooking. In fact, residential use of natural gas accounts for only a fifth of the total US consumption, with more than 60% used by the industrial sector including electric power generation. Until the end of 1980s, high-efficiency fossil fuel (mainly coal) firing or nuclear-powered steam turbine–based power plants supplied the base load electric generation capacity (6,000–8,000 h/year). When newly designed high-efficiency steam plants entered the service, older less-efficient units were relegated to intermediate duty operation (e.g., 4,000 h/year). Early generation, low-efficiency gas turbines, along with old steam plants and hydro pumped-storage plants, were largely limited to short-duration, peak-shaving-type operations. Due to changing economic parameters, introduction of protective environmental regulations, and major technological advances in gas turbine technology, starting in mid-1980s and early 1990s, natural gas became the fuel of choice for new power plants. The steady increase in coal-fired steam plant efficiencies leveled off at the low 40% range, whereas new "F-Class" gas turbines with high firing temperatures, almost exclusively in combined cycle configuration, opened the door for thermal efficiencies easily exceeding 50% and even pushing for 60%. Recently, especially during the so-called energy boom of early 2000s, a large fraction of new capacity addition in the USA has been in natural gas–fired units, e.g., nearly 95% or 22 GW in 2000. Since then, this rate has tapered off significantly. Natural gas–fired power plant capacity addition in terms of construction starts totaled about the same, i.e., 22 GW, in the years 2005–2007 [10]. Scheduled construction start is about 9 GW for 2010, about the same as in 2009 but lower than the peak year of 2008 with nearly 12 GW. As pointed out earlier, natural gas is projected to be the fastest growing fuel for worldwide electric power generation. Thus, the pace of construction and commissioning of plants burning natural gas might be expected to pick up again. Construction of natural gas–fired power plants scheduled to start between 2011 and 2015 was estimated to be 44.5 GW [10]. Even though some projects will undoubtedly be cancelled or postponed, this is a strong positive sign for the future role of natural gas in electric power production.

The driver of the ascendancy of natural gas as the star of fossil fuels is the gas turbine, which also plays a key role in its transportation and distribution via huge pipeline networks. Heavy-duty industrial and/or aeroderivative gas turbines drive the large centrifugal gas compressors, which increase the pressure of natural gas up to 1,500 psia ($\sim$100 bar). In most cases, these gas turbines burn the natural gas taken from the pipeline. Additional compressor stations, placed at regular intervals along the hundreds of miles long pipelines, maintain line pressures reduced via friction between pipe and gas, which can reach flow velocities up to 40 km/h.

## *A Brief History of Gas Turbine*

When lieutenant colonel Heinz Bär died in a civilian aircraft accident in 1956 in (then) West Germany, he was the world's leading jet fighter ace with 16 victories to his credit. Some five decades after his death and more than six decades after his last aerial exploits, he is still unrivaled. The machine that enabled Bär to permanently enter the annals of military history was the world's first mass production jet fighter, the Messerschmitt Me-262. It was powered, remarked another famous German fighter pilot, "as if pushed by angels" – actually two 1984-lb thrust Jumo-004 gas turbine jet engines.

Developed under the leadership of Franz Anselm, the Jumo-004 was essentially the engine that opened up the jet age. Quite naturally, Dr. Anselm did not conceive Jumo-004 in a vacuum. Paraphrasing Sir Isaac Newton, he was standing on the shoulders of giants such as Sir Frank Whittle and Hans von Ohain, who are widely recognized as the pioneers of modern turbojet engine. Sir Frank and Dr. Ohain themselves almost certainly acknowledged early inventors such as Barber (first GT patent in 1791), Stolze (first GT to be built and tried in 1900–1904), Holzwarth, and others, who showed them the path to the ultimate gas turbine engine design through their patents on air and steam turbines.

It is futile to attempt to plot a straight historical line from Hero's turbine (which had no purpose other than entropy generation) to today's modern 2,600°F firing temperature class steam-cooled H-System™. Countless inventors laid down the building blocks that eventually led to the modern gas turbine engine. For an excellent short history of turbomachinery in general and gas turbines in particular, the reader should consult the relevant chapter in Wilson and Korakianitis [11], who highlight the contributions of lesser known (unfairly) inventors such Aegidius Elling and George Jendrassik. Another good narrative with an emphasis on turbojets and German research and development can be found in Hans von Ohain's foreword to Mattingly's book on gas turbine propulsion [12]. For a detailed look at turbojet development activities of Sir Frank Whittle and Hans von Ohain in their respective countries, three articles by Meher-Homji can be consulted [13–15].

Quite obviously, wars or the prewar atmosphere of the western world in the first half of the last century pushed the development of gas turbines primarily as a military aircraft propulsion device. Starting in the late 1940s and early 1950s, engineers in USA and UK managed to derive robust, land-based power generation and, later, marine propulsion engines from these aircraft designs. Claire Soares essentially came to the conclusion that the involvement of so many people from so many different countries with immense pride in their work led to selective "histories" favoring selected milestones [16]. Her brief summary can be consulted for alternate takes on the development history of gas turbines. For a history of Siemens gas turbine development, a good resource is the paper by Leiste [17]. For a US perspective with an emphasis on GE, the reader should consult Miller and Nemec [18] and Brandt [19]. The paper describing the design and development of the GE MS7001F gas turbine by Brandt [20] is a good reference laying out all design aspects of a modern gas turbine and the trade-offs that are unavoidable in bringing a feasible product into the market.

Even though the development of the modern gas turbine was primarily driven by military and, to a lesser extent, civilian aircraft propulsion considerations, the introduction of stationary (land-based) units dedicated to power production did not lag too far behind. In fact, Aurel Stodola credited Hans Holzwarth with "having built the first economically practical gas turbine [21]." Holzwarth built several types of his "explosion" turbine in the first quarter of the twentieth century. Holzwarth's turbine was, strictly speaking, a hybrid construction combining the spark-ignition (constant volume) combustion process of an Otto cycle with the axial expansion process of a Brayton cycle. Stodola calculated a thermal efficiency of 25.6% for the test of a 1,500-rpm experimental turbine in Mühlheim–Ruhr in 1919 [21]. Burning a gas with 434 Btu/ft$^3$ heating value the turbine produced about 725 kW. Air and fuel gas was compressed by steam turbine–driven compressors and sequentially injected into the explosion chamber (gas first) at about 30 psia. Stodola reports an average maximum explosion pressure of 160 psia. The combustion products expanded through a two-stage velocity-compounded turbine. Exhaust gas at about 800°F was recovered in an exhaust heat boiler. In that sense, this unit

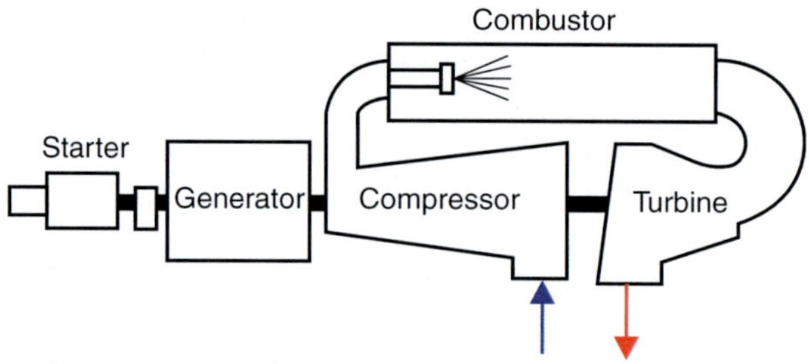


**Fig. 15.4** Schematic diagram of Brown Boveri Co.'s Neuchâtel gas turbine (1939)

could be considered the first combined cycle. Holzwarth designed and developed several different variants of his turbine between 1907 and 1928. From 1928 on, Brown Boveri Company (BBC) took over the development of Holzwarth turbine and in 1933 installed a blast-furnace gas (BFG)–fired unit in a German steel mill, which was destroyed during World War II.

It is generally accepted that the first commercial stationary gas turbine for electric power generation was erected in Neuchâtel, Switzerland, by the former BBC in 1939, 3 years before the first flight of Me-262 powered by Jumo-004. This fuel oil burning 4 MW machine, primarily used for stand-by and peaking duties, was operational for nearly 70 years (see Fig. 15.4). The combustion chamber was derived from the turbocharged *Velox* boiler, which itself resulted from the BBC work done on the Holzwarth turbine. The BBC turbine was tested under the supervision of Aurel Stodola, who reported an overall thermal efficiency of 17.4%, which is less than half of the efficiency of today's advanced F-Class machines [22]. This machine was designated by the ASME as a *Historic Mechanical Engineering Landmark* in 1988.

The first gas turbine installed in an electric utility in the USA (Oklahoma Gas & Electric, Belle Isle Station, Oklahoma) was a 3.5 MW GE Frame 3 unit that entered service in 1949. In addition to generating power, the exhaust gas of this gas turbine was utilized to heat the feed water of a conventional steam plant. In other words, the first US electric utility gas turbine was in a "combined" cycle configuration [23]. In fact the combined cycle concept goes back to Emmet's mercury-vapor process (1925) [24]. For a brief history of the CC power plants and pertinent references, the reader should consult the 1994 Calvin Winsor Rice Lecture by Sir John Horlock [24].

One can look at the technology history of the land-based gas turbine and combined cycle power plants for electric power production in four generations:

1. Generation 1 (1949–1968)
   (a) Smaller than 30 MW GTs (Frames 3 and 5)
   (b) Firing temperatures 1,500–1,800°F
   (c) For repowering and cogeneration
2. Generation 2 (1968–1990)
   (a) B and E Technology GTs (50–120 MW)

(b) Firing temperatures ~2,000°F
(c) $NO_x$ emission control using GT water/steam injection or SCRs
(d) Non-reheat steam cycles

3. Generation 3 (1990–1998)

(a) F Technology GTs (75–260 MW)
(b) Three air-cooled turbine stages
(c) Firing temperatures ~2,400°F
(d) Performance fuel heating (365°F)
(e) DLN combustion system for $NO_x$ control
(f) Three-pressure, reheat (3PRH) steam cycles

4. Generation 4 (1998–present)

(a) H technology GTs (400+ MW in CC)
(b) Closed-loop steam cooling (CL-SC) of the first two turbine stages (both stator and rotor); four turbine stages
(c) Cooling of cooling air (CAC) for turbine wheel spaces and subsequent stages via a heat exchanger (a kettle reboiler) that generates IP steam to be used in the bottoming cycle
(d) Active clearance control (compressor and turbine)
(e) Firing temperatures ~2,600°F
(f) Performance fuel heating (400+°F)
(g) DLN combustion system for $NO_x$ control
(h) Three-pressure, reheat (3PRH) steam cycle integrated with the GT Brayton cycle

The historical advances in GT firing temperature and CC efficiency is summarized in Fig. 15.5. The data illustrates the significant advance in the thermal efficiency of the GT combined cycle power plants, to the tune of nearly 15 percentage points, which is mainly driven by advances in materials, coatings, and cooling techniques that pushed the firing temperatures by almost 800°F (~450°C). In the following chapters, the reader is introduced to the basic thermodynamic principles that govern the performance of natural gas–burning gas turbine power plants. In addition, also provided is a brief overview of state-of-the-art gas turbine power plants, key economic criteria, and operability considerations. Finally, the future directions in gas turbine–based electric power generation systems are briefly elaborated upon.

## Gas Turbine Power Plants

Gas turbine in itself or as a key component of a larger power generation system is a subject that can only be covered in dedicated books. Such books do exist and several selected treatises on the subject, which in the opinion of the authors reflect

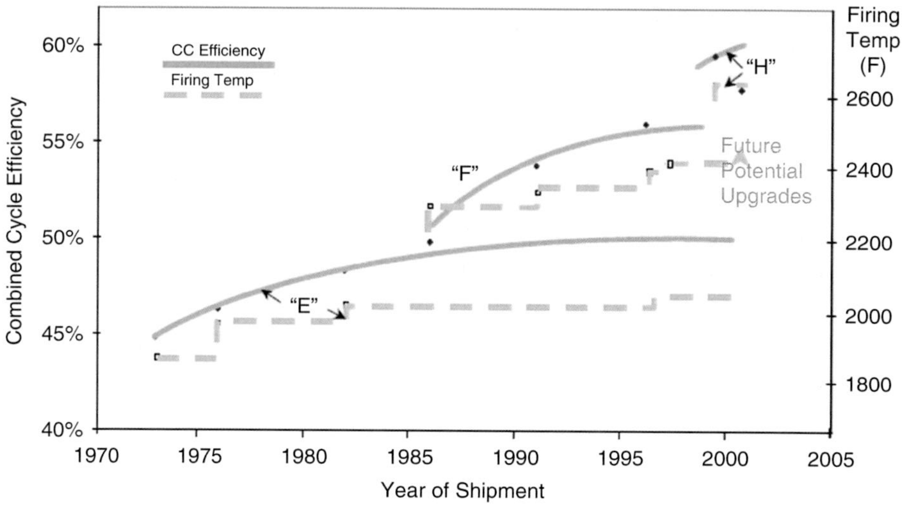


**Fig. 15.5** Incremental evolutions of E-, F- and H-Class GT technology (From Ref. [25])

the best available published work in the field, are listed among the bibliography references. The coverage of the subject in the limited space allotted to the current chapter aims to provide the reader with the fundamental thermodynamic considerations that lead from the Carnot cycle to the present-day advanced gas turbine in a logically coherent manner. There is a reason that the gas turbine combined cycle plant is the most efficient heat engine that is commercially available today and, most likely, in the foreseeable future. That reason is embedded in the laws of thermodynamics. Understanding this will enable someone with a basic knowledge of key engineering concepts to grasp the gas turbine design principles and carry out nearly all calculations using simple formulas and a few technology charts. Furthermore, this will also enable the reader to critically evaluate new developments and claims to future improvements in electric power generation systems that employ turbomachines and utilize fossil fuels such as natural gas.

As Descartes wrote in his *Discourse on Method* "one cannot conceive anything so strange and so implausible that it has not already been said by one philosopher or another." As such, in relation to the treatment of the subject in the paragraphs below, a claim of pure originality would be absurd. Nevertheless, the material is wholly original in the sense that key concepts are developed in a unique manner along with all pertinent formulas, representative design parameters and technology curves so that it forms a coherent and compact reference to be used in basic engineering analysis of gas turbine combined cycles. Even then, it must be pointed out that a similar treatment can be found in the excellent brief chapter on gas turbines written by Haselbacher cited in Ref. [26]. Another valuable short reference

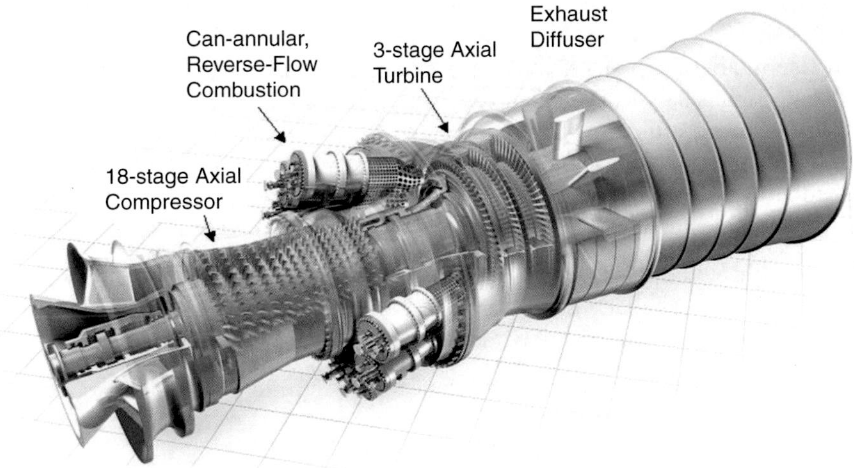


**Fig. 15.6** Modern heavy-duty industrial gas turbine (Courtesy: GE Energy)

incorporating the basic fundamental considerations is cited in Ref. [24]. The treatment herein focuses on land-based heavy-duty industrial gas turbines. For a brief introduction to thermodynamic and economic considerations pertaining to aeroderivative gas turbines and pertinent references, the reader should consult Horlock [27].

## *Basic Thermodynamics*

Gas turbine is a relatively simple turbomachine comprising three key components: Compressor, combustor, and expander (Fig. 15.6). The last one is commonly referred to as a turbine. The compressor and expander are connected through a common shaft. The operation of a gas turbine is described by a thermodynamic cycle comprising four processes: compression, heat addition or combustion, expansion, and heat rejection. A gas turbine can also be classified as an internal combustion engine (just like a car engine), in which the compressed working fluid, i.e., air, mixes with fuel, and the products of the ensuing chemical reaction (i.e., combustion) expand through the turbine. In land-based electric power generation applications, part of the useful mechanical or shaft work, produced by the expanding combustion products (approximately 50% of the total) is consumed by the compressor. The remainder is converted into electric power in a synchronous alternating current (ac) generator, which is connected to the same shaft as the gas turbine. In aircraft propulsion applications, the entire shaft work generated in the turbine is utilized to drive the engine compressor and fan. The thermal energy in the exhaust gas is converted into kinetic energy in the exhaust nozzle, which generates the net engine thrust.

The thermodynamic cycle that describes the gas turbine operation is the *Brayton* cycle. For simple analytical calculations, which can be found in elementary textbooks on thermodynamics, the ideal or *air-standard* Brayton cycle is used. The classical (and, from an engineering perspective, the most logical) representation of the air-standard Brayton cycle is on a temperature–entropy or T-s surface as illustrated in Fig. 15.7. The key assumptions in air-standard cycle analysis are:

1. The working fluid is air, which is a calorically perfect gas (i.e., $c_p$ = constant).
2. Isentropic (i.e., adiabatic and reversible) compression and expansion.
3. Constant pressure heat addition and heat rejection.

The two key nondimensional performance metrics for a gas turbine are specific *net* power output and thermal efficiency:

$$w = \frac{\dot{W}_{net}}{\dot{m} \cdot c_p \cdot T_1} \tag{15.1}$$

$$\eta = \frac{\dot{W}_{net}}{\dot{Q}_{in}} \tag{15.2}$$

The net power output is the difference between the compressor power consumption and the expander power production. In graphical terms, it is exactly equal to the area encompassed by the cycle 1-2-3-4-1 on the T-s diagram in Fig. 15.7. Each component's power can be calculated by the application of the *first law of thermodynamics* for steady-state steady-flow (SSSF) successively to their respective *control volumes*:

$$\dot{W}_{comp} = \dot{m} \cdot c_p \cdot (T_1 - T_2) \tag{15.3}$$

$$\dot{W}_{turb} = \dot{m} \cdot c_p \cdot (T_3 - T_4) \tag{15.4}$$

$$\dot{W}_{net} = \dot{W}_{turb} + \dot{W}_{comp} \tag{15.5}$$

From Fig. 15.7 and (Eqs. 15.3 and 15.4) one can easily see that the compressor power consumption is a *negative* number (i.e., work done *on* the control volume) and expander power generation is a *positive* number (i.e., work done *by* the control volume). Application of the first law for SSSF to the heat addition and heat rejection processes, one gets

$$\dot{Q}_{in} = \dot{m} \cdot c_p \cdot (T_3 - T_2) \tag{15.6}$$

$$\dot{Q}_{out} = \dot{m} \cdot c_p \cdot (T_1 - T_4) \tag{15.7}$$

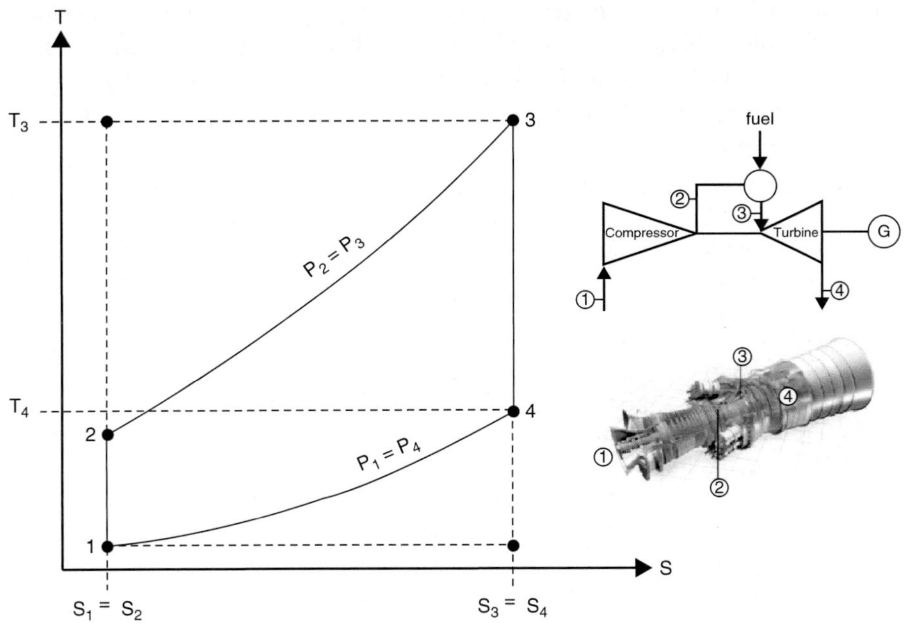


**Fig. 15.7** Temperature–entropy diagram of gas turbine Brayton cycle

From Fig. 15.7 and (Eqs. 15.6 and 15.7), the heat input to the cycle is a *positive* number (i.e., heat supplied *to* the control volume) and heat rejection from the cycle is a *negative* number (i.e., heat taken *from* the control volume).

In addition to the first law of thermodynamics for a SSSF process (or control volume analysis), there is one more key thermodynamic relationship that forms the heart of the GT Brayton cycle calculations: pressure–temperature ratio for an isentropic process:

$$\frac{T_2}{T_1} = \left(\frac{p_2}{p_1}\right)^{\frac{\gamma-1}{\gamma}} \tag{15.8}$$

where $\gamma$ is the ratio of constant pressure and constant volume-specific heats. Combining (Eqs.15.1–15.7), in nondimensional terms and applying the isentropic formula in Eq. (15.8) to compression and expansion processes, one ends up with the following two expressions:

$$w = \left(\pi^k - 1\right) \cdot \left(\frac{\tau_3}{\pi^k} - 1\right) \tag{15.9}$$

$$\eta = 1 - \frac{1}{\pi^k} \tag{15.10}$$

where $\pi = p_2/p_1$ is the cycle pressure ratio, $\tau_3$ is the cycle maximum temperature (nondimensionalized via division by $T_1$) and $k$ is the isentropic exponent, $\gamma - 1/\gamma$. The two simple formulas, Eqs. (15.9) and (15.10), which are strictly valid only for an ideal air-standard cycle, illustrate almost all of the key facts concerning gas turbine performance:

1. Gas turbine or Brayton cycle performance is controlled by two parameters: Cycle pressure ratio, $\pi$, and maximum cycle temperature, $\tau_3$.
2. Cycle efficiency is a function of cycle pressure ratio only; higher cycle pressure leads to higher cycle efficiency.
3. Higher cycle temperature leads to higher cycle specific output.
4. Higher cycle pressure ratio, beyond a certain value, is detrimental to cycle specific output (i.e., $\frac{\partial^2 w}{\partial \pi^2} < 0$ and $\pi = \sqrt[2k]{\tau_3}$ for $\partial w/\partial \pi = 0$).

These observations are also borne out qualitatively by a close examination of the cycle T-s diagram in Fig. 15.7. As the heat addition isobar moves upward (i.e., higher cycle pressure ratio, $\pi$) one can visually verify that the area encompassed by the cycle, 1-2-3-4-1, becomes smaller until, in the limit, it becomes zero. Equation (10) numerically implies that, in the limit when $\pi$ is sufficiently high, the cycle efficiency will approach 100%, or, $\eta \rightarrow 100\%$ as $\pi \rightarrow \infty$. This, of course, is impossible, because, as dictated by the *second law of thermodynamics*, the maximum theoretical efficiency attainable by any cycle is given by the Carnot cycle efficiency. The Carnot cycle is a hypothetical power cycle that is represented by two isentropic processes, i.e., compression and expansion; and two isothermal processes, i.e., heat addition and heat rejection. Graphically, it is the rectangle enveloping the Brayton cycle, which is bounded by $T_1$ and $T_3$ isotherms and $s_1$ and $s_3$ isentropes.

For the cycle in Fig. 15.7, one can easily visualize the limit of $\pi \rightarrow \infty$ as $T_3$ stays constant. In the limiting case, one will end up with a Brayton cycle that is a rectangle with ~0 area (i.e., zero specific output) and heat addition and rejection at constant temperatures $T_3$ and $T_1$, respectively, which essentially defines a Carnot cycle with the efficiency

$$\eta = 1 - \frac{T_1}{T_3} \tag{15.11}$$

Comparing with Eq. (15.10), for a cycle maximum temperature of $T_3$ = 1,500°C and $\gamma = 1.4$, this would correspond to a cycle pressure ratio of $\pi = 577$ and cycle efficiency $\eta = 87.4\%$. In other words, as dictated by the second law of thermodynamics, the maximum theoretical cycle efficiency that is attainable by a gas turbine with a "firing temperature" of 1,500°C is 87.4%. Clearly a cycle pressure ratio of 577 is of little practical interest. For the modern heavy-duty industrial gas turbines, the typical value of $\pi$ is about 18. What does the second law say about a gas turbine for a cycle maximum temperature of $T_3$ = 1,500°C and $\pi = 18$? In order to find that out, consider the isobaric heat addition process 2 → 3 in Fig. 15.7. For a reversible and isobaric process, the modified Gibbs equation is

$$h_3 - h_2 = \int_2^3 T \cdot ds \tag{15.12}$$

One can imagine an (hypothetical) isothermal process that takes place at a constant temperature, whose value lies between $T_3$ and $T_2$, and has the same entropy as the actual end states 2 and 3. This process would have exactly the same numerical value of heat addition as the original process, i.e., $h_3 - h_2$, per unit mass. For that process, Eq. (15.12) can be rewritten as

$$h_3 - h_2 = \bar{T}_H \cdot (s_3 - s_2) \tag{15.13}$$

Using the calorically perfect gas relationships, the *mean-effective* heat addition temperature can be found from Eq. (15.13) as

$$\bar{T}_H = \frac{T_3 - T_2}{\ln\left(\frac{T_3}{T_2}\right)} \tag{15.14}$$

Using the same logic, the mean-effective heat rejection temperature is calculated as

$$\bar{T}_L = \frac{T_4 - T_1}{\ln\left(\frac{T_4}{T_1}\right)} \tag{15.15}$$

These two mean-effective temperatures are graphically depicted in Fig. 15.8. Thus, qualitatively any thermodynamic cycle in general, and the Brayton cycle in particular, 1-2-3-4 in Fig. 15.8, can be represented by its *Carnot-equivalent* cycle, 1′-2′-3′-4′, with mean-effective (constant) heat addition and heat rejection temperatures that are given by Eqs. (15.14 and 15.15). The efficiency of that cycle would be given by the well-known relationship for the Carnot efficiency, i.e.,

$$\eta = 1 - \frac{\bar{T}_L}{\bar{T}_H} \tag{15.16}$$

For an ideal air-standard Brayton cycle, one can show that Eq. (15.16) is the same as Eq. (15.10). Thus, for any gas turbine with a cycle pressure ratio of 18, the maximum theoretical efficiency is 56.2%. Using Eqs. (15.8), (15.14), and (15.15), this would be the efficiency of a Carnot cycle with isothermal heat addition and heat rejection at 852°C and 219°C, respectively. It is interesting to note that, changing the cycle maximum temperature from 1,500°C to 1,400°C would change these temperatures to 814°C and 203°C, respectively, but would *not* change the cycle efficiency, i.e., 56.2%, which is only a function of $\pi$. This is in contrast to the

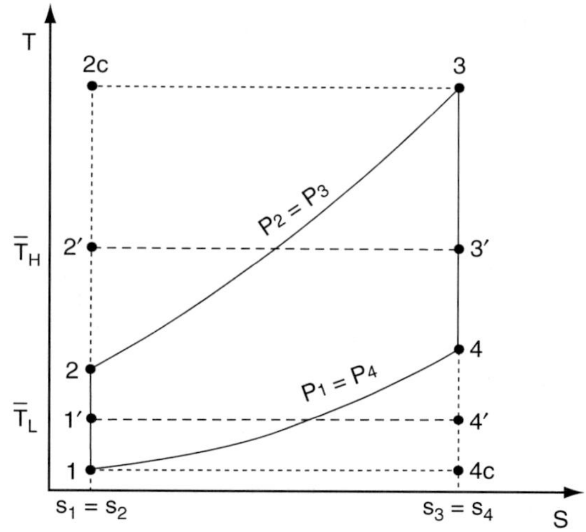


**Fig. 15.8** Temperature–entropy diagrams of gas turbine Brayton cycle and its "Carnot equivalent"

performance trend of the "real" gas turbines, whose thermal efficiency does indeed increase with increasing cycle maximum or *firing* temperature. Why this is so will be understood below.

One key observation from Fig. 15.8 is the large thermal efficiency gap between the ultimate Carnot efficiency given by Eq. (15.11) and the maximum theoretical entitlement level as dictated by the second law, Eq. (115.0) or Eq. (15.16), which is nearly 30 percentage points. Graphically, this gap is represented by the two triangular areas in Fig. 15.8:

1. The upper triangular area 2-2c-3-2, which is equal to the rectangular area 2′-2c-3-3′-2′, and represents the heat addition (i.e., combustion) irreversibility
2. The lower triangular area 1-4-4c-1, which is equal to the rectangular area 1-1′-4′-4c-1, and represents the irreversibility and exergy loss via heat rejection to a heat sink

In terms of cycle temperatures, the upper triangular area is quantified by the gap between the cycle maximum temperature, $T_3$, and the mean-effective heat addition temperature $\bar{T}_H$. Similarly, the lower triangular area is quantified by the gap between the cycle minimum or ambient temperature, $T_1$, and the mean-effective heat rejection temperature $\bar{T}_L$. The latter mechanism of "lost work" associated with heat rejection also points the way to the gas and steam turbine combined cycle power plant concept. This will be explored in some more detail later in the discussion.

The real gas turbine Brayton cycle differs from the ideal, air-standard cycle in the following respects:

1. Non-isentropic compression and expansion (represented by component polytropic or isentropic efficiencies)
2. Pressure losses during heat addition and heat rejection

3. Real gas effects (represented by a suitable equation of state)
4. Change in working fluid composition and properties downstream of combustion process

These nonidealities can be easily incorporated into the simple equations above via polytropic component efficiencies, pressure loss factors, and real gas equation of state. One can refer to one of the available textbooks on the subject for details, e.g., Ref. [29]. For example, Eq. (15.9) can be rewritten as

$$w = \left(1 - \pi_c^{\frac{k_a}{\eta_c}}\right) + (1+f) \cdot \chi_g \cdot \tau_3 \cdot \left(1 - \frac{1}{\pi_t^{\eta_t \cdot k_g}}\right) \tag{15.17}$$

where $\pi_c$ and $\pi_t < \pi_c$ are compressor and turbine pressure ratios, respectively, $k_a$ and $k_g$ are isentropic exponents for air and combustion products, respectively, and $\eta_c$ and $\eta_t$ are compressor and turbine polytropic efficiencies, respectively. The ratio of the specific heats of combustion products and air is $\chi_g$ and $f$ is the fuel–air mass flow ratio. Heat addition can be found from the fuel consumption, i.e.,

$$\dot{Q}_{in} = \dot{m}_f \cdot \mathrm{LHV} \tag{15.18}$$

where LHV is the net or lower heating value of the fuel at the reference temperature, $T_{ref}$. The efficiency of the gas turbine is the ratio of Eqs. (15.17) and (15.18), i.e.,

$$\eta = \frac{\left(1 - \pi_c^{\frac{k_a}{\eta_c}}\right) + (1+f) \cdot \chi_g \cdot \tau_3 \cdot \left(1 - \frac{1}{\pi_t^{\eta_t \cdot k_g}}\right)}{f \cdot \ell} \tag{15.19}$$

where $\ell$ is the nondimensional fuel energy content, $\mathrm{LHV}/c_{p,a} \cdot T_1$. Note that for clarity, mechanical and electric losses are omitted from Eqs. (15.17) and (15.19), which represent the "shaft" performance. Typically quantified via mechanical and generator efficiencies, $\eta_m$ and $\eta_g$, respectively, these losses should be accounted for the gas turbine generator's net electric output. The fuel mass flow rate (as a fraction of turbine airflow) that is requisite for a specified cycle maximum temperature is given by

$$f = \frac{\chi_g \cdot (T_3 - T_{ref}) - (T_2 - T_{ref})}{\ell \cdot T_1 + \chi_f \cdot (T_f - T_{ref}) - \chi_g \cdot (T_3 - T_{ref})} \tag{15.20}$$

where the second term in the denominator is the sensible fuel energy input and $\chi_f$ is the ratio of the specific heats of fuel and air. Comparing Eq. (15.19) (along with Eq. (15.20) that sets the denominator) with Eq. (15.10), one can easily observe that the GT thermal efficiency, when accounting for the "real" cycle effects, is indeed a function of $\tau_3$. While it is difficult to discern the impact of $\tau_3$ on $\eta$ just by looking at Eqs. (15.19) and (15.20), a sample calculation can illustrate that as shown in Fig. 15.9.

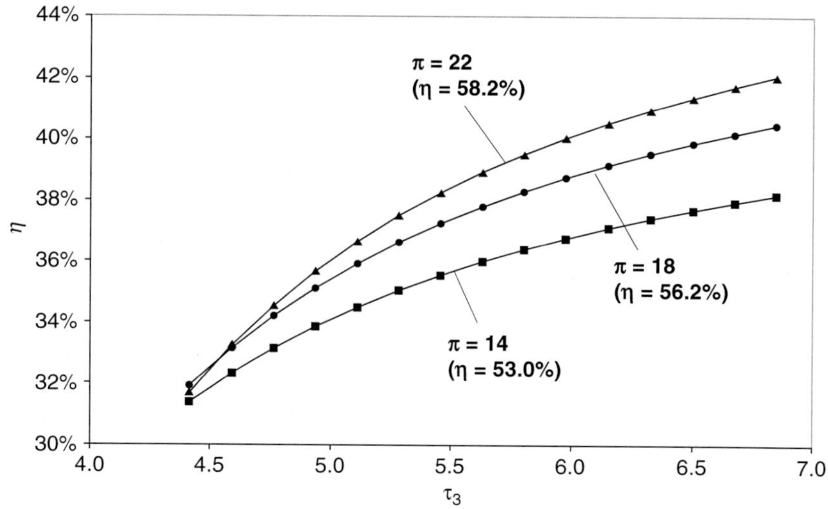


**Fig. 15.9** Real Brayton cycle efficiency from Eqs. (15.19) and (15.20) as a function of cycle pressure ratio ($\pi$) and maximum cycle temperature ($\tau_3$). Also shown (in parentheses) are the ideal, air-standard cycle efficiencies from Eq. (15.10)

Typical values that are reasonably accurate for gas turbines burning natural gas fuel (assumed to be 100% $CH_4$ or methane) are given below (adopted from Ref. [29]) and used in the calculations that resulted in the curves depicted in Fig. 15.9:

LHV = 21,515 Btu/lb
$c_{p,a} = 0.24$ Btu/lb-R (1.005 kJ/kg-K)
$c_{p,g} = 0.274$ Btu/lb-R (1.148 kJ/kg-K) or $\chi_g = 1.1417$
$c_{p,f} = 0.60$ Btu/lb-R (2.5125 kJ/kg-K) or $\chi_f = 2.50$
$\gamma_a = 1.400$
$\gamma_g = 1.333$
$\eta_c = 87.8\%$
$\eta_t = 85\%$
$\eta_m = 99\%$
$\eta_g = 98.9\%$
$k_a = 0.2857$
$k_g = 0.2498$
$T_{ref} = 77°F$ (25°C)
$\ell = 172.728$

Original equipment manufacturer (OEM) data for five 50-Hz (3,000 rpm) heavy-duty industrial gas turbines from a recent trade publication [6] are given in Table 15.2.

For an actual gas turbine, the cycle maximum temperature $T_3$ is equivalent to the turbine inlet temperature (TIT) at the exit of the gas turbine combustor. Another value that is frequently used in the industry is the "firing temperature," which is the temperature at the stage-1 blade row (rotor) inlet. This temperature, also known as

**Table 15.2** OEM data for 50-Hz heavy-duty industrial gas turbines [6]. The maximum cycle temperature, $T_3$, values are estimated using the published output, efficiency, pressure ratio (PR), and exhaust data with Eqs. (15.17–15.20)

| | $T_3$ | | | | | | Exhaust | |
|---|---|---|---|---|---|---|---|---|
| Turbine | (°C) | °F | PR | $PR_t$ | Output (MW) | Eff. | (kg/s) | °C |
| A | 1,415 | 2,579 | 17.7 | 15.9 | 285.0 | 39.6% | 690 | 572 |
| B | 1,443 | 2,629 | 17.0 | 13.5 | 255.6 | 36.9% | 641 | 602 |
| C | 1,501 | 2,734 | 18.3 | 14.5 | 279.2 | 37.8% | 655 | 629 |
| D | 1,435 | 2,616 | 18.2 | 16.3 | 292.0 | 39.8% | 692 | 577 |
| E | 1,476 | 2,688 | 18.0 | 15.5 | 312.1 | 39.3% | 720 | 597 |

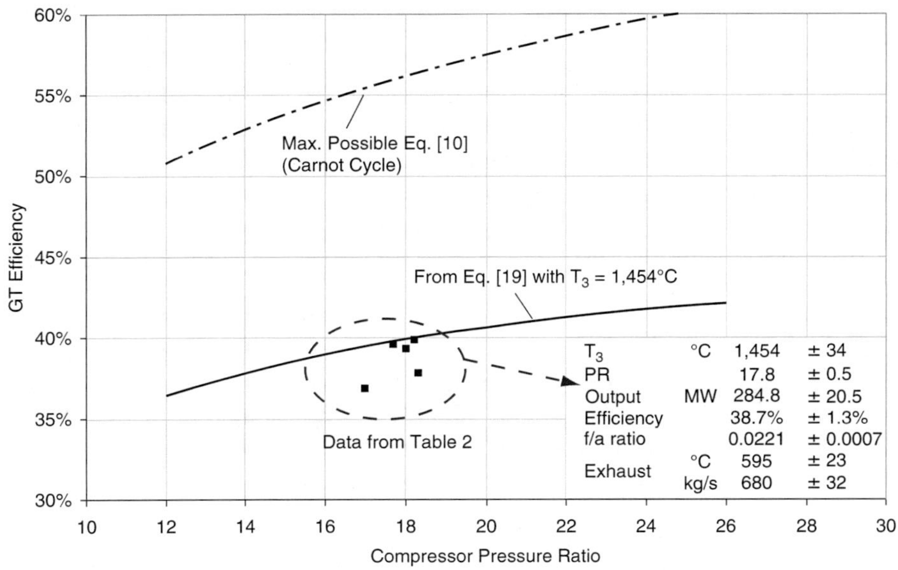


**Fig. 15.10** Gas turbine efficiency from Eq. (15.19) as a function of compressor PR (Data is from Ref. [6] and summarized in Table 15.2)

rotor inlet temperature (RIT) is the maximum temperature at which the combustion products start producing useful expansion work. The difference between TIT and RIT is about 250°F for modern air-cooled gas turbines and reflects the dilution effect of the stage-1 nozzle and forward wheel space cooling flow. One other cycle maximum temperature definition, favored by European OEMs, is the ISO-TIT, which is a fictitious number as defined by ISO-2314 (typically ~180°F lower than RIT). Obviously, for an uncooled turbine all these definitions are the same.

Gas turbine efficiency as a function of PR is shown in Fig. 15.10. Specific output, $w$, in the range covered by the turbines in Table 15.2 is 400–450 kJ/kg. Carbon dioxide emissions are calculated from the airflow and fuel flow using chemical reaction equations for 100% $CH_4$ gaseous fuel assumption as 510 kg/h per MW of output at 39% net efficiency. Each one percentage point improvement in efficiency is worth about 15 kg/h per MW reduction in $CO_2$ emissions. This is equivalent to 27,000 metric tons for a 300 MW GT running at base load for 6,000 h/year.

Figure 15.10 shows that the gap between the current State of the Art (SOA) in gas turbine technology and the maximum theoretical entitlement level, as dictated by the second law, is more than 15 efficiency points. The GT efficiency – compressor pressure ratio (PR) trend in Fig. 15.10 (via Eq. (15.19)) gives the impression that, for a given cycle maximum temperature, the efficiency can be improved indefinitely by increasing the PR. This is *not* true due to a key feature of modern gas turbines that is very difficult to capture in simplified models such as Eqs. (15.17–15.20): turbine hot gas path (HGP) component cooling via air extracted from the compressor discharge and/or interstage locations. Since the early days of GT development, the components exposed to the highest temperature environment have been cooled with air drawn from the compressor. The Junkers Jumo-004 jet engine of the German fighter plane Messerschmitt 262, which entered active service in 1944 during the last stages of the World War II, was the first mass production turbojet engine in the history with a turbine inlet temperature (TIT) of 1,427°F [13]. The hollow turbine blades (manufactured by folding flat sheets of 12% Cr alloy called *Cromadur*) were cooled by air extracted from between the fourth and fifth stages of the compressor and introduced to the blade through the holes drilled into the disk. The cooling technology, although unchanged in basic principles, has advanced greatly over the years with the introduction of complex serpentine cooling passages, impingement, film, and transpiration cooling techniques; paced by concurrent advances in superalloys (e.g., nickel-based directionally solidified or single-crystal), manufacturing techniques (e.g., investment casting), oxidation/corrosion (e.g., MCrAlY), and thermal barrier (e.g., yttria-stabilized zirconia, YSZ) coatings. This development supported the advances in TIT, which nearly doubled the standard set by Jumo-004 over the course of the next 50 years (e.g., see Fig. 15.11 that uses data from Cumpsty [30]). A good overview of the current SOA in advanced GT materials and coatings can be found in Ref. [31].

An alternative to open loop air-cooling (OL-AC) has recently been commercially introduced at the 480-MW Baglan Bay 109H CC power plant in Wales [32]. This machine employs closed-loop steam cooling (CL-SC) for the majority of the turbine HGP components (the first two stages of the four-stage turbine, to be exact). A "lighter" version of the H-System™ is the G-Class turbine with steam cooling limited to the stationary parts upstream of the turbine HGP [33].

A structural refinement to the basic GT Brayton cycle is the inclusion of a second "reheat" combustor within the turbine expansion. Reheat, or *sequential combustion* as it is referred to by the OEM that successfully commercialized the technology increases the mean-effective cycle heat addition temperature and reduces the combustion irreversibility [34] (the upper triangular area 2-2c-3-2 in Fig. 15.3). From a fundamental thermodynamic perspective, lower combustion losses give the GT cycle with reheat combustion a higher performance entitlement than a cycle without it. Similarly, there is no doubt that closed-loop (external) cooling is superior to open-loop (internal) cooling but the exact magnitude of that advantage in real machines is difficult to quantify. Chiesa and Macchi [35] quantified the

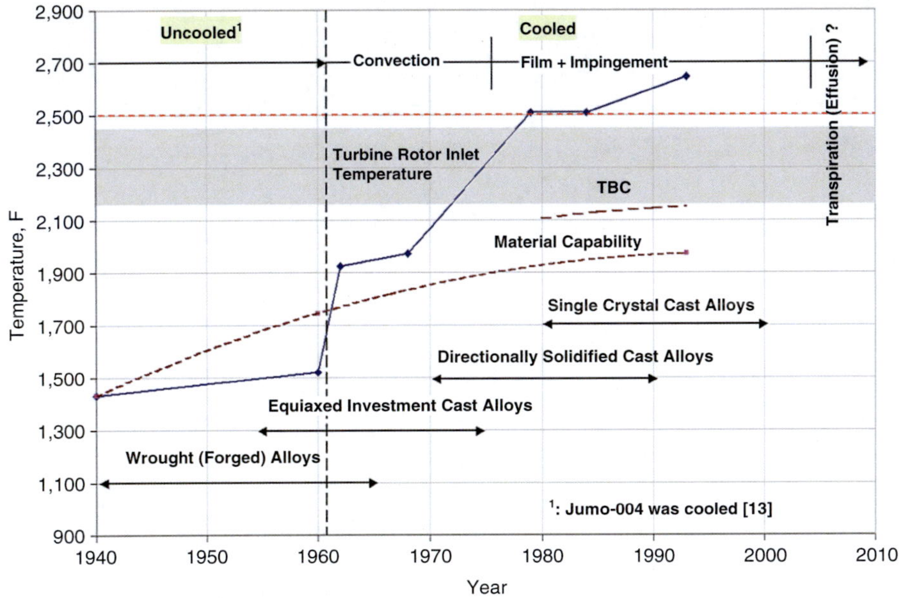


**Fig. 15.11** Turbine RIT and bucket alloy capability history [30]. One can add ~250°F to RIT to get a rough idea about TIT of the advanced F-Class units. The shaded region designates the melting point of nickel-based superalloys

**Table 15.3** Comparison of CC performances based on four major GT technologies [35]: Air-cooled (OL-AC), air-cooled reheat (OL-AC+SC), steam-cooled (CL-SC), and steam-cooled reheat (CL-SC+SC)

| | OL-AC | OL-AC+SC | CL-SC | CL-SC+SC |
|---|---|---|---|---|
| GT PR | 20 | 30 | 23 | 30 |
| HPT RIT (°C) | N/A | 1,290 | N/A | 1,430 |
| HPT RIT (°F) | N/A | 2,354 | N/A | 2,606 |
| RIT (°C) | 1,430 | 1,430 | 1,430 | 1,430 |
| RIT (°F) | 2,606 | 2,606 | 2,606 | 2,606 |
| Net $\eta_{CC}$ (%, points) | Base | +0.81 | +1.77 | +2.74 |

advantages of OL-AC (with and without reheat combustion) and CL-SC technologies via detailed calculations. Using OEM data from GE's 7FB, Siemens V94.3A (now SGT5-4000F), Alstom's GT-26, and MHI's M701G units for calibration, they have developed very detailed cooled GT models and evaluated the CC performances. Their findings are summarized in Table 15.3, which shows that CL-SC has an advantage of ~1.8 points (%) in net $\eta_{CC}$ over OL-AC.

A realistic assessment of the gas turbine requires some means to incorporate the cooling penalty into the GT calculations. Cooled turbine expansion models, no matter how much simplified, involve a rather cumbersome system of equations amenable only to a numerical solution. In the open literature, there are numerous

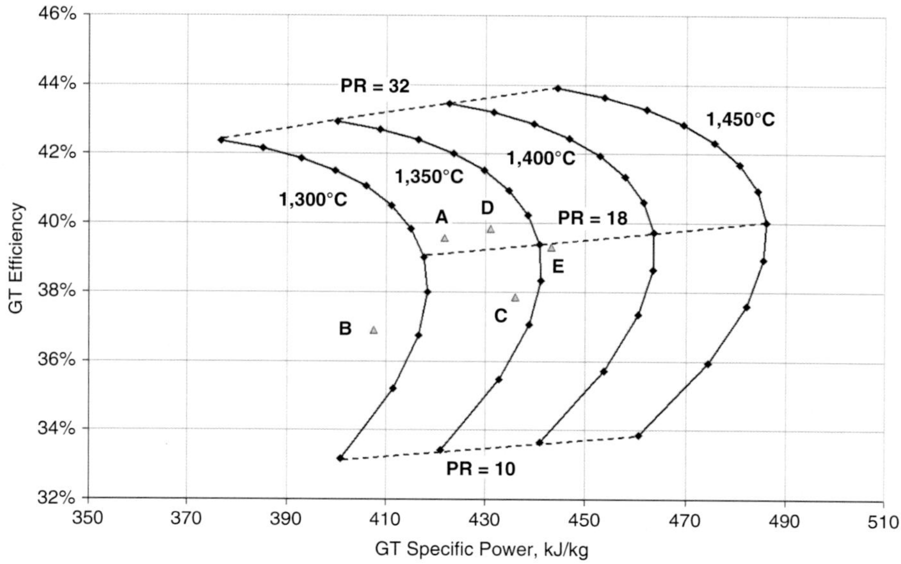


**Fig. 15.12** "Real" gas turbine technology curves for different PR-RIT combinations [46] (Also shown is data from Table 15.2 (as published in Ref. [6]))

published works that range from Holland and Thake [36] and several papers by Elmasri [37–41] during 1980s to the more recent works of Horlock and coworkers [42–45]. Young and Wilcock's comprehensive study [45] and the similar treatment in Ref. [35] incorporate detailed aerothermodynamic calculations. On the business side, obviously not available for public consumption, OEMs have their own in-house proprietary codes containing the field-proven knowledge (instead of theoretical approximations) distilled from decades of experience in manufacturing and testing gas turbines. The results from a more realistic GT calculation [46], which includes the impact of turbine cooling with compressor extraction air, are shown in Fig. 15.12.

Myriad alternate GT design variations have been proposed and entered the literature, some to the extent that they are even covered in elementary textbooks; e.g., compressor intercooling, regeneration (also referred to as recuperation), a combination of both, steam injection (STIG) [47, 48], humid air turbine (HAT) [49], chemical recuperation [50], and closed cycles [51] are among the most well known. For a comprehensive look at those variants and their relative merits, as well as references to the pertinent literature, one can consult the relevant sections in Wilson and Korakianitis [11]. Intercooling and recuperation technologies have been demonstrated successfully in aeroderivative gas turbines [52,53], whereas the combination of both (ICR) has been incorporated into a ship-propulsion unit [54]. Nevertheless, steam-cooled (G- or H-Class) and reheat machines are the only two commercially proven and successful heavy-duty industrial variants of the basic air-cooled GT, which are suitable to large-scale base load electric power generation.

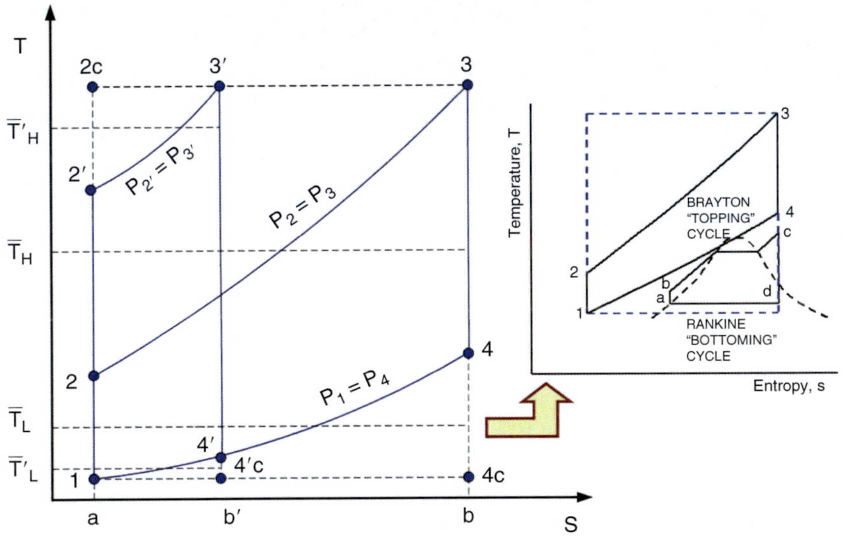


**Fig. 15.13** Comparison of Brayton cycles with different PRs on a T-s diagram. Also illustrated is the "combined" cycle concept, in which the Rankine "bottoming" cycle partially recovers the Brayton "topping" cycle heat rejection "lost" work

So far it has been demonstrated that increasing the compressor PR and/or TIT (i.e., maximum cycle temperature) increases the GT efficiency and specific work. The latter reaches a maximum before further increase in PR decreases it. Furthermore, increasing TIT is favorable to CC-specific work and CC efficiency. These trends can be easily visualized and quantified by the ideal cycle diagrams shown in Fig. 15.13.

The heat rejection "lost work" quantified by the lower triangular area 1-4-4c-1 suggests that a better approximation of the Carnot cycle can be achieved by utilizing the heat source provided by the working fluid at the GT exhaust (i.e., state 4) in a second power generation cycle. At this point, the Rankine cycle presents itself as the ideal choice due to two facts:

1. Good match with GT exhaust (on average about 600°C for SOA F-Class machines) for steam generation at 125–165 bar
2. Constant temperature heat rejection (i.e., in the steam condenser) at a temperature that is much closer to $T_1$ than $\bar{T}_{\mathrm{L}}$

Earlier in the section, it was shown that the Brayton cycle efficiency is given by the mean-effective cycle heat addition and rejection temperatures, $\bar{T}_{\mathrm{H}}$ and $\bar{T}_{\mathrm{L}}$, respectively, e.g., Eq. (15.16). Increasing the cycle maximum pressure from $p_2$ to $p_2{}'$ increases the cycle efficiency by simultaneously increasing $\bar{T}_{\mathrm{H}}$ and reducing $\bar{T}_L$. It is easy to see that the cycle-specific work, quantified by the area 1-2-3-4, would go through a maximum before approaching zero at which point the cycle efficiency is the Carnot efficiency. The ultimately detrimental effect of continuously increasing the Brayton "topping" cycle (BTC) PR on CC efficiency can also be seen from Fig. 15.13. The maximum possible CC "bottoming" cycle work, limited by the second law, is equal to the triangular area 1-4-4c-1, which can be shown to be exactly equal to the GT exhaust gas exergy [55]. It is easy to visually appreciate the steady reduction in this maximum with increasing PR (cf. 1-4-4c-1 and 1-4′-4′c-1). The efficiency of the ideal Rankine bottoming cycle (RBC) is given by the equation below:

$$\eta_{\mathrm{RBC}} = 1 - \frac{T_1}{\bar{T}_{\mathrm{L}}} \tag{15.21}$$

where $\bar{T}_{\mathrm{L}}$ is from Eq. (15.15). The formula in Eq. (15.21) clearly implies that this particular ideal Rankine cycle would be equivalent to a Carnot engine operating between the temperature reservoirs at $\bar{T}_{\mathrm{L}}$ and $T_1$. Furthermore, the product of Eq. (15.21) and the total energy content of the GT exhaust $\dot{Q}_{\mathrm{exh}}$ can be shown to be exactly equal to the total exergy of the exhaust gas [55]. Exergy is a fluid property that can be calculated using a suitable equation of state with two other known properties (e.g., pressure and temperature) and the composition.

By looking at the cycle T-s diagram in Fig. 15.13, the ideal CC efficiency with ideal Brayton and Rankine cycles can be deduced to depend on the ratio of the mean-effective Brayton cycle heat addition temperature given by Eq. (14) and the ambient temperature $T_1$, i.e.,

$$\eta_{\mathrm{CC}} = 1 - \frac{\ln\left(\frac{\tau_3}{\pi^k}\right)}{\tau_3 - \pi^k} \tag{15.22}$$

Ignoring the miscellaneous topping and bottoming cycle losses and minor inputs, a simplified version for the CC shaft efficiency can also be written as

$$\eta_{\mathrm{CC}} = \eta_{\mathrm{BTC}} + (1 - \eta_{\mathrm{BTC}}) \cdot \eta_{\mathrm{RBC}} \tag{15.23}$$

Taking the derivative of both sides with respect to the PR and setting the CC efficiency derivative to zero for the maximum, one finds that (note that the RBC and BTC efficiencies are approximately the same in magnitude)

$$\frac{\partial \eta_{\mathrm{BTC}}}{\partial \mathrm{PR}} \approx -\frac{\partial \eta_{\mathrm{RBC}}}{\partial \mathrm{PR}} \tag{15.24}$$

This relationship states that the maximum *Brayton–Rankine* combined cycle efficiency occurs at the point where the rate of increase of BTC efficiency with PR is the same as the rate of decrease of the RBC efficiency. Using the ideal Brayton cycle formulas, which can be found in any standard thermodynamics textbook such as Ref. [28], the rate of increase of BTC efficiency with increasing PR is

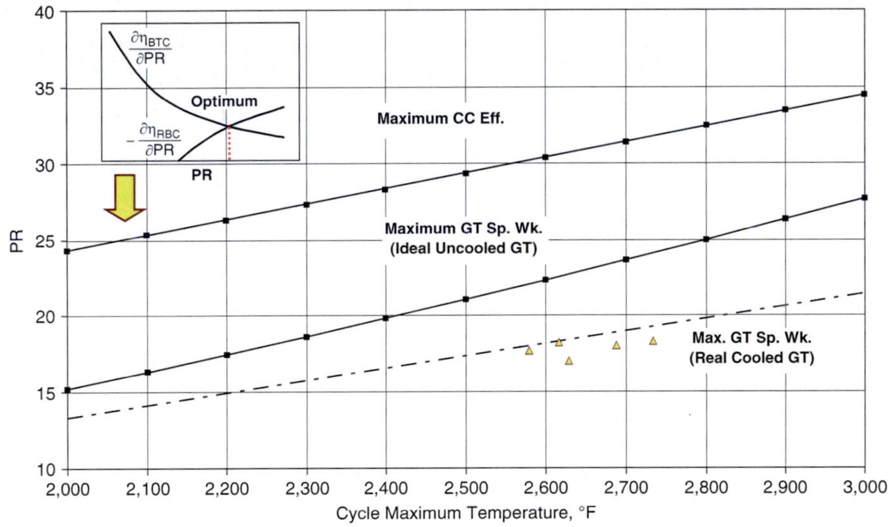


**Fig. 15.14** PR for maximum CC efficiency and maximum BTC specific work. Data points are from Ref. [6]. Cycle maximum temperature or $T_3$ values are from Table 15.2. The inset illustrates the calculation of maximum $\eta_{CC}$ point using Eqs. (15.24–15.26)

$$\frac{\partial \eta_{\mathrm{BTC}}}{\partial \mathrm{PR}} = \frac{\gamma - 1}{\gamma} \cdot \mathrm{PR}^{\frac{1-2\gamma}{\gamma}} \tag{15.25}$$

From Eq. (15.21), once again using the ideal Brayton cycle relationships, the rate of decrease of the RBC efficiency with increasing PR is found as

$$\frac{\partial \eta_{\mathrm{RBC}}}{\partial \mathrm{PR}} = \frac{\gamma - 1}{\gamma} \cdot \frac{\left(\frac{T_1}{T_L}\right) \cdot \left(\frac{T_4}{T_L} - 1\right)}{\mathrm{PR} \cdot \ln\left(\frac{T_4}{T_1}\right)} \tag{15.26}$$

Combining Eqs. (15.21–15.26) with $\gamma = 1.4$ and $T_1 = 85°\mathrm{F}$ (assuming RBC heat rejection at 1.2 in. Hg or 41 mbar of steam condenser pressure) and solving them numerically, the PR of the BTC for maximum CC efficiency can be plotted as a function of $T_3$ (see Fig. 15.14). Also plotted in Fig. 15.14 is the PR corresponding to the maximum BTC specific work, $\pi = \sqrt[2k]{\tau_3}$, with $T_1$=59°F. For comparison, representative OEM data extracted from a recent trade publication [6] is also shown. It is interesting to note the excellent agreement of the PR-TIT data of actual production machines with values corresponding to maximum GT-specific power output predicted using a rigorous cooled gas turbine model.

Using an uncooled *quasi-ideal* GT model with reasonable assumptions outlined earlier in the section to take into account the real turbine losses, CC efficiencies

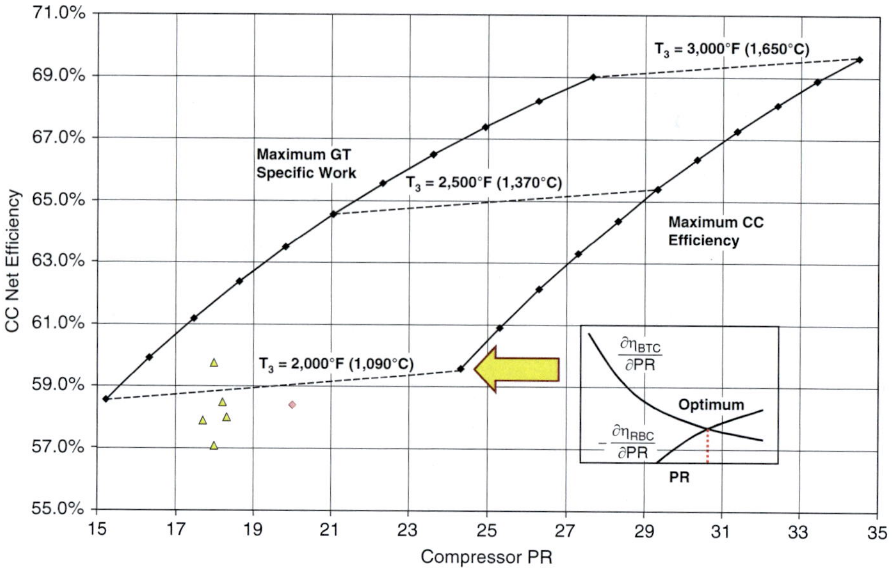


**Fig. 15.15** Uncooled GT-based CC efficiencies (note that $T_3 \equiv$ TIT). The data (symbol Δ) is from Table 15.4. The other data is from Ref. [35] for the SOA air-cooled GT (TIT=2,826°F). The inset illustrates the calculation of maximum $\eta_{CC}$ point using Eqs. (15.24)–(15.26) in the text. Component polytropic efficiencies are assumed to be 92.5% and 90% for the compressor and turbine, respectively

corresponding to the parameters in Fig. 15.14 are calculated and plotted in Fig. 15.15. The key takeaways from the plots in those figures are rather illuminating:

1. Actual OEM designs are at PR values somewhat lower than those for maximum GT-specific output in an uncooled GT but coincide extremely well with the air-cooled GT specific work line, in accordance with the earlier observation made by Horlock [24].
2. This seems to be a reasonable choice from a cost perspective with ~0.50 percentage points sacrifice in performance vis-à-vis the maximum $\eta_{CC}$ that has a ~50% larger PR. In fact, the proximity of the two maxima is even closer in the presence of turbine cooling so that this is a very reasonable trade-off indeed.
3. 65% and 70% net CC efficiencies require uncooled turbines with TIT of 2,500°F and 3,000°F, respectively.

OEM data for combined cycle performances based on five 50-Hz (3,000 rpm) heavy-duty industrial gas turbines from a recent trade publication [6] (see Table 15.2) are given in Table 15.4. The data is for two major combined cycle configurations, which are labeled as 1 × 1 (i.e., one GT and one ST) and 2 × 1 (i.e., two GTs and one ST). Using the GT performance in Fig. 15.12 as a basis, CC performance curves can be developed using the exergy-based bottoming cycle calculations in Ref. [56] (see Fig. 15.16).

**Table 15.4** OEM data for combined cycle performances (1×1 and 2×1) of 50-Hz heavy-duty industrial gas turbines [6] (Gas turbine data is in Table 15.2)

| | | 1×1 | | | 2×1 | | |
|---|---|---|---|---|---|---|---|
| | GT | ST | CC | | ST | CC | |
| Turbine | MW | MW | MW | $\eta$ (%) | MW | MW | $\eta$ (%) |
| A | 285.0 | 141.0 | 415.1 | 57.9 | 282.0 | 830.2 | 57.9 |
| B | 255.6 | 141.8 | 390.8 | 56.7 | 289.2 | 786.9 | 57.1 |
| C | 279.2 | 157.9 | 412.9 | 58.0 | 315.3 | 825.4 | 58.0 |
| D | 292.0 | 144.7 | 423.0 | 58.4 | 290.3 | 848.0 | 58.5 |
| E | 312.1 | 157.3 | 459.0 | 59.5 | 317.7 | 921.1 | 59.7 |

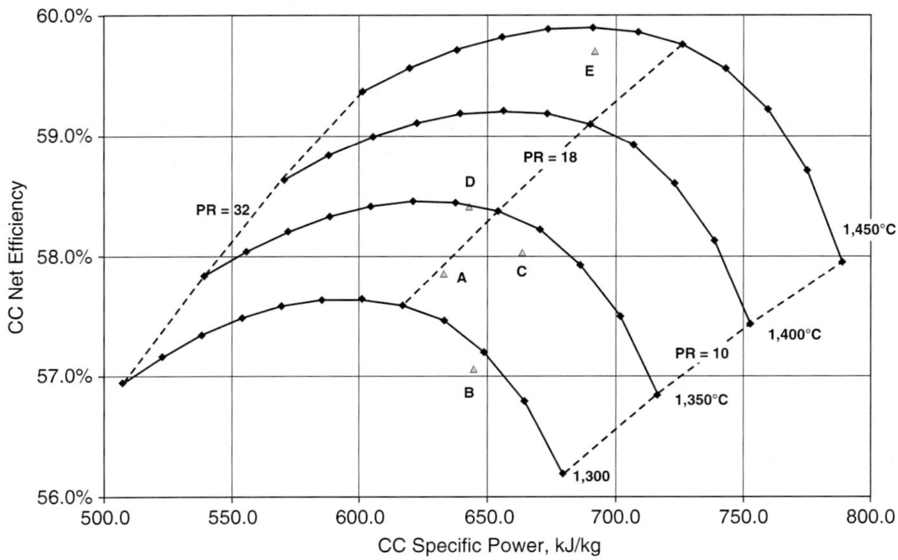


**Fig. 15.16** "Real" combined cycle technology curves for different PR-RIT combinations. They are based on the gas turbine technology curves in Fig. 15.12. (Also shown is data from Table 15.4 (as published in Ref. [6]))

Today's SOA in CC performance with advanced gas turbines corresponds to a range of 57% to ~60% with TIT values in the range of 2,600–2,800°F. This is a far cry from the theoretically possible values in the absence of turbine cooling as shown in Fig. 15.15 (by about 10 percentage points). Sponsored by the DOE's National Energy Technology Laboratory (NETL), the *Vision 21* program has a goal to develop, by 2015, the core modules for a fleet of fuel-flexible, multiproduct energy plants that boost power efficiencies to more than 60% (75% on an LHV basis for gas-fueled plants), emit virtually no pollutants, and with carbon sequestration release minimal or no carbon emissions. Lately, several published claims about the future of the CC plant performance at the end of the first quarter of the twenty-first century stated a goal of 65% with the possibility of over 70% with an air-cooled GT [57]. The uncooled GT–based CC efficiency curves and the current SOA data in Fig. 15.15

underline the significant hurdles to be overcome in terms of cooling flow reduction in pace with increasing firing temperatures in order to achieve those goals. The gap between the uncooled GT TIT (i.e., same as RIT) requirement for 65% $\eta_{CC}$ and the current material capability is nearly 800°F (~450°C). The current trend in superalloy capabilities, considering the competing tendencies in strength (i.e., creep rupture and fatigue limits) and corrosion resistance, does not seem to be likely to alleviate too much of that [58]. A combination of advanced materials including ceramic matrix composites, metal foams (to enable the upper limit of film cooling, i.e., effusion or transpiration cooling) [59] and external cooling (most likely via steam) accompanied by bottoming cycle advances is imperative to even approach the stated efficiency goals with acceptable $NO_x$ and CO emissions. It is thus fairly obvious that a $\eta_{CC}$ goal of 65% and higher in a *Brayton–Rankine* CC power plant is a highly unlikely proposition in near future. This is more or less the conclusion reached by Rao et al. [60] who investigated the plant system configurations that might be able to meet the performance and emission goals of Vision 21 program. They concluded that hybrid systems integrating fuel cells with gas turbines, e.g., SOFC/GT hybrid system [61], were required to meet the program goals. There's little doubt about the very attractive efficiency of the SOFC/GT hybrid systems, e.g., 65–70% for atmospheric cells and 74–76% for pressurized cells [62]. What is really doubtful is whether they will be competitive with gas turbine–based systems for large-scale electric power generation considering their low power density, e.g., less than 5 kW/m$^2$ (cf. ~3,000 kW/m$^2$ for a large GT), and high cost, e.g., about \$1,500/kW to \$3,000/kW in 2010 dollars [63] (cf., \$350/kW to \$500/kW for a GT depending on its size).

## *Today's SOA in CC Power Plants*

A schematic diagram of a CC system is shown in Fig. 15.17, which depicts the four major components: gas turbine, steam turbine, condenser, and the heat recovery steam generator (HRSG). Not shown is the cooling tower (CT), which cools the condenser cooling water. Usually, the heat rejection system (including condenser, CT and circulation pump), and feed pumps along with myriad smaller heat exchangers (such as the GT fuel gas heater in the upper left corner in Fig. 15.17) are collectively referred to as the "balance of plant" (BOP).

The equipment that combines the GT (i.e., Brayton cycle) and ST (i.e., Rankine cycle) is the HRSG. Hot GT exhaust gas (>1,100°F for modern F-Class units) transfers heat to condensate and generates superheated steam, typically at multiple pressure levels, which is used in the ST for power generation. Current SOA is a three-pressure system with reheat (i.e., heating of HP steam turbine exhaust, referred to as "cold reheat," to the same temperature as the main or HP steam) with steam temperatures of 1,050°F or higher. Highest steam pressure at the ST inlet (throttle) is around 1,800–2,500 psia, beyond which the pressures at the generation point in the HRSG exceed the critical pressure and one would require a "once-through" steam generation system as opposed to the existing "drum type"

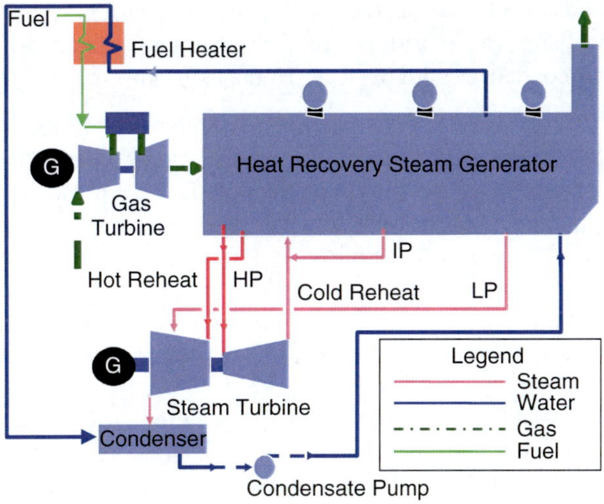


**Fig. 15.17** Three-pressure, reheat (3PRH) combined cycle diagram (Source: GE Energy [64])

system. From the fundamental thermodynamics discussed earlier, the theoretically maximum (net) power output that can be obtained from the Rankine bottoming cycle of the CC plant is exactly equal to the exergy of the GT exhaust gas [55]. Today's most advanced plants with advanced F-, G-, and/or H-Class GTs and 3PRH steam systems, are capable of achieving ~72% of that theoretical maximum, which can only be achieved in a Carnot engine [56]. For rapid estimations, this translates into an ST power output that is about 50% of the combined power output of all GTs in the plant.

As the fundamental CC efficiency relationship, Eq. (15.23), and the graphical representation in Fig. 15.13 make it very clear, the cycle's heat rejection temperature is of prime importance. This temperature is dictated by the steam condenser's operation pressure, which is significantly below the atmospheric pressure. For the best performance, lowest possible steam condensation pressures, around 1.5 in. of Hg or even lower, are imperative. Achieving those low-pressure levels requires striking a delicate balance between heat rejection equipment size, cost, type, and site ambient conditions. The key heat rejection system equipment is the steam condenser. There are many types of condensers that are used to handle the heat rejection from the CC plant. Depending on the site ambient conditions, cooling medium availability, environmental regulations and customer-specific economic criteria, a suitable choice can be made. The major options are listed below.

1. Water-cooled surface condenser
   - Once-through or open-loop (utilizing a naturally available cooling water source such as lake, river, ocean)
   - Closed-loop with a natural (very rare for CC plants) or mechanical (forced or induced draft) cooling tower (CT)

2. Dry, air-cooled (also known as the "A Frame") condenser
3. Air-cooled wet surface (hybrid) or "plume abatement" condenser
4. Direct contact condenser with dry CT (Heller) system

In general, water-cooled systems, in particular open-loop systems near a relatively cold natural source of cooling water (e.g., in coastal locations in the northern hemisphere), offer the best performance, i.e., the lowest condenser pressure. How low in condenser pressure one can go is essentially a question of plant economics. Recently, permitting requirements in conjunction with environmental considerations and water scarcity in many geographic locations render water-cooled heat rejection systems infeasible. Significant size and cost of "dry" or air-cooled systems preclude the optimal condenser pressures and optimistic performance projections displayed in Fig. 15.16.

The key to affordable and sustainable large-scale electric power generation is economies of scale that can be achieved by ever-larger prime movers (gas and steam turbines of 300+ MW) and plant blocks. Modern CC plants include one or more GTs with a single ST; either in a "multi-shaft" or "single-shaft" configuration (e.g., see Figs. 15.18 and Figure 15.19). The most common multi-shaft block is $2\times1$ (i.e., 2 GTs and 1 ST) with power ratings of 600–900 MW. A few larger plants in $3\times1$ or $4\times1$ configuration or in multiple blocks of $2\times1$ are also available (e.g., see Fig. 15.18). Single-shaft (i.e., $1\times1$) plants with power ratings of 300–500 MW per "block" can also be combined into very large generating stations via multiple blocks. For a representative selection of modern CC power plant configuration options and performances, the reader is referred to the white papers listed in [64–68].

Even better economies of scale are achieved via the "reference plant" approach, in which OEMs develop pre-engineered modules that can be shipped and installed anywhere in the world with minimal customization to accommodate different site and customer requirements. This ensures that natural gas burning GT-CC power plants are not only the most efficient means of generating electric power available today, but also the most cost-effective (e.g., at a turnkey capital cost of $600–1,000/kW and cost of electricity of a few cents per kWh)) and fastest to build (e.g., less than 24 months of construction time). On top of all that, GTs with modern "dry low $NO_x$" combustors are also the cleanest fossil fuel burning technology in terms of emissions. This aspect will be explored in detail in the following section.

## *Cogeneration*

Discussion thus far has focused on production of electricity from natural gas. When electricity is the sole product from the thermal cycle the heat rejected to ambient is termed "waste heat" because it is no longer economically useful for generation of additional power. In the larger context of a modern industrial society one finds that low temperature heat of the quality least useful for power generation has many

**Fig. 15.18** Illustration of a 2×207FA (two Frame-7 60-Hz GTs with single ST) NG CC power plant. From *left* to *right*, for each GT-HRSG "block" one can see the GT inlet air filter housing and duct (on top of the GT enclosure), HRSG with drums on the top, and the stack. Two STs are located to the right of the GT-HRSG blocks (cross-over pipe and down-flow LP section are clearly visible). Note the elevation of the ST structure to accommodate the steam condenser. Partially visible to the right is the multicell mechanical cooling tower. In the lower right corner are two distillate oil (GT backup fuel) tanks (Courtesy: GE Energy)

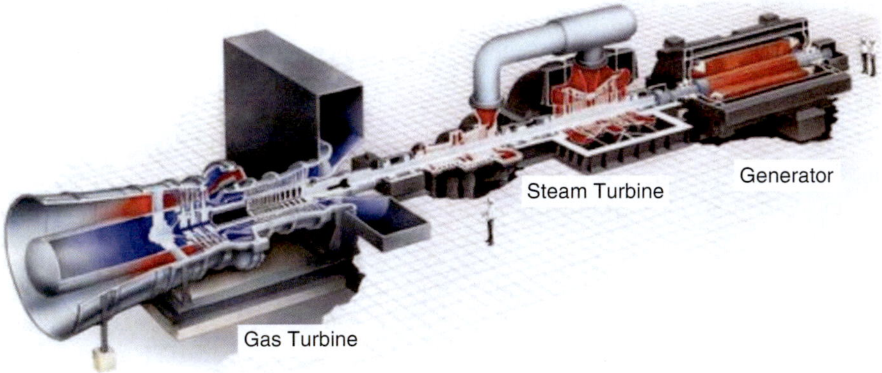


**Fig. 15.19** A single-shaft combined cycle train (Courtesy: GE Energy)

economically important uses. It is thus common in industrial and urban settings to integrate a power plant with another process such that heat energy for that process is provided in whole or in part by the power cycle after the highest quality fuel energy

**Table 15.5** A sampling of natural gas Fired cogeneration opportunities [71]

| Industrial | |
|---|---|
| Paper and board manufacture | Pharmaceuticals and fine chemicals |
| Ceramics | Brewing, distilling, and malting |
| Textile processing | Food processing |
| Oil refineries | Minerals processing |
| Timber processing and paper | Horticulture and glasshouses |
| **Buildings** | |
| Hotels | District heating and cooling |
| Airports | Hospitals |
| Supermarkets and large stores | College campuses and schools |
| Individual houses | Office buildings |

has produced some useful work. Such installations are termed cogeneration systems or Combined Heat and Power (CHP) systems and can be very efficient from an environmental and fuel use perspective. Table 15.5 provides a sampling of cogeneration opportunities with a synergistic combination of heat and power needs.

In the ideal case, the heat energy available from the prime mover (engine or thermal cycle) will always match the needs of the customer process. In practice, this is challenging since many process heat demands are variable (such as building or district heating). The electrical loads can also be out of phase with process heat demands (e.g., an electric grid with high but unsteady wind or solar renewable power contributions). These variations in electrical load vs. process heat loads can be accommodated with appropriate system design and operating strategies. A variety of cogeneration plant options are available between the extremes of a non-cogeneration facility, e.g., fired heater or boiler, with no power generation at all (e.g., a home heating system capturing the heat content of the fuel but wasting its potential to do work [69]) and the aforementioned power generation combined cycle with no process heat demand, which has accordingly been optimized for production of electricity.

Primary considerations in cogeneration system design are the size of the plant, the relative need for heat vs. electrical power, and the electricity and process heat demand variability. The following points summarize the range of system design solutions:

- Large industrial gas turbines ranging from ~20 to 300 MW electrical output are optimized for best combined cycle efficiency, and have exhaust energy of about 60% of heat input with exhaust temperatures between about 1,020°F and 1,160°F (550°C and 625°C). This class of machine is well suited to process heat demands that are large in an absolute sense but small or intermittent in relation to electrical power demand, (0% to ~30% of gas turbine heat input). This level of heat energy supply can readily be accommodated by modifying the power generation combined cycle system to include the capability to extract or divert steam from the bottoming cycle steam turbine to the process. This is a common solution for district heating which may be highly variable and seasonal.

- Aeroderivative gas turbines ranging from ~5 to 60 MW electrical output are optimized for best simple cycle efficiency, and have exhaust energy of about 55% of heat input with exhaust temperatures of about 850°F (455°C). These machines are large enough to support substantial process heat loads and are often configured in a system without a steam turbine to save cost and make a higher portion of gas turbine exhaust energy available to process. In many industrial cogeneration facilities, it is more important to match process heating demand than to meet a specific electric load. The gas turbine load is thus varied to follow process energy demand independent of electric load. Any excess electricity generated is sold to the grid and any power shortfall is purchased. If the design process heat demand is high in relation to electric load, or may occasionally need to be much higher than the unfired gas turbine exhaust energy, the system will need to include a supplemental burner in the HRSG, which can be used to boost process heat export capability. A burner may also be used in combination with part load operation of the gas turbine to maintain process heat export while reducing electricity production to match its load.
- A micro gas turbine ranging from ~0.03 to 0.3 MW electrical output may be selected if the heat load and/or electric load is small. Micro turbines typically employ uncooled regenerative cycles for best simple cycle efficiency, have exhaust energy of about 75% of heat input, and exhaust temperature of about 530°F (275°C). These units are too small to economically justify a bottoming cycle steam turbine so that they fall into the unfired or supplementary-fired heat recovery–type system configuration.
- In the size range between micro turbines and aeroderivatives, a variety of small industrial gas turbines and other prime movers such as a spark ignition reciprocating engine, or fuel cell, may also be considered. In the range from ~10 to 150 MW, direct-fired steam Rankine systems with steam turbine extraction for process heat can also be an attractive configuration that is well suited to highly variable process heat needs. Power can be varied across a wide range with modulated fuel consumption and heat export can also be widely varied via controlled steam extraction. Efficiency at low process demand is inferior to gas turbine–based systems but fuel flexibility is generally better.

Table 15.6 shows some examples of gas turbine–based natural gas–fired cogeneration systems. These are presented to elucidate the behavior of such systems and draw conclusions regarding the role of cogeneration in promoting sustainable use of natural gas. Columns tabulate net electric power export, export heat energy to power ratio, quality of export heat expressed as a ratio of its availability to produce work (exergy), electric generation efficiency, cogeneration efficiency expressed on both an energy and exergy basis, and specific $CO_2$ emissions also expressed on both an energy and exergy basis. Export heat is provided as steam in all cases.

Cogeneration efficiency $\eta_{c,I}$ in Table 15.6 is the straight ratio of the sum of power (kWe) and heat (kWth) products to the fuel burned in the plant on an LHV basis. This is a commonly used (sometimes with fuel input on an HHV basis) measure, which is ultimately misleading because it does not differentiate between the "quality" of different products. In certain cases, especially with supplementary firing, it can reach improbably high values such as 90% or more.

A more rational measure is the cogeneration efficiency per the second law of thermodynamics, $\eta_{c,II}$, which uses the exergy of the thermal product rather than its energy (i.e., enthalpy) [70]. Exergy, as discussed earlier in the present article, is a direct measure of the work generating ability of a material stream via Kelvin–Planck statement of the second law of thermodynamics. Exergy is a property of a given fluid and, as such, for two given properties (e.g., pressure and temperature), it can be calculated using a suitable equation of state such as ASME steam tables. The relationship between $\eta_{c,II}$ and $\eta_{c,I}$ is given as

$$\eta_{\mathrm{c,II}} = \eta_{\mathrm{c,I}} \cdot \frac{1 + \beta \cdot \theta}{1 + \theta}$$

where $\theta$ is the heat-to-power ratio of the cogeneration plant, and $\beta$ is the ratio of the exergy to energy (enthalpy). Since $\beta < 1$, $\eta_{c,II} < \eta_{c,I}$ and has typically a value, which is more in line with standard power plant efficiencies. In passing note that the PURPA (*Public Utility Regulatory Policies Act*, a US federal law enacted in 1978 intended to encourage more energy-efficient and environmentally friendly commercial energy production) efficiency, which is another cogeneration plant efficiency definition, is equivalent to $\eta_{c,II}$ with $\beta = 0.5$. Even $\eta_{c,II}$ is an inflated value because $\beta$ is the efficiency of a hypothetical Carnot engine that converts the thermal energy to useful shaft work.

A realistic turbine that can utilize that thermal efficiency for electric power generation would have a much lower efficiency, which is characterized by the rational efficiency, $\varepsilon$, which is the ratio of actual turbine power output to that of the hypothetical Carnot engine. The most common thermal energy product in cogeneration is steam. Depending on the pressure and temperature of steam, with today's state of the art in ST technology, a reasonably good estimate for $\varepsilon$ would be around 0.7. Thus, a more realistic value for $\eta_{c,II}$ would be given as

$$\eta_{\mathrm{c},II} = \eta_{\mathrm{c,I}} \cdot \frac{1 + \epsilon \cdot \beta \cdot \theta}{1 + \theta}$$

*Case 1* is a micro turbine–based design with no supplemental fuel addition to exhaust heat recovery. Export energy is high in relation to power production since the engine is only 25% efficient but export heat quality is low since exhaust temperature is a modest 530°F (275°C). This low exhaust temperature also compromises the cogeneration energy efficiency in these examples because only a portion of the unfired exhaust energy is recoverable to steam.

*Case 2* adds a supplemental burner fired to 1,600°F (871°C) to boost export energy substantially while also improving its quality. Fuel heat energy is now fully utilized to provide power and process heat (93% cogeneration efficiency, energy

**Table 15.6** Typical heat-to-power ratios and efficiencies for GT-based CHP systems

| | | Net power (MWe) | $\theta$ | $\beta$ | $\eta_{el.}$ (%) | $\eta_{c,I}$ (%) | $\eta_{c,II}$ (%) | Engine+Burner $CO_2$ Emissions | |
|---|---|---|---|---|---|---|---|---|---|
| | | | | | | | | *A* | *B* |
| 1. | Unfired micro GT | 0.1 | 1.74:1 | 0.29 | 25.4 | 69.6 | 38.3 | 284.7 | 516.9 |
| 2. | Supplementary fired micro GT | 0.1 | 10.0:1 | 0.44 | 8.6 | 93.0 | 46.6 | 212.9 | 418.3 |
| 3. | Unfired aeroderivative GT | 40 | 1.13:1 | 0.41 | 41.2 | 87.8 | 60.4 | 225.4 | 327.7 |
| 4. | Supplementary fired Aeroderivative GT | 40 | 2.63:1 | 0.44 | 26.3 | 93.0 | 56.6 | 212.9 | 349.6 |
| 5. | Unfired industrial GT | 100 | 1.50:1 | 0.45 | 35.6 | 89.0 | 59.5 | 222.5 | 332.8 |
| 6. | Supplementary fired Industrial GT | 100 | 2.84:1 | 0.44 | 24.8 | 93.0 | 55.7 | 212.9 | 355.4 |
| 7. | Gas turbine combined cycle (unfired, industrial GT) | 100 | 0.1:1 | 0.29 | 54.3 | 59.7 | 55.9 | 331.8 | 354.4 |
| | | | 0.5:1 | 0.34 | 49.1 | 73.6 | 57.4 | 269 | 344.7 |
| | | | 1.0:1 | 0.38 | 42.4 | 84.9 | 58.7 | 233.3 | 337.2 |
| 8. | Gas turbine combined cycle (industrial GT with supplementary firing) | 100 | 1.5:1 | 0.44 | 35.2 | 88.0 | 58.5 | 225.1 | 338.8 |
| | | | 2.0:1 | 0.44 | 31.6 | 93.0 | 59.4 | 212.9 | 333.2 |

$\theta$ is heat to power ratio in kWth/kWe
$\beta$ is heat quality, which is the ratio of export exergy to export energy
$\eta_{el.}$ is electric generation efficiency
$\eta_{c,I}$ is first law cogeneration efficiency (energy-to-fuel input)
$\eta_{c,II}$ is second law cogeneration efficiency (exergy-to-fuel input)
Emissions are in g/kWh of energy (*A*) and exergy (*B*)

basis), but the performance is still modest on an exergy basis since the engine is inefficient. Note also that specific $CO_2$ emissions (exergy basis) improved because the incremental efficiency due to increased and improved heat export is better than the electrical efficiency of the gas turbine.

*Cases 3* and *4* show the corresponding story for a 40 MW aeroderivative gas turbine while *cases 5* and *6* present a 100 MW industrial gas turbine scenario. These gas turbines both improve exergetic performance over the micro turbine cases but when supplemental firing is added the cogeneration efficiency and specific $CO_2$ emissions (exergy basis) both degrade. The more efficient aeroderivative has slightly higher system performance but lower process heat export capability vs. the industrial gas turbine.

*Case 7* shows the 100 MW industrial gas turbine in a combined cycle configuration with increasing steam extraction to process. Electric generation efficiency falls off as more energy is diverted to process while cogeneration efficiencies improve. *Case 8* begins where case 7 left off with power output maintained as supplemental firing is increased. By the time 1,600°F (871°C) supplemental firing temperature is reached the export heat energy has approximately doubled. As supplementary firing increases the specific $CO_2$ emissions (exergy basis) stays approximately constant.

Note that the supplementary-fired scenarios presented here stop at 1,600°F (871°C) consistent with uncooled metal ductwork. Firing temperature can be pushed further by application of refractory or water-cooled walls until the excess oxygen is nearly expended; or increased even further if supplemental air is also added. High supplementary firing may be required for some applications but is less desirable from a resource conservation perspective than a design with little (or no) supplementary firing.

The following general conclusions can be drawn from this sampling:

1. Cogeneration efficiency should be expressed on an exergy basis to promote the most sustainable use of fossil fuels. Direct use of fuel for low-quality end uses represents a large waste of the thermodynamic potential inherent in the fuel. Cogeneration efficiency as commonly expressed in relation to export energy gives the false impression that supplementary firing improves the environmental ($CO_2$) and resource (natural gas) utilization performance.
2. Cogeneration system configuration will be a function of the quality and quantity of heat energy required, its variability, and the relative heat and power needs. A wide range of applications can be served by appropriate system design.
3. Gas turbine–based systems are most efficient when not supplementary fired while gas turbine combined cycle cogeneration systems maintain high cogeneration exergetic efficiency even with supplemental firing. This is because fuel energy can be recovered to power all the way down to the temperature level required by the process user.

Figure 15.20 shows cogeneration's contribution to electricity production in European Union countries.

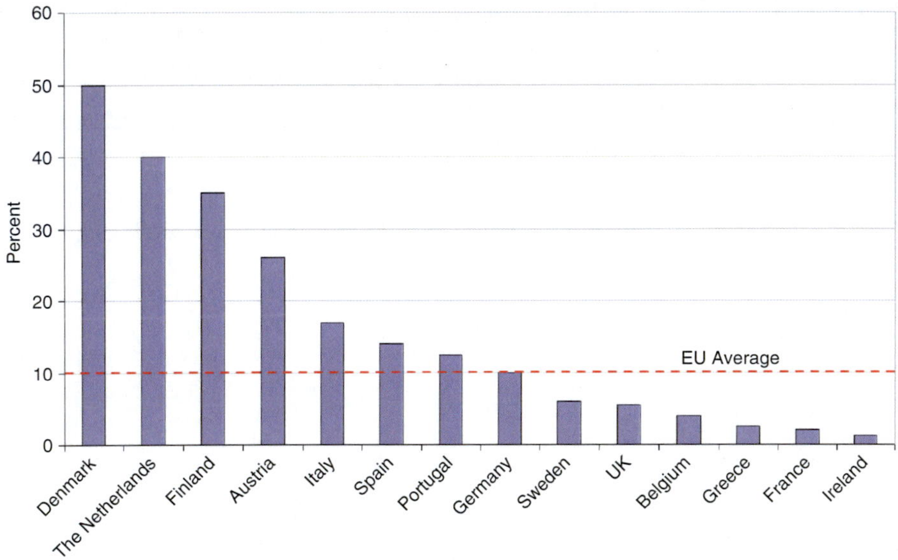


**Fig. 15.20** Cogeneration as a share of national power production (1999) (Source: European Association for the Promotion of Cogeneration [71])

The higher penetration seen in four of these countries is attributed to a combination of favorable economics and supportive government energy policy [71]. In the USA, approximately 4% of electricity production currently comes from cogeneration facilities of which 71% are fueled by natural gas (e.g., see Table 8.2c in [72]).

How far could cogeneration penetrate in the USA and what benefits would accrue? Just as it is possible to look at the utilization rate of renewable energy sources such as wind, hydro, and solar in relation to the size and distribution of these resources, the thermal power plant heat available for low-temperature end users can be quantified to estimate the potential for increased cogeneration and hence its potential to offset other fuel use and emissions.

The annual US thermal power generation totaled 3.7 trillion kWh in 2008 [72]. Factoring in the mix of natural gas–fired systems, nuclear, and direct-fired (mostly coal) plants and their respective thermal efficiencies allow estimation of heat resources available for cogeneration. Table 15.7 shows an estimated breakdown of heat rejection from thermal power plants in the USA.

While cogeneration with direct-fired or nuclear-fueled Rankine systems is certainly possible, the sheer size of these plants makes it difficult to economically match them to end users that themselves are large enough to benefit from the full heat energy available. For this reason, cogeneration systems tend to be relatively small in terms of electrical output, which in turn aligns well with the NG-fired gas turbine and gas turbine combined cycle. Natural gas accounts for 883 billion kWh

**Table 15.7** Annual US thermal energy available supply and rejection, 2008 [72, 73]

| | Coal | Petroleum | Natural gas Combined cycle | Simple cycle | Cogeneration | Nuclear | Total |
|---|---|---|---|---|---|---|---|
| Electricity generation (Billion kWh) | 1,986 | 46.2 | 562 | 122 | 200 | 806 | 3,721 |
| Heat rate (Btu/kWh, HHV) | 9,884 | 9,884 | 7,598 | 11,526 | 11,156 | 10,488 | |
| Heat input (trillion Btu) | 19,628 | 457 | 4,268 | 1,402 | 2,227 | 8,456 | 36,062 |
| Approximate heat rejection to ambient or process (trillion Btu) | 10,889 | 254 | 1,408 | 847 | 1,159 | 5,451 | 19,730 |
| Heat rejection beta (β) | 0.20 | 0.20 | 0.20 | 0.44 | 0.44 | 0.20 | |

or 23.7% of total electricity generation [72, 73]; 199.7 billion kWh or 22.6% of which comes from cogeneration facilities (e.g., see Table 2.1 in [73]). Taking the data in Table 15.6 for cases 5 and 6 as a typical system configuration, it is possible to estimate the current fuel savings from natural gas–fired cogeneration as follows.

## Natural Gas–Fired Cogeneration (GT+Supplemental Firing)

| | | |
|---|---|---|
| 1 | Electricity generation (billion kWh, 2008) | 200 |
| 2 | Combined cycle heat rate (Btu/kWh, HHV) | 8,305 |
| 3 | GT heat input (trillion Btu, 2008) | 1,658 |
| 4 | System heat rate (Btu/kWh, HHV) | 11,156 |
| 5 | Burner supplemental heat input (trillion Btu, 2008) | 569 |
| 6 | System heat input (trillion Btu, 2008) | 2,228 |
| 7 | Heat export to process (trillion Btu, 2008) | 1,159 |
| 8 | Equivalent heat input to meet process (trillion Btu, 2008) | 1,302 |
| | Cogeneration natural gas savings (trillion Btu, 2008) (Line 8 minus Line 5) | 733 |

The fuel savings from application of cogeneration represents ~24.8% reduction in fuel burn vs. making the power with a gas turbine combined cycle and supplying the heat independently with a fired boiler. $CO_2$ emissions reductions follow the fuel saving directly if the fuel saving is booked as natural gas. If booked at the current fuel mix for US electricity production, the $CO_2$ savings are higher by about 33% or higher by 75% if booked as reduced coal fired generation. Looking at the heat

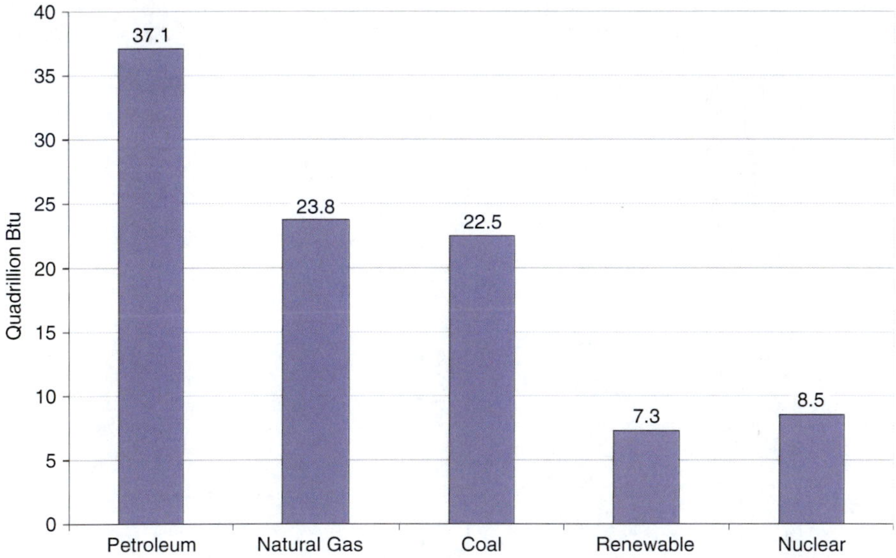


**Fig. 15.21** Primary energy production by source, 2008. Total primary consumption is 99.3 quadrillion Btu. Renewable energy sources are hydroelectric power, geothermal, solar/photovoltaic, wind, and biomass. "Petroleum" does not include the fuel ethanol portion of motor gasoline – fuel ethanol is included in "Renewable Energy." "Natural Gas" excludes supplemental gaseous fuels. "Coal" includes less than 0.1 quadrillion Btu of coal coke net imports (Source: Energy Information Administration (EIA) [72])

rejection from natural gas–fired power generation in the USA in Table 15.7, in comparison to current heat export from natural gas-fueled cogeneration, it appears that the amount of cogeneration could perhaps be doubled without a major shift in fuel use patterns. Reaching the electricity contribution levels seen in some small European countries would likely require more distributed generation with replacement of large coal-fired power plants with more small cogeneration systems (likely gas fired) proximate to potential end users.

Turning now to the potential customers for heat energy, Fig. 15.21 shows the primary energy production by source in the USA for 2008 [72]. Fig. 15.22 shows the incremental retail electric load distributed to each sector and gives a breakdown between residential and commercial sectors. The natural gas contribution to the residential and commercial sectors is 8.2 trillion Btu [72]. Assuming that its primary use is for space heating, it can be seen that it far exceeds the available waste heat from gas-fired power plants tabulated in Table 15.7. It is difficult to make the case that small (micro) cogeneration systems could be aligned to the heat and power needs at the individual house level, though work on fuel cell systems for this purpose is active. The commercial sector and high-density residential (e.g., apartment complex) does offer real potential for expansion. In either case, the performance of the cogeneration system needs to be high enough to improve on the fuel use and environmental performance of an efficient utility scale power plant in combination with direct use of natural gas for heating.

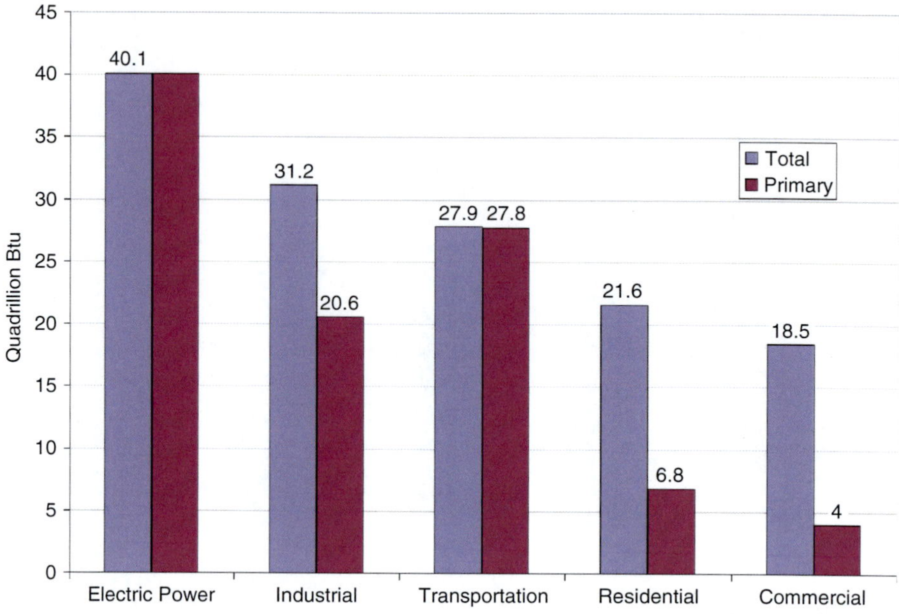


**Fig. 15.22** Primary and total consumption by sector, 2008. See Fig. 15.21 for a breakdown of sources (Source: Energy Information Administration (EIA) [72])

## *Combustion and Emissions*

There are different types of GT combustors that have been used at one time or another by different OEMs in their products:

1. Silo (obsolete now, once used by Siemens and former ABB)
2. Annular (aero-derivatives such as GE LM6000)
3. Can or tubular (heavy-duty industrial GTs by Siemens, GE)

In the following discussion a parameter will appear frequently, i.e., the *equivalence ratio*, $\phi$, which is the ratio of the actual fuel-air ratio, $f$, to the *stoichiometric* fuel air ratio. The latter is the theoretical value requisite for complete combustion. Thus,

$$\phi = f_{\mathrm{act}}/f_{\mathrm{st}}$$

When the equivalence ratio is equal to 1.0, the flame temperature is the highest and the chemical reactions are the fastest. Thus,

1. If $\phi$ is less than 1.0, the mixture that reacts is referred to as "lean" or "fuel lean."
2. If $\phi$ is greater than 1.0, the mixture that reacts is referred to as "rich" or "fuel rich."

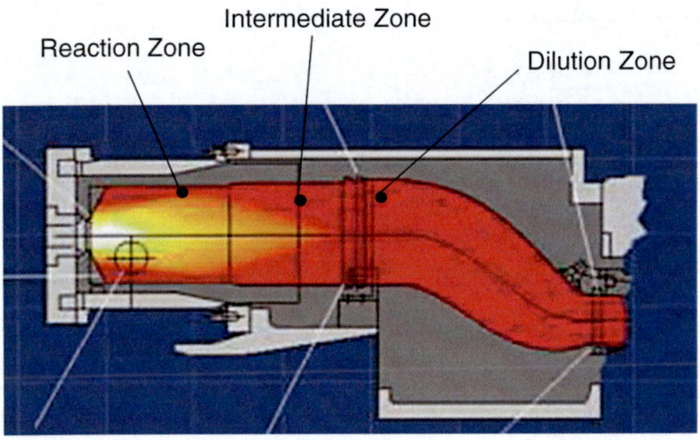


**Fig. 15.23** Single-Nozzle Diffusion (SND) Combustor

In either lean or rich combustion, flame temperature is lower and chemical reaction is slower than their stoichiometric counterparts.

The can-type combustor can be divided into three zones (see Fig. 15.23):

*Primary Zone (PZ)*: This is the combustor *reaction* or flame zone where fuel is injected and mixes with air (about 15–20% of the compressor discharge air is introduced). In order to cap the combustor exit temperatures at a level, which downstream turbine HGP components can tolerate, the requisite value of $\phi$ is typically 0.4–0.5, which is too lean for stable and efficient burning. This is why only a portion of the combustion air is introduced into the primary zone. High-temperature rapid combustion takes place in the PZ. The design of this part of the combustor must facilitate highly turbulent and recirculating flow pattern to anchor the flame at the exit of the fuel injector nozzle. This is critical because of the high-speed of the air stream, even at a fraction of its full value, which is an order of magnitude higher than the turbulent flame speed at which the flame propagates into the fuel-air mixture.

*Secondary Zone (SZ)*: This is the *intermediate* zone where a portion of the remaining combustion air ($\sim$30% of the total) is introduced in the mid-section of the can through holes or cooling slots in the combustor liner. The hot combustion gas from the PZ gets thermally soaked and the combustion is completed. The air that flows between the liner and flow sleeve and enters the can forms a film to cool the liner wall.

*Dilution Zone*: This is where the remaining air enters the can through the metering holes and cools the gases to the desired temperature. A design to promote turbulence and thorough mixing of hot and cold streams is essential to ensure a homogeneous temperature distribution and prevent hot streaks.

A successful combustor design should satisfy the following criteria:

1. Efficient and stable operation, which comprises the following characteristics
2. Complete combustion at or near stoichiometric flame temperature

3. Flame stability and reliability at high turndown
   (a) Ignition
   (b) Crossfire (propagation of flame from fired chambers with spark plugs to the unfired chambers via tubes interconnecting the annular combustion chambers or cans)
   (c) Lean blowout
4. Operational stability and reliability at high turndown
   (a) Dynamics (noise), i.e., large-amplitude pressure oscillations in combustion chamber, driven by heat release oscillations; also known as combustion "humming." These oscillations are destructive to engine hardware.
   (b) Pattern factor, which is a measure of the temperature distribution at the combustor exit defined as $PF = \frac{T_{\text{peak}} - T_{\text{mean}}}{T_{\text{mean}} - T_{\text{inlet}}}$
   (c) Pressure drop
5. Low emissions in compliance with regulatory laws

Achieving these goals simultaneously is very difficult due to the counter-acting impacts of design parameters on each criterion. The design history of the GE heavy-duty industrial gas turbine combustors can be found in Ref. [74]. The design evolution can be summarized in three major designs:

1. Single-nozzle diffusion (SND) combustor
2. Multi-nozzle, quiet combustor (MNQC)
3. DLN or Dry-Low-$NO_x$ combustor

The design evolution was primarily driven by the desire to satisfy the increasingly stringent emission control requirements with improved efficiency via higher combustion and turbine inlet temperatures (TIT). In order to clarify this, one must be familiar with the chemical mechanisms of $NO_x$ production, which is one of the most critical pollutants present in GT exhaust gas. Principal pollutants emitted by a GT are as follows [75]:

1. Oxides of nitrogen ($NO_x$), which are toxic, contribute to chemical smog and lead to depletion of ozone in stratosphere
2. Oxides of sulfur ($SO_x$), which are toxic and corrosive
3. Carbon dioxide ($CO_2$), which contributes to greenhouse effect and global warming
4. Carbon monoxide (CO), which is toxic and asphyxiating
5. Unburned hydrocarbons (UHC), including volatile organic compounds (VOC), which contribute to urban smog
6. Particulate matter (PM), which is suspected to cause respiratory diseases

Detailed information on GT emissions and control can be found in Ref. [76]. Some pollutants, i.e., UHC, PM, and $SO_x$, are generally of no concern for GTs burning gaseous fuel with the exception of coal-gasification SG, which leads to conversion of fuel-contained sulfur to $SO_x$. (Note, however, that in some cases contamination of fuel in the pipeline, e.g., rust particles, can be a problem for HGP components). In any event, currently there is no available GT combustion technology to prevent or control $SO_x$ emissions. Similarly, $CO_2$ is a direct product of the complete combustion process and cannot be controlled or reduced within the GT control volume (except, of course, increasing the thermal efficiency of the GT, which will reduce the amount of fuel consumed and $CO_2$ generated for a given power output). Precombustion or postcombustion capture and storage of carbon dioxide, known as carbon capture and storage (CCS), is currently a hot topic of research and development.

A key pollutant emitted by the GT is $NO_x$. There are two major sources of $NO_x$:

1. Atmospheric $N_2$ in post-flame gases, which leads to formation of NO and $NO_2$ via three principle mechanisms:
   (a) Thermal NO (extended Zel'dovich mechanism)
   (b) Nitrous oxide mechanism
   (c) Prompt NO mechanism
2. Organically bounded nitrogen in fuel or fuel-bound nitrogen (FBN)

The dominant $NO_x$ formation mechanism is the thermal or Zel'dovich mechanism. Although not a significant source of $NO_x$ for NG, FBN in low-quality liquid fuels and LCV gases with hot-gas clean-up (via ammonia) can be high enough to merit attention. $NO_x$ (and CO) formation is a function of three primary parameters:

1. Residence time in the combustion zone
2. Chemical reaction rate
3. Mixing rate

These three reaction parameters can be related to the turbine operating conditions and combustor size [75]. In particular,

$$\text{Residence time} \propto \frac{p \cdot V_c}{\dot{m} \cdot T_{pz}} \tag{15.27}$$

$$\text{Reaction time} \propto p^n \cdot \exp(kT_{st}) \tag{15.28}$$

where

- $T_{st}$ is stoichiometric flame temperature (or, equivalently, $f$ or $\phi$ or firing temperature)
- $T_{pz}$ is the average gas temperature in the reaction or primary zone
- $V_c$ is combustion volume
- $p$ is combustion pressure (raised to a power of $n$)
- $\dot{m}$ is the combustor mass flow rate

Therefore, $NO_x$ production rate can be expressed in an empirical formula that can be written as

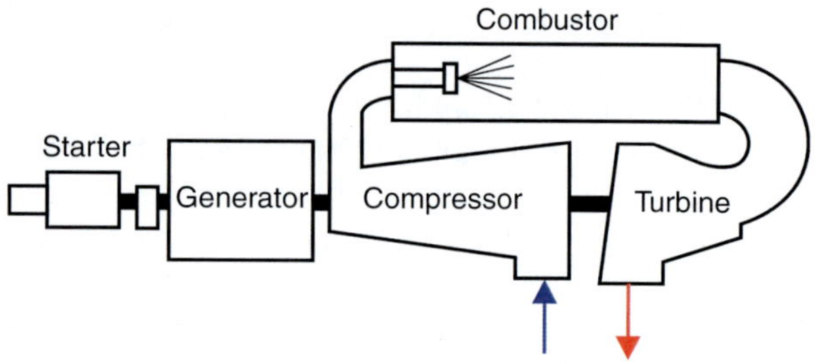


**Fig. 15.24** Effect of pressure and stoichiometric flame temperature on $NO_x$

$$NO_x \propto \frac{V_c}{\dot{m} \cdot T_{pz}} \cdot p^{1+n} \cdot \exp(k \cdot T_{st}) \tag{15.29}$$

The relationship between $NO_x$ production and key parameters per Eq. (15.29) is shown qualitatively in Fig. 15.24. In general, $NO_x$ values are reported in parts per million (ppm) on a dry basis (usually expressed as ppmvd) and corrected to 15% $O_2$ in combustion air. It is extremely important to make sure that the basis of the reported number is clearly defined; otherwise comparisons of data from different sources can be misleading. The expression for correction is as follows:

$$[NO_x]_{15\%O_2} = \frac{[NO_x] \cdot 5.9}{(20.9 - O_{2,meas})} \tag{15.30}$$

The trends in Fig. 15.24 show that the impact of pressure on $NO_x$ formation is stronger at high flame temperatures near the stoichiometric value (i.e., $\phi \sim 1.0$) and negligible at leaner mixtures with lower flame temperatures. Consequently, effective control of the $NO_x$ production in the combustion process is a matter of reducing the flame temperature. There are essentially three ways to accomplish this:

1. Design a primary (reaction) zone with leaner fuel-air mixture
2. Injection of a diluent, i.e., water or steam, into the primary zone of the combustor to act as a "heat sink" and reduce the flame temperature
3. Staged combustion with two distinct reaction zones, which lead to the modern "lean-premixed" designs, which is referred to as *Dry-Low-$NO_x$* (or *DLN*).

The dramatic impact of reducing $\phi$, i.e., leaning out the PZ, on flame temperature and $NO_x$ production can be seen in Fig. 15.25. Leaner fuel-air mixtures lead to lower-value flame and gas temperatures and favorably impact the $NO_x$ production rate via shorter residence and reaction times.

Unfortunately, things get complicated rapidly. For starters, the full operating range of an industrial GT combustor requires significant turn-down ratios, e.g., 40-to-1 in fuel flow, 30-to-1 in air flow, and 5-to-1 in fuel-air ratio. Obviously, the combustor design must ensure stable and efficient combustion across the entire operating range. The early single-nozzle (diffusion flame) combustors, with direct injection of fuel and air into the PZ, were designed for combustion at or near

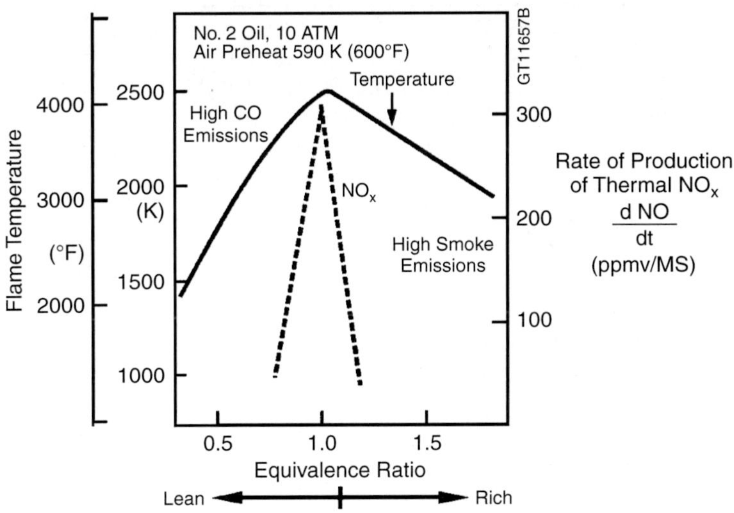


**Fig. 15.25** Thermal $NO_x$ production rate (Source: GE Energy [74])

stoichiometric conditions. For a combustor designed for optimal full load operation at $\phi \sim 1$, combustion at lower loads is bound to be very lean. Thus, further leaning out of the PZ for $NO_x$ reduction is a very limited knob. In fact, initial efforts to reduce $NO_x$ via leaning out reaction zone of an SND combustor did not achieve more than 20% reduction [77].

When it became clear that meeting increasingly stringent pollution-control regulations via leaner reaction zones in SND combustors, other methods to reduce flame temperatures were investigated. Injection of water or steam into the PZ to create a "heat sink" for temperature reduction proved successful. Starting in 1979, when regulations required that $NO_x$ emissions be limited to 75 ppmvd, SND combustors with water or steam injection have been successfully used to meet or exceed these requirements successfully [74]. However, SND combustors that use water and steam injection are limited in their ability to reduce $NO_x$ levels below 42 ppmvd on gas fuel (65 ppmvd on oil fuel). The main reason is the practical limit to the amount of diluent injection into the reaction zone before the increase in dynamic pressure oscillations (i.e., noise) start imposing unacceptable penalties in terms of reduced hardware life and increased maintenance frequencies. High levels of water injection excited discrete dynamic pressure tones within the combustor. Frequencies within the range of the hardware's natural frequency resulted in combustor damage. In terms of dynamic pressure response, steam is a better diluent than water, although a larger mass flow rate is needed. Significant reduction in $NO_x$ production is achievable via steam injection, e.g., ~60% at a steam-to-fuel ratio of unity [77, 78].

The other significant drawback on water or steam injection for $NO_x$ abatement is the overall simple or combined cycle performance penalty. Even though the turbine

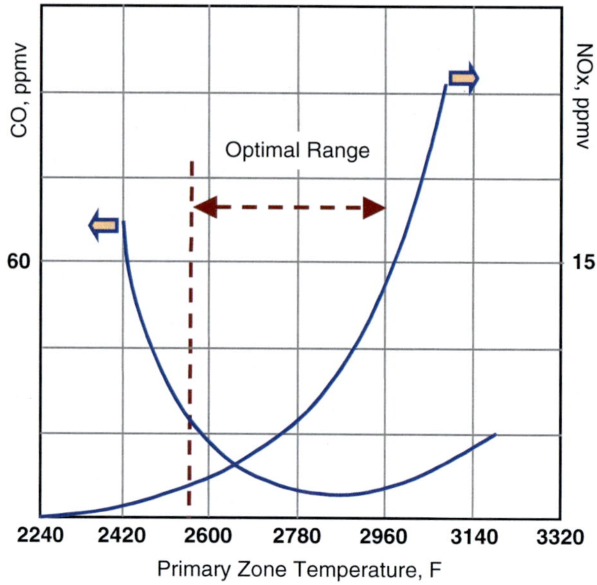


**Fig. 15.26** Influence of PZ temperature on CO and $NO_x$ [75]

output increases via increased mass flow, the associated increase in fuel consumption is detrimental to the heat rate. Furthermore, since the quality of injection water must be high, i.e., of boiler quality, to prevent deposits and corrosion in HGP components, additional *water treatment system* (WTS) cost for simple cycle applications must be considered. In combined cycle applications, where steam from the HRSG or ST (typically *cold reheat* steam) is extracted for injection, increased make-up flow rates and ensuing WTS size increase and water consumption are additional concerns. Even disregarding the combustion noise and life issues, it is easy to imagine that ever-increasing turbine-firing temperatures for improved cycle efficiency would drive the diluent injection flows to such levels where they would literally put the flame out. This is notwithstanding the counterproductive effect of reduced cycle efficiency via higher fuel consumption, which defeats the purpose of increasing the firing temperature in the first place.

Another limit associated with lowering the equivalence ratio and flame temperature for $NO_x$ abatement is the increase in CO emissions. The conflicting emission benefits and harms of lowering or increasing the flame temperature can be seen in Fig. 15.26. Carbon monoxide is a measure of combustion inefficiency, which becomes more pronounced at leaner combustions. This is due to shorter residence time at leaner mixtures, which does not lend itself to complete CO burnout. Thus, the allowable temperature range for emissions control is limited to a relatively narrow band of $\sim 400^{\circ}$F.

When SND combustors hit a practical limit for $NO_x$ emissions reduction, even with diluent injection, designers responded by developing the "*multi-nozzle quiet*

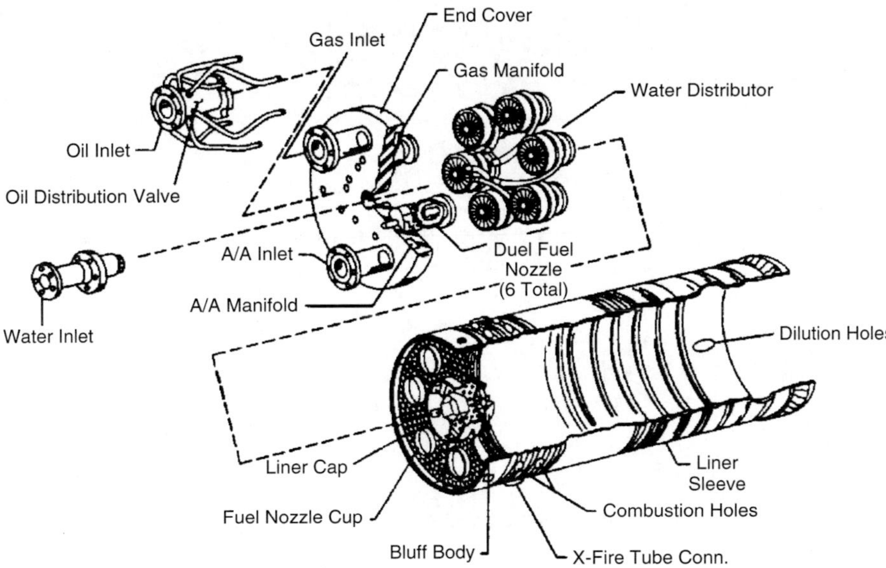


**Fig. 15.27** MNQC Combustor (Source: GE Energy [79])

*combustion*" or MNQC system. For a detailed history of the combustor development, please consult Ref. [74]. Utilizing a multi-nozzle configuration, i.e., six nozzles located peripherally at the combustor head cap, this system eschewed the low-speed ignition and blowout problems associated with excessive leaning out of the PZ in SND systems at low loads (see Fig. 15.27). Furthermore, the MNQC system also helped with the dynamic pressure oscillation (i.e., noise) problem by maintaining a flat dynamic response at increased steam injection levels (that is the reason for the moniker "quiet"), thus enabling the design of combustion systems with higher firing temperatures and compliant $NO_x$ emissions without the problems of increased hardware wear and reduced maintenance intervals. Since 1987, MNQC systems enabled heavy-duty GTs to meet stringent $NO_x$ emission requirements, i.e., 25 ppmvd with gas fuels and 42 ppmvd with liquid fuels. Increase in CO and UHC emissions and combustion stability issues at high injection rates, e.g., steam-to-fuel $>2$, precluded further improvement with this technology [80].

Reduction in $NO_x$ emissions to even lower levels, the need for balancing the low emissions with ever-increasing firing temperatures and the need to avoid the cycle thermal efficiency penalty associated with steam and water injection lead to the development of DLN combustors. DLN combustion system is the natural next-step in combustor development, i.e., a "staged" combustion system with lean-premixed fuel-air mixture. The primary and secondary combustion zones in the staged combustor are operated at the lowest possible $\phi$ at high fuel flow conditions (see Fig. 15.28). During start-up, acceleration, and low-load conditions, fuel is introduced only in the PZ with good ignition and blowout characteristics. At increasing load levels, fuel is introduced into the SZ at a fuel split to balance the CO and $NO_x$ production at an

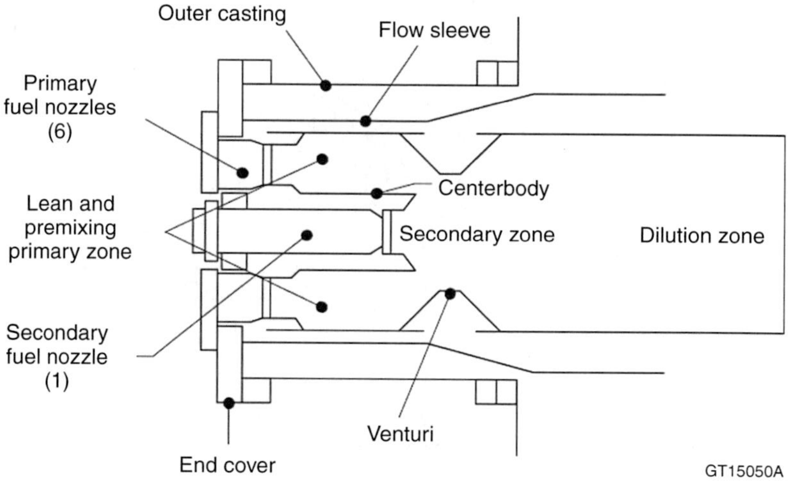


**Fig. 15.28** Dry-Low-$NO_x$ (DLN) combustor (Source: GE Energy [80])

optimal level, i.e., near the middle of the allowable range shown in Fig. 15.25. The first successful DLN system, DLN-I, that went into production incorporated a multi-nozzle system similar to that in MNQC for the PZ and a conventional, large secondary fuel nozzle for the SZ. At normal (50–100% load) operation, fuel and air are "premixed" in the PZ but ignition takes place in the SZ with the flame attached to the secondary nozzle.

DLN technology successfully balanced the requirements for low combustion noise (i.e., dynamic stability) at part-load operation and optimal equivalence ratio for minimum $NO_x$ at acceptable CO levels. DLN achieves this balance without diluent injection (and associated efficiency penalty and cost) while satisfying the requirement for low $NO_x$ emissions, i.e., 25 ppm or lower on gas fuels. A history of the development of DLN-I and DLN-II (for F-Class turbines with high firing temperatures of 2,400+°F) systems can be found in Ref. [74] along with a detailed description of DLN operation modes.

## *Basic Economics*

Natural gas–fired gas turbines and GTCC power plants are the most economic choices for fossil fuel–based electric power generation. This is clearly represented by the data in Table 15.8. The data in the table reflects 2004 prices. The numbers and the relative position of NG-fired plants should be expected to show a certain variation as a function of changing economic climate (i.e., inflation, demand for generating capacity, absolute and relative fuel prices, labor shortage, raw material prices, etc.) of which the recent worldwide depression that started in the last quarter of 2008 is a very good example. Nevertheless, a dramatic shift in the economic advantages of the CC plants vis-à-vis other nonrenewable large central stations or renewable options (i.e., wind, solar, etc.), which cost nearly ten times per kW capacity and available only in small sizes amenable to distributed power generation, should not be expected near term. A very brief introduction to the key aspects of the CC plant economics, i.e., plant capital (turnkey) cost or price and the cost of generating 1 kWh of electric power, is therefore quite useful.

Turnkey combined cycle plant price levels from 2009 Gas Turbine World (GTW) Handbook are shown in Fig. 15.29. Note that listed GTW prices are in 2009 US dollars for a basic natural gas–fired combined cycle with gas turbine generator, unfired multi-pressure heat recovery boiler with no bypass stack, condensing multi-pressure steam turbine generator, step-up transformer, water-cooled heat rejection, standard controls, starting system, and plant auxiliaries. Obviously, depending on the scope of equipment, site-specific requirements, geographic location, and competitive market conditions these prices will vary considerably. Also excluded from the prices in GTW are the indirect costs and other commercial outlays. For a more realistic price level, the data in Fig. 15.29 reflects an adder of about 60% to the listed price, which is based on the assumption that the prices in GTW reflect only 75% of the direct costs, which are 85% of the total installed cost (see Fig. 15.30 and Table 15.9 below).

The data plotted in Fig. 15.29 can be summarized by the power law

$$k = 2,890 \cdot P^{-0.2012} \tag{15.31}$$

where $k$ is the specific price (\$/kW) and $P$ is the net plant output in MW (with $R^2$=0.994). In general, an exponent of 0.2–0.3 is adequate to estimate the capital cost of a CC plant from a known data point (i.e., price and net output). There are several methods and tools to calculate CC power plant cost (from a plant owner's perspective), which are used within OEMs and A&E firms and not available for public consumption. Well-established cost-estimating techniques, especially those that are adopted by the CPI, are readily available in textbooks and scholarly articles. For a discussion of those and the underlying details, the reader is referred to Reference [81]. Another valuable source is Reference [82].

Calculation of the final cost of the CC plant requires consideration of many elements. In fact, the total cost of GT and ST generators and the HRSG, which are the major equipments in the CC power plant, comprises about 50% of the total plant cost. A typical distribution is shown in Fig. 15.30 from Ref. [83]. In addition to these "direct" costs, there are "indirect costs" that can contribute anywhere between 15% and 30% to the final cost of the CC power plant. As an example, consider the representative breakdown in Table 15.9.

The presented data should suffice to give an idea about the scope of the CC plant capital cost and its importance from a design engineer's perspective. All design performance calculations should be evaluated with an eye to the final plant cost. That final plant cost is the money, i.e., the plant "price," that is going to come out of

**Table 15.8** Cost and performance of nonrenewable electric power generation plants

| | | Economic life | Lead time | | Efficiency | | Fuel cost | Fixed | Variable |
|---|---|---|---|---|---|---|---|---|---|
| | Capital cost per kW | years | | Load factor % | % | % (CHP) | $/MMBtu | $/kW-year | $/MWh |
| NG-fired GT | $350–500 | 30 | 1.5 | 5–50 | 38 | | | 9–12 | 4 |
| NG-fired GTCC | $550–900 | 30 | 2 | 60–85 | 55 | | 4–5 | 10–15 | 2 |
| NG-fired GTCC with CHP | $600–700 | 30 | 2 | 50–85 | 40–55 | 79–84 | | 10–15 | 2 |
| US SCPC | $1,300–1,800 | 30 | 4 | 60–85 | 35–42 | | | 20–25 | 3 |
| US IGCC | $1,700–2,100 | 30 | 4 | 60–85 | 39–42 | | 0.75–1.50 | 30–35 | 2 |
| Nuclear | $2,000–3,000 | 40 | 6 | 85–92 | 33 | | 0.5 | 60–70 | 1.5 |

Source: Cambridge Energy Research Associates, MA, USA

Load factor specifies the percent of (equivalent) hours in a year (8,760 h) during which the plant runs at its nominal base load. The data reflects the 2004 prices

*CHP* Combined Heat and Power (also known as Cogeneration), *SCPC* Supercritical Pulverized Coal, *IGCC* Integrated (Coal) Gasification Combined Cycle

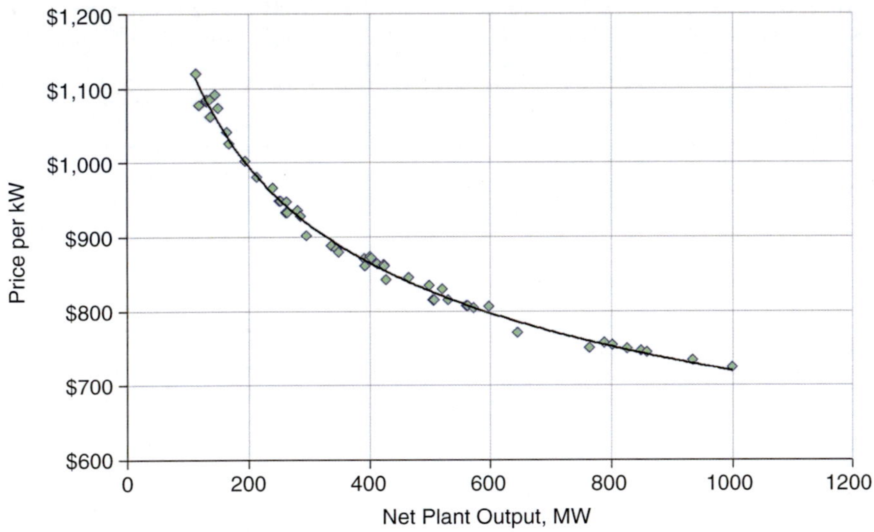


**Fig. 15.29** Turnkey CC plant prices from Gas Turbine World 2009 Handbook. An adder of about 60% is applied to the listed price, which is budgetary turnkey equipment package price, to account for items not covered (see Fig. 15.30 and Table 15.9 below). Provided for reference only

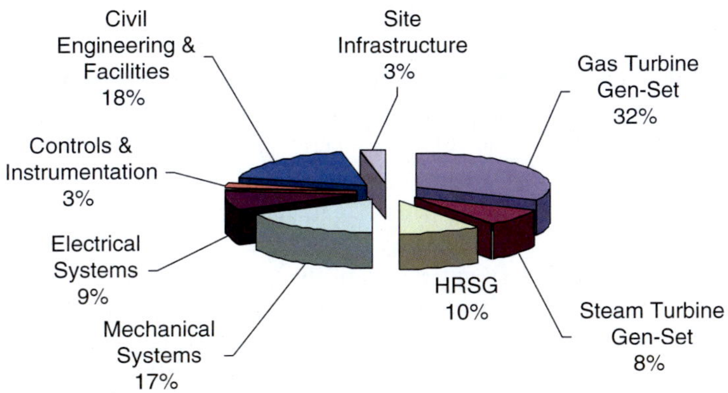


**Fig. 15.30** Direct cost breakdown for a typical 400 MW CCPP (Source: Reference [83], pp 16)

the plant owner's pockets or his credit sources and thus directly impact the generation cost of 1 kWh of electricity.

Probably the most critical customer criterion for choosing an electric power generation product is the life cycle cost of ownership. The metric that is used to quantify this cost is the unit (per kWh) cost of generating electricity. Cost of electricity (COE), in its most commonly used form, is a simple formula that

**Table 15.9** Typical CC capital cost breakdown between direct and indirect costs. From Energy Issues (World Bank), No. 20 June 1999, which is based on a 350/700 MW CC Plant with a Siemens-Westinghouse V94.3A (now SGT5-4000 F) GT

| Integrated services (Indirect costs) | Contribution to total CC cost (%) |
|---|---|
| Project management and subcontracting | 4 |
| Plant and project engineering//software | 2 |
| Plant construction, commissioning, and training | 8 |
| Transport and insurance | 1 |
| **Direct costs** | **Contribution to total CC cost (%)** |
| Civil works | 15 |
| Gas and steam turbine generators | 32 |
| Boiler island | 11 |
| BOP | 16 |
| Electrical systems | 7 |
| Instruments and control | 4 |

combines a power generation system's cost (capital and operating) and thermal performance into a single figure of merit that enables comparison of different systems. It is given as:

$$\mathrm{COE} = \frac{\beta \cdot k}{H} + \frac{L \cdot f}{\eta} + \mathrm{OM} \tag{15.32}$$

where β = Capital charge factor $k$ = Total plant specific cost in \$/kW $H$ = Annual operating hours $f$ = First-year fuel cost in \$/MMBtu(HHV) $L$ = Levelization factor η = Net efficiency of the CC plant OM = Operating & maintenance costs, \$/kWh.

The first term on the right-hand side of Eq. (15.32) represents the carrying charges, the second term is the cost of fuel, and the last term is the cost of maintaining and operating the plant. Fuel cost and operation and maintenance (O&M) costs are the variable components of the COE whereas carrying charge is the fixed component. COE is typically calculated over the economic lifetime of the plant (typically 30 years).

*Levelization factor*, *L*, converts the fuel cost that usually escalates over the plant's economic life due to market conditions and inflation into average annual (constant) values. The formula for *L* can be found in any financial reference and is a function of discount, escalation, and inflation rates.

*The fixed charge rate*, β, represents the plant's financial and tax situation and is a function of the economic life of the power plant, debt–equity ratio used in financing the plant, debt and equity rates, corporate tax rate and tax credits, book depreciation, property tax, and insurance. Various methods are available for calculating β, which can be found in plant economic studies. Typical values of β are 16–18% for *Independent Power Producers* (IPPs) and 12–13% for government-owned utilities, depending on the interest and taxation rates.

If the new and clean performance conditions are used within the calculations, the cost of generation as given by the COE formula should be treated as a figure of merit rather than an accurate assessment of the "real" generating cost. Important drivers – such as seasonal variations in ambient conditions, part load operation, recoverable and unrecoverable component deterioration, uncertainty in fuel prices, start-up and shutdown, and unforeseen events – are not considered. Nevertheless, levelized COE is an important metric that provides insight to performance-cost trade-offs for OEMs (system design) as well as customers (system selection).

The exact distribution of the cost of generating 1 kWh of electric energy varies depending on the financial assumptions. Typically, about two-thirds of the COE for an NG-fired, medium-loaded (~6,000 h/year) modern 3PRH CC plant with an F-Class gas turbine is the fuel cost. About one fourth of the COE is the capital cost. The distribution is markedly different (actually, nearly exactly the opposite of NGCC) for the IGCC plant with very high capital expenditure ($2,000/kW or even higher) utilizing a cheaper fuel, coal ($1–2).

## Future Directions

At the time of writing of this article, natural gas accounts for about 22% of US electricity generation. The share of natural gas in electricity generation in OECD member countries in Europe is about the same. The share of natural gas is expected to increase in the future to the tune of 25% of total world electricity generation by 2030 [2]. However, recent developments suggest that this projection might in fact be a low estimate. One reason is the new drilling technologies that made extraction of natural gas from shale rocks feasible. The technique is called *hydraulic fracturing* and involves injection of high-pressure water into rock formations to destroy them and enable the trapped gas to be released to the surface for capture. So much so that this development resulted in a 35% jump in US natural gas reserves from about 1,500 trillion $ft^3$ in 2006 to more than 2,000 trillion $ft^3$ in 2008 [84]. Note that this number is the total estimated reserves including speculative, possible, and probable sources, and of which only 237 trillion $ft^3$ is considered to be proven by the Energy Department. According to the same source, shale gas constitutes nearly a third of the total reserves.

Another potential source of natural gas is *gas hydrates* (also known as gas *clathrates*), which are "crystalline solids consisting of gas molecules, usually methane, each surrounded by a cage of water molecules." It looks very much like water ice. Methane hydrate is stable in ocean floor sediments at water depths greater than 300 m, and where it occurs, it is known to cement loose sediments in a surface layer several hundred meters thick. The worldwide amount of carbon bound in gas hydrates is conservatively estimated to total *twice the amount of carbon to be found in all known fossil fuels on Earth*. (Source: US Geological Survey Fact Sheet.)

Unfortunately, as of today, commercially feasible production of natural gas from gas hydrates is not a reality.

The dramatic increase in natural gas reserves and actual production via new technology that can extract it from shale rocks, taken at its face value, is a strong driver for its increasing share in electricity generation; at least in the USA. The advantage of natural gas vis-à-vis coal, which roughly accounts for half of US electricity generation, in terms of reduced carbon emissions is thus fortified from lower prices and increased availability. But the situation may not be as simple as it seems. The price of natural gas hit its 7-year low of about \$3/1,000 $ft^3$ in the summer of 2009. This huge drop from a peak of \$13 in the summer of 2008 was a direct result of the *great recession* that led to a decrease in demand, coupled with the increase in production. At the time of writing, NYMEX natural gas futures are trading at about \$4 per million Btu (about \$4.50/1,000 $ft^3$) and they were as high as \$6 per million Btu (about \$6.80/1,000 $ft^3$) at the beginning of 2010. Note that extremely low price levels are not likely to continue for long if they are below the production costs. Extreme price volatility can be expected when demand and supply (or capacity) go neck to neck.

In this context, one should also mention the problems associated with natural gas transportation via pipelines. In January 2009, the price quarrel between Russia and Ukraine ended in Russia cutting off all gas flows through Ukraine for 13 days. This resulted in severe distress in Southeastern Europe, which is nearly completely reliant on Russian gas, and other parts of Europe. While not an issue in North America, political stability of producer and transit countries is a significant concern for European countries that can make them reluctant in investing in natural gas–fired power plants. Alternate routes between Russia, Azerbaijan, and other countries, such as thousands of miles long *South Stream* and *Nabucco* pipelines, are actively pursued projects to provide answers to such concerns. The only other practical way to transport natural gas is to move it by sea by cooling it to $-162°C$ ($-260°F$) and converting it to a liquid (liquefied natural gas or LNG), which reduces its volume by 600 times. Specially built, double-hulled cryogenic ships are used to transport LNG from its port of origin to its port of destination, which are purpose-built facilities used exclusively to export or import LNG. From the receiving port, LNG is shipped to its final destination via tanker trucks, trains, or pipelines (after regasification). Expensive infrastructure investment requirements, safety concerns, large "carbon footprint" during liquefaction, transport, and regasification are key concerns associated with LNG, which limit its share in the overall natural gas consumption (currently less than 10% of US imports).

In terms of cost, regulatory costs, and incentives, e.g., those associated with GHG emission caps and cap-and-trade system requirements (e.g., for manufacturers and utilities to purchase pollution permits), should be considered along with straight dollars per Btu. Coal-fired generation can compete with natural gas in terms of GHG and other pollutant emissions only at a significantly high capital cost and reduced efficiency. Currently, there are two major approaches to accomplish this: Postcombustion carbon capture and storage (CCS) technologies that can clean the stack gas of the existing (via retrofit) and new coal-fired plants and precombustion technologies that are a very good fit with *integrated gasification* CC (IGCC) power plants. Note that carbon emission abatement technologies via stack gas treatment are equally applicable to natural gas–fired CC power plants. In terms of both efficiency and cost, detailed studies indicate that NG-fired CC technology is superior to oil or coal-fired plants as well as IGCC [85].

Objectively viewed, these factors are strong incentives for natural gas burning power plants to replace aging, inefficient, and carbon-emitting, coal-fired power plants in the USA. To a certain extent, this can be accomplished even without building new natural gas-fired CC power plants. A recent report estimated that CC power plants had an average 2007 capacity factor of 42% [86]. In comparison, in the same year, coal-fired plants had a much higher capacity factor of 75%. Clearly, at least on paper, the replacement potential by simply switching generation from coal-fired to natural gas-fired plants is pretty big. In reality, factors such as geographical proximity of the power plants to each other and to the locations of highest demand, transmission system limitations and dispatch issues limited this potential severely; i.e., to about 5–9% of total US coal-based electricity generation [86]. While the proposed switch from coal to natural gas would also require an increase of 5–20% in natural gas supply, this should not be a significant problem in the light of recent developments discussed in the earlier paragraphs. A recent article states that several major US utilities have announced that they will replace at least some of their coal-fired plants slated for decommissioning (total capacity of more than 4.5 GW) [10].

Another factor that favors the natural gas–fired gas turbine plants over the coal-fired steam turbine power plants is operational flexibility. The rapid start-up and good turndown characteristics of gas turbine simple and combined cycle power plants, especially with the increase of renewable energy power plants (mostly solar and wind) connecting to the grid, make them eminently suitable to cyclic on-demand duties to compensate for the variability of the renewable generation. Presently, all major OEMs provide CC plants that can reach base load (several hundred megawatts) within 30–60 min without sacrificing base load efficiency, parts life, and RAM. Aeroderivative gas turbines, such as GE's intercooled LMS-100, can provide 100 MW within 10 min in an emergency or for peak shaving in hot summer months. Especially in Europe and to a certain extent in the USA, large, highly flexible natural gas–fired CC power plants that start and shut down daily for up to 250 starts and a total of 4,000 h of operation per year are expected to supplement renewable power generation. These plants are based on advanced F-Class and H-Class gas turbines that are rated at 58–59% efficiency and expected to reach more than 60% in the near future.

Clearly, then, no matter how regional politics in the USA, Europe, and other parts of the world play out, and what kind of legislative packages eventually emerge from governing bodies to control, cap or regulate carbon and other GHG emissions, natural gas–fired electricity generation with advanced gas turbine–based power plants will be an important part of the overall generating technology portfolio. In all likelihood, natural gas–fired advanced CC power plants will be the bridge technology between today's coal-based power generation and tomorrow's renewable (and probably nuclear) power generation to alleviate the lurking problem of global warming. (In fact, this is pretty much the conclusion reached by a recent MIT report, which was made public after the writing of the present article [87].) The question is what will be the primary technologies in the next, say, quarter century?

Without a doubt, gas turbine–based power plants, in simple cycle (both heavy-duty industrial and aeroderivative variants) and combined cycle configurations will be by far the dominant technology in natural gas electric power generation. Today's advanced air-cooled F-Class and H-Class, steam-cooled G-Class, GE's H-System™, and J-Class engines with firing temperatures up to 1,600°C and PRs up to 23 (up to 35 in reheat machines) are currently capable of 58%, have recorded 59%, and are slated for 60% (or even 61% as claimed by one OEM) in 2011–2012 [88]. These are large units with 500–600 MW per block in $1 \times 1$ configuration designed not only for base load duty but for operational flexibility as described earlier. (In passing, note that the cited efficiencies are typically ISO rating numbers that reflect the most favorable operating assumptions (e.g., once-through open-loop cooling with water drawn from a natural source such as river, ocean, etc). Depending on site and ambient characteristics, performances of actual installed units will be somewhat lower with wide variability.)

Increasing power plant efficiency is the simplest and arguably the easiest way to reduce carbon emissions without breaking the bank or worrying about what to do with all that $CO_2$ or where to put it. As mentioned earlier, ambitious targets as high as 65% or even 70% have been in circulation for some time [57–60]. The discussion in the preceding sections of the article hopefully makes it clear that the key to ever-increasing gas turbine combined cycle efficiencies is increasing firing temperatures and commensurate cycle pressure ratios. The factors that prevent easy realization of that simple goal are primarily related to available component materials that can withstand the extreme pressures and temperatures and limits imposed by pollutant emission regulations. State-of-the-art nickel-based alloys used in turbine HGP components require significant amounts of cooling air supplied from extraction ports in the compressor, which is detrimental to performance. Higher flame zone temperatures in DLN combustors result in excessive $NO_x$ and/or CO in flue gas, which is incompatible with increasingly stringent environmental regulations that limit the emissions of those pollutants to low single-digit numbers.

Single-crystal (SC) and directionally solidified (DS) superalloys with TBCs helped push firing temperatures up to 1,500°C. With closed-loop steam cooling and advanced film cooling techniques, J-Class turbines are rated at 1,600°C. Current research efforts focus on advances in materials, e.g., ceramic matrix composites (CMC) [58] or metal foams [59], which can reduce or even eliminate cooling air requirements at extremely high HGP temperatures. Manufacturing techniques have always been a significant driver in the implementation of aforementioned innovations; e.g., TBC application processes (e.g., electron beam physical vapor deposition), casting and machining technologies to produce SC airfoils. In a similar vein, laser-drilling and casting methods to achieve transpiration (effusion) cooling (the ultimate limit of film cooling) are related areas of research and development to further the goal of increasing turbine inlet temperatures. An associated effort is made on the bottoming cycle side, i.e., development of HRSG heat exchanger tubes, steam headers and piping and steam turbine materials (casing, rotor, and buckets) that can withstand increased steam conditions as high as 180 bar and 600–700°C [89].

Closed-looped steam cooling of the combustor transition piece and stage-1 stators (nozzles) is an existing technique to increase the firing temperature without raising the combustor exit (turbine inlet) or flame temperatures [32, 33]. This enables performance improvement without a concomitant increase in $NO_x$ emissions. Another method to facilitate extreme combustor exit and firing temperatures (up to 1,700°C) is exhaust gas recirculation (EGR), in which a fraction of the stack gas (up to 35%) is recirculated into the compressor or combustor inlet [90]. EGR reduces thermal $NO_x$ production via reduced $O_2$ in combustion air and peak temperatures in flame. EGR's benefit becomes more pronounced at increased flame temperatures and it can reduce $NO_x$ emissions by up to 50% [91]. EGR is also beneficial for postcombustion CCS because it increases the $CO_2$ concentration in stack gas (from ~3–4% to ~8–10% by volume), which leads to a reduction in the capital cost of the separation plant [91].

Lean premix or DLN combustion technology is the industry standard for heavy-duty industrial gas turbines for large-scale electricity production. In terms of the ability to meet the stringent emissions requirements, the technology is nearly at its limit at the current turbine inlet temperatures. EGR is one possibility to extend the applicability of DLN or even diffusion combustors to 1,700°C turbine inlet temperatures with low $NO_x$ emissions. Another way to mitigate the $NO_x$ emissions is selective catalytic reduction (SCR) in the HRSG. SCR is an existing postcombustion $NO_x$ reduction technology. Other combustion technologies are being investigated for deployment in IGCC and other low-Btu syngas and blast-furnace gas applications to extend the fuel flexibility of gas turbines. The reader can consult the relevant sections in *The Gas Turbine Handbook* published by NETL (US DOE, Office of Fossil Energy). Various high-hydrogen and oxyfuel combustion technologies are being actively investigated for $CO_2$-free emissions. Since the subject of the present article is natural gas-burning technologies, the reader is referred to relevant academic journals such as *ASME's Journal of Engineering for Gas Turbines and Power*, which contain a wide selection of recent research papers on the subject.

A potentially promising combustion technology that directly attacks the single largest source of exergy destruction in the GT Brayton cycle is pulse detonation combustion (PDC). Originally proposed for jet engine aircraft propulsion, the technology has been around for more than a half century [92]. Recently, PDC is actively considered for land-based power generation [93]. The technology involves combustion of a fuel-air mixture across detonation waves in tubular devices [94]. Detonation waves are composite waves comprised of a "frozen" shock wave with pressure and temperature jump followed by a relaxation zone where chemical combustion reactions take place. The benefit of the process is temperature *and* pressure rise during the cycle heat addition process with a correspondingly higher mean-effective temperature vis-à-vis the conventional diffusion or DLN combustors with approximately constant pressure (actually a 5–6% pressure loss). Detailed cycle calculations indicate that a PDC gas turbine in CC configuration is potentially 1.5–3 percentage points advantage over standard units at comparable firing temperatures and pressure ratios [95].

SOFC- or MCFC-based hybrid fuel cell plus gas turbine power plants are by far the most efficient future technologies to "burn" natural gas for electric power generation. In these machines, the fuel cell replaces the combustor to oxidize the gaseous fuel. Different configurations are possible. Detailed studies show that fuel cell-GT hybrid plants are capable of thermal efficiencies approaching 70%. Bhargava et al. [96] can be consulted for more information on these and other advanced GT-based power plants such as humid air turbine (HAT) and intercooled recuperated engine (ICR). Almost all variants have been studied in detail and reached various stages of development and commercialization in the past. They have not been able to pass the audition in terms of reliability, operability, capital cost, scalability, and other important considerations (e.g., water consumption) to present viable alternatives to simple Brayton cycle GT in CC configuration. Closed-loop steam cooling and reheat (also known as *sequential combustion*) have been the only successful and commercially accepted variations on the basic cycle. At this time, to the best of the authors' knowledge, there is no reason to expect this situation to change in the foreseeable future; at least in terms of large-scale electric power generation. As always, dual-fuel diesel engines capable of burning gaseous fuels will continue to play a role in small-scale and/or distributed generation applications. The same can be true for very small (several kilowatts) fuel cell–based hybrid units.

## Bibliography

### *Primary Literature*

1. Energy Information Administration (EIA) (2009) Annual energy review 2008. http://www.eia.doe.gov/aer. Accessed 29 June 2009
2. Energy Information Administration (EIA) (2008) International energy outlook 2008. DOE/EIA-0484(2008). www.eia.doe.gov/oiaf/ieo/index.html.
3. United States Environmental Protection Agency (EPA) (2010) Methane. www.epa.gov/methane.
4. Liss WH, Thrasher WR (1992) Variability of natural gas composition in select major metropolitan areas of the United States. Gas Technology Institute, Chicago, GRI-92/013
5. Natural Gas and the Environment (2010) From http://www.naturalgas.org/environment/naturalgas.asp
6. Turbomachinery International (2008) 49(6), 10/2008, Handbook 2009. www.turbomachinerymag.com
7. Wärtsilä 50DF generating set (2010) From www.wartsila.com
8. Khan BH (2006) Non-conventional energy sources. Tata McGraw Hill, New Delhi, India
9. McNeely M (2006) Power generation order survey. Diesel & Gas Turbine Worldwide. Article from Oct 2006 issue

10. Burt B, Mullins S (2010) U.S. gas-fired power development: last man standing, POWER. http://www.powermag.com. Accessed Sept 2010, pp 71–73
11. Wilson DG, Korakianitis T (1998) The design of high efficiency turbomachinery and gas turbines, 2nd edn. Prentice-Hall, Uppersaddle River
12. Von Ohain H (1996) Foreword. In: Mattingly JD (ed) Elements of gas turbine propulsion. Tata McGraw Hill, New Delhi, India, Edition 2005
13. Meher-Homji CB (1997) The development of the Junkers Jumo 004B – the world's first production turbojet. J Eng Gas Turb Power 119:783
14. Meher-Homji CB (1998) The development of the Whittle Turbojet. J Eng Gas Turb Power 120:249
15. Meher-Homji CB (2000) Pioneering turbojet developments of Dr. Hans von Ohain – from the HeS 1 to the HES 011. J Eng Gas Turb Power 122:191
16. Soares C (2006) Gas turbines in simple cycle and combined cycle applications. Section 1.1 in The gas turbine handbook. US DOE, Office of Fossil Energy, NETL. http://www.netl.doe.gov/technologies/coalpower/turbines/refshelf/handbook/TableofContents.html
17. Leiste V (1999) Development of the siemens gas turbine and technology highlights. Siemens Power Generation, Erlangen, Germany
18. Miller H, Nemec T (2006) Gas turbines. In: Myer K (ed) Mechanical engineers' handbook, 3rd edn., Energy and power. Wiley, Hoboken, Chapter 24
19. Brandt D (2007) A brief history of GE energy product lines. General Electric Company, New York
20. Brandt D (1988) The design and development of an advanced heavy-duty gas turbine. J Eng Gas Turb Power 110:243–250
21. Stodola A (1927) Steam & gas turbines. Authorized translation from the 6th German edition: Löwenstein LC. McGraw-Hill, New York
22. Langston LS (2010) World's first gas turbine power plant. ASME Mech Eng 132(4):51
23. Tomlinson LO, Lee DT (1985) Combined cycles. In: Sawyer JW, Japikse D (eds) Sawyer's gas turbine engineering handbook, Chapter 7. Turbomachinery International Publications, Norwalk, Conn., USA
24. Horlock JH (1994) Combined cycle power plants – past, present, and future. J Eng Gas Turb Power 117:608–616
25. Gebhardt E (2000) The F Technology experience story, GER-3950C. http://www.gepower.com/
26. Haselbacher H (1989) Gas turbine fundamentals. In: Elliott TC (ed) Standard handbook of power plant engineering. New York, McGraw-Hill Publishing POWER magazine
27. Horlock JH (1997) Aero-engine derivative gas turbines for power generation: thermodynamic and economic perspectives. J Eng Gas Turb Power 119:119–123
28. Moran MJ, Shapiro HN (1988) Fundamentals of engineering thermodynamics. Wiley, New York
29. Cohen H, Rogers GFC, Saravanamuttoo HIH (1987) Gas turbine theory, 3rd edn. Longman, London
30. Cumpsty N (2003) Jet propulsion, 2nd edn. Cambridge University Press, Cambridge
31. Schilke PW (2004) Advanced gas turbine materials and coatings, GER-3569 G. www.gepower.com
32. Pritchard JE (2003) H-System™ technology update. GT2003-38711, ASME turbo expo – power for land, sea & air, 16–19 June 2003, Atlanta
33. Koeneke C (2006) Steam cooling of large frame gas turbines one decade in operation. VDI Ber Nr 1965:33–42
34. Imwinkelried B (1995) Advanced cycle system gas turbines GT24/GT26: the highly efficient gas turbines for power generation. In: Proceedings of the 21st international congress on combustion engines, CIMAC 1995, Interlaken, Switzerland
35. Chiesa P, Macchi E (2004) A thermodynamic analysis of different options to break 60% electric efficiency in CC power plants. J Eng Gas Turb Power 126:770–785

36. Holland MJ, Thake TF (1980) Rotor blade cooling in high pressure turbines. J Aircr 17:412–418
37. Elmasri MA, Pourkey F (1986) Prediction of cooling flow requirements for advanced utility gas turbines. Part 1: analysis and scaling of the effectiveness curve, 86-WA/HT-43, ASME Winter Annual Meeting, Anaheim, 7–12 Dec 1986
38. Elmasri MA (1986) Prediction of cooling flow requirements for advanced utility gas turbines. Part 2: influence of ceramic thermal barrier coatings. ASME Winter Annual Meeting, Anaheim, 7–12 Dec 1986
39. Elmasri MA (1985) On thermodynamics of gas turbine cycles: part 1 – second law analysis of combined cycles. J Eng Gas Turb Power 107:880–889
40. Elmasri MA (1986) On thermodynamics of gas turbine cycles: part 2 – a model for expansion in cooled turbines. J Eng Gas Turb Power 108:151–159
41. Elmasri MA (1986) On thermodynamics of gas turbine cycles: part 3 – thermodynamic potential and limitations of cooled reheat gas turbine combined cycles. J Eng Gas Turb Power 108:160–170
42. Horlock JH, Watson DT, Jones TV (2001) Limitations on gas turbine performance imposed by large turbine cooling flows. J Eng Gas Turb Power 123:487–494
43. Horlock JH (2001) The basic thermodynamics of turbine cooling. J Eng Gas Turb Power 123:583–591
44. Wilcock RC, Young JB, Horlock JH (2005) The effect of turbine blade cooling on the cycle efficiency of gas turbine power cycles. J Eng Gas Turb Power 127:109–120
45. Young JB, Wilcock RC (2002) Modeling the air-cooled gas turbine: parts 1 and 2. J Turbomach 124:207–222
46. Gülen SC (2010) A simple mathematical model for cooled gas turbines. GT2010-22160, ASME turbo expo – power for land, sea & air, 14–18 June 2010, Glasgow
47. Rice IG (1995) Steam-injected gas turbine analysis: steam rates. J Eng Gas Turb Power 117:347–353
48. Cheng DY, Nelson ALC (2002) The chronological development of the Cheng cycle steam injected gas turbine during the past 25 years. ASME International – IGTI Turbo Expo 2002, GT2002-30119
49. Rao A (1989) Process for producing power. U.S. Patent No. 4,289,763
50. Adelman ST, Hoffman MA, Baughn JW (1995) A methane-steam reformer for a basic chemically recuperated gas turbine. J Eng Gas Turb Power 117:16–23
51. McDonald CF, Boland CR (1981) The Nuclear Closed-Cycle Gas Turbine (HTGR-GT) – dry cooled commercial power plant studies. J Eng Gas Turb Power 103:89–100
52. Reale MJ (2004) New high efficiency simple cycle gas turbine – GE's LMS100™, GER-4222A. www.gepower.com
53. Mercury 50, Recuperated Gas Turbine Generator Set, Solar® Turbines (2010) www.solarturbines.com
54. Cox JC, Hutchinson D, Oswald JI (1995) The Westinghouse/Rolls Royce WR-21 gas turbine variable area power turbine design. ASME Paper 95-GT-54, International Gas Turbine and Aeroengine Congress and Exposition, Houston, TX, 5–8 June 1995
55. Hofer DC, Gülen SC (2006) Efficiency entitlement for bottoming cycles. GT2006-91213, ASME turbo expo – power for land, sea & air, Barcelona, Spain, 8–11 May 2006
56. Gülen SC, Smith RW (2008) Second law efficiency of the Rankine bottoming cycle of a combined cycle power plant. ASME Paper GT2008-51381. ASME turbo expo 2008, Berlin, Germany, 9–13 June 2008
57. Bohn D (2006) SFB 561: Aiming for 65% CC efficiency with an air-cooled GT, Modern Power Systems, pp 26–29, Sept 2006
58. Mutassim Z (2008) New gas turbine materials. Turbomachinery International, Sept/Oct 2008 issue, 38–42
59. Bohn D, Dilthey U, Schubert F (2004) Innovative Technologien für ein GuD-Kraftwerk mit 65% Wirkungsgrad. VDI-Berichte Nr 1857:13–25

60. Rao AD, Robson FL, Geisbrecht RA (2002) Power plant system configurations for the 21st century. In: ASME turbo expo 2002, Amsterdam, the Netherlands, 3–7 June 2002
61. Lundberg WL, Veyo SE, Moeckel MD (2003) A high efficiency solid oxide fuel cell hybrid power system using the mercury 50 advanced turbine system gas turbine. ASME J Eng Gas Turb Power 125:51–58
62. Massardo AF, Lubelli F (2000) Internal Reforming Solid Oxide Fuel Cell – Gas Turbine Combine Cycles (IRSOFC-GT); Part I: cell model and cycle thermodynamic analysis. ASME J Eng Gas Turb Power 122:27–35
63. Massardo AF, Magistri L (2003) Internal Reforming Solid Oxide Fuel Cell – Gas Turbine Combine Cycles (IRSOFC-GT); Part II: exergy and thermoeconomic analyses. ASME J Eng Gas Turb Power 125:67–74
64. Chase DL, Kehoe PT. GE combined-cycle product line and performance, GER-3574 g. GE Energy
65. Chris EM, Leroy OT. GE combined-cycle experience, GER-3651. http://www.gepower.com.
66. Tomlinson LO, McCullough S. Single-shaft combined – cycle power generation system, GER-3767c. http://www.gepower.com
67. Matta RK, Mercer GD, Tuthill RS. Power systems for the 21st century – H GT combined-cycles, GER-3935B. GE Energy
68. Smith RW, Polukort P, Maslak CE, Jones CM, Gardiner BD. Advanced technology combined cycles, GER-3936a. GE Power Systems
69. Phylipsen GJM, Blok K, Worrell E (1998) Handbook on international comparisons of energy efficiency in the manufacturing industry. Department of Science, Technology and Society, Utrecht University, The Netherlands
70. Gülen SC (2010) A proposed definition of CHP efficiency, POWER. http://www.powermag.com, pp 58–63, June 2010
71. European Association for the Promotion of Cogeneration (Mar 2001) A guide to cogeneration. http://www.cogeneurope.eu/wp-content/uploads//2009/02/educogen_cogen_guide.pdf
72. Energy Information Administration (EIA) (2009) Annual energy review 2008. http://www.eia.doe.gov/aer. Accessed 29 June 2009
73. Energy Information Administration (EIA) (2010) Electric power annual. http://www.eia.doe.gov/fuelelectric.html. Accessed 20 Jan 2010
74. Davis LB, Black SH (2000) Dry low nox combustion systems for GE heavy-duty gas turbines, GER-3568 g. http://www.gepower.com
75. Lefebvre AH (1995) The role of fuel preparation in low-emission combustion. J Eng Gas Turb Power 117:617
76. Roointon, Pavri, Moore, Gerald D (2001) Gas turbine emissions and control, GER-4211. http://www.gepower.com
77. Hilt MB, Waslo J (October 1984) Evolution of NOx abatement techniques through combustor design for heavy-duty gas turbines. J Eng Gas Turb Power 106:825
78. Touchton GL (1984) An experimentally verified NOx prediction algorithm incorporating the effects of steam injection. J Eng Gas Turb Power 106:833
79. Davi MA (1994) GE gas turbine combustion flexibility, GER-3946. GE Energy
80. Miller HE (1994) Development of the GE quiet combustor and other design changes to benefit quality, GER-3551. http://www.gepower.com
81. Peters M, Timmerhaus K, West R (2004) Plant design and economics for chemical engineers, 5th edn. McGraw-Hill, London
82. Bejan A, Tsatsaronis G, Moran M (1996) Thermal design & optimization. Wiley, New York
83. Kehlhofer R, Warner J, Nielsen H, Bachmann R (1999) Combined cycle gas & steam turbine power plants, 2nd edn. PennWell Corp, Tulsa
84. As reported in the press per Potential Gas Committee report (2008) Potential supply of natural gas in the United States, Potential Gas Agency, Colorado School of Mines, Golden, 31 Dec 2008

85. Gambini M, Vellini M (Jan 2003) $CO_2$ emission abatement from fossil fuel power plants by exhaust gas treatment. J Eng Gas Turb Power 125:365–373
86. Wagman D (2010) Can natural gas displace coal? Power Eng (Mar 2010 issue): 4
87. The future of natural gas – an interdisciplinary MIT study (2010) Interim report by MIT Energy Initiative, ISBN 978-0-9828008-0-5, Massachusetts Institute of Technology, Boston
88. Robb D (2010) CCGT: breaking the 60 percent efficiency barrier. Power Eng Int 18(3). www.peimagazine.com
89. Review of status of advanced materials for power generation, Technology Status Report, Cleaner Coal Technology Programme, Department of Trade and Industry (Oct 2002) London
90. Tukagoshi K, Muyama A, Uchida S et al (Oct 2005) Latest technology for large capacity gas turbine. MHI Tech Rev 42(3)
91. ElKady AM, Evulet A, Brand A (May 2009) Application of exhaust gas recirculation in a DLN F-class combustion system for postcombustion carbon capture. J Eng Gas Turb Power 131: #034505
92. Kailasanath K (2000) Review of propulsion applications of detonation waves. AIAA J 38(9):1698–1708
93. Goldmeer J, Tangirala V, Dean A (2008) System-level performance estimation of a pulse detonation based hybrid engine. J Eng Gas Turb Power 130:#011201
94. Tangirala VE, Rasheed A, Dean AJ (2007) Performance of a pulse detonation combustor-based hybrid engine, GT2007-28056. ASME turbo expo – power for land, sea & air, Montreal, Canada, 14–18 June 2007
95. Gülen SC (2010) Gas turbine with constant volume heat addition. ESDA2010-24817. ASME 2010 10th biennial conference on engineering systems design and analysis, Istanbul, Turkey, 12–14 July 2010
96. Bhargava R, Bianchi M, Campanari S et al (2010) A parametric thermodynamic evaluation of high performance gas turbine based power cycles. J Eng Gas Turb Power 132:#022001

## *Books and Reviews*

Bejan A (2006) Advanced engineering thermodynamics, 3rd edn. Wiley, New Jersey
Boss M (1996) Steam turbines for STAG™ combined cycle power systems, GER-3582E. http://www.gepower.com
Boyce MP (2006) Gas turbine engineering handbook, 3rd edn. Gulf Professional Publishing, Houston
Chase D (2001) Combined cycle development evolution and future, GER-4206. http://www.gepower.com
Colegrove D, Mason P, Retzlaff K, Cornell D (2001) Structured steam turbines for the combined cycle market, GER-4201. http://www.gepower.com
Constant EW II (1980) The origins of the turbojet revolution. The Johns Hopkins University Press, Baltimore/London
Cotton KC (1998) Evaluating and improving steam turbine performance, 2nd edn. Cotton Fact Inc, Rexford
Denton JD (1993) Loss mechanisms in turbomachines, The 1993 IGTI scolar lecture. J Turbomachinery 115:621–656
Dunn MG (2001) Convective heat transfer and aerodynamics in axial flow turbines. J Eng Gas Turb Power 123:637–686
Elmasri MA (2007) Design of gas turbine combined cycle and cogeneration systems – theory, practice and optimization. Seminar Notes, Thermoflow, Inc., Sudbury, MA. info@thermoflow.com

Han JC, Dutta S, Ekkad SV (2000) Gas turbine heat transfer and cooling technology. Taylor & Francis, New York

Horlock JH (2001) Combined power plants: including Combined Cycle Gas Turbine (CCGT) Plants. Krieger Publishing Company, Malabar

Kehlhofer R, Hannemann F, Stirnimann F, Rukes B (2009) Combined cycle gas & steam turbine power plants, 3rd edn. PennWell Corp, Tulsa

Lakshminarayana B (1996) Fluid dynamics and heat transfer of turbomachinery. Wiley, New York

Lefebvre AH, Ballal DR (2010) Gas turbine combustion: alternative fuels and emissions, 3rd edn. CRC Press/Taylor & Francis, Boca Raton

Nag PK (2006) Power plant engineering, 2nd edn. Tata McGraw-Hill, New Delhi, India

Saravanamuttoo HIH, Rogers GFC, Cohen H, Straznicky PV (2009) Gas turbine theory, 6th edn. Pearson Prentice Hall, England

Traupel W (1977) Thermische Turbomaschinen, Erster Band, Thermodynamisch-strömungstechnische Berechnung, 3, neuarbeitete und erweiterte Auflage. Springer, Berlin/Heidelberg/New York

# .Chapter 16
# $CO_2$ Capture and Sequestration

**S. Julio Friedmann**

## Glossary

| | |
|---|---|
| Carbon capture and sequestration (CCS) | The long-term isolation of carbon dioxide from the atmosphere through physical, chemical, biological, or engineered processes. |
| $CO_2$ capture | The separation and concentration of $CO_2$ from mixed, dilute, or flue gas streams, commonly for the purpose of geological sequestration or EOR (which requires 95% concentrations or higher). |
| $CO_2$ enhanced oil recovery (EOR) | Injection of $CO_2$ into depleted oil fields for the purpose of increasing production. This typically also results in long-term storage of $CO_2$. |
| Geological sequestration | The long-term physical or chemical storage of $CO_2$ in deep geological formations for the purpose of indefinite separation from the atmosphere. |
| Oxyfired combustion | The burning of an hydrocarbon fuel (coal, oil, gas, biofuel) in an oxygen-rich or pure oxygen environment for the purpose of $CO_2$ concentration or capture. |
| Precombustion capture | Concentration of $CO_2$ from mixed gas streams that are the products of fuel conversion (e.g., gasification, methanation, or fermentation) before combustion, commonly through physical solvents. |

This chapter was originally published as part of the Encyclopedia of Sustainability Science and Technology edited by Robert A. Meyers. DOI:10.1007/978-1-4419-0851-3

S.J. Friedmann (✉)
Lawrence Livermore National Laboratory, 7000 East Ave, Livermore, CA 94550, USA
e-mail: friedmann2@llnl.gov

R. Malhotra (ed.), *Fossil Energy: Selected Entries from the Encyclopedia of Sustainability Science and Technology*, DOI 10.1007/978-1-4614-5722-0_16,

| | |
|---|---|
| Post-combustion capture | Concentration of $CO_2$ from mixed, dilute, and flue gas streams that are the products of combustion, commonly through chemical solvents, sorbents, or selective membranes. |

## Definition of Subject

Carbon capture and sequestration (CCS) is the long-term isolation of carbon dioxide from the atmosphere through physical, chemical, biological, or engineered processes. The primary purpose is to reduce the impacts of human-induced climate change associated with increased greenhouse gas concentrations from emissions and land use. CCS can be a major partial solution to climate change, accounting for between 15% and 25% of the needed abatement, and should be considered as part of a portfolio of solutions which include efficiency and conservation, renewable energy, nuclear power, and other options.

## Introduction

Carbon capture and sequestration (CCS) is the long-term isolation of carbon dioxide from the atmosphere through physical, chemical, biological, or engineered processes. This includes a range of approaches including soil carbon sequestration (e.g., through no-till farming), terrestrial biomass sequestration (e.g., through planting forests), direct ocean injection of $CO_2$ either onto the deep seafloor or into the intermediate depths, injection into deep geological formations, or even direct conversion of $CO_2$ to carbonate minerals. Some of these approaches are considered geoengineering (see the appropriate chapter herein). All are considered in the 2005 special report by the Intergovernmental Panel on Climate Change [1].

Of the range of options available, CCS most commonly entails the capture of $CO_2$ from power and industrial plants followed by injection into deep geological formation (geological carbon sequestration, or GCS). While the economics are hotly debated, in many ways this form of CCS appears to be a critical option for major greenhouse gas reduction in the next 10–50 years. The basis for this interest includes several factors:

- There is no obvious immediate technical barrier to deployment. Systems to capture and concentrate $CO_2$ have been in commercial operation for over 80 years (REF), and the oil and gas industry has injected $CO_2$ for enhanced oil recovery for over 35 years.
- The initial estimates of potential geological capacities are enormous (Fig. 16.1). Formal estimates for global storage potential vary substantially, but are likely to be between 800 and 3,300 Gt of C (3,000 and 10,000 Gt of $CO_2$), with significant capacity located reasonably near large point sources of the $CO_2$. (REF)

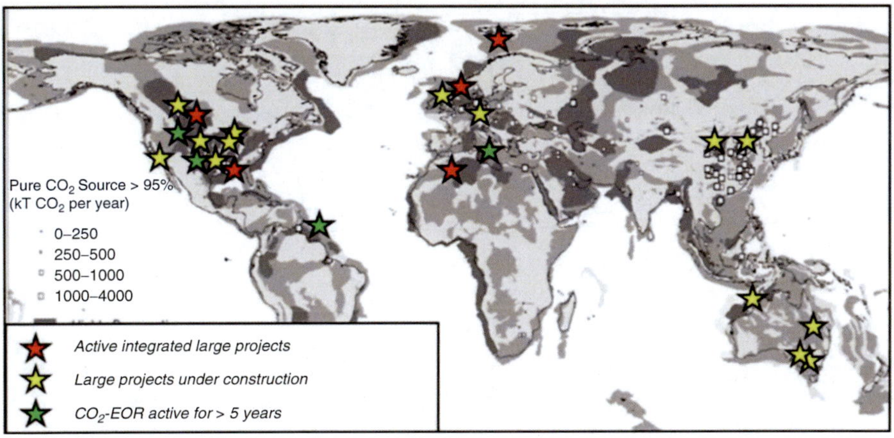


**Fig. 16.1** Map of large CCS projects worldwide. Base map shows highly prospective (*dark shaded*) and prospective (*medium shaded*) areas; squares show size of pure $CO_2$ streams available for sequestration (After Bradford and Dance (2004))

- Testing of large-scale GCS is feasible and has begun (Fig. 16.1). In many industrialized countries, large $CO_2$ sources like power plants and refineries lie near prospective storage sites. These plants could be retrofit today and injection begun (while bearing in mind scientific uncertainties and unknowns). Indeed, some have, and five projects described here provide a great deal of information on the operational needs and field implementation of CCS, with more to follow.

Part of this interest comes from several key documents written in the last few years that provide information on the status, economics, technology, and impact of CCS. These are cited throughout this text and identified as key references at the end of this manuscript.

When coupled with improvements in energy efficiency, renewable energy supplies, fuel switching, and nuclear power, CCS helps dramatically reduce current and future emissions [2–4]. If CCS is not available as a carbon management option, it will be much more difficult and much more expensive to stabilize atmospheric $CO_2$ emissions. Estimates of the cost of carbon abatement without CCS appear 30–80% higher than if CCS were to be available [5].

## Carbon Capture

CCS has two separate but coupled steps. The first is the separation and concentration of $CO_2$ from industrial flue streams, chiefly power plants. This first step is commonly called *carbon capture* and usually includes compression and transportation via pipeline. The second involves the injection of $CO_2$ as a dense, supercritical (liquid-like) phase into deep geological formations. This step is commonly called *geological carbon sequestration*, or GCS. In short, the cost for CCS lies

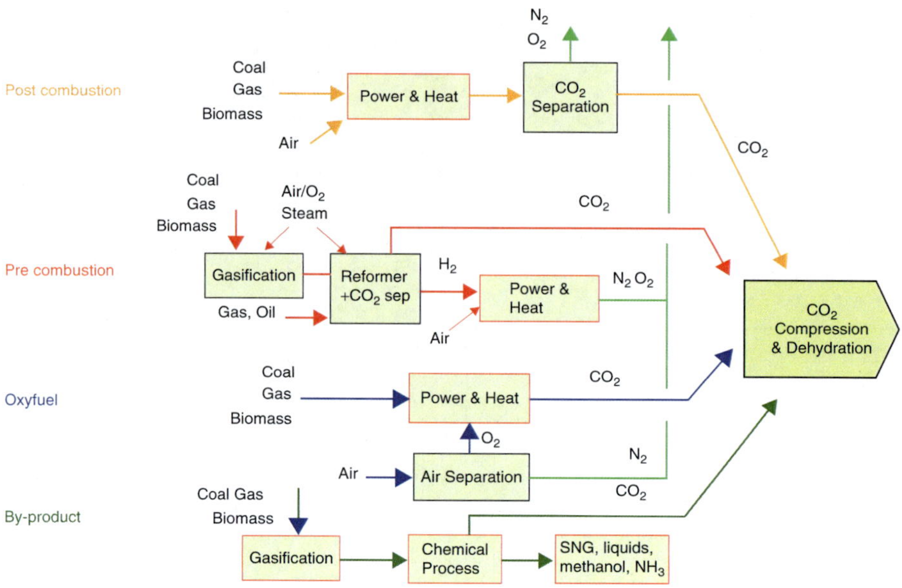


**Fig. 16.2** Schematic diagram of the four main $CO_2$ capture pathways and the main processes involved (After Thambimuthu et al. [6])

mostly in the capture stage. The risk lies mostly in the sequestration stage. Both steps are needed for a successful project.

In the first step, CCS requires the separation of $CO_2$ from industrial flue streams and concentration to $CO_2$ purities of 95% or greater [2, 6]. This limits compression costs and makes effective and efficient use of available sequestration resource (subsurface pore volume). Currently, four technology pathways exist for commercial $CO_2$ capture and separation (Fig. 16.2):

- *Post-combustion capture*: This involves separation of $CO_2$ from nitrogen, commonly with chemical sorbents (e.g., monoethanolamine (MEA)).
- *Precombustion capture*: This involves conversion of fuel feedstocks (e.g., coal) into syngas via gasification, steam reformation, or partial oxidation and then shifting the syngas chemically to hydrogen and $CO_2$, and then separating the $H_2$ from $CO_2$. Currently, this last step is commonly done with physical or chemical sorbents.
- *Oxyfiring combustion*: This involves combustion of fuels in a pure oxygen or $O_2$-$CO_2$ rich environment such that effectively no nitrogen is present in the flue gas. Separation of $O_2$ from air ($N_2$) is required and is the main cost element. Chemical looping [7] is considered a variation of this technology.
- *Direct capture of high-purity streams*: In these cases, $CO_2$ is "pre-captured" and already at or above 95% purity. This requires only compression and transportation before sequestration.

- *Direct air capture*: This involves separation of $CO_2$ directly from the atmosphere. The very low concentrations of $CO_2$ in air (~385 ppm) make this approach both expensive and controversial.

Each of these approaches requires substantial power to run the adsorption and air separation units, raising operating expenses and increasing the amount of $CO_2$ emissions produced simply to drive the sequestration process. They also require more capital in plant construction and have differing operational costs and energy penalties. At present, each technology pathway appears equally viable from an economic and thermodynamic standpoint [2, 6, 8].

Industry has substantial experience with each of these technology pathways, chiefly from operation of hydrogen plants, fertilizer plants, refineries, and natural gas processing facilities. $CO_2$ has been separated from industrial flue streams at scales much greater than 1 MMt $CO_2$/year (270,000 t C/year). Similarly, $CO_2$ has been separated from small-scale power plants, and the technology to scale these operations to plants of 200 MW or greater exists. Large pipelines transport millions of tons of $CO_2$ hundreds of kilometers, and millions of tons of $CO_2$ and other acid gases are compressed and injected into geological formations every year. Thus, a great deal is known about carbon capture, separation, and transportation, and many countries have regulatory frameworks in place to accommodate the permitting of separation facilities and pipelines (in the USA, the Department of Transportation carries this authority).

Cost remains an important barrier to wider commercial deployment [2, 9], especially retrofit costs. However, as the concepts for geological carbon sequestration are proven to be reliable for current power plant technology, improved power plant designs are expected to be able to bring down sequestration costs dramatically. Several large pilot projects are testing pre-, post, and oxyfired combustion tests at the 2–30 MW scale, a necessary precursor to broad commercial deployment. A number of technologies claim to be able to capture and separate $CO_2$ at $9–11/t $CO_2$, less than half the best available technology. As the results from these and future tests are made public, decision makers and investors will be able to plan better for plant economics and design.

## *Post-combustion Capture*

As the name suggests, post-combustion capture (PCC) involves capture and separation of $CO_2$ from flue gas streams after combustion [2, 6]. It can also apply to capture from high- or low-concentration bag-house streams from industrial applications (e.g., cement manufacturing). Most PCC study has focused on separation from coal plants, both because of their central role in global power production and their high concentration of flue gas $CO_2$ (14–20%). However, PCC can also be applied to natural gas plants (4–7% concentration) or other post-combustion streams (e.g., biomass fired power generation).

Typically, $CO_2$ capture involves one of these separation processes:

- *Chemical solvents*: This process operates when flue gas contacts liquid solvents and dissolves into the liquid. The $CO_2$-rich solvent is typically heated to release the $CO_2$ and reconstitute the solvent, which is then recirculated.
- *Membranes*: This process separates $CO_2$ from mixed gases as $CO_2$ preferentially flows through the membrane, which rejects other gases. Membranes can be made of polymers [10], ceramics [11], or more exotic materials (e.g., [12]). Costs and viability are a function of gas selectivity and permeance.
- *Solid sorbents*: In this process, selective solid materials (e.g., zeolites) adsorb $CO_2$ from mixed gas streams. These are released under different physical or chemical conditions, typically through heating.
- *Exotic materials*: Other classes of materials have been proposed and remain promising, including ionic liquids [13, 14] and metal-organic frameworks (MOFs; REF).

The cost and performance of different systems varies as a function of loading, energy costs, capital costs, vapor pressure, process efficiency, and other issues (e.g., Abanades et al. 2009). While some processes appear to be highly efficient, they may have cost or volumetric issues which limit deployment. Many prospective processes remain at the bench-scale and have not been tested in large-scale pilots or commercial plants.

Today, most post-combustion capture is performed by liquid solvents, chiefly amines. This is in large part due to the familiarity of the technology (invented in 1930) and the conventional equipment set used, such as gas stripper towers and thermal regeneration. Rochelle [15] argues that the familiarity and performance of amine-based solvents makes them the class of $CO_2$ capture processes and materials most likely to be widely deployed. Many commercial groups are working to improve the cost and performance of their amine systems today, including Mistubishi, Babcock and Wilcox, Alstom, Aker Clean Carbon, HTC, and Huaneng.

While other capture systems show promise based on bench-scale performance [16], few have been tried. One exception is Alston's chilled ammonia process [17], piloted at the Mountaineer power plant in West Virginia. Work on the pilot project ceased in 2011, and results remain inconclusive.

## *Precombustion Capture*

Precombustion capture involves the removal of $CO_2$ after the coal is converted into syngas, but before combustion in a turbine or boiler. Typically, the first step involves gasifying the coal. This syngas contains $CO_2$, which may be separated at this stage for partial capture (typically around 20% of carbon content). To achieve higher fractional capture, a water-gas shift reactor converts carbon monoxide and steam to $CO_2$ and hydrogen. This increases the concentration of $CO_2$, improving capture efficiency and increasing the amount of carbon (in the form of $CO_2$) that can be removed using this process. The $CO_2$ is then removed using either a chemical or a physical solvent, such as Selexol™ or Rectisol™, and is compressed. This approach is used widely where coal is used for chemical feedstock.

For power production applications, the remnant hydrogen burns in a turbine to generate electricity [2]. The likeliest configuration for these systems is an integrated gasification combined-cycle power plant, or IGCC. While both IGCC and precombustion $CO_2$ capture technologies are available today, the costs limit commercial deployment. Four gigawatts of IGCC power plants have been built worldwide as of the end of 2007 [1], and no existing IGCC plants have $CO_2$ capture. In 2003, the US Department of Energy created the FutureGen project [18], aimed at construction of a 275 MW IGCC + CCS plant capturing 1 million tons $CO_2$/year. FutureGen halted in 2008 [19], and then restarted as an oxyfiring retrofit plant (see below). A similarly scaled project, GreenGen in China [20], is slated for completion in 2012 with $CO_2$ capture in 2014.

Many companies have proposed specific IGCC designs with $CO_2$ capture, including GE, Siemens, and Mitsubishi. Overall, IGCC plans with CCS appear to have higher net efficiencies, lower water consumption, and lower energy penalties than other kinds of new plants with carbon capture [21]. The commercial availability of gasifiers, water-gas shift reactors, solvent towers, and gas turbines also provides some commercial advantages. However, the total capital costs for IGCC systems with CCS remain higher than conventional plants.

Other precombustion plant designs show promise. For example, hydrogen and syngas from gasification can also be used to make chemicals such as ammonia, urea, or olefins. Coal-to-chemical plants can use some hydrogen like an IGCC plant and generate power as well. These *polygeneration* plants show some advantages in terms of economic return (e.g., [22]). Several US projects, including the Texas Clean Energy Project [23] and HECA project [20] are attempts to manage costs and create economic return for precombustion plants through polygeneration.

Given the high efficiency of physical solvents for precombustion, little research has focused on solvent-based precombustion technology improvement. Much research has focused on improving efficiency or reducing costs of gasification or on better plant integration to improve overall efficiency and cost. Additional research has focused on alternative separation technologies, including hydrate formation, ceramic membranes, or exotic materials. Alternative gasification technologies such as molten metal gasifiers [24] or underground coal gasification (Friedmann 2009) remain potential avenues to large cost reductions for precombustion capture–based plant designs.

## *Oxyfired Combustion*

The burning of a hydrocarbon fuel (coal, oil, gas, biofuel) in an oxygen-rich or pure oxygen environment for the purpose of $CO_2$ concentration or capture is called oxyfired combustion (REF). Oxyfiring is premised on this basic chemical reaction:

$$CH_x + O_2 \rightarrow CO_2 + XH_2O$$

where the coefficients x and X vary as a function of fuel type (for methane, x = 4 and X = 2). When oxygen is blended with the fuel stoichiometrically, this reaction balances and water can be separated with compression leaving concentrated, dry $CO_2$.

One of the potential challenges with oxyfiring combustion is that many hydrocarbon fuels will burn extremely hot in a pure oxygen environment, making retrofits difficult and new reactors expensive. One common approach to this problem is to mix oxygen with $CO_2$ at near atmospheric oxygen concentrations (e.g., 25% oxygen and 75% $CO_2$) creating a "synthetic air." The $CO_2$ is recirculated after separation, and acts to moderate the temperature and kinetics of the fuel combustion in the reactor (REF). This approach is considered promising for retrofits, both in the power sector (subcritical and supercritical pulverized coal boilers) and in the industrial sector (e.g., in catalytic crackers common in hydrocarbon refining).

A number of commercial projects have made progress on oxyfiring. In Europe, Vattenfall's Schwartzepumpe pilot project successfully ran a 30 MW thermal boiler with synthetic air. Similarly, Total successfully operates a 50 MW oxyfired boiler in association with injection into a depleted natural gas field in Lacq, France. The largest current program in oxyfiring is the revised FutureGen program (FutureGen 2.0; REF). This US project is attempting to retrofit a 250 MW boiler in Meridosia, Illinois for oxyfired combustion. This project plans to capture 1.3 million tons $CO_2$/year, over 90% of the plant emissions.

The primary cost of oxyfired combustion is the cost of oxygen separation from air (REF). The air separation unit (ASU) requires substantial capital and operating costs. There are also some specific technical considerations associated with the operating pressure of the combustion unit. If the reactor burns its fuel at lower than atmospheric pressure (a common configuration for boilers), then there is a risk of air leaking into the reactor and contaminating the stream with nitrogen.

Chemical looping [7] is considered an oxyfiring variant, although in most ways the technology is radically different from synthetic air combustion. Instead, a metal carrier is oxidized, creating heat which can be used to run a steam cycle. Most commonly, the process is configured with two interconnected fluidized bed reactors: an air reactor and a fuel reactor. The solid oxygen carrier is circulated between the air and fuel reactors, and the metal oxide is reduced in the second stage. This process remains promising, in large part due to the very high theoretical conversion efficiencies (as high as 80%). However, no reactors larger than bench-top have proven viable yet, and questions remain about the operation, longevity, and stability of the metal oxide carriers.

## *Direct Air Capture (DAC)*

Recent carbon cycle research by Solomon et al. [25] and Hansen et al. (2007) has resulted in increased interest in direct capture of carbon dioxide ($CO_2$) from the air (e.g., [26]). Pielke [27] has described the macro-scale economic playing field for air capture, while others ([28], Lackner 2009, [29]) have begun research into specific technologies capable of removing $CO_2$ from the air. Several companies have recently formed to develop and commercialize direct air-capture systems, including Kilimanjaro, Carbon Engineering, and Global Thermostat.

The key technical concern is the low partial pressure of $CO_2$ in air due to its low concentration (<390 ppm). This requires the movement of very large volumes of air through a capture unit, and high expense of energy to execute the separation. The American Physical Society Panel on Public Affairs [30] recently completed a fairly comprehensive assessment of the 2009 state of the art. The key messages of the report include three conclusions:

- DAC is not currently an economically viable approach.
- Generally, low-carbon power and energy is best used directly to minimize losses rather than used for DAC.
- Eventually, DAC could play a role in capture and storage from decentralized sources of $CO_2$, such as vehicles, ships, or planes.

While the situation may change with $CO_2$ demand and price, the consensus today is that the economics and engineering of DAC remain formidable and difficult to overcome.

## Geological Sequestration

A number of geological reservoirs appear to have the potential to store many hundreds to thousands of Gt of $CO_2$ [31]. The most promising reservoirs are porous and permeable rock bodies at depth (Figs. 16.3 and 16.4).

- *Saline formations* contain brine in their pore volumes, commonly with salinities greater than 10,000 ppm. Because $CO_2$ is buoyant in most geological settings, target saline formations require a cap rock or sealing unit above the main injection zone.
- *Depleted oil and gas fields* have some combination of water and hydrocarbons in their pore volumes. In some cases, economic gains can be achieved through enhanced oil recovery or enhanced gas recovery [32–34]. Substantial $CO_2$-enhanced oil recovery already occurs in the USA with both natural and anthropogenic $CO_2$. These fields provide much of the knowledge base available about the potential issues related to $CO_2$ sequestration.
- *Deep coal seams*, often called unmineable coal seams, are composed of organic minerals with brines and gases in their pore and fracture volumes that can preferentially adsorb and bind $CO_2$ as well as store it in pores and minor fractures. These targets present some challenges in that coals are relatively low permeability units, presenting challenges to injection.

Because of their large storage potential and broad distribution, it is likely that most geological sequestration will occur in saline formations. However, initial

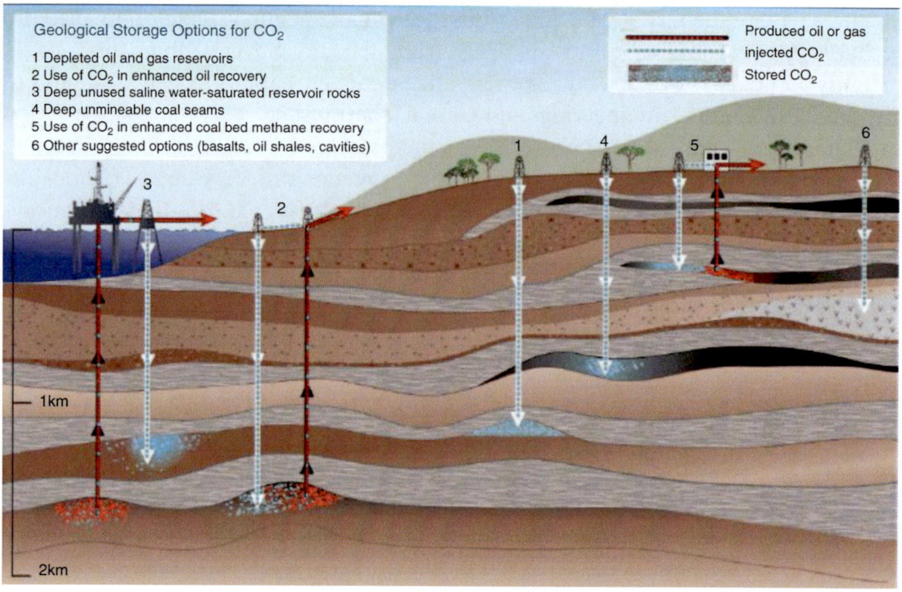


**Fig. 16.3** Options for storing $CO_2$ in underground geological formations (After Benson and Cook [31])

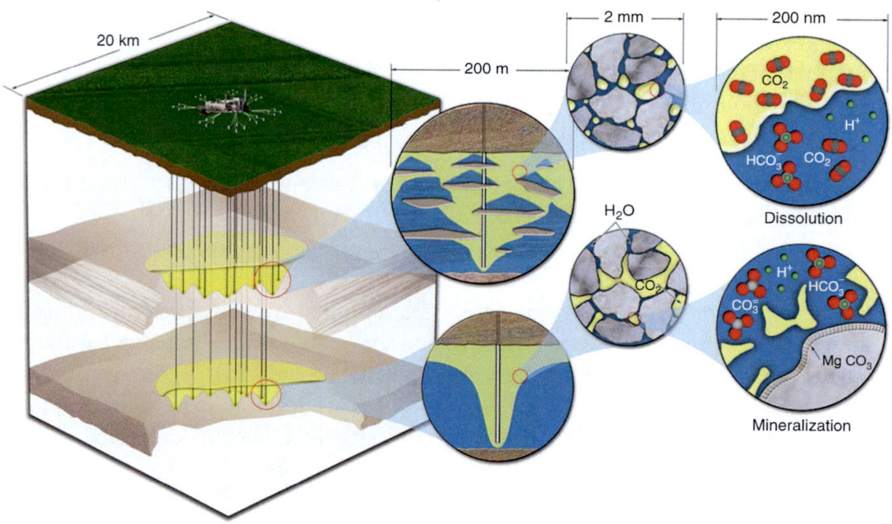


**Fig. 16.4** Schematic diagram of large injection at 10 years time illustrating the main storage mechanisms. All $CO_2$ plumes (*yellow*) are trapped beneath impermeable shales (not shown). The upper unit is heterogeneous with a low net percent usable porosity, whereas the lower unit is homogeneous. Central insets show $CO_2$ as a mobile phase (*lower*) and as a trapped residual phase (*upper*). Right insets show $CO_2$ dissolution (*upper*) and $CO_2$ mineralization (*lower*) (After MIT [2])

projects have been proposed for depleted oil and gas fields, accompanying enhanced oil recovery, due to the high density and quality of subsurface data and the potential for economic return. Although there remains some economic potential for enhanced coal bed methane recovery much less is known about this style of sequestration [2, 31, 35, 36]. Even less is known about sequestration in basalts. As such, many workers are not convinced of the economic viability of sequestration projects in coal, basalts, or oil shales given today's technology and understanding [37].

Storage of large $CO_2$ volumes in geological formations requires that the $CO_2$ be relatively dense, so that storage capacity is efficiently used. Given typical geothermal gradients and hydrostatic loads, $CO_2$ is likely to be in a *supercritical* state at most target sites greater than 800 m depth (e.g., [38]). At the likely range of injection pressures and temperatures for most projects, $CO_2$ would be buoyant and gravitational forces would push $CO_2$ upward from the injection point.

Consequently, trapping mechanisms are needed to store $CO_2$ effectively. For depleted oil and gas fields or for saline formations, $CO_2$ storage mechanisms are reasonably well defined and understood (Fig. 16.4). $CO_2$ sequestration targets will require *physical barriers* to $CO_2$ migration out of the crust to the surface. These barriers will commonly take the form of impermeable layers (e.g., shales, evaporites) overlying the sequestration target. This storage mechanism is highly or directly analogous to that of hydrocarbon trapping, natural gas storage, and natural $CO_2$ accumulations. At the pore scale, *capillary forces* can immobilize a substantial fraction of a dispersed $CO_2$ bubble, commonly measured to be between 5% and 25% of the $CO_2$-bearing pore volume. The volume of $CO_2$ trapped as a residual phase is highly sensitive to pore geometry, and consequently is difficult to predict; however, standard techniques can measure residual phase trapping directly in the laboratory with rock samples.

Once in the pore volume, the $CO_2$ will *dissolve* into other pore fluids, including hydrocarbon species (oil and gas) or brines. Depending on the fluid composition and reservoir condition, this may occur rapidly (seconds to minutes) or over a period of tens to hundreds of years. Once dissolved, the $CO_2$-bearing brines are denser than the original brines, and so the strong buoyant forces of free-phase gas are replaced by small downward forces. Over longer time scales (hundreds to thousands of years) the dissolved $CO_2$ may react with minerals in the rock volume to *dissolve or precipitate* new carbonate minerals. For the majority of the rock volume and major minerals, this process is slow, and may take hundreds to thousands of years to achieve substantial storage volumes. Precipitation of carbonate minerals permanently binds $CO_2$ in the subsurface; dissolution of minerals generally traps $CO_2$ as an ionic species (usually bicarbonate) in the pore fluid.

Although work remains to characterize and quantify these mechanisms, the current level of understanding can be used today to develop estimates of the percentage of $CO_2$ that can be stored over some period of time. Confidence in these estimates is bolstered by studies of hydrocarbon systems, natural gas storage operations, hazardous waste injection, and $CO_2$-enhanced oil recovery (CO2-EOR). Current evaluations of CCS effectiveness based on the current understanding of trapping mechanisms estimate that *more than 99.9% of injected $CO_2$ can be reliably stored over 100 years, and it is likely that 99% of $CO_2$ can be reliably stored for 1,000 years* [31]. These estimates assume careful siting, due diligence before injection, and appropriate management of injection, and reflect the view that the crust contains many sites that are generally well configured to store $CO_2$ effectively.

## Large-Scale Commercial Deployment

In order to achieve substantial GHG reductions, geological storage deployment has several requirements:

- Projects must be large in scale, roughly in the order of 1 Mt/year $CO_2$ or more (Friedmann 2006).
- There must be minimal leakage from the underground storage reservoirs back to the atmosphere.
- There must be minimal impact on other uses of the subsurface environment and the resources it contains.

The issue of scale dominates deployment of GCS ([5, 39]; McFarland 2004; [4]). These volumes would have geological carbon sequestration, providing 25–75 Gt C over 50 years, or 15–43% of emissions reduction needed to stabilize atmospheric $CO_2$ levels at 550 ppm [39].

Today there are five well-established large-scale injection projects with an ambitious scientific program that includes monitoring and verification (Table 16.1): Sleipner in Norway [40], Weyburn in Canada [41], In Salah in Algeria [42]; Snohvit in northern Norway (REF); and Cranfield, Mississippi (REF). Each project has injected $CO_2$ at the rate of ~1 MM t/year (~280,000 t C/year). Each project has had a substantial supporting science program or anticipates one. Recently, the Global CCS Institute in Australia compiled a list of active and near active projects, as well as a roster of announced integrated and pilot projects (REFS). A few are listed here.

These projects have sampled a wide array of geology with varying trapping mechanisms, injection depths, reservoir types, and injectivity. Each of these projects appears to have ample injectivity and capacity for success, and none has detected $CO_2$ leakage of any significance. In addition to the three sequestration projects, many industrial applications have injected large volumes of $CO_2$ into the subsurface. EOR operations in W. Texas, New Mexico, Colorado, Wyoming, Oklahoma, Mississippi, Trinidad, Canada, and Turkey have individual injection programs as large as 3 MM t $CO_2$/year (~820,000 t C/year) and cumulative anthropogenic emission injections of ~10 MM t $CO_2$/year (2.7 MM t C/t) [43]. It should be said that the monitoring and verification program at each site varies substantially [2].

**Table 16.1** Some current and pending large $CO_2$ injection projects

| Site | Location | Reservoir class | Plant/capture type | Permeability | Seal type | Start date[a] |
|---|---|---|---|---|---|---|
| Sleipner | Norway | Offshore Saline Fm. | Gas refining | Very high | Thick shale | 1996 |
| Weyburn | Canada | Onshore EOR | SNG plant/precombustion | Moderate | Evaporite | 2000 |
| In Salah | Algeria | Onshore Sandstone | Gas refining | Low | Thick shale | 2004 |
| Snohvit | Norway | Offshore Saline Fm. | Gas refining | Moderate | Shales | 2008 |
| Cranfield | USA (Mississippi) | Onshore Saline Fm. And EOR | Natural $CO_2$ dome | Very high | Mudstone | 2009 |
| FutureGen 2.0 | USA (Illinois) | Onshore Saline Fm. | IGCC/precombustion | Moderate to very high | Thick shales | 2016 |
| ZeroGen | Australia | Onshore EOR/Saline | IGCC/precombustion | Low to moderate | Shale | 2018 |
| GreenGen | China | Offshore EOR +/− Saline Fm. | IGCC/precombustion | Low to moderate | Shales | 2015 |
| HECA | USA (California) | Onshore EOR | IGCC/precombustion | Moderate to high | Thick shale | 2016 |
| Gorgon | Australia | Offshore Saline Fm | Gas refining | Moderate | Thick shales | 2013 |
| SaskPower Boundary Dam | Canada (Saskatchewan) | Onshore EOR +/− Saline Fm. | SC boiler/Post-combustion | Moderate | Evaporite | 2016 |
| Hauneng Shidongkou | China | Onshore saline Fm. | USC boiler/post-combustion | Unclear | Shales | 2016 |
| Shenhua DCL/Majiata | China | Onshore Saline Fm. +/− EOR | Coal liquefaction/ precombustion | Low to moderate | Shales | 2013 |
| Archer-Daniels Midland | USA (Illinois) | Onshore Saline Fm. | Ethanol by-product | Moderate to high | Shales | 2013 |
| TCEP | USA (Texas) | Onshore EOR | Polygeneration precombustion | Moderate to high | Evaporite | 2017 |

[a] = date of first injection or planned first injection of $CO_2$

In many EOR projects, there is almost no monitoring beyond that required for $CO_2$ flood operations [32].

It is worth noting that many of these projects have come online recently in China, which has the potential to ramp up projects very rapidly ([44, 45]; Liu and Gallagher 2009). It should also be noted that all of the first projects do not involve capture from a power plant, but rather from industrial facilities where $CO_2$ is available at a relatively low cost. While this does improve the knowledge of GCS greatly, some questions will remain about plant integration and economics after these projects which will only be resolved through successive scale-up and demonstration.

## Science and Technology Status

As discussed above, the knowledge of trapping mechanisms and the successes of the three large projects provide substantial information. These are augmented by studies of naturally occurring $CO_2$ systems, [46], natural gas storage facilities, hazardous waste disposal, acid gas injection, and $CO_2$-EOR [31]. This knowledge provides a firm foundation for commercial action and a nascent foundation for the development of regulation, standards, and legal frameworks for sequestration (REFS, including Wilson et al. 2009; CCSREGS project; EPA; EU ruling). GCS itself, however, drives study into specific technical and scientific challenges associated with the central elements of site characterization, selection, operation, and monitoring ([36]; WRI 2009). Forward investigation around these topics will enhance the technical and operational understanding of commercial GCS.

### *Monitoring and Verification (M&V)*

Monitoring and verification must detect and track $CO_2$ in the deep subsurface near injection targets, in the shallow subsurface, and above ground. Monitoring and verification studies are a chief focus of many applied research efforts. The US Department of Energy has defined M&V technology development, testing, and deployment as a key element to their technology roadmap (US DOE 2010). The European Union $CO_2$ReMoVe effort is dedicated to monitoring and verification, and the industry-led $CO_2$ Capture Project continues to study monitoring in commercial settings [47]. Some form of M&V will be required at commercial sites, but the extent of monitoring required by regulators, operators, or financiers remains uncertain. Many geophysical and geochemical methods are sufficiently well understood for them to be used to make reasonable performance predictions at candidate storage sites ([48]; WRI 2009). Testing which of these approaches will be the most valuable for a given geological environment remains to be determined.

## *Key Science and Technology Gaps*

Despite the tremendous amount of applied and basic knowledge, there remain both cross-cutting-specific and site-specific topics for investigation (e.g., [49]). From an applied perspective, the National Energy Technology Laboratory has written several plans to identify and address key technology gaps (e.g., [36]). Recent new R&D projects aimed at specific components of R&D include the National Risk Assessment Program (NRAP), the Carbon Capture Simulation Initiative (CCSI; REF), Weyburn's Final phase project (REF), and portions of China's project 836 and 927. While these efforts and reports are not meant to be comprehensive, they reflect the current state of knowledge and potential to continue scientific investigations in GCS.

## Deployment Challenges

Despite the current gaps in sequestration science and technology, commercial projects have begun and are ready to proceed with confidence in their success. Today, enough is known to safely and effectively execute key tasks around single large-scale injection projects:

- Characterize a site.
- Design and operate the project.
- Monitor the $CO_2$ injection.
- Mitigate problems that might arise.
- Close and abandon the project.

Although this knowledge is currently being brought to bear on specific injection projects around the world, greater work is needed to codify and develop tools, regulations, and standards for deployment of multiple million ton injections in thousands of wells nationwide and worldwide across a range of geological settings. This affects both sides of the deployment rubric over the project life cycle ([50]; WRI 2009). Potential operators must execute a set of tasks to prepare for and execute injection permitting and operation. Similarly, potential regulators, investors, insurers, and public stakeholders require information to make decisions. Part of the challenge is to provide a technical basis for each set of actors to make decisions concerning the minimal amount of information needed to serve all stakeholders.

While many possible goals and terms may be pursued in site characterization, it is difficult to imagine the success of a large-scale injection project without knowledge of three parameters. These are *injectivity*, *capacity*, and *effectiveness* (Friedmann 2006). In general terms, injectivity and capacity may be estimated by conventional means, such as special core analysis, regional and local structural and stratigraphic mapping, and simple multi-phase fluid flow simulations. However, there are explicit standard measures of effectiveness. Ultimately, characterizations must rely on estimates of geomechanical integrity, hydrodynamic stability, and seal continuity for the rock system, fault system, and well system [51–55].

Given this complexity, it is not broadly accepted today what terms constitute effective storage. The FutureGen project's request for proposals [18] laid out a set of minimum criteria for acceptance. These criteria are based on expert opinion, laboratory experimentation, analog studies, and simulations. Similarly, the World Resources Institute (2009) proposed a set of guidelines to potential operators based on a working group of over 40 experts ranging from geoscientists to lawyers and regulators. However, the few active projects today do not constitute a large empirical data set from multiple geological carbon sequestration deployments to provide a definitive standard for characterization of site effectiveness.

## *Hazards Assessment and Risk Management*

Supercritical $CO_2$ is buoyant and will seek the Earth's surface; therefore, $CO_2$ injection carries the possibility of leakage. Importantly, $CO_2$ leakage risk will not be uniform across all sites, thus $CO_2$ storage sites will have to demonstrate minimal risk potential in their site characterization plans [56, 57]. Based on analogous experience in $CO_2$ injection such as acid gas disposal and enhanced oil recovery, these risks appear to be less than those of current oil and gas operations [31].

The direct hazards associated with geologic sequestration fall into three distinct categories:

- Hazards associated with the release of the carbon dioxide to the Earth surface
- Hazards associated with release into groundwater and subsequent degradation
- Hazards associated with Earth movement caused by the injection process itself

The hazards themselves in turn are associated with failure mechanisms and triggers (Friedmann 2007; [56]). Potential triggering of events associated with these hazards could lead to undesired consequences. As such, it is an important goal to identify and understand these hazards in order to avoid triggering hazard events. Identification and characterization of these hazards is the critical first step to managing the risks at a site. They also serve as the basis for a quantitative probabilistic risk assessment (PRA). A robust PRA cannot be made today due to substantial scientific and technical gaps. However, the hazards associated with a site can be identified, mapped, characterized, and parameterized sufficiently to avoid failure or (alternatively) avoid selecting a bad site.

## *Water Issues*

A number of issues arise from CCS regarding water use and quality. These include additional water use due to process steam and additional hear requirements in the plants; displacement of in situ brines; and potential risks to groundwater quality from unplanned $CO_2$ leakage [58]. Additional water needs for surface facilities is highly sensitive to capture process and plant type, ranging from 20% to 110% additional water consumption ([59]; b). $CO_2$ injected into saline formations necessarily interacts with water and brines there. This interaction includes displacement of the brines there, local drying of the reservoir, and dissolution of $CO_2$ into brine forming carbon acid. Early research focused on the potential risks and impacts of acid brine formation, ultimately concluding that these risks are small.

Injection also creates a pressure transient in the reservoir which increases through time. In the right context, this pressure can drive water co-production (enhanced water recovery) and even drive surface desalination processes [60, 61]. Based on preliminary calculations, it appears that water co-production can cut the water demands for an integrated CCS plant by 50%, or produce freshwater for agricultural or civil use. More work is needed to understand the viability of this technical approach.

## Bibliography


1. IPCC (2005) Intergovernmental panel on climate change, IPCC special report on carbon dioxide capture and storage, Interlachen, http://www.ipcc.ch/
2. MIT (2007) Future of coal in a carbon constrained world, MIT Press. http://coal.mit.edu/
3. EPRI REF!! 2007 or 2009
4. US CCTP (2005) US climate change technology program strategic plan, Washington, DC, 256 p. http://www.climatetechnology.gov
5. Edmonds J, Clarke J, Dooley J, Kim SH, Smith SJ (2004) Stabilization of $CO_2$ in a B2 world: insights on the roles of carbon capture and disposal, hydrogen, and transportation technologies. Energy Econ 26(4):517–537, Special Issue EMF 19 Alternative technology strategies for climate change policy, John P. Weyant, ed
6. Thambimuthu K, Soltanieh M, Abanades JC (2005) Chapter 3: capture, IPCC special report on carbon dioxide capture and storage, intergovernmental panel on climate change, Interlachen. www.ipcc.ch, pp 3–1 to 3–114
7. Hossain MM, de Lasa HI (2008) Chemical-looping combustion (CLC) for inherent $CO_2$ separations – a review. Chem Eng Sci 63:4433–4451
8. Rao AB, Rubin ES, Keith DW, Morgan MG (2006) Evaluation of potential cost reductions from improved amine-based $CO_2$ capture systems. Energy Policy 34:3765–3772
9. Al-Juaied M, Whitmore A (2009) Realistic costs of carbon capture, Belfer center discussion paper 2009-08, Harvard University, 73 p. http://belfercenter.ksg.harvard.edu/files/2009_AlJuaied_Whitmore_Realistic_Costs_of_Carbon_Capture_ web.pdf
10. Hiroshi M, Hiroshi H (2000) Development of $CO_2$ separation membranes: (1) Polymer membrane. Sumitomo Electric Tech Rev 157:22–26
11. Meinema HA, Dirrix RWJ, Brinkman HW, Terpstra RA, Jekerle J, Kosters PH (2005) Ceramic membranes for gas separations – recent developments and state of the art. Interceram 54:86–91
12. Holt JK, Park G, Wang Y, Staderman M, Artyukhin AB, Grigoropoulos CP, Noy A, Bakajin O (2006) Fast mass transport through sub-2-nanometer carbon nanotubes. Science 312:1034–1036
13. Blanchard LA, Hancu D, Beckman EJ, Brennecke JF (1999) Green processing using ionic liquids and $CO_2$. Nature 399:28–29

14. Maiti A (2009) Theoretical screening of ionic liquid solvents for carbon capture. Chem Sus Chem 2:628–631
15. Rochelle G (2009) Amine scrubbing for $CO_2$ capture. Science 325:1652–1655
16. Herzog H, Hattan A, Meldon J (2009) An R&D "pipeline" for advanced post-combustion $CO_2$ capture technologies. In: Coal without carbon, Clean Air Task Force. www.coaltransition.org, pp 37–54
17. Kozak F, Petif A, Morris E, Rhudy R, Thimsen D (2009) Chilled ammonia process for $CO_2$ capture. Energy Procedia 1:1419–1426
18. FutureGen Alliance (2006) Final request for proposals for FutureGen facility host site, 54 p. http://www.futuregenalliance.org/news/futuregen_siting_final_rfp_3-07-2006.pdf
19. Scientific American (2008) "Clean" coal power plant cancelled – hydrogen economy, too. http://www.scientificamerican.com/article.cfm?id=clean-coal-power-plant-canceled-hydrogen-economy-too
20. MIT (2011a) GreenGen fact sheet: carbon dioxide capture and storage project, MIT Energy Initiative. http://sequestration.mit.edu/tools/projects/greengen.html
21. DOE-NETL (2007b) U.S. Department of Energy, National Energy Technology Laboratory, Cost and performance baseline for fossil energy plants, vol 1. Bituminous coal and natural gas to electricity, Revision 1, Aug 2007
22. Kreutz T, Williams R, Consonni S, Chiesa P (2005) Co-production of hydrogen, electricity, and $CO_2$ from coal with commercially ready technology, part B: economic analysis. Intl J Hydrogen Energy 30:769–784
23. US DOE (2011) Innovative Texas clean coal project takes major step forward as DOE issues record of decision, press release. http://www.fe.doe.gov/news/techlines/2011/11053-Texas_Clean_Coal_Project_Moves_For.html
24. Redman E, Fennerty K, Fowler M (2009) Mobilizing the next generation coal gasification technology for carbon capture and sequestration. In: Coal without carbon, Clean Air Task Force, www.coaltransition.org, pp 1–16
25. Solomon S et al (2009) Irreversible climate change due to carbon dioxide emissions. Proc Natl Acad Sci 106(6):1704–1709
26. Keith DW (2009) Why capture $CO_2$ from the atmosphere? Science 325:1654–1656
27. Pielke RA (2009) An idealized assessment of the economics of air capture of carbon dioxide in mitigation policy. Environ Sci Policy 12(3):216–225
28. Mahmoudkhani M, Keith DW (2009) Low-energy sodium hydroxide recovery for $CO_2$ capture from atmospheric air–thermodynamic analysis. Int J Greenhouse Gas Control 3(4):376–384
29. Zeman F (2007) Energy and material balance of $CO_2$ capture from ambient air. Environ Sci Technol 41(21):7558–7563
30. American Physical Society (2010) Direct air capture of $CO_2$ with chemicals, APS, 119 p. http://www.aps.org/policy/reports/popa-reports/loader.cfm?csModule=security/getfile&PageID=244407
31. Benson SM, Cook P (2005) Chapter 5: Underground geological storage, IPCC special report on carbon dioxide capture and storage, Intergovernmental Panel on Climate Change, Interlachen, www.ipcc.ch, pp 5–1 to 5–134
32. Jarrell PM, Fox CE, Stein MH, Webb SL (2002) Practical aspects of $CO_2$ flooding. Monograph 22. Society of Petroleum Engineers, Richardson
33. Oldenburg CM, Stevens SH, Benson SM (2004) Economic feasibility of carbon sequestration with enhanced gas recovery (CSEGR). Energy 29:1413–1422
34. Stevens S (1999) Sequestration of $CO_2$ in depleted oil and gas fields: barriers to overcome in implementation of $CO_2$ capture and storage (disused oil and gas fields). IEA Greenhouse Gas R&D Programme, IEA/CON/98/31
35. NPC (2007) Facing hard truths about energy: a comprehensive view to 2030 of global oil and gas, National Petroleum Council, Washington, DC, 442 pp. www.npc.org
36. US DOE (2007a) Carbon sequestration technology roadmap and program plan for 2007, Morgantown, WV, 39 p. www.netl.doe.gov

37. US DOE (2007b) Basic Research needs for geosciences: facilitating 21st century energy systems. Department of Energy Office of Basic Energy Sciences, Washington, DC, 287 p. http://www.sc.doe.gov/bes/reports/list.html
38. Bachu S (2000) Sequestration of $CO_2$ in geological media: criteria and approach for site selection in response to climate change. Energy Convers Manage 41:953–970
39. Pacala S, Socolow R (2004) Stabilization wedges: Solving the climate problem for the next 50 years using current technologies. Science 305:986–999
40. Arts R, Eiken O, Chadwick A, Zweigel P, van der Meer L, Zinszner B (2004) Monitoring of $CO_2$ injected at Sleipner using time-lapse seismic data. Energy 29:1383–1392
41. Wilson M, Monea M (eds) (2004) IEA GHG Weyburn $CO_2$ monitoring & storage project summary report 2000–2004, 273 p
42. Riddiford F, Wright I, Espie T, Torqui A (2004) Monitoring geological storage. In: Salah gas $CO_2$ storage project, GHGT-7, Vancouver
43. Kuuskraa VA, DiPietro P, Koperna GJ (2006) $CO_2$ storage capacity in depleted and near-depleted US oil and gas reservoirs. In: NETL 5th annual conference on carbon sequestration. ExchangeMonitor, Alexandria
44. Asia Society (2009) Roadmap for US-China collaboration on carbon capture and sequestration, Asia Society Press, 41 p. http://asiasociety.org/files/pdf/AS_CCS_TaskForceReport.pdf
45. Dalhousie R, Li X, Davidson C, Wei N, Dooley J (2009) Establishing China's potential for large scale, cost effective, deployment of carbon dioxide capture and storage. Pacific Northwest National laboratory report, PNNL-SA-XXXX, 6 p
46. IEA GHG (2005) A review of natural $CO_2$ emissions and releases and their relevance to $CO_2$ storage. International Energy Agency greenhouse gas R&D programme, report 2005/8. http://www.ieagreen.org.uk/
47. $CO_2$ Capture Project (2009) A technical basis for $CO_2$ storage. In: Cooper C (ed) $CO_2$ capture project. CPL Press, ISBN: 978-1-872691-48-0
48. Benson SM, Hoversten M, Gasperikova E, Haines M (2004) Monitoring protocols and life-cycle costs for geological storage of carbon dioxide. In: 7th international greenhouse gas technology conference, Vancouver
49. Friedmann, SJ, Newmark RA (2009) Commercial deployment of geologic carbon sequestration: technical components of an accelerated U.S. program. In: *Coal without carbon*, Clean Air Task Force. www.coaltransition.org, pp 59–76
50. Wilson EJ, Friedmann SJ, Pollak MF (2007) Research for deployment: incorporating risk, regulation, and liability for carbon capture and sequestration. Environ Sci Technol 41:5945–5952
51. Burton EA, Myhre R, Myer LR, Birkinshaw K (2007) Geologic carbon sequestration strategies for California, The assembly bill 1925 report to the California legislature. California Energy Commission, Systems Office. CEC-500-2007-100-SD. http://www.energy.ca.gov/2007_energypolicy/documents/index.html#100107
52. Chiaramonte L, Zoback M, Friedmann S, Stamp V (2006) $CO_2$ sequestration, fault stability and seal integrity at teapot dome, Wyoming, NETL 5th annual conference on carbon sequestration. ExchangeMonitor Publications, Alexandria
53. Cugini A, DePaolo D, Fox M, Friedmann SJ, Guthrie G, Virden J, 2010, US-DOE's National risk assessment program: bridging the gaps to provide the science base to ensure successful $CO_2$ storage. Energy Procedia
54. Hovorka SD, Benson SM et al (2006) Measuring permanence of $CO_2$ storage in saline formations – the Frio experiment. Environ Geosci 13:103–119
55. Streit JE, Hillis RR (2004) Estimating fault stability and sustainable fluid pressures for underground storage of $CO_2$ in porous rock. Energy 29:1445–1456
56. WRI (2008) Guidelines for carbon capture and sequestration, major contributing author. World Resources Institute, Washington, DC, 103 pp. www.wri.org

57. Bradshaw J, Boreham C, la Pedalina F (2004) Storage retention time of $CO_2$ in sedimentary basins; examples from petroleum systems. Paper presented at the GHGT-7 conference, Vancouver
58. Newmark RA, Friedmann SJ, Carroll SJ (2010) Water challenges for carbon capture and sequestration. Environ Manage. doi:10.1007/s00267-010-9434-1
59. DOE-NETL (2007a) U.S. Department of Energy, National Energy Technology Laboratory, Power plant water usage and loss study, May 2007 revision
60. Aines RD, Wolery TJ, Bourcier WL, Wolfe T, Hausmann C (2010) Fresh water generation from aquifer-pressured carbon storage: feasibility of treating saline formation waters. Energy Procedia 4:2269–2276
61. Buscheck TA, Sun Y, Hao Y, Aines RD, Wolery TJ, Tompson AFB, Jones ED, Friedmann SJ (2010) Combining brine extraction, desalination, and residual-brine injection with $CO_2$ storage in saline formations: implications for pressure management, capacity, and risk management. Energy Procedia 4:4283–4290
62. IEA (2011) Carbon capture and storage: legal and regulatory review. International Energy Agency, Paris, 108 p. www.iea.org
63. Abanades JC, Rubin ES, Anthony EJ (2004) Sorbent cost and performance in $CO_2$ capture systemts. Ind Eng Chem Res 43:3462–3466
64. Bachu S (2003) Screening and ranking of sedimentary basins for sequestration of $CO_2$ in geological media in response to climate change. Environ Geology 44:277–289
65. Birkholzer JT, Zhou Q (2009) Basin-scale hydrogeologic impacts of $CO_2$ storage: capacity and regulatory implications. Int J Greenhouse Gas Control 3:745–756
66. Bradshaw J, Dance T (2004) Mapping geological storage prospectivity of $CO_2$ for the world's sedimentary basins and regional source to sink matching. In: Proceedings of the 7th international conference on greenhouse gas technologies, 2004. Vancouver
67. EPA (2008c) U.S. Environmental Protection Agency, Federal requirements under the Underground Injection Control (UIC) program for carbon dioxide ($CO_2$) geologic sequestration (GS) wells; proposed rule, 40 CFR Parts 144 and 146, Federal register, vol 73, no. 144, July 2008. http://www.epa.gov/EPA-WATER/2008/July/Day-25/w16626.htm
68. Friedmann SJ (2007a) Operational protocols for geologic carbon storage: facility life-cycle and the new hazard characterization approach. In: 6th annual NETL conference on carbon capture and sequestration. ExchangeMonitor, Pittsburgh, Oral 034
69. Friedmann SJ (2007) Geological carbon dioxide sequestration. Elements 3:179–184
70. Friedmann SJ (2009) Emerging technical challenges of 5-million ton/year $CO_2$ injection. In: SPE technical proceedings, SPE-126942-MS, 2009 SPE international conference on $CO_2$ capture, storage, & utilization, San Diego
71. Friedmann SJ (2009) Accelerating development of underground coal gasification: priorities and challenges for U.S. Research and Development. In: Coal without carbon, Clean Air Task Force. www.coaltransition.org, pp 1–16
72. FutureGen Alliance (2011) FutureGen facts, 3p brochure. http://www.futuregenalliance.org/pdf/FutureGenFacts.pdf
73. GCCSI (2011) The global status of CCS, Global Carbon Capture and Storage Institute, Canberra, 204 p. http://www.globalccsinstitute.com/resources/publications/global-status-ccs-2010
74. GCCSI (2011) Status of CCS project database. Global Carbon Capture and Storage Institute, Canberra, http://www.globalccsinstitute.com/resources/data/dataset/status-ccs-project-database
75. Herzog H, Caldeira K, Reilly J (2003) An issue of permanence: assessing the effectiveness of temporary carbon storage. Climatic Change 59(3):293–310
76. IEA GHG (2006) 2nd well bore integrity network meeting, International Energy Agency greenhouse gas R&D, report 2006/12, Sept 2006
77. Ide ST, Friedmann SJ, Herzog, HJ (2006) $CO_2$ leakage through existing wells: current technology and regulatory basis. In: 8th greenhouse gas technology conference, Trondheim

78. Keith DW, Hassanzadeh H, Pooladi-Darvish M (2004) Reservoir engineering to accelerate dissolution of stored $CO_2$ in brines. In: 7th international greenhouse gas technology conference, Vancouver
79. Kharaka YK, Cole DR, Hovorka SD, Gunter WD, Knauss KG, Freifield BM (2006) Gas-water-rock interactions in Frio formation following $CO_2$ injection: Implications for the storage of greenhouse gases in sedimentary basins. Geology 34:577–580
80. Knauss KG, Johnson JW, Steefel CI (2005) Evaluation of the impact of $CO_2$, contaminant, aqueous fluid and reservoir rock interactions on the geologic sequestration of $CO_2$. Chem Geology 217:339–350
81. Lackner KS, Wright AB (2009) USA Patent no. 12/265556
82. Liu H, Gallagher K-S (2010) Catalyzing strategic transformation to a low-carbon economy: a CCS roadmap for China. Energy Policy 38:59–74
83. McFarland JR, Reilly JM, Herzog HJ (2004) Representing energy technologies in top-down economic models using bottom-up information. Energy Econ 26(4):685–707. Special Issue EMF 19 Alternative technology strategies for climate change policy, John P. Weyant, ed
84. MIT (2011b) Hydrogen energy California (HECA) project fact sheet: carbon dioxide capture and storage project, MIT Energy Initiative. http://sequestration.mit.edu/tools/projects/heca.html
85. Morris JP, Detwiler RL, Friedmann SJ, Vorobiev OY, Hao Y (2010) The large-scale effects of multiple $CO_2$ injection sites on caprock and formation stability Intl. J Greenhouse Gas Control 5:69–74
86. Wilson E, Johnson T, Keith D (2003) Regulating the ultimate sink: managing the risks of geologic $CO_2$ storage. Environ Sci Technol 37:3476–3483

# Index




R. Malhotra (ed.), *Fossil Energy: Selected Entries from the Encyclopedia of Sustainability Science and Technology*, DOI 10.1007/978-1-4614-5722-0,

FSC
www.fsc.org
MIX
Papier | Fördert gute Waldnutzung
FSC® C083411

Kolibri 360 GmbH
Ferdinand-Jühlke-Straße 7
99095 Erfurt, Deutschland
produktsicherheit@kolibri360.de